# AN INTRODUCTION TO MASS AND HEAT TRANSFER  ·  Principles of Analysis and Design

# AN INTRODUCTION TO MASS AND HEAT TRANSFER

## PRINCIPLES OF ANALYSIS AND DESIGN

**Stanley Middleman**
*University of California, San Diego*

**John Wiley & Sons, Inc.**

ACQUISITIONS EDITOR Wayne Anderson
MARKETING MANAGER Harper Mooy
SENIOR PRODUCTION MANAGER Lucille Buonocore
SENIOR PRODUCTION EDITOR Monique Calello
SENIOR DESIGNER Karin Kincheloe
COVER DESIGNER Carol C. Grobe
ILLUSTRATION COORDINATOR Jaime Perea
ILLUSTRATION Thunder Graphics, Inc.
MANUFACTURING MANAGER Monique Calello

This book is printed on acid-free paper. ∞

The paper in this book was manufactured by a mill whose forest management programs include sustained yield harvesting of its timberlands. Sustained yield harvesting principles ensure that the number of trees cut each year does not exceed the amount of new growth.

To order books or for customer service please, call 1(800)-CALL-WILEY (225-5945).

*Library of Congress Cataloging-in-Publication Data*
Middleman, Stanley.
    An introduction to mass & heat transfer: principles of analysis and
    design / Stanley Middleman.
        p. cm.
    Includes bibliographical references and index.
    ISBN 0-471-11176-7 (cloth: alk. paper)
    1. Heat—Transmission. 2. Mass transfer. I. Title.
TJ260.M483 1997
621.402—dc21
                                                                97-9531
                                                                CIP

10 9 8 7 6 5 4

This book is dedicated to

Yocheved bat Miriam

*v'Dodi Li*

# Preface

This textbook is the outgrowth of 36 years of teaching this material to students in chemical engineering, environmental engineering, mechanical engineering, bioengineering, and applied mechanics. As such, it represents my ideas regarding how the material should be taught to engineering students. Primary is my belief that the development of a mathematical model is central to the analysis and design of an engineering system or process. Hence the orientation of this text, and the way I handle this material in the classroom, is toward teaching students how to develop mathematical representations (models) of physical phenomena. The key elements in model development involve assumptions about the physics, the application of basic physical principles, the exploration of the implications of the resulting model, and the evaluation of the degree to which the model mimics reality. This latter point—evaluation—is critical. It requires that the model builder have some a priori sense of what is required of the model, in terms of both the specific phenomena that need to be described and the desired degree of correspondence of the predictions to the observations. As a consequence, a great deal of effort has been put forth to provide many examples of experimental data against which the results of modeling exercises can be compared. Students need to see that simple models can often be quite accurate, and that when a model fails to yield the required accuracy, the assumptions of the model need to be explored and reconsidered.

Another goal of this text is to expose students to a wide range of technologies, since they may be called upon to apply their skills in diverse areas. Hence, few of the problems illustrated in this text, or raised in the Homework sections, are sterile mathematical analyses. Where possible, the examples presented are motivated by real engineering applications. Many of the problems are derived from my years of experience as a consultant to companies whose businesses cover a broad spectrum of engineering technologies. My own experience as a teacher is that students are more motivated by problems having their basis in commercial technology than by those having an orientation that is primarily mathematical or analytical.

The text assumes that the reader/student has been exposed to a basic course in ordinary differential equations. My observation is that most students learn their mathematics in an environment that is separated from application. In addition, many students complete required calculus courses early in their academic program, and a course in mass and heat transfer that requires the solution of ordinary differential equations may come more than a year after the completion of the math sequence. This is a serious handicap to students and instructors alike. Remedial instruction may be required in some circumstances, but space has not permitted any significant review of mathematical concepts in this text. Each instructor will need to evaluate the needs of his or her students and proceed accordingly.

This textbook provides more than sufficient material to support a one-semester course in mass and heat transfer; curricula that need to present mass and heat transfer separately in two one-quarter segments are served abundantly. The book is deliberately overwritten. The instructor may choose to illustrate certain concepts with different sections of the text from one year to the next. Sufficient homework problems are offered that it should not be necessary to assign the same problems in successive years. Upon reading the Solutions Manual carefully, the instructor will find that the problems selected can be used to extend the classroom presentation beyond the confines of the text proper. Many of these problems are worked out in considerable detail, and, with accompanying commentary, provide a basis for use by teaching assistants as material in discussion sections.

A comment is in order regarding a common complaint of my students, namely, that the physical property data in my problem statements often are very scanty. This is deliberate, and my goal is to force the student to look outside the textbook for information, or to develop the confidence to estimate physical property values from his or her knowledge of related materials. Parallel to this, I want my students to be troubled enough by the lack of information to investigate the *sensitivity* of their "solution" to errors in physical property data.

*Stanley Middleman*

## ACKNOWLEDGMENTS

I am indebted to many generations of students and colleagues whose questions and criticisms have improved the text. Over the years, many undergraduate students have worked in my laboratory carrying out experiments aimed at illustrating models presented in the classroom. Such examples are typically cited as "unpublished data" throughout the text. A number of colleagues reviewed my manuscript prior to my final revisions, and indeed several taught from preliminary versions of the text. Special acknowledgment must be made of my gratitude to Professors Richard Calabrese of the University of Maryland, Skip Rochefort of Oregon State University, Arup Chakraborty of the University of California, Berkeley, and Martin Wagner of the University of Delaware. The final form of this book is considerably improved in clarity because of the input of these individuals, who also shared input gathered from their students. The inconsistencies, obscurities, and out-and-out errors that remain are all mine to correct in subsequent editions, and the author hopes that his attention will be drawn to these imperfections. I have taught mass and heat transfer from many textbooks over many years, incorporating material from these books into many generations of course notes. Over this time, the origin of this material has often been forgotten. Hopefully, readers will feel free to bring to my attention any examples of material in the text or in the homework problems that warrant acknowledgment and attribution of the original source.

*S. M.*

# Contents

# Part One

# MASS TRANSFER

# Chapter 1

# What Is Mass Transfer?

**W**e begin this book with the study of mass transfer, and state that our goal is to learn how to write mathematical models for a wide variety of important engineering systems in which mass transfer plays a central role. In many respects the study of mass transfer, and the use of the basic principles that describe the rate of mass transfer, provide the features that distinguish chemical engineering from the other engineering disciplines. This introductory chapter indicates the range of problems of interest.

Atoms and molecules are in a constant state of motion. In a crystalline solid the atoms are so tightly constrained by steric hindrance, and by intermolecular/interatomic forces, that their departure from their average sites on the crystal lattice is very small. In a gas, on the other hand, the mean free path between collisions can be of the order of micrometers to centimeters, depending on the mean gas pressure. This distance is many orders of magnitude greater than the size of the molecule.

Thermal (kinetic) energy drives this fluctuating state of motion. In a *homogeneous* fluid or solid (where the concentration is independent of position when averaged over a distance that is large compared to molecular dimensions), this motion brings about no change in mean composition. In an inhomogeneous medium (one with a mean concentration gradient with respect to some chemical species), this fluctuating motion can give rise to a net movement of a particular species, with respect to some coordinate system. *This net molecular motion is what we call diffusion.*

Diffusion is a mode of mass transfer arising from *molecular* motion. *Macroscopic* motion, which we normally describe as "flow," also provides a mechanism of mass transfer. When we stir sugar into tea, or simply pour cream into coffee, a process of homogenization occurs at a rate that depends primarily on the fluid dynamics of the system and is not strongly dependent on diffusion. We refer to this mode of mass transfer as *convection.*

Mass transfer by a combination of diffusion and convection occurs in a wide variety of systems of concern to chemical, biochemical, and environmental engineers. We cite some examples here, all in the context of *design.*

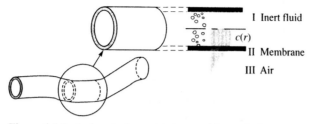

**Figure 1.1.1** A chemical agent is trapped in a tubular membrane. (Its concentration $c(r)$ varies radially across each phase.)

## 1.1 EXAMPLES OF MASS TRANSFER

### 1.1.1 Design of a Sustained Release System for Control of Insects

Consider a biologically active chemical agent dissolved in an inert fluid, and the fluid entrapped within a tubular membrane, as suggested in Fig. 1.1.1.

The rate at which the active agent is released to the surrounding atmosphere depends on the rate at which it passes through a series of resistances to mass transfer: the inert and usually stationary internal fluid, the solid or porous membrane, and the gas external to the membrane. Tubular membranes used in this application are typically "hollow fibers" of inside diameter $O(100\ \mu m)$[1] and having a membrane thickness of 25 $\mu$m. The liquid inside such a small "container" is stationary; hence mass transport within the tube is by diffusion only. Membrane transport is usually regarded as a diffusive process, also. On the exterior of the tube, airflow may promote mass transfer by convection. From a knowledge of the transport resistances of regions I, II, and III, the *design task* is to select a fiber size (diameter and membrane thickness) and agent concentration that produce the desired release rate.

### 1.1.2 Design of a Bubbler for Delivering Dopants to a CVD Reactor

Chemical vapor deposition (CVD) is a process by which solid films are "grown" onto a substrate. In semiconductor fabrication, for example, a thin layer of silicon is grown onto a silicon wafer. However, it is often desired to incorporate minute amounts of "charge carrier" into the film. A typical "dopant" used for this purpose is phosphorus. Doping is accomplished by adding a gaseous compound such as $POCl_3$ to a feed gas of silane ($SiH_4$). The silane decomposes to Si, which is deposited on the wafer surface. The $POCl_3$ decomposes, and phosphorus is incorporated into the deposited film.

$POCl_3$ is a liquid at room temperature. As suggested in Fig. 1.1.2, it can be delivered to the reactor by bubbling an inert gas through liquid $POCl_3$. Some $POCl_3$ then evaporates into the inert gas bubbles and this gas mixture is conveyed to the reactor.

To design such a system, it is necessary to know the rate of transfer of a volatile species into an inert gas bubble rising through a liquid. The flow external to the bubble is sufficient to promote convection *to* the bubble, but this is irrelevant, since the liquid is *pure* (homogeneous) $POCl_3$. The main resistance to mass transfer is due to molecular diffusion of $POCl_3$ vapor into the interior of the gas bubble. (There could be some

---

[1] This notation is read "order of ( )." It refers to the nearest power of 10. For example, 25 is $O(10)$ because it is closer to 10 than to 100. Of what "order of magnitude" is 31.6?

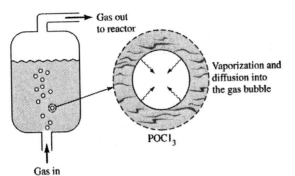

**Figure 1.1.2** Vaporization into bubbles of inert gas.

convective motion within the bubble, but for small bubbles this is normally small and does not contribute significantly to the rate of mass transfer.)

### 1.1.3   Design of a Dryer for a Polymeric Coating

Figure 1.1.3 represents a system in which a polymeric material must be coated onto a thin metallic film to protect the metal from chemical attack. The polymer is applied as a dilute liquid (solution), and the system is then conveyed through a dryer to remove the solvent.

The solvent must diffuse through the solution, and then through the boundary layer of air external to the moving substrate. Since the boundary layer thickness depends on the motion of the substrate, the external mass transfer problem is really one of *convective* diffusion. It is not obvious whether mass transfer through one phase or the other controls the observed rate of transfer without additional information about the operating conditions.

Proper design of this system requires that the mass transfer rate be controlled to prevent bubble formation. The best strategy is not necessarily to dry the coating as quickly as possible. To design a dryer, it is necessary to understand how the rate of drying depends on the film speed, any air motion directed at the film, and the resistance to diffusion within the polymer solution.

### 1.1.4   Design of an Artificial Kidney: An Enzyme/Bead Reactor

The enzyme urease is encapsulated in solid polymeric beads. A reactor is designed in which an agitated slurry of beads is contacted with blood. A membrane permeable to urea separates the blood from the slurry and protects the blood from contamination by viruses in the commercially prepared slurry and its associated tubing and pumps.

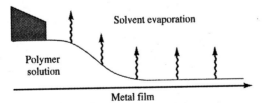

**Figure 1.1.3** Coating a polymer onto a metal film.

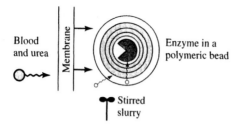

**Figure 1.1.4** Elements of an artificial kidney.

Urea must pass across several resistances before it contacts a reactive enzyme molecule. The design problem requires an estimate of the amount of urease/beads, and the bead size, which will permit the unit to reduce the urea concentration in blood by half, in one hour of operation.

As before, we see that in this system (Fig. 1.1.4), the mass transfer occurs across a series of resistances—diffusive across the membrane, convective to the bead, and diffusive within the bead. Chemical reaction within the bead can profoundly affect the rate of transfer under some conditions of design and be unimportant under other conditions. We will develop the tools to assess the relative roles of the various mass transfer resistances *before* we begin a modeling exercise. That assessment will then serve to guide our model making.

## 1.2 REEXAMINATION AND OVERVIEW

Let's reexamine the example problems of Section 1.1 from several perspectives.

*What are the fields of commerce/science/technology involved?*

    **1.1.1.** *A sustained release system for an insecticide*
        Agricultural products
        Entomology (the science of insects)
        Membrane science and technology
    **1.1.2.** *Chemical vapor deposition of a solid film*
        Microelectronics fabrication
        Solid state physics, but not directly with this problem
        Gas bubble dynamics; two-phase fluid dynamics
    **1.1.3.** *Drying of a polymer film*
        Polymer products—materials science
        Boundary layer flow exterior to a moving surface
        Diffusion in a polymer matrix
    **1.1.4.** *Artificial kidney design*
        Medical products
        Enzymology/biochemistry
        Encapsulation technology
        Polymer materials science

*Can we classify these problems in simple categories, for the purpose of deciding how we will approach the problems? In particular, a categorization can help us decide how to write mathematical models of these systems.*

    **1.1.1.** Transient diffusion
        Cylindrical symmetry
        Exterior resistance (convection) probably negligible

**1.1.2.** Transient diffusion
Spherical symmetry
**1.1.3.** Steady state
Planar geometry, but with no apparent symmetry
Diffusion and exterior convection in series
**1.1.4.** Steady state
Spherical symmetry (for the bead)
Convection (blood), diffusion (membrane), convection (slurry), diffusion
(bead), *chemical reaction*

Clearly, then, to solve problems like these, we must develop an understanding of the *principles of diffusion, convective mass transfer,* and the *interaction of mass transfer* with *chemical reaction.*

We begin this process in Chapter 2 with definitions of various measures of mass transfer (*molar or mass fluxes*) and with a definition of a *diffusion coefficient.* We will learn how to write *Fick's "law,"* which is really a definition of a diffusion coefficient, and we will learn how to estimate the magnitude of these diffusion coefficients for various compounds in solids, liquids, and gases.

Then we will examine the principle of conservation of chemical species and derive a transport equation that expresses that principle. To solve this partial differential equation, we must invoke (assume) boundary conditions that we think match the physics of the problems we are trying to model. Next we examine some particularly simple (steady, one-dimensional) diffusion problems and then some simple transient problems.

Following the development of these basic principles and building blocks, we couple our knowledge of fluid dynamics with mass transfer and develop some models of convective mass transport, especially across interfaces. In particular, we define various mass transfer coefficients for interphase transfer. We will then examine and use available empirical correlations of convective mass transfer coefficients to solve a variety of problems. Finally we illustrate how we use these ideas to design mass transfer equipment.

# Chapter 2

# Fundamentals of Diffusive Mass Transfer

**I**n this chapter we study the fundamental principles of mass transfer by diffusion. We begin with quantitative definitions of the basic characteristics such as concentrations and diffusive fluxes, and then we introduce the diffusion coefficient through Fick's law of diffusion. This permits us to write mathematical models for several very simple, but still interesting mass transfer problems. We then address the question of how diffusion coefficients for various species can be estimated if data are not available. Application of these simple ideas to a design problem is illustrated next. Finally, we introduce the quasi-steady approximation for solving unsteady state mass transfer problems by using steady state models.

## 2.1 CONCENTRATIONS, VELOCITIES, AND FICK'S LAW OF DIFFUSION

Consider a mixture of $i$ distinct chemical species, each of molecular weight $M_{w_i}$, in a volume $\Delta V$. If $m_i$ is the total mass of each species in that volume, then we may define the *(mass) average density* within that volume to be

$$\bar{\rho} = \frac{1}{\Delta V} \sum_i m_i \tag{2.1.1}$$

We may also define $\rho$ locally, that is, at a *point* in the medium, as

$$\rho = \lim_{\Delta V \to 0} \bar{\rho} \tag{2.1.2}$$

The *local mass density of each species* is defined as

$$\rho_i = \lim_{\Delta V \to 0} \frac{m_i}{\Delta V} \tag{2.1.3}$$

Hence

$$\rho = \sum_i \rho_i \tag{2.1.4}$$

We often measure compositions *in molar units*. Then we define

$$C_i = \lim_{\Delta V \to 0} \frac{m_i}{(\Delta V) M_{w_i}} = \frac{\rho_i}{M_{w_i}} \qquad (2.1.5)$$

The *total molar concentration* is just

$$C = \sum_{i=1} C_i \qquad (2.1.6)$$

We may denote the velocity of a species, relative to some *stationary* coordinate system, by the *vector* $v_i$. Then the *mass-average velocity* of the mixture is *defined* as

$$v = \frac{\sum \rho_i v_i}{\rho} = \sum \omega_i v_i \qquad (2.1.7)$$

where $\omega_i$ is the *mass* fraction of the species $i$. This is the velocity we normally use in the equations of fluid dynamics—the *mass*-average velocity.

Note that for a gas mixture at *equilibrium*, all molecules may be in thermal motion, with a *mean speed* determined by the gas temperature but a mean *velocity of zero*, relative to *some* fixed coordinate axes. That fixed coordinate system could be moving, as when the gas is simply being transported as a rigid body.

We may also define a *molar-average* velocity as

$$\frac{\sum C_i v_i}{C} = v^* \qquad (2.1.8)$$

relative to a *fixed* coordinate system.

The model normally adopted for describing diffusion in a *binary* mixture is **Fick's law.** It asserts that the *molar flux* of a species, *relative to the molar average velocity*, is proportional to the concentration gradient

$$J_A^* = -D_{AB} \nabla C_A \qquad (2.1.9)$$

This equation provides a consistent definition of the *binary diffusion coefficient* $D_{AB}$ only for isothermal mixtures that are also isobaric (i.e., under constant pressure). *More generally*, one should define $D_{AB}$ as the coefficient in

$$J_A^* = -C D_{AB} \nabla x_A \qquad (2.1.10)$$

Fick's law

which is not restricted to isothermal isobaric systems, but *is* restricted to *binary* mixtures. Here we use $x_A$ as the mole fraction of species A, that is,

$$x_A = \frac{C_A}{C} \qquad (2.1.11)$$

We have stated that $J_A^*$ is the *molar* flux, relative to $v^*$. *Why is there any $v^*$ in a system in which diffusion occurs?* The answer is that in an isobaric system, diffusion of one species *can* cause motion of the other species, giving rise to a *net* nonzero value of $v^*$. We can see this in the following example.

**EXAMPLE 2.1.1** *Diffusion of Naphthalene in a Narrow Tube*

A tube, open at one end, is surrounded by gaseous nitrogen, $N_2$. Solid naphthalene plugs one end of the tube, as indicated in Fig. 2.1.1. At room temperature, naphthalene

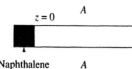

Naphthalene $\quad A$

**Figure 2.1.1** Diffusion from an open tube: naphthalene evaporates and diffuses to the right, toward the open end of the tube.

is a *solid* with a significant vapor pressure. The solid evaporates slowly. Between the closed and open ends of the narrow tube there is a difference in concentration of naphthalene in the gas phase. We expect diffusion to occur. The vapor diffuses toward the open end of the tube.

Across some stationary plane $A—A$, there is a finite molar flux of species A (naphthalene). Given the gradient of A, there must be a gradient of B ($N_2$), since (consistent with the ideal gas law), the total molar concentration remains constant in an isothermal isobaric system:

$$C = C_A + C_B = \text{constant} \qquad \text{at constant } T, P \qquad (2.1.12)$$

It follows then that

$$\nabla C = \nabla C_A + \nabla C_B = 0 \qquad \text{or} \qquad \nabla C_B = -\nabla C_A \qquad (2.1.13)$$

Then, since there is a gradient of composition for species B, there must be diffusion of B, as well; hence, we expect that the flux of B is nonzero.

Suppose $N_2$ is not soluble in the solid naphthalene. Then at the solid naphthalene surface, no B crosses the plane $z = 0$ ($\mathbf{v_B} = 0$). Does this mean $\mathbf{J_B^*} = 0$? No, because $\mathbf{J_B^*}$ is not a flux relative to a *fixed* plane—it is a flux relative to $\mathbf{v^*}$.

We should define another set of fluxes, $\mathbf{N_A}$ and $\mathbf{N_B}$, as *molar* fluxes relative to a *fixed* coordinate system:

$$\mathbf{N_A} = C_A\mathbf{v_A} \qquad (2.1.14)$$

and similarly for component B. The statement that no B ($N_2$) crosses the fixed plane $z = 0$ implies $\mathbf{N_B} = 0$ at that plane. If we assume that the system is in a steady state (no time variations of concentrations or fluxes), then $\mathbf{N_B} = 0$ at $z = 0$ implies $\mathbf{N_B} = 0$ *everywhere.* If this were not so, species B would be accumulating between $z = 0$ and any plane showing nonzero flux of B, and this would violate the steady state assumption. A similar argument tells us that although $\mathbf{N_A}$ may not be zero, it is *constant* with respect to the $z$ coordinate.

The (vector) sum of $\mathbf{N_A}$ and $\mathbf{N_B}$ is what gives rise to a nonzero $\mathbf{v^*}$:

$$C\mathbf{v^*} = \mathbf{N_A} + \mathbf{N_B} = C_A\mathbf{v_A} + C_B\mathbf{v_B} \qquad (2.1.15)$$

Since we normally "measure" the behavior of a system relative to fixed laboratory coordinates, we want to measure the diffusion of A as $\mathbf{N_A}$, and not as $\mathbf{J_A^*}$. The relationship between the two is

$$\mathbf{N_A} \quad = \quad \mathbf{J_A^*} \quad + \quad C_A\mathbf{v^*}$$

Flux of A $\qquad$ Flux of A $\qquad$ Flux of A $\qquad\qquad$ (2.1.16)
relative to fixed axes $\quad$ relative to $\mathbf{v^*}$ $\quad$ associated with $\mathbf{v^*}$

Introducing Fick's "law," we write

$$\mathbf{N_A} = -CD_{AB}\nabla x_A + x_A(\mathbf{N_A} + \mathbf{N_B}) \qquad (2.1.17)$$

Returning to the specific example above, we noted that while $\mathbf{J}_B^* \neq 0$, $\mathbf{N}_B = 0$, since there can be no net flow of $N_2(B)$ across the solid surface. Hence, for *this special case*,

$$\mathbf{N}_A = -CD_{AB}\nabla x_A + x_A\mathbf{N}_A \qquad (2.1.18)$$

or

$$\mathbf{N}_A = \frac{-CD_{AB}}{1-x_A}\nabla x_A \qquad (2.1.19)$$

Suppose the flux is strictly in the $z$ direction (unidirectional diffusion) and is independent of $z$. Then Eq. 2.1.19 becomes

$$\frac{1}{1-x_A}\frac{dx_A}{dz} = -\frac{N_{A,z}}{CD_{AB}} = \text{constant} \qquad (2.1.20)$$

Solving this differential equation for $x_A$ we find, taking $L$ to be the tube length, and $x_{A1}$ and $x_{A2}$ to be the mole fractions of naphthalene at $z = 0$ and $z = L$,

$$\left(\frac{1-x_A}{1-x_{A1}}\right) = \left(\frac{1-x_{A2}}{1-x_{A1}}\right)^{z/L} \qquad (2.1.21)$$

This result shows that the concentration profile is not linear along the length of the tube. It is important to emphasize that diffusion in an apparently "static" system gives rise to velocities. We need to get an idea of the magnitude of these velocities, and especially whether they are at all comparable to what we usually regard as macroscopic velocities in a gas.

---

**EXAMPLE 2.1.2**    *Calculation of Diffusion Velocities in a "Static" System*

Calculate $\mathbf{v}$ and $\mathbf{v}^*$ for the naphthalene/$N_2$ system described in Example 2.1.1. Take $T = 75°F = 24°C$, and $P = 1$ atm. Use a diffusion length of $L = 10$ cm. For this system:

$$M_{w_{naph}} = 128 \qquad P_{vap} = 9.6 \text{ Pa} \qquad D_{AB_{naph/air}} = 8 \times 10^{-2} \text{ cm}^2/\text{s}$$

For an ideal gas,

$$C = \frac{P}{R_G T} = \frac{1}{82\,(297)}\,\text{g}\cdot\text{mol/cm}^3 \qquad (2.1.22)$$

At the naphthalene surface $x_{AS} = P_{vap}/P_{tot}$.[1] Thus we can write

$$x_{AS} = \frac{9.6}{101.3 \times 10^3} = 9.6 \times 10^{-5} \ll 1 \qquad (2.1.23)$$

Hence

$$N_{A,z} \cong -CD_{AB}\frac{dx_A}{dz} \cong CD_{AB}\frac{x_{AS}}{L} \qquad (2.1.24)$$

---

[1] Since $P_{vap}$ is given in the SI units of pascals (newtons per meter squared), we use atmospheric pressure in those same units. The Standard Atmosphere is defined as $1.013 \times 10^5$ Pa. Through most of this book we will approximate this as $10^5$ Pa.

(We have assumed, and we will later show, that $dx_A/dz = \Delta x_A/L$ for a dilute mixture.) Then we find

$$v_z^* = \frac{N_{A,z}}{C} = \frac{D_{AB}x_{AS}}{L} = \frac{8 \times 10^{-2}(9.6 \times 10^{-5})}{10} = 7.7 \times 10^{-7} \text{ cm/s} \qquad (2.1.25)$$

Note that $v_z^*$ is constant within the tube. This is not the case with $v_{A,z}$. From Eq. 2.1.8, with $v_{B,z} = 0$,

$$v_{A,z} = \frac{v_z^*}{x_A} \qquad (2.1.26)$$

Since $x_A$ varies (decreases) along the tube axis, $v_A$ will increase toward the top of the tube. At the midpoint $L/2$ (noting that $x_A$ is linear in $z$), we have

$$v_{A,z=L/2} = \frac{v_z^*}{x_{A,z=L/2}} = \frac{7.7 \times 10^{-7}}{\frac{1}{2}(9.6 \times 10^{-5})} = 1.6 \times 10^{-2} \text{ cm/s} \qquad (2.1.27)$$

Because these velocities arise from molecular motion, their magnitudes, and even their existence, are not intuitive. We make this point in the following example.

*Mass Flow Without Molar Flow*

$N_2$ and $O_2$, initially separated, are allowed to interact through a capillary connection, as in Fig. 2.1.2. Since $\nabla C_A$ and $\nabla C_B$ are nonzero initially, we expect diffusion to occur. For an isothermal isobaric system, the number of moles of gas in each volume is independent of composition, so there can be no net flow (flux) of *moles*:

$$\mathbf{v}^* = 0 \qquad (2.1.28)$$

Then it must follow that

$$\mathbf{N}_A = -\mathbf{N}_B \qquad (2.1.29)$$

since

$$C\mathbf{v}^* = 0 = \mathbf{N}_A + \mathbf{N}_B \qquad (2.1.30)$$

Equation 2.1.17 is valid for binary isothermal isobaric systems, so

$$\mathbf{N}_A = -CD_{AB}\nabla x_A + x_A(\mathbf{N}_A + \mathbf{N}_B) \qquad (2.1.31)$$

$$\mathbf{N}_A = -CD_{AB}\nabla x_A \qquad (2.1.32)$$

This case, where there is no net *molar* flow, is called "equimolar counterdiffusion." Note how the flux expression differs from that for "unidirectional diffusion," Eq. 2.1.19, which is a *nonlinear* flux expression.

What about *mass* flow? While $\mathbf{v}^* = 0$, $\mathbf{v} \neq 0$. Initially there are the same number of moles on each side; *but the masses are different, since the molecular weights are different.* At equilibrium, the same masses are on both sides (equal mixtures with the same

**Figure 2.1.2** Two gases initially separated.

average molecular weight). Therefore there must have been a mass flow. Hence $\mathbf{v}$ is not proportional to $\mathbf{v}^*$, as in Example 2.1.2.

Mass transfer often occurs in the presence of a chemical reaction and can be strongly affected by the rate of the reaction. Example 2.1.4 shows one way in which reaction and diffusion are coupled.

**EXAMPLE 2.1.4**   *Diffusion with Reaction on a Surface*

As indicated in Fig. 2.1.3, oxygen diffuses to a hot carbon surface, on which the following reaction occurs:

$$3C + 2O_2 \rightarrow 2CO + CO_2 \tag{2.1.33}$$

Note that two moles of gas ($O_2$) arrive at the surface, and three moles of gas products leave.

Oxygen must diffuse *counter* to the flow of CO and $CO_2$. From the stoichiometry (note here how the chemical reaction affects a diffusion problem) we find

$$\mathbf{N}_{O_2} = -\mathbf{N}_{CO} \tag{2.1.34}$$

$$\mathbf{N}_{O_2} = -2\mathbf{N}_{CO_2} \tag{2.1.35}$$

$$C\mathbf{v}^* = \mathbf{N}_{O_2} + \mathbf{N}_{CO} + \mathbf{N}_{CO_2} = \mathbf{N}_{O_2}(1 - 1 - \tfrac{1}{2}) \tag{2.1.36}$$

and the molar flux is

$$C\mathbf{v}^* = -\tfrac{1}{2}\mathbf{N}_{O_2} \tag{2.1.37}$$

Can we write Fick's law in this example?

**No! This is *not a binary system.***
A much more complex "law" holds for multicomponent diffusion, and diffusion is described by a *set* of coefficients, rather than by a single binary diffusivity. We will not consider this more complex case in an introductory text. Indeed, this more complex formulation of diffusion is rarely used, and the coefficients that appear in it are rarely measured. Instead, we define an *effective binary diffusivity* as the coefficient in

$$\mathbf{N}_i = -CD_{im}\nabla x_i + x_i \sum_{j=1}^{n} \mathbf{N}_j \tag{2.1.38}$$

(which looks like Eq. 2.1.17). In systems in which species $i$ is very dilute, we treat the rest of the mixture as a single gas with some average properties.

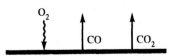

Reaction on this carbon surface

**Figure 2.1.3** Oxidation of a carbon surface.

**EXAMPLE 2.1.5** *Diffusion Across a Thin Barrier Separating Two Fluids*

Suppose a thin immobile barrier of thickness $L$ separates a single fluid into two regions, one to the left and one to the right of the barrier, as in Fig. 2.1.4. There is a species dissolved in the fluid which is capable of dissolving in the material of the barrier. If the concentrations on either side of the barrier are uniform, but at different values that we denote by $C_{1o}$ and $C_{2o}$, respectively, we might expect that diffusion of this species will take place across the barrier.

To describe the rate of diffusion, we must write Fick's law. If we assume that the species A is very *dilute* in either phase (the fluid or the barrier material) and that the species flux is *only* in the $z$ direction, then we may see that Eq. 2.1.20 is valid, and simplifies to

$$\frac{dx_A}{dz} = \text{constant} \qquad (2.1.39)$$

(We use mole fraction $x_A$ here, instead of molar concentration $C_A$.)

Of course the solution of this differential equation is just a linear function, which we write in the form

$$x_A = \alpha + \beta z \qquad (2.1.40)$$

The integration constants $\alpha$ and $\beta$ are to be determined from boundary conditions, which give the values of the concentration of the diffusing species at the planes $z = 0$ and $z = L$. The question at hand is: Do we use the concentrations $C_o$ in the surrounding fluid, or the concentrations $C_i$ inside the barrier? Further, aren't these concentrations the same, since they are defined at the same place?

The answer to the second question is that the concentration of a species at a phase boundary differs on either side of the boundary. The concentrations on either side of a boundary are determined by an equilibrium relationship stating that if the fluid in question had a concentration $C_{1o}$ and were in thermodynamic equilibrium with the barrier of this example, the barrier concentration $C_{1i}$ would be related to the fluid concentration $C_{1o}$ by a relationship of the form

$$C_{1i} = F_{eq}(C_{1o}) \qquad (2.1.41)$$

at boundary 1 ($z = 0$), where $F_{eq}$ is some equilibrium function. It is often the case that equilibrium relationships are *linear*. If that were so, we might write

$$C_{1i} = bC_{1o} \qquad (2.1.42)$$

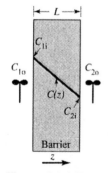

**Figure 2.1.4** Well-mixed fluids are on either side of a permeable barrier.

and a similar relationship at the boundary $z = L$. In that case the appropriate boundary conditions for this example would be

$$C_{1i} = bC_{1o} \quad \text{at } z = 0 \tag{2.1.43}$$

$$C_{2i} = bC_{2o} \quad \text{at } z = L \tag{2.1.44}$$

and the solution for the concentration profile would take the forms

$$C_A = C_{1i} + (C_{2i} - C_{1i})\frac{z}{L} = bC_{1o} + b(C_{2o} - C_{1o})\frac{z}{L} \tag{2.1.45}$$

If we again invoke Fick's law for *dilute* species A, we may write the molar flux across the barrier in either of two forms:

$$N_{A,z} = -D_{AB}\frac{dC_A}{dz} = D_{AB}\frac{(C_{1i} - C_{2i})}{L} \tag{2.1.46}$$

$$N_{A,z} = bD_{AB}\frac{(C_{1o} - C_{2o})}{L} \tag{2.1.46'}$$

When we consider gas permeation through a solid but permeable film or barrier, we usually write the equilibrium relationship in terms of pressure:

$$C_{1i} = b_p p_{1o} \quad \text{at } z = 0 \tag{2.1.47}$$

$$C_{2i} = b_p p_{2o} \quad \text{at } z = L \tag{2.1.48}$$

Then we may go back to Eq. 2.1.40 and write it in the form

$$C_A = C(\alpha + \beta z) = \alpha' + \beta' z \tag{2.1.49}$$

where $C$ is the molar density of the barrier material. In the case of polymeric or glassy materials, or composite materials, the molecular weight is ill defined, and $C$ is not known. From the boundary conditions,

$$b_p p_{1o} = \alpha' \tag{2.1.50}$$

$$b_p p_{2o} = \alpha' + \beta' L = b_p p_{1o} + \beta' L \tag{2.1.51}$$

from which it follows that

$$\beta' = \frac{b_p(p_{2o} - p_{1o})}{L} \tag{2.1.52}$$

so

$$C_A = b_p p_{1o} - \frac{b_p(p_{1o} - p_{2o})}{L}z \tag{2.1.53}$$

We now find the diffusive flux (assuming that $x_A \ll 1$) as

$$N_{A,z} = -D_{AB}\frac{dC_A}{dz} = \frac{b_p D_{AB}(p_{1o} - p_{2o})}{L} \tag{2.1.54}$$

where $N_{A,z}$ is a molar flux. When we deal with a gas permeating through a film, we often measure the flux in *volumetric* units. For example, each mole of gas corresponds to a volume of $R_G T/p_{1o}$ measured at $z = 0$. Keep in mind that $p_{1o}$ is the partial pressure of gas species A. It follows then that the volumetric flux (measured at $p_{1o}$, $T$) is

$$q = \frac{b_p D_{AB} R_G T(p_{1o} - p_{2o})}{L p_{1o}} \tag{2.1.55}$$

We could (and usually do) choose to convert the measured volumetric flux to conditions of standard temperature and pressure (273 K and 1 atm = 101.3 kPa). Then we can replace $R_G T/p_{1o}$ in Eq. 2.1.55 by

$$\frac{R_G T}{p_{1o}} = \frac{8.314 \times 273}{1.013 \times 10^5} = 0.0224 \text{ m}^3/\text{g} \cdot \text{mol} \qquad (2.1.56)$$

(Note that we use 8.314 for the gas constant when we are using *gram*-moles, rather than *kilogram*-moles.) Permeability data are often reported in terms of cubic centimeters as the volume measure. Hence we may write Eq. 2.1.55 in the form

$$q_{STP} \text{ [cm}^3/\text{cm}^2 \cdot \text{s]} = 2.24 \times 10^4 \frac{b_p D_{AB}(p_{1o} - p_{2o})}{L} \qquad (2.1.57)$$

where we now understand that cgs units are in use, except that $b_p$ is often based on a pressure difference in units of atmospheres, so $p_{1o}$ is also in units of *atm* partial pressure difference.

If we were to measure the volumetric flux [converted to standard temperature and pressure (STP)] through an area of 1 cm² of film of thickness $L = 1$ cm, under a partial pressure difference of 1 atm, we would find

$$q_{STP} = 2.24 \times 10^4 \, b_p D_{AB} \, 1 \text{ atm}/1 \text{ cm} = K \qquad (2.1.58)$$

The coefficient $K$ so defined is sometimes called a *gas permeability coefficient*. With this definition, if $K$ is measured, we can find $q_{STP}$ under general conditions as

$$q_{STP} = K \frac{\Delta p}{L} \qquad (2.1.59)$$

where $\Delta p$ is understood to be the partial pressure difference of the diffusing species (atm) and $L$ is the film thickness in centimeters. The units of $K$ are

$$K \, [=] \, \frac{\text{cm}^3_{STP} \cdot \text{cm}}{\text{cm}^2 \cdot \text{s} \cdot \text{atm}} \, [=] \, \frac{\text{cm}^2}{\text{s} \cdot \text{atm}} \qquad (2.1.60)$$

Through this development we see that the permeability coefficient $K$ is related to $D_{AB}$ and $b_p$ by

$$K = 2.24 \times 10^4 \, b_p D_{AB} \qquad (2.1.61)$$

**EXAMPLE 2.1.6**    *Calculation of $b_p$ from Solubility Data*

The solubility of a specific penetrant molecule in a polymeric membrane has been measured. The species has a molecular weight of 75 g/g·mol. The experimental data were obtained by taking a piece of film, 1 cm by 5 cm, of thickness 5 mils, free of this particular solute, and exposing it at 30°C to the pure vapor of this solute. The weight gain was measured until equilibrium (no further weight gain) occurred. At 30°C the vapor pressure of this solute is known to be 76 mmHg. The measured weight gain was $4.76 \times 10^{-3}$ g. The film weight, free of solute, was 0.0572 g. The film density, solute free, is 0.9 g/cm³. Find the solubility coefficient $b_p$ from these data.

At equilibrium, the vapor concentration in the film is

$$C_m = \frac{(4.76 \times 10^{-3})/75}{(5.72 \times 10^{-2})/0.9} = \frac{6.35 \times 10^{-5} \text{ g} \cdot \text{mol}}{6.35 \times 10^{-2} \text{ cm}^3} = 0.001 \text{ g} \cdot \text{mol/cm}^3 \qquad (2.1.62)$$

We will define the solubility coefficient in terms of the pressure, so

$$b_p = \frac{C_m}{p_{lo}} = \frac{0.001 \text{ g} \cdot \text{mol/cm}^3}{76 \text{ mmHg}/(760 \text{ mmHg/atm})} = 0.01 \frac{\text{g} \cdot \text{mol}}{\text{cm}^3 \text{ film} \cdot \text{atm partial pressure}}$$
(2.1.63)

This is a fundamental thermodynamic property of the polymer/solute pair, measured at 30°C.

---

**EXAMPLE 2.1.7**   *Calculation of $D_{AB}$ from Data on Solubility $(b_p)$ and Permeability $(K)$*

The permeability of a polymeric film to a specific solute (the polymer and solute from Example 2.1.6) has been measured, and is given as

$$K' = 10^{-6} \frac{\text{cm}^3_{STP} \cdot \text{cm}}{\text{cm}^2 \cdot \text{s} \cdot \text{cmHg}} \text{ at 30°C}$$
(2.1.64)

If we convert the pressure units from centimeters of mercury to atmospheres we find

$$K = 7.6 \times 10^{-5} \frac{\text{cm}^3_{STP} \cdot \text{cm}}{\text{cm}^2 \cdot \text{s} \cdot \text{atm}}$$
(2.1.65)

What is the value of the diffusion coefficient of the solute through the membrane at 30°C? From Eq. 2.1.61 we find

$$D_{A_m} = \frac{K}{2.24 \times 10^4 b_p} = \frac{7.6 \times 10^{-5}}{2.24 \times 10^4 \times 0.01} = 3.4 \times 10^{-7} \text{ cm}^2/\text{s}$$
(2.1.66)

---

**EXAMPLE 2.1.8**   *Effectiveness of a Vapor Barrier*

The solute described in the two preceding examples sits at the bottom of a glass tube, as sketched in Fig. 2.1.5. The system is maintained at 30°C, and the solute evaporates. We wish to retard evaporation by placing a polymeric barrier film across the end of the tube, as shown. The polymer of the preceding examples is used. If the film thickness is 10 mils (0.0254 cm), and the area is 0.2 cm², what time is required for 10 mg of the solute to pass through the barrier?

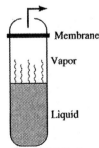

Membrane

Vapor

Liquid

**Figure 2.1.5** A polymeric membrane barrier to evaporation.

The molar flux through the membrane is given by

$$N_A = \frac{b_p D_{AB} p_{1o}}{L} = \frac{0.01[\text{g} \cdot \text{mol/cm}^3 \cdot \text{atm}] \times 3.4 \times 10^{-7}\,[\text{cm}^2/\text{s}] \times 0.1\,[\text{atm}]}{0.0254\,[\text{cm}]}$$
$$= 1.34 \times 10^{-8}\,[\text{g} \cdot \text{mol/s} \cdot \text{cm}^2] \tag{2.1.67}$$

From this, the *mass rate* follows as

$$M_w \times A \times N_A = 75 \times 0.2 \times 1.34 \times 10^{-8} = 2 \times 10^{-7}\,[\text{g/s}] \tag{2.1.68}$$

The time required for the passage of 10 mg = 0.01 g through this barrier is

$$t = \frac{0.01\text{ g}}{2 \times 10^{-7}\,[\text{g/s}]} = 5 \times 10^4\text{ s} \tag{2.1.69}$$

When the flux of some species is directly proportional to a concentration difference, and inversely proportional to the thickness of the region through which diffusion occurs, we speak of the relationship as a *linear transport law*. The important thing to note here is that the proportionality constant is the diffusion coefficient only if the concentration difference is that *within* the medium itself, rather than the concentration difference in the fluid phase *surrounding* the medium. As Example 2.1.9 demonstrates, this is important because systems that obey linear transport laws can give rise to behavior that appears to be *nonlinear*.

---

**EXAMPLE 2.1.9**   *Diffusion of a Diatomic Gas Through a Metal Barrier*

We consider the physical situation as discussed in Examples 2.1.5 through 2.1.8: diffusion through a planar barrier. Now, however, the soluble species that diffuses through the barrier is a diatomic gas, and the barrier is a *metal*. It is known that in such systems the equilibrium relationship is itself nonlinear, and we want to see how this affects the resultant transport model. Under such a nonlinear solubility relationship, Eq. 2.1.42 would be replaced by an expression of the form

$$C_{1i} = bC_{1o}^{1/2} \tag{2.1.70}$$

(This square-root dependence is commonly observed for diatomic gases in metals.) It is not difficult to show that Eq. 2.1.40 still holds for the concentration profile within the metal barrier, but that the imposition of Fick's law for the flux now gives an expression of the form

$$N_A = bD_{AB}\frac{(C_{1o}^{1/2} - C_{2o}^{1/2})}{L} = D_{AB}\left(\frac{C_{1i} - C_{2i}}{L}\right) \tag{2.1.71}$$

Depending on our choice of composition variables, the flux expression appears to be linear, or nonlinear. How do we resolve this apparent confusion?

For dilute systems, Fick's law is a linear rate law, but it is written in terms of the concentrations *in the medium through which diffusion occurs*. Hence Eq. 2.1.71, written in terms of the *interior* concentration difference that we might write as $\Delta C_i$, is a *linear* model. On the other hand, if we choose to write the flux in terms of the outside (i.e., the fluid) concentrations, we observe a *nonlinear* rate law. But it is not the rate law that is nonlinear; it is the *nonlinear equilibrium solubility relationship* that gives rise to this nonlinear *format* for the flux law.

The next obvious question would be: Why would one choose the outside concentrations in writing the rate expression, since their use might give rise to nonlinear expressions? The answer is that we do not normally *measure* concentrations within the barrier, but instead we measure the concentrations in the fluid external to the barrier. Hence data for the relationship of the observed flux to the *measured concentrations* would appear to be nonlinear. An examination of the equilibrium solubility relationship would be required before we knew whether the nonlinearity was in the rate (i.e., diffusive) behavior or in the thermodynamics of the solubility behavior.

## 2.2  ESTIMATION OF DIFFUSION COEFFICIENTS

To use the models of diffusive mass transfer that we will develop, we must be able to estimate values for diffusion coefficients. The diffusion coefficient, or *diffusivity* (as we sometimes call it) is a property of the molecular mixture. Hence we should expect the diffusivity to depend on molecular properties such as molecular weight, molecular size, and the intermolecular forces of attraction among the molecules in the mixture. We will find that a very well-developed theory for diffusion in gases exists, and it is possible to estimate diffusivities very well for *binary gas mixtures*. It is much more difficult to estimate diffusivities for liquid systems, partly because the intermolecular forces are much more difficult to estimate from molecular size, and so on. It is also quite difficult to predict diffusivities in solid systems from basic principles. Consequently, a number of *empirical methods* have been developed for the estimation of diffusivities.

We begin by noting that the dimensions of the diffusion coefficient are length squared divided by time. Most commonly, $D_{AB}$ is reported in cgs units as cm$^2$/s. The SI units are m$^2$/s. In older textbooks one often finds $D_{AB}$ reported in units of ft$^2$/h.

It is important to have in mind the order of magnitude of $D_{AB}$ for common systems of interest to us:

| | |
|---|---|
| Gas *in gas* | 0.1–1 cm$^2$/s = $10^{-5}$–$10^{-4}$ m$^2$/s at one atmosphere |
| Liquid or gas *in liquid* | $10^{-5}$ cm$^2$/s = $10^{-9}$ m$^2$/s |
| Gas in *solid* | $10^{-6}$–$10^{-10}$ cm$^2$/s = $10^{-10}$–$10^{-14}$ m$^2$/s |

While these order-of-magnitude values are useful for quick assessment of the role of diffusion phenomena, diffusivities in specific systems can vary by several orders of magnitude, especially in liquid and solid mixtures. In the following we will present some physical models, and empirical correlations, with which we can relate the diffusion coefficient of a particular species in some medium to the molecular and physical properties of the species and the medium.

### 2.2.1  Binary Diffusion in Gases at Low Pressure

The diffusion coefficient is an inverse measure of the resistance a molecule experiences as it moves through a medium under the influence of some "driving force." In a mixture of gases at low pressure (say, atmospheric pressure and lower), the kinetic theory of gases provides a model of the dynamics of interaction among the gas molecules. The kinetic theory is developed in most standard physical chemistry texts, and the details are not repeated here, but should be reviewed.

In the simplest model of gas dynamics, the molecules are regarded as rigid spheres that interact only upon collision. This implies that there are no intermolecular forces exerted by one molecule on another. Hence the motion of a molecule is uninfluenced by its neighbors *except upon collision*. It is further assumed, in the simplest model, that the collisions are *completely elastic*: that is, no energy is lost by collision. The result of this simple kinetic theory of *rigid noninteracting molecules* gives a diffusion coefficient in the form

$$D_{AB} = \frac{2}{3} \left( \frac{k_B^3}{\pi^3} \right)^{1/2} \left( \frac{1}{2 m_A} + \frac{1}{2 m_B} \right)^{1/2} \frac{T^{3/2}}{p[(d_A + d_B)/2]^2} \qquad (2.2.1)$$

The Boltzmann constant appears here, and has the value $k_B = 1.38 \times 10^{-16}$ erg/(molecule-K) in cgs units. The only characteristics of the molecules are their masses $m_A$ and $m_B$, and their diameters $d_A$ and $d_B$. The diffusivity is predicted to depend on absolute temperature $T$ (to a three-halves power) and on absolute pressure (in an inverse linear fashion). As simple as the kinetic theory model is, it yields quite good estimates of the diffusion coefficient. For greater accuracy, one usually uses a model that accounts for intermolecular forces and is based on an extension of the kinetic theory of gases. This leads to the so-called *Chapman–Enskog* formula for diffusivity, which we write in the form

$$D_{AB} = 1.86 \times 10^{-3} \, T^{3/2} \frac{[(M_{w,A} + M_{w,B})/M_{w,A}M_{w,B}]^{1/2}}{p \sigma_{AB}^2 \Omega_D} \text{ cm}^2/\text{s} \qquad (2.2.2)$$

With the numerical value of the coefficient given above, the pressure must be expressed in units of *atmospheres* and $T$ in kelvins. Notice that molecular weight $M_w$, not molecular mass $m$, appears here. Two new parameters appear in this model. One is $\sigma_{AB}$, a length that may be interpreted as a mean molecular diameter for the gas pair. The other is $\Omega_D$, the *diffusion collision integral,* which is a dimensionless measure of the interaction among the molecules in the system. The collision integral accounts for the fact that real molecules are not billiard balls. With the numerical value of the coefficient given here, $\sigma_{AB}$ must be in angstrom units (Å).

To use Eq. 2.2.2 it is necessary to have values of the $\sigma$ and $\Omega$ parameters for any gas pair of interest. The $\sigma$ values of individual molecules can be found tabulated, and we give some in Table 2.2.1. The $\Omega$ parameter, the collision integral, depends on the potential function selected to describe the intermolecular forces of attraction among the molecules in the gas. One usually works with the *Lennard-Jones 12-6* potential. As a consequence, the $\Omega$ parameter can be determined as a function of the Lennard-Jones energy parameter $\varepsilon$, which may also be found tabulated for various molecular species, usually as the ratio $\varepsilon/k_B$.

**Table 2.2.1** Parameters for the Chapman–Enskog Diffusion Model[a]

|  | Species | $\sigma$ (Å) | $\varepsilon/k_B$ (K) |
|---|---|---|---|
| Ar | Argon | 3.54 | 93.3 |
| He | Helium | 2.55 | 10.2 |
| Air | Air | 3.71 | 78.6 |
| $AsH_3$ | Arsine | 4.15 | 260 |
| $CCl_4$ | Carbon tetrachloride | 5.95 | 323 |
| $CH_4$ | Methane | 3.76 | 149 |
| $C_6H_6$ | Benzene | 5.35 | 412 |
| $H_2$ | Hydrogen | 2.83 | 59.7 |
| $H_2O$ | Water | 2.64 | 809 |
| $N_2$ | Nitrogen | 3.8 | 71.4 |
| $O_2$ | Oxygen | 3.47 | 107 |
| $SiH_4$ | Silane | 4.08 | 208 |
| $Si(OC_2H_5)_4$ | TEOS | 7.6 | 522 |
| $Ga(CH_3)_3$ | TMG | 5.52 | 378 |

[a] For a more extensive tabulation, see Reid, Prausnitz, and Poling, *The Properties of Gases and Liquids,* 4th ed., McGraw-Hill, New York, 1987.

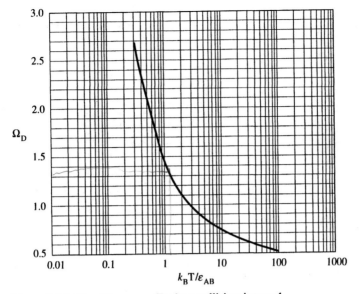

**Figure 2.2.1** The Chapman–Enskog collision integral.

For a binary mixture, we calculate $\sigma_{AB}$ as the *arithmetic* average of the individual $\sigma$s:

$$\sigma_{AB} = \frac{\sigma_A + \sigma_B}{2} \tag{2.2.3}$$

However, $\varepsilon_{AB}$ for the mixture is the *geometric* average:

$$\varepsilon_{AB} = (\varepsilon_A \varepsilon_B)^{1/2} \tag{2.2.4}$$

To find $\Omega_D$ at a given temperature, it is necessary to calculate $\varepsilon_{AB}$ for the pair and then calculate a value of $k_B T/\varepsilon_{AB}$. From Fig. 2.2.1 we then find a value of the collision integral $\Omega_D$.

We illustrate the use of the Chapman–Enskog theory in the following example.

**EXAMPLE 2.2.1**    *Diffusivity of $H_2$ in Argon*

We need an estimate of the diffusivity of hydrogen in argon at a pressure of 200 mtorr and a temperature of 527°C. The $\sigma$ value for the mixture is simply obtained from the individual values for each molecule, given in Table 2.2.1, and Eq. 2.2.3:

$$\sigma_{AB} = \frac{\sigma_A + \sigma_B}{2} = \frac{2.83 + 3.54}{2} = 3.19 \text{ Å} \tag{2.2.5}$$

Likewise, we find the mean $\varepsilon/k_B$ value as

$$\frac{\varepsilon_{AB}}{k_B} = \frac{(\varepsilon_A \varepsilon_B)^{1/2}}{k_B} = 74.6 \text{ K} \tag{2.2.6}$$

At a temperature of 527°C or 800 K, we calculate $k_B T/\varepsilon_{AB} = 10.7$. From Fig. 2.2.1 we find $\Omega_d = 0.74$. We now use Eq. 2.2.2, noting molecular weights of 2 and 18 for hydrogen and argon, and converting the pressure given here in units of millitorr to units of atmospheres [200 mtorr/1000) $\times$ (1/760) = $2.63 \times 10^{-4}$ atm]. The result is

$$D_{AB} = 1.58 \times 10^4 \text{ cm}^2/\text{s} \tag{2.2.7}$$

This is a very large diffusion coefficient! But keep in mind that the calculation is made for a very low pressure and that the diffusion coefficient is an inverse function of pressure. In addition, hydrogen and argon are small molecules—another factor that yields larger diffusion coefficients.

## 2.2.2 Diffusion in Liquids

In general the theory of the liquid state is less well developed than that of the gas state, and as a consequence we do not have diffusion models, such as the Chapman–Enskog theory, that permit accurate estimation of diffusion coefficients strictly from well-defined molecular parameters. A very idealized model is based on the hydrodynamic idea that the diffusivity of a molecule through a liquid can be regarded as a measure of the *viscous* resistance to its motion through the liquid, as if the liquid were a continuum. The resultant model is the *Stokes–Einstein equation,* which is written in the form

$$D_{AB} = \frac{k_B T}{6\pi\mu r_d} \tag{2.2.8}$$

where $k_B$ is the Boltzmann constant, $\mu$ is the liquid viscosity, and $r_d$ is the radius of the diffusing molecule. This *hydrodynamic* theory is not quantitatively accurate, but it does indicate with reasonable accuracy the functional dependence of liquid diffusivities on temperature, viscosity, and molecular size. Ordinarily one would not use Eq. 2.2.8 for molecules of molecular weight less than about 1000.

A reasonably accurate model for liquid diffusivities that requires minimal material property data is due to Wilke and Chang. It takes the form

$$D_{AB} = 7.4 \times 10^{-8} \frac{(\phi M_{w,B})^{1/2} T}{\mu_B V_A^{0.6}} \text{ cm}^2/\text{s} \tag{2.2.9}$$

where $M_{w,B}$ and $\mu_B$ are the molecular weight and viscosity (in *centipoise* units) of the *solvent,* and $T$ is, as usual, the absolute temperature. This model includes an important

Table 2.2.2 Molal Volumes at the Normal Boiling Point for Use in the Wilke–Chang Equation

|  | Molecule | $V_A(\text{cm}^3/\text{g} \cdot \text{mol})$ |
|---|---|---|
| $H_2$ | Hydrogen | 14.3 |
| $O_2$ | Oxygen | 25.6 |
| $CO$ | Carbon monoxide | 30.7 |
| $CO_2$ | Carbon dioxide | 34 |
| $NH_3$ | Ammonia | 25.8 |
| $H_2S$ | Hydrogen sulfide | 32.9 |
| $SO_2$ | Sulfur dioxide | 44 |
| $CH_4$ | Methane | 37.7 |
| $C_6H_6$ | Benzene | 96.5 |
| $C_6H_{14}$ | Hexane | 141 |
| $C_7H_8$ | Toluene | 118 |
| $C_3H_6O$ | Acetone | 74 |
| $C_8H_{10}$ | m-Xylene | 140 |
| $C_{10}H_8$ | Naphthalene | 148 |
| $C_9H_{12}$ | 1,3,5-Trimethylbenzene | 163 |

**Table 2.2.3** Atomic Volumes for Use in Estimating the Molal Volume at the Normal Boiling Point

| | Atom | $v_a$ (cm$^3$/g·mol) |
|---|---|---|
| C | Carbon | 14.8 |
| H | Hydrogen | 3.7 |
| O, except for: | Oxygen | 7.4 |
| C—O—C | as in esters and ethers$^a$ | 11.0 |
| O—H | In acids | 12.0 |
| O—N | | 8.3 |
| Cl | Chlorine | 24.6 |
| S | Sulfur | 25.6 |
| N | in NH | 12.0 |
| N | in NH$_2$ | 10.5 |

In ring compounds, the addition of the contributions of the individual atoms overestimates the volume, so an amount is *subtracted* from the sum to account for this. For example:

| | | |
|---|---|---|
| C$_6$ | Benzene ring, as part of a larger compound | −15.0 |

$^a$ But in methyl esters and ethers, use 9.1, and in ethyl esters and ethers, use 9.9.

*thermodynamic* parameter related to the size of the diffusing solute, $V_A$, which is the *molal volume* of solute at its normal boiling temperature, in units of cubic centimeters per gram-mole. The parameter $\phi$ is a dimensionless *solvent association factor*. For water the recommended value of $\phi$ is 2.6. For unassociated (i.e., non-hydrogen-bonded) solvents such as organic liquids, one uses $\phi = 1$. The values 1.9 and 1.5 are recommended for methanol and ethanol. Table 2.2.2 gives a few values of molal volumes of simple compounds.

For compounds for which no measurements are available, the molal volume at the normal boiling point can be estimated by the additive method of Le Bas, which assigns certain numerical values of an "atomic volume" to atoms in a molecule. The atomic volumes, weighted by the number of such atoms in the molecule, are then added. Sometimes, the atomic volume depends on the binding of that atom in the molecule. For example, the atomic volume of oxygen will depend on whether it is bonded to a hydrogen atom as in the OH group of an acid, or joins two carbons as in the COC group of an ether or an ester. Table 2.2.3 gives some values for use in estimating the molal volume at the normal boiling point. See Reid, Prausnitz, and Poling for additional values and discussion.

We emphasize that the Wilke–Chang equation is valid only for highly *dilute* solutes. Methods for the prediction of diffusion coefficients in nondilute solutions are available, but no single method has proven to be accurate for a wide range of solvent-solute pairs. An extensive discussion of some of these issues may be found in Reid, Prausnitz, and Poling, *The Properties of Gases and Liquids.*

**EXAMPLE 2.2.2** *Estimate the Diffusivity of Benzoic Acid in Water at 25°C*

Benzoic acid (C$_7$H$_6$O$_2$) consists of a benzene ring to which is attached an acid group (COOH). Using the additive method of Le Bas and Table 2.2.3 we find

$$V_A = 7(14.8) + 6(3.7) + 7.4 + 12 - 15 = 130.2 \ (\text{cm}^3/\text{g·mol}) \qquad \textbf{(2.2.10)}$$

**Table 2.2.4** Limiting Ionic Conductances in Water at 25°C

| Cation | $\lambda_+^\circ$ | Anion | $\lambda_-^\circ$ |
|--------|-------|-------|-------|
| $H^+$ | 350 | $OH^-$ | 198 |
| $Na^+$ | 50 | $Cl^-$ | 76.3 |
| $K^+$ | 73.5 | $NO_2^-$ | 71.4 |
| $NH_4^+$ | 73.4 | $HCO_2^-$ | 44.5 |

Notice that we add the atomic volumes from the table, weighted by the number of atoms of each kind in the compound, and then subtract 15 from the sum to account for the benzene ring.

For the solvent (water) we use $\phi = 2.6$ and $M_w = 18$. The solvent viscosity at 25°C is 0.89 cP. From the Wilke–Chang equation we find

$$D_{AB} = 7.4 \times 10^{-8} \frac{(2.6 \times 18)^{1/2} \, 298}{0.89(130.2)^{0.6}} = 0.91 \times 10^{-5} \, cm^2/s \qquad (2.2.11)$$

The measured value is reported as $1.21 \times 10^{-5} \, cm^2/s$. This is a larger error than usual, as discussed in Reid, Prausnitz, and Poling.

An important case of interest, the diffusion of *ionic* species in water, arises commonly with dissolved salts. For dilute solutions the Nernst–Haskell equation is recommended, which has the form

$$D_{AB} = \frac{R_G T}{\mathbf{F}^2} \frac{1/n_+ + 1/n_-}{1/\lambda_+^\circ + 1/\lambda_-^\circ} \, cm^2/s \qquad (2.2.12)$$

where $R_G$ is the gas constant (8.314 J/g·mol K), and $\mathbf{F}$ is the Faraday constant (96,500 C/g-equiv). (Remember—a coulomb is an ampere-second.) The $n$'s are the valences of the cation ($+$) and anion ($-$), and the $\lambda$s are the limiting ionic conductances [with cgs units of $(A/cm^2)$ $(V/cm)$ $(g\text{-equiv}/cm^3)$]. Table 2.2.4 gives some values for these coefficients in water at 25°C.

## 2.3 APPLICATION OF THESE PRINCIPLES TO ANALYSIS AND DESIGN OF A SUSTAINED-RELEASE HOLLOW FIBER SYSTEM

In Chapter 1 we introduced the notion of designing a system for control of insect pests by means of a chemical agent encapsulated in a hollow fiber membrane. The fiber is simply a permeable tube that permits us to control the rate of release of the agent. An important design parameter is the diffusion coefficient of the active agent across the membrane wall. Since it is very unlikely that fundamental parameters of the active agent have been measured and tabulated, we must resort to experimental methods for the measurement of the diffusion coefficient. Let's see how we can use the information put forth in this chapter to facilitate this measurement.

We should begin with a clear statement of our goal.

*It is to develop an experimental protocol that permits measurement of the diffusion coefficient of the active chemical agent across any candidate membrane from which we can fabricate hollow fibers.*

There are many ways to achieve this goal. This is one of the characteristics of engineering practice. What we must learn to do is to propose several possible solutions to our problem, and then evaluate these solutions with respect to criteria such as economics, environmental impact, and safety, which are outside of the scope of a course in mass transfer fundamentals, *per se*.

Here is one way to achieve the goal just stated. With reference to Fig. 2.3.1, we will take a candidate hollow fiber and pump a solution of the active chemical agent through it. We will use the same solvent as that proposed in the static, encapsulated formulation of this device.

Our first task is to develop a mathematical model of the process shown in Fig. 2.3.1. This means that we must understand the physics taking place, and the principles with which we describe the physics in quantitative language. Thus we must combine our knowledge with our expectations (which may be intuitive, but incorrect to some degree), and if there are any remaining "holes" in the description of the physics, we must fill them with our assumptions about the behavior of the system. If we do not learn how to make intelligent assumptions, we will fall into the holes and be buried by invalid assumptions. The result will be a model that reflects the real physics so poorly that we will not be able to achieve our stated goal.

We should understand enough about diffusion at this point to realize that we have at our disposal a model that relates the rate at which a chemical species diffuses (across a barrier such as a membrane) to the concentration gradient across that barrier. Of course, we are talking about Fick's law, in the form, for example of Eq. 2.1.19. If we assume that the diffusion is strictly in the radial direction across the membrane, we may write Eq. 2.1.19 in the form

$$N_{A,r} = -\frac{CD_{A,m}}{1 - x_A}\frac{dx_A}{dr} \tag{2.3.1}$$

We have also assumed that the membrane material itself is fixed and does not diffuse, so that we set $N_{B,r} = 0$. We will assume that the active chemical agent is only slightly soluble in the membrane, which permits us to neglect $x_A$ in the denominator of Eq. 2.3.1. Then we may write this as

$$N_{A,r} = -CD_{A,m}\frac{dx_A}{dr} \tag{2.3.2}$$

We are going to assume that we can design the experiment in such a way that the concentration of the active species within the solution flowing along the fiber axis is independent of axial position.

> *We are free to assume anything that we wish. At issue is whether this statement about concentration is a good assumption—Does it reflect observable physics?*

As we often do when we make assumptions, we will put off an assessment of the validity of the assumption until we have evaluated the resulting model. We will find that we

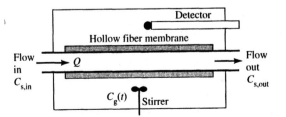

**Figure 2.3.1** System for determining membrane diffusivity.

can then often define a criterion, based on the operating parameters of the experiment, that tells us the conditions under which the assumption will be reasonable. Then we can examine the possibility of actually operating under the conditions specified by this criterion.

We must now think about what is going on *outside* the hollow fiber. According to our sketch in Fig. 2.3.1, the active chemical species will first dissolve in and then diffuse across the membrane, finally evaporating into the surrounding gas volume V, which we take to be well mixed. By this we mean that the concentration of the active species is uniform throughout the external volume V. The concentration of this species will build up in the surrounding volume as more and more of the species crosses the membrane. If this is the case, we could eventually saturate the volume V, in the sense that no further evaporation of the species could occur from the membrane surface. This idea suggests, in turn, that the rate of mass transfer eventually will decrease as we approach saturation. In particular, if we permit (or better, cause) the system to operate in this way, the system does not operate at steady state conditions. Our analysis would be easier if we could ensure steady state behavior, but clearly that is not possible if the volume V is closed, since the active agent must build up in concentration in a closed volume. If the external volume is large enough, however, the species will be so diluted in the surrounding gas space that even if its concentration is increasing, it is still a very small concentration. In other words, we might assume that the concentration $C_g(t)$ is large enough to be detected by a sensitive analytical instrument, but small enough (compared to what?) that we could assume that the system is very far from saturation.

If we return to the mathematical formulation, these ideas can be clarified now. Equation 2.3.2 gives the active species flux, normal to any cylindrical surface of radius r within the membrane—that is, for $R_{in} < r < R_{out}$, where the subscripts *in* and *out* refer to the inside and outside radii of the hollow fiber membrane. Since at steady state the species is not accumulating within the membrane region, the *rate* of diffusion across any surface at r must be independent of r. Note that it is not the flux, but the *rate*, that is independent of r. The mathematical formulation of this statement is just

$$2\pi r L N_{A,r} = -2\pi r L C D_{A,m} \frac{dx_A}{dr} = \text{constant} \qquad (2.3.3)$$

where L is the axial length of the fiber. More simply,

$$r\frac{dx_A}{dr} = \text{constant} = A \qquad (2.3.4)$$

We may now solve this first-order ordinary differential (ODE) equation and find

$$x_A = A \ln r + B \qquad (2.3.5)$$

where A and B are constants of integration that must be specified through boundary conditions on the system. Now we return to the physics of the system. What do we state about the mole fraction of species A on the boundaries of the membrane? We know, or have at least assumed, that the concentration of the active species in the solution flowing down the inside of the hollow fiber is uniform. Hence the concentration at the inner surface is $C_{s,in}$. But this is the concentration in the *solution*—not in the membrane! What is the connection between the two? Aren't the concentrations *continuous* across the boundary between the liquid and the solid (membrane)?

The answer is no. It is not the concentrations that are continuous, but the free energies. The active species may have a very different solubility in the membrane, with respect to its concentration in the solution. What we do know, from thermodynamics, is that if the solution were in equilibrium with the membrane, the solute (the active chemical species in this case) would partition itself between the two phases, and the concentrations in the two phases would be related by an equilibrium expression. This is often observed to be a linear relationship of the form

$$C_{m,A} \equiv C x_A = \alpha_{m/s} C_s \qquad \text{at } R_{in} \qquad (2.3.6)$$

We call $\alpha_{m/s}$ the solubility or partition coefficient, or simply the solubility. It is a thermodynamic property. Equation 2.3.6 is the equilibrium relationship at the solution/membrane interface. What is the situation at the membrane/gas interface? The usual observation is that there is also a linear relationship between the concentrations in the membrane and gas phases, but with a different solubility coefficient. Hence we expect to find

$$C_{m,A} = \alpha_{m/g} C_g \qquad \text{at } R_{out} \qquad (2.3.7)$$

where $\alpha_{m/g}$ is the equilibrium coefficient for this case. (This is more commonly written in the form of Henry's law,

$$p_{g,A} = H'_A C_{m,A} \qquad (2.3.8)$$

in terms of the *partial pressure* of the active species in the gas that is in equilibrium with the membrane, where $H'$ is the Henry's law constant).[1]

We now make the further assumption that these equilibrium relationships hold even when the system is not at equilibrium. In other words, we assume that the concentrations of some species on the two sides of a given interface satisfy an equilibrium relationship even when mass is being transferred (i.e., when the system is *not* in equilibrium). We make this assumption for two very good reasons:

1. We cannot think of an alternative.
2. It leads to results that are in close correspondence with observation.

Even if we feel some uncertainty about one or more assumptions that have been introduced into the mathematical model, we should move ahead. At least we will obtain a *testable* model, which is better than no model at all.

We now introduce boundary conditions based on these ideas. They take the forms

$$\text{at } r = R_{in} \qquad x_A = \frac{\alpha_{m/s}}{C} C_s \qquad (2.3.9)$$

and

$$\text{at } r = R_{out} \qquad x_A = \frac{\alpha_{m/g} C_g}{C} \qquad (2.3.10)$$

But we earlier remarked that we will hope to get away with the approximation that $C_g$ is vanishingly small with respect to the equilibrium or saturation value. Hence we write Eq. 2.3.10 in the form

$$\text{at } r = R_{out} \qquad x_A = 0 \qquad (2.3.11)$$

---

[1] Often we find Henry's law in the form

$$p_{g,A} = H_A x_{m,A} \qquad (2.3.8')$$

in terms of the mole fraction $x_{m,A}$ of A in the membrane. Mass fraction units are also in common use, so one must be vigilant in looking at solubility data.

The two boundary conditions of Eqs. 2.3.9 and 2.3.10 permit us to find the constants of integration $A$ and $B$, with the result that the concentration profile of the active species in the membrane is

$$\frac{x_A}{(\alpha_{m/s}C_s/C)} = \frac{\ln(r/R_{out})}{\ln \kappa} \qquad (2.3.12)$$

where

$$\kappa \equiv R_{in}/R_{out} \qquad (2.3.13)$$

Now we must use this result to obtain a model that relates what we *want* to measure—the diffusion coefficient of the active species across the membrane—to what we *can* measure—the concentration $C_g$ of the active species in the gas volume $V$ outside the membrane. To do this we go back to Eq. 2.3.3 and find the rate at which moles of the active species cross the outer surface of the hollow fiber membrane. The result is simply a statement that the rate at which the number of moles of the volatile species increases in the volume surrounding the membrane fiber is equal to the rate of transport of the species across the membrane:

$$\frac{\text{moles}}{\text{time}} = \frac{d}{dt}(C_g V) = -\frac{2\pi L D_{A,m}\alpha_{m/s}C_s}{\ln \kappa} \qquad (2.3.14)$$

(This result is obtained by performing the indicated differentiation in Eq. 2.3.3, using Eq. 2.3.12, and evaluating the derivative at $r = R_{out}$). Based on the assumptions we have made, everything in this equation is constant except the measured gas concentration $C_g$. Hence if we measure the concentration $C_g$ as a function of time, we can write the following model:

$$\frac{dC_g}{dt} = -\frac{2\pi L D_{A,m}\alpha_{m/s}C_s}{V \ln \kappa} \qquad (2.3.15)$$

This result suggests that by taking data on $C_g$ as a function of time and then taking the slope of the data, we will have the right-hand side of Eq. 2.3.15. Everything is known on the right-hand side except the product $\alpha_{m/s}D_{A,m}$. Thus we actually get this product, rather than the diffusion coefficient itself. Since this is a material property of the membrane and the active species, it will suffice to have this property as the product of the solubility and the diffusivity.

## 2.3.1   Assessment of Some Assumptions of the Model

We have illustrated the application of some simple principles of mass transfer by diffusion to the design of an experiment for obtaining an important characteristic parameter of the membrane. Once the model has been obtained, we can assess some of the assumptions that were part of the modeling procedure. One assumption was that the concentration $C_s$ was uniform along the fiber axis. Let's establish a criterion for the validity of this assumption.

Equation 2.3.14 gives the rate at which moles cross the hollow fiber boundary. Since these moles come out of the solution traveling down the fiber axis, we should be able to write a mole balance on the liquid stream in the form

$$Q(C_{s,in} - C_{s,out}) = -\frac{2\pi L D_{A,m}\alpha_{m/s}C_s}{\ln \kappa} \qquad (2.3.16)$$

where $Q$ is the volume flowrate of solution through (i.e., *inside*) the hollow fiber. We have assumed $C_s$ constant along the axis. Equivalent to this assumption, a suitable *definition* of a negligible change in $C_s$ is

$$\frac{C_{s,in} - C_{s,out}}{C_{s,in}} = -\frac{2\pi L D_{A,m}\alpha_{m/s}}{Q \ln \kappa} \ll 1 \qquad (2.3.17)$$

Upon obtaining an estimate of $\alpha_{m/s}D_{A,m}$, we can evaluate this criterion and decide whether the experimental conditions are consistent with the use of this model. A suitable choice of parameters $Q$ and $L$ should permit us to design an appropriate experimental procedure. If, however, $\alpha_{m/s}D_{A,m}$ is so large that no practical choice is possible, a more complicated mathematical model will have to be developed.

A second assumption of this model was a very small gas phase concentration. We used this assumption in setting the concentration at the outer radius equal to zero in the boundary condition of Eq. 2.3.11. If we permit the experiment to go on for a long time, eventually this assumption will cease to be valid. Rather, we would observe a gradual decrease in the slope $dC_g/dt$ from its initial value, given by the right-hand side of Eq. 2.3.15. Hence we would not use the data on $C_g(t)$ beyond the region where $C_g(t)$ is linear in time. We could use the model developed here to estimate the time at which nonlinearity would appear, if we have estimates of the product $\alpha_{m/s}D_{A,m}$ and the additional coefficient $H'$ of Eq. 2.3.8, or $\alpha_{m/s}$ of Eq. 2.3.10.

Now that we have a method for evaluating the fundamental thermodynamic and transport properties of this system, we could go on and develop a model to aid in the design of a system that meets certain criteria in terms of the rate of sustained release of the active agent. We will return to this design problem, but not until Chapter 4, where we establish some additional background in mass transfer.

## 2.4  THE QUASI-STEADY APPROXIMATION

Consider the following diffusion problem. A long, thin-walled tube contains a binary gas mixture. Species A is not soluble in the wall of the tube, and so there is no mechanism for A to leave the system. The amount of A within the tube is constant. Species B is soluble in the wall of the tube and diffuses across the wall to the exterior medium. The exterior medium is so large, and well mixed, that $C_B$ is essentially zero outside the tube. The amount of species B inside the tube changes with time. How would we model the depletion of species B from the interior of tube?

Since this is an unsteady-state problem, we should begin by writing appropriate equations that describe the time dependence of concentrations within the tube interior, as well as within the wall of the tube. A little thought indicates that the concentration of species B within the tube wall $C_{w,B}$ is a function of time as well as radial position. With reference to Fig. 2.4.1, we can sketch the concentration profiles of species B over time, as the concentration of B falls in the tube interior. With the assumption that $C_{i,B}$ within the *interior* of the tube ($0 < r < R_i$) is related to the concentration at the inner surface of the tube wall $C_{w,B}$ by a boundary condition of the form of Eq. 2.3.9, but written now in the form

$$\text{at } r = R_i \qquad C_{w,B} = \alpha C_{i,B} \qquad (2.4.1)$$

we see that as B is depleted from the interior of the tube, the boundary condition above changes with time. As a consequence of these comments, we conclude that we cannot use the analysis of Section 2.3, in which the interior concentration $C_s$ was assumed to be maintained at a *constant* level, subject to the restriction that for this to be a good approximation, Eq. 2.3.17 had to be satisfied. At successive times, the interior concentration $C_{i,B}$ falls and the concentration profile in the tube wall changes, always satisfying Eq. 2.4.1.

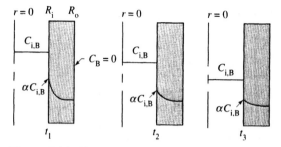

**Figure 2.4.1** Time-dependent concentration profiles in the wall of a tube. Note that $C_{i,B}$ decreases with time.

It turns out that the unsteady-state analysis of this type of problem is very complicated mathematically. An alternative, an approximate approach that we call a "quasi-steady" analysis, begins with the following assumption:

*We assume that although the concentration in the interior of the tube is changing with time, we can still use the steady state model for the concentration profile in the wall of the tube,* with the modification that the concentration at the interior surface of the tube is written as a function of time.

Hence we write Eq. 2.3.12, but in the current notation of this problem:

$$C_{w,B}(r,t) = \left[ \frac{\ln(r/R_o)}{\ln \kappa} \right] \alpha C_{i,B}(t) \tag{2.4.2}$$

The concentration within the tube wall $C_{w,B}$ is written in a notation that expresses the expectation that it is a function of both radial position $r$ and time $t$.

We have written the interior concentration $C_{i,B}$ as if it were a function only of time. Wouldn't $C_{i,B}$ also vary across the radius in the interior region: $(0 < r < R_i)$? It could; but the diffusion coefficient of B in the gas phase would be very high compared to that in the tube wall, and so we could expect that for small-diameter tubes, diffusion in the tube interior would be very efficient in keeping the interior concentration of B very uniform spatially. We will make that assumption. In later chapters we will learn how to evaluate it quantitatively.

The next part of the quasi-steady analysis is the *unsteady* part. We write a species balance on B, stating that the rate at which B leaves the exterior of the tube must equal the rate at which it disappears from the interior.

The rate at which species B leaves the exterior of the tube is just the flux at the outer surface (i.e., at $r = R_o$) times the surface area over an axial length $L$. Application of Fick's law to Eq. 2.4.2 yields

$$\left( -D_{B,m} \frac{\partial C_{w,B}(r,t)}{\partial r} \right)_{R_o} 2\pi R_o L = -\alpha D_{B,m} C_{i,B}(t) \frac{2\pi R_o L}{R_o \ln \kappa} = -\frac{\partial(\pi R_i^2 L C_{i,B})}{\partial t} \tag{2.4.3}$$

The term on the right-hand side is simply the time rate of change of the amount of species B in the tube interior. The minus sign on the right reflects the fact that if the flux is positive, the time derivative term must be negative.

Now we may rearrange Eq. 2.4.3 into the form of an easily solved differential equation:

$$\frac{\partial C_{i,B}}{\partial t} = \frac{2\alpha D_{B,m}}{R_i^2 \ln \kappa} C_{i,B}(t) \tag{2.4.4}$$

An appropriate initial condition is

$$\text{at } t = 0 \qquad C_{i,B} = C_{i,B}^{\circ} \tag{2.4.5}$$

The solution for the interior concentration then follows and is

$$C_{i,B} = C_{i,B}^{\circ} \exp\left[\left(\frac{2\alpha D_{B,m}}{R_i^2 \ln \kappa}\right) t\right] \tag{2.4.6}$$

Note that since $\ln \kappa < 0$, the concentration decreases with time, as expected.

Equation 2.4.6 is our quasi-steady model. It combines a steady state analysis of the concentration profile inside the tube wall with an unsteady mass balance on the interior of the tube. The connection is through the boundary condition at the interior surface. The justification of this approach awaits our later study of unsteady-state diffusion.

## SUMMARY

In a binary system the diffusive flux of some species is described by Fick's law—Eq. 2.1.10. In the form given, this is the molar flux relative to the molar average velocity of the binary mixture. Even in a mixture that is static in the sense that we do not induce any flow in it through external forces, diffusion itself can create a flow. If we want to write the diffusive flux relative to *fixed* coordinates, as we usually do in engineering analysis, we must use Fick's Law in the form of Eq. 2.1.17. For multicomponent systems we make the assumption that the diffusive flux can be approximated by a relationship of the *form* of Fick's law—Eq. 2.1.38—which uses an *effective* diffusivity.

Methods exist for the estimation of the diffusion coefficients of various species in gases and liquids. For the gas phase, the *Chapman–Enskog* formula is recommended (Eq. 2.2.2). For the liquid phase, the Wilke–Chang equation (Eq. 2.2.9) is most often used.

With these basic tools, we are able to illustrate, in Sections 2.3 and 2.4, methods by which we can model several simple mass transfer situations of interest to engineers.

## PROBLEMS

**2.1** A long glass capillary tube, of diameter 0.01 cm, is in contact with water at one end and dry air at the other. Water vapor evaporates at the wet end within the capillary, and the vapor diffuses through the capillary toward the dry end. How long is required for one gram of water to evaporate through this sytem? The vapor pressure of water is 17.5 mmHg at 20°C, the temperature at which the entire system is maintained. Take the diffusivity of water vapor in air at 20°C to be 0.3 cm²/s, and assume that the dry air is at a pressure of 760 mmHg. Assume that the distance from the wet interface within the capillary to the dry end is always 10 cm.

**2.2** Give the derivation of Eq. 2.1.21. State clearly the boundary conditions imposed that lead to that solution. If species A was so dilute that $x_A$ was negligible compared to unity, what form would Eq. 2.1.20 take, and what would the solution be? Does Eq. 2.1.21 reduce to this solution in the limit as $x_A$ becomes very small? Show this.

**2.3** Go back to Example 2.1.4. Assume that even though the system is ternary in the gas phase, the diffusion of oxygen may be treated by using a *binary* diffusion model (i.e., Fick's law). Let $D_O$ be the diffusivity of oxygen in the gas mixture. Derive a differential equation for the concentration profile of oxygen in the direction normal to the surface. Assume as one boundary condition that the concentration of oxygen is constant outside a thin layer of gas of thickness $\delta$ adjacent to the surface. We want the concentration distribution in the region $[0, \delta]$.

**2.4** Estimate the binary diffusion coefficient of methane ($CH_4$) in nitrogen ($N_2$) at 1000 K and $10^{-5}$ atm.

**2.5** Calculate the diffusivity of water vapor in air at 25°C and one atmosphere pressure.

**2.6** An organometallic compound is known to have a diffusivity in nitrogen of 0.25 $cm^2$/s at 600 K and atmospheric pressure. Estimate its diffusivity in nitrogen at 950 K and a pressure of 2 atm, absolute.

**2.7** Begin with Eq. 2.1.20 and derive Eq. 2.1.21. Show that when $x_A \ll 1$, Eq. 2.1.20 leads to an approximate flux of the form

$$N_{A,z} = \frac{CD_{AB}(x_{A1} - x_{A2})}{L} \quad \text{(P2.7.1)}$$

Show that Eq. 2.1.20 leads to an exact flux expression of the form

$$N_{A,z} =$$

$$\frac{CD_{AB}(x_{A1} - x_{A2})}{L} \left[ \frac{\ln\left[(1 - x_{A2})/(1 - x_{A1})\right]}{x_{A1} - x_{A2}} \right]$$
$$\text{(P2.7.2)}$$

When the approximation $x_A \ll 1$ is made, is the approximate flux (Eq. P2.7.1) less than or greater than the exact flux? For the case $x_{A1} = 0.1$ and $x_{A2} = 0$, what percentage error is made in using Eq. P2.7.1 as an approximation to the flux?

**2.8** Predict the diffusion coefficient of naphthalene ($C_{10}H_8$) in ethanol at 52°C, using the additive method of Le Bas for the atomic volume, and the equation of Wilke–Change. The viscosity of ethanol at 52°C is 0.68 cP.

**2.9** Predict the diffusion coefficient of behenic acid [$CH_3(CH_2)_{20}COOH$] in ethanol at 52°C, using the additive method of Le Bas for the atomic volume, and the equation of Wilke–Chang. The viscosity of ethanol at 52°C is 0.68 cP.

**2.10** A process is carried out under *ultrahigh vacuum* in a reactor whose walls are a glass tube of inside diameter 10 cm and wall thickness 1 cm. We do not want $H_2$ to "leak" across the tube wall by diffusion and thereby reduce the quality of the vacuum environment for the process. The diffusivity of $H_2$ in glass at the process temperature, 700 K, is $10^{-8}$ $cm^2$/s. The molar solubility ratio of $H_2$ in glass is 0.2 ($mol/cm^3$) $H_2$ in glass/($mol/cm^3$) $H_2$ in air. Suppose the tube is one meter long, and the end plates and fittings of the tube are impermeable to hydrogen. Sup-

pose further that at some instant of time the vacuum pump had brought the system to a pressure of $10^{-6}$ torr. The glass tube is surrounded by atmospheric air, and the partial pressure of hydrogen in air is $3.8 \times 10^{-4}$ torr. Present a plot of the pressure within the process tube as a function of time, over the interval [0, 1] h. Use a *quasi-steady approximation,* assuming that the steady state concentration profile has been established instantly within the wall of the glass tube.

**2.11** Data for permeation of a gas species through a solid are often presented in terms of a so-called *permeation constant K*, which is defined as the product of the solubility constant and the diffusivity: $K = bD_{AB}$. It is commonly expressed as the amount of gas [$cm^3$/s at standard temperature and pressure (STP): 273 K and 1 atm] permeating through a 1 $cm^2$ cross section of solid of 1 cm thickness, for a partial pressure difference of 1 atm. In a textbook on vacuum technology we find a value for $K$ for hydrogen through Pyrex glass at 1000 K given as $1.2 \times 10^{-8}$ $cm^2$/s · atm.

If the diffusion coefficient measured under these conditions is known to be $2 \times 10^{-8}$ $cm^2$/s, find the solubility of hydrogen in glass, in *mass fraction units* (i.e., grams of $H_2$ per gram of glass per atmosphere partial pressure). Take the density of the glass to be 2.4 $g/cm^3$.

**2.12** The permeation constant $K$, defined as in Problem 2.11, has been measured for hydrogen through a special glass. The value given at 500 K is $10^{-8}$ $cm^2$/s · atm. Suppose a glass vessel is pumped down to a pressure of $10^{-6}$ torr. The vessel is surrounded by air at 1 atm total pressure. How long will the pressure inside the vessel remain within 10% of the initial pumpdown value? Take the vessel to be a sphere of diameter 20 cm, with a wall thickness of 0.5 cm. The temperature is 500 K. Assume that only hydrogen permeates the glass wall.

**2.13** In the analysis of Section 2.3 we assumed that the partial pressure of the volatile species in the volume $V$ surrounding the hollow fiber membrane was negligible compared to the equilibrium value. Let's relax that assumption. Then we will use Eq. 2.3.10 as the second boundary condition on Eq. 2.3.5.

**a.** Find the solution for $x_A$ that replaces Eq. 2.3.12 in this case.

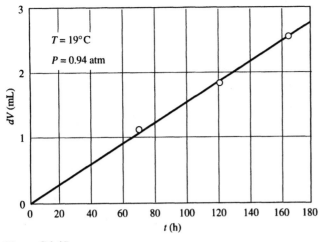

**Figure P2.15**

**b.** Find and solve the equation that replaces Eq. 2.3.15.

**c.** Define a time scale for the achievement of equilibrium in the volume $V$ outside the hollow fiber membrane, and give an expression for the time scale.

**d.** Show in a sketch how you could use the measured $C_g(t)$ data to determine a value for the parameter $\alpha_g D_{A,m}$ of the model.

**2.14** In the design of an experiment along the lines of the discussion of Section 2.3, we expect to have to pump a liquid with a viscosity of 0.001 Pa·s through a hollow fiber of inside diameter 100 $\mu$m. Suppose the fiber length is 10 cm, and the outside diameter is 200 $\mu$m. We expect that the $\alpha D_{A,m}$ product has a value of the order of $10^{-7}$ cm²/s. Using Eq. 2.3.17 as a criterion for selection of a flowrate, and choosing

$$\frac{C_{s,in} - C_{s,out}}{C_{s,in}} < 0.01 \qquad \text{(P2.14)}$$

estimate the expected pressure drop required to convey this liquid through the inside of the hollow fiber.

**2.15** Liquid hexane is placed in a vertical glass tube of inside diameter 0.95 cm. The tube is closed at the bottom and open, at the top, to air flowing at a uniform pressure of 0.94 atm. The tube assembly is immersed in a constant temperature bath maintained at 19°C. The hexane evaporates and diffuses through the air column in the tube. The loss in volume of liquid hexane is recorded as a function of time, and a set of such

data is shown in Fig. P2.15. The initial length of air column above the hexane surface is 15.6 cm. Calculate the diffusion coefficient of hexane through air from these data. The vapor pressure of hexane at 19°C is $1.5 \times 10^4$ Pa, and the density of the liquid is 0.66 g/cm³.

**2.16** Show that the molar average velocity $\mathbf{v}^*$ and the mass average velocity $\mathbf{v}$ are identical in a binary system of equal molecular weight species.

**2.17** Find the initial evaporation rate of water, in units of centimeters per second of surface velocity, for the Arnold cell operating as shown in Fig. P2.17. The entire system is maintained at 25°C and 1 atm. The saturated humidity of water in air at 75°F is 0.0189 lb H₂O/lb dry air.

**2.18** Silane gas [diluted to 1% (molar) in inert helium] is pumped through the inside of a small glass hollow "fiber." At the temperature and pressure of the operation, the silane pyrolyzes at the inner glass surface, and a film of solid silicon is deposited. This technique is used for the coating of optical fibers used in telecommuni-

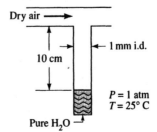

**Figure P2.17**

cations. The coating thickness is always a small fraction of the inside diameter of a fiber.

The (heterogeneous) pyrolysis of silane may be written stoichiometrically as

$$SiH_4 \rightarrow Si + 2H_2$$

Assume that the gas flow inside the fiber is steady and laminar, and the pressure drop across the ends of the fiber is small in comparison to the average pressure within the gas. The average axial velocity of the feed gas is 1 cm/s at the entrance to the fiber.

The overall goal of this problem is the development of the differential equation and boundary conditions that describe this system.

**a.** Keeping in mind that the molar flux of silane is a vector quantity, write the axial and radial components of Fick's law of diffusion for silane. Then impose any additional physical statements that permit you to write these fluxes in terms of the mole fraction of silane gas and the axial gas velocity. State clearly any assumptions you make in simplifying the flux expressions, and offer supporting arguments for your assumptions based on your physical insight, intuition, and/or knowledge.

**b.** Examine a fixed differential element of volume within the flow, and write a species balance for silane entering and leaving that volume element. Assume that silane does not react in the gas phase—it just reacts on the glass surface. After all your analysis and discussion, present the final result as the diffusion equation, a partial differential equation for silane concentration within the gas phase, and boundary conditions, the solution of which would give the concentration (mole fraction) of silane everywhere within the gas phase within fiber. Do not attempt to solve the equation.

**c.** Suppose you had the solution $x_{SiH_4}(r, z)$ to the equation in part b. Show how you would calculate the rate of growth of solid silicon film at the inner wall of the fiber from x.

**d.** Does the concept of a quasi-steady analysis enter anywhere in this problem? Explain your answer. How, if at all, did you use the statement that the silane is highly diluted within the gas phase?

**2.19** As a test of a hollow fiber device for controlling the release of a volatile sex pheromone, the fibers are filled with a test liquid, and data are obtained on the rate of release of the liquid to the ambient medium, which is air. The fiber is sealed at one end and open at the other, as shown in Fig. P2.19. The wall of the fiber is impermeable to diffusion.

Derive a mathematical model for the rate of release (the evaporation rate). The model should give an explicit prediction of $L(t)$, and of the mass evaporation rate $W(t)$. The vapor pressure $P_v$ and the diffusivity $D$ should appear in your model, along with other relevant parameters that you define.

**2.20** The Southern California Florists Cooperative has contracted with us to design a system for control of insects in the event of sudden infestation in a greenhouse. An outline of design constraints, and available physical property data, follow.

The insecticide is available in the form of a liquid solution containing 1% (molar) of the active agent in water. Our plan is to encapsulate this liquid in hollow polysulfone fibers. These fibers are available commercially, with outside diameters in the range 100 to 300 $\mu$m and wall thicknesses in the range 20 to 40 $\mu$m. The "partition coefficient" of the active agent between the solution and the fiber wall is defined by

$$\frac{C_A(\text{wall})}{C_A(\text{solution})} = \alpha \qquad (C_A \text{ in molar units})$$

A value of $\alpha = 5 \times 10^{-4}$ has been determined. The diffusion coefficient of the active agent in the solution is $10^{-5}$ cm$^2$/s at 25°C. The diffusion

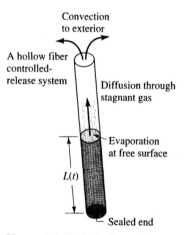

**Figure P2.19** A hollow fiber, controlled-release system.

coefficient in the fiber wall is $10^{-7}$ cm$^2$/s at 25°C. The "half-life" of the system is to be 2 days (i.e., half the active agent should remain encapsulated after 2 days of exposure to the environment).

The active agent has a molecular weight $M_w$ of 100. Its equilibrium solubility in water is 3% (molar). The air in the greenhouse is circulated by overhead fans.

Toxicity tests indicate that the active agent should be kept below 150 parts per million by weight (ppmw) in the greenhouse air, so that employees will not be at risk. Good insect control requires an air concentration in excess of 30 ppmw. The customer wants this level to be achieved within one day after activation of the product.

The standard greenhouse contains 20,000 ft$^3$ of air, and there is negligible exchange of air with the external environment.

**a.** Recommend a fiber diameter and wall thickness for this application.

**b.** How long can the product remain active in the greenhouse, with respect to employee safety?

**c.** It appears that it will cost about $400 per pound of packaged solution to put this product in the hands of customers, and make an acceptable profit. Customers indicate that they are looking for a product that will cost them $15 for a single application. Does this appear to be a good business venture?

**2.21** A liquid A evaporates at one end of a capillary tube and the vapor diffuses toward the other end, which is open to a large gas space B that is essentially free of A. (See Fig. P2.21.)

Assume that the gas B is insoluble in liquid A, and work with the following conditions and properties:

Total pressure        $P = 1$ atm
Vapor pressure of A   $P_{vap} = 550$ torr

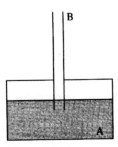

**Figure P2.21**

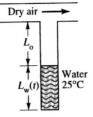

Dry air →

$L_o$

$L_w(t)$    Water 25°C

**Figure P2.22**

| Temperature | $T = 25°C$ |
| Capillary length | $L = 10$ cm |
| Molecular weights | $M_{w,A} = 100, M_{w,B} = 28$ |
| Binary diffusivity | $D_{AB} = 0.25$ cm$^2$/s |

**a.** Find the magnitude of the molar average velocity $v_z^*$.

**b.** What is the factor by which you would be in error if you assumed that $x_A \ll 1$?

**c.** Is $v_z^*$ constant along the capillary axis? Explain your answer.

**d.** Is $v_z$ (the *mass*-average velocity) constant along the capillary axis? Explain your answer.

**2.22** Water evaporates at the surface of a column confined to a long narrow capillary tube, closed at the bottom, as shown in Fig. P2.22. Derive and solve the equations that describe the length $L_w(t)$. For the case that $L_o = 10$ cm and $L_{w,o} = 10$ cm, how long will it take for the column of water to vanish?

**2.23** The vapor pressure of solid naphthalene at 23°C is 0.07 mmHg. A naphthalene sphere of diameter 1 cm is suspended in still air. How long is required for the sphere to lose 87.5% of its initial mass? Use $\rho_{solid} = 1150$ kg/m$^3$, $M_{w,naph} = 128$ g/g·mol, and $D_{naph/air} = 8 \times 10^{-6}$ m$^2$/s.

**2.24** Estimate the diffusion coefficient of benzene ($C_6H_6$) in water at 20°C. The measured value is $10^{-5}$ cm$^2$/s at that temperature. Do the same for tetrachlorobenzene ($C_6H_2Cl_4$). Use the method of Le Bas, as described in Section 2.2, for the calculation of the molar volume at the boiling point, $V_A$.

**2.25** Estimate the diffusion coefficients for the following organic compounds in water at 20°C. Plot your results against molecular weight, and comment on the applicability of Graham's law, which states that the diffusion coefficient of a series of similar compounds in a solvent varies as the inverse square root of the molecular weight of the compound.

Benzene ($C_6H_6$) Chlorobenzene ($C_6H_5Cl$)
Dichlorobenzene ($C_6H_4Cl_2$)
Tetrachlorobenzene ($C_6H_2Cl_4$)
Phenol ($C_6H_5OH$)
Diphenol ($C_{10}H_7OH$) Naphthalene ($C_{10}H_8$)

**2.26** Funazukuri et al. [*J. Chem. Eng. Data*, **39**, 911 (1994)] present the following data for the diffusion coefficients of several solutes in hexane at 313.2 K and a pressure of 16 MPa. Test the data against the Wilke–Chang equation. For the solvent, hexane, $M_w$ = 86.17 g/mol, and $\mu$ = 0.304 cP.

| Solute | $10^5 D_{AB}$ (cm²/s) |
|---|---|
| Acetone | 5.29 |
| Benzene | 4.48 |
| Toluene | 4.32 |
| $m$-Xylene | 3.63 |
| 1,3,5-Trimethylbenzene | 3.55 |
| Naphthalene | 3.70 |

**2.27** Snidjer et al. [*J. Chem. Eng. Data*, **40**, 37 (1995)] measured the diffusivity of carbon monoxide (CO) in liquid toluene and curve-fit the data with the following expression:

$$D_{AB} = 2.77 \times 10^{-3} \exp\left(\frac{-1164.2}{T}\right) \text{cm}^2/\text{s}$$

**(P2.27.1)**

when $T$ is in kelvins. Independent measurements of the viscosity of toluene as a function of absolute temperature can be curve-fit by

$$\mu = 0.0152 \exp\left(\frac{1072}{T}\right) \text{cP} \quad \textbf{(P2.27.2)}$$

Use the Wilke–Chang equation to predict $D_{AB}(T)$ and compare the predicted model to the empirical model (Eq. P2.27.1).

**2.28** Snidjer et al. [*J. Chem. Eng. Data*, **40**, 37 (1995)] measured the diffusivity of carbon dioxide ($CO_2$) in liquid toluene and curve-fit the data with the following expression:

$$D_{AB} = 2.96 \times 10^{-3} \exp\left(\frac{-1224.5}{T}\right) \text{cm}^2/\text{s}$$

**(P2.28.1)**

where $T$ is in kelvins. Independent measurements of the viscosity of toluene as a function of absolute temperature can be curve-fit by

$$\mu = 0.0152 \exp\left(\frac{1072}{T}\right) \text{cP} \quad \textbf{(P2.28.2)}$$

Use the Wilke–Chang equation to predict $D_{AB}(T)$ and compare the predicted model to the empirical model (Eq. P2.28.1).

**2.29** Predict the diffusion coefficient of vitamin E in hexane at 313.2 K. (The measured value is $1.52 \times 10^{-9}$ m²/s.) Vitamin E has the formula $C_{29}H_{50}O_2$. Its molecular weight is $M_w$ = 430.7 g/mol. Its molar volume at the normal boiling point is estimated to be 603 cm³/mol. For the solvent, hexane, $M_w$ = 86.17 g/mol, and $\mu$ = 0.304 cP at 313.2 K.

**2.30** Diffusion coefficients of naphthalene in supercritical carbon dioxide have been measured

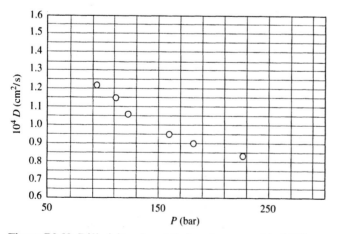

**Figure P2.30** Diffusivity of naphthalene in supercritical $CO_2$ at 308 K.

[Liong et al., *Ind. Eng. Chem. Res.*, **30**, 1329 (1991)] and are shown in Fig. P2.30. A supercritical fluid has some properties of a gas and some of a liquid. Predict the diffusion coefficient using the Wilke–Chang model for a liquid. What conclusion do you draw?

For naphthalene, use $V_A = 147.6$ cm$^3$/mol, which is calculated using the method of Le Bas (Chapter 3 of Reid, Prausnitz, and Poling). The viscosity of supercritical $CO_2$ as a function of pressure, at 308 K, is 0.0566 cP at $P = 98.7$ bar, 0.072 cP at $P = 138.5$ bar, and 0.0814 cP at $P = 181.3$ bar.

**2.31** Funazukuri et al. [*Ind. Eng. Chem. Res.*, **30**, 1325 (1991)] give a value of $5.8 \times 10^{-9}$ m$^2$/s for the diffusion coefficient of linoleic acid methyl ester in supercritical $CO_2$ at 308.2 K and 19 MPa. The molecular weight of this compound is 294.5. The molar volume at the boiling point for this compound is estimated to be $4.25 \times 10^{-4}$ m$^3$/mol. The viscosity of the solvent is described in Problem 2.30. Use the Wilke–Chang equation to estimate a value for $D$, and compare it to the measured value.

**2.32** Funazukuri et al. [*AIChE J.*, **38**, 1761 (1992)] give the following values of diffusion coefficients of a series of organic compounds in supercritical $CO_2$ at $T = 313$ K and $P = 16$ MPa. Plot the data as diffusion coefficient against molecular weight and comment on the applicability of Graham's law, which states that the diffusion coefficient of a series of similar compounds in a solvent varies as the inverse square root of the molecular weight of the compound.

**2.33** One often finds expressions for the diffusivity of a gas pair in the form

$$D_{AB} = \frac{AT^n}{P} \qquad \text{(P2.33)}$$

where $A$ and $n$ are constants, and $T$ and $P$ are absolute temperature and pressure; $n$ is often found to be in the range of 1.5 to 1.7. Use the Chapman–Enskog formula (Eq. 2.2.2) to calculate the diffusivity of silane in hydrogen at $P = 1$ atm, and plot your results as a function of temperature in the range 300 to 1000 K. How good a model is Eq. P2.33?

**2.34** The binary diffusion coefficient of a species in a gas is inversely proportional to the absolute pressure, according to the Chapman–Enskog theory. In liquids, the diffusion coefficient is usually assumed to be independent of pressure, according to the Wilke–Chang equation, unless the viscosity is a function of pressure. Normally we think of the viscosity of a liquid as independent of pressure. This may not be the case at very high pressures, however, because of compressibility effects. This is especially true in the important case of supercritical fluids.

Suarez et al. [*Chem. Eng. Sci.*, **48**, 2419 (1993)], who measured the diffusion coefficient of ethylbenzene in supercritical $CO_2$, obtained the following data, which relate to parts (a) and (b).

**a.** At a fixed temperature, determine the dependence of the diffusion coefficient on pressure.

**b.** Test the prediction of the Wilke–Chang equation, that the $\mu D_{AB}$ product is a constant.

| Compound | | $M_w$ | $10^9 D_{AB}$ (m$^2$/s) |
|---|---|---|---|
| Myristoleic acid methyl ester | $C_{15}H_{28}O_2$ | 240 | 7.2 |
| *cis*-11-Eicosenoic acid methyl ester | $C_{21}H_{40}O_2$ | 324.6 | 6.2 |
| Erucic acid methyl ester | $C_{23}H_{44}O_2$ | 352.6 | 5.9 |
| Vitamin K$_3$ | $C_{11}H_8O_2$ | 172.2 | 9.3 |
| Vitamin E | $C_{29}H_{50}O_2$ | 430.7 | 5.5 |
| Vitamin A acetate | $C_{22}H_{32}O_2$ | 328.5 | 6.0 |
| Vitamin K$_1$ | $C_{31}H_{46}O_2$ | 450.7 | 5.4 |
| DL-Limonene | $C_{10}H_{16}$ | 136.2 | 10.9 |
| *cis*-Jasmone | $C_{11}H_{16}O$ | 164.3 | 8.8 |
| Indole | $C_8H_7N$ | 117.2 | 10.4 |
| Benzene | $C_6H_6$ | 78 | 12.4 |
| Naphthalene | $C_{10}H_8$ | 128 | 11.2 |

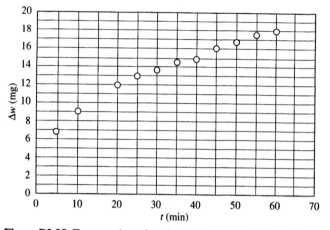

**Figure P2.35** Evaporation of methanol from a vertical capillary.
(Oliver and Clarke, *Chem. Eng.*, p. 58, February 1971.)

| $T$ (°C) | $P$ (bar) | $10^6\mu$ (Pa·s) | $10^9 D_{AB}$ (m²/s) |
|---|---|---|---|
| 40 | 150 | 67.2 | 12.15 |
| 40 | 200 | 77.2 | 11.02 |
| 40 | 250 | 85.0 | 9.59 |
| 40 | 300 | 93.1 | 9.04 |
| 40 | 350 | 102.3 | 8.79 |
| 50 | 150 | 57.1 | 14.68 |
| 50 | 200 | 68.8 | 12.56 |
| 50 | 250 | 77.0 | 11.26 |
| 50 | 300 | 85.1 | 10.40 |
| 50 | 350 | 91.5 | 9.79 |
| 60 | 150 | 47.6 | 17.15 |
| 60 | 200 | 59.8 | 14.53 |
| 60 | 250 | 68.7 | 12.56 |
| 60 | 300 | 73.8 | 11.59 |
| 60 | 350 | 83.9 | 10.83 |

**2.35** Methanol initially fills a vertical capillary of inside diameter $D = 0.17$ cm. The tube is closed at the lower end and the entire system is maintained at 50°C. Data for the cumulative mass loss from the capillary are shown in Fig. P2.35. Develop and test a model to describe these data. The physical properties are $\rho_L = 0.79$ g/cm³, $p_{vap} = 450$ mmHg, and $D_{AB} = 0.186$ cm²/s, all at 50°C.

**2.36** Tamimi, Rinker, and Sandall give the data shown in Fig. P2.36 for the diffusivity of $CO_2$ in water as a function of temperature. Test the ability of the Wilke–Chang equation to predict these data.

**2.37** Repeat Example 2.2.2, but use the method of Hayduk and Laudie described in Appendix C, Eq. C8.1.

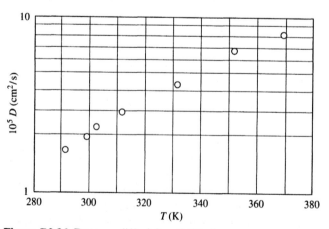

**Figure P2.36** Data on diffusivity of $CO_2$ in water.
Tamimi et al., *J. Chem. Eng. Data*, **39**, 330 (1994).

# Chapter 3

---

# Steady and Quasi-Steady Mass Transfer

**W**e begin this chapter with a derivation of the unsteady state species balance equation for a mixture that may include convective mass transfer, diffusion, and chemical reactions on the boundaries as well as in the mixture. Then through a long series of examples of steady state systems we show how this equation often can be simplified to the piont of yielding an analytical solution. Examples drawn from several areas of technology include the interactive roles of convection, or reaction, with diffusion. These introductory examples are followed by several sections illustrating the application of mass transfer analysis to the design of engineering systems and the interpretation of data required for design.

---

## 3.1 MASS BALANCE ON A SPECIES

In this chapter we present the partial differential equation (PDE) that describes the concentration of a chemical species at some point in space, as a function of position and time. We often refer to this equation as the *species balance equation*, or the *species conservation equation*, or simply as the *diffusion equation*. It is a mass balance, but specifically it provides an equation for *each* chemical species in the system of interest. In fluid dynamics we encounter the continuity equation, which is truly a *mass* balance equation that does not differentiate among the various chemical species in the system. In a sense, the species balance equation is the continuity equation for *individual chemical species*. It is derived in a manner very similar to that used to obtain the continuity equation, except that the balances (the bookkeeping parts) are written on each chemical species in the system.

A mass balance is a bookkeeping procedure on a small volume element *fixed in space*, which we call a *control volume*. We pick out a particular chemical species A, and then account for the transport of A across imaginary surfaces that surround this volume element. We keep account of the rate at which the mass of species A changes within the volume element. (We could also do this in *molar* units, but we choose mass units here.) Species A crosses surfaces of the control volume at a rate determined by the *mass* flux $\mathbf{n}_A$ relative to a fixed coordinate system. ($\mathbf{n}_A$ is the sum of a diffusive flux and convective flux. For the moment, we simply assume that there *is* a mass flux, and we name it $\mathbf{n}_A$.) We pick a volume element bounded by the surfaces of a rectangular parallelepiped of dimensions $\Delta x\, \Delta y\, \Delta z$, as depicted in Fig. 3.1.1. The procedure is very much like that used in deriving the continuity equation, which should be reviewed (see Chapter 4, Section 4.1, of Middleman, *An Introduction to Fluid Dynamics,* or any standard fluid dynamics textbook).

The rate at which a species crosses the surface of area $\Delta y\, \Delta z$, oriented normal to the $x$ axis, is the product of the $x$ component of the flux, $n_{A,x}$, times that area, or $n_{A,x}\Delta y\, \Delta z$, with $n_{A,x}$ evaluated at the plane $x$. Note that if $n_{A,x}$ is negative, as drawn in Fig. 3.1.1, this term represents an efflux *from* the volume—not an influx *into* the volume—across that boundary. Across the parallel face we would write exactly the same expression, with the understanding that $n_{A,x}$ is evaluated at $x + \Delta x$. Hence the *net* rate of mass *influx* across the pair of faces separated by $\Delta x$ is

$$\{n_{A,x}|_x - n_{A,x}|_{x+\Delta x}\}\Delta z\, \Delta y$$

For the other two pairs of faces that define the rest of this volume element, the net rate of *influx* would be

$$\{n_{A,y}|_y - n_{A,y}|_{y+\Delta y}\}\Delta x\, \Delta z$$

across the surface normal to the $y$ axis, and

$$\{n_{A,z}|_z - n_{A,z}|_{z+\Delta z}\}\Delta x\, \Delta y$$

across the surface normal to the $z$ axis.

Accumulation, by which we mean either an increase or decrease of the amount of A in the element over time, is measured by $\partial(\rho_A\, \Delta x\, \Delta y\, \Delta z)/\partial t$, where $\rho_A$ is the mass density of species A.

Species A may be created by chemical reaction. If $r_A$ is the (*mass*) rate of production (not *disappearance*) of species A, per unit volume, then the reaction term is $r_A\, \Delta x\, \Delta y\, \Delta z$.

In this control volume, then, the amount of a particular species may change as a result of convective or diffusive fluxes across the boundaries, as described by the vector $\mathbf{n}_A$, and through chemical reactions. Hence our species balance takes the form

$$\frac{\partial}{\partial t}(\rho_A\Delta x\, \Delta y\, \Delta z) = \{n_{A,x}|_x - n_{A,x}|_{x+\Delta x}\}\, \Delta y\, \Delta z + \{n_{A,y}|_y - n_{A,y}|_{y+\Delta y}\}\, \Delta x\, \Delta z$$

$$+ \{n_{A,z}|_z - n_{A,z}|_{z+\Delta z}\}\, \Delta x\, \Delta y + r_A\, \Delta x\, \Delta y\, \Delta z \qquad \textbf{(3.1.1)}$$

Now we divide both sides of this equation by the differential volume and examine the limit of $(\Delta x\, \Delta y\, \Delta z) \to 0$. As a result, we obtain a partial differential equation, which

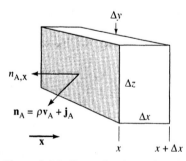

**Figure 3.1.1** Control volume for species balance.

we may write in a compact vector notation as

$$\nabla \cdot \mathbf{n}_A + \frac{\partial \rho_A}{\partial t} - r_A = 0 \qquad (3.1.2)$$

At this point we can introduce the statement that the mass flux $\mathbf{n}_A$ is the sum of the diffusive and convective fluxes

$$\mathbf{n}_A = \mathbf{j}_A + \rho \omega_A \mathbf{v} \qquad (3.1.3)$$

where $\mathbf{j}_A$ is the mass flux relative to the mass average velocity $\mathbf{v}$, $\omega_A$ is the mass fraction of species A, and $\rho$ is the mass density of the mixture. This equation is identical in meaning to Eq. 2.1.16, except that we are now working in *mass*, rather than *molar*, units. We may use this in Eq. 3.1.2, along with the continuity equation for a variable density fluid[1] to find (see Problem 3.1)

$$\rho \left( \frac{\partial \omega_A}{\partial t} + \mathbf{v} \cdot \nabla \omega_A \right) + \nabla \cdot \mathbf{j}_A - r_A = \rho \frac{D\omega_A}{Dt} + \nabla \cdot \mathbf{j}_A - r_A = 0 \qquad (3.1.4)$$

This is the desired *mass* balance on species A. We have introduced the derivative notation $D/Dt$, the so-called Stokes derivative, as a shorthand notation for the terms in parentheses on the left-hand side of Eq. 3.1.4.

If we go back to Eq. 3.1.2 and divide all terms by the molecular weight $M_{w,A}$, we may write this species conservation equation in *molar units*, obtaining

$$\frac{\partial C_A}{\partial t} + \nabla \cdot \mathbf{N}_A - R_A = 0 \qquad (3.1.5)$$

In this form, the reaction rate term is written in molar units, consistent with the use of molar concentration and molar flux in the equation. Now we introduce Fick's law for a *binary* mixture (Chapter 2, Eq. 2.1.17) in the form

$$\mathbf{N}_A = -CD_{AB} \nabla x_A + x_A (\mathbf{N}_A + \mathbf{N}_B) \qquad (3.1.6)$$

We most frequently write a diffusion equation in terms of the molar concentration $C_A$. In most problems of interest we may *assume constant $\rho$ and $D_{AB}$*. Then, after some algebraic manipulation (see Problem 3.2), we find

$$\frac{\partial C_A}{\partial t} + \mathbf{v} \cdot \nabla C_A = D_{AB} \nabla^2 C_A + R_A \qquad (3.1.7)$$

Equation 3.1.7 is a vector equation, written in a format that is not specific to any choice of coordinate system. The specific form of the Laplacian operator $\nabla^2$ in Cartesian, cylindrical, and spherical coordinates is given later (Chapter 4, Table 4.2.1).

Simplified forms of Eq. 3.1.7 follow for steady state and stationary media: for example, setting $\mathbf{v} = 0$, as well as the time derivative, we find

$$D_{AB} \nabla^2 C_A + R_A = 0 \qquad (3.1.8)$$

for steady diffusion with a reaction distributed homogeneously throughout the material. Note that $R_A$ is the rate of reaction that occurs *in* the fluid or solid phase. If a reaction

---

[1] In vector format, the continuity equation is simply:

$$\frac{\partial \rho}{\partial t} = -\nabla \cdot \rho \mathbf{v}$$

occurs only on a bounding surface of the region of interest (we call that a *heterogeneous* reaction), the rate does not appear in the species balance equation. But it has to appear somewhere. This leads us naturally to the next topic.

## 3.2 BOUNDARY CONDITIONS

A specific number of boundary conditions is required to match the order of the differential equation for species conservation. (An initial condition is needed for an unsteady state problem, as well.)

Boundary conditions express, in mathematical form, what we *know* or *assume* about the physical phenomena that occur, or are imposed, at the boundaries. "Boundary" refers to the ends of the *domain* over which the diffusion equation is written. Thus, for diffusion *within* a spherical bubble of radius $R$, the *domain* is $0 \le r \le R$. The point $r = 0$ is a *boundary*, even though it is interior to the sphere.

*Symmetry* that arises from the geometry often provides one boundary condition (Fig. 3.2.1a). There is no flux across a plane, line, or point of symmetry; therefore the gradient of concentration vanishes

$$\nabla C \equiv 0 \qquad (3.2.1)$$

Often the *composition* at a boundary is fixed by the physics. For example, if *a species reacts very rapidly* at a surface, its composition will be essentially zero at that surface. If the fluid exterior to a surface is *very well mixed*, then the concentration at the surface, *on the side of the surface where mixing occurs*, will be nearly equal to the bulk concentration in the mixed region (Fig. 3.2.1b).

We often *assume thermodynamic equilibrium at the interface* between two media. Then the compositions of a species, on either side of the interface, are related by (Fig. 3.2.1c)

$$C_+ = \alpha C_- \qquad (3.2.2)$$

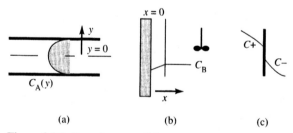

(a)               (b)               (c)

**Figure 3.2.1** Boundary conditions on concentration are determined by the geometry, symmetry, fluid dynamics, and thermodynamics. (a) Symmetry about the plane $y = 0$ implies that $\partial C_A/\partial y = 0$. (b) The efficiency of mixing of the fluid near the boundary $x = 0$ determines how close the concentration at the boundary is to the concentration $C_B$ in the well-mixed region away from the boundary. (c) Two phases may be in thermodynamic equilibrium across a phase boundary, in which case the concentrations of the same species on either side of the boundary are related through a thermodynamic rule, such as Eq. 3.2.2.

Examples are vapor/liquid equilibria

$$P_{vap} = HC_L \tag{3.2.3}$$

liquid/liquid equilibria such as

$$C_+ = \alpha_{sol}C_- \tag{3.2.4}$$

where $\alpha_{sol}$ is a solubility relationship, or partition coefficient. And of course there is the possibility that *nonlinear* relationships exist among the compositions on either side of the boundary, rather than the linear relationships just given.

Sometimes we can *specify a flux at a surface*. If a surface is impermeable to a given species, there is *no* flux across the interface: $\nabla C = 0$. If a reaction occurs at an interface (heterogeneous reaction), the flux must balance the reaction rate per unit of surface area

$$-D_{AB}\frac{\partial C_A}{\partial z} = R'_A \quad \text{for} \quad x_A \ll 1 \tag{3.2.5}$$

diffusive flux = rate of reaction per unit of surface area

On the other hand, if a convective process occurs at the interface, the convective mass transfer coefficient $k$ is defined in such a way that, setting the diffusive flux equal to the convective flux,

$$-D_{AB}\frac{\partial C_A}{\partial z}\bigg|_{z=0} = k(C|_{z=0} - C_b) \tag{3.2.6}$$

diffusive flux = convective flux

where the bulk concentration $C_b$ is specified. Equation 3.2.6 simply says that the rate at which a species gets *to* the surface must balance the rate at which it *leaves* the surface by convection, if no reaction occurs *on* the surface.

We can now illustrate and explain some of these ideas through a series of examples.

**EXAMPLE 3.2.1** *Mass-Transfer Controlled by External Diffusion Resistance*

Suppose a liquid drop contains a *supersaturated* solution of some species. The solute could exist either as a solid or as a liquid immiscible phase. The liquid concentration of the solute is $C_{A,sat}$. The solute species is volatile, and it has a vapor pressure $p_{A,sat}$.

The saturated drop in Fig. 3.2.2 is surrounded by a large body of stagnant gas. Analyze the steady diffusion *into the gas*. Assume the solution remains supersaturated throughout the process. We can initiate the model by two routes:

(a) For this problem:

$$\mathbf{v} = 0 \tag{3.2.7a}$$

(In the dilute approximation, $\mathbf{v}^* = 0$.)

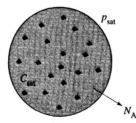

**Figure 3.2.2** A drop saturated with solute.

We assume steady state and no reaction, so

$$R_A = 0 \quad \text{and} \quad \frac{\partial C_A}{\partial t} = 0$$

Then all that remains of Eq. 3.1.6 is

$$\nabla^2 C_A = 0 \tag{3.2.8a}$$

In spherical coordinates, assuming spherical symmetry we write

$$\frac{1}{r^2} \frac{d}{dr} \left( r^2 \frac{dC_A}{dr} \right) = 0 \tag{3.2.9a}$$

or

$$r^2 \frac{dC_A}{dr} = a \tag{3.2.10}$$

**(b)** Or, begin with the mass balance: the rate of transfer across any spherical surface at radius $r$ is constant, so

$$r^2 N_{A,r} = \text{constant} \tag{3.2.7b}$$

Assume that species A is a dilute component: Then $v_r = 0$, and with Fick's law we find

$$D_{AB} r^2 \frac{dC_A}{dr} = \text{constant} \tag{3.2.8b}$$

or

$$r^2 \frac{dC_A}{dr} = a \tag{3.2.9b}$$

Both routes lead to the same differential equation for $C_A$:

$$\frac{dC_A}{dr} = \frac{a}{r^2} \tag{3.2.11}$$

and the solution is

$$C_A = \frac{a}{r} + b \tag{3.2.12}$$

We need two boundary conditions to evaluate the integration constants $a$ and $b$. What are they, and where are they imposed?

The domain of interest is not $0 \le r \le R$; it is the semi-infinite region $r \ge R$, *outside* the sphere. At $r = R$, does $C_A = C_{A,\text{sat}}$? Yes, but at $r = R^-$. This is in the liquid drop, which is *not in the domain* in which diffusion is occurring. What is the concentration at $r = R^+$? In the gas phase, at $r = R^+$, we have $p_A = p_{A,\text{sat}}$. But we want the concentration in *molar* units, not in partial pressure. The relationship between these units follows from the relationship for an ideal gas,

$$C = \frac{p}{R_G T} \tag{3.2.13}$$

Then we may write one boundary condition as

$$C_A = \frac{p_{A,\text{sat}}}{R_G T} = C_R \quad \text{at } r = R^+ \tag{3.2.14}$$

This equation defines $C_R$.

Vapor/liquid equilibrium data will give us

$$p_{A,sat} = H_A C_{A,sat} \tag{3.2.15}$$

where $H_A$ is some constant, usually called Henry's constant.[1] Equation 3.2.15 is an assumption, but a commonly made assumption, generally valid for dilute concentrations of the solute.

Where is the other boundary? It is far removed from the sphere, so we write

$$C_A = 0 \quad \text{at } r = \infty \tag{3.2.16}$$

on the assumption that species A evaporates from the drop into a gas initially free of this species, and that far from the drop, species A is so diluted by the large surrounding volume of gas that we may take its concentration as effectively zero.

Then it follows that $b = 0$ and

$$a = -RC_R \tag{3.2.17}$$

so

$$C_A = C_R \frac{R}{r} \tag{3.2.18}$$

We usually want more than the concentration profile $C_A(r)$, however.

What is the *rate of mass transfer* (evaporation of A) for this system? The radial component of the flux at the surface of the drop is given by

$$N_{A,r}|_R = -D_{AB} \frac{dC_A}{dr}\bigg|_R \tag{3.2.19}$$

Note that we ignore any motion of the interface—there is no $x_A(N_{A,r} + N_{B,r})$ term. Using Eq. 3.2.18 we find (we stop writing the subscript r, denoting that we are looking at the radial flux)

$$N_A|_R = D_{AB} \frac{C_R}{R} \tag{3.2.20}$$

The *rate* of evaporation is the product of the flux $N_A$ and the transfer area, so

$$W_A = 4\pi R^2 N_A|_R = 4\pi R C_R D_{AB} \tag{3.2.21}$$

We often *define* a surface mass transfer coefficient $k$ as the proportionality constant between the flux and the concentration difference between the surface and the surroundings

$$N_A = k\,\Delta C = k(C_R - 0) = kC_R \tag{3.2.22}$$

Hence we find, for this specific case, that

$$k = \frac{N_A}{C_R} = \frac{D_{AB}}{R} \tag{3.2.23}$$

---

[1] More commonly, Henry's law is written in terms of mole fraction as the concentration unit

$$p_{A,sat} = H'_A x_A \tag{3.2.15'}$$

or

$$\frac{2kR}{D_{AB}} = 2 \qquad (3.2.24)$$

We arrange the result in this form because the *dimensionless* mass transfer coefficient, $2kR/D_{AB}$, has a special importance, as we will see later. (It is called the Sherwood number, denoted Sh.) Hence we find that

$$Sh = 2 \qquad \text{for } \textit{steady diffusion external} \text{ to a } \textit{sphere} \qquad (3.2.25)$$

Note that the Sherwood number is a dimensionless *convective* mass transfer coefficient, while the example given here refers to transfer solely by diffusion, not by convection. Although $k$ is properly defined by Eq. 3.2.22, it is not really a *convective* coefficient in this example, except in the limit of no external flow.

**EXAMPLE 3.2.2** *Diffusion Controlled by External Convection*

We consider a problem here that is nearly the same as in that posed in Example 3.2.1, but the exterior fluid is not stationary. Now the release is into a *well-stirred* region of fluid, within which convection aids the transport. We will represent the *convective resistance* as if it were due to diffusion through a boundary layer of known thickness $\delta$ (Fig. 3.2.3). If the exterior region is large enough to dilute the diffusant to a very small concentration, the appropriate boundary conditions are

$$C_A = C_R \qquad \text{at } r = R \qquad (3.2.26)$$

and

$$C_A = 0 \qquad \text{at } r = R + \delta \qquad (3.2.27)$$

Now the *domain* of interest is $R \leq r \leq R + \delta$. We still get

$$C_A = -\frac{a}{r} + b \qquad (3.2.28)$$

as in Example 3.2.1. When we apply the boundary conditions

$$C_A = C_R = -\frac{a}{R} + b \qquad \text{at } r = R \qquad (3.2.29)$$

$$C_A = 0 = -\frac{a}{R + \delta} + b \qquad \text{at } r = R + \delta \qquad (3.2.30)$$

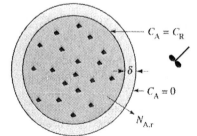

**Figure 3.2.3** Boundary layer surrounding a drop.

we find

$$b = \frac{a}{R + \delta} \quad \text{and} \quad \frac{a}{R} = b - C_R = \frac{a}{R + \delta} - C_R \qquad (3.2.31)$$

Therefore

$$a = \frac{-RC_R}{1 - \frac{1}{1 + \delta/R}} = \frac{-RC_R}{\frac{\delta/R}{1 + \delta/R}} \qquad (3.2.32)$$

Note that in the limit as $\delta/R \gg 1$ we recover the earlier result. Hence the boundary condition at infinity is a good *approximation* as long as $\delta \gg R$.

The radial flux at $R$ is still given by

$$N_{A,r}|_R = -D_{AB} \frac{dC_A}{dr} \bigg|_R \qquad (3.2.33)$$

but now

$$\frac{dC_A}{dr} \bigg|_R = \frac{a}{r^2} \bigg|_R = \frac{a}{R^2} = -\frac{C_R(1 + \delta/R)}{\delta} \qquad (3.2.34)$$

Hence the flux is

$$N_A = \frac{D_{AB}C_R}{R} f\left(\frac{\delta}{R}\right) \qquad (3.2.35)$$

where $f(\delta/R) = (1 + \delta/R)/(\delta/R)$, and we may define a mass transfer coefficient, as before, as

$$k = \frac{N_A}{\Delta C} = \frac{N_A}{C_R} \qquad (3.2.36)$$

Now we find

$$\text{Sh} = \frac{k(2R)}{D_{AB}} = 2f\left(\frac{\delta}{R}\right) = 2\left(1 + \frac{R}{\delta}\right) \qquad (3.2.37)$$

We described $\delta$ as a boundary layer thickness. Its magnitude must depend on the fluid dynamics external to the sphere. We will need to learn how to estimate $\delta$ from hydrodynamic parameters. Obviously we can promote mass transfer by controlling the hydrodynamics of the system.

**EXAMPLE 3.2.3**  *Respiration of a Spherical Cell*

In this example we consider a spherical particle within which a *homogeneous* reaction occurs, involving a single species A (Fig. 3.2.4). The concentration of A is somehow maintained at the level $C_R$ at $r = R^-$. If an *average* rate of consumption of A *must be maintained,* denoted by $\bar{R}_A$ [mol/cm$^3 \cdot$ s], what level of $C_R$ must be provided? The species balance takes the form

$$D_{AB} \nabla^2 C_A + R_A = 0 \qquad (3.2.38)$$

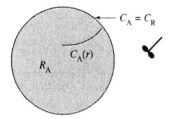

**Figure 3.2.4** A spherical cell in which a reaction occurs.

assuming steady diffusion, homogeneous reaction, and no convection. The domain is the region $0 \leq r < R$.

The average rate of consumption is defined as

$$\overline{R}_A = \frac{\int_0^R 4\pi r^2 R_A(r)dr}{\frac{4}{3}\pi R^3} \qquad (3.2.39)$$

In spherical coordinates, the diffusion equation takes the form

$$D_{AB}\frac{1}{r^2}\frac{d}{dr}\left(r^2\frac{dC_A}{dr}\right) + R_A = 0 \qquad (3.2.40)$$

We have to introduce some assumption about the kinetics. *If we assume* zero-order kinetics

$$R_A = \text{constant} = -R_A^o = -\overline{R}_A^o \qquad (3.2.41)$$

$$\uparrow$$

consumption has a *minus* sign

then the diffusion equation is

$$\frac{d}{dr}\left(r^2\frac{dC_A}{dr}\right) = +\frac{\overline{R}_A^o}{D_{AB}}r^2 \qquad (3.2.42)$$

With one integration this becomes

$$r^2\frac{dC_A}{dr} = \frac{\overline{R}_A^o r^3}{3D_{AB}} + a \qquad (3.2.43)$$

For $dC_A/dr = 0$ at $r = 0$ (*a symmetry condition*), we must have $a = 0$. Then another integration yields

$$C_A = +\frac{\overline{R}_A^o r^2}{6D_{AB}} + b \qquad (3.2.44)$$

For

$$C_A = C_R \qquad \text{at } r = R \qquad (3.2.45)$$

we find

$$C_A(r) - C_R = -\frac{\overline{R}_A^o(R^2 - r^2)}{6D_{AB}} \qquad (3.2.46)$$

As expected, if species A is being consumed by the reaction, then $C_A < C_R$ at $r = 0$.

Since $C_A$ must be positive, we see that at a fixed value of $C_R$ there is a limiting value of $R_A^o$ that satisfies this equation. For any $R_A^o$ larger than this limiting value, we cannot supply sufficient A to the interior, by diffusion, to maintain a positive $C_A$.

Note that $C_R$ is an *unknown* parameter. Our goal is to specify a value of $C_R$ that will provide a specified *average* rate of consumption, $\overline{R}_A$. Since we have selected zero-order kinetics, we have $R_A = \text{constant} = \overline{R}_A = -R_A^\circ$. How do we relate $\overline{R}_A$ to $C_R$? We must write a material balance on species A, that states that the rate of delivery of A to the cell exactly balances the rate of consumption

$$4\pi R^2 N_A|_R = \tfrac{4}{3}\pi R^3 \overline{R}_A \qquad (3.2.47)$$

or

$$R^2 N_A|_R = -\tfrac{1}{3}R^3 R_A^\circ \qquad (3.2.48)$$

if we note that $\overline{R}_A = -R_A^\circ$.

From the $C_A(r)$ result (Eq. 3.2.46), we find

$$N_A|_R = -D_{AB}\frac{dC_A}{dr}\bigg|_R = -\frac{R_A^\circ R}{3} \qquad (3.2.49)$$

or

$$R^2\left(-\frac{R_A^\circ R}{3}\right) = -\frac{R^3 R_A^\circ}{3} \qquad (3.2.50)$$

This leads us to $1 = 1$!

What happened? We are apparently not able to solve for the $C_R$ level that will maintain the specified reaction rate. This is because we have specified our way into a nonsolvable—nonsensical—problem. If, as assumed, the reaction is really zero-order, the consumption rate is *independent of concentration,* and *any* value of $C_R$ will maintain the required reaction rate. In short, we should not be surprised that the analysis failed to answer the question posed at the beginning. We imposed this burden on the analysis by the assumption of zero-order kinetics. A related example will make this clearer.

**EXAMPLE 3.2.4** *Life Support for a Spherical Organism: First-Order Kinetics for Respiration*

As an extension of Example 3.2.3, we suppose that an average consumption rate (of oxygen) must be maintained, but now we take the kinetics (the respiration rate) to be first-order in A (oxygen). The diffusion equation and boundary conditions are

$$D_{AB}\frac{1}{r^2}\frac{d}{dr}\left(r^2\frac{dC_A}{dr}\right) - kC_A = 0 \qquad \text{on } 0 \le r \le R \qquad (3.2.51)$$

$$C_A \text{ is finite at } r = 0 \qquad (3.2.52)$$

$$C_A = C_R \qquad \text{at } r = R \qquad (3.2.53)$$

Some mathematical manipulation allows us to solve this equation easily. The steps are as follows:

$$\frac{1}{r^2}\frac{d}{dr}\left(r^2\frac{dC_A}{dr}\right) = \frac{k}{D_{AB}}C_A \qquad (3.2.54)$$

For simplicity, let's drop the subscript A on the concentration variable $C$. We now transform the differential equation into an easily solved form by following a series of simple steps:

$$\frac{1}{r}\frac{d}{dr}\left(r^2\frac{dC}{dr}\right) - \frac{k}{D_{AB}}rC = 0 \tag{3.2.55}$$

$$2\frac{dC}{dr} + r\frac{d^2C}{dr^2} - \frac{k}{D_{AB}}rC = 0 \tag{3.2.56}$$

$$\frac{dC}{dr} + \frac{dC}{dr} + r\frac{d^2C}{dr^2} - \frac{k}{D_{AB}}rC = 0 \tag{3.2.57}$$

$$\frac{d}{dr}\left(C + r\frac{dC}{dr}\right) - \frac{k}{D_{AB}}rC = 0 \tag{3.2.58}$$

$$\frac{d}{dr}\frac{d}{dr}(rC) - \frac{k}{D_{AB}}rC = 0 \tag{3.2.59}$$

$$\frac{d^2}{dr^2}(rC) - \frac{k}{D_{AB}}(rC) = 0 \tag{3.2.60}$$

By forcing the equation into this form, we see that we can define a new variable: $rC = y$. This simplifies Eq. 3.2.60 to the form

$$\frac{d^2y}{dr^2} - \alpha^2 y = 0 \tag{3.2.61}$$

where we define $\alpha^2 = k/D_{AB}$. Equation 3.2.61 has the solution (see Example 3.2.8 for details)

$$y = a \sinh \alpha r + b \cosh \alpha r$$

or (recalling that $y = rC$)

$$rC = a \sinh\left(\frac{k}{D_{AB}}\right)^{1/2} r + b \cosh\left(\frac{k}{D_{AB}}\right)^{1/2} r \tag{3.2.62}$$

To satisfy the boundary conditions, we must find the integration constants $a$ and $b$. The final result for $C_A$ is then found to be expressible in the form

$$C_A = C_R \frac{R}{r}\frac{\sinh(k/D_{AB})^{1/2}r}{\sinh(k/D_{AB})^{1/2}R} \tag{3.2.63}$$

The consumption rate is equal to the diffusion *rate* across the surface of the sphere, which is

$$+4\pi R^2 N_a|_R = -4\pi R^2 \frac{dC_A}{dr}\bigg|_R D_{AB} = -4\pi R^2 D_{AB}\frac{C_R}{R}(\phi\coth\phi - 1) \tag{3.2.64}$$

where a dimensionless parameter

$$\phi = \left(\frac{kR^2}{D_{AB}}\right)^{1/2} \tag{3.2.65}$$

appears, which is called the Thiele modulus.

Now we find, using Eq. 3.2.39 for $\overline{R}_A$, and the statement that consumption = diffusion rate,

$$\tfrac{4}{3}\pi R^3\overline{R}_A = 4\pi R^2 D_{AB}\frac{C_R}{R}(\phi\coth\phi - 1) \tag{3.2.66}$$

To support respiration at the specified rate $\overline{R}_A$, we conclude that we must have a $C_R$

value of

$$C_R = \frac{1}{3}\left(\frac{R^2}{D_{AB}}\right)\overline{R}_A(\phi \coth \phi - 1)^{-1} \tag{3.2.67}$$

The modeling task is complete, but now we must examine the implications of the model. For example, we want to find the parametric dependence of $C_R$ on $R$, $D_{AB}$, and $\phi$. We can rewrite Eq. 3.2.67 in the form

$$C_R = \frac{1}{3}\frac{R^2 k}{D_{AB}}\frac{\overline{R}_A}{k}(\phi \coth \phi - 1)^{-1} = \frac{1}{3}\frac{\overline{R}_A}{k}\phi^2(\phi \coth \phi - 1)^{-1} \tag{3.2.68}$$

For small $\phi$ we may simplify this expression:

$$\coth \phi = \frac{e^\phi + e^{-\phi}}{e^\phi - e^{-\phi}} = \left[\phi - \frac{\phi^3}{3} + \frac{2\phi^5}{15} - \cdots\right]^{-1} \tag{3.2.69}$$

and so

$$\phi \coth \phi - 1 = \frac{\phi}{\phi - \phi^3/3 + \cdots} - 1 = \frac{(\phi^3/3 - \cdots)}{\phi - \cdots} \approx \frac{\phi^2}{3} \tag{3.2.70}$$

For large $\phi$ another approximation is possible:

$$\phi \coth \phi - 1 = \phi\frac{e^\phi + e^{-\phi}}{e^\phi - e^{-\phi}} - 1 = \phi - 1 \approx \phi \tag{3.2.71}$$

Going back to Eq. 3.2.68, which we rewrite slightly as

$$C_R = \frac{1}{3}\frac{\overline{R}_A}{k}\frac{\phi^2}{\phi \coth \phi - 1} \tag{3.2.72}$$

we now find the following results:

***For Small $\phi$***

$$C_R = \frac{1}{3}\frac{\overline{R}_A}{k}\frac{\phi^2}{\phi^2/3} = \frac{\overline{R}_A}{k} \tag{3.2.73}$$

In the case of small values of the parameter $\phi$, $C_R$ is independent of $R$ and $D_{AB}$; moreover, it is controlled by reaction rate, not by diffusion.

***For Large $\phi$***

$$C_R = \frac{1}{3}\frac{\overline{R}_A}{k}\frac{\phi^2}{\phi} = \frac{1}{3}\frac{\overline{R}_A}{k}\phi = \frac{1}{3}\frac{\overline{R}_A}{k}R\left(\frac{k}{D_{AB}}\right)^{1/2} \tag{3.2.74}$$

Our goal in looking at these *asymptotic limits* (this is what we call these expressions for large and small values of a parameter) is to obtain algebraic expressions that tell us how the variable of interest, in this case $C_R$, depends on system parameters and properties. For example, for small $\phi$, from Eq. 3.2.73:

$$kC_R = \overline{R}_A \tag{3.2.75}$$

We conclude that at a specified $\overline{R}_A$, $C_R$ is independent of $R$ and $D_{AB}$, and depends only on $k$. On the other hand, for large $\phi$, Eq. 3.2.74 gives (using the definition of $\phi$)

$$kC_R = \overline{R}_A\frac{\phi}{3} \tag{3.2.76}$$

In this limit of behavior we find that at a specified $\overline{R}_A$, $C_R$ increases with increasing $\phi$, as expected. That is, $C_R$ depends on $k$, even at large $R$.

An important feature of mathematical modeling is the examination of the manner in which the solution depends on the physical parameters that appear in the problem. It is often very useful to know how certain parameters affect the solution, especially in the limits of very large and very small values of certain parameters. Putting the solution into a dimensionless format aids that process. This is one of the lessons of Example 3.2.4.

**EXAMPLE 3.2.5** *Control of Organism Growth*

As a consequence of a search for intelligent life in Urey Hall at the University of California, San Diego, a small spherical organism is discovered. It has a respiration rate of 7 g $O_2$/L · h, and this respiration rate appears to be independent of external oxygen concentration in the surrounding air. Controlled tests indicate that the organism will not reproduce as long as an organ located at the center of the organism is exposed to a concentration of oxygen below $10^{-7}$ mol/cm³. Assume that the diffusion coefficient of $O_2$ through the organism is $D = 10^{-5}$ cm²/s. The organism has a radius of 100 $\mu$m ($10^{-2}$ cm).

   **a.** What surface concentration of oxygen (exterior to the organism) must be maintained to permit reproduction? Give your answer in units of atmospheres of oxygen (i.e., as a partial pressure).
   **b.** What happens if the surface concentration is reduced to 85% of the critical value found in part (a), *but* the organism somehow divides into two equal spheres? (Will they reproduce?)

A respiration rate independent of external oxygen concentration implies that the kinetics are of zero order. Apparently the limitation on reproduction is independent of respiration. From a result given earlier (Eq. 3.2.46) for zero-order kinetics, setting $r = 0$, we find the center concentration to be

$$C(r = 0) = C_R - \frac{R_A^o R^2}{6D_{AB}} \tag{3.2.77}$$

We convert the respiration rate to molar units

$$R_A^o = 7 \text{ g/L} \cdot \text{h} = \frac{7}{32} \frac{10^{-3}}{3600} = 6 \times 10^{-8} \text{ mol/cm}^3 \cdot \text{s} \tag{3.2.78}$$

and then we find, from Eq. 3.2.77,

$$C_R^{\text{crit}} = C^{\text{crit}}(r = 0) + \frac{R_A^o R^2}{6D_{AB}} = 10^{-7} + \frac{6 \times 10^{-8}(10^{-2})^2}{6 \times 10^{-5}} = 2 \times 10^{-7} \text{ mol/cm}^3 \tag{3.2.79}$$

This concentration is in the *interior* of the organism, just inside its boundary. To connect this value with the required *exterior* concentration (in the gas phase), we need solubility data for oxygen in the organism. This would be in the form of a relationship between $C$ and $p$ (e.g., Henry's law: Eq. 3.2.15). Suppose we use the properties of water as characteristic of a cell. We know that at 25°C the solubility of oxygen in water that is in equilibrium with normal air with 21% oxygen is $C = 2.7 \times 10^{-7}$ mol/cm$^3$. We conclude that a partial pressure of oxygen of [2]

$$p_{crit} = \left(\frac{0.21}{2.7 \times 10^{-7}}\right) \times (2 \times 10^{-7}) = 0.156 \text{ atm} \qquad (3.2.80)$$

would support respiration and reproduction. Normal air at 0.21 atm oxygen is more than sufficient to support respiration.

Suppose now that the oxygen in the ambient air is reduced to

$$C_R = 0.85 C_R^{crit} = 1.7 \times 10^{-7} \text{ mol/cm}^3 \qquad (3.2.81)$$

and the organism splits so that $R = 80 \times 10^{-4}$ cm. Then our model says that

$$C(r = 0) = C_R - \frac{R_A^o R^2}{6 D_{AB}} = 1.7 \times 10^{-7} - \frac{6 \times 10^{-8}(0.8 \times 10^{-2})^2}{6 \times 10^{-5}} \qquad (3.2.82)$$
$$= 1.7 \times 10^{-7} - 0.64 \times 10^{-7} = 1.06 \times 10^{-7} \text{ mol/cm}^3$$

This value just *exceeds* the level stated earlier ($10^{-7}$ mol/cm$^3$) at which reproduction takes place. The organism can breed by fission, but it would fail to breed without fission since, if $R = 100$ $\mu$m were maintained, Eq. 3.2.82 would yield a concentration below the critical value of $10^{-7}$. [It would give $C(r = 0) = 0.7 \times 10^{-7}$.]

Example 3.2.5 illustrates one use of a mathematical model: it provides a means of interpretation of physical phenomena for the purpose of enhancing our understanding of physical processes and providing a means of predicting the behavior of the system under some set of imagined or proposed conditions. Another important use of mathematical models is in design, by which we mean the specification of operating conditions, geometry, sizes, and materials to facilitate the performance of a proposed system according to some desired specifications. The next two example problems lead us in that direction.

**EXAMPLE 3.2.6**  *A Steady State Convection/Diffusion Problem*

Two gases are initially separated into two bulbs connected by a narrow diameter capillary. Pure $N_2$ then flows from right to left, as shown in Fig. 3.2.5. To what degree can $O_2$ diffuse *upstream*, against the $N_2$ flow?

We *assume* that the system has achieved some steady state.[3] We begin the analysis with the diffusion equation in the vector form

$$\mathbf{v} \cdot \nabla C_A = D_{AB} \nabla^2 C_A + R_A \qquad (3.2.83)$$

The domain is the connecting tube $z = [0, L]$. We choose the oxygen concentration $C_A$ to be the dependent variable. (We drop the subscript A in subsequent equations.)

---

[2] First we find the Henry's law constant as $H = p/C = 0.21$ atm/($2.7 \times 10^{-7}$ mol/cm$^3$). Then $p_{crit} = HC_{crit}$ yields Eq. 3.2.80.

[3] Might this be another example of an assumption that leads to a nonsensical result? Yes—it might! Such an early warning is not always provided, but if the assumption is misleading, we should be able to tell when we examine the resulting model and consider its predicted behavior.

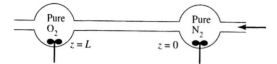

**Figure 3.2.5** Diffusion of oxygen counter to a flow of nitrogen.

Assume that the volumes at the two ends are so large that each remains nearly pure, even though some exchange takes place. This will be a good approximation in the early stages of this process, before much $N_2$ has gone into the left-hand bulb. The early stage model will be sufficient for us to make the necessary point.

We assume that no reactions occur:

$$R_A = 0 \tag{3.2.84}$$

We also assume that the velocity vector has only an axial component and that there is diffusion only in the axial direction. Then the convective diffusion equation reduces to (we drop the subscript AB on $D$),

$$v\frac{dC}{dz} = D\frac{d^2C}{dz^2} \tag{3.2.85}$$

Boundary conditions are

$$C = C_o \text{ at } z = L \tag{3.2.86}$$

$$C = 0 \text{ at } z = 0 \tag{3.2.87}$$

Let us define a new variable $p$ as

$$p \equiv \frac{dC}{dz} \tag{3.2.88}$$

Then Eq. 3.2.85 becomes

$$\frac{dp}{dz} - \frac{v}{D}p = 0 \tag{3.2.89}$$

and the solution is easily found as

$$p = a'e^{vz/D} = \frac{dC}{dz} \tag{3.2.90}$$

After putting the definition of $p$ back in, and integrating once more, we find $C$ as

$$C = ae^{vz/D} + b \tag{3.2.91}$$

One of the assumed boundary conditions is Eq. 3.2.87. From this it follows that

$$0 = a + b \tag{3.2.92}$$

The solution now has the form

$$C = a(e^{vz/D} - 1) \tag{3.2.93}$$

From the first boundary condition (Eq. 3.2.86) we may write

$$C_o = a(e^{vL/D} - 1) \tag{3.2.94}$$

After solving for the constant $a$ we may write the solution in the form

$$\frac{C}{C_o} = \frac{e^{vz/D} - 1}{e^{vL/D} - 1} \tag{3.2.95}$$

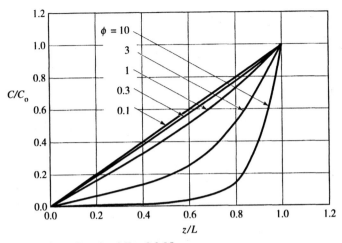

**Figure 3.2.6** Graph of Eq. 3.2.95.

The concentration distribution along the connecting tube is sketched in Fig. 3.2.6. We note that in the limit of very small values of $vL/D$ the solution becomes linear:

$$\text{for } \frac{vL}{D} \ll 1, \qquad \frac{C}{C_o} = \frac{vz/D}{vL/D} = \frac{z}{L} \tag{3.2.96}$$

An important feature of the solution is the flux at the end of the tube—$z = L$. Hence we need to differentiate the concentration function, and we find

$$\left.\frac{dC}{dz}\right|_{z=L} = C_o \frac{v}{D} \frac{e^{vL/D}}{e^{vL/D} - 1} \tag{3.2.97}$$

It is useful to nondimensionalize the solution to this problem. For example, we may introduce a dimensionless concentration $C/C_o$ and a dimensionless space variable $z/L$. Then a dimensionless oxygen flux becomes

$$S \equiv \frac{dC/C_o}{dz/L} = \frac{\phi e^{\phi}}{e^{\phi} - 1} \qquad \text{at } z = L \tag{3.2.98}$$

where

$$\phi = \frac{vL}{D} \tag{3.2.99}$$

We have noted here that $vL/D$ is dimensionless, and so we change its name to $\phi$.

The behavior of the flux at very large, and very small, values of the parameter $\phi$ is interesting. For large values, for example, it is easy to show that

$$\lim_{\phi \to \infty} S = \phi \tag{3.2.100}$$

Now, suppose that the practical engineering design issue is that we wish to prevent the intrusion of oxygen into the right-hand chamber of this system. We have available a *mathematical model* with which we can explore the effect of various design and operating conditions on the rate of intrusion of oxygen into that chamber. *That is why we have displayed the solution in terms of the flux of oxygen at z = L.*

As an example we can pose the following question:

*How large must $\phi = vL/D$ be to obtain $C/C_o = 0.01$ at $z/L = 0.99$?*

Inserting these conditions into the solution for $C(z)$ we find

$$0.01 = \frac{e^{(z/L)\phi} - 1}{e^\phi - 1} = \frac{e^{0.99\phi} - 1}{e^\phi - 1} \qquad \textbf{(3.2.101)}$$

We wish to solve for $\phi$. But Eq. 3.2.101 is not explicit in $\phi$, and a trial-and-error solution, or a numerical solution, would be required. Since $\phi$ must be large, try an approximation in which the $-1$ terms are dropped in both the numerator and denominator of Eq. 3.2.101. Then

$$0.01 = e^{-0.01\phi} \qquad \textbf{(3.2.102)}$$

Taking logarithms of both sides we find

$$0.01\phi = 4.6 \qquad \textbf{(3.2.103)}$$

and

$$\phi = 4.6 \times 10^2 = 460 \qquad \textbf{(3.2.104)}$$

Hence, taking $C/C_o = 0.01$ as a criterion, diffusion against the flow can be effectively stopped if

$$\frac{vL}{D} = 460 \qquad \textbf{(3.2.105)}$$

or

$$v = \frac{460D}{L} \qquad \textbf{(3.2.106)}$$

For $D = 0.1$ cm$^2$/s, $L = 100$ cm, $v = 0.46$ cm/s is sufficient.

It should be apparent that $\phi = vL/D$ plays an important role here. Note that it is dimensionless. We call $\phi = vL/D$ a *Peclet number*, written as Pe. Note also that $\phi$ may be written (upon introducing the kinematic viscosity $\nu = \mu/\rho$)

$$\phi = \frac{vL}{D} = \frac{vL}{\nu}\frac{\nu}{D} = \mathrm{Re_L}\frac{\nu}{D} = \mathrm{Re_L Sc} = \mathrm{Pe} \qquad \textbf{(3.2.107)}$$

where $\mathrm{Re_L} = vL/\nu$ is a Reynolds number (but based on the length $L$, rather than on the tube *diameter*) and $\mathrm{Sc} = \nu/D$ is called the Schmidt number. We see from this model that the physics of intrusion of a gas phase species, diffusing against a gas flow, is determined by the values of certain dimensionless parameters—in this case the Peclet number $\phi$. *It appears that the Peclet number is an important measure of the relative importance of diffusion and convection.* We will see numerous confirmations of this observation as we continue modeling.

---

**EXAMPLE 3.2.7**   *Design of an Outlet for a Reactor*

Now we illustrate the use of a mathematical model of a particular physical process to explore the implications of that model with a view to making some *engineering design* decisions.

A chemical reaction is carried out within a heated, closed reactor. When the reaction is completed, the contents of the reactor are cooled by blowing cold inert gas through

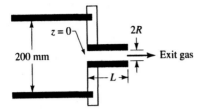

**Figure 3.2.7** End cap of a reactor.

the reactor. Of course there must be an exit for the inert gas (Fig. 3.2.7), and this opens the possibility of outside air diffusing into the reactor. Such contamination must be prevented by proper design of the gas exit port.

We have a 10-liter vessel, and the inert gas is nitrogen, which is supplied at 100 *standard cubic centimeters per minute (sccm)*. By "standard" we mean that the volume is referred to one atmosphere pressure and 273 K. The gas leaves the reactor at an elevated temperature, but below 300°C. We have to design an exit port for this system, along the lines suggested in Fig. 3.2.7.

*Our primary design task is to specify suitable values of L and R.* The *design* constraints are as follows: the pressure within the reactor is not to exceed 800 torr, and the oxygen concentration in the reactor is not to exceed one part per million by volume.

We begin by following along the lines of Example 3.2.6. If we assume a flat velocity profile in the nozzle, we must solve

$$v\frac{dC}{dz} = D\frac{d^2C}{dz^2}$$

(3.2.108)

where $C$ is the molar concentration of oxygen. We will convert the concentration units to partial pressure units, noting that for an ideal gas

$$C = \frac{p}{R_G T}$$

(3.2.109)

Hence

$$v\frac{dp}{dz} = D\frac{d^2p}{dz^2}$$

(3.2.110)

and the boundary conditions are $p = 0.2$ atm at $z = L$ (assuming the ambient air is 20% oxygen) and $p = 10^{-6}$ atm at $z = 0$ (assuming near-atmospheric pressure at $z = 0$, and an oxygen concentration of one part per million by volume).

In Example 3.2.6 we found

$$C = ae^{vz/D} + b$$

(3.2.111)

We replace $C$ by $p$ and impose the boundary conditions (which are in terms of $p$) and find

$$p = \frac{(0.2 - 10^{-6})\, e^{vz/D} + 10^{-6}\, e^{vL/D} - 0.2}{e^{vL/D} - 1}$$

(3.2.112)

Now we want to find the velocity $v$ that is high enough to ensure that at the entrance to the exit port (at $z = 0$) $p = 10^{-6}$. But if we substitute $z = 0$ into Eq. 3.2.112 we find that $p = 10^{-6}$ *for any v!* Does this make sense physically?

We have again made a very common mistake in problem solving: we overspecified the problem. As a result, the role of the key design variable (in this case, the exit velocity $v$) actually was eliminated from the problem because we *asserted*, in writing the boundary conditions, that $p = 10^{-6}$ at $z = 0$ *regardless of $v$*. Hence the result makes sense physically only because we have imposed some unknown physics on the problem. We stated that we have a mechanism for keeping $p = 10^{-6}$ at $z = 0$, but we neglected to describe how the mechanism operates.

> *We have not formulated this problem properly. We must go back to the physics and talk about what is going on here.*

When the reactor is first "opened up" to exterior diffusion, we may have some initial concentration of oxygen within the reactor. We will assume that we can take this initial value to be essentially zero compared to $10^{-6}$. Otherwise we probably have a problem even before we begin to flush and cool the reactor. By controlling the outlet design and the flowrate of the flushing gas, we can retard the intrusion of any additional oxygen into the reactor. The mechanism of intrusion is diffusion against the outlet flow. There will be some finite *rate* at which oxygen diffuses against the efflux from the reactor. The key word, and the key to the physics of the problem, is *rate*. We cannot control the concentration of oxygen at $z = 0$ because that concentration is dependent on the concentration of oxygen *within* the reactor. There is a *transient* process whereby oxygen diffuses into the reactor and is diluted within the reactor volume. The dynamics *within* the reactor, and the diffusion upstream *through the exit port*, are *coupled processes*.

We must write a species balance on the reactor itself. *We will assume that the reactor contents are well mixed in the gas phase.* This means that the concentration of oxygen within the reactor volume $V_R$ is spatially uniform. In particular, the "well mixed" assumption implies that if the concentration at some time, within the reactor, is $C_o$, then the concentration right at the position $z = 0$, and at that particular time, is also $C_o$. But $C_o$ is unknown—we cannot fix its value through any physical argument and still have control over $C_o$ by controlling the velocity $v$ at the exit.

Now we go back to the diffusion/convection problem for the exit tube. We still solve

$$v \frac{dC}{dz} = D \frac{d^2C}{dz^2} \tag{3.2.113}$$

but now the boundary conditions are $C = C_o$ at $z = 0$ and $C = 0.2$ at $z = L$. (We will use partial pressure units for the concentration variable in what follows. However, we have changed the notation from $p$ to $C$.)

The solution remains

$$C = ae^{vz/D} + b \tag{3.2.114}$$

but now we find

$$C = \frac{(0.2 - C_o)e^{vz/D} + C_o e^{vL/D} - 0.2}{e^{vL/D} - 1} \tag{3.2.115}$$

Next, we must write a species balance on the reactor. It is simply a statement that the rate of increase of oxygen *within* the reactor volume $V_R$ must arise from the *net* effect of a diffusive inflow and a convective outflow

$$V_R \frac{dC}{dt} = \left[ +D \frac{\partial C}{\partial z}\bigg|_{z=0} - vC|_{z=0} \right] \pi R^2 \tag{3.2.116}$$

Note the sign on the diffusive term. A positive diffusive flux would be in the $+z$ direction, which is *out* of the reactor. Hence the diffusive *inflow* to the reactor is the negative of the flux

$$-\text{flux} = -\left(-D\frac{\partial C}{\partial z}\right)$$

Now we need $\partial C/\partial z$ at $z = 0$, which is obtained from Eq. 3.2.115 as

$$\frac{\partial C}{\partial z}\bigg|_{z=0} = \frac{(0.2 - C_0)v/D}{e^{vL/D} - 1}$$ (3.2.117)

The balance on the reactor takes the form

$$\frac{V_R}{\pi R^2 v}\frac{dC}{dt} = \frac{0.2 - C_o}{e^{vL/D} - 1} - C_o = \frac{0.2 - C_o e^{vL/D}}{e^{vL/D} - 1}$$ (3.2.118)

with an initial condition $C = 0$ at $t = 0$.

Recall the discussion at the end of Chapter 2, on the *quasi-steady approximation*. We have just carried out another quasi-steady analysis, in which we used the steady state solution, Eq. 3.2.115, coupled with the unsteady reactor balance, to yield Eq. 3.2.118. We examine first the case of no flushing.

Let's think about this special case for a moment. Suppose the reactor were opened but no flushing flow occurred. Then we should be able to set $v = 0$ in Eq. 3.2.118. To avoid problems, we must write

$$e^{vL/D} = 1 + \frac{vL}{D} + \frac{1}{2}\left(\frac{vL}{D}\right)^2 + \cdots$$ (3.2.119)

and then take the limit as $v \to 0$. If we do this, after multiplying both sides of Eq. 3.2.118 by $v$, we find

$$\frac{V_R}{\pi R^2}\frac{dC}{dt} = \frac{(0.2 - C_o)D}{L}$$ (3.2.120)

This is exactly what we should expect, since in the absence of flow, the diffusion flux down the tube of length $L$ (for dilute diffusion) is just $D\Delta C/L$, and $\Delta C = 0.2 - C_o$.

If we take the reactor to be well mixed, so that we can set $C_o = C$, we must solve

$$\frac{V_R}{\pi R^2}\frac{dC}{dt} = \frac{(0.2 - C)D}{L}$$ (3.2.121)

No flushing

subject to the initial condition

$$C = 0 \quad \text{at } t = 0$$

Define a dimensionless time as

$$\frac{\pi R^2 D}{L V_R}t = \tau$$ (3.2.122)

Then Eq. 3.2.121 is

$$\frac{dC}{d\tau} = 0.2 - C$$ (3.2.123)

which is the same as

$$-\frac{d(0.2 - C)}{d\tau} = 0.2 - C$$ (3.2.124)

Letting

$$0.2 - C = C' \tag{3.2.125}$$

we find

$$-\frac{dC'}{d\tau} = C' \tag{3.2.126}$$

subject to the initial condition

$$C' = 0.2 \quad \text{at } \tau = 0 \tag{3.2.127}$$

The solution is

$$C' = 0.2e^{-\tau} \tag{3.2.128}$$

or

$$0.2 - C = 0.2e^{-\tau}$$

Hence we find

$$C = 0.2(1 - e^{-\tau}) \tag{3.2.129}$$

We can now find how long it will take for $C$ to reach $10^{-6}$ *in the absence of any inert flushing.* Set $C = 10^{-6}$ and solve for $\tau$.

$$10^{-6} = 0.2(1 - e^{-\tau}) \tag{3.2.130}$$

$$e^{-\tau} = 1 - \frac{10^{-6}}{0.2} = 1 - 5 \times 10^{-6} \tag{3.2.131}$$

The solution is

$$\tau = 5 \times 10^{-6} \tag{3.2.132}$$

In *real* time this is

$$t = 5 \times 10^{-6} \frac{LV_R}{\pi R^2 D} \tag{3.2.133}$$

No flushing

Using $V_R = 10 \text{ L} = 10^4 \text{ cm}^3$, $D = 0.3 \text{ cm}^2/\text{s}$ (an order-of-magnitude estimate), and $R = 1$ cm, we find

$$t = 0.053 \, L \text{ seconds} \tag{3.2.134}$$

when $L$ is in centimeter units. We see that unless $L$ is $O(10^3)$ cm the reactor quickly exceeds the $10^{-6}$ limitation. *This is why we need to flush the reactor.*

Alternatively we could use a very small value of $R$. For example, for $R = 10^{-2}$ cm we find $t = 530 \, L$ seconds and an $L$ of 10 cm gives us over an hour during which $C$ is less than $10^{-6}$.

We should remind ourselves at this point that this is a mathematical model of this system, and it is subject to a number of approximations that permit us to get a simple analytical result. One key issue here is whether the "well mixed" assumption makes any sense for the no-flushing case, where the fluid is not in forced motion. Some thought suggests that if the reactor is not well mixed, the rate of intrusion of contaminant will be retarded. Hence this result (Eq. 3.2.134) is conservative.

We must examine another feature of this model, namely, the pressure distribution along the length of the efflux tube. By implication, we have neglected any variation of pressure in this system. The only reason to use an efflux tube at all is to flush the reactor. If $R$ is too small, the pressure required to flush the reactor would be excessive.

For example, we can calculate the pressure required to flow 100 sccm through a tube of radius $R = 0.01$ cm and length $L = 10$ cm at 300°C. *If we ignore the volume change with pressure,* to get a first approximation, we use Poiseuille's law (review your fluid dynamics)

$$\Delta P = \frac{8\mu LQ}{\pi R^4} \qquad (3.2.135)$$

Estimate $\mu = 3 \times 10^{-4}$ poise at 300°C. Take $Q$ as

$$100 \text{ sccm} = \frac{100}{60}\left(\frac{273 + 300}{273}\right) \text{cm}^3/\text{s} = 3.5 \text{ cm}^3/\text{s} \qquad (3.2.136)$$

Then

$$\Delta P = \frac{8(3 \times 10^{-4})10(3.5)}{\pi(0.01)^4} = 2.67 \times 10^6 \text{ dyn/cm}^2 \qquad (3.2.137)$$

$$= 2.67 \text{ atm} = 2035 \text{ torr}$$

This $\Delta P$ exceeds the design constraint. Keep in mind, however, that this calculation assumes an incompressible fluid, so that $Q$ is independent of axial position. We must use a modified Poiseuille's law for a compressible fluid in which $Q$ varies along the axis of the tube. We will not follow up on this here. (See Problem 3.5.)

We have examined behavior in a limiting case—no flushing. Now we turn to an examination of the effect of flushing. For a finite efflux velocity $v$ we define

$$\frac{vL}{D} = \text{Pe} \qquad (3.2.138)$$

and, to simplify the writing,

$$e^{\text{Pe}} = \Phi \qquad (3.2.139)$$

Then we may write the quasi-steady dynamic equation for the transient oxygen concentration as

$$\frac{dC}{d\tau} = \text{Pe}\frac{0.2 - C\Phi}{\Phi - 1} \qquad (3.2.140)$$

With flushing

(We have continued to use the "well mixed" approximation.)

We can solve this equation, as before, but some valuable information may be available just from the equation itself. For instance, a *dynamic steady state* will eventually be reached, where the efflux balances the diffusive influx, and $C$ will reach a constant value. When this occurs, $dC/d\tau = 0$ (steady state occurs and $C$ is not a function of time). By inspection of the dynamic equation we see that when $dC/d\tau$ becomes zero we must have

$$0.2 - C\Phi = 0 \qquad (3.2.141)$$

or

$$C = \frac{0.2}{\Phi} \qquad (3.2.142)$$

is the steady state value.

Thus the *maximum* concentration of oxygen reached in the reactor is given by this value. If we demand that $v$ be large enough that $C$ *always* will be below $10^{-6}$ atm, we solve for

$$\Phi = e^{Pe} = e^{vL/D} \tag{3.2.143}$$

from

$$10^{-6} = \frac{0.2}{\Phi} \tag{3.2.144}$$

We find

$$e^{Pe} = 2 \times 10^5 \tag{3.2.145}$$

or

$$Pe \geq 12.2 \tag{3.2.146}$$

If we ignore any compressibility effect of $\Delta P$ on the volumetric flow (note the comments just following Eq. 3.2.137) and write

$$v = \frac{Q}{\pi R^2} \tag{3.2.147}$$

then we require that

$$Pe = \frac{vL}{D} = \left(\frac{Q}{\pi D}\right)\frac{L}{R^2} \geq \frac{3.5}{\pi(0.3)}\frac{L}{R^2} \tag{3.2.148}$$

(We have used an order-of-magnitude estimate for $D$ from $D \approx v$ for an ideal gas, a result that comes out of the simplest form of the kinetic theory of gases.) Inverting Eq. 3.2.148, we conclude that we require

$$\frac{L}{R^2} = \frac{0.3\pi}{3.5}(12.2) = 3.3 \text{ cm}^{-1} \tag{3.2.149}$$

If $R$ were $O(0.1)$ cm or less, we would find that $L/R < 1$ unless a much larger Pe than the minimum given by Eq. 3.2.146 were used. In this case the pressure drop $\Delta P$ would be calculated for flow through an orifice or abrupt contraction, rather than with Poiseuille's law (i.e., for $L/R < 1$).

*We may choose* $R = 0.33$ cm and $L = 5$ cm. This will give Pe > 12.2 (conservative design). Now it is possible to use Poiseuille's law to estimate pressure loss (since $L/R = 15$ which is tubelike rather than orifice-like). From Eq. 3.2.135 we easily find that $\Delta P$ is very small. This fact supports the use (about which we expressed appropriate caution) of Poiseuille's law for a compressible gas, for this choice of parameters.

What is the Reynolds number under this design? Using the more standard definition based on tube diameter ($2R$), we find

$$Re = \frac{2Rv\rho}{\mu} = \frac{2Q}{\pi Rv} = \frac{2(3.5)}{\pi(0.33)(0.3)} = O(10) \tag{3.2.150}$$

Flow is laminar in the exit tube, and likely so in the entrance tube. This raises a suspicion about the assumption of a well-mixed reactor, since there is little convective mixing at such a low Reynolds number. But poor internal mixing would retard the inflow by diffusion *(Why? Think about this.),* and so we probably have a conservative design.

**EXAMPLE 3.2.8**  *Diffusion Through a Film Within Which There is a Homogeneous Reaction*

We consider a thin static film of material that separates two gaseous regions of different concentrations. Suppose the concentrations are denoted by partial pressures $p_{A,o}$ and $p_{A,L}$ in the gas on each side of the film. We will take $p_{A,L} = 0$. We suppose that the two interfaces at $z = 0$ and $z = L$ are in equilibrium, with the result that the concentrations *within* the film, at the interface, satisfy a relationship such as Henry's law

$$C_A = Hp_A \qquad (3.2.151)$$

with $H$ constant.

We seek an expression for the flux of species A across the film, and specifically we want to know how the reaction affects the flux.

For this simple geometry, and *assuming steady state*, we may begin with the diffusion equation in the form

$$0 = D\frac{d^2C_A}{dz^2} + R_A \qquad (3.2.152)$$

where $R_A$ is a volumetric rate of reaction of species A within the film (in units of moles/time-volume). We will assume a first-order reaction and write

$$R_A = -kC_A \qquad (3.2.153)$$

where the minus sign corresponds to the *consumption* of species A, rather than to its *production* by reaction. Thus we must solve

$$D\frac{d^2C_A}{dz^2} - kC_A = 0 \qquad (3.2.154)$$

Since the equation is *second*-order with respect to $z$, we need *two* boundary conditions. We will imagine that the gas phases are so well mixed that the partial pressures are

$$p_A = p_{A,o} \qquad \text{at } z = 0 \qquad (3.2.155)$$
$$p_A = 0 \qquad \text{at } z = L \qquad (3.2.156)$$

right at the interfaces. Then, with Eq. 3.2.151, boundary conditions on $C_A$ are

$$C_A = Hp_{A,o} \equiv C_{A,o} \qquad \text{at } z = 0$$
$$C_A = 0 \qquad \text{at } z = L \qquad (3.2.157)$$

A lot of algebraic manipulation is going to follow. We are less likely to make mistakes if we simplify the notation. One way to do this is to introduce dimensionless combinations of the many parameters that appear in the equations. By nondimensionalizing at this point we can also see what dimensionless groups are important in this problem.

If we define

$$\zeta = \frac{z}{L} \qquad (3.2.158)$$

and (cf. Eq. 3.2.65) a parameter called the Thiele modulus

$$b_1 = \left(\frac{kL^2}{D}\right)^{1/2} \qquad (3.2.159)$$

then we may write Eq. 3.2.154 as

$$\frac{d^2C_A}{d\zeta^2} - b_1^2 C_A = 0 \qquad (3.2.160)$$

which must now be solved subject to the boundary conditions

$$C_A = C_{A,o} \qquad \text{at } \zeta = 0 \qquad\qquad (3.2.161)$$

$$C_A = 0 \qquad \text{at } \zeta = 1 \qquad\qquad (3.2.162)$$

We might recognize Eq. 3.2.160 as a standard form of second-order ordinary differential equation whose solution may be written in terms of hyperbolic functions (sinh and cosh). Otherwise we may derive the solution as follows: If we substitute

$$C_A = ae^{\alpha\zeta} \qquad\qquad (3.2.163)$$

into Eq. 3.2.160 we find

$$a\alpha^2 e^{\alpha\zeta} - b_1^2 a e^{\alpha\zeta} = 0 \qquad\qquad (3.2.164)$$

or

$$\alpha^2 - b_1^2 = 0 \qquad\qquad (3.2.165)$$

Hence

$$\alpha = \pm b_1 \qquad\qquad (3.2.166)$$

and $C_A$ is given by

$$C_A = a_1 e^{b_1\zeta} + a_2 e^{-b_1\zeta} \qquad\qquad (3.2.167)$$

and $a_1$ and $a_2$ must be found from Eqs. 3.2.161 and 3.2.162:

$$C_{A,o} = a_1 + a_2 \qquad\qquad (3.2.168)$$

$$0 = a_1 e^{b_1} + a_2 e^{-b_1} \qquad\qquad (3.2.169)$$

We find

$$a_1 = C_{A,o} - a_2 \qquad\qquad (3.2.170)$$

$$0 = (C_{A,o} - a_2)e^{b_1} + a_2 e^{-b_1} \qquad\qquad (3.2.171)$$

or

$$a_2 = \frac{C_{A,o} e^{b_1}}{e^{b_1} - e^{-b_1}} \qquad\qquad (3.2.172)$$

and

$$a_1 = C_{A,o} - a_2 = C_{A,o}\left(1 - \frac{e^{b_1}}{e^{b_1} - e^{-b_1}}\right)$$

$$= C_{A,o}\frac{-e^{-b_1}}{e^{b_1} - e^{-b_1}} \qquad\qquad (3.2.173)$$

Hence the solution is

$$\frac{C_A}{C_{A,o}} = \frac{-e^{-b_1}e^{b_1\zeta} + e^{b_1}e^{-b_1\zeta}}{e^{b_1} - e^{-b_1}}$$

$$= \frac{\exp[b_1(1-\zeta)] - \exp[-b_1(1-\zeta)]}{e^{b_1} - e^{-b_1}} \qquad\qquad (3.2.174)$$

$$= \frac{\sinh b_1(1-\zeta)}{\sinh b_1}$$

The molar flux of A at $\zeta = 0$ is

$$N_{A,z}|_{z=0} = -D\frac{dC_A}{dz}\bigg|_{z=0} = -\frac{DC_{A,o}}{L}\frac{d(C_A/C_{A,o})}{d\zeta}\bigg|_{\zeta=0} \tag{3.2.175}$$

and we find

$$N_{A,z}|_{z=0} = \frac{DC_{A,o}}{L}\left[\frac{b_1 \cosh b_1(1-\zeta)}{\sinh b_1}\right]_{\zeta=0} = \left(\frac{DC_{A,o}}{L}\frac{b_1}{\tanh b_1}\right) \tag{3.2.176}$$

It is easy to show that if there were no reaction at all, the flux would be

$$N_{A,z}^o = \frac{DC_{A,o}}{L} \tag{3.2.177}$$

(We can either solve Eq. 3.2.154 with $k = 0$ or examine Eq. 3.2.176 as $b_1$ vanishes. In the latter case we note that

$$\lim_{b_1 \to 0} \tanh b_1 = b_1 \tag{3.2.178}$$

and Eq. 3.2.177 follows directly from Eq. 3.2.176.)

We may define a dimensionless flux as the ratio

$$\frac{N_{A,z}|_{z=0}}{DC_{A,o}/L} = \varphi_0 \tag{3.2.179}$$

In this case $\varphi_0$ is a measure of the *enhancement* of the flux of A, at $z = 0$, due to reaction within the film. Hence

$$\varphi_0 = \frac{b_1}{\tanh b_1} \tag{3.2.180}$$

It is easy to confirm that $\varphi_0$ has the behavior

$$\lim_{b_1 \to 0} \varphi_0 = 1 \tag{3.2.181}$$

$$\lim_{b_1 \to \infty} \varphi_0 = b_1 \tag{3.2.182}$$

We present a plot of $\varphi_0(b_1)$ in Fig. 3.2.8. Note that for a very fast reaction (i.e., for $b_1 \gg 1$), we find (since $\tanh b_1 \to 1$)

$$N_{A,z}|_{z=0} = \frac{DC_{A,o}}{L}b_1 = C_{A,o}(Dk)^{1/2} \tag{3.2.183}$$

while for very slow reaction

$$N_{A,z}|_{z=0} = \frac{DC_{A,o}}{L} \quad \text{for} \quad b_1 \ll 1 \tag{3.2.184}$$

Note also that for a fast reaction the flux $N_{A,z}|_{z=0}$ depends on $D$ and $k$. For a slow reaction this flux is independent of $k$. It is diffusion controlled.

Now let us look at the flux *out* of the film, which is (note Eq. 3.2.176 but put in $\zeta = 1$)

$$N_{A,z}|_{z=L} = \frac{DC_{A,o}}{L}\frac{b_1}{\sinh b_1} \tag{3.2.185}$$

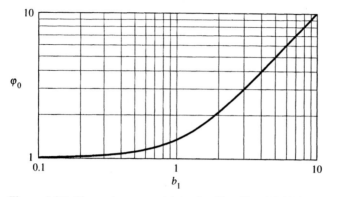

**Figure 3.2.8** Flux enhancement into the film (Eq. 3.2.180).

and let us define an "enhancement" factor

$$\varphi_1 = \frac{N_{A,z}|_{z=L}}{DC_{A,o}/L} = \frac{b_1}{\sinh b_1} \tag{3.2.186}$$

Note that

$$\lim_{b_1 \to 0} \varphi_1 = 1 \tag{3.2.187}$$

and

$$\lim_{b_1 \to \infty} \varphi_1 = \frac{2b_1}{e^{b_1}} = 0 \tag{3.2.188}$$

*Hence the flux leaving the film is not enhanced—it is reduced.* This makes sense, of course, since the faster the reaction, the less A can diffuse across the entire film without disappearing through reaction.

It is interesting to contrast this problem to a related problem. Suppose species A enters the film *but does not leave it,* since the boundary at $z = L$ is impermeable. This will require a change in boundary conditions, in comparison to the model developed above. You should carry out the complete analysis, and show that the dimensionless flux at $z = 0$ may be written as (cf. Eq. 3.2.180)

$$\varphi_0 = b_1 \tanh b_1 \tag{3.2.189}$$

In this case, $\varphi_0$ is not an *enhancement* factor, because the term to which the flux is normalized, $DC_{A,o}/L$, is not the flux in the absence of reaction. This term is simply an arbitrary but appropriate scale factor for the definition of $\varphi_0$.

It is not difficult to show that the asymptotic behavior of $\varphi_0$ is

$$\lim_{b_1 \to 0} \varphi_0 = b_1^2 \tag{3.2.190}$$

and

$$\lim_{b_1 \to \infty} \varphi_0 = b_1 \tag{3.2.191}$$

Hence we find a plot of Eq. 3.2.189 as shown in Fig. 3.2.9.

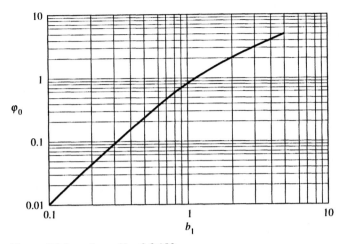

**Figure 3.2.9** $\varphi_0$ from Eq. 3.2.189.

In carrying out the model for the case that the lower boundary is *closed*, you will find, from Eq. 3.2.190, that

$$N_{A,z}|_{z=0} = \frac{DC_{A,o}}{L} b_1^2 = C_{A,o}Lk \qquad (3.2.192)$$

for slow reaction ($b_1 \ll 1$) and, from Eq. 3.2.191,

$$N_{A,z}|_{z=0} = \frac{DC_{A,o}}{L} b_1 = C_{A,o}(kD)^{1/2} \qquad (3.2.193)$$

for fast reaction ($b_1 \gg 1$). Hence we see that for slow reaction the flux is independent of $D$ and depends on $k$ (cf. Eqs. 3.2.192 and 3.2.184), while for a fast reaction (as in Eq. 3.2.183) the flux depends on $D$ as well as $k$. Note also how the parametric dependence on $L$ differs between Eqs. 3.2.184 and 3.2.192. This example permits us to appreciate the importance of boundary conditions in model formulation. *The boundary conditions are an essential part of the physics of the model; they are not just an afterthought that modifies the coefficients of the model.*

We look next at a diffusion reaction problem featuring fairly complicated kinetics.

**EXAMPLE 3.2.9**   *Diffusion with Second-Order Reaction*

Imagine a film of stagnant liquid in which a chemical species of concentration $C_2$ is dissolved. At the surface $z = 0$ a concentration $C_{1o}$ of another species is maintained. As sketched in Fig. 3.2.10, species 1 and 2 will diffuse, and profiles $C_1(z)$ and $C_2(z)$ will be established. Note that we *assume* that there is some external mechanism that permits us to maintain $C_{1o}$ at $z = 0$ and $C_{2L}$ at $z = L$.

Now suppose further that species 1 and 2 react at a rate

$$R_{12} = -kC_1C_2 \qquad (3.2.194)$$

($R_{12}$ is the rate of disappearance of 1 and of 2.) This reaction occurs homogeneously throughout the liquid film.

Species 1 and 2 diffuse through the surrounding liquid according to different diffusion

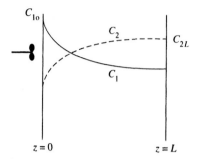

**Figure 3.2.10** Reaction inside a film.

coefficients, which we will denote here as $D_1$ and $D_2$. We assume that the system is somehow maintained at a steady state.

For one-dimensional steady diffusion we must solve

$$0 = D_1 \frac{d^2C_1}{dz^2} - kC_1C_2 \tag{3.2.195}$$

$$0 = D_2 \frac{d^2C_2}{dz^2} - kC_1C_2 \tag{3.2.196}$$

with boundary conditions that reflect the following physical ideas:

At $z = 0$ we can maintain $C_1$ at $C_{1o}$, and species 2 does not cross the plane at $z = 0$.

$$C_1 = C_{1o} \quad \text{at } z = 0 \tag{3.2.197}$$

$$\frac{dC_2}{dz} = 0 \tag{3.2.198}$$

At $z = L$ we can maintain $C_2$ at $C_{2L}$, and species 1 is depleted by reaction before that plane

$$C_2 = C_{2L} \quad \text{at } z = L \tag{3.2.199}$$

and

$$C_1 = 0 \tag{3.2.200}$$

Note that Eqs. 3.2.195 and 3.2.196 are nonlinear in the concentrations $C_1$ and $C_2$. An analytical solution is not possible, although a numerical solution would not be difficult to achieve.

A simple approximation is possible if we assume that the rate of reaction is very fast. Imagine, first, that we keep track of two individual molecules—one of species 1 and one of species 2. Each diffuses toward the other, according to Fick's law, down its concentration gradient, as in Fig. 3.2.11. When the molecules meet they react and disappear (as a third species—the product of reaction). We call the plane at which they meet $z = L_R$, the "reaction front." If the reaction is sufficiently fast, the profiles of $C_1$ and $C_2$ will appear as sketched (we assume no excess of species 2).

In the region $0 \leq z \leq L_R$ there is no species 2. Hence the $C_1$ profile satisfies

$$\frac{d^2C_1}{dz^2} = 0 \tag{3.2.201}$$

subject to

$$C_1 = C_{1o} \quad \text{at } z = 0 \tag{3.2.202}$$

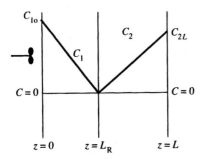

Figure 3.2.11 Concentration profiles with fast reaction.

$$C_1 = 0 \qquad \text{at } z = L_R \text{ (but } L_R \text{ is unknown at this point in the analysis)} \qquad \textbf{(3.2.203)}$$

By a similar argument, the $C_2$ profile satisfies

$$\frac{d^2 C_2}{dz^2} = 0 \qquad \textbf{(3.2.204)}$$

subject to

$$C_2 = C_{2L} \qquad \text{at } z = L \qquad \textbf{(3.2.205)}$$

$$C_2 = 0 \qquad \text{at } z = L_R \qquad \textbf{(3.2.206)}$$

The solutions are

$$C_1 = a_1 + b_1 z \qquad \textbf{(3.2.207)}$$

$$C_2 = a_2 + b_2 z \qquad \textbf{(3.2.208)}$$

and $a_1$, $a_2$, $b_1$, $b_2$ are found by using the boundary conditions of Eqs. 3.2.202, 3.2.203, 3.2.205, 3.2.206. The result is

$$a_1 = C_{1o} \qquad\qquad b_1 = -\frac{C_{1o}}{L_R}$$

$$a_2 = -\frac{C_{2L} L_R}{L - L_R} \qquad b_2 = \frac{C_{2L}}{L - L_R} \qquad \textbf{(3.2.209)}$$

*But $L_R$ is unknown.* To find $L_R$, we need an additional physical statement—but not a boundary condition. Let us look at the individual species fluxes:

$$N_1 = -D_1 \frac{dC_1}{dz} = -D_1 b_1 = \frac{C_{1o} D_1}{L_R} \qquad \textbf{(3.2.210)}$$

$$N_2 = -D_2 \frac{dC_2}{dz} = -D_2 b_2 = -\frac{C_{2L} D_2}{L - L_R} \qquad \textbf{(3.2.211)}$$

Since one mole of species 1 reacts with one mole of 2, at steady state the two fluxes must be equal in magnitude:

$$N_1 = -N_2 \qquad \textbf{(3.2.212)}$$

This permits us to solve for $L_R$:

$$L_R = \frac{L}{1 + \dfrac{C_{2L} D_2}{C_{1o} D_1}} \qquad \textbf{(3.2.213)}$$

and we find $N_1$ as

$$N_1 = \frac{C_{10}D_1}{L}\left(1 + \frac{C_{2L}D_2}{C_{10}D_1}\right)$$

(3.2.214)

Note that the rate of transfer of species 1 into the film is enhanced by reaction. One result of this mathematical model is the suggestion that chemical reaction can serve to enhance the rate of absorption of a gas by a liquid. This possibility has very important implications for the design of gas absorbers as antipollution devices, as well as for many other mass transfer devices. An example with implications for biomedical device design follows.

The analysis above is simplified by the assumption that the reaction is instantaneous. In the case of a finite reaction rate, Eqs. 3.2.195 and 3.2.196 must be solved numerically. A nondimensional formulation of these equations introduces some new dimensionless parameters.

**EXAMPLE 3.2.10** *A Design for Enhanced Oxygen Transfer*

In many areas of technology we need to transfer oxygen to a fluid. Examples include biochemical fermentations, sewage treatment, and blood oxygenation during surgery. Mass tranfer is limited by a number of factors, which include fluid dynamics, convective mass transfer across interfaces, and the limiting solubility of oxygen in the fluid.

The solubility of oxygen in water can be calculated if we can find a value for Henry's constant. For example from Levine (*Physical Chemistry*, McGraw-Hill, 1978 p. 221), we find Henry's law written in the form

$$P = k_i x$$

(3.2.215)

where $P$ is the partial pressure of the species in the gas phase that is in equilibrium with that species in the liquid phase, where the mole fraction is $x$. At 25°C, the $k_i$ value for $O_2/H_2O$ is given as[4]

$$k_i = 3.3 \times 10^7 \text{ torr}$$

(3.2.216)

In normal air at one atmosphere, the partial pressure of oxygen is (using air as 21% $O_2$ by volume)

$$P = 0.21\,(760) \text{ torr}$$

(3.2.217)

Hence the solubility of $O_2$ in $H_2O$ at 25°C and one atmosphere is (in mole fraction units)

$$x = \frac{0.21(760)}{3.3 \times 10^7} = 4.9 \times 10^{-6} \text{ mol } O_2/\text{mol } H_2O$$

(3.2.218)

In molar concentration units we find that the oxygen solubility in water at 25°C and one atmosphere is (using 18 for the molecular weight of water, and taking the density of water as 1 g/cm²)

$$C_1 = \frac{4.9 \times 10^{-6}}{18} = 2.7 \times 10^{-7} \text{ mol/cm}^3$$

(3.2.219)

---

[4] We get nearly the same value, but in different units, if we use Fig. C2.1 (Appendix C).

Certain solutes in aqueous liquids have much greater solubilities for oxygen than does water. Hence we have the prospect of using such a liquid as a "carrier" of oxygen. An example would be a protein that "complexes" with oxygen according to a second-order reaction, as in Eq. 3.2.194. Suppose we have a protein, of molecular weight 1000, that binds one mole of oxygen per mole of protein. If we want to increase the solubility of oxygen in the aqueous solution by an order of magnitude, then we want a value of $C_1$ of

$$C_1 = 2.7 \times 10^{-6} \text{ mol/cm}^3 \qquad \textbf{(3.2.220)}$$

This in turn requires (essentially) the same molar concentration of the protein. (We ignore the oxygen dissolved in the pure water fraction of the solution.) With a molecular weight of 1000, this corresponds to

$$2.7 \times 10^{-6} \times 1000 = 2.7 \times 10^{-3} \text{ g/cm}^3$$

which is a very dilute solution.

The diffusivity $D_1$ of oxygen in water at 25°C is about

$$D_1 = 2 \times 10^{-5} \text{ cm}^2/\text{s} \qquad \textbf{(3.2.221)}$$

and we will use the same value in this dilute aqueous solution. We have no experimental information on which to estimate a value for $D_2$. We certainly expect it to be much less than $D_1$, since the solute has a molecular weight of 1000. We do know that raffinose, with a molecular weight of 504, has a diffusivity of $4.3 \times 10^{-6}$ cm$^2$/s. We can expect a protein of molecular weight 1000 to have a diffusivity near to that of raffinose (but less). A crude estimate of diffusion coefficients can be based on Graham's law, which states that the diffusivity of similar molecules in a given solvent varies inversely as the square root of the molecular weight of the solute. On this basis we *estimate*

$$D_2 = 3 \times 10^{-6} \text{ cm}^2/\text{s} \qquad \textbf{(3.2.222)}$$

If we now impose Eq. 3.2.214, we find that the presence of this solute enhances the *rate* of transfer of oxygen from air into the liquid by a factor of

$$1 + \frac{C_{2L}D_2}{C_{1,o}D_1} = 1 + \frac{2.7 \times 10^{-6}}{2.7 \times 10^{-7}} \frac{3 \times 10^{-6}}{2 \times 10^{-5}} = 2.5 \qquad \textbf{(3.2.223)}$$

It appears that we have the possibility of achieving two important goals through the use of a solute that has an affinity for oxygen. One is that we increase the *solubility* of oxygen in the liquid, thus increasing the *carrying capacity* of the liquid for oxygen. This property has obvious implications in the design of a blood oxygenator or a fermentation broth for a biochemical reactor. The other effect of this solute is that we increase the *rate* at which transfer takes place. Hence a smaller amount of liquid has a capacity to deliver oxygen to the target.

The analysis presented involved the assumption of very rapid reaction. It would be necessary to examine this assumption, and modify the analysis if the reaction is not fast enough. A mathematical model could aid in establishing a criterion for determining when a reaction is fast enough to give validity to the assumption made here.

Other simplifying assumptions were made. Note that $C_{1,o}$ was taken as the solubility of oxygen in *protein-free* water. If we assume a rapid reaction then, according to our sketch of the concentration profiles (Fig. 3.2.11), there is *no uncomplexed protein* in the region $z < L_R$. We used the protein-free solubility in that region, therefore, since uncomplexed protein is not available to increase the solubility.

For $C_{2L}$ we used the value of $C_1$ from Eq. 3.2.220, since the protein concentration must match the desired increased oxygen concentration.

## 3.3 DESIGN OF AN ARTIFICIAL KIDNEY UTILIZING UREASE IN POLYMERIC BEADS

Let's take a look at the problems encountered in the design of an artificial kidney, with the design based on the concept described in Section 1.1.4. There are a number of issues associated with the design of such a system, and we are able to discuss some of them at this point. In particular, we want to develop a model that guides the decision regarding the appropriate size for the polymeric bead that encapsulates the enzyme. Is there an optimum size, and if so, what factors determine the optimum? This turns out to be a complex problem (at this early stage of our knowledge of mass transfer), so we look first at a simpler problem. Our goal is to develop a model that permits us to answer the following questions:

What type of experimental system would facilitate our obtaining fundamental data on the kinetics and mass transfer for this system?
Given the constraint of fixed total amount of enzyme in the reactor, is there a bead diameter that maximizes the rate of conversion of urea?

The polymeric beads will be spherical and will contain a uniform distribution of the enzyme urease. The beads will be surrounded by a solution of urea. The urea must dissolve in the bead and diffuse into the interior, where reaction occurs. If diffusion is inefficient in some sense, the amount of the enzyme that is available to act on the urea will be greatly restricted. The relative roles of chemical kinetics and diffusive mass transfer are at issue here. We want to design a system that will maximize the utilization of the enzyme and permit the conversion of urea to proceed at a rate that is not limited by mass transfer.

Some thought suggests that we have already solved a problem that has the elements of this reactor concept, and will permit us to achieve our goal. It is the problem described in Example 3.2.3, which involves reaction and diffusion within a spherical particle. If the kinetics of the conversion of urea by urease are first-order, we may use the model developed in that section directly. We will make that assumption, while noting that under some circumstances the kinetics differ from first-order in the urea/urease system.

The molar rate of conversion of urea is given by Eq. 3.2.64, which we write in the form

$$\text{rate} = 4\pi R^2 D_{AB} \frac{C_R}{R}(\phi \coth \phi - 1) \tag{3.3.1}$$

(in units of moles per cubic centimeter-second), where

$$\phi = \left(\frac{kR^2}{D_{AB}}\right)^{1/2} \tag{3.3.2}$$

is the Thiele modulus for a first-order reaction. The rate written in Eq. 3.3.1 is *per bead*. If we have $N$ beads of identical radius $R$, the total rate is just

$$\text{total rate} = 4\pi R^2 N D_{AB} \frac{C_R}{R}(\phi \coth \phi - 1) \tag{3.3.3}$$

If the total amount of enzyme is fixed (e.g., by economic considerations), and if the concentration of enzyme in each bead is independent of the choice of bead radius, then

$$\tfrac{4}{3}\pi R^3 N = \text{constant} = M \tag{3.3.4}$$

and

$$\text{total rate} = 3M D_{AB} \frac{C_R}{R^2}(\phi \coth \phi - 1) \tag{3.3.5}$$

This may be written as

$$\Phi \equiv \frac{\text{total rate}}{MkC_R} = 3\frac{\phi \coth \phi - 1}{\phi^2} \tag{3.3.6}$$

We introduce the function $\Phi$ for two reasons. First of all, it is convenient to cast mathematical models in dimensionless formats, since this indicates the minimum number of dimensionless parameters that affect the physics. This knowledge is useful in guiding the design of experiments and in the correlation of data. Second, dimensionless formats of models often have simple interpretations. For example, it is not difficult to argue that $\Phi$, as defined above, is the ratio of the expected rate of conversion of urea to the maximum possible rate that would occur if there were no mass transfer limitations to accessibility of the enzyme. Because of diffusion resistance within the bead, the rate of reaction is slowed, and eventually limited by diffusion, regardless of the kinetics of the reaction. We will refer to $\Phi$ as an effectiveness factor. *Keep in mind that it is based on the constraint of constant amount of enzyme.*

We can now address the second of the two questions raised above. We should choose a value of $R$ that maximizes $\Phi$. Figure 3.3.1 plots $\Phi$ versus $\phi$. We see that as soon as $\phi$ is less than unity, the system has reached its maximum efficiency. Hence there is no need to reduce $R$ any further once we have satisfied $\phi = 1$, or

$$R_{\min} = \left(\frac{D_{AB}}{k}\right)^{1/2} \tag{3.3.7}$$

Hence this is the minimum $R$ we would use. This result provides the beginning of the answer to the first question raised above. We wish to design an experiment by which we can determine the kinetic coefficient $k$ and the diffusivity $D_{AB}$ of urea within the bead matrix. According to our model, if we carry out experiments with beads smaller than $R_{\min}$ we will find a rate of conversion that is independent of $R$, and independent of $D_{AB}$ as well. We see this directly from Eq. 3.3.6 since, with $\Phi = 1$,

$$\text{total rate} = \phi MkC_R = MkC_R \tag{3.3.8}$$

We may also test the assumption of linear kinetics in this way, since if we change the concentration of enzyme $C_R$ but keep $M$ constant, the observed rate of conversion

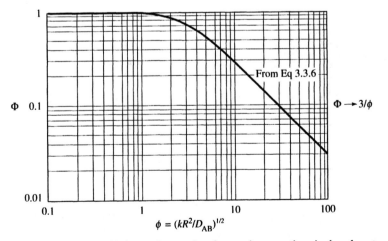

**Figure 3.3.1** The efficiency factor for first-order reaction in beads, at constant amount of enzyme.

should be a linear function of $C_R$. We conclude that we can most easily obtain information about the kinetics, as well as the value of the rate constant $k$, through experiments using beads smaller than $R_{min}$.

Once the rate constant $k$ has been established, we may find $D_{AB}$ by performing experiments with larger beads, for which there is a significant diffusion resistance. For example, if we select values of $R$ such that $\phi > 10$ we find

$$\frac{\text{total rate}}{MkC_R} = \frac{3}{\phi} = \frac{3}{R}\left(\frac{D_{AB}}{k}\right)^{1/2} \tag{3.3.9}$$

All parameters in this equation are known, once $k$ has been established, so a measurement of the total rate will yield a value for the diffusivity. Again, keep in mind that these experiments are performed at a *constant total amount of enzyme* in the reactor. Hence, as we change $R$, and assuming that $C_R$ is held constant, we must change the number of beads $N$ to maintain the total enzyme loading at a constant value.

There is an implicit assumption here that requires comment. We assume that there is no resistance to transfer of urea *to* the beads in the suggested experiments. This is not necessarily the case, and care must be taken to reduce any *external* resistance factors. In a typical reactor the beads would be surrounded by liquid within which the urea is dissolved. We must control the flow of this solution in a way that promotes very efficient convective mass transfer to the bead surface. In a stirred reactor, for example, we would operate at a stirring rate high enough to ensure that the observed conversion of urea becomes independent of stirring rate and is limited by factors internal to the beads—diffusion and reaction.

Now that we know one way to find $k$ and $D_{AB}$, let's assume some values for these parameters and examine an artificial kidney design problem. Suppose that a specific formulation of enzyme/polymer beads is known to correspond to the following parameter values

$$k = 10^{-3}\,\text{s}^{-1} \quad \text{and} \quad D_{AB} = 0.9 \times 10^{-5}\,\text{cm}^2/\text{s}$$

First we will specify the bead radius $R_{min}$. From Eq. 3.3.7 we find

$$R_{min} = \left(\frac{0.9 \times 10^{-5}}{10^{-3}}\right)^{1/2} = 0.095\,\text{cm} \tag{3.3.10}$$

We will use 1 mm for the bead radius in this system.

Next we have to design a system that will permit us to reduce the urea concentration in blood at an acceptable rate. A rough design is suggested in Fig. 3.3.2. Blood is pumped at a volumetric flowrate $Q$ from the body into a membrane system, and back to the body. The membrane system promotes the transfer of urea out of the blood into a reservoir of volume $V_R$, where the urea is diluted. In this reservoir the enzyme/polymer bead system is used to keep the concentration of urea so low that the rate of membrane transport out of the blood is maximized.

We will make several assumptions here, which simplify the analysis without destroying its usefulness.

1. There is no loss of fluid volume across the membrane, so $Q$ returning to the body is the same as that leaving the body.
2. The blood volume in the body, and the fluid volume in the reactor/reservoir, are well mixed and so of uniform concentrations throughout their respective volumes.

To minimize the blood volume within the membrane system, we will specify the use of a hollow fiber membrane unit consisting of a large number $N_f$ of parallel hollow fibers,

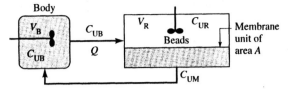

**Figure 3.3.2** Schematic of a blood dialysis system, where $C_{UB}$, $C_{UR}$ and $C_{UM}$ are, respectively, the concentrations of urea in the body, the reservoir, and in the blood leaving the membrane unit.

each of axial length $L$ and inner and outer radii $R_{in}$ and $R_{out}$. Our first task is to develop for the hollow fiber unit a "performance equation," that is, a model that relates the concentration of urea in the blood leaving the unit, denoted $C_{UM}$, to the inlet concentration $C_{UB}$ and the characteristics of the membrane. The flowrate through the unit will also affect the performance of the unit. Figure 3.3.3 will aid the analysis.

We will assume that transfer of urea across the membrane wall is governed by simple Fickian diffusion. We already have a model for this geometry—it is Eq. 2.3.14—which gives the rate of mass transfer across a hollow fiber. From that model we may write the "local" flux, which is the flux at any axial position $z$ where the axial concentration of urea is $C_U(z)$, simply by dividing the rate by the area, $2\pi R_{in} L$. The result is

$$\text{flux} = -\frac{\alpha_m D_{U,m} C_U(z)}{R_{in} \ln \kappa} \tag{3.3.11}$$

where $\alpha_m$ is the solubility of urea in the membrane and $D_{U,m}$ is the diffusivity of urea in the membrane. We have introduced the assumption here that operating conditions serve to reduce the urea concentration $C_{UR}$, through the reaction promoted by the urease beads, to a value that is very small compared to $C_U(z)$. We will have to come back to this point and define the operating range in which this assumption is valid.

Since we used the inside area of the fiber in writing Eq. 3.3.11, this is the flux at the *inner* surface. As before, we define $\kappa = R_{in}/R_{out}$. Now we write a simple mass balance (really, a species balance on urea) across a differential axial length $dz$, with the result

$$Q_f[C_U(z) - C_U(z + dz)] = -\frac{\alpha_m D_{U,m} C_U(z)}{R_{in} \ln \kappa} 2\pi R_{in}\, dz \tag{3.3.12}$$

where $Q_f$ is the volume flowrate through each fiber. We assume that the hollow fiber unit is a collection of $N_f$ hollow fibers in parallel, each performing according to Eq.

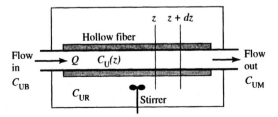

**Figure 3.3.3** Schematic of a single fiber in a blood dialysis system.

3.3.12. After dividing by $dz$ and taking the limit as $dz \to 0$, we find

$$\frac{dC_U(z)}{dz} = -\beta C_U(z) \tag{3.3.13}$$

where a parameter $\beta$ is defined as

$$\beta \equiv -\frac{2\pi \alpha_m D_{U,m}}{Q_f \ln \kappa} \tag{3.3.14}$$

($\beta$ is not dimensionless.) The solution to Eq. 3.3.13 is easily found in the form

$$\frac{C_U(z)}{C_{UB}} = e^{-\beta z} \tag{3.3.15}$$

where we have used the entrance condition that

$$C_U(z) = C_{UB} \quad \text{at} \quad z = 0 \tag{3.3.16}$$

Our goal is a model for the exit concentration, which follows from Eq. 3.3.16 upon setting $z = L$:

$$\frac{C_{UM}}{C_{UB}} = e^{-\beta L} \tag{3.3.17}$$

(Note that $\beta L$ is dimensionless.) Then the change in concentration of urea after one pass of the blood through the fiber is

$$C_{UM} - C_{UB} = C_{UB}(e^{-\beta L} - 1) \tag{3.3.18}$$

Thus we see that $\beta L$ is an important *dimensionless* parameter, related to the efficiency of the system.

Now we must write a transient balance around the blood volume of the body. This takes the form

$$-V_B \frac{dC_{UB}}{dt} = Q(C_{UB} - C_{UM}) \tag{3.3.19}$$

where we now write the total volume flowrate through the hollow fiber unit as

$$Q = N_f Q_f \tag{3.3.20}$$

When this relation is coupled to Eq. 3.3.18, we find that the blood urea concentration changes with time according to

$$\frac{-V_B}{Q} \frac{dC_{UB}}{dt} = (1 - e^{-\beta L}) C_{UB} \tag{3.3.21}$$

The solution of this equation is easily found to be

$$\frac{C_{UB}}{C_{UBo}} = \exp\left(-\frac{\lambda'}{\tau_B} t\right) \tag{3.3.22}$$

where

$$\lambda' \equiv 1 - e^{-\beta L} \tag{3.3.23}$$

and

$$\tau_B \equiv \frac{V_B}{Q} \tag{3.3.24}$$

is a time scale for the system.

Before proceeding further, we should evaluate the assumption, used in writing Eq. 3.3.11, that

$$C_{UR} \ll C_{UM} \tag{3.3.25}$$

We can write a quasi-steady mass balance on the membrane system of the form

$$k\alpha_b C_{UR} f_v V_R = Q(C_{UB} - C_{UM}) \tag{3.3.26}$$

where $\alpha_b$ is the solubility of urea in the beads and $f_v$ is the volume fraction of beads in $V_R$. This is simply a statement that the rate of reaction promoted by the enzyme/polymer beads balances the loss of urea across the membrane. When we use Eq. 3.3.18 we find

$$\frac{C_{UR}}{C_{UB}} = \frac{Q(1 - e^{-\beta L})}{k\alpha_b f_v V_R} \tag{3.3.27}$$

This is a good point at which to pause and recall what we are attempting to achieve here. Our goal is the development of a model that will aid in the design of a hollow fiber dialysis unit. As is usually the case, our model has been obtained through the imposition of several assumptions. Since this is a design problem, in which several design and operating variables are not yet specified, we may have the freedom to select some of these parameters in such a way as to yield operation under conditions consistent with the assumptions. For example, if we want Eq. 3.3.25 to hold, we could select parameters to give

$$\frac{C_{UR}}{C_{UB}} = \frac{Q(1 - e^{-\beta L})}{k\alpha_b f_v V_R} = 0.05 \tag{3.3.28}$$

The choice of 0.05 is arbitrary, but it represents a reasonable criterion of $C_{UR} \ll C_{UM}$.

At this point, two key design and operating parameters are unspecified. They are the total flowrate $Q$ and the volume $V_R$. Equation 3.3.28 provides a constraint on these parameters such that one of the key assumptions of the analysis is valid. What other constraint might there be on the design?

An essential design criterion addresses the speed with which the dialysis unit can bring the blood urea level down to a physiologically acceptable level. Let us suppose that we are told that the urea concentration is supposed to be reduced by half after 2 hours of operation (i.e., $t = 7200$ s). Then, from Eq. 3.3.22 we must select parameters such that

$$\frac{7200\lambda'}{\tau_B} = 0.693 \tag{3.3.29}$$

Suppose the following physical properties are available to us:

$$\alpha_m D_{U,m} = 10^{-5} \text{ cm}^2/\text{s} \quad R_{out} - R_{in} = 50 \ \mu\text{m} \quad R_{in} = 50 \ \mu\text{m}$$
$$A_{in} = 500 \text{ cm}^2 \quad k = 10^{-3} \text{ s}^{-1} \quad \alpha_b = 1$$

We begin with Eq. 3.3.23 and calculate $\beta L$:

$$\beta L \equiv -\frac{2\pi R_{in} L \alpha_m D_{U,m}}{R_{in} Q_f \ln \kappa} = -\frac{A_{in} \alpha_m D_{U,m}}{R_{in} Q \ln \kappa} = \frac{1.443}{Q} \tag{3.3.30}$$

with $Q$ in units of cubic centimeters per second.

We will take the blood volume to be

$$V_B = 10 \text{ L} = 10^4 \text{ cm}^3 \tag{3.3.31}$$

Then Eq. 3.3.29, with the aid of Eqs. 3.2.23 and 3.2.24, becomes

$$\left[1 - \exp\left(-\frac{1.443}{Q}\right)\right] = \frac{0.9625}{Q} \tag{3.3.32}$$

or

$$Q = 1.65 \text{ cm}^3/\text{s, or } 5.94 \text{ L/h} \tag{3.3.33}$$

Since this result is much less than the flowrate through a single kidney in a normal human, it seems to be an acceptable value from a physiological perspective. Now we must return to Eq. 3.3.28 and find a reactor volume large enough to ensure that the concentration of urea on the reservoir side of the membrane unit will be very small. With the parameters specified already, we find

$$f_v V_R = 19.3 \text{ L} \tag{3.3.34}$$

If the volume fraction of polymer beads is too large, it will be difficult to agitate the reactor slurry sufficiently to promote good external mass transfer to the beads. Hence we should expect the bead volume fraction to be limited by mass transfer considerations. If we select a value of $f_v = 0.2$, we will require a reactor volume of

$$V_R = 97 \text{ L} \tag{3.3.35}$$

This is really quite large, and we should explore other design parameters that will permit us to carry out the dialysis at the required rate, but with a much smaller reactor.

## 3.4 A MODEL TO AID THE INTERPRETATION OF DATA ON THE DISSOLUTION OF "NUCLEAR WASTE" GLASS

A key issue in nuclear waste management is the development of suitable materials for the immobilization of radioactive species, to prevent their release to the environment. One proposal for waste management is the long-term burial of radioactive waste that has been mixed in a molten glassy matrix, poured into metal cylinders, and stored in an underground repository. Over long periods of time there is some probability of leakage of underground water into the repository, as well as corrosion of the metal casing to expose the glassy matrix. A specific concern is the difficulty of predicting the rate of dissolution of the glassy matrix and the subsequent release, over "geologic" time scales, of the radioactive material. A significant literature is available, but its reading requires a firm basis in the analysis of diffusion/reaction systems. In this section we apply our knowledge of mass transfer, gained to this point, to the development of a simplified model of the rate of dissolution of glass. This provides a starting point for anyone interested in following the literature in this important field, with a view to contributing to the development of more secure storage technology.

The dissolution of glass is really a complex process. Chemically, glass consists of a solid solution of numerous oxides, the main constituents of a typical glass being $SiO_2$, $B_2O_3$, $Al_2O_3$, $Na_2O$, and $CaO$. (If the silicon and boron oxides constitute the major fraction of the species, it is called a *borosilicate* glass.) The dissolution kinetics are usually followed by measuring the rate of release of specific atoms such as boron or silicon into the surrounding environment.

An example of such data is shown in Fig. 3.4.1. A powdered glass was placed in a magnesium chloride rich brine at 190°C. This corrosive brine solution was analyzed periodically and the boron concentration was determined as a function of time. From

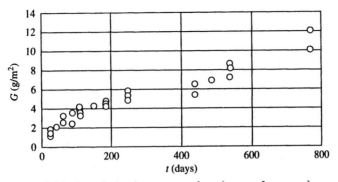

**Figure 3.4.1** Cumulative boron mass loss (per surface area).
Data of Grambow, Lutze, and Muller, *Mater. Res. Soc. Symp.
Proc.,* **257,** 143, (1992).

this, the total boron (mass) flux across the glass/brine interface was calculated and used
as a measure of dissolution of the glass. Note that Fig. 3.4.1 plots the *cumulative* mass
(per surface area) $G$, not the flux. $G$ is just the integral of the flux over time. There
are several interesting features of these data. One is that the experiments run over a
period of nearly 3 years. By most standards, this is a long experiment, but it is very
short in terms of geologic time. It is necessary to develop an understanding of the
dissolution process that will support extrapolation of the data to times of the order of
many hundreds, and preferably thousands of years.

Another feature of these data is the nonlinearity of the rate of dissolution. The initial
rate is high, but it falls off toward an apparently linear behavior. When the data are
replotted on log–log coordinates (Fig. 3.4.2), however, a different picture emerges. The
dissolution rate is not linear at all. Over the entire time of measurement, the data follow

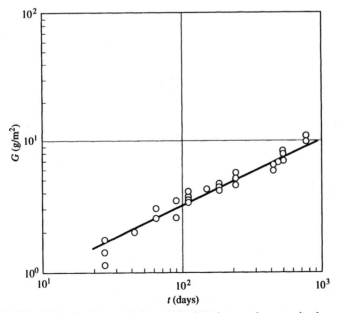

**Figure 3.4.2** Cumulative boron mass loss (per surface area), plotted as a test of a diffusion model.

a relationship of the form

$$G = Kt^{1/2} \tag{3.4.1}$$

(The straight line through the data in Fig. 3.4.2 has a slope of 0.5.) We shall see that this is a clue to understanding the mechanism of dissolution.

With this introduction behind us, let's examine some mechanisms of dissolution that lead to models that can be compared to these data. One possible mechanism is the following:

*The corrosive solution (most likely, water containing a variety of salts) reacts with the glass, breaking silicon and boron bonds in the respective oxides, and producing soluble silicon and boron complexes.*

*As the silicon and boron complexes dissolve in the surrounding solution, they leave behind a glass whose structure has been degraded to a more porous or gel-like state. If this glass contained nuclear waste, it would be possible for "hot" uranium, cesium, and strontium compounds to leach out of the matrix.*

*As reaction continues at the glass/gel interface (see Fig. 3.4.3) the mobile silicon and boron complexes must diffuse through the gel layer to reach the external solution.*

*The rate of removal of the boron and silicon compounds is determined by the reaction rate(s) of the silicon and boron oxides, and by the rates of diffusion of the boron and silicon compounds through the growing gel layer.*

With this model as a basis, we would expect that as the gel layer grows thicker, there will be a reduction in the observed rate of reaction (dissolution). Let's turn this physical picture into a mathematical model.

We will denote the concentration of the species that reacts with boron to yield a diffusible boron complex by $C_R$. The boron complex concentration will be denoted by $C_B$. These concentrations exist and are defined *in the gel phase*. The molar rate of reaction at the glass/gel interface is denoted $\mathscr{R}$. It is a *surface* reaction.

The reactant gets to the glass/gel interface by diffusion. Hence its concentration must satisfy the following diffusion equation:

$$0 = D_R \frac{d^2 C_R}{dz^2} \tag{3.4.2}$$

with boundary conditions

$$C_R = C_o \quad \text{on } z = 0 \text{ (the exposed surface)} \tag{3.4.3}$$

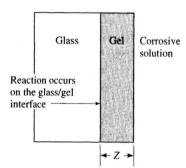

**Figure 3.4.3** Schematic model for glass dissolution.

and

$$-D_R \frac{dC_R}{dz} = \mathscr{R} \qquad \text{on } z = Z(t) \text{ (the reacting surface)} \qquad \textbf{(3.4.4)}$$

These simple equations imply several things about the physics. First, Eq. 3.4.2 is a steady state diffusion equation, and we are applying it to an unsteady problem where $Z = Z(t)$. This would be a good approximation if we knew that $Z$ changes very slowly. In fact, measurements of the thickness of the gel layer suggest that $dZ/dt$ is of the order of micrometers per day.

Equation 3.4.3 implies that at the interface of the solid with the corrosive solution, the concentration of the active species is constant in time. This would be a reasonable approximation as long as the glass is surrounded by a relatively large body of liquid.

Equation 3.4.4 is simply a balance between the rate of diffusion of the reactant to the reacting plane and the rate of its disappearance ($\mathscr{R}$) due to reaction.

We will adopt a simple kinetic model for the dissolution reaction. We assume that the rate of reaction is linearly proportional to the concentration of the reactant $C_R$ in the gel, at the reacting plane. Then we write Eq. 3.4.4 as

$$-D_R \frac{dC_R}{dz} = kC_R \qquad \text{on } z = Z(t) \qquad \textbf{(3.4.5)}$$

With this boundary condition, the solution to Eq. 3.4.2, in the *quasi-steady* approximation where $Z$ is regarded as constant, is simply

$$C(z) = C_o - \frac{kC_o}{kZ + D_R} z \qquad \textbf{(3.4.6)}$$

From this we find the reaction rate as

$$\mathscr{R} = kC_R(Z) = \frac{kC_o D_R}{kZ + D_R} \qquad \textbf{(3.4.7)}$$

The essential idea of any quasi-steady model is that we begin with a steady state analysis—in this case the diffusion model for constant $Z$—and we introduce some dynamic model for the unsteady state feature of the physics. In this case, we must introduce a model for $Z(t)$. We do so with the following assumption:

The rate of growth of the gel layer is directly proportional to the molar rate of reaction $\mathscr{R}$.

Keep in mind that $\mathscr{R}$ refers to a *surface* reaction, and is a molar rate of reaction *per unit of reactive surface*. Hence we write

$$\frac{dZ}{dt} = \frac{\nu kC_o D_R}{kZ + D_R} \qquad \textbf{(3.4.8)}$$

where $\nu$ is the molar volume of the gel layer. The initial condition on this equation is simply

$$Z = 0 \qquad \text{at} \quad t = 0 \qquad \textbf{(3.4.9)}$$

The solution to Eq. 3.4.8 may be written in the form

$$t = \frac{Z}{\nu kC_o} + \frac{Z^2}{2\nu C_o D_R} \qquad \textbf{(3.4.10)}$$

In the early stages of the corrosion process, when $Z$ is very small, the second term on the right-hand side of Eq. 3.4.10 is negligible, and we see that the rate of reaction controls the growth of the gel layer and the diffusive resistance is unimportant. As $Z$ grows large enough to retard the availability of the reactant at the gel/glass boundary, the second term in Eq. 3.4.10 dominates, and we expect to find $Z$ to grow as $t^{1/2}$.

Our model is not finished, because we need an expression for the rate of release from the solid matrix of some constituent, such as boron. However, if we assume that the rate of release is controlled by diffusion through the gel layer, then the rate of release will be inversely proportional to the gel layer thickness, which is growing as $t^{1/2}$, after the initial stage of dissolution. Hence we would expect to observe the *cumulative* release rate (per unit of surface area) to have the form

$$G(t) = \int_0^t N_R \, dt = \int_0^t \frac{\alpha}{t^{1/2}} \, dt = 2\alpha t^{1/2} \tag{3.4.11}$$

where $N_R$ is the diffusive flux through the growing gel layer. The coefficient $\alpha$ is a combination of parameters that characterize diffusion through the gel layer. Equation 3.4.11 is consistent with the data shown in Fig. 3.4.2.

## 3.5  ANALYSIS OF "BARRIER FILMS" FOR PACKAGING

The plastic films used as packaging materials are designed to serve many purposes. In this section we will apply our knowledge of mass transfer to the analysis and design of films used to control the permeation of vapors into or out of a container of some kind. While such systems operate under unsteady conditions because the concentration of at least one species in the container is changing with time, the "dynamics" are usually slow enough to permit a *quasi-steady* approximation to work quite well. We select some examples from the food industry, but it is clear that similar problems and opportunities arise in a variety of technologies.

**EXAMPLE 3.5.1**  *Prevention of Water Intrusion into a Food Package*

We wish to evaluate the use of a plastic (polymer) film for the prevention of water intrusion into a packaged food. Figure 3.5.1 will serve as an aid to the discussion. A food product is packaged in a two-piece container: the lower supporting container is a thick molded plastic "box," open at the top; the top is sealed with a thin transparent plastic film. The top film must enable customers to see the product inside the package. The thicker the film, the poorer the clarity. The "box" portion of the package is made of a thick opaque plastic material that is chosen to serve as an effective barrier to the transport of water vapor into the package. Choices of transparent films for the cover are more limited. Some compromise must be made between visual clarity and water protection.

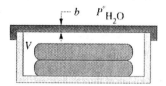

**Figure 3.5.1** Schematic for analysis of a barrier film.

In this example, the design task is to specify a film thickness that will provide both the clarity and the "shelf life" dictated by the marketing experts. We have available the following information, which provides the basis for design.

In the worst-case scenario the package will be on the shelf in an environment at 90°F and a relative humidity of 90%. The barrier film will have a surface area of 100 cm². The interior of the package has a free volume of 400 cm³ and is initially free of water vapor; product quality is compromised if the interior relative humidity reaches 50%. A shelf life of 90 days is required. Assume, for the sake of this design calculation, that the food product is always in equilibrium with the water vapor that intrudes into the package and that the rate of absorption of the water vapor by the food product is rapid in comparison to transmission into the package. Assume further that the amount of water vapor absorbed by the food product is small in comparison to the amount in the free volume of the package, at equilibrium.

Data are available for water vapor transmission rates through a variety of polymers (Gerlowski, in *Barrier Polymers and Structures*, Koros, Ed., ACS Symposium Series 423, 1990). We select the following values:

Permeability $P_{90}$ = 0.0066 (g · mil/100 in.² · d) at 100°F and 90% relative humidity differential

Solubility for water $S$ = 0.0071 (g/100 g polymer) at 25°C and 100% relative humidity

An initial task is to interpret these data, given as they are in mixed units. It is very common to find permeability data presented in this form and requiring conversion to a consistent set of units. The first step, however, is to connect the property called "permeability" to our understanding of mass transfer fundamentals. When we have done so, we will find that the units, as stated in the reference cited, are not correct for a permeability coefficient, as we define it.

To begin, let's consider the transfer of some species across a thin film, under steady state conditions. We will assume that the transport is by Fickian diffusion, so that the *mass* flux of water may be written in the form

$$n_{H_2O} = \frac{D_{H_2O/film} \rho_{film} \Delta\omega_{int}}{b} \tag{3.5.1}$$

In this expression we find two significant permeation parameters: the diffusion coefficient $D_{H_2O/film}$ of water vapor through the polymeric film and the film thickness, $b$. We must be careful about the concentration driving force $\Delta\omega_{int}$, which represents the concentration difference across the film, but *in the interior* of the film—not the difference in values across the film in the air outside, exterior to the film. The concentration difference $\Delta\omega_{int}$ used here is a mass fraction, in units of mass of water per unit mass of solid film. Hence we need the mass density of the polymer, $\rho_{film}$, to convert the concentration to a volumetric basis (g/cm³).

The connection of Fick's law to a relationship involving permeability is through the solubility of water vapor in the film. We assume that there is a linear solubility relationship, and that Eq. 3.2.2 holds. (This is a reasonable assumption because this polymer is hydrophobic, the water solubility is very low, and a linear relationship is often a good approximation at high degrees of dilution.) For this example we write the solubility relationship in the form

$$S_C \equiv \frac{\omega_{int}}{C_{ext}} = \frac{\text{g water /g polymer}}{\text{g water /cm}^3 \text{ air}} \tag{3.5.2}$$

with the choice of concentration units shown. The solubility information provided by the data presented above gives the concentration of water in the polymer film in units of grams of water per 100 g of polymer, while the air concentration is given in relative humidity units. Since the data are available in this form it makes sense to define the solubility parameter as

$$S_{RH} \equiv \frac{\omega_{int}}{RH} = \frac{g\,water\,/g\,polymer}{relative\,humidity\,of\,air} \tag{3.5.3}$$

We must use mass, rather than molar, units on concentration ($\omega_{int}$) because the molecular weight of the polymer is unspecified. (Actually most polymeric films will have a mixture of molecular weights, and this information is not often available.) Hence in this analysis we will use mass units (i.e., mass flux and mass concentrations in the film) in working with Eq. 3.5.1. We have quickly come to realize that the most tedious part of this analysis will be keeping the units straight and consistent.

The data for the barrier film are given in terms of permeability. Fick's law (Eq. 3.5.1) does not have this parameter. What is the connection? In reference works, the permeability is often defined in such a way that the flux may be written in the form

$$n_{H_2O} = P\frac{\Delta(RH)}{b} \tag{3.5.4}$$

where $P$ is the permeability of the film for the penetrant (water).[1] The concentration driving force is the *external* difference across the barrier, in the surrounding medium, in some set of convenient units. We will use *percentage* of relative humidity as the concentration unit in this discussion. If we introduce the film solubility relationship we can write Fick's law (Eq. 3.5.1) as

$$n_{H_2O} = D_{H_2O/film}\,\rho_{film}\,S_{RH}\frac{\Delta(RH)}{b} \tag{3.5.5}$$

Hence the permeability coefficient, using *percentage* of relative humidity as the driving force for transfer, may be expressed as

$$P = D_{H_2O/film}\,\rho_{film}\,S_{RH} \tag{3.5.6}$$

Since the coefficient $S_{RH}$ is dimensionless, the units of the permeability (as we define it through Eq. 3.5.4) are

$$P\,[=]\,\frac{n_{H_2O}\,b}{\Delta(RH)}\,[=]\,\frac{g\cdot cm}{cm^2\cdot s\cdot \%RH} \tag{3.5.7}$$

How do we reconcile this relation with the data given earlier? The permeability was given in the reference cited as $P_{90} = 0.0066$ (g $\cdot$ mil/100 in.$^2$ $\cdot$ d) at 100°F and 90% relative humidity differential. We use the notation $P_{90}$ here because the value, as stated, is not really a permeability. It is actually a measured flux per mil ($10^{-3}$ in.) of film thickness, at a concentration difference corresponding to a relative humidity difference of 90%.

---

[1] This definition of "permeability" is not agreed on by all who use the term. Just as frequently, one finds a "permeability" defined as $P/b$ of Eq. 3.5.4. This is a useful definition when the film thickness is unknown. The reader must be very careful when applying "permeability" values taken from various literature sources.

Hence the actual permeability is given by

$$P = \frac{P_{90}}{90} = \frac{0.0066}{90\%RH} \frac{g \cdot mil}{100 \, in.^2 \cdot day} \qquad (3.5.8)$$

$$= \frac{0.0066}{90 \times 100 \, in.^2 \times [(2.54)^2 \, cm^2/in.^2] \cdot day \times (3600 \times 24) \, s/day}$$

$$= 3.3 \times 10^{-15} \frac{g \cdot cm}{cm^2 \cdot s \cdot \%RH}$$

at a temperature of 100°F. This definition of the permeability takes the observed permeation rate through a film of a given thickness, under a specific concentration differential, and yields a coefficient that is independent of the film thickness and concentration (RH) differential. The result is a property of the specific polymer film and the penetrant (water vapor).

It is important to recognize what we have done so far: we have managed to use our knowledge of diffusive mass transfer to define and calculate a permeability coefficient for water through a polymer film, given a measured value of a mass flux through the film, under a specific concentration (relative humidity) difference.

We can now illustrate the use of this coefficient in the design of a barrier film to meet the specifications cited at the beginning of this section. The package volume is given as 400 cm³ and the temperature for which we are designing the barrier is 90°F. (In the absence of additional data, we will use the 100°F permeability value at the design temperature of 90°F.) The equilibrium vapor pressure of water at 90°F (= 305 K) is 0.698 psi = 0.0475 atm (see Fig. P3.18). Using the ideal gas law, we find that this corresponds to a mass density at equilibrium of

$$\rho_{H_2O}^{eq} = \frac{p_{vap} M_w}{R_G T} = \frac{0.0475 \times 18}{82 \times 305} = 3.42 \times 10^{-5} \, g/cm^3 \qquad (3.5.9)$$

or 0.0137 g in 400 cm³. We want to find the time required to transport enough water vapor across the barrier to raise the relative humidity within the package to 50%. Hence we want to find the time required to transport half this amount of water, or 0.0137/2 = 0.0068 g.

First we must develop a model for the rate of transport of water vapor across the barrier film. Isn't Eq. 3.5.4 that model? Not quite! In the case of transport across a film into a *closed* region, the interior relative humidity is changing with time. Hence $\Delta(RH)$ is changing. We need an additional equation that couples the rate of change of $\Delta(RH)$ to the flux. This follows from a simple mass balance on water that states that the rate of increase of water mass within the closed package is equal to the mass rate of transport into the package, or

$$\frac{d}{dt}(\rho_{H_2O} V) = n_{H_2O} A = PA \frac{\Delta(RH)}{b} \qquad (3.5.10)$$

If we divide both sides of this equation by $\rho_{H_2O}^{eq}$ as given in Eq. 3.5.9, it becomes

$$\frac{d}{dt}(RH) = \frac{100 \, PA}{bV\rho_{H_2O}^{eq}} \Delta(RH) = \beta \, \Delta(RH) = \beta[90 - (RH)] \qquad (3.5.11)$$

In writing this result we recognize that the mass density of water within the package at any instant of time, divided by the equilibrium mass density $\rho_{H_2O}^{eq}$, is the relative humidity as a *fraction,* or RH(%)/100. To simplify the form of Eq. 3.5.11, we introduced

$\beta$ to represent a combination of the membrane properties. For $\Delta$(RH) we have assumed that the exterior relative humidity remains constant at 90% and that the exterior temperature remains constant as well. (Otherwise $\rho_{H_2O}^{eq}$ would change as well.) Finally, in this case we are assuming that the food product inside does not absorb a significant fraction of the water vapor that intrudes.

We may solve Eq. 3.5.11 by integration, with the initial condition that (RH) = 0 at $t = 0$, and the result may be written in the form

$$\frac{(RH)}{90} = 1 - e^{-\beta t} \tag{3.5.12}$$

The time for (RH) to reach 50% is then found from

$$t_{50} = -\frac{1}{\beta} \ln \left( 1 - \frac{50}{90} \right) = \frac{0.811}{\beta} \tag{3.5.13}$$

Since everything in this expression is specified except the barrier film thickness $b$, we may solve for $b$ to find the film thickness that will yield 50% relative humidity at a time corresponding to 90 days, the required shelf life of the package. The result is

$$b_{50} = \frac{100 \, PA \, t_{50}}{0.811 \, V\rho_{H_2O}^{eq}} = \frac{100(3.3 \times 10^{-15})(100)(90 \times 24 \times 3600)}{0.811(400)(3.42 \times 10^{-5})} \tag{3.5.14}$$

$$= 0.023 \, cm = 9.1 \times 10^{-3} \, in. = 9.1 \, mils$$

**EXAMPLE 3.5.2**    *Retention of Carbonation in a Soda Bottle*

Carbonated beverages are packaged in plastic containers that must retain the carbonation over a long shelf life. A commercial polymeric material has been developed and its permeability to several vapors has been measured. Our task is to evaluate a proposed experiment for rating the performance of this material as a bottle for a carbonated beverage. The bottle has an internal volume of 320 cm³, a surface area (excluding the neck and closure of the bottle) of 225 cm², and a wall thickness of 30 mils (0.03 in. = 0.26 mm). To perform the evaluation, we will fill the empty bottle with a gas mixture containing equal volumes of air and carbon dioxide, and seal it at a total pressure of 2 atm absolute. The initial partial pressure of carbon dioxide in the bottle is 1 atm absolute. The partial pressure of air inside matches that of the ambient air, and so no oxygen and nitrogen will permeate the bottle wall. The internal and external air will be dry. A pressure sensor, sealed inside the bottle, provides a continuous readout of the internal pressure, thus making available information about the rate of loss of carbon dioxide.

The manufacturer of the polymer presents us with permeation data. In particular, we are told that at 25°C the permeability of this material to carbon dioxide is $P = 0.02$ barrer. Obviously, our first task is to discover what a "barrer" is. We find that a "barrer" is both a *who* and a *what*. R. M. Barrer was one of the pioneering investigators of diffusion through solids.[2] In his honor, a unit of permeability carries his name. It is defined as

$$1 \, barrer = \frac{10^{-10} \, cm^3 \, (STP) \cdot cm}{cm^2 \cdot s \cdot cmHg} \tag{3.5.15}$$

---

[2] He authored the classic monograph *Diffusion in and Through Solids,* Cambridge University Press, 1951.

The permeation is measured in units of gas volume (corrected to standard temperature and pressure, that is, $cm^3_{STP}$), and the pressure driving force for permeation is given in units of centimeters of mercury (cmHg). From its units we can see that this permeability is defined consistent with Eq. 3.5.4, but with pressure as the driving force

$$\text{volumetric flux} = \frac{cm^3}{cm^2 \cdot s} = P \frac{\Delta p}{b} \tag{3.5.16}$$

Additional information on this barrier polymer is provided to us. The diffusivity of carbon dioxide through the polymer has been measured and found to be

$$D = 6 \times 10^{-10} \ cm^2/s \text{ at } 273 \text{ K} \tag{3.5.17}$$

The solubility of $CO_2$ in the polymer is also available, and is given as

$$S = 0.25 \frac{cm^3 \ CO_2 \ (STP)}{cm^3 \ polymer \cdot atm} \text{ at } 273 \text{ K} \tag{3.5.18}$$

The definition of permeability, as Eq. 3.5.16 implies, is relevant only to a quasi-steady or truly steady process, since it assumes that the concentration gradient is represented by the concentration difference (whatever the units used for concentration) divided by the thickness of the film being permeated. The first thing we must recognize is that Eq. 3.5.16 does not hold during the transient portion of the permeation process, when the linear concentration profile across the film is being developed. In Chapter 4 (Section 4.2), we will discover that this start-up transient requires a time interval that is well approximated by

$$t^* = \frac{b^2}{D} \tag{3.5.19}$$

For the parameters that characterize this polymeric film, we find

$$t^* = \frac{b^2}{D} = \frac{(0.026 \ cm)^2}{(6 \times 10^{-10}) \ cm^2/s} = 1.13 \times 10^6 \ s = 13 \text{ days} \tag{3.5.20}$$

This discussion implies that we cannot use Eq. 3.5.16 to predict the loss of carbonation until the quasi-steady regime has been achieved, and this will require nearly 2 weeks of permeation time. Until we have the transient solution available, we cannot estimate the loss of carbonation during this initial period. Hence we will return to this problem in Chapter 4. What we can calculate is the rate of loss during the quasi-steady period. From Eq. 3.5.16 this is found to be

$$\text{volumetric rate} = AP \frac{\Delta p}{b} = 225 \ [cm^2] \times 0.02 \times 10^{-10} \left[ \frac{cm^3 \ (STP) \cdot cm}{cm^2 \cdot s \cdot cmHg} \right]$$
$$\times \frac{76 \ [cmHg]}{0.026 \ cm} = 1.3 \times 10^{-6} \ cm^3/s \tag{3.5.21}$$

where we assume that the carbon dioxide level remains approximately at its initial value during this period. The initial volume (at STP) of carbon dioxide is 160 $cm^3$ (half the bottle volume). At this rate, the time to reduce the carbonation by 5% of the initial charge is 71 days:

$$\frac{0.05 \times 160 \ cm^3}{1.3 \times 10^{-6} \ cm^3/s} = 6.2 \times 10^6 \ s = 71 \text{ days} \tag{3.5.22}$$

We cannot proceed with further calculations until Chapter 4, where we will learn more about modeling the transient portion of the loss (see Problem 4.85).

## 3.6 MASS TRANSFER ISSUES IN THE PRODUCTION OF FUEL PELLETS FOR A CONTROLLED FUSION REACTOR

One approach to the design of a controlled fusion reactor is called "inertial confinement." The fuel is a mixture of isotopes of hydrogen (deuterium and tritium) contained under high pressure in microscopic pellets—thin-walled glass or plastic microspheres. The pellets can be rapidly heated by laser light. The pellets are designed to implode rapidly, compressing the gas inside and raising its temperature until a fusion reaction can occur. Each pellet is a mini hydrogen "bomb," and the released energy is absorbed by a heat exchange medium and transferred out of the reactor, where it can be converted to a useful form for power generation. Various design configurations are under study. We will consider a specific case of design and operating parameters.

Glass microspheres can be produced in a variety of sizes. We consider a sphere with an outside diameter of 100 $\mu$m and a wall thickness of 1 $\mu$m. The fuel is produced by "loading" the spheres with the gas mixture through a process of high temperature/high pressure permeation. The spheres, *initially containing no fuel*, are placed in a chamber containing the gaseous fuel. The temperature and pressure are then raised (e.g., to 700 K and 200 atm), and the gas enters the interior of each fuel pellet. When equilibrium has been reached, the system is rapidly cooled with liquid nitrogen and the chamber is vented. At low temperatures, including room temperature, the permeability of the gas across the pellet wall is so small that the gas is effectively trapped in the cold pellet. Our design and analysis issue here is the dynamics of penetration. We need to be able to estimate the time required to fill the pellets under some given set of operating conditions. Alternatively, we must be able to estimate how long it will be before significant *leakage* of the fuel from the pellets occurs.

Data are available for the permeability of glass to hydrogen and its isotopes [Tsugawa et al., *J. Appl. Phys.*, **47**, 1987 (1976)]. A set of such data is shown in Fig. 3.6.1, for hydrogen itself. Before proceeding, we should clarify the definition of permeability used by these authors. It is based on the *molar* flux of hydrogen across a glass film, and it

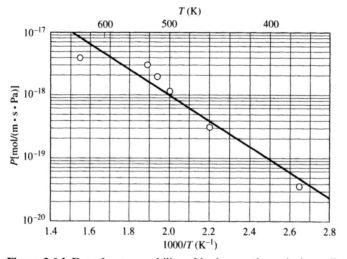

**Figure 3.6.1** Data for permeability of hydrogen through the wall of glass microspheres; the units of permeability are moles per meter-second-pascal.

is the coefficient $P$ in the expression

$$j_{H_2} = P\frac{\Delta p}{B} \tag{3.6.1}$$

where $\Delta p$ is the difference in partial pressure of hydrogen in the gas phases inside and outside the microsphere (Pa), $B$ is the thickness of the glass wall of the microsphere (m), and $j_{H,2}$ is the molar flux (moles of hydrogen/s · m$^2$) across the glass wall. Equation 3.6.1 is essentially the statement that hydrogen diffuses across the wall according to Fick's law. The permeability coefficient is a product of the diffusion coefficient of hydrogen through glass (in this specific case) and the solubility of hydrogen in glass. The development of Eq. 3.6.1 from Fick's law parallels the analysis in Section 3.5 that yields Eq. 3.5.4 from Eq. 3.5.1. The permeability coefficient is all that is required to permit us to characterize the rate of transport across the glass film, given the film thickness and the partial pressure difference driving the transfer. This is illustrated in the following example.

**EXAMPLE 3.6.1**   *Leakage from Glass Microspheres in Storage*

Fuel pellets have been prepared from glass microspheres of radius 50 $\mu$m and a wall thickness of 1 $\mu$m. The interior hydrogen pressure is 100 atm ($\approx$ 10 MPa). If the microspheres are exposed to an ambient medium essentially free of hydrogen, and at a temperature of 300 K, how much (%) of the fuel will be lost in 3 days?

First we need the permeability at 300 K. A simple exponential curve-fit of the data in Fig. 3.6.1 gives the equation

$$P = Ge^{-H/T} \tag{3.6.2}$$

and the parameter values[1]

$$G = 1.1 \times 10^{-14} \text{ mol/(m} \cdot \text{s} \cdot \text{Pa)} \quad \text{and} \quad H = 4660 \text{ K}$$

This yields an estimated permeability at 300 K of

$$P = 2 \times 10^{-21} \text{ mol/(m} \cdot \text{s} \cdot \text{Pa)}$$

To solve this example problem we first need a model for permeation across the wall of the sphere. Since the wall thickness is only 2% of the sphere radius, we may use a planar film model. That is essentially how the permeability is defined in Eq. 3.6.1, so that expression yields the flux directly:

$$j_{H_2} = P\frac{\Delta p}{B} = [2 \times 10^{-21} \text{ mol/(m} \cdot \text{s} \cdot \text{Pa)}] \left(\frac{10^7 \text{ Pa}}{10^{-6} \text{ m}}\right) = 2 \times 10^{-8} \text{ mol/(m}^2 \cdot \text{s)} \tag{3.6.3}$$

with the assumption that $\Delta p$ is constant. This is the molar flux from a single sphere. The molar *rate* of loss is

$$4\pi R^2 j_{H_2} = 4\pi(50 \times 10^{-6})^2(2 \times 10^{-8}) = 6.3 \times 10^{-16} \text{ mol/s} \tag{3.6.4}$$

---

[1] These parameters are based on a least-squares fit of the data shown. They differ slightly from those parameters as presented in the original reference. This discrepancy is due to small errors in reading values from the original published graph, rather than from the original data set, which was not available.

Three days corresponds to $2.6 \times 10^5$ s. Hence the fuel loss is

$$(6.3 \times 10^{-16}\,\text{mol/s}) \times (2.6 \times 10^5\,\text{s}) = 1.6 \times 10^{-10}\,\text{mol} \qquad \textbf{(3.6.5)}$$

How many moles of hydrogen were initially in the sphere? We will assume that the ideal gas law holds[2] and find

$$n = \frac{pV}{R_G T} = \frac{10^7\,\text{Pa}}{8.314[\text{J/(g}\cdot\text{mol}\cdot\text{K)}] \times 300\,\text{K}}\,\frac{4\pi}{3}\,(50 \times 10^{-6}\,\text{m})^3 = 2.1 \times 10^{-9}\,\text{mol} \quad \textbf{(3.6.6)}$$

Hence the fractional loss is $1.6 \times 10^{-10}/(2.1 \times 10^{-9}) = 0.076 = 7.6\%$. The fuel loss is small, as desired under storage conditions. (The small loss also validates the assumption that $\Delta p$ is constant.) Colder storage conditions would extend the "shelf life" of the fuel.

Now we can proceed to develop a model for the time required to fill the fuel pellets. Suppose filling takes place at 750 K with a constant external hydrogen pressure of $p^\circ = 100$ atm. As filling proceeds, the flux is given by Eq. 3.6.1 (we take the flux as positive inward):

$$j_{H_2} = P\frac{\Delta p}{B} = P\frac{p^\circ - p(t)}{B} \qquad \textbf{(3.6.7)}$$

where $p(t)$ is the hydrogen pressure within the sphere at any time. This pressure will be increasing with time, of course. Since we wish to have a model that holds right up to equilibrium, we cannot neglect $p(t)$ with respect to $p^\circ$. The change in internal pressure is obtained from a mass (mole) balance of the form

$$\frac{dp(t)}{dt} = \frac{R_G T}{V}\frac{dn(t)}{dt} = \frac{R_G T}{V}4\pi R^2 j_{H_2} = \frac{R_G T}{V}4\pi R^2 P\frac{p^\circ - p(t)}{B} \qquad \textbf{(3.6.8)}$$

or

$$\frac{dp(t)}{dt} = \beta[p^\circ - p(t)] \qquad \textbf{(3.6.9)}$$

where

$$\beta \equiv \frac{3R_G T}{BR}P \qquad \textbf{(3.6.10)}$$

The solution of Eq. 3.6.9, subject to the initial condition

$$p(t) = 0 \qquad \text{at } t = 0 \qquad \textbf{(3.6.11)}$$

is

$$p(t) = p^\circ(1 - e^{-\beta t}) \qquad \textbf{(3.6.12)}$$

---

[2] Examination of a compressibility factor chart for gases shows that for hydrogen at these conditions, the compressibility factor is nearly unity (i.e., the gas is nearly ideal). Hence the error arising from this assumption will be small. Note also that we use the SI unit for $R_G$, but with gram-moles, not kilogram-moles, since the permeability coefficient is in gram-mole units. When kilogram-moles are in use, the SI value is 8314 J/(kg·mol·K).

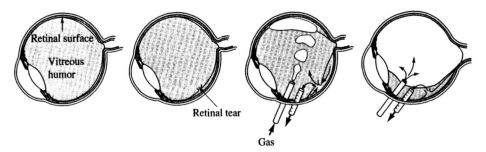

**Figure 3.7.1** The human eye.

The time required to achieve 95% of the equilibrium fill is found by setting $p(t)/p°$ = 0.95 in Eq. 3.6.12. The result is

$$t_{0.95} = \frac{3}{\beta} = \frac{BR}{PR_GT} \tag{3.6.13}$$

At 750 K we find, from Eq. 3.6.2, $P = 2.2 \times 10^{-17}$ mol/(m·s·Pa). For the case of a sphere with an outside diameter of $2R = 100\ \mu$m and a wall thickness of $B = 1\ \mu$m, we find the time required to (nearly) fill the fuel pellets from

$$t_{0.95} = \frac{10^{-6}\,\text{m}\,(50 \times 10^{-6}\,\text{m})}{\left(2.2 \times 10^{-17}\,\dfrac{\text{g}\cdot\text{mol}}{\text{m}\cdot\text{s}\cdot\text{Pa}}\right)\left(8.314\,\dfrac{\text{Pa}\cdot\text{m}^3}{\text{g}\cdot\text{mol}\cdot\text{K}}\right)750\,\text{K}} = 364\,\text{s} \tag{3.6.14}$$

By operating at such high temperatures, it is possible to fill the spheres very quickly.

## 3.7 THE USE OF A SIMPLE MASS TRANSFER MODEL TO GUIDE A STRATEGY FOR A MEDICAL PROCEDURE: THE REPAIR OF A RETINAL DETACHMENT[1]

A simplified sketch of the human eye is shown in Fig. 3.7.1. The retina is a thin layer of light-sensitive cells that lines a section of the interior of the eye. If a hole develops in the retina, some of the fluid inside the eye—the *vitreous humor*—can flow through the hole and cause a separation of the retina from the inner surface of the eye. Unless this retinal *detachment* is repaired, blindness can result. In one method of repair a gas mixture is injected into the vitreous humor to raise the pressure in the interior of the eye. This pushes the retinal layer back against the eye wall, forcing fluid out from the region between the retina and the eye wall. Complete replacement of the vitreous fluid by gas is required to keep the fluid from the retinal defect. Retinal readhesion approaches its full strength if the pressure can be maintained for about 10 days. Proper treatment requires that the pressure be controlled within specific limitations (upper and lower bounds) and ultimately return to the normal interior pressure of the eye.

After the vitreous fluid has been replaced by the gas mixture, the pressure within the eye changes in response to mass transfer between the gas and the capillaries that perfuse the retina. A commonly used procedure involves a gas mixture consisting of

---

[1] This problem is motivated by the following reference, from which much of the physical property data are obtained: R. Tyson, W. T. Welch, J. Berreen, C. Wells, and C. E. Pearson, *J. Theor. Biol.,* **164**, 15 (1993).

nitrogen, an inert gas such as sulfur hexafluoride ($SF_6$), which is not normally found in the blood, and a mixture of gases including $O_2$, water vapor, and $CO_2$ in concentrations that are in *equilibrium* with the blood that perfuses the retina. As a consequence, no exchange occurs among the equilibrium gas species, but nitrogen and the inert gas can diffuse across the barrier that separates the retinal capillaries from the interior of the eye. By controlling the amounts of the nitrogen and inert gas, the time history of the interior pressure can be controlled.

It is no simple task to develop a model of this exchange process, the details of which are poorly understood. We can appreciate this fact better if we go immediately to a possible model of gas exchange within the eye. Figure 3.7.2 shows the main features of the model. The eye is modeled as a closed unit containing a fixed volume of gas (i.e., the vitreous fluid has been replaced by the gas mixture), and the partial pressures of the various species are denoted $p_i(t)$ as a function of time.

At this stage, some assumption (model) must be introduced regarding the exchange process. It is known that a bed of capillaries perfuses one section of the retinal surface. We will assume that any gas diffuses across a tissue barrier into the capillary bed and thereby changes the concentration of that species within the blood. Since the capillary bed is small, it seems likely that we can use an average concentration of each species in the whole of the capillary bed in writing a mass transfer model. Hence we take both the gas phase and the blood region to be well mixed and separated by a permeable barrier of area $A$ and thickness $B$. With Fig. 3.7.2 as a guide, the mass transfer model takes the form

$$Q(\overline{C}_i - C_i^o) = \frac{AD_{B,i}}{B}(C_{g,i}^B - C_{b,i}^B) = -\frac{d}{dt}\left(\frac{VM_{w,i}p_i}{R_G T}\right) \qquad (3.7.1)$$

where

$Q$ = volume flowrate of blood through the capillary bed
$A$ = exchange area for transfer
$V$ = volume of the interior of the eye
$B$ = equivalent thickness of the diffusion barrier
$D_{B,i}$ = diffusion coefficient of the $i$th species through the barrier
$T$ = absolute temperature
$M_{w,i}$ = molecular weight of the diffusing species
$R_G$ = ideal gas constant

and the relevant mass concentrations are, for each species $i$,

$\overline{C}_i$      in the blood region (with entering value $C_i^o$)
$C_{g,i}^B$      within the barrier, at the *gas*/barrier interface
$C_{b,i}^B$      within the barrier, at the *blood*/barrier interface

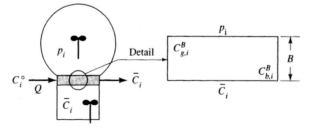

**Figure 3.7.2** A model for gas exchange within the eye.

We need two equilibrium relationships for the solubilities of each gas in the barrier and the blood. We will write these in the forms

$$C_{g,i}^B = \alpha_{g,i} p_i \quad \text{and} \quad C_{b,i}^B = \alpha_{b,i} \overline{C}_i \tag{3.7.2}$$

Between Eqs. 3.7.1 and 3.7.2, we may eliminate the barrier concentrations and derive a differential equation for the partial pressure of each species that takes the form

$$\frac{d(p_i - \alpha_{bg,i} C_i^o)}{d\tau} = -K_i(p_i - \alpha_{bg,i} C_i^o) \tag{3.7.3}$$

where

$$\tau \equiv \frac{Qt}{V} \qquad \alpha_{bg,i} \equiv \frac{\alpha_{b,i}}{\alpha_{g,i}} \qquad K_i \equiv \frac{\dfrac{R_G T}{\alpha_{bg,i} M_{w,i}}}{1 + \dfrac{BQ}{AD_{B,i}\alpha_{b,i}}} \tag{3.7.4}$$

We will assume that so little exchange takes place that $C_i^o$ is constant for all species. This is equivalent to the assumption that within the total blood volume of the body the concentration changes only slightly, since the gas volume within the eye is so small compared to the total blood volume of the body.

The solution to Eq. 3.7.3 takes the form

$$p_i(t) = \alpha_{bg,i} C_i^o + (p_i^o - \alpha_{bg,i} C_i^o)e^{-K_i \tau} \tag{3.7.5}$$

for each of the $i$ species that transfers across the exchange surface A. Here, $p_i^o$ is the *initial* partial pressure of each species after all the vitreous fluid has been displaced by the gas mixture.

The pressure of physiological significance is the so-called intraocular pressure (IOP), which is simply the sum of the partial pressures of all species in the eye:

$$\begin{aligned} \text{IOP} \equiv \sum_i p_i(t) &= \alpha_{bg,N_2} C_{N_2}^o + (p_{N_2}^o - \alpha_{bg,N_2} C_{N_2}^o) \exp(-K_{N_2}\tau) \\ &+ \alpha_{bg,SF_6} C_{SF_6}^o + (p_{SF_6}^o - \alpha_{bg,SF_6} C_{SF_6}^o) \exp(-K_{SF_6}\tau) + p_{equi}^o \end{aligned} \tag{3.7.6}$$

Here $p_{equi}^o$ is the partial pressure of all the gases ($O_2$, water vapor, and $CO_2$) in concentrations that are in *equilibrium* with the blood that perfuses the retina. The equilibrium relationships may be used to simplify this expression further.

At equilibrium we know that (i.e., Eqs. 3.7.2 imply that)

$$\overline{C}_i^o \alpha_{b,i} = C_{b,i}^{B\,equi} = C_{g,i}^{B\,equi} = \alpha_{g,i} p_i^{equi} \tag{3.7.7}$$

from which it follows that

$$\overline{C}_i^o \alpha_{bg,i} = p_i^{equi} \tag{3.7.8}$$

This permits us to write Eq. 3.7.6 in the form

$$\text{IOP} = p_{N_2}^{equi} + (p_{N_2}^o - p_{N_2}^{equi}) \exp(-K_{N_2}\tau) + p_{SF_6}^o \exp(-K_{SF_6}\tau) + p_{equi}^o \tag{3.7.9}$$

where $p_{N_2}^{equi}$ is the normal (equilibrium) value of $N_2$ in the blood. Since $SF_6$ is not normally in the blood, $p_{SF_6}^{equi} = 0$.

At this point our model is complete, and we can examine some of its features. To do so, we adopt the following values for the parameters of the model.

$$T = 37°C = 310 \text{ K} \qquad B = 60 \ \mu\text{m} \qquad A = 1 \text{ cm}^2 \qquad V = 4 \text{ cm}^3 \qquad Q = 0.002 \text{ cm}^3/\text{s}$$

Solubility data are available at 37°C:

| | |
|---|---|
| $N_2$ in blood, in equilibrium with air | 0.0124 cm³ $N_2$/cm³ blood |
| $N_2$ in tissue, in equilibrium with air | 0.014 cm³ $N_2$/cm³ tissue |
| $SF_6$ in blood, in equilibrium with pure $SF_6$ gas at 1 atm | 0.0072 cm³ $SF_6$/cm³ blood |
| $SF_6$ in tissue, in equilibrium with pure $SF_6$ gas at 1 atm | 0.015 cm³ $N_2$/cm³ tissue |

The diffusion coefficient of $N_2$ in tissue is $2.2 \times 10^{-5}$ cm²/s at 37°C.

The diffusion coefficient of $SF_6$ in tissue is $10^{-5}$ cm²/s at 37°C. This value is not from measurement, but from the assumption of the validity of Graham's law (see Problem 2.25), using the molecular weight of $SF_6$ of 146.

From the solubility data we can calculate the $\alpha$ parameters. First we find[2]

$$\overline{C}_{N_2} = \frac{0.0124}{22,400} \text{ mol/cm}^3 \text{ tissue} \times 28 \text{ g/mol} \times \frac{273}{310} = 1.4 \times 10^{-5} \text{ g/cm}^3 \quad \textbf{(3.7.10)}$$

From the equilibrium expressions given above we find

$$\overline{C}_{N_2} = \frac{C^B_{b,N_2}}{\alpha_{b,N_2}} = \frac{\alpha_{g,N_2} p_{N_2}}{\alpha_{b,N_2}} = \frac{p_{N_2}}{\alpha_{bg,N_2}} \quad \textbf{(3.7.11)}$$

The solubility value given is with respect to air at 1 atm pressure, or 0.8 atm $N_2$ partial pressure. Hence

$$\alpha_{bg,N_2} = \frac{p_{N_2}}{\overline{C}_{N_2}} = \frac{0.8}{1.4 \times 10^{-5}} = 5.7 \times 10^4 \text{ atm/g/cm}^3 \quad \textbf{(3.7.12)}$$

In the tissue phase

$$C^B_{N_2} = \frac{0.014}{22,400} \text{ mol/cm}^3 \text{ tissue} \times 28 \text{ g/mol} \times \frac{273}{310} = 1.6 \times 10^{-5} \text{ g/cm}^3 = \alpha_{g,N_2} \times 0.8 \quad \textbf{(3.7.13)}$$

and we find

$$\alpha_{g,N_2} = \frac{1.6 \times 10^{-5}}{0.8} = 2 \times 10^{-5} \text{ g/cm}^3 \cdot \text{atm} \quad \textbf{(3.7.14)}$$

and

$$\alpha_{b,N_2} = \alpha_{g,N_2} \alpha_{bg,N_2} = 2 \times 10^{-5} (5.7 \times 10^4) = 1.1 \quad \textbf{(3.7.15)}$$

For the sulfur hexafluoride the results are

$$\overline{C}_{SF_6} = \frac{0.0072}{22,400} \text{ mol/cm}^3 \text{ tissue} \times 146 \text{ g/mol} \times \frac{273}{310} = 4.1 \times 10^{-5} \text{ g/cm}^3 \quad \textbf{(3.7.16)}$$

for $p_{SF_6} = 1$ atm. Hence

$$\alpha_{bg,SF_6} = \frac{p_{SF_6}}{\overline{C}_{SF_6}} = \frac{1}{4.1 \times 10^{-5}} = 2.4 \times 10^4 \text{ atm/g/cm}^3 \quad \textbf{(3.7.17)}$$

---

[2] The factor 22,400 is the volume (cm³) of 1 mole of gas at 273 $K$ and 1 atm. The factor 273/310 accounts for the fact that the solubility data are at 310 $K$.

and in the tissue phase

$$C_{SF_6}^B = \frac{0.015}{22,400} \text{mol/cm}^3 \text{ tissue} \times 146 \text{ g/mol} \times \frac{273}{310} = 8.5 \times 10^{-5} \text{ g/cm}^3 = \alpha_{g,SF_6} p_{SF_6} \quad \textbf{(3.7.18)}$$

Since this value is given with respect to pure $SF_6$ at one atmosphere, we find

$$\alpha_{g,SF_6} = 8.5 \times 10^{-5} \text{ g/atm} \cdot \text{cm}^3 \quad \textbf{(3.7.19)}$$

and

$$\alpha_{b,SF_6} = \alpha_{g,SF_6} \alpha_{bg,SF_6} = 8.5 \times 10^{-5} (2.4 \times 10^4) = 2 \quad \textbf{(3.7.20)}$$

Finally, we can calculate the $K$ parameters

$$K_{N_2} \equiv \frac{\dfrac{82.06 \times 310}{5.7 \times 10^4 (28)}}{1 + \dfrac{60 \times 10^{-4}(0.002)}{(1)(2.2 \times 10^{-5})}} = 0.01 \quad \textbf{(3.7.21)}$$

and

$$K_{SF_6} \equiv \frac{\dfrac{82.06 \times 310}{2.4 \times 10^4 (146)}}{1 + \dfrac{60 \times 10^{-4}(0.002)}{(1)(10^{-5})(2)}} = 0.0045 \quad \textbf{(3.7.22)}$$

We will take the gas mixture injected into the eye to have the following composition:

$$p_{N_2}^o = 504 \text{ mmHg} \qquad p_{SF_6}^o = 84 \text{ mmHg} \qquad p_{equi}^o = 187 \text{ mmHg}$$

Note that the total pressure, initially, is the sum of these partial pressures—775 mmHg. We also can find that the *equilibrium* partial pressure of $N_2$ in blood is 573 mmHg, which is the pressure $p_{N_2}^{equi}$.

Finally we find that the IOP varies with time as

$$\text{IOP [mmHg]} = 573 + (504 - 573)e^{-0.01\tau} + 84e^{-0.0045\tau} + 187 \quad \textbf{(3.7.23)}$$

For this specific case, the IOP history is shown in Fig. 3.7.3. The medical literature suggests that we want the pressure to be above 775 mmHg but below 795 mmHg for about 8 to 10 days. Figure 3.7.3 shows this behavior. With this model we could now explore the sensitivity of the model to changes in the initial mixture composition, as well as to the sensitivity to the physical properties that are assumed or calculated. In

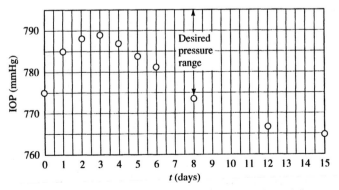

**Figure 3.7.3** Intraocular pressure (IOP) as a function of time.

addition, it would be useful to explore other physical models of the transfer process. These exercises are left to the homework problems.

## SUMMARY

The basic species balance equation in vector form is derived and presented as Eq. 3.1.7. We *assume constant* $\rho$ *and* $D_{AB}$, which are not serious restrictions for the problems of interest to us in an introductory treatment. The equation allows for unsteady behavior, convection, and reaction, in addition to diffusion. Boundary conditions, an essential feature of the procedure of model formulation, must include what we know or assume about reactions, convection, and equilibrium at the boundaries. In the course of these analyses we are naturally led to discover important dimensionless groups such as the Thiele modulus (Eq. 3.2.65) and the Peclet and Schmidt numbers (Eq. 3.2.107).

Through the rest of this chapter we examined a series of relatively complicated physical problems to which we can apply the knowledge gained thus far. The result was the production of testable models of the behavior of these systems. We restricted the examples to steady state phenomena, or to situations in which a quasi-steady analysis is appropriate.

## PROBLEMS

**3.1** Give the derivation of Eq. 3.1.4 in detail. Do not assume that the medium is incompressible.

**3.2** Give the derivation of Eq. 3.1.7 in detail. Point out where the assumptions of constant $\rho$ and $D_{AB}$ enter.

**3.3** In Example 3.2.4 we find the surface concentration $C_R$ required to maintain an average respiration rate $\overline{R}_A$. Define a nondimensional critical concentration $\Theta = \overline{R}_A/kC_R$ and plot $\Theta$ versus $\phi$. From this plot give definitions to the terms "large" and "small" $\phi$.

**3.4** Extend Example 3.2.4 to the case of an organism that is in the shape of a long cylinder. Define a variable $\Theta = \overline{R}_A/kC_R$ and plot $\Theta$ versus $\phi$. From this plot give definitions to the terms "large" and "small" $\phi$. With respect to maintenance of respiration *via* diffusion, which shape has greater "survival" value: spherical or cylindrical?

**3.5** In Example 3.2.7 we give an estimate for the pressure drop across a reactor efflux tube of dimensions $R = 0.01$ cm and $L = 10$ cm. The result given in Eq. 3.2.137 is incorrect, because Eq. 3.2.135 does not hold for a compressible fluid. For a compressible fluid in laminar flow through a long cylindrical tube, the analog to Poiseuille's law is

$$Q_o = \frac{\pi D^4}{128\,\mu L}\frac{P_o^2 - P_L^2}{2\,P_o} \qquad \text{(P3.5)}$$

where $Q_o$ is the volumetric flowrate evaluated at the upstream pressure $P_o$. The pressure drop is $\Delta P = P_o - P_L$. For a flowrate of 100 sccm, and with a downstream pressure of $P_L = 1$ atm, find $\Delta P$ and compare it to the result given in Eq. 3.2.137. What is the Reynolds number under these conditions? Define two values of Re for this problem, one based on downstream conditions and the other on upstream conditions. What conclusion do you draw regarding your solution for $P_L$?

**3.6** With reference to Example 3.2.8, how large must $b_1$ be to yield a significant degree of enhancement of the flux?

**3.7** With reference to Example 3.2.8, suppose the boundary at $z = L$ is impermeable. Carry out the analysis of this problem, and verify Eq. 3.2.189, when $\varphi_0$ is defined as in Eq. 3.2.179.

**3.8** In Example 3.2.9, suppose species 2 is in such a degree of stoichiometric excess that $C_2$ can be regarded as uniform throughout the film. Solve Eq. 3.2.195 for $C_1$ under conditions of a finite rate of reaction. Plot the concentration profile across the film. Define the plane $L_R$ as the position where $C_1$ falls to zero. Nondimensionalize the equations and plot a nondimensional $\Lambda = L_R/L$ as a function of some dimensionless kinetic parameter. Does your solution behave like Eq. 3.2.213 when the rate of reaction gets very fast? Explain your result.

**3.9** With the geometry chosen at the end of Example 3.2.7 ($R = 0.33$ cm and $L = 5$ cm), what flowrate $Q$ will just suffice to yield the critical Peclet value of 12.2?

**3.10** Return to the dialysis design problem in Section 3.3. There are two free parameters, the operating parameter $Q$ and the design parameter $V_R$. They are fixed by imposing two constraints: Eq. 3.3.28 is a constraint that yields a very small urea concentration in the reactor volume outside the membrane module, and Eq. 3.3.29 is the requirement on the rate of dialysis. Prepare plots of $Q$ and $V_R$ as a function of total area $A_{in}$ for the range [200, 1000] cm$^2$. Then do the same for $k$ in the range [0.1, 0.5] s$^{-1}$, for $A_{in} = 500$ cm$^2$. In each case, keep all the other parameters of the design constant, at the values given in the presentation in Section 3.3.

**3.11** For the parameters specified in Section 3.3, plot the half-time for dialysis as a function of $A_{in}$ and $Q$. Maintain Eq. 3.3.28 as a constraint on the design.

**3.12** Return to the dialysis design problem in Section 3.3. Relax the assumption of Eq. 3.3.25, and instead develop a model that includes the response of the urease/polymer side of the system. For $A_{in} = 5000$ cm$^2$, which is 10 times that of the example in Section 3.3, plot the half-time of the dialysis system as a function of $Q$, for $V_R = 10$ L and $f_v = 0.10$. Keep all the other parameters of the design constant, at the values given in the presentation in Section 3.3.

**3.13** Show that in a dialysis system of the type defined in Section 3.3, the minimum $Q$ is determined only by the specified half-time and the blood volume $V_B$. For the parameters selected in that section, what membrane area is required to yield the minimum $Q$?

**3.14** Complete the model expressed by Eq. 3.4.11. Specifically, make a reasonable set of assumptions consistent with this model and present an expression for the coefficient $\alpha$ in terms of the parameters that characterize diffusion and reaction in the glass and gel.

**3.15** A composite film is composed of a pair of parallel films (call them A and B) that share a common interface. The composite film is supposed to separate two regions differing in humidity and to retard the passage of water. Film A has a higher solubility and a higher diffusivity for water than film B. Does it matter which film, A or B, is in contact with the high humidity region? Answer by developing a model for the water flux in the two cases and comparing the fluxes under conditions of equal humidity difference. The following notation and definitions will help.

| | |
|---|---|
| Solubility of water in film A | $S_A$ (kg water/kg film) |
| Solubility of water in film B | $S_B$ (kg water/kg film) |
| Partition coefficient of water between film A and B | $\alpha_{A/B}$ (kg water in film A/kg water in film B at equilibrium) |
| Diffusion coefficient of water in film A | $D_{w,A}$ (m$^2$/s) |
| Diffusion coefficient of water in film B | $D_{w,B}$ (m$^2$/s) |
| Concentration of water in air on the A side of the film | $\rho_A$ (kg/m$^3$) |
| Concentration of water in air on the B side of the film | $\rho_B$ (kg/m$^3$) |

All the above are assumed to be constant for the analysis that you do. Assume that Fick's law holds for diffusion of water through the individual films.

**3.16** Data indicate that the flux of oxygen across a 75-mil-thick polypropylene film at 30°C is $35 \times 10^{-9}$ mol/m$^2 \cdot$ s per atmosphere of oxygen partial pressure difference. A device is packaged in a closed volume of 125 cm$^3$. (This is the volume available to gas within the package.) The package is initially flushed with nitrogen and has no detectable oxygen. The internal pressure is one atmosphere, and the package is exposed to atmospheric air at 30°C. The package walls are impermeable except for a "window" of polypropylene, of thickness 25 mils and exposed area 5 cm$^2$. Find the time required for the oxygen partial pressure to rise to 0.01 atm within the package. State clearly what assumption you make about the transport of nitrogen across the polypropylene film.

**3.17** Repeat Problem 3.16, but find the time required for the oxygen partial pressure to rise to 95% of equilibrium with the surrounding air.

**3.18** For the package described in Problem 3.16, find the time required for the water vapor content to rise to 0.1 mol %. Assume that the ambient air is at 50% relative humidity regardless of temperature, and find and plot the required time as a function of temperature in the range 20–30°C. Assume that the permeability coefficient depends on temperature according to

$$P = P_o \exp\left(\frac{-E_a}{R_G T}\right) \qquad \textbf{(P3.18)}$$

where $R_G$ is the gas constant [using units of $R_G$ = 0.001987 kcal/(mol · K)] and the activation energy is $E_a$ = 9 kcal/mol. The vapor pressure of water as a function of temperature is given in Fig. P3.18. Use the water permeability data given in Section 3.5.

**3.19** For the package described in Problem 3.16, find the ratio of water vapor concentration to oxygen concentration when the oxygen partial pressure inside the package reaches 0.01 atm. Assume that the ambient air is at 30°C and the relative humidity is 75%.

**3.20** Water transport is measured across a 1.64 mm thick polystyrene film at 100°F and 100% relative humidity difference across the film. The result, stated as a water flux, is $3.45 \times 10^{-14}$ mol/m² · s. What is the permeability coefficient of polystyrene at this temperature? Answer in cgs units, as well as in units of g · mil/100 in.² · d · %RH.

**3.21** When a 2000 Å film of glass is deposited on the polypropylene film described in Section 3.5, the water flux is reduced by a factor of 200,

all other conditions being constant. Find the permeability coefficient for this glass film. The following data are available:

Diffusion coefficient of water in glass    $D_{H_2O,glass} = 10^{-19}$ cm²/s

Diffusion coefficient of water in polypropylene    $D_{H_2O,PP} = 10^{-7}$ cm²/s

**3.22** A polypropylene test tube has an inside diameter of 1 cm and a length of 5 cm. It has a wall thickness of 1 mm. It initially contains 1 mL of water, and the system is maintained at 30°C. The internal pressure is the same as ambient. What is the rate of water loss, in percentage per month? Assume no loss through the stopper. The ambient air is at 50% RH.

**3.23** Return to the model that led to Eq. 3.6.14. Find the corresponding fill time if the ambient temperature is only 400 K.

**3.24** With reference to the problem described in Section 3.6, suppose that the filling of the fuel pellets takes place from a *closed* pressurized vessel. Specifically, suppose that $3.4 \times 10^{-5}$ kg of pellets with an outside diameter of $2R = 42.5$ μm and a wall thickness of $B = 1$ μm is placed in a pressure vessel with an inside diameter of 3.5 mm and an axial length of 50 mm. The pellets are glass, and the mass density of the glass itself is 2400 kg/m³. The pellets are initially free of hydrogen, as is the surrounding vessel. The interior of the vessel is suddenly charged with pure

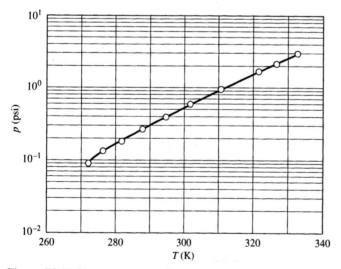

**Figure P3.18** Vapor pressure of water versus temperature.

hydrogen at 100 atm (10 MPa) and $T = 400$ K.
**a.** Derive a model for the rate at which equilibrium is approached under these conditions. Neglect any hydrogen dissolved within the pellet walls, in your model. Provide an expression for the equilibrium pressure of hydrogen in the pellets when filling stops. Plot the hydrogen pressure inside the pellets, normalized to the equilibrium value, as a function of time. The filling is taking place from a closed vessel of finite volume. State clearly the points at which you must account for this condition.
**b.** Modify the analysis for part (a) by accounting for the finite solubility of hydrogen within the glass walls of the pellets. Suppose that the solubility relationship is

$$C_{H_2,glass} \, [g \cdot mol/m^3] = \alpha(T) C_{H_2,gas} \, [g \cdot mol/m^3]$$

$$\text{(P3.24)}$$

and $\alpha(T)$ is a function only of temperature. Plot the hydrogen pressure inside the pellets, normalized to the equilibrium value, as a function of time, for values of $\alpha(T)$ of 0.01, 0.3, and 1.0.
**3.25.** With reference to Example 3.6.1, find the storage temperature that reduces the fuel loss to less than 5% over 15 days.
**3.26** Read the paper by Peppas and Khanna [*Polym. Eng. Sci.*, **20**, 1147 (1980)] on water transport across food packaging; it provides an excellent discussion of the relationship of permeability models to Fickian diffusion. Look carefully at the definition of the "water permeability" $P'_w$ given by *these authors*. Is their Eq. 13 dimensionally consistent with the units stated for the various parameters that appear in their analysis?
**3.27** Solubility and permeation data have been obtained (Barrie, Nunn, and Sheer, in *Permeability of Plastic Films and Coatings to Gases, Vapors, and Liquids*, Hopfenburg, Ed., Plenum, 1974, p. 167) for water vapor through a polyurethane elastomer. Figure P3.27.1, which shows water solubility at two temperatures, plots the equilibrium concentration of dissolved water as a function of the external partial pressure of water. The pressure axis is normalized by dividing by the vapor pressure of water at the temperature of absorption.

Figure P3.27.2 shows the volumetric water vapor flux through the film as a function of the pressure difference of water vapor across the film. Again, the pressure of water vapor is normalized to the vapor pressure at the temperature of the experiment. The volumetric flux is multiplied by the film thickness, which was $b = 0.08$ cm.

From these data, find values for the solubility coefficient, the permeability coefficient, and the diffusion coefficient, at temperatures of 35.8 and

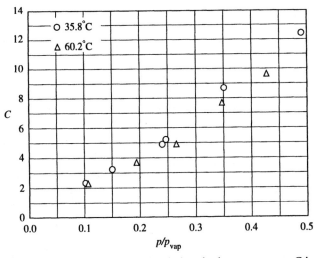

**Figure P3.27.1** Concentration of absorbed water vapor: $C$ is in units of cubic centimeters of water vapor (at STP)/per cubic centimeter of polymer.

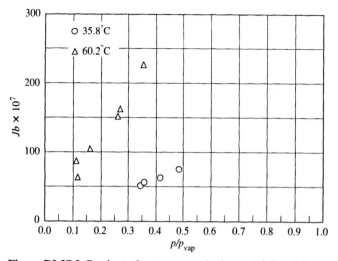

**Figure P3.27.2** Product of water vapor (volumetric) flux $J$ times film thickness $b$, as a function of partial pressure difference of water vapor across the film (normalized to the vapor pressure at each temperature). The product $Jb$ has units of $cm^3_{STP} \cdot cm/cm^2 \cdot s$.

60.2°C. Figure P3.18 gives the required vapor pressure data. Assume that the temperature dependence of the permeability coefficient may be expressed in the form

$$P = P_0 \exp\left(\frac{-B}{T}\right) \quad \textbf{(P3.27.1)}$$

with $T$ in kelvins. Give the value of $B$ for this polymer film.

**3.28** An airborne spherical cellular organism, 0.015 cm in diameter, utilizes $4.5 \ g \cdot mol \ O_2$ per hour, per kilogram of cell mass. Assume $Sh = 4$ for external convective resistance to $O_2$ transfer *to* the cell. ($Sh = kd/D_{AB}$ is based on $D_{AB}$ in the *gas* phase.) Assume zero-order kinetics for respiration.

What is the concentration of $O_2$ at the center of the cell? Use a diffusion coefficient for $O_2$ through the cellular material of $10^{-5} \ cm^2/s$. Take the solubility of oxygen in the cellular material to be $1.4 \times 10^{-6} \ mol/cm^3$, in equilibrium with air at 25°C and one atmosphere total pressure.

**3.29** A drop of pure liquid A ($\rho_L = 0.85 \ g/cm^3$) is attached (by surface tension) to a fine wire, as shown in Fig. P3.29. The vapor pressure of the liquid is $P_A$. The molecular weight is $M_{w,A} = 150$. The surrounding gas is still, and the partial pressure of A is essentially zero far from

the drop. The gas is at 25°C and 1 atm pressure. The diffusion coefficient of the vapor A in the ambient gas is $D_{AB} = 0.2 \ cm^2/s$.

Find the time required for the drop radius to be reduced to half of its initial value if $R_0 = 500$ $\mu m$ and $P_A = 0.25$ atm.

**3.30** A silica/alumina material is used as a catalyst in a reaction of interest. The material (in a powdered form initially) can be fabricated into small beads to be used as packing in a reactor. It is known that regardless of the bead size, the available area per volume of the beads is $a = 350 \ cm^2/cm^3$ of bead. The beads have a density of $1.14 \ g/cm^3$ of bead. A chemist runs a reaction, using beads of diameter $2R = 0.43 \ cm$, and assumes that the effectiveness factor is unity. She reports a first-order rate constant as $k''_1 = 0.7$ $cm^3/s \cdot cm^2$. (The area in the denominator refers to the total available catalyst area—not the geometrical surface area of the outside of the bead.) Assume an effective diffusivity of the reactant within the bead of $D_{AB} = 0.2 \ cm^2/s$.

**Figure P3.29**

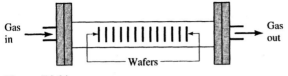

**Figure P3.31**

Correct the reported rate constant, and give the true value of $k_1''$ (cm$^3$/s · cm$^2$) and the effectiveness factor $\eta$.

**3.31** A diffusion furnace is a tube (see Fig. P3.31) containing a stack of silicon wafers, which are heated and exposed to a flow of gas that contains a dopant compound. When the doping process is completed, the downstream end of the tube is opened to the atmosphere, and an inert gas enters the upstream end to prevent the intrusion of room air into the furnace tube. The inert flow cools the wafers—a process that takes a half-hour. The flowrate of the entering gas (nitrogen) is 20 standard cubic centimeters per minute (sccm), and the gas temperature at the end of the diffusion tube is 300°C. The tube is one meter long.

Design an end cap for the diffusion tube. The geometry must be as sketched in Fig. 3.2.7, and you are to specify $L$ and $R$. The design constraints are as follows: the pressure within the diffusion furnace must not exceed 10 psi above atmospheric, and the oxygen concentration in the furnace must not exceed one part per million (by volume).

**3.32** A porous ceramic sphere of radius $R_1$ is kept saturated with a pure component liquid A. The vapor pressure of A is 50 torr. This sphere is surrounded by a concentric solid spherical surface of radius $R_2$. Species A reacts at the surface $r = R_2$ according to A → B (s). Species B is deposited as a *solid* film by this reaction, which is first-order.

Assume that this system is in a steady state, and derive an expression for the partial pressure of species A at the reacting surface $r = R_2$. Assume that the space between the two spheres is isothermal and at a uniform pressure of one atmosphere.

**3.33** Compound semiconductors are grown in a reactor by the process of chemical vapor deposition. For example, gallium arsenide films (GaAs) are commonly grown by mixing arsine (AsH$_3$) with trimethylgallium [TMG: Ga(CH$_3$)$_3$]. An excess of hydrogen is used as a carrier gas for arsine and TMG. The feed gases are fed separately, as sketched in Fig. P3.33.

The arsine is supplied from a liquid arsine source in a bubbler with hydrogen as the carrier gas. The concentration of arsine in the vapor space above the liquid surface is fixed at the value $C_{AS}$ by controlling the temperature and pressure in the bubbler, as well as the hydrogen flowrate. The volumetric flowrate out of the bubbler is $q$ and constant.

Tubing of total length $L$ and uniform cross-sectional area $A$ connects the arsine bubbler outlet to the mixing chamber $M$ through a valve $V$, which is normally open. The valve is a distance $L_1$ from the mixing chamber.

Hydrogen and arsine mix in the chamber $M$ and then flow toward the reactor. The main hydrogen flowrate (see Fig. P3.33) is $Q \gg q$.

**a.** Derive an expression for the steady state concentration profile of arsine along the delivery tube from the bubbler ($z = L$) to the mixing chamber ($z = 0$).

**b.** Give the expression for the arsine concentration at the valve ($z = L_1$). Assume $L_1 \ll L$. Assume that all parts of the system illustrated here are at constant temperature and pressure.

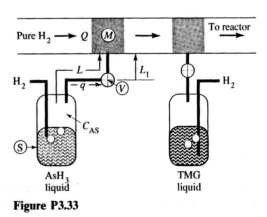

**Figure P3.33**

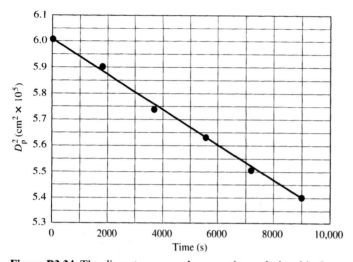

**Figure P3.34** The diameter squared-versus-time relationship for a di-*n*-butyl phthalate droplet evaporating into still air at 40°C. The initial diameter is 77.5 $\mu$m.

**3.34** Given the data shown in Fig. P3.34, find the vapor pressure of di-*n*-butyl phthalate at 40°C. The diffusion coefficient is $D = 0.035 \text{ cm}^2/\text{s}$, and the molecular weight is 278 g/g · mol. Liquid density is 1.05 g/cm$^3$.

**3.35** A spherical catalyst particle is surrounded by a uniform film of stationary liquid, of thickness $\delta$, as sketched in Fig. P3.35. A gaseous species A with partial pressure $p_A$ is in the gas surrounding the particle and film. Species A is slightly soluble in the liquid film but does not react with any species dissolved in the liquid; it does, however, react on the catalytic surface. The reaction follows Michaelis–Menten kinetics, according to which the molar rate of reaction per unit of catalytic surface area is of the nonlinear form

$$R = \frac{kC_A}{1 + KC_A} \qquad \textbf{(P3.35)}$$

The catalytic surface is impermeable to species A.

Write the differential equation and boundary conditions that determine the concentration profile of A *in the liquid film*. Do not attempt to solve the equation.

**3.36** A biochemical reactor (see Fig. P3.36) is designed in the following manner. Cells are immobilized in a gelatin film that is surrounded by a nutrient solution. The solution is well stirred and has a steady concentration $C_{N,o}$ of a key nutrient species N. The nutrient species N diffuses through the gelatin film with a diffusivity $D_{N,g}$. In each cell there is a chemical reaction that produces a product species P and water W. Both P and W pass through the gelatin film and into the nutrient solution. P diffuses through the gelatin with a diffusivity $D_{P,g} < D_{N,g}$.

The molar stoichiometry of the cellular reaction is given by

$$N + E \rightarrow P + 2W + E \qquad \textbf{(P3.36.1)}$$

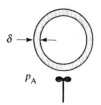

**Figure P3.35**

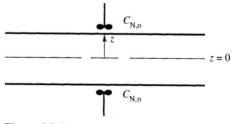

**Figure P3.36**

where E represents the enzymatic "apparatus" within the cells.

The kinetics of the reaction is first-order in N:

$$R_P = k_E C_N \frac{\text{moles of P produced}}{\text{film volume} - \text{time}} \quad \textbf{(P3.36.2)}$$

Note that the reaction rate is based on *film* volume. From the stoichiometry note also that

$$R_P = \frac{R_W}{2} = -R_N \quad \textbf{(P3.36.3)}$$

The cells are uniformly distributed throughout the film thickness $2L$, and the cell size is very small compared to $L$. Although the reaction occurs at discrete "points" within the gelatin film (i.e., only inside the cells), assume that the cells may be thought of as if they were spread uniformly in the gel and that the reaction occurs *homogeneously throughout the gel,* at a molar rate given by Eq. P3.36.2.

Find the concentration profile $C_N(z)$ within the film. Derive a model for the rate of "expression" of the product P from the film.

**3.37** Nitrogen is used to flush a reactor (see Fig. P3.37) during cooldown to prevent the intrusion of air. What volumetric flowrate is required if the concentration of oxygen at point $A$ is to remain below a millionth of its value in the external air? Take the gas temperature throughout the reactor to be 500 K.

**3.38** Write the differential equation and boundary conditions for the steady state concentration profiles in the following system:

A slightly soluble species A, at partial pressure $P_A$ in a well-mixed gas, dissolves in and diffuses through a thin liquid film that separates the gas from a planar solid surface. A reaction occurs on the solid, impermeable surface at $z = 0$. The rate of reaction is first order in $C_A$. The reaction product, B, is soluble in the liquid, does not deposit as a solid on the surface $z = 0$, and is not volatile.

**3.39** Begin with

$$\frac{\partial C_A}{\partial t} + \nabla \cdot N_A - R_A = 0 \quad \textbf{(P3.39.1)}$$

and show that for a binary mixture of species A and B, in the absence of reactions, if the total concentration of the mixture C and the diffusion coefficients are independent of position, the diffusion equation takes the form

$$\frac{\partial C_A}{\partial t} + \mathbf{v}^* \cdot \nabla C_A = D_{AB} \nabla^2 C_A \quad \textbf{(P3.39.2)}$$

where $\mathbf{v}^*$ is the *molar*-average—not mass-average—velocity.

**3.40** An enzyme is uniformly dissolved in a spherical droplet of liquid, which is immersed in a second immiscible liquid. The droplet remains spherical, and no flow is induced inside it. In the surrounding liquid there is a dissolved species that is soluble in the droplet and reacts with the enzyme. The kinetics of the reaction are first-order with respect to the molar concentration of the soluble species in the droplet. Assume that the concentration of the soluble species *in the surrounding liquid* is maintained at a constant level $C_L$. The equilibrium partition coefficient of the solute between the droplet and the surrounding liquid is given by $\alpha = C_D/C_L$. Write, but do not attempt to solve, the diffusion equation and boundary conditions from which it would be *possible* to solve for the *steady state* radial concentration distribution of the solute within the droplet.

State clearly any assumptions you introduce in writing this mathematical model. Make a sketch that defines clearly any variables that appear in your model.

**3.41** Develop a quasi-steady mathematical model of the system sketched in Fig. P3.41. Specifically, derive a relationship for the half-time $t_{1/2}$ at which the concentration of a dilute diffusible species in the spherical volume enclosed by

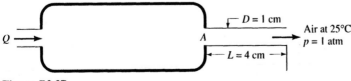

**Figure P3.37**

A solute within the spherical shell leaves by diffusion across the shell wall

**Figure P3.41**

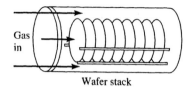

Gas in

Wafer stack

**Figure P3.44.1**

the shell is reduced to half its initial value. Your final result should be in a dimensionless format.

Assume that the gas inside the spherical shell is well mixed and that the region outside the shell is characterized by a convective mass transfer coefficient $k_c$. Assume that the shell thickness is much less than the radius of the spherical volume it encloses. Define clearly any notation you introduce in writing your model.

**3.42** Experiments with transfer of a pure gas by diffusion across a membrane film indicate that the molar flux of the gas is proportional to the partial pressure difference of the gas across the membrane:

$$N_A \text{ (mol/cm}^2 \cdot \text{s)} = K_p \, \Delta p \text{ (atm)} \quad \textbf{(P3.42.1)}$$

This equation defines the *permeability coefficient* $K_p$, based on a pressure driving force for transport.

We assume that the gas is soluble in the membrane. Define a solubility coefficient as

$$b \equiv \frac{C(\text{membrane})}{C(\text{external gas})} \quad \textbf{(P3.42.2)}$$

where both the concentrations are in units of moles per cubic centimeter.

Derive a relationship between the permeability coefficient and the membrane thickness $h$, the solubility coefficient $b$, and $D_{A,m}$, the Fickian diffusion coefficient of the gas species through the membrane.

**3.43** A membrane permeability for a particular gas species is defined as in Eq. P3.42.1 above. We wish to measure $K_p$ in the following experiment. A thin-walled hollow fiber is fabricated from the membrane material. The wall thickness is $h$, and $h$ is very small compared to the inside radius $R$ of the fiber. Air flows through the inside of the hollow fiber, at a volume flowrate $Q$, and at atmospheric pressure. The airflow is along the fiber axis. The air does not permeate across the wall of the fiber. We assume that the flowrate is small enough to prevent any significant pressure

drop along the axial length $L$ of the hollow fiber. The region outside the fiber is surrounded by air, at atmospheric pressure, containing a small mole fraction $x_o$ of the gas of interest; $x_o$ is kept constant during an experiment. We measure the mole fraction $x_L$ of the gas of interest in the airflow coming from the exit of the hollow fiber. None of this gas is in the entrance flow $Q$ into the hollow fiber. The following data are available:

$$x_o = 0.1 \qquad x_L = 0.01 \qquad Q = 0.2 \text{ cm}^3/\text{s}$$
$$R = 1 \text{ mm} \qquad L = 10 \text{ cm}$$
$$T = 400 \text{ K} \qquad \text{Gas constant} = R_G = 82.06$$
$$\text{cm}^3 \cdot \text{atm/(g} \cdot \text{mol} \cdot \text{K)}$$

Calculate a value for $K_p$ from these data. Give the units of $K_p$.

**3.44** One of the steps in semiconductor processing involves the placement of a stack of silicon wafers into a tube through which a reactant species is conveyed, usually highly diluted in an inert carrier gas. The wafers are stacked coaxial with the tube, as shown in Fig. P3.44.1. A detail of the interwafer region is shown in Fig. P3.44.2.

Formulate the diffusion equation and boundary conditions for the case of a single reactant species in the interwafer region. Assume that

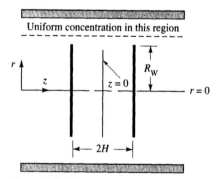

Uniform concentration in this region

**Figure P3.44.2** Detail of the interwafer region. Wafers have a radius $R_w$, and the interwafer spacing is $2H$. The midplane separating each pair of wafers is $z = 0$.

the concentration of the reacting species is constant in the annular space just outside the wafer stack. The reactant mole fraction is small compared to unity. A first-order reaction occurs on the surfaces of the wafers. The system is isothermal and at steady state. Assume that the system is axisymmetric, *but allow for diffusion in both the radial and axial directions in the region between wafers.* Do not attempt to solve the resulting partial differential equation.

**3.45** A spherical droplet of a pure liquid fuel is held stationary in a large volume of still air. The liquid evaporates, and the drop radius $R$ decreases with time. Develop a quasi-steady model from which the radius may be found as a function of time. Assume that the external phase controls the mass transfer (i.e., the evaporation rate) and that the Sherwood number is given by Eq. 3.2.25 (Sh = 2).
**a.** First present a differential equation for $R(t)$. State clearly any assumptions you made to obtain this equation.
**b.** Solve the equation and prove that the radius satisfies

$$\left(\frac{R}{R_o}\right)^2 = 1 - \alpha\left(\frac{D_{vap}t}{R_o^2}\right) \quad \textbf{(P3.45)}$$

and give an expression for the constant $\alpha$ in this equation.

**3.46** The interior gas space of a hollow fiber membrane is filled with pure oxygen. Conditions are such that the oxygen concentration in the gas phase in the fiber interior is radially and axially uniform, and independent of time. Oxygen dissolves in the membrane material and diffuses radially across the wall of the membrane. At the outer surface of the membrane a first-order reaction occurs that consumes oxygen. Derive a mathematical model for the concentration profile of oxygen across the wall of the membrane. Define carefully each symbol that appears in your model.

**3.47** A biocatalytic absorber is designed as shown in Fig. P3.47. Derive an expression for the flux of a species A across the surface of this system. The species A has the same solubility in each layer, and its diffusion coefficient is the same in the reactive layer as it is in the encapsulating layer. A first-order reaction occurs *homogeneously* throughout the reactive layer. Assume

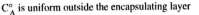
$C_A^\circ$ is uniform outside the encapsulating layer

Impermeable film

**Figure P3.47**

that the system is at steady state. The encapsulating layer and the reactive layer are attached to an impermeable film, as shown.

**3.48** A thin-walled, spherical plastic container is initially filled with pure nitrogen at $T = 300$ K and $P = 1$ atm. The interior volume is $V = 1$ L and the wall thickness of the container is $h = 2$ mm. The wall of the container is permeable to oxygen, but not to nitrogen. The contents of the container are well mixed. The container is initially stored in a pure nitrogen environment.

An experiment is performed in which the container is removed from the nitrogen environment, and the *exterior* of the container is suddenly surrounded by a large volume of pure oxygen at $T = 300$ K and $P = 1$ atm. The concentration of oxygen inside the container is measured continuously; it reaches 500 parts per million by volume (ppmv) after 2.7 h and 1000 ppmv after 5 h.

What is the value of the permeability coefficient of oxygen in the plastic wall of the container?

The permeability coefficient is defined here as

$$K_p = \frac{\beta D_{O,P}}{h} \quad \textbf{(P3.48.1)}$$

where

$$\beta = \frac{C_O}{p_O} \quad \textbf{(P3.48.2)}$$

Here, $\beta$ is the ratio of the molar concentration of oxygen in the plastic ($C_O$) divided by the partial pressure of oxygen ($p_O$) in the surrounding gas, at thermodynamic equilibrium, and $D_{O,P}$ is the diffusivity of oxygen through the plastic. Use a quasi-steady model in solving this problem.

**3.49** Equilibrium data are important in determining the permeability of membranes to various penetrants. Best and Moylan [*J. Appl.*

*Polym. Sci.,* **45,** 17 (1992)] present the data given in Fig. P3.49 for the water vapor sorption isotherm (21°C) for a thin film of polymer. The density of the polymer is 1.224 g/cm³. The equilibrium relationship may be written in various forms:

$$C_p \text{ [mol water/cm}^3 \text{ polymer]}$$
$$= \alpha C_v \text{ [mol water/cm}^3 \text{ vapor]}$$

$$C_p \text{ [mol water/cm}^3 \text{ polymer]}$$
$$= H p_v \text{ [atm water vapor pressure]}$$

$$x \text{ [mass fraction water in polymer]}$$
$$= H'y \text{ [mass fraction water in vapor]}$$

$$C_p \text{ [mg water/cm}^3 \text{ polymer]}$$
$$= H''p_v \text{ [torr water vapor pressure]}$$

Give values for $\alpha$, $H$, $H'$, and $H''$.

**3.50** An artificial kidney is designed along the following lines:

One of the major toxic wastes, urea, is to be removed from the blood by means of the enzyme urease, which catalyzes the reaction

$$\text{urea} \rightarrow \text{ammonia} + \text{carbon dioxide}$$

The enzyme is trapped in gelatin beads of diameter $d = 0.1$ cm. The urea must cross several mass transfer resistances to get to the bead surface. None of these resistances is significant in comparison to the diffusion resistance inside the gelatin bead.

Laboratory data have been obtained with gelatin beads of diameters in the range 10 to 50 $\mu$m. Per unit mass of beads (and assuming the mass fraction of enzyme in gelatin is held constant), a specific rate of conversion of urea is observed, and the following conclusions are drawn:

Conversion is independent of bead size in the range studied.

Conversion occurs by first-order reaction, with a rate constant $k_1 = 0.01$ s⁻¹, where $k_1$ is defined as follows: rate of conversion of urea (mol/cm³·s) $= k_1 C_A$ when $C_A$ is the concentration of urea in moles per cubic centimeter *in the fluid surrounding the beads.*

The diffusivity of urea in gelatin has been determined to be $D = 10^{-6}$ cm²/s, at the temperature of interest.

Per unit mass of beads of diameter 0.10 cm, find the expected rate of conversion of urea, relative to the laboratory data. In other words, find the value of $K$, defined by

$$K = \frac{\begin{array}{c}\text{rate of conversion of urea/}\\\text{unit mass of 0.1 cm beads}\end{array}}{\begin{array}{c}\text{rate of conversion of urea/}\\\text{unit mass of 10 }\mu\text{m beads}\end{array}}$$

What does this result imply about transforming laboratory data to the design process?

**3.51** Return to Section 3.7 and examine the sensitivity of the model to changes in the initial gas mixture. Keep the total initial pressure constant at 775 mmHg, and $p^o_{\text{equi}} = 187$ mmHg. Do the cases $p^o_{SF_6} = 30$ mmHg, and $p^o_{SF_6} = 120$ mmHg, and plot the results on a copy of Fig. 3.7.3 for comparison.

**3.52** What is the sensitivity of the result given in Fig. 3.7.3 to the assumed value for the area of transfer, $A$? Plot results for $A = 0.5$ cm² and 2 cm².

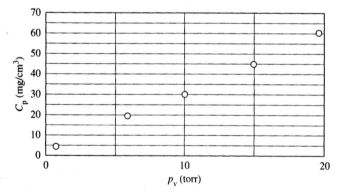

**Figure P3.49** Water vapor sorption isotherm (21°C) for a thin film of polymer.

**3.53** What is the sensitivity of the result given in Fig. 3.7.3 to the assumed value for the thickness $B$ of the diffusion barrier? Plot results for $B = 30$ $\mu$m and 120 $\mu$m. Do these results make sense to you? How does the model developed in Section 3.7 behave in the limit of vanishing diffusion resistance ($B \to 0$)? If there is no diffusion resistance, what controls the dynamics of the model?

**3.54** Derive Eq. 3.7.3.

**3.55** Show that if the species balance equation in mass units (Eq. 3.1.2) is summed over all species, the result is the continuity equation

$$\frac{\partial \rho}{\partial t} + \nabla \cdot \rho \mathbf{v} = 0 \qquad \text{(P3.55.1)}$$

On the other hand, show that if we sum the species balance equation in *molar* units (Eq. 3.1.5) over all species, we obtain

$$\frac{\partial C}{\partial t} + \nabla \cdot C\mathbf{v}^* = \sum_{i=1}^{n} R_i \qquad \text{(P3.55.2)}$$

Why do reactions not occur in the *mass* form of the balance?

**3.56** A fermentation broth consists of an aqueous solution of nutrients and cells. As the cells grow, they cluster into spherical pellets of radius $R(t)$. On average, the cell density inside a pellet is 0.02 mg of cell mass per cubic millimeter of pellet volume. The dissolved oxygen concentration in the broth is 5 $\mu$g/cm$^3$. The cells utilize oxygen at a rate of 1.2 mmol of oxygen per hour per gram of cell mass, via a zero-order reaction.

Assume that the diffusion coefficient $D_{AB}$ of oxygen within the pellet is $1.8 \times 10^{-5}$ cm$^2$/s. How large can $R(t)$ become before the oxygen concentration becomes zero at the center of a pellet? Assume that the broth external to the pellets is well mixed.

**3.57** Repeat Problem 3.56, but allow for the presence of a finite convective resistance to mass transfer to the surface of the pellet, such that the flux to the surface is given by

$$\text{molar flux} = k_c(C_P - C_B) \qquad \text{(P3.57)}$$

where $C_B$ is the broth oxygen concentration and $C_P$ is the (unknown) concentration of oxygen at the pellet surface. Calculate the "critical" value of $R(t)$ (at which the oxygen concentration becomes zero at the center of a pellet) for two cases: $2k_c R/D_{AB} = 2$, and 20.

**3.58** Problems 3.3 and 3.4 explore the behavior of the dependence of a dimensionless parameter $\Theta$, which is a measure of the efficiency of reaction within a solid, on a Thiele modulus $\phi$ defined in Eq. 3.2.65. Show that for large values of $\phi' = (k/D_{AB})^{0.5} R^*$, where

$$R^* = \frac{\text{volume}}{\text{surface area}} \qquad \text{(P3.58)}$$

the $\Theta$ versus $\phi'$ relationship is identical for spheres and long cylinders. Is this true also for a thin slab of thickness $2B$, whose width $W$ and length $L$ satisfy $W \gg B$ and $L \gg B$?

**3.59** A thin-walled glass capillary is filled with a pure gas at an absolute pressure of 5 atms. The gas permeates the wall of the capillary and is rapidly diluted in the ambient medium, which is essentially free of this gas. The wall thickness is 20 $\mu$m, and the inside diameter is 1250 $\mu$m. The permeability coefficient $P$ is defined as in Eq. 3.6.1. At the temperature of interest the value of $P$ is $10^{-15}$ mol/m·s·Pa. How long a time is required for 90% of the gas to "leak" from the capillary? By what factor is this time reduced if the ambient medium is a vacuum at absolute zero pressure?

**3.60** A homogeneous first-order chemical reaction occurs within a spherical particle. When the particle radius is too large, the reaction rate at the center of the particle falls below 10% of its maximum (surface) value, because of diffusion limitations. Find that radius. At the temperature of interest, the internal diffusion coefficient is $D = 0.02$ cm$^2$/s, and the kinetic rate constant is found to be 0.08 s$^{-1}$.

**3.61** The dominant life form on the planet $\beta$-Wally is a telepathic worm (see Fig. P3.61) with a length of 100 norblicks. [The norblick (nb) is the engineering unit of length on $\beta$-Wally.] Life support is maintained primarily through a metabolic process located in an organ (the *fenster*) that lies along the longitudinal axis of the worm. The organ itself is 1 norblick in diameter and stretches for nearly the entire length of the worm. The fenster is surrounded by fleshy tissue, which is homogeneous right to the surface of the worm.

The major metabolic "fuel" of the worm is methane. The metabolic rate is such that the methane concentration is essentially zero at the surface of the fenster. The metabolic rate (per

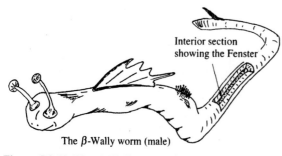

Interior section
showing the Fenster

The $\beta$-Wally worm (male)

**Figure P3.61** The $\beta$-Wally worm (male).

total body volume) required to sustain viability is $\mathscr{R} = 2.5$ mol/nb³-kipper.

The methane concentration in the atmosphere of $\beta$-Wally is 0.31 mol/nb³. The (molar) solubility ratio of methane in worm flesh is

$$\alpha = 0.1 = C_{\text{methane}}(\text{flesh})/C_{\text{methane}}(\text{atmosphere}) \quad \textbf{(P3.61.1)}$$

If the diffusion coefficient of methane through the body of the worm is known to be

$$D_{\text{methane}} = 0.3 \text{ nb}^2/\text{kipper} \quad \textbf{(P3.61.2)}$$

what is the maximum diameter that the worm can attain?

# Chapter **4**

# Unsteady State (Transient) Mass Transfer

In this chapter we examine unsteady state diffusion problems. To simplify matters as much as possible we consider one-dimensional diffusion—diffusion in only a single coordinate direction—but we still must deal with *partial* differential equations by which such problems are modeled. Fortunately, the equations that arise are classical mathematics problems, in the sense that the solutions have been worked out and presented analytically and—often—graphically. Hence one of our tasks here is to learn how to recognize that a specific physical problem leads to one of these classical models. Having accomplished this, we must find out how to use the solutions that are available for analysis and design of systems in which transient diffusion controls the behavior.

## 4.1 UNSTEADY STATE MASS TRANSFER

By definition, steady state mass transfer means that the concentration variable(s), and the rate of mass transfer, are *independent* of time. Many systems of importance operate under conditions of steady state behavior. Other systems exhibit *transient* (i.e., time-dependent behavior), and the time dependence may be the key issue in developing a mathematical model of the system. Yet other systems may be truly unsteady state, but the essential physics may be describable through the use of steady state equations for at least a part of the model. When this part is coupled with the key unsteady feature of the system, we obtain a "quasi-steady" model.

The assumption that a mass transfer system is in the steady state, so that all time derivatives vanish, is not always consistent with the physics. We can illustrate this with a simple example (see Fig. 4.1.1). Consider the hollow fiber sustained-release concept described earlier. For simple diffusion *in the core* of the fiber (i.e., in the region $0 \leq r \leq R_i$), with no convection and no chemical reactions, the diffusion equation in cylindrical coordinates reduces to the form

$$\frac{d}{dr} r \frac{dC_A^s}{dr} = 0 \quad \text{at steady state} \tag{4.1.1}$$

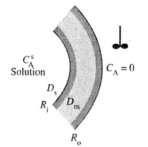

**Figure 4.1.1** A hollow fiber: diffusion in the core.

and the general solution is easily found to have the form

$$C_A^s = a \ln r + b \tag{4.1.2}$$

in the region $0 \leq r \leq R_i$.

Since $C_A^s$ is bounded everywhere, including $r = 0$, which is in the domain $[0, R_i]$, it follows that $a = 0$. Otherwise the "$\ln r$" term blows up at $r = 0$. Hence the only possible *steady state* solution is

$$C_A^s = b = \text{constant in the core} \tag{4.1.3}$$

This solution does not permit any depletion of the active species from within the core of the fiber. Hence there could be no evaporative loss from the exterior of the fiber without a reduction in $C^s$ in the interior. This would violate a simple mass balance.

There is a way the steady state assumption could make sense, but it calls for a specific assumption about the physics. Namely, if the encapsulated solution were *supersaturated*—for example, by including a solid precipitate of the active species—then the concentration $C^s$ would remain constant as long as the solid phase existed. The mass balance would be satisfied because any evaporation would be balanced by dissolution of the solid form of the active species, which would maintain the concentration of $C^s$ at the equilibrium saturation value. We will not consider that case here.

Still, our goal is to discuss a nonsteady mass transfer process. We must write the *transient diffusion equation*—the partial differential equation that describes the unsteady state behavior of the system—to get an exact solution for this transient process. Unfortunately this gives us a complicated equation to solve. (See Problem 4.1.) Instead we will use a *quasi-steady* analysis. It is based on the idea that diffusion in the core might be rapid compared to diffusion in the membrane surrounding the core. In a sense, we are assuming that the core is well mixed.

Now the *diffusion* problem that must be solved is in the annular membrane surrounding the core, that is, on the domain $[R_i, R_o]$. The steady diffusion equation is still Eq. 4.1.1, and Eq. 4.1.2 is still the general solution. However, since we must consider diffusion in the membrane surrounding the core (Fig. 4.1.2), we write the solution for the concentration profile *in the membrane* as

$$C_A^m = a \ln r + b \quad \text{on} \quad R_i \leq r \leq R_o \tag{4.1.4}$$

with the following boundary conditions:

1. The active species evaporates at $R_o$ and is instantly swept away by convection into an infinite exterior volume. Hence the partial pressure of A *exterior* to $R_o$ vanishes.

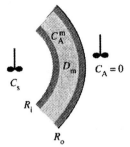

**Figure 4.1.2** A hollow fiber: diffusion in the membrane only.

We expect the concentration of A in the membrane, at $r = R_o^-$, to be proportional to the partial pressure at $R_o^+$. Hence we conclude that

$$C_A^m = 0 \qquad \text{at} \quad r = R_o^- \tag{4.1.5}$$

2. The active species has the concentration $C_A^s$ on the region $0 \le r < R_i$. There is a thermodynamic relationship between concentration of the active species in the liquid ($r = R_i^-$) and in the solid ($r = R_i^+$) (membrane) at the interface. We assume a linear relationship and write

$$C_A^m = \alpha C_A^s \qquad \text{at} \quad r = R_i \tag{4.1.6}$$

as the second boundary condition. We call the thermodynamic parameter $\alpha$ a *partition coefficient*, or a *solubility coefficient*. With these two boundary conditions it now follows that

$$C_A^m = \alpha C_A^s \frac{\ln(r/R_o)}{\ln(R_i/R_o)} \tag{4.1.7}$$

and the molar *flux* out of the fiber (this is the evaporative flux) is

$$N_A\big|_{r=R_o} = -D_m \frac{dC_A^m}{dr}\bigg|_{r=R_o} = -\frac{\alpha C_A^s D_m}{R_o \ln(R_i/R_o)} \tag{4.1.8}$$

The molar *rate* of evaporation, per unit of fiber length $L$, is

$$\frac{\dot{M}_A}{L} = 2\pi R_o N_A\bigg|_{r=R_o} = \frac{2\pi \alpha C_A^s D_m}{\ln(R_o/R_i)} \tag{4.1.9}$$

But in reality the interior concentration is unsteady [i.e., $C_A^s = C_A^s(t)$], while the solution presented here is valid only in the steady state: $C_A^s = $ constant. In the *quasi-steady approximation* we assume that the rate of transfer is given by the expression above, *even in the unsteady state*, if $C_A^s$ changes slowly in some sense.

To find $C_A^s(t)$ we now must introduce an overall mass (actually, mole) balance that equates the change in $C_A^s$ to the evaporative loss. This takes the form

$$\frac{\dot{M}_A}{L} = -\frac{d}{dt}(\pi R_i^2 C_A^s) = \frac{2\pi \alpha C_A^s D_m}{\ln(R_o/R_i)} \tag{4.1.10}$$

If we define a time scale $\tau$ as

$$\tau = \frac{R_i^2 \ln(R_o/R_i)}{2 D_m \alpha} \tag{4.1.11}$$

we can write Eq. 4.1.10 in the simpler format

$$-\frac{dC_A^s}{dt} = \frac{C_A^s}{\tau}$$

(4.1.12)

The solution is

$$\frac{C_A^s}{C_A^{s,o}} = e^{-t/\tau}$$

(4.1.13)

where $C_A^{s,o}$ is the initial value of $C_A^s$.

After an elapsed time $T$, the total amount evaporated is

$$\frac{M_A}{L} = \int_0^T \frac{\dot{M}_A}{L} dt = \pi R_i^2 C_A^{s,o}(1 - e^{-T/\tau})$$

(4.1.14)

This mathematical model is much simpler in format than the more exact unsteady state analysis would yield. If we were to discuss this model further at this point, it would be appropriate to provide some guidelines that indicate when the quasi-steady analysis can be expected to yield a good approximation to the more exact model. The guidelines would be in terms of the value of some dimensionless parameters that appear naturally in the problem. (See Problem 4.1.) Instead we turn to a discussion of the transient diffusion equation, and the properties of its solutions.

### 4.1.1  Some Examples of Unsteady State Mass Transfer Problems

There is a large body of interesting and important mass transfer problems that involve transient diffusion. We pose some of these problems here, and later we will develop mathematical models of some these problems, as well as others.

**Example A** *Unsteady Diffusion Across a Membrane*  A membrane test cell consists of two compartments separated by a polymeric membrane. In this example, we suppose that the membrane is permeable to olefin vapors, like ethylene, but impermeable to aliphatic hydrocarbon vapors. The top compartment is initially evacuated, but the lower one is filled with ethylene gas. We measure the ethylene concentration subsequently appearing in the vapor in the top compartment as a function of time.

We will carry out an analysis of this problem, and show that we can use the resulting mathematical model to determine the solubility and the diffusion coefficient of ethylene in the film.

**Example B** *A Dissolving Particle*  The dissolution of a particle is often controlled by diffusion into the surrounding liquid. A transient analysis can tell us how much time is required to reach a *steady* rate of dissolution, following some initial transient period. In addition, we can find the time for the particle to "disappear" through dissolution.

**Example C** *Evaporation of Solvent from a Film*  Polymer coatings are often applied from solution, and it is necessary to "dry" the resulting film by promoting the evaporation of solvent. Because diffusion is relatively slow within the viscous polymeric matrix, evaporation is controlled by the diffusion process within the film.

**Example D** *Stroke and Cell Death*  Capillaries supply oxygen to tissue, and diffusion through the tissue region is rate limiting. If blood flow is stopped, cell respiration depletes the dissolved oxygen content until the oxygen level is too low to support cell function. A transient analysis can indicate how long a cell region will remain viable under these conditions.

**Example E** *Doping of a Semiconductor Film*  To control the conductivity of a silicon film in a region of a semiconductor device, very small amounts of a specific atom, such as boron, are caused to diffuse into the region near the surface of the solid film. This transient process, called "doping," is central to control of the electronic properties of a device. Careful control of the concentration profile of the dopant (e.g., boron atoms) is achieved through an understanding of transient diffusion in the solid.

Before we can model any of the foregoing problems, we will have to look at the mathematics of transient diffusion. We begin with a general analysis in which we find *dimensional analysis* to be very useful. We then examine the mathematical solutions to several simple problems, and the use of graphical forms of these solutions (such as the Gurney–Lurie charts).

## 4.2 GENERAL TRANSIENT DIFFUSION: NO REACTION OR INTERNAL CONVECTION

We use as a starting point Eq. 3.1.7 from the preceding chapter:

$$\frac{\partial C_A}{\partial t} = D_{AB} \nabla^2 C_A \tag{4.2.1}$$

in some domain $V$. The form of the Laplacian operator (the $\nabla^2$ term) depends on the coordinate system. Table 4.2.1 gives these forms for planar, cylindrical, and spherical coordinates. For some arbitrary-shaped volume, as in Fig. 4.2.1, there is no obvious or natural choice of a coordinate system. Note that in the absence of convection, there is no velocity term on the left-hand side of the diffusion equation.

An initial condition is required, which we write as

$$C_A = C_A^o(\mathbf{r}, 0) \quad \text{at} \quad t \le 0 \tag{4.2.2}$$

While generally the *initial* state of the system could be spatially nonuniform, rendering $C$ a function of the vector position $\mathbf{r}$, we will almost always consider simple problems here, where the initial concentration distribution is spatially uniform. Boundary conditions can take any of a number of forms depending on the physics of the problem at hand.

**Table 4.2.1** The Laplacian Operator $\nabla^2$

| Cartesian Coordinates |
|---|
| $\nabla^2 = \dfrac{\partial^2}{\partial x^2} + \dfrac{\partial^2}{\partial y^2} + \dfrac{\partial^2}{\partial z^2}$ |

| Cylindrical Coordinates |
|---|
| $\nabla^2 = \dfrac{1}{r}\dfrac{\partial}{\partial r}\left(r\dfrac{\partial}{\partial r}\right) + \dfrac{1}{r^2}\dfrac{\partial^2}{\partial \theta^2} + \dfrac{\partial^2}{\partial z^2}$ |

| Spherical Coordinates |
|---|
| $\nabla^2 = \dfrac{1}{r^2}\dfrac{\partial}{\partial r}\left(r^2\dfrac{\partial}{\partial r}\right) + \dfrac{1}{r^2 \sin\theta}\dfrac{\partial}{\partial \theta}\left(\sin\theta\dfrac{\partial}{\partial \theta}\right) + \dfrac{1}{r^2 \sin^2\theta}\dfrac{\partial^2}{\partial \phi^2}$ |

For example,

$$C_A = C_A^s(t) \text{ on the surface S}$$

Specified $C_A$

(4.2.3)

Often the concentration on the surface of the volume is known because it is maintained by some external conditions over which we have control. In simple problems this concentration is independent of both time and position on the surface. An alternative to Eq. 4.2.3 is

$$-D_{AB}\nabla C_A = \mathbf{N}_A \text{ on the surface S}$$

Specified flux

(4.2.4)

The most common form of control of the surface flux is through making the surface, or at least some portion of it, impermeable, so that the flux $\mathbf{N}_A$ is zero on those portions.

Yet another boundary condition is

$$\mathbf{n} \cdot (-D_{AB}\nabla C_A) = k(C_A^s - C_A^\infty)$$

Convective transfer

(4.2.5)

This is a very common form of boundary condition, since so many problems of interest to us involve transfer from the interface into a fluid in which there is controlled dynamics. The right-hand side of Eq. 4.2.5 represents an assumption of a *linear* relationship between the flux and the *difference* between the concentration in the domain $V$—but right at the surface—$C_A^s$, and the concentration $C_A^\infty$ that the surface would have *if it were in thermodynamic equilibrium with the surrounding medium.* This assumption is borne out by experience. The coefficient $k$ that appears in this boundary condition is called the "convective mass transfer coefficient" and is assumed to be known. We will later address the issue of how we can estimate values of $k$ from a knowledge of characteristics of the external flow field. Note that in Eq. 4.2.5 the left-hand side is a *scalar* quantity; it is the component of the flux in the direction normal (perpendicular) to the surface. (We will take a more careful look at Eq. 4.2.5 in Section 4.3. This form, and these comments, will suffice for now.)

For a volume of *arbitrary shape* (Fig. 4.2.1), it is impossible to write a simple and general mathematical solution to the diffusion equation with its initial and boundary conditions. The form of the solution depends on the type of initial and boundary conditions that describe the specific physics of interest to us. We can learn something through dimensional arguments, however. In the following we analyze the equations and boundary conditions, coming up with possible presentations of the solutions to transient diffusion problems in a compact dimensionless format. The results of this analysis also provide guidelines for the design of experiments.

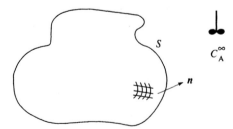

**Figure 4.2.1** An arbitrary volume within which transient diffusion occurs.

We will begin by defining a dimensionless concentration variable. The form of the definition depends in part on what we know, or assume, about conditions external to the body. If *convection* is taking place between the body (the domain $V$) and an exterior fluid at some known concentration equivalent to $C_A^\infty$, we usually choose to define $Y$ with $C_A^\infty$ as a reference concentration:

$$Y \equiv \frac{C_A - C_A^\infty}{C_A^0 - C_A^\infty} \qquad (4.2.6)$$

This type of definition is useful when there is *external convection* to a region of *known* concentration equivalent to $C_A^\infty$ far from $V$. Hence the choice of a reference concentration in the definition of $Y$ depends on what we know about the physics of the system. Since we may not know some features of the system with certainty, we are really making an *assumption* about the behavior, and consequently we obtain a model based on that assumption. Ultimately, experience is a valuable guide to this aspect of the modeling.

Note (from inspection of Eq. 4.2.5) that if convection is very effective (large $k$), then $C_A^s \to C_A^\infty$, and the surface concentration would be specified as in Eq. 4.2.3. In that case we would define $Y$ as

$$Y \equiv \frac{C_A - C_A^s}{C_A^0 - C_A^s} \qquad (4.2.7)$$

However, this choice of definition for $Y$, specifically the use of $C_A^s$ in the definition, is useful only if $C_A^s$ is known on the surfaces (the case of specified $C_A$ as in Eq. 4.2.3). We now have, in Eq. 4.2.7, a transformed variable $Y$ that takes on the value of zero on the surface. Note that with either of these definitions of $Y$, $Y = 1$ at time $t = 0$. This is why we defined $Y$ in this way.

Next, we make *each* space variable dimensionless using *some* characteristic length scale $L$. Although $L$ is undefined at this stage, we can say that it is a parameter with the units of length and is related to the size of the body. For a simply shaped body, $L$ could be one of the dimensions. For an odd-shaped body of volume $V$, $L$ could be $V^{1/3}$, which would just be a measure of the size of the body. Another possible choice would be the ratio of volume to surface area $A : L = V/A$. Whatever the definition, we define

$$x/L = \tilde{x} \qquad y/L = \tilde{y} \qquad z/L = \tilde{z} \qquad \tilde{\mathbf{r}} = (\tilde{x}, \tilde{y}, \tilde{z}) \qquad (4.2.8)$$

The expression for the vector position $\mathbf{r}$ is just a shorthand vector notation. In Cartesian coordinates, for example, we would write the Laplacian operator as

$$\nabla^2 = \frac{\partial^2}{\partial x^2} + \frac{\partial^2}{\partial y^2} + \frac{\partial^2}{\partial z^2} = \frac{1}{L^2}\left(\frac{\partial^2}{\partial \tilde{x}^2} + \frac{\partial^2}{\partial \tilde{y}^2} + \frac{\partial^2}{\partial \tilde{z}^2}\right) = \frac{\tilde{\nabla}^2}{L^2} \qquad (4.2.9)$$

Then

$$\frac{\partial Y}{\partial t} = \frac{D_{AB}}{L^2} \tilde{\nabla}^2 Y \qquad (4.2.10)$$

Next we define a dimensionless time variable as

$$X_D = \frac{D_{AB} t}{L^2} \qquad (4.2.11)$$

so that the differential equation becomes

$$\frac{\partial Y}{\partial X_D} = \tilde{\nabla}^2 Y \tag{4.2.12}$$

Note that no parameters appear in this differential equation. Boundary conditions now take the following forms:

Eq. 4.2.2 gives      $Y = 1$    at   $X_D = 0$     (the initial condition)    **(4.2.13)**

Eq. 4.2.5 gives      $\mathbf{n} \cdot (-\tilde{\nabla} Y) = \dfrac{kL}{D_{AB}} Y$    on S    **(4.2.14)**

We would use this boundary condition in a problem in which transfer to the ambient medium is influenced by convection. In that case $Y$ would not be known on the surface S. If the surface concentration $C_A^s$ were specified, we would use Eq. 4.2.6 to define $Y$. Then the boundary condition simplifies to the form

$$Y = 0 \quad \text{on} \quad S \tag{4.2.15}$$

On inspecting Eq. 4.2.14 we see that we could define a dimensionless "transport number" Bi, called the Biot number, by

$$Bi = \frac{kL}{D_{AB}} \tag{4.2.16}$$

where $D_{AB}$ is in the domain $V$, not in the convecting exterior fluid. Then for a *specified* surface concentration the boundary condition is still

$$Y = 0 \tag{4.2.17}$$

and Bi does not appear, of course. On the other hand, for a general convective boundary condition we would find

$$-\mathbf{n} \cdot \tilde{\nabla} Y = Bi \; Y \tag{4.2.18}$$

What can we say about the solution to the partial differential equation (PDE), (Eq. 4.2.12), with its initial and boundary conditions? For an arbitrarily shaped body, or even for a simple shape, but one without the geometrical simplicity and symmetry of bodies such as spheres or long cylinders or very thin planar slabs, the only statement we can make is the fairly obvious one that

$$Y = Y(\tilde{\mathbf{r}}, X_D; Bi) \tag{4.2.19}$$

Equation 4.2.19 is really just the obvious statement that the solution to Eq. 4.2.12 with initial condition 4.2.13 and boundary condition 4.2.18 is a function of whatever variables and parameters appear in these equations. This does not seem like a very powerful or remarkable result. On the other hand, it really does include some significant information:

**a.** A natural definition of a dimensionless time is $X_D$. This choice removes length scales and the diffusion coefficient from the differential equation.

**b.** A dimensionless convective coefficient Bi appears in the boundary condition.

For one-dimensional problems with geometrical symmetry (the concentration depends on only one component of $\tilde{\mathbf{r}}$) we have available analytical solutions that can be presented in a very compact graphical format known as the Gurney–Lurie charts, three examples of which are presented in Fig. 4.2.2. Hence there is a class of simple transient diffusion problems that have already been solved, and the solutions are available in a graphical format. We will see that many simple, but important and interesting, problems can be analyzed using the Gurney–Lurie charts.

We may summarize by observing that by defining the dimensionless variables in the manner shown above, we can present the analytical solutions to *certain one-dimensional* problems with known/uniform external concentration or convective transfer in a com-

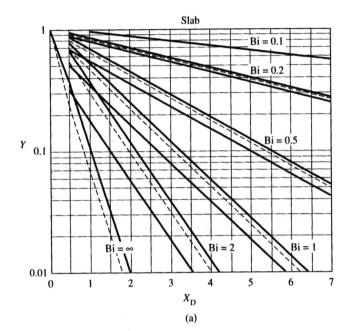

(a)

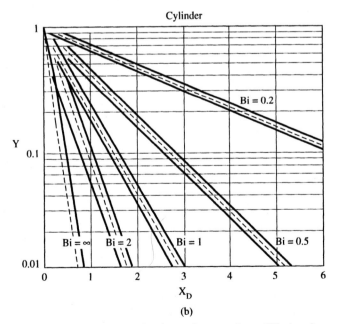

(b)

**Figure 4.2.2** Gurney–Lurie charts for transient diffusion in (a) a slab, (b) a long cylinder, and (c) a sphere. For each value of the Biot number Bi, a set of three lines is shown. The upper line is the value of $Y$ at the center of the body. The dashed

Sphere

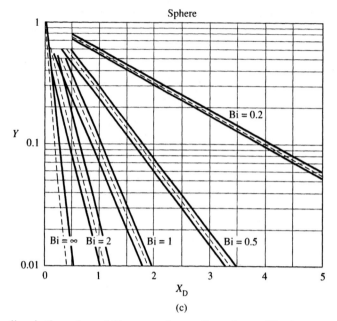

(c)

line is the value of $Y$ *averaged* over the volume. The lower line is the value of $Y$ at the surface. In the case $Bi = \infty$, the surface is at $Y = 0$ for all time, so no line is shown. (It would be the vertical $Y$ axis.)

pact format. The graphs replace the need to solve the PDE, or to manipulate the analytical functions that appear in the solutions. We are using existing known solutions for *special* simple cases.

What do we do if the geometry is *not simple one-dimensional,* or if the physics leads to *nonsymmetry in the geometry*? Let's look at some specific examples corresponding to Examples A, B, and C in Section 4.1.1. We will see that these problems do not fall into the class for which the Gurney–Lurie charts can be used, primarily because of the boundary conditions.

**EXAMPLE 4.2.1**  *Unsteady Diffusion Across a Membrane Test Cell*

With reference to Fig. 4.2.3, ethylene diffuses across a membrane that separates two compartments. The partial pressure (or, equivalently, the molar concentration) of ethylene is assumed to be zero in the upper compartment. This is equivalent to saying that the ethylene is extremely diluted in the upper compartment, a result that could be achieved through careful design of the experimental system.

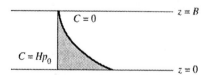

**Figure 4.2.3** Diffusion profile across a membrane. Concentrations are maintained constant on the boundaries.

The partial pressure of ethylene is fixed at the value $p_0$ in the lower compartment (i.e., for $z \leq 0$). Keep in mind that this region is *outside* the membrane. The diffusion is assumed to be one-dimensional. Hence only one space variable, $z$, need be considered. In this example we are interested in the transient part of the problem, and so we seek a solution to the unsteady diffusion equation,

$$\frac{\partial C}{\partial t} = D \frac{\partial^2 C}{\partial z^2} \tag{4.2.20}$$

where $C$ is the molar concentration of ethylene in the *membrane*. Initial and boundary conditions are assumed, and take the forms

$$C = 0 \qquad \text{at} \quad t \leq 0 \qquad 0 \leq z \leq B \tag{4.2.21}$$

$$C = Hp_0 \qquad \text{at} \quad z = 0^+ \qquad t > 0 \tag{4.2.22}$$

$$C = 0 \qquad \text{at} \quad z = B^- \qquad t > 0 \tag{4.2.23}$$

Note that there is no symmetry about any plane in the membrane in this problem. Figure 4.2.2a is valid only for the case of such symmetry. Hence we cannot use the Gurney–Lurie chart for this geometry because of the form of the boundary conditions. Equation 4.2.21 reflects our *assertion* that the membrane is initially free of the penetrant. This is our claim of how this system is set up. It is *physically* possible for the membrane to be saturated with ethylene initially. If that were the case, we would write a different initial condition.

Equation 4.2.22 reflects our assertions that equilibrium exists at the lower boundary and that a Henry's law type of equilibrium relationship holds. While this is a common pair of assumptions, and one that is often consistent with physical observation, other boundary conditions would be possible.

Equation 4.2.23 reflects our earlier statement that the ethylene is extremely diluted in the upper compartment.

If we now define a modified concentration variable $Y'$ as

$$\frac{Hp_0 - C}{Hp_0} \equiv Y' \tag{4.2.24}$$

we will find that the boundary conditions will take a simpler form. In addition, if we define a dimensionless space variable $Z$, and a dimensionless time variable $X$, as

$$\frac{z}{B} = Z \tag{4.2.25}$$

$$\frac{Dt}{B^2} = X \tag{4.2.26}$$

the differential equation will take a form free of any parameters

$$\frac{\partial Y'}{\partial X} = \frac{\partial^2 Y'}{\partial Z^2} \tag{4.2.27}$$

Furthermore, with the choices of dependent variables we have just made, the boundary conditions become independent of parameters also:

$$Y' = 1 \qquad \text{at} \quad X = 0 \qquad 0 \leq Z \leq 1 \tag{4.2.28}$$

$$Y' = 0 \qquad \text{at} \quad Z = 0 \qquad X > 0 \tag{4.2.29}$$

$$Y' = 1 \qquad \text{at} \quad Z = 1 \qquad X > 0 \tag{4.2.30}$$

The advantage of this nondimensionalization is twofold:

1. The equation and boundary conditions are simpler to write, and when there are fewer parameters, we are less likely to make errors as we write.
2. With some experience we can recognize that the differential equation and boundary conditions are in a standard form, for which a solution has already been worked out, and we need only find that solution in a reference somewhere, rather than developing it ourselves. From a practical, *engineering*, perspective, the best way to solve a differential equation is to find that someone else has already solved it and copy this solution!

From the physics we know that at steady state the concentration profile will be linear. (This follows by setting the left-hand side of Eq. 4.2.27 to zero and solving the resulting ordinary differential equation for $Y'$.) Hence it follows that (in the dimensionless variables)

$$Y' = Z \quad \text{as } X \to \infty \tag{4.2.31}$$

Now we define $Y = Y' - Z$ to be a new variable, and we find that Eq. 4.2.27 becomes

$$\frac{\partial Y}{\partial X} = \frac{\partial^2 Y}{\partial Z^2} \tag{4.2.32}$$

subject to boundary conditions

$$Y = f(Z) = 1 - Z \quad \text{at} \quad X = 0 \tag{4.2.33}$$

$$Y = 0 \quad \text{at} \quad Z = 0 \tag{4.2.34}$$

$$Y = 0 \quad \text{at} \quad Z = 1 \tag{4.2.35}$$

We have manipulated the differential equation into this form (especially now, $Y = 0$ on *both* surfaces) because it is easy to solve in this form. The solution technique, called "separation of variables," can be found in any applied math book that deals with partial differential equations.[1] The result can be written in an infinite series (this is a Fourier series), where we sum over the "counter" $n$:

$$Y = \frac{2}{\pi} \sum_{n=1}^{\infty} \frac{\sin n\pi Z}{n} \exp\left(-n^2\pi^2 X\right) \tag{4.2.36}$$

Now we must work our way back toward the solution in terms of the actual dependent variable of interest, the molar concentration $C$. Using the definition (Eq. 4.2.24)

$$Y' = 1 - \frac{C}{Hp_0} = Z + Y \tag{4.2.37}$$

we find

$$\frac{C}{Hp_0} = 1 - Z - Y = 1 - Z - \frac{2}{\pi} \sum_{n=1}^{\infty} \frac{\sin n\pi Z}{n} \exp(-n^2\pi^2 X) \tag{4.2.38}$$

But the *concentration profile* in the membrane is not really what we want. We want the *rate of transfer* of ethylene into the upper compartment. For a membrane area $A$,

---

[1] A good reference is Rice and Do, *Applied Mathematics and Modeling for Chemical Engineers*, Wiley, New York, 1995.

this rate is

$$-AD\frac{\partial C}{\partial z}\bigg|_{z=B} = \frac{HADp_0}{B}\frac{\partial Y'}{\partial Z}\bigg|_{Z=1} \qquad (4.2.39)$$

The number of moles of ethylene $n_E$ in the upper compartment can be expressed in terms of the partial pressure of ethylene. If the upper compartment is initially at zero pressure, the partial pressure of ethylene subsequently is the absolute pressure in the upper compartment. If the upper compartment has a fixed volume $V$, the number of moles is

$$n_E = \frac{pV}{R_G T} \qquad (4.2.40)$$

and

$$-AD\frac{\partial C}{\partial z}\bigg|_{z=B} = \frac{dn_E}{dt} = \frac{V}{R_G T}\frac{dp}{dt} = \frac{HADp_0}{B}\frac{\partial Y'}{\partial Z}\bigg|_{Z=1} \qquad (4.2.41)$$

We find $(\partial Y'/\partial Z)$ from Eqs. 4.2.37 and 4.2.38 to be

$$\frac{\partial Y'}{\partial Z} = 1 + \frac{\partial Y}{\partial Z} = 1 + 2\sum_{n=1}^{\infty}\cos n\pi Z \exp(-n^2\pi^2 X) \qquad (4.2.42)$$

and at $Z = 1$

$$\frac{\partial Y'}{\partial Z}\bigg|_{Z=1} = 1 + 2\sum_{n=1}^{\infty}(-1)^n\exp(-n^2\pi^2 X) \qquad (4.2.43)$$

Then

$$\frac{dp}{dt} = \frac{HADp_0 R_G T}{VB}\frac{\partial Y'}{\partial Z}\bigg|_{z=1} = \frac{HADp_0 R_G T}{VB}\left[1 + 2\sum_{n=1}^{\infty}(-1)^n\exp(-n^2\pi^2 X)\right]$$
$$(4.2.44)$$

Equation 4.2.44 is a first-order ordinary differential equation for $p(t)$. With the initial condition

$$p = 0 \qquad \text{at} \quad t = 0 \qquad (4.2.45)$$

the solution is obtained directly by integrating the right-hand side of Eq. 4.2.44 term by term. The result is

$$p = \frac{HADp_0 R_G T}{VB}\left\{t + \frac{2B^2}{D\pi^2}\sum_{n=1}^{\infty}\frac{(-1)^n}{n^2}[1 - \exp(-n^2\pi^2 X)]\right\} \qquad (4.2.46)$$

Note that in this solution the time dependence appears in two places: there is a simple linear term $t$, to which is added an infinite series of exponential functions of $t$, in terms of the dimensionless time $X = Dt/B^2$. It is possible to simplify this solution and obtain a useful approximation. For a long time, defined as $X > 1$, the exponential terms starting with $n = 1$ are small compared to unity, and we find

$$p = \frac{HADp_0 R_G T}{VB}\left[t + \frac{2B^2}{D\pi^2}\sum_{n=1}^{\infty}\frac{(-1)^n}{n^2}\right] \qquad (4.2.47)$$

From a decent set of math tables we can find that

$$\sum_{n=1}^{\infty}\frac{(-1)^n}{n^2} = -\frac{\pi^2}{12} \qquad (4.2.48)$$

so an approximate solution for the pressure history $p(t)$ is

$$p = \frac{HADp_0 R_G T}{VB}\left[t - \frac{B^2}{6D}\right]$$

(4.2.49)

We may write this in the form

$$p = \frac{HADp_0 R_G T}{VB}t\left[1 - \frac{B^2}{6Dt}\right]$$

(4.2.50)

Since we have already assumed that

$$X = \frac{Dt}{B^2} > 1$$

(4.2.51)

in going from Eq. 4.2.46 to 4.2.47 we should be consistent and write

$$p = \frac{HADp_0 R_G T}{VB}t \qquad \text{for} \quad X > 1$$

(4.2.52)

It would not be difficult to show that we would get this result exactly if we carried out a quasi-steady solution to this problem. 'Now we have the additional insight that the quasi-steady solution is valid after a period of time defined by $X > 1$.

If we were to plot $p(t)$ it would look like the sketch in Fig. 4.2.4. (See Problem 4.3.) Hence we can take data on $p(t)$ and find values for $D$ from the intercept and $H$ from the slope (assuming $B$ is known).

If you were paying attention to the **physics,** instead of to the **mathematics,** you should have been uncomfortable with the solution procedure and the resulting model. Go back to the boundary condition at the upper surface, Eq. 4.2.23. This condition followed from an earlier assumption that the ethylene partial pressure in the upper compartment satisfied $p = 0$. Having made that assumption, we end up with a solution for $p(t)$ in the upper compartment, which is an obvious violation of the assumption that $p$ vanishes there. How do we resolve this?

We conclude that the solution (Eq. 4.2.50, or Eq. 4.2.52, or even the more exact result, Eq. 4.2.46) might be a good approximation as long as $p(t)$ is very small. What is an appropriate criterion of small $p$? It seems reasonable to say that the solution we have derived is a good approximation to the physics of the problem as long as $p/p_0$ is much smaller than 1. We can ensure this in the design of the experimental system described by making the volume of the upper compartment so great that any ethylene that diffuses into the volume $V$ is so highly diluted that $p/p_0$ is small compared to unity (see Problem 4.5).

In working this example it has been assumed that the membrane is initially free of

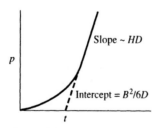

**Figure 4.2.4** Pressure rise in the closed volume above a membrane test cell.

the diffusing species, as the initial condition (Eq. 4.2.21) shows. Under other starting conditions it might be possible for the membrane to be equilibrated with the gas on the high pressure side. Then the initial condition would take the form

$$C = Hp_0 \quad \text{at} \quad t \leq 0 \tag{4.2.53}$$

For this initial condition, the solution for $C$ would be

$$\frac{C}{Hp_0} = 1 - Z - Y = 1 - Z - \frac{2}{\pi} \sum_{n=1}^{\infty} \frac{(-1)^n \sin n\pi Z}{n} \exp(-n^2\pi^2 X) \tag{4.2.54}$$

(There is a subtle difference between this and Eq. 4.2.38. Look carefully!) The change in pressure would then take the form

$$\frac{dp}{dt} = \Theta \left[ 1 + 2 \sum_{n=1}^{\infty} \exp(-n^2\pi^2 X) \right] \tag{4.2.55}$$

where $\Theta$ is the coefficient

$$\frac{HADp_0 R_G T}{VB}$$

The solution for $p(t)$ for long time (Eq. 4.2.49) would be changed accordingly, but we leave this part of the problem as an exercise (see Problem 4.6.).

## EXAMPLE 4.2.2   *A Dissolving Particle (an External Diffusion Problem)*

We want to develop a mathematical model of an important process: the dissolution of a particle in a fluid. As usual we introduce a number of simplifying assumptions. With respect to geometry, we assume that the particle is spherical, as in Fig. 4.2.5, and that the dissolution process is spherically symmetric. With regard to the mass transfer conditions in the fluid surrounding the dissolving particle, we assume that a *stagnant* liquid is in contact with the particle. (Later we will examine dissolution controlled by convection, rather than diffusion.)

In this example we will develop another *quasi-steady* model: that is, we will assume that although the particle is dissolving, its radius remains constant. This probably appears to make no sense physically, but it is the first step toward the development of an approximate but useful model of the process. This raises an important question, which we will not answer at this point: *How can it make sense to claim that a mathematical model based on apparently nonphysical assumptions is useful for the description of a real process of concern to us?* This is a central question in mathematical modeling, and we will have to discuss it.

**Figure 4.2.5** A solid particle from which transient diffusion occurs.

The transient diffusion equation in spherical coordinates, with only radial diffusion, is written as

$$\frac{\partial C}{\partial t} = D \frac{1}{r^2} \frac{\partial}{\partial r} \left( r^2 \frac{\partial C}{\partial r} \right) \tag{4.2.56}$$

and the initial and boundary conditions are

$$
\begin{array}{llll}
\text{at } t = 0 & C = 0 & r > R_0 & \text{(4.2.57)} \\
\text{at } r = R_0^+ & C = C_{\text{sat}} & t > 0 & \text{(4.2.58)} \\
\text{at } r \to \infty & C = 0 & t > 0 & \text{(4.2.59)}
\end{array}
$$

Notice the *domain* in which the diffusion equation is written. It is in the fluid surrounding the particle, *not within the particle* itself. We are *assuming* here that the major resistance to mass transfer is in the fluid surrounding the particle and that there is no significant resistance to diffusion inside the solid particle itself. This is an assumption that would have to be argued with more information than is presented here. We assume that such evidence exists, and it has dictated the way we choose to write the physics in this problem formulation.

We can now define a new variable $u = rC$ and show that Eq. 4.2.56 takes the simpler form

$$\frac{\partial u}{\partial t} = D \frac{\partial^2 u}{\partial r^2} \tag{4.2.60}$$

and that the boundary conditions become

$$\text{at } t = 0 \qquad u = 0 \tag{4.2.61}$$

$$\text{at } r = R_o^+ \qquad u = C_{\text{sat}} R_o = U_o \tag{4.2.62}$$

$$\text{at } r \to \infty \qquad u = 0 \tag{4.2.63}$$

Here is another partial differential equation, identical in form to Eq. 4.2.32, but with a different notation. The solution is not the same, because the boundary conditions are different. This is an important point to appreciate; *two identical partial differential equations can have different solutions depending on the form of the boundary conditions.* A key difference in this case lies in the extension to infinity of the domain for solution. This specific boundary value problem[2] is solvable by a technique known as "combination of variables." Again, this is a standard textbook example in applied mathematics. The reader not interested in the mathematics can skip directly to Eq. 4.2.97.

We define a new variable that combines both $r$ and $t$ in the form

$$\eta = \frac{r - R_o}{2\sqrt{Dt}} \tag{4.2.64}$$

and then we convert the derivatives $\partial/\partial t$ and $\partial/\partial r$ to derivatives $\partial/\partial \eta$. We do this as follows:

$$\frac{\partial}{\partial t} = \frac{\partial \eta}{\partial t} \frac{\partial}{\partial \eta} \tag{4.2.65}$$

$$\frac{\partial \eta}{\partial t} = \frac{\partial}{\partial t} \left( \frac{r - R_o}{2\sqrt{D}} t^{-1/2} \right) = \frac{r - R_o}{2\sqrt{D}} \left( -\frac{t^{-3/2}}{2} \right) = -\frac{1}{2} \frac{r - R_o}{2\sqrt{Dt}} t^{-1} = -\frac{\eta/t}{2} \tag{4.2.66}$$

---

[2] We use the term "boundary value problem" as a reminder that the solution to the partial differential equation is strongly dependent on the boundary conditions.

Since

$$\frac{\partial}{\partial r} = \frac{\partial \eta}{\partial r} \frac{\partial}{\partial \eta} \qquad (4.2.67)$$

we find

$$\frac{\partial \eta}{\partial r} = \frac{1}{2\sqrt{Dt}} = \frac{\eta}{r - R_o}$$

and

$$\frac{\partial^2}{\partial r^2} = \frac{\partial}{\partial r}\frac{\partial}{\partial r} = \frac{\partial \eta}{\partial r}\frac{\partial}{\partial \eta}\left[\frac{\partial \eta}{\partial r}\frac{\partial}{\partial \eta}\right]$$

$$= \frac{\eta}{r - R_o}\frac{\partial}{\partial \eta}\left(\frac{\eta}{r - R_o}\right)\frac{\partial}{\partial \eta}$$

$$= \frac{\eta}{r - R_o}\left[\frac{\eta}{r - R_o}\frac{\partial^2}{\partial \eta^2} + \frac{1}{r - R_o}\frac{\partial}{\partial \eta} - \frac{\eta}{(r - R_o)^2}\frac{\partial r}{\partial \eta}\frac{\partial}{\partial \eta}\right] \qquad (4.2.69)$$

$$= \frac{\eta}{r - R_o}\left[\frac{\eta}{r - R_o}\frac{\partial^2}{\partial \eta^2} + \frac{1}{r - R_o}\frac{\partial}{\partial \eta} - \frac{\eta}{(r - R_o)^2}\left(\frac{r - R_o}{\eta}\right)\frac{\partial}{\partial \eta}\right]$$

$$= \frac{\eta^2}{(r - R_o)^2}\frac{\partial^2}{\partial \eta^2}$$

With these results, Eq. 4.2.60 becomes

$$-\frac{\eta/t}{2}\frac{\partial u}{\partial \eta} = D\frac{\eta^2}{(r - R_o)^2}\frac{\partial^2 u}{\partial \eta^2} \qquad (4.2.70)$$

or

$$\frac{\partial^2 u}{\partial \eta^2} + \frac{(r - R_o)^2}{2\eta tD}\frac{\partial u}{\partial \eta} = 0 \qquad (4.2.71)$$

which is simplified to the form

$$\frac{\partial^2 u}{\partial \eta^2} + 2\eta\frac{\partial u}{\partial \eta} = 0 \qquad (4.2.72)$$

since $\eta$ was defined such that

$$\eta^2 = \frac{(r - R_o)^2}{4Dt} \qquad (4.2.73)$$

Note that the original differential equation for $u(t, r)$ has been *transformed* to an equation for $u(\eta)$. The two independent variables have *merged* and only one *new* independent variable, which is $\eta$, appears in the differential equation. This merging of two independent variables into one gives the method the name "combination of variables."

Now we must look at the boundary conditions.

$$u = 0 \text{ at } t = 0 \qquad \text{becomes} \qquad u = 0 \text{ at } \eta = \infty \qquad (4.2.74)$$

$$u = 0 \text{ at } r = 0 \qquad \text{becomes} \qquad u = 0 \text{ at } \eta = \infty \qquad (4.2.75)$$

These two boundary conditions in $t$ and $r$ collapse to a single boundary condition on $\eta$. The third (or, now, second) boundary condition becomes

$$u = U_o \quad \text{on} \quad \eta = \frac{R_o - R_o}{2\sqrt{Dt}} = 0 \qquad (4.2.76)$$

To summarize, we have transformed the PDE and boundary conditions to the form

$$\frac{\partial^2 u}{\partial \eta^2} + 2\eta \frac{\partial u}{\partial \eta} = 0 \qquad (4.2.77)$$

$$u = 0 \quad \text{at} \quad \eta = \infty \qquad (4.2.78)$$

$$u = U_o \quad \text{at} \quad \eta = 0 \qquad (4.2.79)$$

Note that since $\eta$ is the only independent variable to appear in the PDE *and* the boundary conditions, we conclude that

$$u = u(\eta) \quad \text{only} \qquad (4.2.80)$$

and Eq. 4.2.77 is actually an ODE:

$$\frac{d^2 u}{d\eta^2} + 2\eta \frac{du}{d\eta} = 0 \qquad (4.2.81)$$

$$u = 0 \quad \text{at} \quad \eta = \infty \qquad (4.2.82)$$

$$u = U_o \quad \text{at} \quad \eta = 0 \qquad (4.2.83)$$

We may solve this second-order linear ordinary differential equation by defining a variable

$$\omega = \frac{du}{d\eta} \qquad (4.2.84)$$

so that the differential equation becomes

$$\frac{d\omega}{d\eta} + 2\eta\omega = 0 \qquad (4.2.85)$$

We rewrite this as

$$\frac{d\omega}{\omega} = -2\eta \, d\eta \qquad (4.2.86)$$

and integrate to find

$$\ln \omega = -\eta^2 + \text{constant} \qquad (4.2.87)$$

or

$$\ln \omega = -\eta^2 + \ln \text{constant}$$
$$= -\eta^2 + \ln \omega_o \qquad (4.2.88)$$

so that

$$\ln \frac{\omega}{\omega_o} = -\eta^2 \qquad (4.2.89)$$

Hence

$$\frac{\omega}{\omega_o} = e^{-\eta^2} \qquad (4.2.90)$$

or

$$\frac{du}{d\eta} = \omega_o e^{-\eta^2} \qquad (4.2.91)$$

One more integration gives

$$u - U_o = \omega_o \int_0^\eta e^{-\eta^2} d\eta \qquad (4.2.92)$$

(We have used Eq. 4.2.83 to introduce $U_o$. This corresponds to the lower limit $\eta = 0$ on the integral sign.)

The boundary condition at $\eta = \infty$ gives

$$0 - U_o = \omega_o \int_0^\infty e^{-\eta^2} d\eta \qquad (4.2.93)$$

The *definite* integral above is some pure number. In fact, its value is $\sqrt{\pi}/2$. Hence we find

$$\omega_o = -\frac{2U_o}{\sqrt{\pi}} \qquad (4.2.94)$$

so we obtain as the solution for $u(\eta)$

$$\frac{u}{U_o} = 1 - \frac{2}{\sqrt{\pi}} \int_0^\eta e^{-\eta^2} d\eta \qquad (4.2.95)$$

This integral is available in graphical or tabular form, and is called the *error function* (see Fig. 4.2.6). Hence we write

$$\frac{u}{U_o} = 1 - \text{erf}(\eta) = \text{erfc}(\eta) \qquad (4.2.96)$$

where erf is the error function and we define the *complementary error function* erfc as $1 - \text{erf}$.

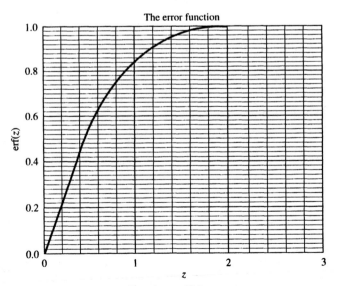

**Figure 4.2.6** The error function erf($z$).

Recovering the definitions of $u = rC$ and $\eta$ (Eq. 4.2.64) we find

$$\frac{C}{C_{\text{sat}}} = \frac{R_o}{r}\left(1 - \text{erf}\frac{r - R_o}{2\sqrt{Dt}}\right) \tag{4.2.97}$$

As is often the case, it is the flux and not the concentration profile itself that is of primary interest. Hence we need to calculate the molar flux $N$ at the surface $R_o$ from (see Appendix B, Eq. B26)

$$N = -D\left.\frac{\partial C}{\partial r}\right|_{R_o^+} = \frac{DC_{\text{sat}}}{R_o}\left(1 + \frac{R_o}{\sqrt{\pi Dt}}\right) \tag{4.2.98}$$

(We use the notation $R_o^+$ as a reminder that the flux is at the surface, but on the *outside* of the particle.) Note that the steady solution (which we can see is equivalent to Sh = 2, a result we derived earlier) is reached when

$$\frac{R_o}{\sqrt{\pi Dt}} \ll 1 \tag{4.2.99}$$

or

$$\frac{\pi Dt}{R_o^2} \gg 1 \tag{4.2.100}$$

Hence a criterion for *steady dissolution* is given by Eq. 4.2.99 or 4.2.100, and if values for $D$ and $R_o$ were available, we would be able to estimate the transient time prior to steady dissolution. Two important issues are not addressed here. One is the validity of the assumption that there is no internal diffusion resistance, with its corollary that only the external (fluid) domain requires consideration. The other is the assumption that although the particle is dissolving, we may use $R_o$ = constant in the boundary condition. Instead of following up on these points we go on to another problem.

**EXAMPLE 4.2.3** *Transient Evaporation of Solvent from a Sheet of Polymer*

Find the time required to remove 95% of the solvent that is in a polymeric film attached to an *impermeable* substrate; the film has a thickness of 2 mm (Fig. 4.2.7). The solvent diffusivity within the polymer is $D = 4 \times 10^{-7}$ cm²/s. The film initially has 2000 ppm (by weight) of solvent uniformly distributed throughout the polymeric phase.

We recognize this as a problem in transient diffusion, with no convection within the film, and no reaction. We assume that the film is thin in the sense that its dimensions in the plane of the surface (the $x$ and $z$ directions) are large compared to $B$. Then the diffusion is one-dimensional, to a good approximation.

**Figure 4.2.7** Solvent dissolved in a polymeric film. Evaporation is from the upper surface; the lower surface is impermeable.

Hence we must solve

$$\frac{\partial C}{\partial t} = D \frac{\partial^2 C}{\partial y^2} \qquad (4.2.101)$$

with the initial condition

$$C = C_o = 2000 \text{ ppm} \qquad \text{at} \quad t = 0 \qquad (4.2.102)$$

At the boundary $y = 0$ we assume that there is no penetration of solvent

$$\frac{\partial C}{\partial y} = 0 \qquad \text{at} \quad y = 0 \qquad (4.2.103)$$

At the boundary $y = B$ we assume that there is rapid exterior convective removal of solvent vapor (Bi = $\infty$), and thus a reasonable approximation is

$$C = 0 \qquad \text{at} \quad y = B \qquad (4.2.104)$$

This problem is not exactly of the form of the diffusion equation/boundary conditions that lead to the Gurney–Lurie charts. *Note specifically that there is no symmetry about the central plane of the sheet.* But note further that the boundary condition at $y = 0$ (no flux) looks like a symmetry boundary condition. Hence the problem at hand is equivalent to evaporation from *both* surfaces of a sheet of thickness $2B$. Figure 4.2.8 shows this relationship. Here we are stating our recognition that since a line or plane of symmetry corresponds to zero flux, this is equivalent to an impermeable plane. Although the plane of symmetry is open and permeable, it is because of symmetry that there is no net flux across the plane.

It follows that the graphical solution in the Gurney–Lurie chart for a slab of *half-thickness B* is usable for the problem of transfer from *one side only* of a slab of *total thickness B*.

The initial concentration is $C_o = 2000$ ppm. The final *average* concentration is specified in the problem statement to be 5% of this, or $\overline{C} = 100$ ppm. Thus

$$\overline{Y} = \frac{C_s - \overline{C}}{C_s - C_o} = \frac{0 - 100}{0 - 2000} = 0.05 \qquad (4.2.105)$$

$$X_D = \frac{Dt}{B^2} = \frac{4 \times 10^{-7} t}{0.2^2} = 10^{-5} t \qquad (4.2.106)$$

Using the dashed line on Fig. 4.2.2a for Bi = $\infty$, we see that $X_D$ must be 1.12 for $\overline{Y} = 0.05$.
Hence

$$10^{-5} t = 1.12 \qquad (4.2.107)$$

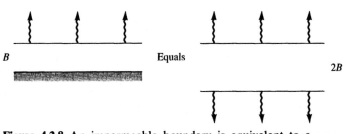

**Figure 4.2.8** An impermeable boundary is equivalent to a plane of symmetry.

or

$$t = 1.12 \times 10^5 \, \text{s} \qquad \qquad \textbf{(4.2.108)}$$

This very long time (about 30 hours) is a consequence of the very small diffusion coefficient of the solvent in the polymeric film.

Note that we have assumed that the mass transfer process is limited by diffusion within the film. According to this model, there is no advantage to improving the external mass transfer process. Any improvement in drying time will have to come about by means of alteration of conditions internal to the film. This includes the possibility of a more diffusive solvent, a higher temperature of operation (which will increase the diffusion coefficient), or the use of a thinner film. We could explore the impact of several of these changes with this mathematical model. In other words, we could do "experiments" on the system by using the mathematical model in place of the real physical system. Obviously this potential flexibility places a premium on the development of a realistic model of the process by which solvent leaves.

## 4.2.1   Transient Diffusion for Very Short Times

Under some circumstances we are interested in transient diffusion for times so short that very little mass transfer has occurred. To be specific, let's consider the preceding problem (diffusion-limited evaporation of solvent from a sheet) and examine its mathematical formulation and solution. We may begin with Eq. 4.2.12, which we rewrite slightly in the form

$$\frac{\partial Y}{\partial X_D} = \nabla^2 Y = \frac{\partial^2 Y}{\partial \bar{y}^2} \qquad \qquad \textbf{(4.2.109)}$$

with an initial condition

$$Y = 1 \qquad \text{at } X_D = 0 \qquad \qquad \textbf{(4.2.110)}$$

and a "no-flux" boundary condition at the impermeable surface

$$\frac{\partial Y}{\partial \bar{y}} = 0 \qquad \text{at } \bar{y} = 0 \qquad \qquad \textbf{(4.2.111)}$$

We consider the case of one-dimensional diffusion in the $\bar{y}$ direction through a slab, so the Laplacian and gradient operators simplify as indicated. At the exposed surface we will use the convective boundary condition in the form

$$-\frac{\partial Y}{\partial \bar{y}} = \text{Bi } Y \qquad \text{at } \bar{y} = 1 \qquad \qquad \textbf{(4.2.112)}$$

Here Bi is defined as in Eq. 4.2.16, and the length scale $L$ is $L = B$.

This boundary value problem is solved by a method similar to that used in the solution illustrated in Example 4.2.1. In fact the problems are quite similar, but with some differences in the boundary conditions. As a consequence, the form of the solution is similar, but not identical, to that given in Eq. 4.2.36, and in this case we find

$$Y = \sum_{n=1}^{\infty} \frac{4 \sin \lambda_n}{2\lambda_n + \sin 2\lambda_n} \exp(-\lambda_n^2 X_D) \cos(\lambda_n \bar{y}) \qquad \qquad \textbf{(4.2.113)}$$

where the eigenvalues $\lambda_n$ are defined as the infinite set of roots of the equation (see Fig. A1.3)

$$\lambda_n \tan \lambda_n = \text{Bi} \qquad (4.2.114)$$

In principle, Eq. 4.2.113 is the solution to our problem. In practice we find that if we wish to evaluate the solution for very small time (by which we mean $X_D \ll 1$), it is necessary to carry out the indicated summation over a large number of terms of the infinite series. This form of the solution does not lend itself to easy and accurate computation for short times.

This is true as well in the special case of very efficient external convection: $\text{Bi} = \infty$. In that case the surface boundary condition becomes simply

$$Y = 0 \qquad \text{at} \quad \tilde{y} = 1 \qquad (4.2.115)$$

and the solution for $Y$ takes the form

$$Y = \frac{4}{\pi} \sum_{n=0}^{\infty} (-1)^n \frac{\cos\left(\dfrac{2n+1}{2}\right)\pi\tilde{y}}{2n+1} \exp\left[-\frac{(2n+1)^2}{4}\pi^2 X_D\right] \qquad (4.2.116)$$

in place of Eq. 4.2.113. The same objection regarding the difficulty of using this infinite series when $X_D$ is very small holds here, as well. Hence we seek an alternative form for the solution that is accurate and lends itself to simple computation when we want to study the behavior of a solution at short times.

To reformulate the solution just obtained, we go back to the diffusion equation in the form of Eq. 4.2.1:

$$\frac{\partial C_A}{\partial t} = D_{AB} \nabla^2 C_A = D_{AB} \frac{\partial^2 C_A}{\partial y^2} \qquad (4.2.117)$$

We continue to consider the case of one-dimensional diffusion through a slab, so the Laplacian takes the form indicated. The initial condition is

$$C_A = C_A^o \qquad \text{at} \quad t \le 0 \qquad (4.2.118)$$

To simplify matters, we will assume that convective transfer at the exposed surface is very efficient ($\text{Bi} = \infty$), and so we write the boundary condition as

$$C_A = C_A^s = \beta C_A^{f\infty} \qquad \text{at } y = B \qquad (4.2.119)$$

where $C_A^{f\infty}$ is the concentration of volatile solvent in the ambient gas phase, and $\beta$ is the solubility ratio for the solvent between the two phases.

As before, we need a second boundary condition, and we might expect to be able to repeat our use of Eq. 4.2.111, the statement that the boundary $y = 0$ is impermeable. Instead, however, we introduce a different physical statement—one that reflects our interest in the behavior of the solution for very short times. We do this in three steps.

First, we state that for very short times the concentration in the near neighborhood of $y = 0$ will not change measurably. This seems reasonable. We are saying that for short times the diffusion is confined to a region very near the exposed surface. Hence we write this statement in the form

$$C_A = C_A^o \qquad \text{at} \quad y = 0 \qquad (4.2.120)$$

The next step seems trivial. We shift the coordinate system to place the exposed surface at $y = 0$ and the plane of symmetry at $y = B$. Equations 4.2.117 and 4.2.118

are unchanged in format by this shift, but Eqs. 4.2.119 and 4.2.120 become

$$C_A = C_A^s = \beta C_A^{f\infty} \quad \text{at} \quad y = 0 \tag{4.2.121}$$

and

$$C_A = C_A^0 \quad \text{at} \quad y = B \tag{4.2.122}$$

In the third step we assert that if the concentration has not changed at the distance $y = B$ from the exposed surface, then the evaporation from the exposed surface would be unchanged if the slab were thicker than $B$, since all the "action" is occurring in a thin region near the exposed surface. If the slab can be thicker, why not set the impermeable boundary at $y = \infty$? If we do this, the boundary condition given in Eq. 4.2.122 becomes

$$C_A = C_A^0 \quad \text{at} \quad y \rightarrow \infty \tag{4.2.123}$$

Now we can ask the question: *What have we achieved?* The answer is that this simple change in boundary condition makes it possible to write a solution to Eq. 4.2.117 in a completely different form, and it is a form that lends itself to easy computation for short times, which was our initial goal.

To summarize, then, our mathematical problem has taken the form

$$\frac{\partial C_A}{\partial t} = D_{AB} \frac{\partial^2 C_A}{\partial y^2} \tag{4.2.124}$$

$$C_A = C_A^0 \quad \text{at} \quad t \leq 0 \tag{4.2.125}$$

$$C_A = C_A^s = \beta C_A^{f\infty} \quad \text{at} \quad y = 0 \tag{4.2.126}$$

$$C_A = C_A^0 \quad \text{at} \quad y \rightarrow \infty \tag{4.2.127}$$

Now we can define a dimensionless concentration variable, as before:

$$Y \equiv \frac{C_A - \beta C_A^{f\infty}}{C_A^0 - \beta C_A^{f\infty}} \tag{4.2.128}$$

and the mathematical formulation becomes

$$\frac{\partial Y}{\partial t} = D_{AB} \frac{\partial^2 Y}{\partial y^2} \tag{4.2.129}$$

$$Y = 1 \quad \text{at} \quad t \leq 0 \tag{4.2.130}$$

$$Y = 0 \quad \text{at} \quad y = 0 \tag{4.2.131}$$

$$Y = 1 \quad \text{at} \quad y \rightarrow \infty \tag{4.2.132}$$

This problem contains no *natural* length scale, like the thickness of the slab, since the solid slab is now regarded as unbounded in the positive $y$ direction. Hence we do not define a dimensionless space variable.

We can show that if we define a new variable as

$$\eta \equiv \frac{y}{2\sqrt{D_{AB}t}} \tag{4.2.133}$$

which is a combination of the two independent variables $y$ and $t$, the *partial* differential equation becomes an *ordinary* differential equation of the form

$$\frac{d^2Y}{d\eta^2} + 2\eta \frac{dY}{d\eta} = 0 \tag{4.2.134}$$

with boundary conditions

$$Y = 0 \quad \text{for} \quad \eta = 0 \tag{4.2.135}$$

and

$$Y = 1 \quad \text{for} \quad \eta = \infty \tag{4.2.136}$$

(Note how two of the boundary conditions given earlier—the initial condition and the condition far from the surface—collapse to a single boundary condition for $\eta = \infty$.)

This is almost (but not quite) exactly the differential equation and boundary conditions given in Eqs. 4.2.77 to 4.2.79. The solution of this differential equation is expressible in the form

$$Y = \text{erf} \; \eta = \text{erf}\left(\frac{y}{2\sqrt{D_{AB}t}}\right) \tag{4.2.137}$$

where erf is the error function introduced earlier (Eqs. 4.2.95 and 4.2.96) and plotted in Fig. 4.2.6. Since our interest is in short times, for which $\eta$ is large, it is useful to have Fig. 4.2.9 in addition to Fig. 4.2.6. Actually, what we show is the complementary error function: $\text{erfc}(x) = 1 - \text{erf}(x)$. For values of the argument $x$ exceeding $x = 3$, the error function equals unity for any practical purpose, and we do not bother to plot $\text{erfc}(x)$ for larger arguments.

Equation 4.2.137 gives the concentration distribution within the slab. We are also (and sometimes, more) interested in the *average* concentration within the slab, at any time after the beginning of evaporation. This is most easily found from a simple material balance. The amount of solvent within the slab initially is

$$M_o = C_A^o B \tag{4.2.138}$$

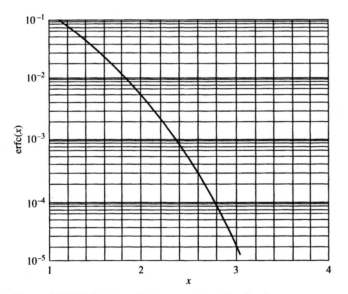

**Figure 4.2.9** Behavior of the error function for large arguments (short times).

per unit of surface area normal to the $y$ axis. In the interval of time $t$ the amount of solvent that has crossed the free surface is given by

$$M = \int_0^t + D_{AB} \left( \frac{\partial C_A}{\partial y} \right)_{y=0} dt \tag{4.2.139}$$

(Again, this is per unit of surface area. Hence $M$ is just the diffusive flux to the surface integrated over the time interval. Note the positive sign on the diffusive flux. Why positive?) To connect Eq. 4.2.139 to the solution given in the form of Eq. 4.2.137, we must do some calculus. First of all, we write $C_A$ as

$$C_A = \beta C_A^{f\infty} + (C_A^0 - \beta C_A^{f\infty}) Y = \beta C_A^{f\infty} + (C_A^0 - \beta C_A^{f\infty}) \mathrm{erf}\, \eta$$
$$= \beta C_A^{f\infty} + (C_A^0 - \beta C_A^{f\infty}) \frac{2}{\sqrt{\pi}} \int_0^\eta e^{-\eta^2} d\eta \tag{4.2.140}$$

Now we have to differentiate this expression with respect to $y$. The $y$ dependence lies in the character of the upper limit $\eta$ of the integral: namely, it is a function of $y$ (and $t$). Hence we need to recall a rule for differentiation of an integral, in the case that one of the limits of integration is a function of the variable with respect to which we are differentiating. This rule is called the Leibniz rule, and it takes the form

$$\frac{\partial}{\partial y} \int_0^{\eta(y)} f(\eta)\, d\eta = \int_0^{\eta(y)} \frac{\partial f(\eta)}{\partial y}\, d\eta + f(\eta) \frac{\partial \eta}{\partial y} \tag{4.2.141}$$

In the case at hand, the function $f(\eta)$ is

$$f(\eta) = \frac{2}{\sqrt{\pi}} e^{-\eta^2} \tag{4.2.142}$$

Next we note that

$$\frac{\partial f(\eta)}{\partial y} = \frac{df}{d\eta} \frac{\partial \eta}{\partial y} \tag{4.2.143}$$

From Eq. 4.2.133 it follows that

$$\frac{\partial \eta}{\partial y} = \frac{1}{2\sqrt{D_{AB}t}} \tag{4.2.144}$$

From Eq. 4.2.142 we find

$$\frac{df(\eta)}{d\eta} = -\frac{4}{\sqrt{\pi}} \eta e^{-\eta^2} \tag{4.2.145}$$

and so

$$\frac{\partial f}{\partial y} = -\frac{2\eta}{\sqrt{\pi D_{AB}t}} e^{-\eta^2} \tag{4.2.146}$$

Then the integral on the right-hand side of Eq. 4.2.141 becomes

$$\int_0^{\eta(y)} \frac{\partial f(\eta)}{\partial y}\, d\eta = \int_0^{\eta(y)} \frac{-2\eta e^{-\eta^2}}{\sqrt{\pi D_{AB}t}}\, d\eta = \frac{1}{\sqrt{\pi D_{AB}t}} (1 - e^{-\eta^2}) \tag{4.2.147}$$

With reference to Eq. 4.2.139, we see that we evaluate this function at $y = 0$, which is also $\eta = 0$. Hence this term vanishes. The second term of Eq. 4.2.141 becomes

$$f(\eta) \frac{\partial \eta}{\partial y} = \frac{e^{-\eta^2}}{\sqrt{\pi D_{AB}t}} \tag{4.2.148}$$

Again we evaluate this at $\eta = 0$ and find

$$f(\eta)\frac{\partial \eta}{\partial y} = \frac{1}{\sqrt{\pi D_{AB}t}} \tag{4.2.149}$$

Finally Eq. 4.2.139 becomes

$$M = D_{AB}(C_A^o - \beta C_A^{f\infty})\int_0^t \frac{1}{\sqrt{\pi D_{AB}t}}\,dt = 2(C_A^o - \beta C_A^{f\infty})\left(\frac{D_{AB}t}{\pi}\right)^{1/2} \tag{4.2.150}$$

The amount of solvent remaining after time $t$ is simply the difference between the amounts given in Eqs. 4.2.138 and 4.2.139; hence the mean concentration remaining after time $t$ is

$$\frac{M_o - M(t)}{B} = C_A^o - \frac{M}{B} = C_A^o - 2(C_A^o - \beta C_A^{f\infty})\left(\frac{D_{AB}t}{\pi B^2}\right)^{1/2} \tag{4.2.151}$$

It is important to keep in mind that Eq. 4.2.151 is valid only for a short time and holds for evaporation from only one surface of the film. The first point should be obvious from inspection of the expression, which predicts a negative mean concentration at long times. This physical impossibility is an "artifact" of the solution method, which replaces the finite slab by a slab of infinite extent in the $+y$ direction.

We can provide a simple (though arbitrary) criterion of "short time" by the following argument. We will consider the boundary between short and long time as the time at which the concentration at the *impermeable* surface falls from its initial value of $Y = 1$ to a value of $Y = 0.99$. Then from Eq. 4.2.137 we solve for the time $t$ at which $Y = 0.99$ at $y = B$. Hence we want the value of $\eta$ at which $\text{erf}(\eta) = 0.99$, or $\text{erfc}(\eta) = 0.01$. From Fig. 4.2.9 we see that $\eta = 1.8$. It follows then that

$$\eta \equiv \frac{y}{2\sqrt{D_{AB}t}} = 1.8 \tag{4.2.152}$$

or, at $y = B$

$$\frac{D_{AB}t}{B^2} \le \left(\frac{1}{3.6}\right)^2 = 0.077 \tag{4.2.153}$$

Equation 4.2.153 is the (arbitrary) criterion we use to define a "short time." It is arbitrary because we impose an arbitrary, though reasonable, definition of how we understand this term.

With this model in hand, let's ask the following question: For an evaporative process described by the model, and for the special case that $C_A^{f\infty} = 0$, how long can evaporation proceed before the average concentration of solvent has changed by 1% of its initial value?

The problem statement implies that we want the time at which the fraction remaining is

$$\frac{C_A^o - M/B}{C_A^o} = 0.99 = 1 - 2\left(\frac{D_{AB}t}{\pi B^2}\right)^{1/2} \tag{4.2.154}$$

It follows that

$$\frac{D_{AB}t}{B^2} = 7.854 \times 10^{-5} \tag{4.2.155}$$

Notice how much shorter a time this is than the time given in Eq. 4.2.153. These times refer to two very different stages in the evaporation process. For example, at the time corresponding to Eq. 4.2.155, the concentration of solvent at the impermeable boundary is nowhere near $Y = 0.99$. Instead, we find

$$Y = \operatorname{erf} \frac{1}{2\sqrt{D_{AB}t}} = \operatorname{erf} \frac{1}{2\sqrt{7.854 \times 10^{-5}}} = \operatorname{erf} 56.42 \qquad (4.2.156)$$

From inspection of Fig. 4.2.9 we conclude that $Y$ is negligibly different form $Y = 1$.[3]

## EXAMPLE 4.2.4   *A Device for Treatment of Glaucoma*

Glaucoma can be treated on a continuous basis with a biomedical device developed by Alza Corporation. The device, called Ocusert, can be thought of as a membrane that has been soaked in a solution of pilocarpine until equilibrium is achieved within the membrane. The membrane is placed into the eye (much like a contact lens, but smaller, so that it fits into the corner, or *cul-de-sac*, of the eye), and the continuous release of pilocarpine to the eye begins. The rate of release is controlled by diffusion through the membrane. The fluid dynamics arising from blinking provides effective convection.

In designing such a system, it is important to measure the diffusion coefficient of the therapeutic agent, here pilocarpine, through the membrane. This can be done by equilibrating a membrane with pilocarpine (or, as in the case of the data shown in Fig. 4.2.10, pilocarpine nitrate) and measuring the rate of release of the compound to a surrounding solution under conditions designed to ensure that diffusion in the membrane controls the release rate. For the data to be examined, a 70 mg Hydron contact lens was soaked in a 4% solution of pilocarpine nitrate until it came to equilibrium. Then the rate of release was measured, as shown in Fig. 4.2.10.

If we assume that the data were obtained under conditions promoting release from *both sides* of the membrane, and if we regard the membrane as a slab of thickness $L$, then the fractional amount released as a function of time may be written as (see Problem 4.9)

$$\frac{M}{M_\infty} = 1 - \sum \frac{8 \exp(-D[2n+1]^2\pi^2 t/L^2)}{(2n+1)^2\pi^2} \qquad (4.2.157)$$

where $M_\infty$ is the amount of releasable compound. From this solution we may calculate the fraction remaining in the membrane at any time, and the rate of release at any time. Because the infinite series expression above is tedious to use for computation, we take note of two approximations that can be used for short and long times. These are, for the final stage of release:

$$\frac{M}{M_\infty} = 1 - \frac{8 \exp(-\pi^2 Dt/L^2)}{\pi^2} \qquad \text{for } 0.4 \le M/M_\infty \le 1 \qquad (4.2.158)$$

which simply truncates the series solution (Eq. 4.2.157) after the first term, and for the early stage of release:

$$\frac{M}{M_\infty} = 4 \left[ \frac{Dt}{\pi L^2} \right]^{1/2} \qquad \text{for } M/M_\infty \le 0.4 \qquad (4.2.159)$$

---

[3] For the curious, $Y = 1 - \dfrac{1}{56.42\sqrt{\pi}\exp(3183)}$

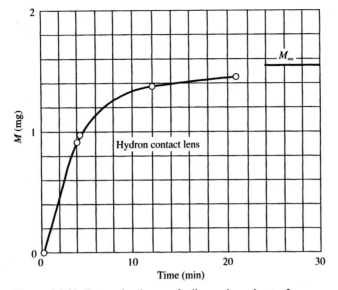

**Figure 4.2.10** Rate of release of pilocarpine nitrate from a Hydron contact lens; $M_\infty$ is the total amount of the compound available for release.

which does not follow by inspection or approximation to Eq. 4.2.157. (See Problem 4.13.) From these expressions, we may also calculate approximations for the release rate $dM/dt$ at any time, but we do not show these results here. (See Problem 4.9.)

Using the data of Figure 4.2.10 we may plot $\overline{Y} = (1 - M/M_\infty)$ as a test of Eqs. 4.2.158 and 4.2.159. From Fig. 4.2.11 we see that 90% of the drug release occurs in accordance

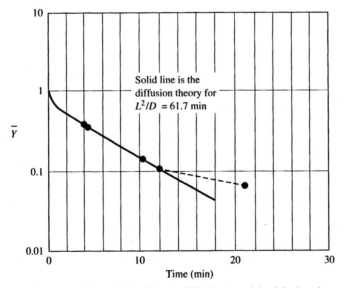

**Figure 4.2.11** Test of the simple diffusion model with the data of Figure 4.2.10: $\overline{Y}$ is the fraction of pilocarpine remaining in the membrane.

with the prediction of this simple diffusion model. The line drawn through the data (excepting the lowest value of $\overline{Y}$) corresponds to a value of $L^2/D = 61.7$ min. From this value, and a measurement of the wet thickness of the membrane, we could calculate the diffusivity of pilocarpine.

Our next example corresponds to Example E in Section 4.1.1.

**EXAMPLE 4.2.5**    *Doping a Semiconductor Film*

To control the electrical conductivity in a region of a semiconductor device, it is necessary to introduce very small amounts of specific atoms into the base material (e.g., silicon) of the semiconductor. These "impurity" atoms are commonly referred to as "dopants." For the device to function properly, it is necessary to control the spatial distribution of the dopant atoms very precisely. The simplest doping system is one in which the solid film is exposed to a gas phase, which is held at a constant concentration of dopant atoms over some period of time. This is known as "constant source" diffusion. Assuming that the diffusivity of the dopant species in the solid is constant, we may write the transient diffusion equation for the concentration $C(z, t)$ of the dopant as

$$\frac{\partial C}{\partial t} = D \frac{\partial^2 C}{\partial z^2} \tag{4.2.160}$$

and impose the initial condition

$$C = C_o \quad \text{at} \quad t = 0, z \geq 0 \tag{4.2.161}$$

This assumes that there is an initial uniform distribution of dopant atoms at some concentration $C_o$, which could be zero in some cases. Dopant diffusion is permitted to occur until the dopant atoms have penetrated a very small distance into the solid, relative to the thickness of the solid. Hence we will regard the solid to be semi-infinite in extent. At the boundary $z = 0$ we impose the condition

$$C = C_e \quad \text{at} \quad z = 0, t \geq 0 \tag{4.2.162}$$

Here $C_e$ is the solid phase concentration of the dopant species that is in equilibrium with the gas phase concentration. For the second boundary condition we assume that far from the exposed surface ($z = 0$), no dopant penetration has occurred:

$$C = C_o \quad \text{at} \quad z = \infty, t \geq 0 \tag{4.2.163}$$

This set of equations has an analytical solution in terms of the error function

$$\frac{C - C_0}{C_e - C_0} = \text{erfc} \frac{z}{2\sqrt{Dt}} = 1 - \frac{2}{\sqrt{\pi}} \int_0^{z/2\sqrt{Dt}} \exp(-s^2) \, ds \tag{4.2.164}$$

The error function is plotted in Fig. 4.2.6. The complementary error function (erfc) is just $1 - \text{erf}$.

Dopants such as arsenic and phosphorus are called *n-type* because they contribute free electrons (hence *negative* charge carriers) to the crystalline solid within which they are diffused. A common *p-type* dopant is boron, which "removes" electrons from the atoms of the crystal structure (such as silicon); hence charge is carried by the movement of "holes," which have a *positive* charge.

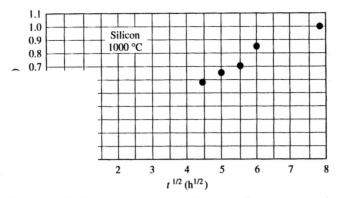

**Figure 4.2.12** Data on junction depth versus time as a test of Eq. 4.2.165.

Often doping of an n-type atom is performed on a solid that already has a concentration of the opposite (p-type) carrier. An important characteristic is the depth at which the concentrations of the n- and p-carriers are equal. This is called the "junction depth." Suppose a substrate (e.g., a silicon wafer) has a uniform concentration of a p-dopant, $C_p$, and we diffuse an n-type dopant into the solid using the constant source method. We assume that $C_{n,o} = 0$ in the initial condition (Eq. 4.2.161). Then Eq. 4.2.164 (with $C$ replaced by $C_n$) gives the profile of $C_n (z)$, and the junction depth $Z_J$ is the solution to (setting $C_n = C_p$).

$$\frac{C_p}{C_{n,e}} = \text{erfc} \frac{Z_J}{2\sqrt{D_n t}} \qquad (4.2.165)$$

Note that the diffusivity appearing here is that of the n-type dopant, since that is the species being diffused into the solid.

This simple model predicts that the junction depth is a linear function of the square root of time. Data are shown in Fig. 4.2.12 for phosphorus (n-type) doping of a silicon wafer that had an initial uniform doping of boron (p-type) of $6.1 \times 10^{15}$ atoms/cm$^3$. Conditions corresponded to constant source doping with a surface concentration of phosphorus of $7.5 \times 10^{20}$ atoms/cm$^3$. The data confirm the expected $t^{1/2}$ behavior. From these data it is possible to determine the diffusion coefficient of phosphorus in silicon (see Problems 4.12 and 4.30).

---

**EXAMPLE 4.2.6**  *Comparison of Two Methods for Measurement of the Diffusivity of a Species Through a Membrane*

In carrying out the task of development of new membranes for specific applications, it is necessary to characterize the membranes with respect to the diffusivity of various species in the membrane, under different conditions of temperature, total pressure, and concentration of the penetrant species. The models we have described to this point are used in this kind of characterization.

In one common method of measurement, a membrane is exposed to a change in concentration of some species that is soluble within the membrane. The species is absorbed at the exposed surface(s) of the membrane and then diffuses into the membrane interior. The process continues until the equilibrium solubility of the diffusant within the membrane has been achieved. A typical data set is shown in Fig. 4.2.13: diffusion of a soluble gas, $CO_2$ in this case, within a polyimide membrane. Such membranes are important for the development of gas separation systems.

In Section 4.2.1 we developed a model for evaporative loss of a solvent from a film, one side (surface) of which was impermeable to the solvent. It is not difficult to demonstrate that if we considered uptake (absorption) of the solvent, instead of release (evaporation), we would have obtained Eq. 4.2.150, with a few small modifications. First, we would set $M_o = 0$, and $C_A^o = 0$, to correspond to an initially "empty" membrane. Second, $M(t)$ would be the amount of the diffusant within the membrane at any time $t$. (Actually, it is the amount per unit of area normal to the direction of diffusion.) When equilibrium with the outside concentration $C_A^{f\infty}$ is achieved, the amount (again, per area) of the diffusing species within the membrane will be $M_\infty = L\beta C_A^{f\infty}$, where $\beta$ is the partition coefficient of $CO_2$ between the gas and polymer phases and $L$ is the total thickness of the membrane (not the half-thickness!). Finally, if absorption is across both faces of the membrane, we multiply the rate of absorption given by the one-sided evaporative model by a factor of 2. As a consequence, Eq. 4.2.150 takes the form

$$\frac{M}{M_\infty} = 4\left(\frac{D_{AB}t}{\pi L^2}\right)^{1/2} \tag{4.2.166}$$

Keep in mind that this equation is a suitable model only for short times, defined such that $M/M_\infty < 0.4$.

Note that the data of Fig. 4.2.13 have been plotted as total concentration $C_t$ against $t^{1/2}$, anticipating that the short-time data will be used to determine the diffusion coefficient. Notice, as well, that the predicted linearity is observed up to about half the total uptake. From the data, we may find the equilibrium (long-time) concentration $C_{eq}$. Hence, as plotted, the slope of the short time portion of the data is

$$\text{slope} = \frac{dC_t}{dt^{1/2}} = 4C_{eq}\left(\frac{D_{AB}}{\pi L^2}\right)^{1/2} \tag{4.2.167}$$

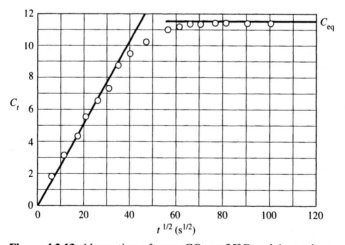

**Figure 4.2.13** Absorption of pure $CO_2$ at 25°C and 1 atm by a polyimide membrane. Toi et al., *J. Poly. Sci.: B: Polymer Phys.*, **30**, 549 (1992).

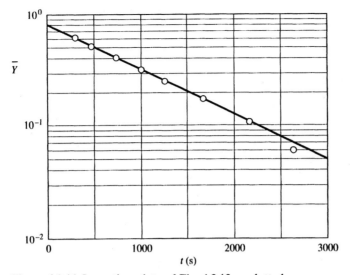

**Figure 4.2.14** Long-time data of Fig. 4.2.13, replotted.

With $C_{eq} = 11.5$ and a slope estimated from Fig. 4.2.13 as 0.26, and using the stated value of membrane thickness of 32 $\mu$m, we calculate a diffusion coefficient of $D = 1.04 \times 10^{-9}$ cm$^2$/s.[4]

We may also calculate a diffusion coefficient from the long-time data as equilibrium is approached. To do so we have to replot the data as $\overline{Y} = 1 - C_t/C_{eq}$ against time, as shown in Fig. 4.2.14. From the plot we see that $\overline{Y} = 0.05$ at $t = 3000$ s. From Fig. 4.2.2a, we see that $\overline{Y} = 0.05$ at $X_D = 1.12$ for the assumed value of Bi $= \infty$. Hence, noting that the half-thickness of the membrane is used in the definition of $X_D$,

$$D_{AB} = \frac{1.12 \, (L/2)^2}{t} = \frac{1.12 \times 0.016^2}{3000} = 9.5 \times 10^{-10} \, \text{cm}^2/\text{s} \qquad \textbf{(4.2.168)}$$

which is in excellent agreement with the value obtained from the short-time data.

## 4.3  INTERNAL VERSUS EXTERNAL RESISTANCES

In general the rate of mass transfer from (or to) a solid depends on some combination of *internal* diffusion (i.e., internal to the solid) and *external* convection (i.e., convective transfer at the surface generated by flow external to the solid). For simple geometries *and symmetries,* the Gurney–Lurie charts show the transient solutions, and the *internal* diffusion coefficient appears in the dimensionless time $X_D$. The parameter Bi contains the convective coefficient $k$ *and* the internal diffusion coefficient. In some sense, Bi is a ratio of external (convective) to internal (diffusive) mass transfer resistances.

Since we are often interested primarily in the *average* concentration inside a body, we can present graphical representations of the analytical solutions that are available for simply shaped bodies. These are shown on Figs. 4.2.2 a, b, c, as the dashed lines.

---

[4] As long as we use the same units for $C_t$ and $C_{eq}$, it is not necessary to convert the concentration units, which in the original reference are cubic centimeters of $CO_2$ (at standard temperature and pressure) per cubic centimeter of polymer.

Keep in mind that

$$\text{Bi} = \frac{kR}{D_{AB}} \tag{4.3.1}$$

When we consider transport across a *phase boundary*, we must be careful about the definition of the mass transfer coefficient that appears in the boundary condition. It turns out that, as we have written Eq. 4.2.14, our $k$ includes the *equilibrium partition coefficient* between the two phases. We can see that more clearly in the following analysis.

We *should* define a convection boundary condition in the following way. We let $C_A^s$ represent the concentration of the species of interest to us, *within the solid or stagnant phase*, and we let $C_A^f$ be the concentration in the *fluid or convecting* phase. A proper definition of a convective transfer coefficient is (cf. Eq. 4.2.5)

$$(-D_{AB}\nabla C_A^s) \cdot \mathbf{n} = k_c(C_A^f - C_A^{f\infty}) \tag{4.3.2}$$

where $C_A^{f\infty}$ is the concentration of species A in the fluid phase far from the boundary that separates the two phases. In Eq. 4.3.2 the gradient of $C_A^s$ is evaluated *at the boundary, but within the solid or stagnant phase*, and $C_A^f$ is the concentration of species A *at the boundary, but in the fluid (convecting) phase.*

Now we define the partition or solubility coefficient between the two phases. We write[1]

$$C_A^f = \alpha C_A^s \tag{4.3.3}$$

and define $\alpha$ as the ratio of concentrations of species A in the fluid (convecting) and solid (diffusing) phases, *at equilibrium*. Equation 4.3.2 now becomes (compare this to Eq. 4.2.5)

$$(-D_{AB}\nabla C_A^s) \cdot \mathbf{n} = \alpha k_c(C_A^s - C_A^{s\infty}) \tag{4.3.4}$$

In this form, all the concentrations refer to the *internal* phase within which only diffusion occurs.

The concentrations on the right-hand side of Eq. 4.3.4 are not those that would be measured in the external convecting phase. They are the concentrations in the *solid phase* that would be in equilibrium with the fluid phase concentrations $C_A^f$, right at the surface, and $C_A^{f\infty}$, far from the surface.

Another way of looking at the need to introduce $\alpha$ emphasizes that when equilibrium occurs (at steady state) the concentration $C_A^s$ in the solid does *not* approach the concentration $C_A^{f\infty}$ in the fluid. Instead,

$$C_A^s \rightarrow \frac{1}{\alpha} C_A^{f\infty} \tag{4.3.5}$$

or

$$(C_A^s - C_A^{s\infty}) \rightarrow 0 \qquad \text{at equilibrium} \tag{4.3.6}$$

since, from Eq. 4.3.3

$$C_A^{s\infty} = \frac{1}{\alpha} C_A^{f\infty} \tag{4.3.7}$$

---

[1] $\alpha$ is just the inverse of the coefficient $\beta$ introduced in Eq. 4.2.119.

From this discussion we can see that the $k$ that we used in Eq. 4.2.5 was really the product of a convective mass transfer coefficient and a solubility coefficient. We will have to be careful about nomenclature when we evaluate Bi, since mass transfer correlations usually provide $k_c$, as it is defined in Eq. 4.3.2, and not $k$, and we must introduce $\alpha$ for the system of interest. We can illustrate this with an example.

**EXAMPLE 4.3.1**   *Decaffeination of Coffee Beans*

Coffee beans are treated to release the caffeine *within*. Whole beans are contacted with an organic solvent into which the caffeine is leached. The primary resistance to transfer, in a well-agitated solvent, is the diffusion of caffeine within the solid bean. We have the following information:

A bean is equivalent to a sphere of diameter 0.6 cm.
The diffusion coefficient at the operating temperature is estimated to be $D_{AB} = 1.8 \times 10^{-10}$ m²/s. *(Note units.)*

Assuming no external convective resistance (i.e., assuming Bi $= \infty$), we want to find the time required to reduce the caffeine content to 3% of the initial value.

For the case Bi $= \infty$, we define the dimensionless concentration as

$$\overline{Y}_A = \frac{\overline{C}_A - C_A^{S\infty}}{C_A^o - C_A^{S\infty}} \tag{4.3.8}$$

[This is Eq. 4.2.6, using the average $\overline{Y}$ instead of the local $Y(r/R)$.] For Bi $= \infty$ and a nearly caffeine-free external solvent, we assume that the *equilibrium* concentration of caffeine within the bean will be

$$C_A^{S\infty} \approx 0 \tag{4.3.9}$$

We want a specific value of $\overline{Y}$, which we denote $\overline{Y}^*$, to satisfy

$$\overline{Y}_A^* = \frac{\overline{C}_A^*}{C_A^o} = 0.03 \tag{4.3.10}$$

From Fig. 4.2.2c we find

$$\frac{D_{AB}t}{R^2} = 0.3 \tag{4.3.11}$$

and therefore

$$t = \frac{0.3R^2}{D_{AB}} = \frac{0.3(0.3 \text{ cm})^2}{(1.8 \times 10^{-10} \times 10^4)\text{cm}^2/\text{s}} \tag{4.3.12}$$

$$= 1.5 \times 10^4 \text{ s} = 4.2 \text{ h}$$

This is a long time. Can it be reduced? What happens if we halve the size of the coffee beans, so that they have an equivalent diameter of 0.3 cm?

Under the assumption that Bi $= \infty$ (no convective resistance), we still use Eqs. 4.3.9, 4.3.10, and 4.3.11, and since $t$ is proportional to $R^2$, the time is reduced by a factor of $0.5^2$ to $t = 1$ h.

Now suppose we switch to a process that introduces a finite convective resistance. We can no longer assume $C_A^{S\infty} = 0$ on $r = R$ as a boundary condition. We must use one of the curves in Fig. 4.2.2c, but for a finite Bi value. Note that we define Bi as (cf. Eq. 4.2.16)

$$\text{Bi} = \frac{kR}{D_{AB}} = \frac{\alpha k_c R}{D_{AB}} \tag{4.3.13}$$

and we now need two pieces of data that did not enter into the Bi $= \infty$ analysis above. The equilibrium solubility coefficient is known to be

$$\alpha = \frac{\text{caffeine conc. in solvent}}{\text{caffeine conc. in wet bean}} = 0.10 \tag{4.3.14}$$

From independent studies of this new process, we are told to expect a convective coefficient of

$$k_c = 9 \times 10^{-4}\,\text{cm/s} \tag{4.3.15}$$

This leads to (assuming $R = 0.3$ cm)

$$\text{Bi} = \frac{0.10\,(9 \times 10^{-4})(0.3)}{1.8 \times 10^{-6}} = 15 \tag{4.3.16}$$

We will interpolate crudely between the time values we predict at Bi $= \infty$ and Bi $= 10$:

$$X_D = \frac{D_{AB}t}{R^2} = \begin{cases} 0.3 \text{ at Bi} = \infty \\ 0.4 \text{ at Bi} = 10 \end{cases} \tag{4.3.17}$$

Then, at Bi $= 15$ we estimate

$$X_D = \frac{D_{AB}t}{R^2} = 0.35 \tag{4.3.18}$$

Because of convective resistance, the time required to reduce the caffeine content to 3% of its original value will be increased by a factor of

$$\frac{0.35}{0.3} = 1.2 \tag{4.3.19}$$

times the value in the absence of external convective resistance, or

$$t = 1.2(4.2) = 5\,\text{h} \tag{4.3.20}$$

Now what happens if we halve the bean size, *in the presence of convective resistance?* *The Biot modulus will decrease,* for smaller $R$, since Bi $\sim R$ (see Eq. 4.3.13). However, we will find later in our study of convective mass transfer that $k_c$ also depends on $R$, approximately as

$$k_c \sim R^{-0.4} \tag{4.3.21}$$

Hence $k_c$ will *increase* for smaller $R$, and

$$\text{Bi} \sim R \times R^{-0.4} \sim R^{0.6} \tag{4.3.22}$$

will *decrease* if $R$ is halved, by a factor of approximately

$$2^{0.6} = 1.5 \tag{4.3.23}$$

This will give Bi $= 10$ and

$$X_D = 0.4 \tag{4.3.24}$$

Then, for the small beans with convective resistance, we find

$$t = \frac{R^2}{D_{AB}} X_D = \frac{0.15^2}{1.8 \times 10^{-6}} 0.4 = 1.4 \, \text{h} \qquad \textbf{(4.3.25)}$$

while for the large beans with convective resistance we require 5 h (Eq. 4.3.20). In the presence of convective resistance, halving the bean size does not reduce the required time by a factor of $0.5^2$ but by a somewhat lower factor.

One of the key points of Example 4.3.1 is that we usually need equilibrium data for the analysis of the nonequilibrium (transient) behavior of a system undergoing mass transfer. In practice, such equilibrium data are not always available, but if they are, the data are often presented in a form that requires further manipulation. We illustrate this point in Example 4.3.2.

**EXAMPLE 4.3.2** *Drying of Paper*

A commercial paper is being dried in a "convection oven," which is shown schematically in Fig. 4.3.1. The heated air is at 35°C, and independent measurements have yielded a convective mass transfer coefficient of $k_c = 0.5$ cm/s. The relative humidity of the air at 35°C is 10%. The "wet" paper is initially at 18% moisture, by weight, and we must dry it to 8%.

Wet paper is a complex material, but studies of its drying behavior in this range of moisture content indicate that the internal resistance to transport of the water to the surface of the sheet can be considered diffusive, with an effective diffusion coefficient at 35°C of $10^{-5}$ cm$^2$/s. Experiments have been carried out in which this paper has been equilibrated with air at various controlled humidities. The data were presented in the form of the moisture content, at thermodynamic equilibrium, as a function of the partial pressure of water vapor in the air. The results have been presented as a Henry's law coefficient, defined as

$$p(\text{atm}) = Hx(\text{ppmw}) \qquad \textbf{(4.3.26)}$$

where $x$ is the moisture content in *parts per million by weight*. (Hence 1% water content by weight is 10,000 ppmw. We multiply mass *fraction* by $10^6$ to get ppmw.) The value obtained for $H$ at 35°C is found to be

$$H = 4.2 \times 10^{-7} \, \text{atm/ppmw} \qquad \textbf{(4.3.27)}$$

The paper thickness $b = 3$ mils ($= 7.62 \times 10^{-3}$ cm), is assumed to remain constant over this range of drying.[2]

To model the dynamics of drying, we must begin by writing the diffusion equation with an appropriate set of initial and boundary conditions that are consistent with the foregoing physical statements. We will use the weight fraction (but as ppmw) as the concentration unit within the paper phase. To define the dimensionless concentration variable $Y$, we must first determine the equilibrium state of this system. Keep in mind that we are not drying the paper to equilibrium. The final state we are designing for,

---

[2] In fact, paper shrinks as it dries, and the drying dynamics are more complicated than suggested by this simplified example.

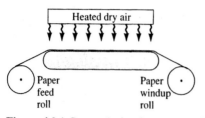

**Figure 4.3.1** Paper drying in a convection oven.

8% water, is well above the final equilibrium state that the paper would achieve if it were allowed sufficient time to equilibrate with 10% relative humidity air at 35°C.

To determine the equilibrium state, we first need the partial pressure of water vapor in air at 10% relative humidity. From a curve of vapor pressure versus temperature for water (see Chapter 3, Fig. P3-18), we find

$$p_{\text{vap}} = 0.85 \text{ psi} = 0.058 \text{ atm at } 35°C \tag{4.3.28}$$

Since relative humidity (RH) is just the fraction of the *equilibrium* water content in the air, the air used in the convective drier (RH = 10%) has a partial pressure of a tenth of this value, or

$$p^\infty = 0.0058 \text{ atm} \tag{4.3.29}$$

Now we can use the Henry's law relationship to find the moisture content in the paper at equilibrium with 10% RH air:

$$x^\infty(\text{ppmw}) = \frac{p(\text{atm})}{H} = \frac{0.0058}{4.2 \times 10^{-7}} = 1.38 \times 10^4 \text{ ppmw} = 1.38\% \tag{4.3.30}$$

We will be interested in the average concentration variable $\overline{Y}$ instead of the spatial distribution of water within the paper. Using $x$ as the concentration variable, we find

$$\overline{Y} = \frac{\bar{x} - x^\infty}{x^\circ - x^\infty} = \frac{8 - 1.38}{18 - 1.38} = 0.4 \tag{4.3.31}$$

(Notice that we did not need to write the common factor of $10^4$ for each term in this fraction.)

The next step in the analysis requires the calculation of a Biot number from the data given. We will define the Biot number as

$$\text{Bi} = \frac{\alpha k_c b}{D_{\text{AB}}} \tag{4.3.32}$$

Notice that we are using the total thickness of the paper for the length scale $b$. As in the preceding example, with convective loss from only one face, the film behaves the same as a film of thickness $2b$, losing mass by convection from both faces. Hence, if we use the Gurney–Lurie charts, $b$ is the proper length scale in Bi and $X_D$.

Everything is known except the solubility coefficient $\alpha$, but this can be obtained from the Henry's law constant. We define $\alpha$ as

$$\alpha = \frac{C_{\text{water}}^{\text{air}}}{C_{\text{water}}^{\text{paper}}} = \frac{\text{moles water/cm}^3 \text{ air}}{\text{moles water/cm}^3 \text{ paper}} \tag{4.3.33}$$

Now the task is to convert the molar concentrations in each phase to units that appear in Henry's law, namely, pressure in the air phase and weight fraction in the paper phase.

The air phase is easy: we use the ideal gas law and find

$$C_{\text{water}}^{\text{air}} = \frac{p_{\text{water}}}{R_{\text{G}}T} \tag{4.3.34}$$

The conversion for the paper phase is a little more tedious, but we proceed step by step as follows:

$$18 C_{\text{water}}^{\text{paper}} = \text{g water/cm}^3 \text{ paper} \tag{4.3.35}$$

where 18 is the molecular weight (g/mol) of water. Now we introduce the mass density of the *dry* paper:

$$\frac{18 C_{\text{water}}^{\text{paper}}}{\rho_{\text{paper}} \, [\text{g paper/cm}^3 \text{ paper}]} = \frac{18 C_{\text{water}}^{\text{paper}}}{\rho_{\text{paper}}} \, [\text{g water/g paper}] = x \tag{4.3.36}$$

This gives us the mass fraction $x$. To convert to $x$ in units of ppmw we multiply by $10^6$. (Specifically, $10^6 \times (\text{wt fraction}) = x$ (ppmw).) Solving for the molar concentration from Eq. 4.3.36, we find

$$C_{\text{water}}^{\text{paper}} = \frac{\rho_{\text{paper}}}{18 \times 10^6} x [\text{ppmw}] \tag{4.3.37}$$

Now, from Eqs. 4.3.33 and 4.3.34, and using Eq. 4.3.26, we find

$$\alpha = \frac{18 \times 10^6}{R_{\text{G}}T\rho_{\text{paper}}} \frac{p}{x[\text{ppmw}]} = \frac{18 \times 10^6}{R_{\text{G}}T\rho_{\text{paper}}} H \tag{4.3.38}$$

This is the sought-after relationship between the required $\alpha$ and the provided $H$.

With pressure in atmospheres, the value of $R_{\text{G}}$ is $R_{\text{G}} = 82.06 \text{ cm}^3 \cdot \text{atm/mol} \cdot \text{K}$. We will take the paper density as $\rho_{\text{paper}} = 1.1 \text{ g/cm}^3$. At the operating temperature of $T = 35°\text{C} = 308 \text{ K}$, we find

$$\alpha = 2.7 \times 10^{-4} \frac{\text{moles water/cm}^3 \text{ air}}{\text{moles water/cm}^3 \text{ paper}} \tag{4.3.39}$$

We are interested in the *average* water content as a function of time. For the slab geometry, Fig. 4.2.2a presents this information. Since drying is from one side only, as in Example 4.3.1, the length scale that appears in both $X_{\text{D}}$ and Bi is taken to be the thickness $b$, and not the half-thickness, as would be the case if drying were from both sides.

We calculate the Biot number and find

$$\text{Bi} = \frac{\alpha k_{\text{c}}b}{D_{\text{AB}}} = \frac{2.7 \times 10^{-4}(0.5 \text{ cm/s}) \, 7.62 \times 10^{-3} \text{ cm}}{10^{-5} \text{ cm}^2/\text{s}} = 0.1 \tag{4.3.40}$$

We want the time at which $\bar{Y}$ takes on the value $\bar{Y} = 0.4$. From Fig. 4.2.2a, extrapolating along the line for Bi = 0.1, we estimate $X_{\text{D}} = 9.25$. This gives us an estimated time of

$$t = \frac{9.25 b^2}{D_{\text{AB}}} = \frac{9.25(7.62 \times 10^{-3} \text{ cm})^2}{10^{-5} \text{ cm}^2/\text{s}} = 57 \text{ s} \tag{4.3.41}$$

With this value of a "residence time" inside the drier, we may determine the required path length for the paper, in terms of the speed of travel of the paper through the oven.

---

**EXAMPLE 4.3.3**   *Evaporation of a Solute from a Small Water Droplet*

A small drop of water initially contains $SO_2$ with a *mole* fraction of $x = 0.002$. The drop, which has a diameter of 2 mm, is suspended in an airstream, and the system is isothermal at 35°C. Find the time for the average mole fraction of $SO_2$ in the drop to fall by a factor of 10. Assume that the convective mass transfer coefficient at the drop interface is $k_c = 20$ cm/s. At 35°C the vapor/liquid equilibrium is described by Henry's law in the form

$$p_{vap} \text{[atm]} = Hx \qquad (4.3.42)$$

where $x$ is the mole fraction of $SO_2$ in the water phase and $H = 50$ atm/mol fraction.

This looks like a relatively straightforward problem involving the use of the Gurney–Lurie charts. The major task is to find a value for the Biot number, using the information given. We can start with a quick estimate of the behavior by assuming that a convective coefficient of 20 cm/s is, in some sense, large enough to allow us to regard the external gas as well mixed (i.e., $Bi = \infty$). We make this assumption not because we believe it is true, but because it gives us a quick lower bound on the required time. Any finite external mass transfer resistance (i.e., a smaller coefficient) will raise the time required for the drop to lose 90% of the dissolved vapor.

We will assume that the ambient airstream is free of $SO_2$ and that the evaporation from this small drop does not add any significant $SO_2$ to the ambient medium. Then a 10-fold drop in average concentration of $SO_2$ in the liquid corresponds to a value of $\overline{Y} = 0.1$. From Fig. 4.2.2c, at $\overline{Y} = 0.1$, we find $X_D \approx 0.2$. Then

$$t = \frac{R^2}{D_{AB}} X_D \qquad (4.3.43)$$

We need an estimate of the diffusion coefficient $D_{AB}$ of $SO_2$ in the drop—water. Since we are only looking for an approximate estimate here, let's use a nominal value, typical of gas/liquid diffusion, of $D_{AB} = 2 \times 10^{-5}$ cm²/s. Then we find

$$t = \frac{0.1^2}{2 \times 10^{-5}} 0.2 = 100 \text{ s} \qquad (4.3.44)$$

This quick estimate is based on the assumption that there is no external mass transfer resistance. We should evaluate this assumption next.

The Biot number is given by Eq. 4.3.40, with the drop radius $R$ as the length scale

$$Bi = \frac{\alpha k_c R}{D_{AB}} \qquad (4.3.45)$$

Hence the next step is calculation of the thermodynamic coefficient $\alpha$, which we define in general as

$$\alpha = \frac{C_A^f}{C_A^s} \qquad (4.3.46)$$

which is the ratio of concentration in the fluid (convecting) phase (the external air) to that in the stationary (diffusive) phase (the aqueous drop).

The molar density (concentration) of the drop, which is nearly all water, is simply

$$C^s = \frac{1 \text{ g/cm}^3}{18 \text{ g/mol}} = 0.055 \text{ mol/cm}^3 \qquad (4.3.47)$$

The molar density of $SO_2$ in the liquid is just the product of this number times the mole fraction $x$:

$$C_A^s = 0.055x \ \text{mol/cm}^3 \tag{4.3.48}$$

In the external air phase, the molar concentration of $SO_2$ in that phase is (using Eq. 4.3.42)

$$C_A^f = \frac{p_{vap}}{R_G T} = \frac{Hx}{R_G T} \ \text{mol/cm}^3 \tag{4.3.49}$$

As a result, we find

$$\alpha = \frac{C_A^f}{C_A^s} = \frac{Hx/R_G T}{0.055x} = \frac{H}{0.055 R_G T} = \frac{50}{0.055(82.06)(308)} = 0.036 \tag{4.3.50}$$

and

$$\text{Bi} = \frac{0.036(20)(0.01)}{2 \times 10^{-5}} = 360 \tag{4.3.51}$$

This large value for Bi confirms the assumption of negligible external resistance.

## 4.4 NEGLIGIBLE INTERNAL RESISTANCE

Sometimes internal diffusion is so rapid that there is no significant resistance to mass transfer within the solid or stagnant fluid. All the resistance is *external* to the stationary medium and arises from convection at the surface. In such a case we may regard the "solid" as *well mixed* and at some uniform concentration $C_A^s$ with respect to a particular species A. Now a transient mass balance takes the form

$$-\frac{d}{dt}(VC_A^s) = Ak_c(C_A^f - C_A^{f\infty}) = A\alpha k_c(C_A^s - C_A^{S\infty}) \tag{4.4.1}$$

where $C_A^{S\infty}$ is defined by Eq. 4.3.7. In this equation $V$ is the volume of the "solid," and $A$ is its surface area across which convective transfer occurs. (We put quotation marks on the word "solid" because this analysis would apply as well to transfer from a fluid droplet, or a bubble, as long as internal resistance was much less than external resistance to transport across the interface.) An initial condition is

$$C_A^s = C_A^0 \qquad \text{at} \quad t \leq 0 \tag{4.4.2}$$

It is not difficult to see that we can write Eq. 4.4.1 in the form

$$-\frac{d\overline{Y}}{dX_C} = \overline{Y} \tag{4.4.3}$$

where

$$\overline{Y} = \frac{\overline{C}_A^S - C_A^{S\infty}}{C_A^0 - C_A^{S\infty}} = \frac{\overline{C}_A^S - C_A^{f\infty}/\alpha}{C_A^0 - C_A^{f\infty}/\alpha} \tag{4.4.4}$$

(note Eq. 4.2.6) and a new dimensionless time variable is defined as

$$X_c = \frac{\alpha k_c A}{V} t \tag{4.4.5}$$

If we define a characteristic length scale as[1]

$$L' = \frac{3V}{A} \tag{4.4.6}$$

then we find

$$X_C = \frac{3\alpha k_c}{L'} t \tag{4.4.7}$$

We can define a dimensionless convective mass transfer coefficient, the Sherwood number, based on this length scale $L'$ as

$$Sh' = \frac{k_c L'}{D_{AB}^*} \tag{4.4.8}$$

where we use the diffusion coefficient $D_{AB}^*$ *in the convecting medium*—not within the solid or stagnant phase. Note that $k_c$ (defined by Eq. 4.3.2) appears in $Sh'$.

Notice that the Biot modulus and the Sherwood number look alike, but they have different diffusion coefficients. Note also that $\alpha$ does not appear in the definition of the Sherwood number. Finally, be careful about which length scale ($L$ or $L'$) is used in the definition of either.

Now we can show that

$$X_C = 3\alpha X_D' Sh' \frac{D_{AB}^*}{D_{AB}} = 3X_D' Bi' \tag{4.4.9}$$

where

$$X_D' = \frac{D_{AB} t}{L'^2} \quad \text{and} \quad Bi' = \frac{\alpha k_c L'}{D_{AB}} \tag{4.4.10}$$

As before, these groups are defined using $D_{AB}$ *within the solid* or stagnant volume. Note the use of the prime mark to alert us to the use of $L'$ as the length scale. $X_C$ and $X_D'$ differ in two significant respects: $X_C$ has $k_c$ in it ($X_D'$ does not); and $X_D'$ has a diffusion coefficient in it (and $X_C$ does not).

The solution to Eq. 4.4.3 is

$$\bar{Y} = \exp(-X_C) = \exp(-3X_D' Bi') \tag{4.4.11}$$

Recall that a key assumption that leads us to this model is that all the resistance to mass transfer lies in the external—convecting—phase. How can we obtain a criterion for neglecting internal resistance? For the case of a sphere, for example, the simplest approach is to take the graph of $\bar{Y}(X_D, Bi)$ presented earlier (Fig. 4.2.2c), and replot it as $\bar{Y}(X_D Bi, Bi)$, as shown in Fig. 4.4.1. We observe that for $Bi < 0.2$ the lines for $Bi = 0.2, 0.1, \ldots$ collapse to a single line independent of $Bi$, and furthermore, that line is identical to Eq. 4.4.11. (Similar plots are possible for the cylinder and slab geometries.) *We conclude that the Biot modulus determines the relative roles of internal*

---

[1] Be careful of length scales! In the definitions of $X_D$ and $Bi$ given in Section 4.2, we used a length scale $L$ which is, for simple shapes such as the slab, cylinder, or sphere, the half-thickness of the slab or the radius of the cylinder or sphere. The length scale $L'$ defined here is different from the $L$ used in Eq. 4.2.11. Here we find that Eq. 4.4.6 yields $L' = R$ for a sphere, $L' = 1.5R$ for a cylinder, and $L' = 3b$ where $b$ is the half-thickness for a slab.

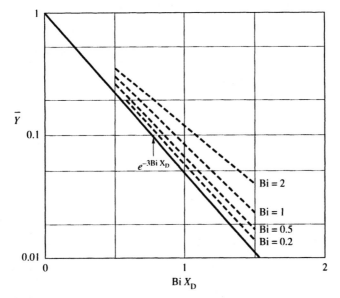

**Figure 4.4.1** A replot of Fig. 4.2.2c. For Bi ≤ 0.2 the lines correspond to Eq. 4.4.11. For a sphere, $L = L' = R$, $X_D = X'_D$, Bi' = Bi.

*diffusion and external convection.* For the case of the sphere, when[2]

$$\text{Bi}' = \frac{\alpha k_c L'}{D_{AB}} = \text{Bi} = \frac{\alpha k_c R}{D_{AB}} < 0.2 \tag{4.4.12}$$

*internal diffusion resistance is negligible,* and transient mass transfer is controlled by external convection.

From this discussion, the following problem-solving strategy emerges. In examining a problem in convective diffusion, if the convective coefficient $k_c$ is available, and if we have sufficient equilibrium data to yield a value of $\alpha$, we should first evaluate Bi. In the case of a sphere, if the value of Bi is large (or small) compared to Bi = 0.2, we can use either of two simple transient models. For large Bi we use the transient solution with constant surface concentration equal to the solubility value that is in thermodynamic equilibrium with the ambient. For small Bi we use the "well-mixed interior" model— Eq. 4.4.11.

The foregoing discussion is specific to the comparison of the two solutions (no internal resistance *vs.* finite internal resistance) for the case of a *spherical* geometry. However, the same ideas hold for a slab or a long cylinder, with appropriate changes in the plots that would be analogous to Fig. 4.4.1, and in the equations analogous to Eq. 4.4.12. (See Problem 4.14.)

**EXAMPLE 4.4.1** *Kinetics of a Sustained-Release Drug Delivery System*

Some mass transfer systems do not fit neatly into the categories we have considered so far. This example exhibits some of the features of a system that is free of internal

---

[2] Since $L' = R$ in this case, Bi = Bi' and $X_D = X'_D$.

diffusive resistance. Other aspects of the system, however, correspond to our earlier discussion of diffusion across a film. Let us see how a model of such a system is developed.

The problem we wish to illustrate comes up in the context of analysis of a biomedical device—a drug delivery system. When the drug is dispersed uniformly as a *solid* within a polymeric material (but not in a discrete reservoir *surrounded* by a membrane), the release behavior is somewhat complex. For a drug solution in a *liquid* phase reservoir, the diffusivity of the drug in the liquid phase is high enough that the drug reservoir may be regarded as well mixed, and release is controlled by diffusion through the surrounding membrane. If the drug were dispersed as a solid, *throughout* the polymer matrix, the behavior would be as in the case of the Ocusert device in Example 4.2.4 above, and transient diffusion would control release. Often, however, the solid is dispersed at a concentration well above its solubility in the polymer, and in such cases another type of release kinetics is attainable. Figure 4.4.2 shows the expected behavior, in terms of the concentration profile $C(x)$ within the matrix.

We may carry out an analysis of the release kinetics for this system, under the assumption that a quasi-steady state exists in the polymeric matrix. We define $M_t$ as the amount delivered from (not remaining in) the device at some time $t$ after the start of delivery. Then the rate of delivery (per unit width) is given by

$$\frac{dM_t}{dt} = \frac{AD(C_s - 0)}{x} \tag{4.4.13}$$

where $A$ is the *total* exposed area (i.e., the area of *both* sides of the polymeric film). Note that as the drug leaves the matrix, the boundary at $x$ is a function of time: $x$ increases as the boundary moves toward the center of the matrix. In addition, $x$ is the thickness of the undersaturated region (on each side), and $x = 0$ is measured from the exterior surface.

If we denote the initial amount of drug in the matrix by $M_\infty$, then a simple material balance (obtained essentially by inspection of the geometry in Fig. 4.4.2) gives the amount that has been released over the time interval $t$ as

$$M_t = M_\infty \left(\frac{2x}{L}\right) - C_s \left(\frac{Ax}{2}\right) \tag{4.4.14}$$

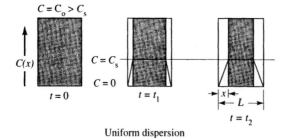

Uniform dispersion

**Figure 4.4.2** Delivery from a solid drug in a polymeric matrix. The drug is initially well above its solubility limit $C_s$ in the matrix. At later times, dissolution of the drug into the exterior fluid occurs, and the drug is replenished by diffusion from the supersaturated core. The supersaturated region becomes thinner with time. There is no concentration gradient within the core. In the regions of thickness $x$ adjacent to the exterior surfaces, there is a concentration gradient.

We may solve for $x$ from this expression and substitute that into Eq. 4.4.13 to yield

$$\frac{dM_t}{dt} = \frac{ADC_sM_\infty}{M_t}\left(\frac{2}{L} - \frac{AC_s}{2M_\infty}\right) \tag{4.4.15}$$

We may now integrate this expression to yield the solution for $M_t(t)$ in the form

$$M_t^2 = ADC_sM_\infty\left(\frac{2}{L} - \frac{AC_s}{2M_\infty}\right)t \tag{4.4.16}$$

or, substituting

$$M_\infty = \frac{AC_oL}{2} \tag{4.4.17}$$

we find

$$M_t = A[DC_s(2C_o - C_s)t]^{1/2} = A(2DC_oC_st)^{1/2} \tag{4.4.18}$$

(The far-right expression follows on the assumption that $C_o \gg C_s$.)

Equation 4.4.18 predicts that this type of sustained-release system will show a square root of time dependence of the total amount of drug released and that the amount released will depend on the square root of $C_o$, which is called the "loading." Data are shown in Fig. 4.4.3a, for the release of chlormadinone acetate from a silicone elastomer into water. Are the data consistent with this model?

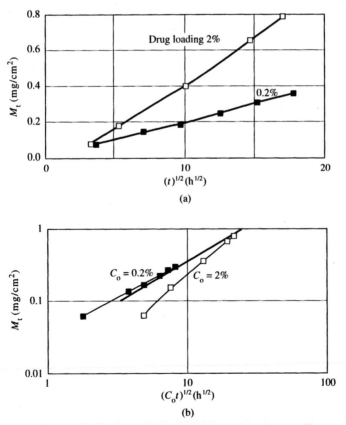

**Figure 4.4.3** (a) Release of chlormadinone acetate from a silicone elastomer. (b) Replot of the data in (a).

We do observe an approximately linear relationship of $M_t$ with $\sqrt{t}$, consistent with Eq. 4.4.18. If we replot the data as in Fig. 4.4.3b, however, we see that the model largely fails to describe the data. First, note that we replot the data as the square root of the product $C_o t$. According to Eq. 4.4.18, both sets of data should fall on a single curve. They don't! By plotting on logarithmic coordinates we can see that the data form separate curves that are about a factor of 2 apart. This is a large deviation. We have placed a straight line with a slope of unity (the solid line) on Fig. 4.4.3b. The small-loading data and some of the high-loading data might be regarded as falling about a single line with the expected slope. However, the original reference gives no reason to discount the short-time, high-loading data. It is possible that the approximation just noted (below Eq. 4.4.18) is not obeyed by these data; but again, we do not have enough evidence to make a determination.

We have chosen to show an example of a partial failure of a simple model to describe experimental data because it is important to understand that such failure is commonplace. Many systems of real commercial interest are much more complex than any simple model might pretend. In many cases, however, simple models mimic observations quite well, and this fact should encourage us to explore simple models as a first step in the analysis of complex systems. See Problem 4.25 for an example.

### EXAMPLE 4.4.2 *Dissolution of Solid Particles in Liquids*

A common chemical process involves the dissolution of a mass of discrete solid particles in an agitated liquid. The nature of the dissolution process may differ from one system to another. We examine the case in which the rate-controlling step is the mass transfer of the dissolved solute (the solid material in solution) into the bulk of the agitated liquid medium. We assume that there is no internal resistance to mass transfer to the surface of the particle. This would be the case, for example, for a pure solid or for a solid mixture of components all equally soluble in the surrounding liquid. (In either of these cases there would be no internal concentration gradients, hence no internal diffusion process to retard the transport to the surface of the particle.) Figure 4.4.4 suggests an approach to modeling this problem.

The simplest model, and the one we begin with, assumes the following:

1. The solid particle is *pure solute*, not a solid mixture. The concentration of this solute at the solid/liquid boundary [at $r = R(t)$] is the saturation value $C_s$.
2. The solid particle remains spherical.
3. Convective mass transfer into the liquid phase controls the rate of dissolution, and

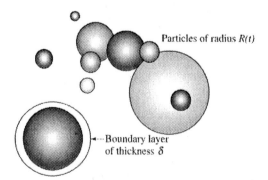

Particles of radius $R(t)$

Boundary layer of thickness $\delta$

**Figure 4.4.4** Solid particles of radius $R(t)$ in an agitated liquid.

this may be represented by a "film" model. This means that dissolution takes place by diffusion of the solute across a film, concentric with the spherical particle, of uniform thickness $\delta$. We will later address the issue of how $\delta$ is to be estimated.

4. $\delta$ is small compared to $R(t)$, and a quasi–steady state approach may be used to describe diffusion within the film.

In spherical coordinates the *steady* diffusion equation takes the form

$$D \frac{d}{dr}\left(r^2 \frac{dC}{dr}\right) = 0 \quad \text{in} \quad R(t) \leq r \leq R(t) + \delta \tag{4.4.19}$$

Keep in mind that in this example $D$ is the diffusivity of the solute in the surrounding liquid film. Appropriate boundary conditions are

$$C = C_s \quad \text{at} \quad r = R(t) \tag{4.4.20}$$

$$C = C_b \quad \text{at} \quad r = R + \delta \tag{4.4.21}$$

where $C_b$ is the bulk concentration in the agitated liquid. $C_b$ may be constant if the volume $V$ of the bulk phase is large enough. Otherwise we must add a *transient* species balance equation for the bulk region as well.

For a single sphere, the diffusive transfer into the film must balance the loss of material from the particle, hence the decrease in its radius. Thus we write

$$-4\pi R^2 D \left(\frac{dC}{dr}\right)_{r=R} = -\frac{d}{dt}\left(\frac{4}{3}\pi R^3 \frac{\rho_s}{M_w}\right) \tag{4.4.22}$$

or

$$D\left(\frac{dC}{dr}\right)_{r=R} = \frac{\rho_s}{M_w}\frac{dR}{dt} \tag{4.4.23}$$

Here we denote the mass density of the solid species by $\rho_s$ and its molecular weight by $M_w$.

Equation 4.4.23 embodies the quasi-steady part of the analysis, because we are going to use a *steady state* analysis (Eqs. 4.4.19–4.4.21) to find the concentration gradient for constant radius $R$, and then put this steady state result into the transient balance— Eq. 4.4.23.

We may integrate Eq. 4.4.19 for the concentration profile in the film surrounding the drop and find the concentration gradient at $R$ in the form

$$\left(\frac{dC}{dr}\right)_{r=R} = -\frac{(1 + \delta/R)(C_s - C_b)}{\delta} \tag{4.4.24}$$

Upon combining this expression with Eq. 4.4.23, we find a differential equation for the radius as a function of time:

$$\frac{dR}{dt} = -\frac{DM_w}{\rho_s \delta}\left(1 + \frac{\delta}{R}\right)(C_s - C_b) \tag{4.4.25}$$

We begin with the case that $C_b$ is constant. Keep in mind that the saturation concentration $C_s$ is also constant.

## $C_b$ Constant

We may solve Eq. 4.4.25 and find

$$t^* = \frac{\left[(1 - R^*) + \delta^* \ln\left(\frac{R^* + \delta^*}{1 + \delta^*}\right)\right]\delta^*}{C^*} \tag{4.4.26}$$

where we have defined the following dimensionless variables and parameters:

$$t^* = \frac{4\pi t D R_o N_v}{3} \qquad \delta^* = \frac{\delta}{R_o}$$

$$R^* = \frac{R}{R_o}$$

$$C^* = \frac{3 M_w (C_s - C_b)}{4\pi \rho_s R_o^3 N_v}$$

(4.4.27)

The initial radius of the particles is $R_o$, and $N_v$ is the number of identical particles *per total volume of bulk liquid*. Hence $\frac{4}{3}\pi \rho_s R_o^3 N_v$ is the mass density of solids loading in the vessel. Equation 4.4.26 gives an implicit solution for $R(t)$.

## $C_b$ Not Constant

If the volume of the bulk liquid is not large enough, dissolution will increase the concentration $C_b$, and a transient balance on the bulk liquid must be written. This takes the form

$$\left(\frac{dC_b}{dt}\right) = -\frac{4\pi \rho_s N_v}{M_w} R^2 \frac{dR}{dt}$$

(4.4.28)

which has the solution (with $C_b = C_o$ at $t = 0$, $R = R_o$)

$$C_b = C_o + \frac{4\pi \rho_s N_v}{3 M_w} (R_o^3 - R^3)$$

(4.4.29)

When this expression for $C_b$ is substituted into Eq. 4.4.25 we obtain a new differential equation for $R(t)$. The solution now takes a very cumbersome algebraic form, which we do not display here (see Problem 4.15).

## Dissolution Time

We may calculate the dimensionless time $T^*$ required to dissolve a particle by setting $R^* = 0$ in Eq. 4.4.26. The result is

$$T^* = \frac{\delta^* \left[1 + \delta^* \ln \left(\frac{\delta^*}{1 + \delta^*}\right)\right]}{C^*}$$

(4.4.30)

The corresponding result for the case of variable $C_b$ (Eq. 4.4.29) may be written in the form

$$T^* = -\delta^* \left\{ a_1 \ln \left(\frac{\delta^*}{1 + \delta^*}\right) + a_2 \ln \left(\frac{\alpha}{1 + \alpha}\right) \right.$$

$$+ b_2 \ln \left[(0.5)\left(\frac{\alpha^2}{(1 + \alpha/2)^2 + 3\alpha^2/4}\right)\right]$$

$$\left. - \left(\frac{b_2\alpha + 2c_2}{\sqrt{3}\alpha}\right) \left[\tan^{-1}\left(\frac{1}{\sqrt{3}}\right) + \tan^{-1}\left(\frac{2 - \alpha}{\sqrt{3}\alpha}\right)\right] \right\}$$

(4.4.31)

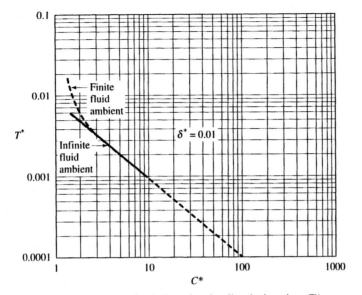

**Figure 4.4.5** Theoretical solutions for the dissolution time $T^*$.

where

$$\alpha^3 = C^* - 1 \qquad\qquad a_1 = -\frac{\delta^*}{\alpha^3 - \delta^{*3}} = -b_1$$

$$a_2 = \frac{b_1\alpha^2 - c_1\alpha + d_1}{3\alpha^2} \qquad b_2 = \frac{2b_1\alpha^2 + c_1\alpha - d_1}{3\alpha^2} \qquad (4.4.32)$$

$$c_2 = \frac{-b_1\alpha^2 + c_1\alpha + 2d_1}{3\alpha^2} \qquad c_1 = -\frac{\delta^{*2}}{\alpha^3 - \delta^{*3}}$$

$$d_1 = \frac{\alpha^3}{\alpha^3 - \delta^{*3}}$$

Figure 4.4.5 shows these solutions (Eqs. 4.4.30 and 4.4.31) for the case $\delta^* = 0.01$. As expected, the dissolution time is greater for the case of the finite ambient medium because as dissolution occurs, the driving "force" for mass transfer decreases. From the figure we see that for $C^*$ greater than about 4, the ambient medium can be regarded as infinite.

After wading through this swamp of algebra, we must recall that the convective transport is encoded in the parameter $\delta^*$ ($= \delta/R_o$). Hence we must address the question: How do we obtain estimates of $\delta$ under various conditions of agitation of a slurry of solid dissolving particles? This topic will be taken up in a later chapter.

**EXAMPLE 4.4.3**    *Control of a Toxic Gas Release in a Closed Space*

In this example we look at a physical system having some features similar to those of Example 4.4.2. In particular, the presence of a bounded external medium exerts a retarding influence on the rate of approach to equilibrium. However, in *this* example the transport resistance is *interior* to the particle—in this case a liquid droplet.

Arsine ($AsH_3$) is a highly toxic compound used commonly in the semiconductor fabrication industry as a means of supplying arsenic atoms for doping of semiconductor films. The potential for release of arsine to the laboratory environment must be considered, and a means of rapidly removing arsine vapor from the air must be developed. The following facts about arsine are relevant.

It is a gas at room temperature. It boils at $-55°C$.
Its solubility in water at 20°C is 0.07 g/100 $cm^3$.
Its Henry's law constant at 20°C is 6191 atm/mole fraction in water.
Its molecular weight is 78.
Its vapor density is 2.7 times that of air.
Its toxicity is rated as 6 ppmv; this is the so-called IDLH value, which is the concentration immediately dangerous to life or health. An IDLH value represents the maximum level at which escape could be made within 30 minutes without escape-impairing symptoms and without irreversible health effects.

We want to examine some proposals for removal of arsine from air in the event of an accidental release. Imagine the following scenario:

*An accidental release of arsine has occurred, with the result that the concentration of vapor in the room air is 20 ppmv. It is proposed that the immediate response to this is to spray the room with a water mist.*

The concentration of 20 ppmv in air implies that the partial pressure of the arsine will be

$$p_{As} = 20 \times 10^{-6} \, atm = 2 \times 10^{-5} \, atm \qquad (4.4.33)$$

Let us suppose that the room is closed (no air circulation) and is of dimensions 2 m $\times$ 4 m $\times$ 3 m. This volume of 24 $m^3$ contains 480 $cm^3$ of arsine vapor when the concentration is 20 ppmv. This corresponds (at 20°C) to

$$\frac{480}{22400 \times 293/273} = 0.02 \, \text{mol arsine} = 78 \times 0.02 = 1.6 \, \text{g arsine} \qquad (4.4.34)$$

We might start by determining how much water is needed to completely absorb all this arsine. From the equilibrium solubility data, we know that the solubility of arsine in water is 0.07 g/100 $cm^3$ water. Hence we need

$$\frac{100}{0.07} \times 1.6 = 2300 \, cm^3 \, \text{water} \qquad (4.4.35)$$

Thus it appears that we need only spray a little more than 2 L of water to absorb all the arsine. However, this conclusion is based on the assumption that a state in which air *free of arsine* is in contact with 1.6 g of arsine dissolved in 2300 $cm^3$ of water is an equilibrium state. This is clearly not the case. A calculation of the partial pressure of arsine in equilibrium with arsine-saturated water (using the Henry's law constant given earlier) shows that the partial pressure is, in fact, one atmosphere! This is exactly the expected result, since the solubility value given refers to the concentration of arsine in water that is in equilibrium with *pure* arsine gas at one atmosphere and 293 K.

If arsine is transferred into the mist of water, equilibrium requires that the concentration of arsine in the water correspond to the partial pressure of arsine in the air. Hence our calculation of the amount of water that will hold *all* the arsine is correct, but irrelevant. This misleading number follows from an act of unclear thinking. We must ask a different question.

For example, let us find how much water is required to ensure that, *at equilibrium*, 90% of the arsine has been absorbed by the water and that the water is then in equilibrium with the remaining 10% of the arsine that is in the air. The equilibrium condition is that the partial pressure of arsine in air must equal the vapor pressure of arsine in the water/arsine solution. If 10% of the arsine remains in the air, then the partial pressure is

$$0.1 \times 20 \times 10^{-6} \times 1 \text{ atm} = 2 \times 10^{-6} \text{ atm} \qquad (4.4.36)$$

Using Henry's law, the mole fraction of arsine in the water is then

$$x = \frac{p}{H} = \frac{2 \times 10^{-6} \text{ atm}}{6191 \text{ atm/mole fraction}} = 3.23 \times 10^{-10} \qquad (4.4.37)$$

Since the water now contains 90% of the original 0.02 mol of arsine, or 0.018 mol, the number of moles of water is

$$\frac{0.018}{3.23 \times 10^{-10}} = 5.6 \times 10^{7} \text{ mol} = 10^{9} \text{ g of water} \qquad (4.4.38)$$

or *one million liters!* Hence a thermodynamic analysis tells us quite clearly that we cannot remove arsine from the room air by simply spraying the room with water droplets.

Let us suppose that we *could* pick a system that was thermodynamically more favorable. This implies that we could find a liquid whose solubility for arsine is enormously higher than that of water. There are still several design issues here. One is the need to mix the water spray efficiently throughout the air space of the room. (A well-designed ceiling spray system could achieve this.) In addition, the equilibrium absorption must occur rapidly. Finally, there must be a means of disposing of the arsine/liquid solution that results from the absorption, since this substance is itself hazardous.

Let us examine the rate at which water drops will absorb arsine vapor. We suppose that the water is sprayed as drops that are on average 100 $\mu$m in diameter. We ask the question: How long is required for such a drop to achieve 90% of its equilibrium concentration of arsine if it is surrounded by a 20 ppmv arsine/air mixture? If we assume that the air is well mixed, so that the Biot modulus is large, we can use Fig. 4.2.2c to find that when $\overline{Y}$ is 0.1 (which corresponds to 90% absorption) the dimensionless time is $X_D = 0.18$. If we set $Dt/R^2 = 0.18$, and assume a diffusion coefficient of about $10^{-6}$ cm$^2$/s for arsine in water (it may be larger), we find $t = 4.5$ s to achieve (near) equilibrium. On the basis of this calculation, we would conclude that this system would very rapidly reduce the arsine concentration in the air below the toxic level, *assuming there was enough water* ($10^{9}$ g).

On what basis, if any, can we argue that the Biot modulus is large? Isn't it likely, in fact, that there is a significant resistance to mass transfer in the vapor phase? The answer is available to us through a model that we developed and presented in Example 3.2.1. In the worst case, if there is no convection to aid transfer in the gas phase, diffusion will give mass transfer at a rate that corresponds to a convective coefficient $k_c$ that satisfies Eq. 3.2.23[3]:

$$k_c = \frac{D_{AB}}{R} \qquad (4.4.39)$$

*The diffusivity that appears in this equation is that in the fluid (air) outside the sphere.* For arsine in air, we might expect $D_{AB} = 0.1$ cm$^2$/s. For the case cited, this leads to $k_c = 0.1/50 \times 10^{-4} = 20$ cm/s. Then we find a Biot modulus of

$$\text{Bi} = \frac{\alpha k_c R}{D_{AB}^s} = \frac{20 \times 50 \times 10^{-4}}{10^{-6}} \alpha = 10^{5} \alpha \qquad (4.4.40)$$

---

[3] Although written $k$, it is the coefficient $k_c$ that is defined in Eq. 3.2.23.

*Note here that (as usual) the Biot modulus is defined with the diffusivity of the interior phase,* which we estimate to be of the order of $10^{-6}$ cm²/s. We see then that the Biot modulus is indeed very large, unless $\alpha$ is a very small number. (From the equilibrium data above, it is not difficult to show—see Eq. 4.4.53 below—that $\alpha$ (= $1/\beta$) is about 5.) It seems likely then that we could achieve (near) equilibrium very quickly by spraying small drops of water into the room. (Note that we have used rough order-of-magnitude estimates for the diffusivities in these calculations. Even order-of-magnitude errors will not change the conclusion: the Biot modulus is large compared to unity.)

This estimate of a time scale might be poor for another reason: we want to remove 90% of the arsine from the air; yet as absorption occurs, the concentration of arsine that is "driving" the mass transfer across the phase boundary is getting smaller, thereby retarding mass transfer. Thus the problem is more complex than this first, simple, approach would suggest. What we need is the solution to a more difficult, but still solvable, problem. Figure 4.4.6 suggests the model. Each water droplet is regarded as being surrounded by a spherical volume of air. The volume ratio is that of water to room volume. (Keep in mind that we have already rejected water as the appropriate liquid absorbant, on thermodynamic grounds. We continue the analysis of the *dynamics* of absorption, ignoring this point.) We must solve the diffusion equation in the water drop and couple that equation to a mass balance for the surrounding air. Since we have seen that transfer is controlled by the internal resistance in the drop (based on the value of Bi given in Eq. 4.4.40), we might regard the air region to be well mixed. In that case the mathematical model takes the form of the following set of equations:

$$\frac{\partial C_d}{\partial t} = \frac{D_{AB}}{r^2} \frac{\partial}{\partial r}\left[ r^2 \frac{\partial C_d}{\partial r}\right] \tag{4.4.41}$$

with an initial condition

$$C_d = 0 \quad \text{for } t = 0 \tag{4.4.42}$$

and boundary conditions

$$\frac{\partial C_d}{\partial r} = 0 \quad \text{at} \quad r = 0 \tag{4.4.43}$$

and

$$C_d = \beta C_f \quad \text{at } r = R \tag{4.4.44}$$

Be careful of the notation here: the concentration of arsine in the droplet is $C_d$ and in the surrounding air is $C_f$. The solubility relationship between the two phases appears in the coefficient $\beta$, which we may calculate from the Henry's law constant $H$ given earlier. (We must transform the concentration units from partial pressure to moles per cubic centimeter. In addition, note that this definition of $\beta$ is inverse to H, and is not

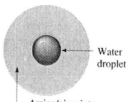

**Figure 4.4.6** A small water droplet absorbs gas from a surrounding finite volume.

the same as the $\alpha$ defined in some earlier analyses. The difference lies in whether we choose to write concentration ratios as liquid-to-vapor or vapor-to-liquid ratios. The $\alpha$ in Eq. 4.4.40 is $1/\beta$.)

In this unsteady system the concentrations at the interface are functions of time. We must find $C_f$ from a balance on the air region, which takes the form

$$V_f \frac{\partial C_f}{\partial t} = -4\pi R^2 D_{AB}^d \left[ \frac{\partial C_d}{\partial r} \right]_{r=R} \tag{4.4.45}$$

where $V_f$ is the volume of the air surrounding one drop. We emphasize that the diffusivity that appears here is that of the drop by writing it as $D_{AB}^d$.

An initial condition on $C_f$ is

$$C_f = C_{f,o} \quad \text{at} \quad t = 0 \tag{4.4.46}$$

The solution to this problem may be written in the following form[4]:

$$\frac{C_f}{C_{f,o}} = \frac{B}{1+B} + 6B \sum_{k=1}^{\infty} \left[ \frac{\exp(-b_k^2 \tau)}{B^2 b_k^2 + 9(1+B)} \right] \tag{4.4.47}$$

The dimensionless time is defined as

$$\tau = \frac{D_{AB}^d t}{R^2} \tag{4.4.48}$$

The parameter $B$ is given by

$$B = \frac{V_f}{\beta V_d} \tag{4.4.49}$$

The drop volume is $V_d$.

The eigenvalues $b_k$ are the nonzero roots of the equation

$$\tan b = \frac{3b}{3 + Bb^2} \tag{4.4.50}$$

Of particular importance to us is the time required to achieve a specified approach to equilibrium, such as $C_f/C_{f,o} = 0.1$. The results are shown in Fig. 4.4.7, in the form of a plot of $\tau_{0.1}$ as a function of $B$. The effect of depletion of the arsine from the gas phase is clearly seen. We note, as well, that $\tau_{0.1}$ cannot be achieved at *any* time unless $B$ satisfies the constraint that $B/(1 + B)$ be less than 0.1. If we select $B$ such that $B/(1 + B) = 0.1$, we can find the *minimum* required volume of water drops from $B = 0.11$ and Eq. 4.4.49. This requires a value for $\beta$, which we can find from Henry's law, or from the solubility data, both given above. However, an infinite time would be required to bring the ambient to exactly this condition.

From solubility data we find that at equilibrium, $C_d = 0.07$ g of arsine per 100 mL of water at 20°C. We assume that this solubility statement is based on the amount of arsine absorbed by water when it reaches equilibrium with *pure* arsine gas at one atmosphere pressure. Based on that assumption, we find the molar concentration of pure arsine at atmospheric pressure

$$C_f = \frac{p}{R_G T} = \frac{1}{82 \times 293} = 4.2 \times 10^{-5} \, \text{mol/cm}^3 \tag{4.4.51}$$

---

[4] It's not likely that you have learned how to solve this kind of partial differential equation at this stage. Just take the result as a gift!

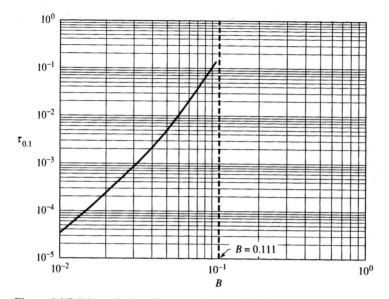

**Figure 4.4.7** Dimensionless time to reduce ambient concentration by 90%.

In a water drop in equilibrium with this gas, using the molecular weight of arsine as 78, we have

$$C_d = \frac{0.07/78}{100} = 9 \times 10^{-6} \, \text{mol/cm}^3 \tag{4.4.52}$$

This yields

$$\beta = \frac{C_d}{C_f} = \frac{9 \times 10^{-6}}{4.2 \times 10^{-5}} = 0.21 \tag{4.4.53}$$

Then from Eq. 4.4.49 we find

$$B = 0.11 = \frac{V_f}{\beta V_d} = \frac{V_f}{0.21 V_d} \tag{4.4.54}$$

or, with $B = 0.11$,

$$V_d = 43 V_f = 43 \times 24 \, \text{m}^3 \approx 10^6 \, \text{L} \tag{4.4.55}$$

This is the same result obtained earlier in Eq. 4.4.38 and of course it reflects the fact that the water requirement far exceeds the volume of the room!

The conclusion that removal of 90% of the arsine from the air into the water requires a value of $B$ of 0.11 or less has an interesting implication. It says that the minimum amount of water required to achieve this task is independent of the amount of arsine initially in the air! This seems counterintuitive at first; we might expect more water to be required if there is more arsine in the ambient medium to be removed. We can resolve this point with some thought. (See Problem 4.24.)

This particular illustrative example may seem somewhat foolish at the end, since the proposed method to remove the toxic gas is found to be so clearly ineffective. With a little thought we could have drawn this conclusion from a qualitative inspection of the equilibrium thermodynamics, and it would not have been necessary to deal with the dynamics of the transient diffusion problem at all. Among the goals of the example are to demonstrate the quantitative analysis of the equilibrium requirements, and to show how that analysis might prevent us from expending much effort on the mass transport dynamics of this system. But aside from that point, there could be examples of release of a toxic vapor for which the water solubility is much greater than that of arsine, such that rapid absorption by water droplets *would* provide a feasible emergency relief. (See Problem 4.52.) In addition, the ideas described in Example 4.4.3 can be modified to treat the problem of control of an *unconfined* release of a toxic vapor, as might occur, for example, if there were a break in a pipeline or pressurized storage vessel *exterior* to a chemical processing plant.[5]

## 4.5   DIFFUSION LIMITATIONS IN THE DECONTAMINATION OF SOIL

When liquid organic pollutants are released to soil, they can become physically and chemically bound within the soil phase, as well as within the pore spaces, both wet and dry, that separate the actual soil particles from one another. Soil is a complex medium, with both organic and mineral phases mixed together within the solid phase. The design of decontamination procedures calls for some knowledge of the limiting factors that control the rate of release of the contaminant to the environment. Diffusion is often a limiting factor, and here we introduce some ideas based on our developing knowledge, in this chapter, of diffusion under unsteady state conditions.

Figure 4.5.1 suggests a possible model of an individual soil particle. It is composed of an aggregation of small solid particles surrounded by a void space. Alternatively, we may think of the individual soil particle as a consolidated solid permeated by internal voids, much like an individual catalyst particle.[1]

The *intraparticle* fractional void space is described by a porosity or void fraction denoted by $\varepsilon$. In the simplest model, the solid phase is itself regarded as impermeable, but the organic contaminant can be immobilized *on* the solid phase by adsorption. It is assumed that there is an equilibrium between the concentration of the contaminant on the solid and the concentration in the surrounding void volume. Because of the multiphase character of the medium, we have to be careful about our definitions of concentrations. Let's introduce the following notations:

$C_f$ = concentration [moles per volume of the fluid (void) phase] of the contaminant in the void volume

$C_s$ = concentration (moles per mass of the solid phase) of the contaminant associated with the solid phase

$\rho_s$ = mass density of the solid phase (mass of solid per volume of solid)

Then in a representative volume *within* a porous soil particle, the apparent concentration of contaminant (moles per volume of the mixed phases) may be written as

$$C = \varepsilon C_f + (1 - \varepsilon)\rho_s C_s \qquad (4.5.1)$$

---

[5] In this regard, see the interesting paper by Fthenakis [*Chem. Eng. Commun.*, **83**, 173 (1989)].
[1] We make a distinction here between the individual particle (think of a single grain of sand) and a collection of such particles (a pile of sand). A single particle, if it is truly porous, has an intraparticle porosity that characterizes its internal structure. A collection of particles (a pile of sand or a reactor tube packed with catalyst particles) has an additional porosity characteristic of the packing of the individual particles.

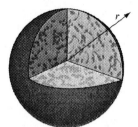

**Figure 4.5.1** A soil "particle," modeled as a porous sphere.

The equilibrium relationship may be written in the form

$$C_s = K_p C_f \tag{4.5.2}$$

This is an assumed linear relationship between the concentrations in the two phases. On the basis of experimental studies, we might find that some nonlinear relationship is more appropriate.

In the simplest model of diffusion of the contaminant through the individual porous particle, we ignore the detailed geometrical structure of the particle and imagine instead that the particle is a homogeneous sphere, of radius $R$; diffusion is envisioned as taking place homogeneously through the spherical volume, but with a correction that accounts for the fraction $(1 - \varepsilon)$ of the volume that is impermeable, hence unavailable for diffusion. We can derive the appropriate diffusion equation by a procedure that parallels the development given in Chapter 3. Figure 4.5.2 will help.

In a volume element between two spheres of radius $r$, and $r + dr$, the time rate of change of contaminant may be written *in terms of the concentration C* as

$$\frac{\partial}{\partial t} 4\pi r^2 dr C(r, t) \tag{4.5.3}$$

We assume spherical symmetry, ignore chemical reactions, neglect any convective flow through the void space, and write the net diffusive flow of contaminant in the form

$$\left[ -D_f \frac{\partial C_f(r, t)}{\partial r} 4\pi r^2 \varepsilon_A \right]_r - \left[ -D_f \frac{\partial C_f(r, t)}{\partial r} 4\pi r^2 \varepsilon_A \right]_{r+dr} \tag{4.5.4}$$

In writing the diffusive terms, we account for the fact that only a fraction $\varepsilon_A$ of the spherical surface area is open to diffusion—the rest being occupied by the impermeable solid phase. The balance on the contaminant now takes the form of a differential equation upon dividing by the volume element $4\pi r^2 dr$ and taking the limit as $dr$ vanishes.

$r$     $r + dr$

**Figure 4.5.2** Volume element in a soil particle.

The result is

$$\frac{\partial C}{\partial t} = \frac{1}{r^2}\frac{\partial}{\partial r}\left\{\varepsilon_A D_f\left[r^2\frac{\partial C_f(r,t)}{\partial r}\right]\right\} \tag{4.5.5}$$

Now we impose Eqs. 4.5.1 and 4.5.2 and obtain, after some rearrangement,

$$\frac{\partial C}{\partial t} = \left[\frac{\varepsilon_A D_f}{\varepsilon + (1-\varepsilon)K_p\rho_s}\right]\frac{1}{r^2}\frac{\partial}{\partial r}\left(r^2\frac{\partial C}{\partial r}\right) \tag{4.5.6}$$

It is commonly assumed that the area fraction $\varepsilon_A$ and the void fraction $\varepsilon$ are the same. With that assumption we may write Eq. 4.5.6 as a diffusion equation with an apparent diffusion coefficient written in the form $D_{eff}/R_K$:

$$\frac{\partial C}{\partial t} = \frac{D_{eff}}{R_K}\frac{1}{r^2}\frac{\partial}{\partial r}\left(r^2\frac{\partial c}{\partial r}\right) \tag{4.5.7}$$

Here we choose to define an effective diffusion coefficient as

$$D_{eff} = \varepsilon D_f \tag{4.5.8}$$

Note that the effective diffusion coefficient of the contaminant through the individual soil particle is less than that for diffusion through the void fluid itself by an amount that depends on the geometrical structure (the parameter $\varepsilon$) of the particle. The factor $R_K$ accounts for an apparent retardation of diffusion that arises from the absorption of the solute on the surface of the particle. It depends on the thermodynamics of equilibrium of the contaminant between the fluid and solid phases, $K_p$. By inspection of Eq. 4.5.6 we see that

$$R_K = \varepsilon + (1-\varepsilon)K_p\rho_s \tag{4.5.9}$$

Observations suggest that this model overestimates the effective diffusion coefficient because it fails to account for the occurrence of diffusion over a tortuous path that is considerably longer than the radius of the particle. An empirical relationship that improves upon Eq. 4.5.8 is given by introducing another factor of $\varepsilon$, with the result

$$\frac{D_{eff}}{R_K} = \frac{\varepsilon^2 D_f}{\varepsilon + (1-\varepsilon)K_p\rho_s} \tag{4.5.10}$$

Equation 4.5.10 gives only an order-of-magnitude estimate of the apparent diffusion coefficient. One could use Eq. 4.5.7 (or more specifically, a solution to that equation) as part of an experimental program for the measurement of an apparent diffusion coefficient for a specific soil/solute pair.

**EXAMPLE 4.5.1**   *Desorption of Tetrachlorobenzene from a River Sediment: Intraparticle Diffusion Controlling*

If a good estimate of the apparent diffusion coefficient of some contaminant is available, we can apply this model to the estimation of the rate of desorption of an organic pollutant from an aquatic environment. In this example we suppose that an inadvertent release of tetrachlorobenzene ($C_6H_2Cl_4$) has led to contamination of the sedimental soil on the bottom of a river. A study indicates that the soil particles are porous, and approximately spherical, with an average diameter of 35 $\mu$m. Additional studies have provided the following parameters:

$$\varepsilon = 0.13 \qquad K_p = 418 \text{ cm}^3/\text{g} \qquad \rho_s = 2 \text{ g/cm}^3$$

We need a diffusion coefficient for tetrachlorobenzene in water. For the sake of this example, we will use a value of (see Problem 2.24)

$$D_f = 9.6 \times 10^{-6} \text{ cm}^2/\text{s at } 20°\text{C}$$

This leads to an apparent diffusion coefficient (using Eq. 4.5.10) of

$$\frac{D_{\text{eff}}}{R_K} = \frac{(0.13)^2 \times (9.6 \times 10^{-6})}{0.13 + (0.87)(418)2} = 2.3 \times 10^{-10} \text{ cm}^2/\text{s}$$

Our goal now is to use the model above for an estimation of the time required for desorption of the contaminant from the sediment. Hence we seek a solution to Eq. 4.5.7.

To solve this diffusion equation, we must specify initial and boundary conditions that reflect the physical situation of interest to us. One possible scenario is the following. We may assume that the sediment phase is initially saturated with respect to the tetrachlorobenzene in the surrounding water and that the initial concentration $C$ satisfies

$$C(r, 0) = C_o = \text{constant} \tag{4.5.11}$$

Thus our initial condition is that of uniform concentration of tetrachlorobenzene within the intraparticle void volume of the individual particles that make up the sediment. The first boundary condition is simple: we assume spherical symmetry within each particle, which requires that

$$\frac{\partial C(r, t)}{\partial r} = 0 \qquad \text{at } r = 0 \tag{4.5.12}$$

For the second boundary condition we will assume that after some initial time, the fluid phase exterior to the individual sediment particles, that is, in the region $r > R$ for each sediment particle, is essentially free of the tetrachlorobenzene. The boundary condition corresponding to this assumption would take the form

$$C(R, t) = 0 \qquad \text{on } r = R \tag{4.5.13}$$

This latter condition might occur when the fluid outside the sediment particles is agitated sufficiently to remove the contaminant efficiently from each particle surface and dilute it into a large volume of surrounding water. There is still a release of the tetrachlorobenzene to the ambient water, with the potential for harmful biological consequences. All we are assuming here is that the rate of release is limited by the diffusion within the sediment particles and that the concentration of tetrachlorobenzene in the surrounding water is so low that the release occurs as if the surrounding water is at zero concentration.

The physical conditions described here would not correspond to a situation in which the sediment particles made up a thick stagnant layer of sediment. In such a case, it is likely that diffusion within the fluid phase surrounding the particles would also hinder the rate of transfer of the contaminant. Under some physical conditions, however, a small but effective flow of water may permeate the sediment bed, whereupon the diffusion within the particulate phase becomes the controlling factor in contaminant release, as assumed in this example.

We could achieve this set of initial and boundary conditions *in the laboratory* by the following experimental protocol. We could first saturate the sediment by agitating the particles in an aqueous solution of the contaminant. This would, after some equilibrium time, bring the sediment particles to the uniform concentration $C_o$. Then we could drain the interstitial water from the sediment mass and dump the partially wetted sediment into a large volume of agitated pure water. Although these conditions do not necessarily mimic what would actually occur in a natural environment, we could achieve them in a laboratory setting, where our goal would be to confirm the applicability of this simple diffusion model to the release of the contaminant from the sediment. Let's suppose that this is in fact the case, and our goal is to examine the correspondence of a set of desorption data to the diffusion model. Hence our task is to use the solution to Eq. 4.5.7, subject to the conditions given in Eqs. 4.5.11 to 4.5.13 to predict the rate of loss of contaminant from the sediment.

We may solve Eq. 4.5.7 by the method called "separation of variables." To begin, we introduce the substitution

$$u = rC \qquad (4.5.14)$$

Next we note that

$$\frac{\partial u}{\partial t} = r \frac{\partial C}{\partial t} \qquad (4.5.15)$$

and for the spatial derivatives

$$\frac{\partial u}{\partial r} = r \frac{\partial C}{\partial r} + C$$

$$\frac{\partial^2 u}{\partial r^2} = \frac{\partial C}{\partial r} + r \frac{\partial^2 C}{\partial r^2} + \frac{\partial C}{\partial r} = 2 \frac{\partial C}{\partial r} + r \frac{\partial^2 C}{\partial r^2} = \frac{1}{r} \frac{\partial}{\partial r} \left( r^2 \frac{\partial C}{\partial r} \right) \qquad (4.5.16)$$

It follows that if $C$ obeys Eq. 4.5.7, then $u$ is a solution to

$$\frac{\partial u}{\partial t} = \frac{D_{\text{eff}}}{R_K} \frac{\partial^2 u}{\partial r^2} \qquad (4.5.17)$$

The initial and boundary conditions transform to

$$u = RC_o \qquad \text{at } t = 0 \qquad (4.5.18)$$
$$u = 0 \qquad \text{at } r = 0 \qquad (4.5.19)$$

and

$$u = 0 \qquad \text{at } r = R \qquad (4.5.20)$$

Applying the method of separation of variables to this boundary value problem, we find the solution in the form

$$\frac{C}{C_o} = \frac{2R}{\pi r} \sum_{n=1}^{\infty} \frac{(-1)^{n+1}}{n} \sin \frac{n\pi r}{R} \exp\left( -\frac{(D_{\text{eff}}/R_K)\, n^2\pi^2 t}{R^2} \right) \qquad (4.5.21)$$

The amount of contaminant remaining in the spherical particle at any time $t$ is found from

$$\frac{M(t)}{M_o} = \int_0^R \frac{C}{C_o} 4\pi r^2\, dr = \frac{6}{\pi^2} \sum_{n=1}^{\infty} \frac{1}{n^2} \exp\left( -\frac{(D_{\text{eff}}/R_K)\, n^2\pi^2 t}{R^2} \right) \qquad (4.5.22)$$

Except for very short times, the first term of this infinite series solution is adequate for evaluation of the history of desorption. Thus, if we want to find the half-time for desorption, under the specified boundary conditions, we must solve for the time $t$ at which $M(t)/M_o = 0.5$, or (using just the dominant $n = 1$ term)

$$\frac{(D_{\text{eff}}/R_K)\pi^2 t}{R^2} = -\ln\left(\frac{\pi^2/6}{2}\right) = 0.1954 \qquad (4.5.23)$$

Hence, for the values of $D_{\text{eff}}$, $R_K$, and $R$ given in this example, we expect that the half-time is

$$t_{1/2} = \frac{0.1954 R^2}{\pi^2 (D_{\text{eff}}/R_K)} = \frac{0.1954(17.5 \times 10^{-4})^2}{\pi^2 \times 2.3 \times 10^{-10}} = 263.6 \text{ s} \qquad (4.5.24)$$

Example 4.5.1 deals with how we would use our diffusion model to interpret data, or estimate the rate of release of a contaminant, *if the release were truly described by this simple diffusion model.* A number of references in the literature on environmental contamination/decontamination[2] indicate that this model is applicable to some situations of interest, but not to others. Hence one must study the literature carefully. A firm knowledge of diffusion modeling is a valuable aid to reading this sometimes confusing material.

## 4.6   THE DYNAMICS OF TRANSPORT THROUGH LANDFILL BARRIERS

Current regulations on the construction of municipal landfills require the preparation of a barrier that separates the fill region from the groundwater supply. The barrier, usually a polymeric film with a low permeability to toxic solutes, is placed on top of a thick layer of clay that prevents the vertical passage of water that could carry additional contaminants into the underground water supply. Because the clay has a high resistance to flow (a low hydraulic permeability), the clay layer is usually designed (i.e., specified) to have a thickness that reduces the vertical flow velocity below some minimal level. However, this concept, rooted in *geohydraulics,* ignores the possibility of diffusion as a parallel and important path by which toxins can move in the vertical direction. As a consequence, to predict the rate at which contaminants move toward the subterranean groundwater supply we must understand the hydraulics of the water phase, as well as the diffusive transport of any solutes in the water phase and through the clay barrier. We examine some features of this problem here.

Figure 4.6.1 shows the regions through which transport of solutes from the landfill takes place, and the mechanisms of transport. In our analysis, we will take the landfill to be a well-mixed reservoir with a concentration $C_i$ of any of several toxic species. Analysis of the barrier properties of the polymeric film is along the lines of the discussion

---

[2] The following papers, all from the journal *Environmental Science and Technology*, are a good test of your ability to use your knowledge of thermodynamics and transport phenomena as an aid toward further learning:

Wu, S-C, and P. M. Gschwend, "Sorption Kinetics of Hydrophobic Organic Compounds to Natural Sediments and Soils" [**20**, 717 (1986)].

Formica, S. J., J. A. Baron, L. J. Thibodeaux, and K. T. Valsaraj, "PCB Transport into Lake Sediments. Conceptual Model and Laboratory Simulation" [**22**, 1435 (1988)].

Grathwohl, P. and M. Reinhard, "Desorption of Trichloroethylene in Aquifer Material: Rate Limitation at the Grain Scale" [**27**, 2360 (1993)].

Farrell, J., and M. Reinhard, "Desorption of Halogenated Organics from Model Soils, Sediments, and Soil Under Unsaturated Conditions" [**28**, 53 (1994)].

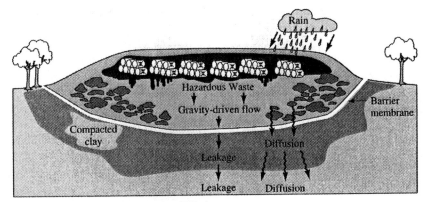

**Figure 4.6.1** Schematic of a landfill site.

in Section 3.5. It is necessary to know the solubility and diffusivity/permeability of various species in the polymeric film. The same is true of the clay barrier, with the added complication that the clay is a composite medium characterized by a porosity, a tortuousity, and a retardation coefficient (defined in Eq. 4.5.9) that depends on the ability of the clay to absorb various species.

Here we want to develop means by which to characterize some of these properties, using the knowledge that we acquired to this point. We begin with the characterization of the clay barrier as a diffusive medium. The most direct approach is to take a sample of the clay medium and set up a transient diffusion experiment in which the transport is controlled and measured. As an alternative, if there is access to the landfill region, or to a region that is geologically similar to the landfill region, a field study may be possible. Such studies have been carried out, and an example is provided.

Data obtained in a field study of a Canadian landfill site were presented and analyzed by Johnson et al. [*Environ. Sci. Technol.*, **23**, 340 (1989)], who removed core samples from the natural clay underlying the landfill. The clay layer was estimated to be about 25 to 40 m thick. Samples were taken from the bottom of the landfill reservoir and down into the first meter or so of the clay. Hence the data are all from the region that is relatively near the reservoir/clay interface.

Figure 4.6.2 plots data on chloride ion concentration as a function of distance below the reservoir. The time after "start-up" of the landfill was 5 years (1825 days). Since the data are confined to a region near the clay/landfill interface, we will model this system as diffusion into a semi-infinite planar region with an initial concentration $C_o = 0$ for $Cl^-$, and a constant concentration $C_s$ on the landfill/clay boundary. The solution to this problem is given in Example A1.3 Appendix A, Eq. A1.8.

$$Y = \text{erf}\left(\frac{y}{2\sqrt{D_{AB}t}}\right) \tag{4.6.1}$$

where "erf" is the error function introduced earlier and plotted in Fig. 4.2.6 and $Y$ is defined as in Appendix A, Eq. A1.2. In terms of the measured concentrations we may write this as

$$\frac{C(x,t)}{C_s} = \text{erfc}\left(\frac{y}{2\sqrt{D_{AB}t}}\right) \tag{4.6.2}$$

in terms of the complementary error function. Figure 4.6.2 shows a fit of the data, using Eq. 4.6.2 with an effective diffusivity value of $D_{\text{eff}} = 5.2 \times 10^{-6}\ \text{cm}^2/\text{s}$. Since the diffusivity

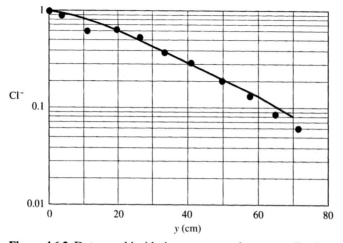

**Figure 4.6.2** Data on chloride ion concentration, normalized to the value at the landfill/clay interface. Data are fit with Eq. 4.6.2, using $D_{AB} = 5.2 \times 10^{-6}$ cm$^2$/s.

of chloride ion in water (at the site temperature of 10°C) is about $1.5 \times 10^{-5}$ cm$^2$/s, which is quite a bit larger than the effective diffusivity, the clay obviously retards the diffusion relative to that through free water. Since chloride ion is not believed to be absorbed by the organic or mineral content of the clay, the retardation must be associated with the porosity and tortuousity of the clay medium. Hence the geometry of the clay layer gives rise to a retardation factor of about 3.

With this estimate of the effective diffusion coefficient of chloride ion through the clay layer, we may estimate the time required for the chloride to penetrate or "break through" the clay barrier region and begin to contaminate the aquifer below. If we select a critical concentration level of $C(t)/C_s = 0.001$, and take the clay layer thickness to be 25 m, we find, from Eq. 4.6.2, that this level of concentration will be reached in a time that satisfies

$$\text{erfc}\left(\frac{25 \times 100}{2\sqrt{5.2 \times 10^{-6}\,t}}\right) = 0.001 = \text{erfc}(2.3) \tag{4.6.3}$$

or $t = 5.7 \times 10^{10}$ s = 1800 years. We see that a 25 m clay barrier provides a long delay before Cl$^-$ would begin to appear in the aquifer. Note that this time is directly proportional to the *square* of the liner thickness.

Effective diffusion coefficients for soil are very difficult to estimate from first principles. We need to know not only the detailed "architecture" of the soil, but also the kinetics and thermodynamics of adsorption of various organic species into the particulate matter that makes up the soil. If possible, field studies should be carried out to provide a more reliable means of estimation of the transport characteristics of the soil than a theoretical analysis can give.

Unfortunately, data from field studies are compromised to the extent that sampling is normally done from a very small fraction of the site, and inhomogeneities in the landfill and the underlying geological features give rise to scatter in the concentration measurements. With only a small amount of data, this statistical scatter makes it difficult to obtain reliable physical property values by curve-fitting the data. For example, Fig. 4.6.3 presents data from the same study that yielded Fig. 4.6.2. No attempt was made

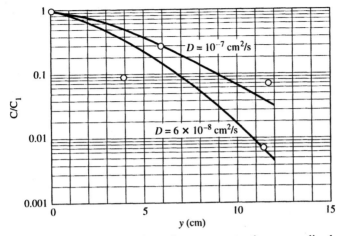

**Figure 4.6.3** Data on trichloroethene concentration, normalized to the value at the landfill/clay interface.

to fit these data statistically, and it is clearly impossible to select a single diffusion coefficient that fits all the data with only a small deviation. The two diffusion coefficients shown were selected just to illustrate the difficulty in picking a "best value" from these data.

**EXAMPLE 4.6.1**  *Diffusion Across a Clay Liner*

We want to characterize some features of the diffusion of a contaminant (say, benzene) across a clay liner under a landfill. We imagine that the landfill is well mixed, and at an average concentration of the contaminant $C_1 = 1$ g/L. We assume that the partition coefficient of the species between the landfill medium and the void space of the clay liner medium is 1.0. We take the average temperature within the liner to be 10°C over an annual basis. The liner is taken to have a uniform thickness of 10 m. We begin with the following questions:

**a.** How many years must elapse before the concentration of benzene at the bottom of the liner reaches 1% of the concentration at the landfill/liner interface? We can regard this as the time at which the contaminant breaks through the barrier.

**b.** How many years will pass between the opening of the landfill site and the achievement of the steady state?

To begin, we need the diffusion coefficient of benzene in water at 10°C. Using the Wilke–Chang equation (see Chapter 2: Section 2.2.2), we find $D_{AB} = 7.1 \times 10^{-6}$ cm²/s. This is the unimpeded diffusion coefficient through water. We will use Eq. 4.5.10 for the *apparent* diffusion coefficient through the clay medium, accounting for two effects that slow diffusion. One is the tortuosity that arises because the diffusant can only move along a path that circumvents the impermeable solid material within the clay medium. The other effect is due to the possibility that the diffusant species may be soluble in some fraction (usually organic) of the clay medium. When this is the case, there is a kinetic process associated with the adsorption of the diffusant, and the movement of the diffusant species appears to be retarded because some fraction of the species is removed from the pathway available for diffusion. Returning to Eq. 4.5.10, we find

$$\frac{D_{\text{eff}}}{R_K} = \frac{\varepsilon^2 D_{AB}}{\varepsilon + (1 - \varepsilon) K_p \rho_s} \tag{4.6.4}$$

Suppose that in the case at hand we have previously determined values of $\varepsilon = 0.4$ and $K_p \rho_s = 50$. Then we find an apparent diffusion coefficient

$$D_{\text{app}} = \frac{D_{\text{eff}}}{R_K} = 3.7 \times 10^{-8} \text{ cm}^2\text{/s} \tag{4.6.5}$$

Now we have to select a model for diffusion through the clay liner. For part (a) we require the time at which the concentration at the lower surface of the liner rises to 1% of the value on the upper surface. This is a "short-time" diffusion problem, in the sense that over the time in question the diffusing species has barely penetrated to the other side of the barrier. Then it will be easier to use Eq. 4.6.2 than the infinite series solution, and hence we want the time at which

$$\text{erfc}\left(\frac{10 \times 100}{2\sqrt{3.7 \times 10^{-8} t}}\right) = 0.01 = \text{erfc}(1.8) \tag{4.6.6}$$

The solution is

$$t = 2.07 \times 10^{12} \text{ s} = 65{,}000 \text{ years!} \tag{4.6.7}$$

We see that there will be no appreciable breakthrough of the benzene across the bottom of the liner, except over geologically long times. The sensitivity of this conclusion is not strongly dependent on the criterion for breakthrough. If, for example, we ask for the time at which the concentration of benzene will be $10^{-6}$ of the landfill value, the time is reduced only by a factor of about 4, which is still in the range of geological times. If the apparent diffusion coefficient is really larger by an order of magnitude, the time calculated would be reduced only by a factor of about 3 ($\sqrt{10} = 3.16$).

The time at which the steady state is achieved is clearly much longer than 65,000 years, and we will not even bother with its calculation here. An example with time scales of more practical interest is given in Problem 4.44.

**EXAMPLE 4.6.2** *Diffusion Across a Clay Liner: Near Steady State*

Let's change the design of the clay liner so that the steady state is achieved over times of more realistic concern to us. Suppose the liner is only a meter thick, all the other conditions of Example 4.6.1 being unchanged. A 10-fold decrease in liner thickness will decrease the time for breakthrough, defined here as 1% of the landfill concentration, to 650 years, since the time to achieve a specified concentration scales with the *square* of the liner thickness. Now we return to the question:

> How many years will pass between the opening of the landfill site
> and the achievement of the steady state?

Once this has been established we will plot the flux of benzene as a function of time (in years) over the period from 10% to 100% of the steady state time.

To find the steady state time, we have to go to the solution for transient diffusion across a slab of *finite* thickness. Hence we seek a solution to the unsteady diffusion equation

$$\frac{\partial C}{\partial t} = D \frac{\partial^2 C}{\partial y^2} \tag{4.6.8}$$

where $C$ is the mass concentration of benzene in the *barrier*. We drop the subscript on $D$ at this point, but we understand that it is the *apparent* diffusion coefficient through the clay barrier. Initial and boundary conditions are assumed, and they take the forms

$$C = 0 \quad \text{at} \quad t \le 0 \quad 0 \le y \le B \tag{4.6.9}$$

$$C = C_s \quad \text{at} \quad y = 0 \quad t > 0 \tag{4.6.10}$$

We need a boundary condition at the lower end ($y = B$) of the clay liner. We will assume that this lower end is the interface with an aquifer through which the flow of water is sufficient to reduce the concentration at $y = B$ to $C = 0$. Then

$$C = 0 \quad \text{at} \quad y = B \quad t > 0 \tag{4.6.11}$$

is the second boundary condition. (See Problem P4.46.)

Note that there is no symmetry about any plane within the membrane in this problem. Hence we cannot use the Gurney–Lurie chart for a slab geometry because of the form of the boundary conditions. We must first obtain the solution to this diffusion problem.

We have already solved an identical problem, but in a different physical context. It was the problem given in Example 4.2.1. All we need to do is rename the concentration $Hp_0$ of that problem to $C_s$ in this one. The mathematical formulations are then identical, and so we may use the solution (Eq. 4.2.38) for this example. We write the solution as

$$\frac{C}{C_s} = \left(1 - \frac{y}{B}\right) - \frac{2}{\pi} \sum_{n=1}^{\infty} \frac{\sin n\pi(y/B)}{n} \exp(-n^2\pi^2 X_B) \tag{4.6.12}$$

where

$$X_B = \frac{Dt}{B^2} \tag{4.6.13}$$

The steady state solution is reached when all the terms of the infinite series become vanishingly small compared to the linear term outside of the summation. (The linear term is the steady state profile.) An arbitrary but suitable definition of "small" leads us to

$$X_B^s \approx 0.4 \tag{4.6.14}$$

or

$$t_{ss} = \frac{0.4B^2}{D} \tag{4.6.15}$$

For the data of this example we find

$$t_{ss} = \frac{0.4 \times 100^2}{3.7 \times 10^{-8}} = 1.1 \times 10^{11} \, \text{s} = 3430 \, \text{years} \tag{4.6.16}$$

This is still a very long time, but we need to see how large the benzene flux is at times prior to the steady state. Hence we need to derive the flux expression from Eq. 4.6.12 for the concentration profile. But, again, this is already available to us. Look at Eq. 4.2.43. This is the dimensionless flux for this problem. It is not difficult to see that in the notation of this example we have

$$\frac{-D\frac{\partial C}{\partial y}\Big|_{y=B}}{\dfrac{DC_s}{B}} = 1 + 2\sum_{n=1}^{\infty}(-1)^n \exp(-n^2\pi^2 X_B) \qquad \text{(4.6.17)}$$

The steady state flux is just

$$\frac{DC_s}{B} = \frac{3.7\times10^{-8}\,\text{cm}^2/\text{s}\times 1\,\text{g/L}\times10^{-3}\,\text{L/cm}^3}{100\,\text{cm}} = 3.7\times10^{-13}\,\text{g/cm}^2\cdot\text{s} \qquad \text{(4.6.18)}$$

Now let's find the flux at 10% (and then 30% and 50%) of the time to achieve steady state. At 10% of 3430 years we find

$$\frac{-D\frac{\partial C}{\partial y}\Big|_{y=B}}{\dfrac{DC_s}{B}} = 1 + 2\sum_{n=1}^{\infty}(-1)^n \exp(-0.04n^2\pi^2) \approx 0.01 \qquad \text{(4.6.19)}$$

In the same way we find, at subsequent times, that the benzene flux is 0.4 of the steady state flux at 1030 years, and 0.72 of the steady state flux at 1715 years. While it seems unlikely that a landfill would operate without change over a millennium, we will examine the impact of the calculated behavior after 1000 years of transport of benzene into an aquifer underlying the landfill.

First, let's see what this flux at about 1000 years implies. At 1030 years the flux is 40% of the steady value, or

$$\text{flux} = 0.4(3.7\times10^{-13})\,\text{g/cm}^2\cdot\text{s} = 1.5\times10^{-13}\,\text{g/cm}^2\cdot\text{s} \qquad \text{(4.6.20)}$$

We will suppose that the landfill has a surface area of $10^4$ m². Then the rate of transport, at this point in time, would be

$$\begin{aligned}
\text{rate} &= 1.5\times10^{-13}\,\text{g/cm}^2\cdot\text{s}\times10^8\,\text{cm}^2\times10^{-3}\,\text{kg/g}\\
&= 1.5\times10^{-8}\,\text{kg/s} = 0.47\,\text{kg/yr}
\end{aligned} \qquad \text{(4.6.21)}$$

It is a simple matter to plot the rate of leakage of benzene from the landfill as a function of time, with the results shown in Fig. 4.6.4.

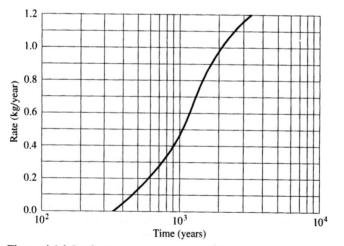

**Figure 4.6.4** Leakage rate as a function of time.

Now that we have a set of results with respect to diffusion across a clay liner at near steady state, we may address the primary issue—What does this flux imply for contamination of the environment surrounding the landfill? To answer this question, we must have some information regarding the pathway by which the contaminant puts the surroundings at risk. One such pathway is considered in the next example.

**EXAMPLE 4.6.3**    *Contamination of an Aquifer by a Landfill*

Suppose the landfill of Example 4.6.2 lies above an aquifer. What information do we need to assess the impact of leakage of contaminants from the landfill on the aquifer? At the least, we must have some information regarding the flow of water through the aquifer—in terms of the flowrate of the water, as well as with regard to the geometrical interface of the aquifer with the landfill. Figure 4.6.5 represents the situation that might occur under some circumstances. For the sake of this example, some typical dimensions and characteristics of an aquifer/landfill system have been selected.

We imagine that once the toxic contaminant has crossed the lower boundary of the landfill barrier, it passes into the aquifer with no significant resistance. We also assume that by the time the aquifer flow has worked its way farther downstream—away from the landfill—the contaminant is well mixed within the aquifer flow. This is a reasonable assumption, since flow through a porous medium creates efficient mixing over a sufficiently long length of flow. The contamination model is very simple. It states that the contaminant transferred into the aquifer remains in the aquifer flow:

$$\begin{aligned} \text{landfill flux} \times \text{area of interface with the aquifer} = \\ \text{contaminant concentration in the aquifer} \times \text{flowrate of the aquifer} \end{aligned} \quad \textbf{(4.6.22)}$$

The flowrate through the aquifer is found from the flux stated on Fig. 4.6.5 as

$$Q_{aq} = q_{aq}A_{c,aq} = 0.2 \text{ cm}^3/\text{s} \cdot \text{cm}^2 \times 20 \text{ m}^2 \times 10^4 \text{ cm}^2/\text{m}^2 \times 10^{-3} \text{ L/cm}^3 = 40 \text{ L/s} \quad \textbf{(4.6.23)}$$

For the landfill flux we will use the value found at 1000 years (Eq. 4.6.20). The interfacial area between the landfill and the aquifer is 2000 m². Then Eq. 4.6.22 yields

$$C_{aq} = \frac{(1.5 \times 10^{-13} \text{ g/cm}^2 \cdot \text{s})(2000 \text{ m}^2 \times 10^4 \text{ cm}^2/\text{m}^2)}{40 \text{ L/s}} = 7.5 \times 10^{-8} \text{ g/L} \quad \textbf{(4.6.24)}$$

This concentration is two orders of magnitude below the EPA limit on benzene in water.

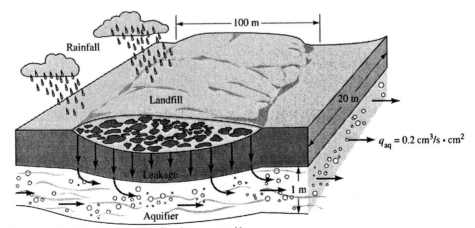

**Figure 4.6.5** A landfill contaminating an aquifer.

## 4.7 TRANSIENT DIFFUSION WITH A REACTION ON THE BOUNDARIES

Let's suppose that we have a gas confined between large parallel planes. The gas is initially binary—some species A that is diluted in species B. The initial concentration of A is $C_A^o$. The planar surfaces at $y = \pm B$ are coated with a catalyst that promotes a first-order reaction

$$A \to P \tag{4.7.1}$$

with a rate

$$R_A = k_s C_A \tag{4.7.2}$$

$R_A$ represents the rate of a surface reaction, and it has the units of moles of A reacted, per second, per centimeter squared of surface area. At the time $t > 0$ the catalytic reaction is "switched on," and A disappears through this reaction. Figure 4.7.1 shows the geometry, and the transient diffusion equation is

$$\frac{\partial C_A}{\partial t} = D_{AB} \frac{\partial^2 C_A}{\partial y^2} \quad \text{on} \quad -B < y < +B \tag{4.7.3}$$

with initial condition

$$C_A = C_A^o \quad \text{at} \quad t < 0 \tag{4.7.4}$$

and boundary conditions

$$-D_{AB} \frac{\partial C_A}{\partial y} = k_s C_A \quad \text{on} \quad y = B \tag{4.7.5}$$

and

$$\frac{\partial C_A}{\partial y} = 0 \quad \text{on} \quad y = 0 \tag{4.7.6}$$

The way to proceed will be clearer if we nondimensionalize this boundary value problem. Define

$$Y = \frac{C_A}{C_A^o} \qquad \eta = \frac{y}{B} \qquad X_D = \frac{D_{AB} t}{B^2} \tag{4.7.7}$$

and we see that the equations transform to

$$\frac{\partial Y}{\partial X_D} = \frac{\partial^2 Y}{\partial \eta^2} \tag{4.7.8}$$

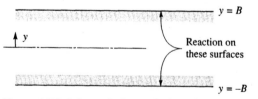
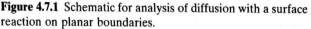

**Figure 4.7.1** Schematic for analysis of diffusion with a surface reaction on planar boundaries.

$$Y = 1 \quad \text{at} \quad X_D < 0 \tag{4.7.9}$$

$$-\frac{\partial Y}{\partial \eta} = GY \quad \text{on} \quad \eta = 1 \tag{4.7.10}$$

$$\frac{\partial Y}{\partial X_D} = 0 \quad \text{on} \quad \eta = 0 \tag{4.7.11}$$

where

$$G = \frac{k_s B}{D_{AB}} \tag{4.7.12}$$

Although the physics is different, this is exactly the *mathematical* problem we obtain for diffusion within a slab with finite convection on the planar boundaries. Instead of the group $G$, we found a dimensionless group Bi (the Biot number, see Eq. 4.2.16). Thus we should be able to use the Gurney–Lurie charts (in particular, Fig. 4.2.2a for the planar geometry) with Bi replaced by $G$. This happens because in each case, the flux of A to the surface at $y = B$ is directly proportional to the concentration of A at that surface. The convective flux and the reaction rate per unit of surface area are both given by linear models. If the kinetics were other than first-order, this would not be so.

The dimensionless group $G$ tells us something about the relative roles of reaction and diffusion in controlling the rate of reduction of the concentration of species A. If $G$ is large enough, reaction is so fast that the reduction of A is limited by the diffusion of A. The system is said to be "diffusion-limited." If $G$ is very small the rate-limiting step is reaction, and diffusion is rapid across the region between the planes. In fact, for small $G$, we would say that there is no internal resistance (in the spirit of the discussion in Section 4.4) and we could write a very simple model for the reduction in A.

We begin with the assumption that for rapid diffusion the concentration $C_A$ is uniform across $y$, and write

$$-\frac{d}{dt}(VC_A) = k_s C_A A \tag{4.7.12}$$

Hence we may use Eq. 4.4.11, but we replace $\alpha k_c$ by $k_s$ and write

$$\overline{Y} = \frac{C_A}{C_A^o} = \exp(-GX_D) \tag{4.7.13}$$

Now let's examine the half-time for disappearance of A, in the limits of large and small $G$. For small $G$, Eq. 4.7.13 holds, and

$$\overline{Y} = 0.5 = \exp(-GX_{D,1/2}) \tag{4.7.14}$$

leads to

$$X_{D,1/2} = \frac{0.69}{G} \quad (G \ll 1) \tag{4.7.15}$$

For very large $G$ we simply examine the Bi $= \infty$ line on Fig. 4.2.2a, and we find

$$X_{D,1/2} = 0.2 \quad (G \gg 1) \tag{4.7.16}$$

These two "asymptotic" results are shown plotted in Fig. 4.7.2, along with additional points for several intermediate values of $G$. From this figure we can infer that $G <$

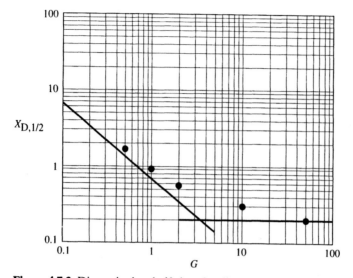

**Figure 4.7.2** Dimensionless half-time for disappearance of A.

0.5 corresponds to reaction-limited behavior, while $G > 50$ corresponds to diffusion-limited behavior.

## SUMMARY

Transient diffusion problems are examined in this chapter, leading to partial differential equations. Of special importance is the definition of a dimensionless time variable (Eq. 4.2.11) which, in a sense, scales the transient behavior to a characteristic diffusion time $L^2/D_{AB}$. When convective boundary conditions apply, we discover the importance of another dimensionless group Bi (Eq. 4.2.16), which expresses the relative roles of diffusion to a surface and convection from the surface in controlling the rate of transfer of some species across an interface between two phases.

For one-dimensional transient diffusion in simple bodies (thin planar slabs, long solid cylinders, spheres) with spatially uniform initial conditions and spatially uniform surface conditions independent of time, the basic partial differential equation for diffusion has been solved and features of the solution are available to us—for example, through the Gurney–Lurie charts (Fig. 4.2.2). In more complex situations, which do not have these features necessary for use of the Gurney–Lurie charts, mathematical techniques are available for obtaining the desired solutions. Appendix A presents some examples of useful solutions.

An important problem-solving (modeling) strategy involves the estimation of the relative roles of internal (diffusive) and external (convective) resistances to mass transfer. We discover that the relevant dimensionless parameter is the Biot modulus (Eq. 4.3.1), which appeared in the earlier nondimensionalization of the diffusion equation and its boundary conditions. Criteria based on the value of Bi tell us when (if at all) we can neglect one resistance relative to the other. A closely related dimensionless group is the Sherwood number (Eq. 4.4.8) which, we will find in subsequent chapters, appears in empirical correlations of convective mass transfer coefficients. It is important to understand the distinction between Bi and Sh. The key lies in the manner in which the convective mass transfer boundary condition is written (see the discussion in Section 4.3).

## PROBLEMS

**4.1** Equation 4.1.13 gives the *quasi-steady* solution for the *rate* of mass transfer from a hollow fiber system. Write the *transient* equations that describe the physics of this system: that is, the diffusion equation(s) and the initial and boundary conditions that define the full unsteady-state behavior. Then identify the dimensionless parameters that define the conditions under which the quasi-steady model is a good approximation. Give a criterion for validity of the quasi-steady model.

Assume that the fiber wall thickness is much less than the core radius. Assume that the interior of the fiber is well mixed and the transient behavior is in the membrane, which is initially in equilibrium with the core. Calculate the rate of mass loss.

**4.2** Confirm the statement following Eq. 4.2.52 by deriving the quasi-steady solution to Example 4.2.1.

**4.3** With reference to Eq. 4.2.46, define a dimensionless pressure and plot it as a function of the dimensionless time $X$. Define a time $t^*$ long enough that the linear portion of the $p(t)$ curve is a good approximation to the behavior, and express $t^*$ in terms of appropriate parameters of the problem.

**4.4** Relax the assumption (Eq. 4.2.23) that the upper compartment in Fig. 4.2.3 is always free of ethylene, and assume instead that the upper compartment is well mixed.

**a.** Write the equations that define the solution to this problem.

**b.** Find the solution to this problem, plot $p(t)$ as a dimensionless pressure against the dimensionless time $X$, and compare this solution to the quasi-steady solution. Define a time $t^{**}$ such that the assumption that the upper compartment is free of ethylene is no longer valid.

**4.5** We want to determine the diffusivity of oxygen in a new polycarbonate membrane. Studies with similar, but not identical, polycarbonate formulations have indicated that the diffusivity is in the range of 4 to $6 \times 10^{-8}$ cm$^2$/s and that the Henry's law constant $H$ (see Eq. 4.2.22) is of the order of $10^{-5}$ mol/cm$^3 \cdot$ atm. If you want to use the linear portion of Fig. 4.2.4 to determine the actual $D$ and $H$ for this particular polymeric film, what volume of test chamber will you use, and when will the linear $p(t)$ data occur? Assume the polycarbonate film is available with a thickness of 10 mils (0.01 inch).

**4.6** Beginning with Eq. 4.2.55, plot the pressure–time history as a dimensionless pressure (give your definition of this pressure) versus $X$, and compare this history to that for the case of an initially "empty" film—Eq. 4.2.46.

**4.7** RTV silicone rubber is a common gasketing material used in systems subjected to high vacuum conditions—reactors that operate at very low pressure, space shuttle systems, and so on. The performance of these systems is strongly dependent on the degree to which trace amounts of volatile materials are released from the solid rubber in which they are dissolved. This process of release is known as *outgassing*.

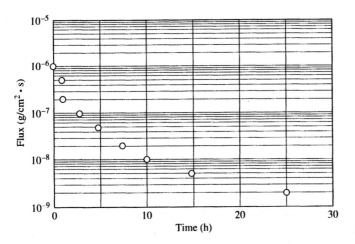

**Figure P4.7**

To predict the performance of such systems, it is necessary to have estimates of the diffusion coefficients of various gases within the rubber. The data can be obtained by analyzing the rate of outgassing from a rubber disk of known geometry. The solution to the transient diffusion equation is used for interpretation of the data.

Some outgassing data of Liu and Glassford are shown in Fig. P4.7 [*J. Vac. Sci. Technol.*, **15**, 1761 (1978)]. The authors interpret the results as corresponding to the release of *two* species (which they call $M_2$ and $M_3$) that diffuse independently, with diffusion coefficients $D_2$ and $D_3$. Species $M_3$ is believed to be very dilute relative to $M_2$ and to have a very low diffusion coefficient with respect to $M_2$.

From the data given, calculate the two diffusion coefficients, as well as the initial concentrations of each species within the rubber sample. The sample thickness is $L = 0.5$ cm, and the disk diameter is 2.25 cm. The sample temperature was 398 K. The outgassing was from only *one* face of the rubber disk.

**4.8** A solid slab of thickness $2B$, having a uniform initial concentration $C_0^s$ of a diffusible species, is suddenly exposed to a well-stirred ambient medium of concentration $C^{f\infty}$. A solubility coefficient $\beta$ relates the solid and fluid concentrations at equilibrium, in the form $\beta = C^s/C^f$. The analytical solution for the concentration profile in the slab takes the following form:

$$\frac{C^s - \beta C^{f\infty}}{C_0^s - \beta C^{f\infty}}$$

$$= 4 \sum_{n=0}^{\infty} (-1)^n \frac{\exp[-(2n+1)^2\pi^2Dt/4B^2]}{(2n+1)\pi} \quad \textbf{(P4.8.1)}$$

$$\cos\left[\left(n + \frac{1}{2}\right)\frac{\pi y}{B}\right]$$

Using this result, write an expression for the average concentration in the slab at any time. Put your result into a dimensionless format. Do the case $C^{f\infty} = 0$.

Present an expression for the time required to reduce the average concentration to a level such that 90% of the potential loss of diffusible species has occurred. Do the same for 99% and 99.9%. Give a general result for the case in which the amount of diffusible species remaining is $10^{-n}$ of the initial amount.

**4.9** A thin film of polymer of thickness $L$ contains an amount $M_\infty$ of a dissolved drug that is uniformly distributed within the film initially. If diffusion within the film controls the rate of release of the drug to the surrounding environment, then the fractional amount released (not remaining) as a function of time may be written as

$$\frac{M}{M_\infty} = 1 - \sum_{n=0}^{\infty} \frac{8 \exp[-D(2n+1)^2\pi^2 t/L^2]}{(2n+1)^2\pi^2}$$
$$\textbf{(P4.9.1)}$$

where $M$ is the amount released at time $t$.

Derive this result, starting with Eq. P4.8.1 in Problem 4.8. Assume $C^{f\infty} = 0$. Show that a good approximation for the half-time (i.e., the time at which half the drug has been released) is

$$t_{1/2} = \frac{0.049\, L^2}{D} \qquad \textbf{(P4.9.2)}$$

Find the rate of release ($dM/dt$) at the half-time. This is often used as a nominal measure of the performance of a delivery system.

**4.10** A sphere of polymer of radius $R$ contains a dissolved drug that is uniformly distributed within the sphere initially. If diffusion within the polymer controls the rate of release of the drug to the surrounding environment, then the fractional amount released as a function of time may be written as

$$\frac{M}{M_\infty} = 1 - \frac{6}{\pi^2} \exp\left(\frac{-\pi^2 Dt}{R^2}\right) \text{ for } 0.5 \le M/M_\infty \le 1$$
$$\textbf{(P4.10.1)}$$

Find an expression for the half-time (i.e., the time at which half the drug has been released) in the form

$$\frac{Dt_{1/2}}{R^2} = \text{constant} \qquad \textbf{(P4.10.2)}$$

Find an expression for the rate of release ($dM/dt$) at the half-time. This is often used as a nominal measure of the performance of a delivery system.

**4.11** Consider the sandwich-type drug delivery system, as shown in Fig. P4.11. Suppose the reservoir (considered a large slab of thickness $2B$) is supersaturated, causing the device to act as a constant-activity delivery system. Assume that after the drug has equilibrated within the surrounding membranes (each of thickness $L$), the

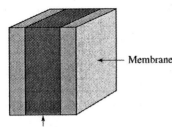

Membrane

Drug reservoir

**Figure P4.11** A sandwich-type drug delivery system.

where

$$Y = \frac{C^s - C^{f\infty}/\alpha}{C^s_o - C^{f\infty}/\alpha} \qquad X_D = \frac{D_{AB}t}{b^2}$$

$$\eta = \frac{y}{b} \qquad \alpha = \frac{C^{f\infty}}{C^s_{eq}}$$

and the eigenvalues $\lambda_n$ are the roots to

$$\lambda_n \tan \lambda_n = Bi \qquad \text{(P4.14.2)}$$

with Bi defined as $Bi = \alpha k_c b/D_{AB}$. The first two roots $\lambda_1$ and $\lambda_2$ are plotted as functions of Bi in Appendix A (Fig. A1.3).

device is implanted in an environment that efficiently removes the escaped drug from the surface of the device. Find the half-time for this device, assuming that there is no significant initial transient period due to establishment of the drug concentration profile in the membrane. Then find the time required to establish the steady delivery regime (i.e., the time for the transient period associated with the membrane transfer to disappear) and compare the transient time to the half-time. Draw some conclusions regarding proper design of the system.

Find the half-time for the case $B = 0.1$ cm, $L = 0.01$ cm, $D_{membrane} = 3 \times 10^{-8}$ cm$^2$/s, and

$$\alpha = \frac{C_{membrane}}{C_{reservoir}} = 0.3 \qquad \text{(P4.11)}$$

when concentrations are in mass per volume units.

**4.12** Use the data in Fig. 4.2.12 (Example 4.2.5) and calculate the diffusivity of phosphorus in silicon.

**4.13** Derive Eq. 4.2.159. Begin with Eq. 4.2.151.

**4.14** The following analytical solutions are available for transient diffusion from a slab or cylinder, initially at a uniform concentration, with symmetric transfer to (or from) a medium with a finite convective resistance (i.e., a finite Biot number).

### For a Slab of Half-Thickness b:

$$Y = \sum_{n=1}^{\infty} \frac{4 \sin \lambda_n}{2\lambda_n + \sin 2\lambda_n} \cos(\lambda_n \eta) \exp(-\lambda_n^2 X_D)$$

$$\text{(P4.14.1)}$$

### For a Cylinder of Radius R:

$$Y = 2Bi \sum_{n=1}^{\infty} \frac{J_0(\lambda_n s)}{(\lambda_n^2 + Bi^2)J_0(\lambda_n)} \exp(-\lambda_n^2 X_D)$$

$$\text{(P4.14.3)}$$

where $Y$ and $\alpha$ are defined as in the case of the slab, above, and

$$X_D = \frac{D_{AB}t}{R^2} \qquad \text{and} \qquad s = \frac{r}{R}$$

and the eigenvalues $\lambda_n$ are the roots to

$$\lambda_n J_1(\lambda_n) - Bi J_0(\lambda_n) = 0 \quad \text{(P4.14.4)}$$

with Bi defined as $Bi = \alpha k_c R/D_{AB}$. The first two roots $\lambda_1$ and $\lambda_2$ are plotted as functions of Bi in Appendix A (Fig. A2.1).

Prepare plots analogous to Fig. 4.4.1 and establish criteria analogous to that of Eq. 4.4.12. It is useful to know that

$$\int sJ_0(\lambda s)\, ds = \frac{s}{\lambda} J_1(\lambda s) \qquad \text{(P4.14.5)}$$

**4.15** Present the detailed solution to Eq. 4.4.25 with Eq. 4.4.29 for $C_b$.

**4.16** Plot curves similar to those of Fig. 4.4.5, for $\delta^*$ values of $10^{-3}$ and 0.1. Instead of using $T^*$ for the y coordinate, plot $T^*/\delta^*$ versus $C^*$. Why is this a useful method of plotting?

**4.17** The data shown in Fig. P4.17 were obtained for dissolution of a powder (human tooth enamel) with an initial radius of 90 $\mu$m. Fit the model of Example 4.4.2 to the data and determine a value of $\delta$. Use the following data [see Thomann *et al.*, *J. Colloid Interface Sci.*, **132**,

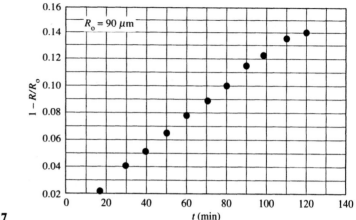

**Figure P4.17**

403 (1989)]:

> 10 mg powder in 50 mL of solvent
> $M_w = 502$    $\rho_s = 2.9$ g/cm$^3$
> $(C_s - C_b) = 1.9 \times 10^{-6}$ mol/cm$^3$
> $D = 4.4 \times 10^{-7}$ cm$^2$/s

**4.18** Using the model of dissolution developed in Example 4.4.2, find the time $T'$ for 90% of the solid particle to dissolve. Do the constant $C_b$ case. What is the particle radius at this point in time? Use the parameters given in Prob. 4.17.

**4.19** Go back to Eq. 4.4.25, and assume that $\delta/R$ is very small compared to unity for all time. What is the resulting $t^*(\delta^*)$ relationship? Compare the results to those of Fig. 4.4.5. What do you conclude from your observations?

**4.20** Nondimensionalize the mathematical model (see Eqs. 4.4.41–4.4.46) for the transfer of a soluble gas into a sphere, from a surrounding concentric sphere of finite volume, and show that

the dimensionless parameter $B$ appears naturally. What is the physical interpretation of $B$?

**4.21** Plot the smallest nonzero eigenvalue $b_1$ defined by Eq. 4.4.50 as a function of the parameter $B$. By analogy to Fig. 4.4.7 for $\tau_{0.1}$ versus $B$, plot $\tau_{0.01}$ versus $B$.

**4.22** Return to the problem defined by Fig. 4.4.6. Suppose that instead of efficient convection (Bi = ∞) from the surrounding gas volume to the interface of the droplet, the physics are such that *diffusion* in the surrounding gas is the only mode of mass transfer to the interface. What form do the boundary conditions at the interface take in this case? Write the full set of equations and boundary conditions that define this model, and nondimensionalize the formulation. Do not attempt to solve the equations.

**4.23** A pressurized cylinder of arsine gas contains 10 L at a pressure of 200 psia. What is the mass of arsine in the cylinder? If all that arsine

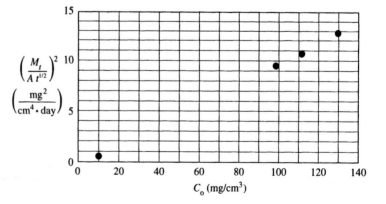

**Figure P4.25** Steroid release from a silicone rubber matrix. Chien et al., *J. Pharm. Sci.*, **63**, 365 (1974).

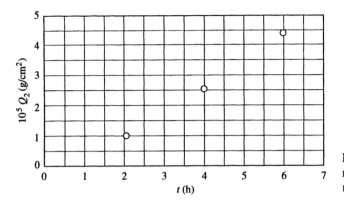

**Figure P4.26** Cumulative dye mass (per membrane area) on the dilute side of a membrane test cell.

were suddenly released into a 24 m$^3$ enclosed space, what would the concentration of arsine be, in ppmv?

**4.24** Resolve the issue of the apparently counterintuitive finding mentioned just after Eq. 4.4.55.

**4.25** A sustained-release device was created by dissolving a steroid (ethynodiol diacetate) in a silicone elastomer matrix. The solubility of the steroid in the rubber (at 37°C) was $C_s = 1.48$ mg/cm$^3$. The cumulative amount of drug released (per exposed area $A$) $M_t/A$ was measured for various loadings ($C_o$ in Eq. 4.4.18). The data are plotted in Fig. P4.25. Do the data agree well with Eq. 4.4.18? If so, calculate a value of the diffusion coefficient of the steroid in the rubber.

**4.26** Paul and McSpadden [*J. Membrane Sci.*, **1**, 33 (1976)] present the data shown in Fig. P4.26 for the transport of a dye (Sudan III) through a vulcanized silicone rubber membrane. Their experiment is along the lines described in Example 4.2.1. They cite values for the diffusion coef-

ficient ($D = 2.68 \times 10^{-6}$ cm$^2$/s) and the solubility of the dye in the membrane ($\rho_s = 0.274$ g/L) for a membrane in equilibrium with the lower (high and constant concentration) reservoir containing the dye solution. What is the membrane thickness? Calculate the thickness from both the slope and the intercept of the data. Discuss any inconsistency in the two numbers obtained thereby.

**4.27** Modify the analysis given in Example 4.2.1 and allow for a finite mass transfer coefficient on the dilute side of a membrane test cell.

**4.28** The data shown in Fig. P4.28 are for the cumulative uptake of water that is dissolving in and diffusing into thin glass samples exposed on both faces to humid air. $C - C_o$ is the molar concentration change of dissolved water in the glass, from an initially uniform concentration $C_o$. The data were obtained for various glass thicknesses $d$ in the range of 30 to 50 $\mu$m, at 400°C and 355 mmHg pressure. Estimate the diffusivity of water through this glass, from these data of Tomozawa and Tomozawa [*J. Non-Cryst. Solids*,

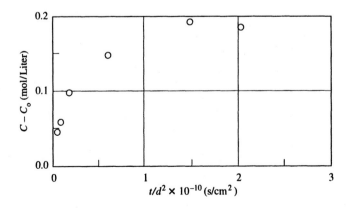

**Figure P4.28** Cumulative uptake of water by a glass film.

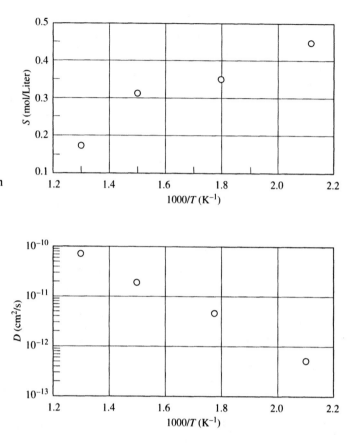

**Figure P4.29.1** Solubility of water in glass under 355 mmHg water vapor.

**Figure P4.29.2** Diffusivity of water in glass under 355 mmHg water vapor.

**109**, 311 (1989)]. Use two methods—one based on the entire set of data shown, and one based on the "short-time" portion of the data.

**4.29** Data are shown in Figs. P4.29.1 and P4.29.2 for the solubility and diffusivity of water in glass at various temperatures. From these data, find the following:

**a.** The time required for a glass film, 50 $\mu$m in thickness, initially "dry," to reach equilibrium with water vapor at 355 mmHg at 350°C.

**b.** The time required to reduce the water vapor concentration in a glass film, 50 $\mu$m in thickness, from 3000 ppm (weight) to 300 ppm (weight) when in contact with dry air at 350°C.

**4.30** Gallium is a common dopant atom used in the manufacture of germanium-based semiconductors. It is important to measure the diffusion coefficient of the dopant through germanium at high temperature. One method for doing this is suggested in Fig. P4.30.1. Two long solid cylinders of germanium, one doped uniformly, and the other undoped, are butted together at one pair of faces. The two cylinders are rapidly heated to the temperature of interest, and diffusion proceeds. Periodically the system is rapidly quenched, and the concentration profile of the dopant in the cylinders is measured.

**a.** Write, but do not attempt to solve, the bound-

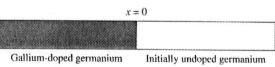

Gallium-doped germanium     Initially undoped germanium
$C = C_0$

**Figure P4.30.1** Schematic of the gallium diffusion experiment.

ary value problem that defines this diffusion problem. Keep in mind that initial and boundary conditions are part of the formulation of a boundary value problem.

**b.** It can be shown that the solution for the gallium concentration distribution, corresponding to the physics of this experiment, takes the form

$$C(x, t) = \frac{C_o}{2}\left(1 + \text{erf}\frac{x}{2\sqrt{Dt}}\right) \quad \textbf{(P4.30)}$$

Some experimental data, obtained by Bourret et al. [*J. Electrochem. Soc.*, **128**, 2437 (1981)] are shown in Fig. P4.30.2. Give the value for the diffusion coefficient.

**4.31** With the model that leads to Eq. 4.4.47 as a basis, calculate and plot the dimensionless half-time for release of a contaminant from a spherical particle, as a function of the parameter $B$.

**4.32** An experiment is planned to determine some of the characteristics of decontamination of a sedimental soil. An organic pesticide is the solute of interest, and the following estimates are available for the characteristic parameters of the solute/soil system.

Internal porosity of the sediment grains
  $\varepsilon = 0.15$
Sediment grain size (diameter)
  $D_p = 100 \ \mu m$
Equilibrium partition coefficient of the solute between water and the solid (20°C)
  $K_p = 1200 \ \text{cm}^3/\text{g}$
Equilibrium solubility of the solute in water (20°C)
  $2 \times 10^4 \ \mu\text{g/L}$
Diffusivity of the solute in water (20°C)
  $8 \times 10^{-6} \ \text{cm}^2/\text{s}$

The following experiment is planned. Into 2 L of water, just saturated with the pesticide and well stirred, is dumped 10 g (on a dry basis) of wet sedimental soil. The vessel is continuously agitated.

**a.** How long is required for the system to reach equilibrium with respect to the partitioning of the pesticide into the internal pores of the soil particles?

**b.** At equilibrium, what is the concentration $C_f$ of the pesticide in the aqueous phase?

**c.** At equilibrium, what is the concentration $C_s$ of the pesticide in the solid phase?

**d.** Repeat parts a, b, and c for 500 g (dry) of wet sedimental soil dumped into the agitated vessel. Assume that the only water brought into the agitated vessel with the wet soil is that in the pore spaces of the solid particles.

**4.33** The bottom of a lake is suddenly exposed to a release of 2,4,6-trichlorophenol (TCP). Immediate action removes the TCP source from the lake bottom, but it is determined that the sediment is contaminated to a depth of 0.2 m with a uniform concentration, measured in the interstitial water of the sediment, of 250 mg/L. We need an estimate of the time required for the diffusive release of the TCP into the lake water.

A sample of contaminated bottom sediment is taken to the laboratory, and an experiment is performed on the rate of release of the contaminant. In the laboratory experiment, a layer of sediment, 1.5 cm thick, is placed on the bottom of a glass aquarium filled with 25 gallons of water. The water is mildly agitated to provide some movement of water across the sediment/water interface. The water temperature is fixed at that of the lake bottom, which is 15°C. The data of

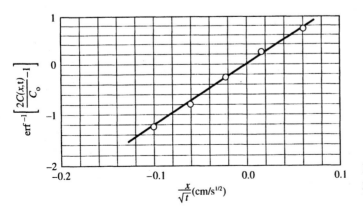

**Figure P4.30.2** Data from the gallium diffusion experiment at 1219 K.

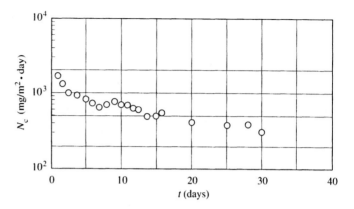

**Figure P4.33** Laboratory data on the release of TCP from a lake sediment.

Thoma et al. [*Environ. Sci. Technol.*, **27**, 2412 (1993)] for the TCP flux are given in Fig. P4.33. What is the expected half-time for release from the lake sediment? State clearly the physical assumptions about the mechanism of release your answer depends on.

**4.34** Equation 4.5.10 gives only an order-of-magnitude estimate of the apparent diffusion coefficient. Another empirical model often seen in the literature is the following:

$$\frac{D_{eff}}{R_K} = \frac{\varepsilon^{4/3} D_f}{\varepsilon + (1-\varepsilon)K_p \rho_s} \qquad \textbf{(P4.34)}$$

Rework Example 4.5.1 using this model.

**4.35** Foam in carbonated beverages has a significant effect on the "feel" and flavor release of the drink. The data shown in Fig. P4.35 are taken from a paper by Bisperink and Prins [*Colloids Surf: A*, **85**, 237 (1994)]. The authors, who are food physicists in the Netherlands, formed a hemispherical bubble of gas on the end of a small capillary submerged in a saturated aqueous solution of $CO_2$, at atmospheric pressure. The ambient pressure and the pressure inside the bubble were then suddenly decreased to 0.36 atm, and as a consequence the solution was supersaturated with $CO_2$, with respect to the bubble. Hence $CO_2$ was transported into the undersaturated bubble, causing the bubble to grow.

The following data are available:

Initial bubble radius
   $R_o = 0.058$ mm
Molar volume of pure $CO_2$ at 1 atm and 20°C,
   *in the gas phase*
   $V_m = 0.0245$ m$^3$/mol
Henry's constant
   $H = 2.97 \times 10^6$ Pa/(mol/kg) for $CO_2$ in water
   at 20°C

Use a quasi-steady model to predict the bubble radius versus time, and compare the prediction to the data shown. What physical phenomenon does your model neglect that might explain the failure of the model?

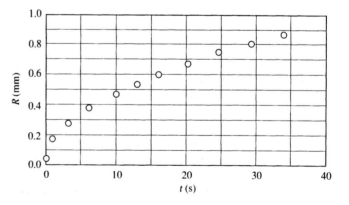

**Figure P4.35** Data on growth of $CO_2$ bubbles in a supersaturated solution.

**4.36** Equation 4.2.38 gives the transient solution for the concentration buildup within a film, under the initial and boundary conditions stated in Eqs. 4.2.21 to 4.2.23. Find an expression for the difference between the flux of the diffusing species into the film, and out of the film. Aren't the fluxes supposed to be independent of position within the film, in the planar case? How can there be a difference? Show that the difference vanishes at large times, and recommend a definition of "large time," in terms of a value of $X$, based on the vanishing of the flux difference at steady state.

Suppose that we want to calculate the total *amount* of the diffusing species (per unit area) that has crossed the boundary at $Z = 0$, up to the time $X$, and the same for the boundary $Z = 1$. Define a dimensionless amount as a function of time, and present expressions for the amounts passed in and out of the film, as a function of time. Show that for large times these amounts differ by a constant value, and explain what that constant corresponds to physically. Plot these amounts as a function of time over the interval $[0, 1]$ in $X$.

**4.37** Return to Example 4.3.2. Suppose that the drying temperature is increased to 45°C, in such a way that there is no change in the molar density of water vapor in the incoming air. What is the relative humidity of the incoming air?

Assume that the diffusion coefficient $D_{AB}$ is larger by a factor of 2, by virtue of this increase in temperature. Suppose that the Henry's law constant $H$, given by Eq. 4.3.28, increases with temperature in the same way that the pure water vapor pressure increases with temperature. To what extent is the required drying time reduced?

**4.38** Return to Example 4.3.2. Suppose that the oven is redesigned so that the paper dries from both sides. By what factor is the drying time reduced?

**4.39** Return to Example 4.3.2. Suppose that the drying temperature is decreased to 25°C, in such a way that there is no change in the molar density of water vapor in the incoming air. What is the relative humidity of the incoming air?

**a.** Assume that the diffusion coefficient $D_{AB}$ is smaller by a factor of 2, by virtue of this decrease in temperature. Suppose that the Henry's law constant $H$, given by Eq. 4.3.28, decreases with temperature in the same way that the pure water vapor pressure decreases with temperature. To what extent is the required drying time increased?

**b.** Suppose that the oven is redesigned so that the paper dries from both sides. By what factor is the drying time reduced, in comparison to the value you obtained from part a?

**4.40** The solution to Problem 4.14 yields a criterion for assuming that Eq. 4.4.11 can be used as an approximation for the transient loss of a diffusible substance from a slab geometry. Return to Example 4.3.2. What is the value of Bi′ for this example? Use Eq. 4.4.11 to find the required drying time, and compare the result to that given in Eq. 4.3.41. Is your conclusion compatible with your expectation based in Problem 4.14?

**4.41** In the field of nuclear reactor design it is necessary to understand and control the diffusion of hydrogen isotopes through metals used in the reactor construction. One case of interest is that of tritium diffusing within zirconium at high temperatures. A method used to make this measurement is referred to as "limited source" diffusion. (This is different from "constant source" diffusion illustrated in Example 4.2.5.) A very thin layer of material containing the diffusant molecule is deposited onto the surface of the solid of interest. The solid is then raised to the temperature at which the diffusion measurement is to be made, and the diffusant diffuses into the solid. An oxide layer on the other side of the deposited film prevents loss from that surface. Figure P4.41.1 shows an idealization of this system.

A mathematical model of diffusion under these idealized conditions consists of the one-dimensional planar diffusion equation

$$\frac{\partial C}{\partial t} = D \frac{\partial^2 C}{\partial x^2} \qquad \textbf{(P4.41.1)}$$

and initial and boundary conditions. The barrier oxide layer is in contact with the plane $x = 0$. Hence, at the boundary $x = 0$ the concentration is assumed to satisfy the condition

$$\frac{\partial C}{\partial x} = 0 \qquad \text{at } x = 0 \qquad \textbf{(P4.41.2)}$$

If the diffusant penetrates over a distance that is large compared to the initial deposited film thickness, but small compared to the thickness of

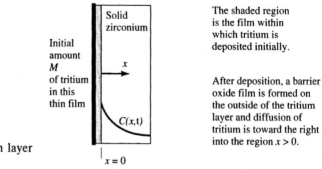

**Figure P4.41.1** Diffusion from a thin layer into the solid at $x > 0$.

the solid sample, a second boundary condition is

$$C = 0 \qquad \text{at } x \to \infty \qquad \textbf{(P4.41.3)}$$

The initial condition is one we haven't seen before. The physical statement is that all the diffusant species, an amount $M$ (per unit of area), is confined to an infinitesimally thin region at the plane $x = 0$. The mathematical "device" that mimics this physical idea is the so-called Dirac delta function $\delta(x)$, which has the properties

$$\delta(x) = 0 \qquad \text{for } x \neq 0$$
$$\int_{-\infty}^{\infty} M\,\delta(x)\,dx = M \qquad \textbf{(P4.41.4)}$$

Hence Eq. P4.41.4 is the initial condition.

The solution for the concentration profile may be found in the form

$$C(x, t) = \frac{M}{\sqrt{\pi D t}}\exp\left(-\frac{x^2}{4Dt}\right) \quad \textbf{(P4.41.5)}$$

The data shown in Fig. P4.41.2 can be analyzed according to this model. Do so, and give a value for the diffusion coefficient of tritium in zirconium.

**4.42** Ants, and some other insects, apparently communicate a wide range of messages (Danger! Food over here! Let's start a family!) by releasing and detecting volatile substances called *pheromones* that diffuse from the source insect to the recipient(s) through diffusion and convection. E. O. Wilson, an entomologist at Harvard University, has spent his life studying the behavior of ants. One of his experiments provides an opportunity for us to apply diffusion theory.

The harvester ant communicates danger through an alarm substance, a volatile liquid, secreted from a gland in its head. While convection can disperse this substance, in certain circumstances diffusion is the only mechanism of transmittal of the message. Wilson [*J. Theor. Biol.*, **5**, 443 (1963)] devised a series of experiments for measuring the diffusion coefficient of the alarm pheromone. In one such experiment (see Fig. P4.42) a group of harvester ants were acclimated to a tube of inside diameter 2.6 cm and length approximately 1 m. At some initial instant of time a small vial containing the crushed heads of several ants was coupled to one end of

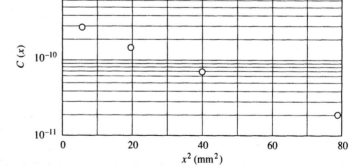

**Figure P4.41.2** Diffusion profile of tritium in zirconium after 8 h at 723 K. $C(x)$ is in relative units, proportional to the actual concentration $C(x, t)$, as explained by Isobe and Sadaoka, *J. Nuclear Sci. Technol.*, **31**, 1290 (1994).

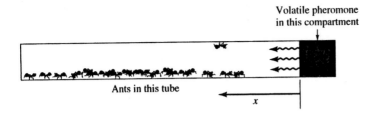

Volatile pheromone
in this compartment

Ants in this tube

$x$

**Figure P4.42** Experimental system for detecting the onset of the alarm reaction in ants.

the main tube. Sufficient time prior to this had been allowed to permit evaporation of the volatile substance into the small vial. Hence it was presumed that the vapor within the vial was at some uniform concentration of the pheromone.

The assumption was made that the source was "limited," in the sense described in Problem 4.41, rather than "constant," in the sense described in Example 4.2.2. On the basis of this assumption, the concentration profile of the pheromone, in the absence of convection, should be given by Eq. P4.41.5 above. Wilson's observations consisted of the determination of the time $t_1$ at which the ants at a distance $x_1$ from the pheromone source began to display behavior characteristic of alarm (running in circles).

The following set of data is available:

| | | |
|---|---|---|
| Time of observation of alarm response, $t_1$ (s) | 60 | 360 |
| Distance from source of alarmed ants, $x_1$ (cm) | 16 | 32.5 |

To analyze these data, we must begin with the assumption that "alarm" occurs when the ant senses a critical or threshold level of concentration, $C^*$. Begin with Eq. P4.41.5, and show that if $(x_1, t_1)$ and $(x_2, t_2)$ are two pairs of (distance, alarm time) measurements, the diffusion coefficient is given by

$$D = \frac{x_1^2/t_1 - x_2^2/t_2}{2 \ln(t_2/t_1)} \quad \text{(P4.42.1)}$$

Calculate the diffusion coefficient from these data.

**4.43** The following data (Belton et al., Chapter 25, in *Polymeric Materials for Electronics Packaging and Interconnection,* Lupinski and Moore, Eds., ACS Symp. Ser. 407, 1989) are available for the uptake of water by an epoxy film immersed on both surfaces in liquid water at 25°C. Values of $f$ are the ratios of mass of water ab-

sorbed at any time to the final or equilibrium mass of water that will be absorbed. The film thickness is 0.07 inch. Find the diffusion coefficient of water in the polymer.

| $t$(h) | $f$ |
|---|---|
| 22.7 | 0.086 |
| 49 | 0.15 |
| 68 | 0.21 |
| 214 | 0.37 |
| 306 | 0.48 |
| 470 | 0.59 |
| 650 | 0.65 |

**4.44** The following statement is found in a textbook on the design of hazardous waste systems. "Employing data from a landfill study for a 1 meter thick liner, an assumed landfill area of $10^4$ m², and a benzene concentration of 1 g/L in the landfill, a steady-state transport rate of 19 kg/yr is computed. This transport rate has the potential to contaminate 3.8 billion liters of water at the EPA drinking water limit of 0.005 mg/L."

This is a very misleading statement. Show why, by making the following calculations:

**a.** At what time (in years) is the steady state achieved, following the opening of the landfill site?

**b.** At what time does the concentration at the bottom of the liner rise to 1% of the concentration at the landfill/liner interface?

**c.** Plot the flux of benzene as a function of time (in years) over the period from 3% to 30% of the steady state time.

A free-solution diffusion coefficient for benzene through the aqueous phase that saturates the clay liner is $7 \times 10^{-6}$ cm²/s. The effective diffusion coefficient (actually, $D_{eff}/R_K$ of Eq. 4.5.10) for benzene in clay has been measured and found to be of the order of $10^{-7}$ cm²/s [Johnson et al., *Environ. Sci. Technol.,* **23,** 340 (1989)].

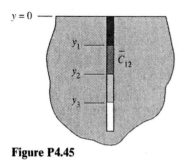

$y = 0$

$y_1$

$\bar{C}_{12}$

$y_2$

$y_3$

**Figure P4.45**

For a partition coefficient of benzene between the landfill medium and the clay liner, use $\alpha = 0.33$.

**4.45** The data shown in Figs. 4.6.2 and 4.6.3 were obtained from so-called core samples, which are obtained by removing long cylindrical plugs of clay and analyzing sections periodically along the length of the core, as suggested in Fig. P4.45. Common practice is to obtain the average concentration of the species of interest in a core section, simply by mixing the contents of the core section prior to analysis. Then this concentration is associated with the average depth of that section of the core. Does a simple arithmetic average depth make sense? To answer this question, we could consider a section in which the concentration profile satisfies Eq. 4.6.2 between two points $y_1$ and $y_2$. Our goal is to find the average concentration in that section, and then find the position in the region $y_1 < y < y_2$ where the concentration has that average value. The issue is whether that position is anywhere near the midpoint of the section.

Because the complementary error function is difficult to work with algebraically, let's make up a simpler analytical representation of the concentration profile through a core sample. Take the concentration to satisfy the following expression:

$$\frac{C(y)}{C_s} = A \exp\left(-\frac{y}{b}\right) \qquad \textbf{(P4.45.1)}$$

with $y$ in centimeters; $b$ is some parameter with units of length, also in centimeters.

Find the average value of $C(y)/C_s$ in a core sample between the levels $y_1 = 4$ cm and $y_2 = 8$ cm. At what value of $y$ is $C(y)/C_s$ equal to this average value? Is this near the mean position $y = 6$ cm? Do two cases: $A = 1$, $b = 0.5$ cm; $A = 1$, $b = 0.1$ cm.

**4.46** Example 4.6.1 describes one way to define the time at which "breakthrough" occurs, for diffusion through a finite planar barrier of some kind. We used Eq. 4.6.2 and a definition of breakthrough based on a small but finite concentration at the lower boundary of the barrier. We then applied this idea to a situation in which the concentration at the lower boundary was maintained at zero concentration. There seems to be the possibility of inconsistency in this approach. Find the breakthrough times for diffusion across a planar barrier of finite thickness $B$ by two methods, and compare the dimensionless times that you obtain.

**Method A:** Use Eq. 4.6.2, and define breakthrough as the time $X_B = Dt/B^2$ at which the concentration at the lower boundary $y = B$ rises to $0.01C_s$, when the boundary and initial conditions are as stated in Eqs. 4.6.9 to 4.6.11.

**Method B:** Begin with Eq. 4.6.12, and define breakthrough as the time $X_B = Dt/B^2$ at which

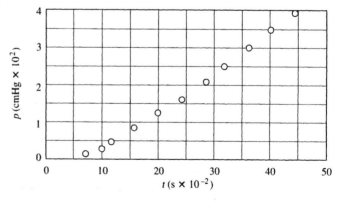

**Figure P4.47** Data for permeation of oxygen across a polyimide membrane.

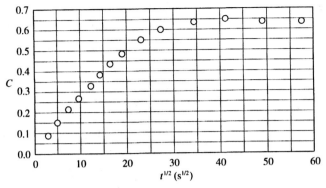

**Figure P4.48** Data for absorption of oxygen into a polyimide membrane. Concentration $C$ is in units of cubic centimeters of oxygen at standard pressure and temperature, per cubic centimeter of polymer membrane.
Toi et al., *J. Polym. Sci., B: Polym. Phys.*, **33**, 777 (1995).

the *flux* at the lower boundary $y = B$ rises to $0.01(DC_s/B)$, when the boundary and initial conditions are as stated in Eqs. 4.6.9 to 4.6.11.

**4.47** Toi et al. present the data shown in Fig. P4.47. The membrane was a polyimide of thickness 25 $\mu$m. The temperature was 25°C. The high pressure side of the membrane was exposed to oxygen at a pressure of 99 cmHg. The data give the oxygen pressure on the low pressure side of the membrane as a function of time. Using the discussion in Section 4.2 as a guide, find the diffusion coefficient of oxygen through the membrane.

**4.48** Toi et al. present the data shown in Fig. P4.48 for the absorption of oxygen at 25°C by a membrane initially free of any oxygen. Such data are sometimes referred to as "sorption" data. The membrane was a polyimide of thickness 25 $\mu$m. The data show the average concentration of oxygen in the polymer as a function of time of exposure to pure oxygen at a pressure of 76

cmHg. The average concentration of oxygen is given in cubic centimeters of oxygen at standard pressure and temperature, per cubic centimeter of polymer membrane. Absorption was across both sides of the membrane. Find the diffusion coefficient of oxygen through the membrane and the equilibrium solubility of oxygen in the membrane.

**4.49** The data shown in Fig. P4.49 are for the solubility of $CO_2$ in a polyimide membrane at 25°C. This type of curve is called an "equilibrium isotherm." The membrane was exposed to pure $CO_2$ at the pressures shown. The units on concentration are given in cubic centimeters at standard pressure and temperature of $CO_2$ per cubic centimeter of polymer membrane. Convert the concentration units to moles of $CO_2$ per cubic centimeter of polymer membrane, and plot $\alpha$ as a function of $CO_2$ pressure. Define $\alpha$ by

$$C_{CO_2}^{\text{membrane}} \text{ [mol/cm}^3 \text{ membrane]} = \alpha C_{CO_2}^{\text{gas}} \text{ [mol/cm}^3 \text{ gas]} \quad \text{(P4.49)}$$

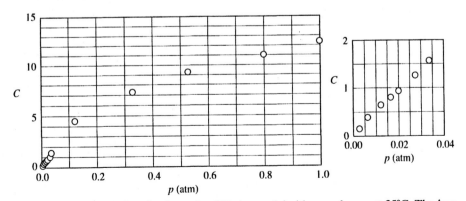

**Figure P4.49** Equilibrium isotherm for $CO_2$ in a polyimide membrane at 25°C. The low pressure region is expanded in the plot at the right.
Toi et al., *J. Polym. Sci. B: Polym. Phys.*, **30**, 549 (1992).

Notice that while the data are linear at small partial pressures, over the range of pressures studied the relationship is quite nonlinear.

**4.50** A polymeric material is being considered as a coating film to protect an optical storage disk from water vapor damage. Suppose we need to protect a disk surface from an ambient medium that will be at 40% relative humidity at 21°C. How thick must the film be if the concentration at the disk/protective film interface is to be below 0.03 wt % water in the film, for a period of 5 years? The film density is 1.224 g/cm³. At 100% relative humidity (at 21°C) the equilibrium uptake of water in the film is 5.4 wt %. The absorption isotherm for water in this polymer is linear. The diffusion coefficient of water in the film is $8.2 \times 10^{-10}$ cm²/s.

**4.51** Consider a polymeric film with properties of the order of those described in Problem 4.50. We wish to make precise measurements of the diffusion coefficient of water vapor in the polymer film, as well as measurements of the equilibrium uptake of water (the absorption isotherm). A sensor capable of continuously measuring the total weight of the polymeric film is available. The sensor is a disk of diameter 1 cm, and one side of it will be coated with a 5 μm thick layer of the polymer. How long will it be before the data on weight gain as a function of time yield a diffusion coefficient? For the parameters given in Problem 4.50, plot weight gain versus time up to equilibrium.

**4.52** Following the analysis of Example 4.4.3, examine the feasibility of controlling a sudden release of HCl gas within a confined space. The water solubility (at 0°C) is 82 g/100 cm³. The Henry's law constant at that temperature is 0.015 atm/mole fraction. The IDLH concentration is 100 ppm. The initial concentration of HCl in air, following the release, is 200 ppmv. All other conditions are as stated in Example 4.4.3. Use twice the minimum amount of water required, and state the time required for the concentration to fall below 10 ppmv in the ambient air.

**4.53** On a copy of Fig. 4.4.7, draw lines corresponding to $\tau_{0.01}$, and $\tau_{0.001}$.

**4.54** A sustained-release system for an antinausea drug (Dramamine) is to be designed along the following lines. The drug is dissolved in a thin gelatin film and applied as a "patch" to the skin, much like a Band-Aid (see Fig. P4.54). The

Film →
Skin

**Figure P4.54**

gelatin is sandwiched between two very thin polymeric membranes that offer no resistance to passage of the drug. The drug is not volatile in the surrounding air. The diffusivity of Dramamine in the gelatin is $3 \times 10^{-7}$ cm²/s. Find the gelatin film thickness such that 90% of the Dramamine is delivered to the skin over a 3-hour period. Assume that the skin region is so highly vascularized that the Dramamine concentration is practically zero in the skin and tissue beneath the patch.

**4.55** A gas bubble is retained (by the force of surface tension) on the end of a very fine hypodermic tube, as shown in Fig. P4.55. The gas is a binary mixture, only one component of which is soluble in the surrounding liquid. As the soluble species dissolves and diffuses away from the bubble, a small flow of (only) that species into the bubble is imposed through the hypodermic tube, in such a way as to keep the bubble diameter constant. Derive a relationship between the volumetric flowrate (cm³/s) of added gas and the diffusion coefficient (cm²/s) of the dissolved gas in the surrounding liquid. Assume that the liquid exterior to the bubble is quiescent.

**4.56** One of the steps in semiconductor processing involves the placement of a stack of silicon wafers into a tube through which a reactant gas is conveyed. The wafers are stacked coaxial with the tube, and the interwafer spacing is 0.5 cm. The wafer radius is 7.5 cm. (See Fig. P4.56.)

When the tube is loaded with a stack of wafers, room air occupies the interwafer space. After the reactor has been loaded, an inert gas is pumped

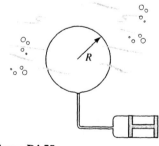

**Figure P4.55**

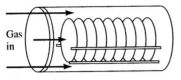

**Figure P4.56**

through the reactor. Exchange between the flowing inert gas and the interwafer gas occurs only by diffusion in the interwafer region.

The purpose of the inert gas flow is to assist the removal of water vapor from the interwafer space. If the inert gas is dry, how long will it be before the water vapor concentration has been reduced to $10^{-5}$ of its initial value? Assume that operation is at atmospheric pressure and 300°C. Assume that there is no adsorbed water on the wafers.

**4.57** A "cast acrylic" artwork (see Fig. P4.57) is to be produced by polymerization of the methacrylate monomer in a mold. When the polymerization is complete, the mold can be removed from around the piece, and the polymeric structure will be stiff enough to stand on its own. However, it cannot be shipped until the residual monomer concentration has fallen below 0.01% (by weight) throughout the piece. (When polymerization is stopped, the residual monomer concentration is uniform, at a level of 0.5%.) This process is known as "curing."

Design an experiment using a scale (smaller) model, to estimate the required time for curing of the full-scale piece. The following information is available:

The "curing room" is a large warehouse. The piece will stand on an open floor in front of a bank of large fans, under conditions such that the mean air velocity impinging on the piece is 10 cm/s.

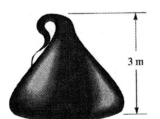

**Figure P4.57** Work of art for Problem 4.57.

This curing system was used to cure two *spherical* acrylic test castings, with the following results:

| | | |
|---|---|---|
| Casting A: | diameter = 6 inches | cure time = 56 hours |
| Casting B: | diameter = 12 inches | cure time = 201 hours |

In your design, specify the dimensions of the model and your method of estimation of the cure time for the model and for the large piece of art. The model cure should occur in 6 hours, for convenience.

**4.58** A valuable hydrocarbon is diluted in air at 25°C and 1 atm. The hydrocarbon has a molecular weight of 44 and its mole fraction in the gas phase is $y = 0.01$. This gas is contained in a closed tower of diameter $D = 2$ feet and height $H = 20$ feet. A proposal has been made to recover the hydrocarbon by absorption into a nonvolatile oil ($M_w = 300$). The oil will be sprayed into the top of the tower and the falling droplets will absorb the hydrocarbon. The oil has a density of 1.2 g/cm³. The diffusion coefficient of the dissolved vapor in the oil is $10^{-6}$ cm²/s.

Design a spray system that will operate in such a way that the oil droplets reaching the bottom of the tower will be 90% saturated with the hydrocarbon. The first step in this design is to specify the required drop size.

Equilibrium data are available, and in the operating range of interest may be expressed as $y = 0.1x$, where $y$ and $x$ are mole fractions of the hydrocarbon in the vapor and liquid, respectively.

For this preliminary design, assume that the vapor space is well mixed throughout.

**4.59** A bubble of pure oxygen rises through a column of water. The system is isothermal at 25°C, and at atmospheric pressure. The bubble diameter is 3 mm. How far must the bubble rise to become saturated with water vapor? The vapor pressure of water at 25°C is 23.8 mmHg.

**4.60** A dilute dye solution in water is being pumped into a turbulent flow of water in a pipe, as sketched in Fig. P4.60. At some instant of time a valve $V$ is shut, trapping the dye in the length $L$ of tube between $V$ and the pipe. The tube has a diameter $d = 1$ mm, and the length $L$ is 10 cm. Assume that the dye concentration is uniform along the length $L$ at the instant the valve is shut.

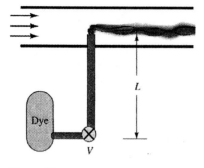

**Figure P4.60**

After the valve has been shut, dye will continue to pass into the turbulent pipe flow, but only by diffusion. If the diffusion coefficient of the dye in water is $D = 5 \times 10^{-6}$ cm$^2$/s, estimate the time required for 95% of the dye to be removed by diffusion from the length $L$ of the tube.

**4.61** An oxygen electrode is used to measure the diffusivity of oxygen through a liquid film. Figure P4.61.1 shows the geometry. The test

layer has thickness $L$. The experimental procedure is simple. The liquid is saturated with oxygen at some uniform partial pressure $P_o$. Then the gas above the test liquid is suddenly reduced in oxygen partial pressure to the value $P_1$. The oxygen sensor records partial pressure $P_b(t)$ on the surface $x = 0$.

Assume that the oxygen concentration gradient set up by the cathodic reaction does not disturb the concentration profile near $x = 0$ and that the system behaves as if the plane $x = 0$ were impermeable to oxygen.

The data shown in Fig. P4.61.2 were obtained at 23.3°C in distilled water. The liquid film thickness was $L = 0.0961$ cm. Give a value for the diffusivity of oxygen in water.

**4.62** Tetraethoxysilane (TEOS) is a liquid source used in the semiconductor industries to grow silicon dioxide. One method of delivery of the TEOS to a reactor involves bubbling of an inert gas through a container of pure liquid TEOS (Fig. P4.62). The TEOS evaporates at the

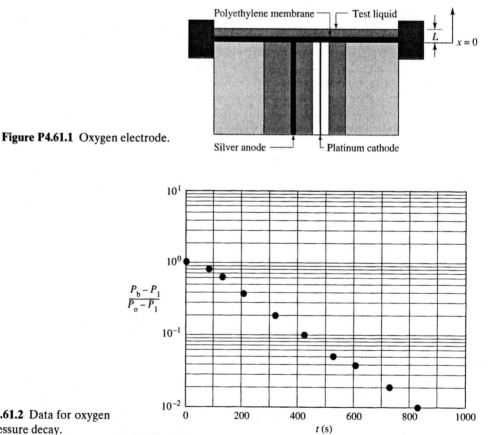

**Figure P4.61.1** Oxygen electrode.

**Figure P4.61.2** Data for oxygen partial pressure decay.

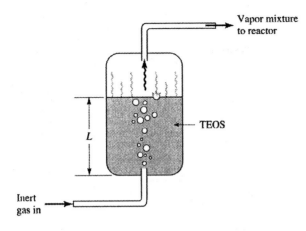

**Figure P4.62** Schematic of a TEOS bubbler.

gas/liquid interface and diffuses into the inert gas bubble. Suppose bubbles of uniform size are created in the bubbler. We want to know how long a liquid path in the bubbler must be if the bubble is to be saturated with TEOS when it leaves the liquid. In other words, how tall must the liquid height in the bubbler be? We have the following data for TEOS, all at 60°C.

Vapor pressure, $P_{vap} = 16$ torr.
Viscosity of the liquid TEOS is $\mu = 0.03$ poise, and the liquid density is 0.95 g/cm³.
Bubbles are introduced to the TEOS bubbler at a pressure of 160 torr, and they have a diameter of 1 mm. The chemical formula for TEOS is $Si(OC_2H_5)_4$.

**4.63** Sustained-release drug delivery systems are often created in a "sandwich form," as shown in Fig. P4.63.1. The drug reservoir is a liquid between two membranes of thickness $L$.

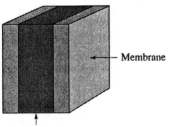

Drug reservoir

**Figure P4.63.1** A sandwich-type drug delivery system.

If a drug is encapsulated in the membrane sandwich under conditions that allow its thermodynamic activity to remain constant despite the loss (release) of the drug, then a steady state can be achieved in which the rate of delivery is constant in time. This is sometimes called "zero-order release," by analogy to chemical kinetics, where the rate of conversion is independent of concentration. One way to achieve constant activity is to encapsulate a *supersaturated* solution of the drug. Of course, at some point enough release will have occurred to render the drug undersaturated. After that point the release rate will fall exponentially, and the release is referred to as "first-order." These comments are strictly valid in the situation of a drug that is so diluted upon release to the surrounding fluid or tissue that its concentration on the exterior surface of the sandwich is essentially zero.

Some data are shown in Fig. P4.63.2 for the release of chloramphenicol from a membrane sandwich. The slope of the curve is constant, corresponding to zero-order release, for about five days. There is a falling rate of release for another day and a half, after which the drug is exhausted, and there is no further release.

Write a mathematical model of this sandwich system and examine how the initial or zero-order release rate $(dM/dt)_o$ depends on the diffusivity of the drug *in the membrane* $D_m$, the membrane thickness $L$, the saturation concentration of the drug *in the reservoir* $C_s$, and the solubility of the drug in the membrane $\alpha = C_s/C_m$.

**4.64** Ozone can be used to sterilize an aqueous solution and prepare it for subsequent pro-

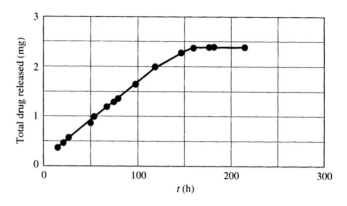

**Figure P4.63.2** Milligrams of chloramphenicol released from a sandwich device.

cessing. It is necessary to design an ozonation process that quickly transfers the ozone gas into solution, and a mathematical model of a small part of this process must be developed.

Suppose a short *transient* burst of pure ozone ($O_3$) is bubbled into a stirred vessel at atmospheric pressure under conditions that include a mean bubble diameter $d_b$ of 4 mm. We know from related studies of ozone transfer into water that the Sherwood number [Sh = $k_c(d_b/D_{AB})$] has a value of Sh = 10 for 4 mm bubbles when the system is stirred in the same manner as the process we are considering.

You have been requested to demonstrate (by a simple model) that 90% of the ozone can be quickly transferred into the surrounding liquid. State clearly any assumptions you make in order to give a rational response to this request.

**4.65** A process is carried out under *ultrahigh vacuum* in a reactor whose walls are a glass tube of inside diameter 10 cm and wall thickness 1 cm. We are concerned with the possibility that $H_2$ can "leak" into the reactor across the tube wall by diffusion and thereby reduce the quality of the vacuum environment for the process. The diffusivity of $H_2$ in glass at the process temperature, 700 K, is $10^{-8}$ cm$^2$/s. Suppose the tube is one meter long, and its end plates and fittings are impermeable to hydrogen. Suppose further that at some instant of time ($t = 0$) the vacuum pump has brought the system to a pressure of $10^{-6}$ torr. The glass tube is surrounded by atmospheric air, and the partial pressure of hydrogen in that air is $3.8 \times 10^{-4}$ torr. The solubility $b$ of hydrogen in glass, defined as the ratio of the molar concentration of hydrogen in the glass to

the molar concentration in the gas, *at equilibrium,* is 0.2 at 700 K.

**a.** Present a mathematical model that relates the pressure in the tube at any time to the parameters of this system. Do not use a quasi-steady analysis.

**b.** Find the time $t_2$ required for the pressure to double from its initial value of $10^{-6}$ torr at $t = 0$.

**c.** Is the quasi-steady approximation reasonable for this problem? Why or why not? Over what fraction of the time $t_2$ calculated above would the system be in an unsteady mode of behavior?

**d.** Does the quasi-steady approximation represent an *over*estimate or an *under*estimate of the time $t_2$? Explain your answer. Think, particularly, about the initial condition that you assume.

**4.66** An initially pure single component gas (A) is contained in a long cylindrical pipe. The gas is stationary, and the initial molar concentration is $C_{A,o}$. At some instant of time a chemical reaction is initiated that occurs only on the inner surface of the pipe. The reaction is first-order in the concentration of A, and the stoichiometry is A → B. The reaction product (B) is a gas.

**a.** Write, but do not attempt to solve, the diffusion equation and boundary conditions from which it would be possible to solve for the *unsteady state* radial concentration distribution of the species A. State clearly any assumptions you introduce in writing this mathematical model. Make a sketch that defines clearly any variables that appear in your model.

**b.** Give the solution for the *average* concentration of species A in the pipe, as a function of time. Plot the half-time for decrease in $C_A$ as a function of some appropriate dimensionless

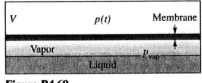

**Figure P4.69**

group. Make use of the information given in Problem 4.14.

**4.67** Liquid containing a soluble gas is in a cylindrical "bottle." The bottle radius is 10 cm, and the axial length is 1 m. The concentration of the gas in the liquid is initially 0.2 mol/L. The bottle wall is a polymeric material of unspecified thickness $L$ cm. The diffusivity of the gas in the liquid is $D_L = 2 \times 10^{-4}$ cm$^2$/s at the temperature of interest. The diffusivity of the gas in the polymer is $D_P = 2 \times 10^{-6}$ cm$^2$/s at the temperature of interest. The solubility ratio of the gas in the polymer, with respect to the liquid is 0.03 mol/L (polymer) per mole/liter (liquid). Assume that the surrounding air is free of the gas in question. Assume that the contents of the bottle are well mixed because of frequent shaking of the bottle.

Prepare a quick but reasonable estimate of the polymer wall thickness required to result in a gas concentration that falls to half its initial value in 67 days.

**4.68** A droplet of a viscous liquid, at a uniform temperature of 300 K, and with a uniformly distributed initial concentration of a solute, is suddenly exposed to a large volume of air containing none of the volatile solute. The droplet has an initial diameter $D$ of 173 $\mu$m, and it is suspended in the air under such conditions that the Sherwood number satisfies

$$Sh \equiv \frac{k_c D}{D^{air}} = 2$$

The diffusion coefficient $D^{air}$ of the volatile species in the surrounding air is 0.3 cm$^2$/s. The diffusion coefficient of the volatile species in the viscous liquid is $3 \times 10^{-5}$ cm$^2$/s. How long is required for the concentration of the solute, at the center of the droplet, to fall to 1% of its initial value?

**4.69** A membrane test cell is designed and operates along the lines suggested in Fig. P4.69. A pure liquid is under its own vapor pressure in the closed container below the membrane. The upper volume is initially pumped down to (nearly) zero absolute pressure, after which vapor from the lower compartment dissolves in the membrane and diffuses across the membrane. Assume that both vapor phases are well mixed. Given the membrane area $A$ normal to the diffusion flux, the membrane thickness $H$, the volume $V$ of the upper compartment, and the vapor pressure $p_{vap}$ of the liquid, derive an expression from which you can calculate the product of solubility and diffusivity of the vapor in the membrane, from data on the time $t_{0.5}$ at which the pressure in the upper compartment rises to the value $0.5p_{vap}$. Do not assume that $p(t) \ll p_{vap}$.

**4.70** A binary gas mixture is contained in a pressurized spherical vessel of radius $R = 10$ cm. Component A is an inert gas, and component B is initially present at a mole fraction of 0.01. B reacts on the inside wall of the vessel by a first-order reaction with a rate constant $k = 0.1$ cm/s. The diffusion coefficient of B in A/B is $D_{AB} = 0.2$ cm$^2$/s. At what time will the concentration of B be reduced to 1% of its initial value? At what time will the concentration of B be reduced to 0.01% of its initial value?

**4.71** A sustained-release system for an antinausea drug (Dramamine) is designed along the following lines (see Fig. P4.71). The drug is dissolved in a thin gelatin film 1 mm thick and applied as a "patch" to the skin, much like a Band-Aid. The gelatin is sandwiched between two very thin polymeric membranes that offer no resistance to passage of the drug. The drug is not volatile in the surrounding air.

The diffusivity of Dramamine in the gelatin is $3 \times 10^{-7}$ cm$^2$/s. Estimate the time required for 90% of the drug to be delivered to the skin. Assume that the Dramamine is transferred into the circulatory system through a diffusive resistance that is equivalent to a thin (0.25 mm) film of gelatin. Take the concentration of Dramamine to be essentially zero in the circulatory system.

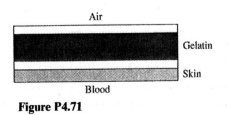

**Figure P4.71**

Take the solubility ratio of Dramamine in gelatin and skin to be $\alpha = 1$.

**4.72** You have been asked to design an experiment to test a series of controlled-release drug delivery systems for steroids. The active drug agent, ethynodiol diacetate, is dissolved in a silicone elastomer and polymerized in a mold into the shape of a thin circular disk (see Fig. P4.72). The polymer disk is placed in a large reservoir of an aqueous solution, which is very well stirred. Before you can design the experimental program, you must provide an estimate of the time required for half the steroid to be released from the polymeric matrix. The following data are available:

The solubility limit of the steroid in the polymer is $C_s = 1.5$ mg/mL.

The diffusion coefficient of the steroid in the polymer is $D_p = 3.8 \times 10^{-7}$ cm²/s at 37°C.

The polymeric disk has a thickness of 6 mm and a radius of 50 mm.

The disk is attached on one side to a supporting surface that is impermeable.

**4.73** An investigator wishes to measure the permeability of a thin gelatin membrane to a specific molecule that is in a gas phase in contact with the membrane. A membrane test cell is used, designed as suggested in Fig. P4.73. The membrane is stretched across a chamber to divide the chamber into two sections. In the lower section the partial pressure of species A is maintained constant, with the value $p^o_{A,g}$, and that partial pressure is uniform throughout the volume $V_1$. The upper region is initially free of species A, is well stirred and closed. It has a fixed volume $V_2$. We can measure the partial pressure $p_{A,g}(t)$ in the volume $V_2$ as a function of time.

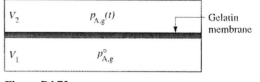

**Figure P4.73**

You can assume that species A is dilute in both gas phases, but you cannot make the assumption that $p_{A,g}(t) \ll p^o_{A,g}$ is always true. We expect that Henry's law holds for the equilibrium of the species A between the gas and gelatin phases. Hence we write $p_{A,g} = H' C_{A,m}$ where $C_{A,m}$ is the molar concentration of species A in the gelatin phase. We assume that $H'$ is constant. The membrane thickness $B$ is uniform and is known, as is the membrane area A. The system is isothermal.

**a.** Show that the solution for $p_{A,g}(t)$ may be written in the following form:

$$\varphi(X_D, S) = \frac{p_{A,g}(t)}{p^o_{A,g}}$$

$$= 1 - \sum_{n=1}^{\infty} \frac{2 \sin \beta_n (\beta_n^2 + S^2)}{\beta_n (\beta_n^2 + S^2 + S)} \exp(-\beta_n^2 X_D)$$

$$\text{(P4.73.1)}$$

where

$$\beta_n \tan \beta_n = S \qquad \text{(P4.73.2)}$$

and

$$X_D = \frac{D_{A,m} t}{B^2} \quad \text{and} \quad S = \frac{A B R_G T}{H' V_2} \quad \text{(P4.73.3)}$$

(See Fig. A1.3 for the first two roots to Eq. P4.73.2.)

**b.** Plot several curves of $p_{A,g}(t)/p^o_{A,g}$ versus $X_D$ for values of $S = 0.5, 1,$ and $2$.

**c.** The following set of data, obtained at 273 K, is available for the system described above. Find the diffusion coefficient of the penetrant through the membrane.

| $t$ (s) | 5000 | $10^4$ | $2 \times 10^4$ | $3 \times 10^4$ | $4 \times 10^4$ |
|---|---|---|---|---|---|
| $p_{A,g}(t)$ (atm) | 0.23 | 0.47 | 0.75 | 0.88 | 0.94 |

For this experimental system the following data are available:

$B = 0.1$ cm    $A = 10$ cm²    $V_2 = 100$ cm³
$H' = 224$ atm·cm³/mol    $p^o_{A,g} = 1$ atm

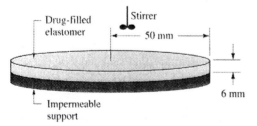

**Figure P4.72** Drug-filled polymer (silicone elastomer) film. The surrounding solution is well stirred.

**4.74** A chemical reaction occurs at the interface between a liquid and a planar horizontal solid surface, leading to the release of reaction products as small gas bubbles. The following information is available to us:

The bubble radius is $R = 1$ mm. The gas within the bubble is mostly hydrogen, but it also contains 5000 ppm (by volume) of arsine ($AsH_3$, $M_w = 78$). The entire system is under a pressure of 100 atm. The temperature is 100 K. The reacting surface is covered with a thick layer of an arsine-absorbing liquid. Arsine is soluble in this liquid, and arsine reacts rapidly with this liquid to produce liquid-soluble products. This liquid is saturated with respect to hydrogen. The terminal velocity of the gas bubbles in this liquid is 10 cm/s.

The absorbing liquid layer must be thick enough to ensure that when the bubbles reach the surface, they have only 5 ppm arsine. What is the required thickness?

**4.75** A glass "bubble," with an outside diameter of 0.005 m and a wall thickness of 50 $\mu$m, is filled with air at one atmosphere absolute pressure. It is suddenly exposed to a gaseous medium that is pure hydrogen gas at a pressure of 100 atm absolute. Estimate the time required for the internal hydrogen partial pressure to rise to a value of 50 atm. The following data are available:

The solubility ratio of hydrogen in glass, relative to pure hydrogen gas, is

$$\alpha = \frac{\text{mol/vol hydrogen in glass}}{\text{mol/vol hydrogen in gas}} = 0.01$$

The diffusion coefficient for hydrogen through glass is $D = 2 \times 10^{-6}$ cm$^2$/s at the temperature of the system (600 K).

**4.76** The following data (Belton et al., Chapter 25 in *Polymeric Materials for Electronics Packaging and Interconnection,* Lupinski and R. S. Moore, ACS Symp. Ser. 407, 1989) are available for the uptake of water by an epoxy film completely immersed in liquid water at 25°C. Values of $f$ are the ratios of mass of water absorbed at any time to the final or equilibrium mass of water that will be absorbed. The film thickness is 0.07 inch. Find the diffusion coefficient of water in the polymer by two methods—one using the short-time data and one using the long-time data.

| $t$(h) | $f$ |
|---|---|
| 22.7 | 0.086 |
| 49 | 0.15 |
| 68 | 0.21 |
| 214 | 0.37 |
| 306 | 0.48 |
| 470 | 0.59 |
| 650 | 0.65 |

**4.77** A binary gas mixture is contained in a pressurized cylindrical vessel of radius $R = 1$ cm and length $L = 20$ cm. Component A is an inert gas, and component B is initially present at a mole fraction of 0.01. B reacts on the inside wall of the vessel by a first-order reaction with a rate constant $k = 0.1$ cm/s. The diffusion coefficient of B in A/B is $D_{AB} = 0.2$ cm$^2$/s. At what time will the concentration of B be reduced to 1% of its initial value? Suppose the system is held at a lower temperature, such that $k = 0.001$ cm/s and $D_{AB} = 0.02$ cm$^2$/s. At what time will the concentration of B be reduced to 1% of its initial value?

**4.78** A liquid drop evaporates into still air. The liquid is pure and has a vapor pressure $p_{vap} = 0.5$ psi at 300 K. The diffusivity of the vapor in air is $D_{AB} = 0.2$ cm$^2$/s at 300 K. The initial radius of the drop is $R_o = 1$ mm. The liquid density is 95 g/cm$^3$. The molecular weight of the vapor is 80. Derive a model for the fractional evaporation as a function of time. After what time is the evaporation quasi-steady? What fraction has evaporated at that time, and does that amount suggest whether the quasi-steady model is accurate?

**4.79** A small liquid droplet of species B with constant radius $R$ absorbs a gas species A from a surrounding ambient medium. Conditions are such that the concentration of A on the surface of the drop, in the liquid phase, is constant at a value $C_{A,s}$. There is no A initially within the drop. The dissolved species A reacts homogeneously within the liquid phase via an irreversible first-order reaction of the form A + B → C. The mole fraction of species A is always small.

Show that the solution to this unsteady state diffusion–reaction problem may be written in the form

$$C_{A,k}(r, t) = k \int_0^t C_{A,0}(r, t')e^{-kt'} \, dt' + C_{A,0}(r, t)e^{-kt} \tag{P4.79.1}$$

where $C_{A,0}(r, t)$ is the solution to the unsteady-state diffusion problem, *in the absence of reaction*, with the same initial and surface conditions. The parameter $k$ is the first-order rate constant for the conversion of A to C that is,

$$R_A = \frac{dC_A}{dt} = -kC_A \qquad \text{(P4.79.2)}$$

**4.80** A long solid cylinder contains a highly diluted solvent A at an initially uniform concentration $C_{A,o}$. The cylinder is immersed in a stirred aqueous bath containing none of species A, and the solvent, which is also soluble in the aqueous phase, is removed from the cylinder into the bath. Transfer across the solid/fluid interface is controlled by convection. Assume that $C_A$ in the external fluid remains nearly zero. The diffusion resistance within the solid cylinder is small. Show that the fraction $f$ of solvent remaining within the solid at any time $t$ may be expressed in the form

$$f = \exp(-2\text{Bi}\,X_D) \qquad \text{(P4.80)}$$

if a particular choice of length scale $L$ is made in defining $\text{Bi} = k_c \alpha L/D$ and $X_D = Dt/L^2$. Here, $\alpha$ is the solubility coefficient in the equilibrium expression $C_A(\text{fluid}) = \alpha C_A(\text{solid})$.

Write the definition of the length scale $L$ in terms of the surface area $A$ and the volume $V$ of the cylinder.

**4.81** The following data (Fig. P4.81) are available for the sorption of $CO_2$ into a polyimide membrane. The ambient medium is at a constant partial pressure of $CO_2$ of 2.5 cmHg. Sorption is from both sides of the membrane, which has a thickness of 32 $\mu$m. At equilibrium, with $CO_2$ at 2.5 cmHg the membrane can absorb 1.56 cm$^3$ (STP) of $CO_2$ per cubic centimeter of polymer. All data are at 25°C. Calculate the diffusion coefficient of $CO_2$ within the membrane by two methods: the short-time portion of the data, and the time at which the system reaches 90% of equilibrium.

**4.82** An implant device is designed for delivery of a drug into a localized region of the brain. The device is in the shape of a cylinder, 2 mm in diameter by 1 cm in length. The device is molded from an ethylene–vinyl acetate copolymer (EVAc) that is biocompatible in brain tissue. The drug dopamine is dissolved within the EVAc phase, uniformly, at an initial concentration of 40% by weight. Provide an estimate of the half-time for delivery of the drug into the surrounding tissue. The diffusion coefficients of dopamine in the polymer and in brain tissue are $3.5 \times 10^{-8}$ cm$^2$/s and $3 \times 10^{-6}$ cm$^2$/s, respectively, all at the body temperature of 37°C. State clearly the assumptions you make in order to model this system. [See Saltzman and Radomsky, *Chem. Eng. Sci.*, **46**, 2429 (1991) for a detailed discussion of the complexities that arise in considering the diffusion and elimination of a drug in brain tissue.]

**4.83** A small liquid droplet is surrounded by a closed but well-mixed volume of gas 10 times the volume of the droplet. In the surrounding

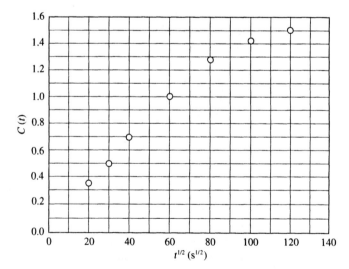

**Figure P4.81** Sorption data of $CO_2$ into a polyimide membrane. $C(t)$ is in units of cm$^3$ (STP) $CO_2$ per cubic centimeter of polymer. Toi et al., *J. Polym. Sci.: B: Polymer Phys.*, **30**, 549 (1992).

gas there is a species that is very dilute and dissolves in the droplet, reacting within it homogeneously according to first-order kinetics with respect to the soluble species. Derive an expression for the time to reduce the external gas phase concentration of the soluble species to a value $1/e = 0.368$ of its initial value. You will need to apply the principle stated in Problem P4.79.

**4.84** A spherical pressure vessel contains a binary gas mixture with a very small mole fraction of a species B. The inner surface of the vessel is coated with a reactive material, and species B reacts on that surface at a rate (per unit surface area) proportional to $C_B$. The reaction is very slow, in the sense that diffusion of B to the reactive surface does not limit the rate of conversion of B.

**a.** Write and solve a model for the fractional disappearance of B as a function of time.

**b.** Give an expression for the half-time of disappearance of B, and calculate the half-time for the case that the sphere radius is 5 cm, the rate constant for conversion of B is $k = 0.002$ cm/s, and $D_{BA} = 5$ cm²/s.

**4.85** In Example 3.5.2 of Chapter 3 we found that 5% of the carbonation was lost in 71 days. This calculation was based on a steady state analysis. Take the initial transient period into account, and give a more accurate estimate of the time. Assume that the wall of the container is initially free of dissolved $CO_2$.

**4.86** Data given in Fig. P4.86 are for the normalized weight gain of three different thicknesses ($B$) of a polymeric film exposed on both surfaces to water vapor at 21°C. (See Fig. P3.18 for vapor pressure of water vs. temperature.) What is the

diffusion coefficient of water vapor through the film?

**4.87** The film described in Problem 4.86 is used as a barrier to water intrusion into a container of volume $V = 1$ L. The film is 5 $\mu$m thick and has an exposed surface area of 10 cm². The external medium is at 21°C and has a relative humidity of 60%. The interior of the container is at 1 atm total pressure and is initially free of water vapor. Find the time for the relative humidity within the container to rise to 30%.

**4.88** Data given in Fig. P4.88 are for the normalized weight gain of three different thicknesses ($B$) of a polymeric film exposed on both surfaces to *liquid* water at 21°C. What is the diffusion coefficient of water vapor through the film?

**4.89** The film described in Problem 4.88 is used as a barrier to water intrusion into a container of volume $V = 1$ L. The film is 75 $\mu$m thick and has an exposed surface area of 50 cm². The container is immersed in water at 21°C and the interior of the container is initially free of water vapor. Find the time for the relative humidity within the container to rise to 10%.

**4.90** In most technologically important interphase transfer problems there are convective and/or diffusive resistances on both sides of the interface. Sometimes one or the other side of the interface controls the transfer. In some cases we have one chemical species being transferred in one direction across the interface, and a different species being transferred in the other direction. If both species are dilute, the transfer of each may be independent of the other.

A gas bubble is formed at a submerged orifice. When it is released from the orifice, it has a

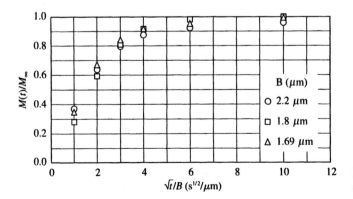

**Figure P4.86** Normalized weight gain of three films exposed to water vapor.

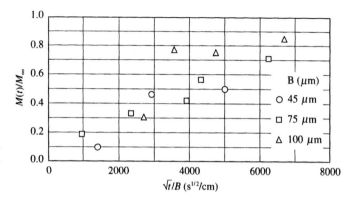

**Figure P4.88** Normalized weight gain
of three films exposed to liquid water.
Best and Moylan, *J. Appl. Polym. Sci.,*
**45,** 17 (1992).

diameter of $d_p = 0.05$ cm and an internal pressure of 15 psia. The gas is pure oxygen initially, and the bubble rises through water at 20°C. The height of the water above the orifice is $H =$ 10 cm.

**a.** *What fraction of the oxygen initially in the bubble is transferred into the water?*

**b.** *What is the mole fraction of water vapor in the bubble after 10 cm of rise?*

Use the following data to calculate your answers.

Vapor pressure of water at 20°C is 17.5 mmHg.

Diffusivity of $O_2$ in liquid water at 20°C is $D^\infty = 2 \times 10^{-5}$ cm²/s.

Diffusivity of $H_2O$ vapor in $O_2$ at 20°C is $D = 0.3$ cm²/s.

Solubility of $O_2$ in $H_2O$ at 20°C and 1 atm is $1.4 \times 10^{-6}$ mol $O_2$/cm³ $H_2O$, which corresponds to

$$\alpha = \frac{\text{mol } O_2/\text{cm}^3 \text{ } H_2O}{\text{mol } O_2/\text{cm}^3 \text{ } O_2} = 0.034 \quad \textbf{(P4.90.1)}$$

If Re for bubble rise is not small, a recommended Sherwood number correlation is

$$\text{Sh} \equiv \frac{k_c d_p}{D^\infty} = 2 + 0.6 \text{ Re}_{d_p}^{1/2} \text{ Sc}^{1/3} \quad \textbf{(P4.90.2)}$$

where the Schmidt number is defined here as $\text{Sc} = \nu/D^\infty$, and $\nu$ is the kinematic viscosity of water.

**4.91** A gas bubble rises through an aqueous solution. The bubble is saturated with water vapor, with respect to the solution. The bubble initially has 20 mol% oxygen in addition to the water vapor. The bubble rises at a steady velocity of 20 cm/s. The bubble has an initial diameter of 3 mm, and its diameter does not change signifi-

cantly as it rises. What height of liquid above the point of introduction of the bubble must be provided if the goal is to permit sufficient time for practically all the oxygen in the bubble to be delivered to the solution? A dissolved chemical agent in the solution consumes the oxygen very rapidly as it is transferred into the solution, and as a result there is no dissolved oxygen in the solution itself. Data for the water/oxygen system can be found in Problem 4.90. (Take a reduction by a factor of $10^6$ as a definition of "practically all." How does the answer change if we use a reduction by a factor of $10^5$ as a definition of "practically all"?)

**4.92** A small water droplet at 35°C is suspended in a large volume of a well-stirred gas mixture of air and $SO_2$. The mole fraction of $SO_2$ in the air is 0.05. The drop has a diameter of 2 mm. Find the time for the drop to become saturated with $SO_2$. At 35°C the vapor/liquid equilibrium is described by Henry's law in the form

$$p_{\text{vap}} [\text{atm}] = Hx \quad \textbf{(P4.92)}$$

where $x$ is the mole fraction of $SO_2$ in the water phase, and $H = 50$ atm/mole fraction. At equilibrium, give the concentration of $SO_2$ in the droplet, in parts per million by weight.

**4.93** Rework Example 4.3.3, but calculate the diffusion coefficient of $SO_2$ in water from the Wilke–Chang equation (Chapter 2).

**4.94** Rework Example 4.3.3, but do the case that the solute is carbon monoxide (CO) in water, for which the Henry's law constant is $H = 8 \times 10^4$ atm/mole fraction. Comment on the relative effects of convection and solubility on whether external mass transfer resistance is significant.

**4.95** For the acetone/water system at low mole

fractions $x$ of acetone in water, the vapor/liquid equilibrium with air at one atmosphere total pressure and 20°C is described by the relationship

$$y = 2x \qquad \text{(P4.95.1)}$$

where $y$ is the mole fraction of acetone in the vapor phase. Give the numerical value of the coefficient $\alpha$ defined as

$$\alpha \equiv \frac{C_{\text{acetone}}^{\text{air}}[\text{mol/cm}^3]}{C_{\text{acetone}}^{\text{water}}[\text{mol/cm}^3]} \qquad \text{(P4.95.2)}$$

**4.96** A long hollow cylindrical tube absorbs a vapor from a gas stream flowing in the tube as well as outside it. The internal and external flows are sufficient to yield very large values of the Biot number. The tube wall is initially free of dissolved vapor. How long will be required to bring the concentration at the mean radius to within 1% of its ultimate equilibrium value? The inner and outer radii are 2 and 4 mm, respectively. The diffusion coefficient is $10^{-6}$ cm²/s. This is not a problem we have solved in the text, but it is one that others have solved. All the required information can be found in Appendix A.

What error would you make if you neglected the curvature and used the solution for diffusion in a planar system?

**4.97** The Gurney–Lurie charts (Figs. 4.2.2) show lines for $\overline{Y}$, the average value of $Y$ over the volume of the body. Show that for a slab, $\overline{Y}/Y$ (midplane) $= (\sin \lambda_1)/\lambda_1$ for $X_D > 0.5$. (See Eq. A1.3.) What is the corresponding result for a sphere? (See Eq. A3.1.) For a long cylinder? (See Eq. A2.1).

**4.98** Repeat Problem 4.77 for the case of a *second-order* reaction on the inner surface of the cylinder, that is,

$$R_B = k_s C_B^2 \qquad \text{(P4.98)}$$

where $R_B$ is the molar rate of disappearance of species B, per unit of surface area. Do two cases: very large $k_s$ and very small $k_s$. Suggest a suitable definition of large and small $k_s$.

**4.99** An ultra-high purity gas A used in semiconductor manufacturing is shipped from the producer to the semiconductor manufacturer in pressurized spherical vessels. It is discovered that an impurity in the vessel wall acts as a catalyst for the conversion of A to a product B, which behaves as an undesired contaminant in the semiconductor manufacturing process. The manufacturer can tolerate concentrations of B up to 10 ppmw (parts per million by weight). We have the following information.

The inner radius of the storage vessel is $R = 30$ cm. The initial internal pressure in the vessel is $P = 1$ atm, at $T = 300$ K, and the vessel is maintained at 300 K throughout shipping and storage. The reaction of A proceeds according to

$$2A \rightarrow A_2 \qquad \text{(P4.99.1)}$$

that is, B is a dimer of A. The molecular weight of A is 60. The reaction is second-order in A, and

$$R_A = k_s C_A^2 \; [=] \; \frac{\text{mol of A reacted}}{\text{cm}^2 \, \text{surface} \cdot \text{s}} \qquad \text{(P4.99.2)}$$

and $k_s$ has the value $k_s = 10^{-6}$ cm⁴/mol·s at the storage conditions. Find the storage time available before the concentration of B (i.e., $A_2$) reaches 10 ppmw.

# Chapter 5

# Diffusion with Laminar Convection

$M$ass transfer often occurs in the presence of convective flow, and often the mass transfer is dominated by the details of the flow field. Another way to look at this point is to emphasize that we have the potential to control mass transfer by exercising some control over the fluid dynamics. This is a very powerful design tool, then, and so we need to learn something about the connection of fluid dynamics to mass transfer. Often the flow field is so complex—turbulent flows are a good example—that we cannot find an analytical solution for the fluid dynamics. As a consequence, we are then unable to find a solution to the convective diffusion equation and hence we cannot produce a simple mathematical model of the system of interest. Laminar flows, while usually much simpler to model, may also have sufficient complexities in their geometry to prevent us from finding a useful analytical approximation to the velocity field. In this chapter we present a series of examples of convection in the presence of laminar flow.

## 5.1 THE FALLING FILM EVAPORATOR

A vertical surface is contacted with a continuous flow of liquid flowing downward under the action of gravity. Evaporation of some volatile species occurs at the liquid/air interface. We want to calculate the rate of evaporation and develop a model of how much evaporation occurs over a length of film, as a function of the liquid film velocity. Figure 5.1.1 shows the geometry.

We assume that the liquid film is planar, very long and wide compared to the film thickness $\delta$. We assume $\delta$ is constant, independent of $x$. A velocity profile is established, given by (review your fluid dynamics)

$$v_x = \frac{\rho g \delta^2}{2\mu}\left[1 - \left(\frac{y}{\delta}\right)^2\right] \tag{5.1.1}$$

The surface ($y = 0$) velocity is

$$v_m = \frac{\rho g \delta^2}{2\mu} \tag{5.1.2}$$

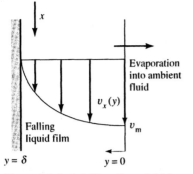

**Figure 5.1.1** A falling film of thickness $\delta$ and width $W$ normal to the page.

The volumetric flowrate is obtained by integrating the velocity profile across the film, and

$$\frac{Q}{W} = \frac{\rho g \delta^3}{3\mu} = \frac{2 v_{\mathrm{m}} \delta}{3} \tag{5.1.3}$$

The convective diffusion equation for the steady state takes the form

$$\mathbf{v} \cdot \nabla C = D\nabla^2 C \tag{5.1.4}$$

where $C$ (with no subscript) is the concentration of some volatile solute in the liquid film. Since the velocity field is one-dimensional, we write

$$\mathbf{v} = (v_x, 0, 0) \tag{5.1.5}$$

and find that the convective term simplifies to

$$\mathbf{v} \cdot \nabla C = v_x \frac{\partial C}{\partial x} \tag{5.1.6}$$

The diffusion equation takes the form

$$v_x \frac{\partial C}{\partial x} = D \left( \frac{\partial^2 C}{\partial x^2} + \frac{\partial^2 C}{\partial y^2} \right) \tag{5.1.7}$$

We *assume* that convection in the $x$ direction is faster than diffusion in the $x$ direction:

$$v_x \frac{\partial C}{\partial x} \gg D \frac{\partial^2 C}{\partial x^2} \tag{5.1.8}$$

We can show that this would be a good approximation as long as the Peclet number is large:

$$\mathrm{Pe} = \frac{v_{\mathrm{m}} \delta}{D} \gg 1 \tag{5.1.9}$$

This assumption on the relative rates of transfer in the $x$ direction (convection dominates diffusion in that direction, because there is macroscopic flow in that direction) leads to a simplification of the diffusion equation. We now want to solve

$$v_x \frac{\partial C}{\partial x} = D \frac{\partial^2 C}{\partial y^2} \tag{5.1.10}$$

with the following boundary conditions:

$$\text{at } x = 0, \quad C = C_0 \tag{5.1.11}$$

$$\text{at } y = 0, \quad C = C_a \tag{5.1.12}$$

$$\text{at } y = \delta, \quad \frac{\partial C}{\partial y} = 0 \tag{5.1.13}$$

Before leaping too deeply into the mathematics, let us review the *physical meaning* of the boundary conditions on this problem. Keep in mind that the boundary conditions are essential to the definition of the model. Equation 5.1.11 states that the solute is uniformly distributed across the film at $x = 0$. Equation 5.1.12 states that the external gas is well mixed and the solute has a fixed and known concentration at the liquid/gas interface. We could use Henry's law, for example,

$$C_a = \frac{p_a}{H} \tag{5.1.14}$$

assuming that the ambient partial pressure of the solute is known. Finally, Eq. 5.1.13 states that the solid boundary is impermeable to the solute.

Note that because of Eq. 5.1.1 the coefficient of the $\partial C/\partial x$ term in Eq. 5.1.10 is a function of $y$. This complicates the solution procedure, and we choose to introduce an approximation that gives rise to a simpler (but approximate) solution. The approximation is based on the following idea (see Fig. 5.1.2). We assume that the region over which solute is depleted by evaporation from the liquid film is confined to a *thin* layer near the liquid/gas interface. If this region is thin enough we may approximate the velocity $v_x(y)$ by its surface value $v_m$, *everywhere in this thin depletion layer.* We will make this approximation and assess the limitations it imposes on our model later.

The diffusion equation now becomes

$$v_m \frac{\partial C}{\partial x} = D \frac{\partial^2 C}{\partial y^2} \tag{5.1.15}$$

Equations 5.1.11 and 5.1.12 still serve as useful boundary conditions. Now, since our interest is confined to a thin layer near the surface, we replace Eq. 5.1.13 with the statement that there is no depletion at a distance $y$ *far* from the liquid/gas surface. We write this as

$$C = C_0 \quad \text{at} \quad y \to \infty \tag{5.1.16}$$

In effect, we are assuming that if the depletion layer is thin compared to $\delta$, then the physics of evaporation is independent of how thick the liquid layer is. Hence we can

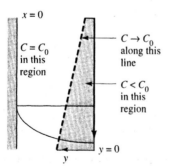

**Figure 5.1.2** Evaporation from a thin layer near the surface.

replace the boundary at $y = \delta$ by $y = \infty$. We do this because the mathematical character of the problem changes if the domain in the $y$ direction stretches to infinity, and the resulting diffusion equation has a solution in a form that is convenient for computational purposes. We dealt with a similar problem in Chapter 4 (Section 4.2.1).

Next we define a dimensionless concentration variable

$$u = \frac{C - C_0}{C_a - C_0} = 1 - Y \tag{5.1.17}$$

where $Y = (C - C_a)/(C_0 - C_a)$ would be the fractional approach to equilibrium, and Eq. 5.1.15 becomes

$$v_m \frac{\partial u}{\partial x} = D \frac{\partial^2 u}{\partial y^2} \tag{5.1.18}$$

with boundary conditions

$$u = 0 \quad \text{at} \quad x = 0 \tag{5.1.19}$$

$$u = 1 \quad \text{at} \quad y = 0 \tag{5.1.20}$$

$$u = 0 \quad \text{at} \quad y \to \infty \tag{5.1.21}$$

Now, if we define a new independent variable $\eta$ as

$$\eta = \frac{y}{(4Dx/v_m)^{1/2}} \tag{5.1.22}$$

we can show that Eq. 5.1.18 transforms to an ODE for $u(\eta)$. The details are essentially identical to the treatment of the transient diffusion problem in Section 4.2.1.

The solution is expressible in terms of the complementary error function

$$u = \text{erfc } \eta \tag{5.1.23}$$

or

$$\frac{C - C_0}{C_a - C_0} = \text{erfc} \frac{y}{(4Dx/v_m)^{1/2}} \tag{5.1.24}$$

The local evaporative flux at the surface $y = 0$ is

$$N_y(x) = -D \left.\frac{\partial C}{\partial y}\right|_{y=0} = -(C_0 - C_a) \left(\frac{Dv_m}{\pi x}\right)^{1/2} \tag{5.1.25}$$

(You will probably need to review some calculus you thought you would never see again! Check out the Leibniz rule for differentiating an integral of two variables, with the limits dependent on one of the variables. See Eq. 4.2.141.)

The total rate at which moles are evaporated over a length $L$ and width $W$ of a surface is

$$\mathsf{M} = W \int_0^L N_y(x)\, dx = -WL(C_0 - C_a) \left(\frac{4Dv_m}{\pi L}\right)^{1/2} \tag{5.1.26}$$

Equation 5.1.26 provides a design equation that relates the evaporative loss to the velocity $v_m$ and the exposed length of the surface, $L$. We should now review some of the limitations of this model. *The primary limitation arises from the assumption that the depletion layer is thin.* What do we mean by this?

We require $C(y)$ to undergo most of its change, from $C_a$ to $C_0$ (see Fig. 5.1.3), over a thin region such that $v_x \approx v_m$ within that region. From Eq. 5.1.1, if we require that

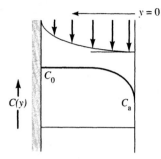

**Figure 5.1.3** The depletion layer must lie in the region of constant velocity.

$v_x$ satisfy

$$0.9 \leq \frac{v_x}{v_m} \leq 1 \tag{5.1.27}$$

then this occurs over a distance

$$0 \leq \frac{y}{\delta} \leq 0.316 \tag{5.1.28}$$

If we require that over this thin region

$$\frac{C - C_a}{C_0 - C_a} = 0.90 \quad \text{or} \quad \frac{C - C_0}{C_a - C_0} = 0.10 \tag{5.1.29}$$

then, from a table or plot of the error function, we find

$$\text{erfc}(1.15) = 0.10 \tag{5.1.30}$$

and this leads to

$$\frac{y}{(4Dx/v_m)^{1/2}} = 1.15 \tag{5.1.31}$$

Setting

$$\frac{y}{\delta} = 0.316 \tag{5.1.32}$$

we find that the depletion region is thin, as we have defined "thin," over a distance $X$ such that

$$\frac{y/\delta}{(4DX/\delta^2 v_m)^{1/2}} = \frac{0.316}{2(DX/\delta^2 v_m)^{1/2}} = 1.15 \tag{5.1.33}$$

Hence we find

$$\frac{DX}{\delta^2 v_m} = \left(\frac{0.316}{2.30}\right)^2 = 0.019 \tag{5.1.34}$$

or

$$\frac{X}{\delta} = 0.019 \frac{\delta v_m}{D} \tag{5.1.35}$$

This is a criterion for application of this *thin depletion film* approximation. The approximation holds over a distance in the direction of flow, $X$, such that Eq. 5.1.35 is valid. We may write Eq. 5.1.35 in the form

$$\frac{X}{\delta} = 0.019 \, \text{Re} \, \text{Sc} = 0.019 \, \text{Pe} \qquad (5.1.36)$$

where the *Peclet number* is Pe = Re Sc and the Reynolds number and Schmidt number for this flow are defined as

$$\text{Re} = \frac{v_m \delta}{\nu} \quad \text{and} \quad \text{Sc} = \frac{\nu}{D} \qquad (5.1.37)$$

We have assumed that the falling liquid film is laminar and that its interface with the surrounding gas is smooth, without waves or ripples. Laminar flow persists out to Reynolds numbers of about 1000, but some waviness can be observed at Reynolds numbers as low as of order 10. Waviness will enhance mass transfer to some extent, so use of this model under those circumstances would give a conservative result for the required transfer length—that is, the required length could be shorter than calculated. For the sake of the argument here, which has to do with limitations in the length of a film over which this model might give useful predictions, we will take Re = 1000 as the upper limit. Most aqueous systems would likely operate in this regime. Schmidt numbers would be of the order of 1000, so we find that an upper limit on $X/\delta$ is about $2 \times 10^4$ and is probably at least $2 \times 10^3$.

Suppose, for example, that we have an aqueous film of thickness $\delta = 0.05$ cm and a viscosity of 0.01 poise. From Eq. 5.1.2 we find (assuming $\rho = 1 \, \text{g/cm}^3$)

$$v_m = \frac{(1) \, 980 \, (0.05)^2}{2(0.01)} = 123 \, \text{cm/s} \qquad (5.1.38)$$

and we find Re = 615. For a diffusivity of $D = 10^{-5} \, \text{cm}^2/\text{s}$ we find a Schmidt number ($\nu/D$) of approximately 1000. Then from Eq. 5.1.36,

$$\frac{X}{\delta} = 0.019 \times 615 \times 1000 = 1.15 \times 10^4 \qquad (5.1.39)$$

or

$$X = 585 \, \text{cm} \qquad (5.1.40)$$

This does not look like a serious limitation.

But let us examine another aspect of the model. What fraction of the incoming solute evaporates over a given length $L$? The volatile species *enters* at the rate

$$\text{M}_0 = W C_0 \bar{v} \delta = 2/3 \, C_0 v_m \delta \, W \qquad (5.1.41)$$

(We use the fact that for a laminar falling film attached to a vertical wall, the average film velocity is 2/3 $v_m$). The fraction evaporated is

$$f = \frac{-\text{M}}{\text{M}_0} = \frac{WL(C_0 - C_a)(4Dv_m/\pi L)^{1/2}}{(2/3)C_0 v_m \, \delta W} = \frac{3}{\sqrt{\pi}} \left(1 - \frac{C_a}{C_0}\right) \left(\frac{DL}{v_m \delta^2}\right)^{1/2} \qquad (5.1.42)$$

Since the model is limited in the length $L$ over which it can be applied (see Eq. 5.1.34) we find that the model should not be used for $f$ larger than

$$f = \frac{3}{\sqrt{\pi}} \left(1 - \frac{C_a}{C_0}\right) \sqrt{0.019} = 0.23 \left(1 - \frac{C_a}{C_0}\right) \qquad (5.1.43)$$

Since $C_a/C_0$ is often small in practical applications, Eq. 5.1.43 does not represent such a serious limitation, especially in light of Eq. 5.1.40.

For an *absorber*, rather than an *evaporator*, we would write the maximum possible rate of removal of absorbable species *from* the gas phase (using $C_a$ as the equilibrium concentration of the species in the liquid with respect to the ambient gas from which absorption takes place, and $C_0$ as the entering amount) as (cf. Eq. 5.1.41)

$$M_a = 2/3\, C_a v_m \delta W \tag{5.1.44}$$

and the fractional approach to the maximum possible absorption (which would occur only for $L \to \infty$) would be

$$f = \frac{WL(C_a - C_0)(4Dv_m/\pi L)^{1/2}}{(2/3)C_a v_m \delta W} = \frac{3}{\sqrt{\pi}}\left(1 - \frac{C_0}{C_a}\right)\left(\frac{DL}{v_m \delta^2}\right)^{1/2} \tag{5.1.45}$$

**EXAMPLE 5.1.1**  *Design of a Falling Film Absorber for Oxygen*

One of the characteristics of a design problem is that very little is specified in terms of details of the operation. For example, suppose our task is to design a falling film absorber for transferring oxygen into a process stream. Aside from the identification of the system as a falling film absorber, we are told only that we will be treating 100 L/min of an aqueous solution whose density and viscosity may be taken to be those of water at 20°C and that we are to bring the solution to (near) equilibrium with respect to air at one atmosphere.

How do we begin such a task?

We could assume, for example, that the entire process stream is delivered to the top of a tall pipe in such a way that the stream overflows into the inside of the pipe and falls as a film along that inner surface. If we select what seems like a reasonable pipe diameter, 30 cm, we quickly find through the use of Eqs. 5.1.3 and 5.1.37, and the knowledge that $Q = (2/3)\,v_m \delta W$, that the Reynolds number is much too high for application of this laminar film model. We would be better off to design the system, if at all practical, to operate in a range that offers a fairly reliable design model. We choose to distribute the process stream into a set of parallel surfaces, as suggested in Fig. 5.1.4.

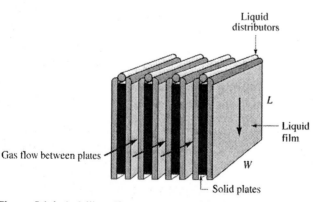

**Figure 5.1.4** A falling film oxygenator. Liquid is distributed to fall down a set of parallel plates. Gas flows horizontally between the plates.

Suppose we design a system having five such plates in parallel. (Only four are shown in the figure.) This distributes the process stream into 10 parts, each with a flowrate of only 10 L/min. We select values of $L = 50$ cm and $W = 30$ cm. Let's use our model, now, to get an idea of how this system would perform.

We can calculate the film thickness from Eq. 5.1.3, and we find

$$\delta = \left(\frac{3\mu Q}{\rho g W}\right)^{1/3} = \left[\frac{3 \times 0.01 \times (10 \times 1000/60)}{1 \times 980 \times 30}\right]^{1/3} = 0.056 \text{ cm} \qquad \textbf{(5.1.46)}$$

We should check the Reynolds number. (We find $v_m = 154$ cm/s from Eq. 5.1.2.)

$$\text{Re} = \frac{\delta v_m}{\nu} = \frac{0.056 \times 154}{0.01} = 840 \qquad \textbf{(5.1.47)}$$

This is on the high side for use of the laminar film model, primarily because of surface waves and ripples, rather than turbulence. At this stage our goal is to get a rough idea of how a usable system would operate. We will continue with the calculation, with the understanding that there will be some error associated with applying the laminar film model at a Reynolds number of order 1000.

We next calculate the extent of absorption, using Eq. 5.1.45:

$$f = \frac{3}{\sqrt{\pi}}\left(\frac{2DLW}{3Q\delta}\right)^{1/2} = 1.69\left(\frac{2 \times 2 \times 10^{-5} \times 50 \times 30}{3 \times 167 \times 0.056}\right)^{1/2} = 0.078 \qquad \textbf{(5.1.48)}$$

Note that we are assuming that the liquid enters with no oxygen ($C_0 = 0$). This result suggests that we will achieve less than 10% saturation of the liquid after one pass through the system. Assuming that the gas is air, we are not close to our goal of near equilibrium. We could seek a value of $L$ that would raise $f$ closer to unity, but we have two problems. One is that the model itself will not be very reliable as we approach equilibrium. Note the comments leading to Eq. 5.1.43. (Should we choose to move in that direction for our design, we would probably want to have a model that is valid beyond the limitations indicated by Eq. 5.1.43. We could certainly obtain such a model.)

The second point follows from inspection of Eq. 5.1.48. Since $f$ increases only with the square root of $L$, an impractical length would be required to bring the system to equilibrium in one pass. Another approach would be to reduce $Q$ further by adding more plates. But again Eq. 5.1.48 tells us that we would have to add a large number, perhaps as high as 100. The film thickness in this case would be so small that we might have difficulty maintaining a smooth wetting film. The flow might be more in the nature of a set of "rivulets" falling along the solid surface. This is an interesting fluid dynamics problem, but we will not address it here.

Instead, let's look at another route to design of this system. Suppose, as suggested in Fig. 5.1.5, that we take a fraction $\varepsilon$ of the stream coming to the bottom of the absorber and pump it back to the top of the system. This is called "recycle." We must now write some simple material balances so that we can determine the degree of absorption that occurs in this system.

The concentration entering the absorber is the result of mixing the feed stream, at concentration $C_f$, with the recycled output stream from the absorber, at concentration $C_1$. A balance on the total flow into the absorber gives us

$$Q = Q_f + \varepsilon Q \qquad \textbf{(5.1.49)}$$

where $Q$ is the total flow through the absorber, while the species balance yields

$$QC_0 = Q_f C_f + \varepsilon Q C_1 \qquad \textbf{(5.1.50)}$$

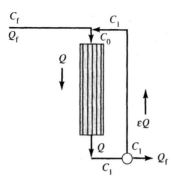

**Figure 5.1.5** An absorber with recycle.

The result is

$$C_0 = (1 - \varepsilon)C_f + \varepsilon C_1 \tag{5.1.51}$$

In the following we will assume that $C_f = 0$, so

$$C_0 = \varepsilon C_1 \tag{5.1.52}$$

There is a second relationship between these two concentrations, based on the performance of the absorber. This follows from Eq. 5.1.45 (or Eq. 5.1.48), since $f$ really represents the fractional approach to equilibrium the liquid undergoes after passing through the absorber. Hence we write

$$C_1 = C_0 + f(C_{eq} - C_0) \tag{5.1.53}$$

The result of these balances is that we can write

$$\frac{C_1}{C_{eq}} = \frac{f}{1 - \varepsilon(1 - f)} \tag{5.1.54}$$

[When $\varepsilon = 0$ there is no recycle, and we get the expected result. When $\varepsilon = 1$ there is complete recycle and $f = 1$, *but there is no output stream.* Thus the only *steady state* solution that makes physical sense for $\varepsilon = 1$ is one in which there is no external feed stream to the top of the absorber ($Q_f = 0$) and the liquid exiting the bottom of the absorber is sent right back to the top. At steady state this liquid will eventually achieve equilibrium, and continued operation will bring no further change.]

For the specific example at hand we have calculated a value of $f = 0.078$ (see Eq. 5.1.48), but that value is based on a feed of $Q_f = 10$ L/min (167 cm$^3$/s). The flow through the absorber is actually (from Eq. 5.1.49)

$$Q = \frac{Q_f}{1 - \varepsilon} \tag{5.1.55}$$

Hence Eq. 5.1.48 becomes

$$f = \frac{3}{\sqrt{\pi}} \left( \frac{2DLW(1 - \varepsilon)}{3Q_f\delta} \right)^{1/2} = 0.078 \, (1 - \varepsilon)^{1/2} \tag{5.1.56}$$

Now we must solve Eq. 5.1.54 for the value of the recycle ratio $\varepsilon$, using Eq. 5.1.56 for $f$.

The process specification was to achieve equilibrium with the ambient air. Of course if we set $C_1/C_{eq} = 1$ we would find that complete recycle ($\varepsilon = 1$) would be required, and there would be no continuous product stream. The question here is whether we can even *approach* equilibrium (e.g., $C_1/C_{eq} = 0.9$) with this absorber, at some reasonable value of $\varepsilon$. When Eq. 5.1.56 is substituted into Eq. 5.1.54 we find a value of $\varepsilon$ of $\varepsilon = 0.993$. We conclude that recycle is not a viable approach to meeting the specifications of this design. The fractional absorption per pass through the absorber is simply too low to permit this design goal to be met. Recycle yields marginal improvement in this particular system because the recycle stream is flowing *faster* than the original feed to the unit, and the higher velocity of the film (reduced residence time) produces an even less efficient absorber, since $f$ decreases with increasing $v_m$. Equation 5.1.56 shows the same thing: as $\varepsilon$ approaches unity, the value of $f$ decreases.

Let's examine one more strategy for reaching our design goal. Suppose we put a series of absorbers together, as shown in Fig. 5.1.6. Each absorber is identical to the one previously described: each has an $f$ value of 0.078. The outflow from one unit becomes the inflow to the next. Ultimately we bring the liquid close to equilibrium. The design issue here is: How many such units in series do we require?

Our basic design equation is Eq. 5.1.53, which we now apply sequentially to the output of each absorber:

$$C_1 = C_0 + f(C_{eq} - C_0) \tag{5.1.57}$$

or, subtracting $C_{eq}$ from both sides and reversing the signs,

$$C_{eq} - C_1 = C_{eq} - C_0 - f(C_{eq} - C_0) \tag{5.1.58}$$

For the second stage

$$C_2 = C_1 + f(C_{eq} - C_1) \tag{5.1.59}$$

or

$$
\begin{aligned}
C_{eq} - C_2 &= C_{eq} - C_1 - f(C_{eq} - C_1) \\
&= C_{eq} - [C_0 + f(C_{eq} - C_0)] - f\{C_{eq} - [C_0 + f(C_{eq} - C_0)]\} \\
&= C_{eq} - C_0 - 2f(C_{eq} - C_0) + f^2(C_{eq} - C_0) \\
&= (C_{eq} - C_0)(1 - f)^2
\end{aligned}
\tag{5.1.60}
$$

and generally, for the $n$th unit,

$$C_{eq} - C_n = (C_{eq} - C_0)(1 - f)^n \tag{5.1.61}$$

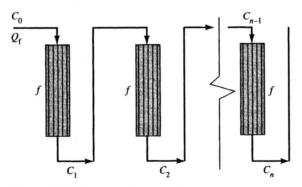

**Figure 5.1.6** Identical absorbers in series.

Thus, if we wish to raise the final output to the level $C_n = \kappa C_{eq}$, where $\kappa$ is the fractional approach to equilibrium, it follows that the number of required units in series is

$$n = \frac{\ln[(1 - \kappa)/(1 - C_0/C_{eq})]}{\ln(1 - f)} \tag{5.1.62}$$

which in our example here gives us $n = 28$. Economic factors will dictate whether the use of this many units in series is an appropriate solution to this design problem.

We can explore, as well, an alternative concept based on the idea of units in series. The key characteristic of the improved performance of a sequence of units is that we periodically mix the film. The average concentration, which is well below that of the surface, is then exposed to the ambient medium, and the driving force for mass transfer — the concentration difference between the surface and the ambient — is greatly increased. Figure 5.1.7 suggests how this might be done if we had a single unit, but the falling film was periodically interrupted and mixed. Here the inside of a large pipe becomes the surface along which the film falls, and an internal rotating set of wiper blades, much like the windshield wipers on an automobile, interrupts the film as it falls. An element of liquid would experience absorption over an interval of time corresponding to the distance between successive wipes. The film would then be mixed and would continue to fall and absorb, but with a new surface concentration that has been much reduced by the mixing. Such a unit is called a "wiped film" absorber.

The performance of this system may be modeled with Eq. 5.1.61 above. The length between wipes is $L_f$, and this can be approximated as $v_m t_w$, where $v_m$ is the *surface* velocity of the film (not the average velocity) and $t_w$ is the time between wipes. This time is determined roughly by the rotational speed of the wiper blade,

$$t_w = \frac{1}{n_B \Omega} \tag{5.1.63}$$

where $\Omega$ is the rotational speed in revolutions per time and $n_B$ is the number of blades around the circumference of the pipe.

We can reexamine our design study, and determine the performance of this system as a function of the characteristics of the wiper. We begin by writing Eq. 5.1.45 (with $C_0 = 0$) in the form

$$f = 1.69 \left(\frac{D}{n_B \Omega \delta^2}\right)^{1/2} \tag{5.1.64}$$

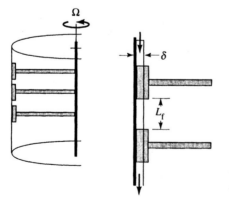

**Figure 5.1.7** A wiped film absorber. A rotating shaft with wiper blades periodically interrupts and mixes the falling film.

We continue to use the values

$$D = 2 \times 10^{-5} \, \text{cm}^2/\text{s} \quad \text{and} \quad \delta = 0.056 \, \text{cm}$$

and we select $n_B = 6$. We want to know the required height of a wiped film unit that will yield liquid with an average concentration of 90% of the equilibrium value it *could* attain with respect to the gas that it is in contact with.

In the absence of wiping the required length follows from Eqs. 5.1.61 (with $n = 1$) and 5.1.48, and the result is

$$L = 4934 \, \text{cm} \tag{5.1.65}$$

By comparison, for rotation at 1 rev/s, with $n_B = 6$, we find that the film falls for approximately 20 cm between wipes (mixes), and the total required length is 834 cm. This is a considerable improvement over the unwiped case.

We leave our example at this point, noting as always that the model used is based on a number of assumptions that are questionable and would have to be examined. For example, we assume that the film is in fully developed flow between wipes. In practice, however, there would be some length over which the velocity field is being developed, and this would affect the rate of transfer into the film. The film would move more slowly until achieving a velocity consistent with the force of gravity acting on the film. Thus the residence time would be increased by these periodic regions of acceleration, and the transfer would be enhanced over that calculated.

In addition, we neglected the effect of the wiper blades on the surface of the film, hence on the transfer across the surface. It seems likely that transfer would be enhanced by any disturbance to the film surface, and so with respect to this specific issue, our result would overestimate the required length. Thus the model put forth here would seem to be conservative in its prediction of the required length.

## 5.2 DISSOLUTION OF A SOLID FILM BY A FLOWING SOLVENT

Many of the characteristics of the falling film evaporator are found in the case of a solid surface that is soluble in a liquid flowing parallel to the surface. Figure 5.2.1 shows the physical situation of interest. A laminar flow is moving parallel to a solid planar surface at $y = 0$. This flow could be confined within a tube or parallel plate channel, or it could be a liquid film of thickness $\delta$ falling under gravity along the solid surface. The analysis that follows is valid for any of these cases.

Beyond the point $x = 0$, the surface is soluble in the liquid, and a concentration profile develops as the material dissolves and diffuses into the flowing stream. If the solid is only slightly soluble, then at least near $x = 0$ the extent of penetration of the

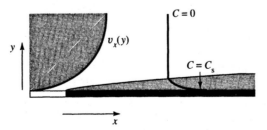

**Figure 5.2.1** Dissolution of a solid by a flowing solvent.

solute into the flow may be small. But what do we mean by "small"? A thing is small only in comparison to something else. In this case, we wish to assume that the concentration field is confined to a region close enough to the solid surface that the velocity profile may be approximated as a linear function of $y$. We will see why we want to say this in a moment.

The velocity field is assumed to be one-dimensional:

$$\mathbf{v} = (v_x, 0, 0) \tag{5.2.1}$$

The convective diffusion equation for the steady state takes the form

$$\mathbf{v} \cdot \nabla C = D_b \nabla^2 C \tag{5.2.2}$$

where $C$ (with no subscript) is the concentration of the solute in the liquid, and $D_b$ is the diffusivity of the solute in the external (bulk) liquid.

The convective term simplifies to

$$\mathbf{v} \cdot \nabla C = v_x \frac{\partial C}{\partial x} \tag{5.2.3}$$

and the diffusion equation takes the form

$$v_x \frac{\partial C}{\partial x} = D_b \left( \frac{\partial^2 C}{\partial x^2} + \frac{\partial^2 C}{\partial y^2} \right) \tag{5.2.4}$$

As in the previous example we *assume* that *convection* in the $x$ direction is faster than *diffusion* in the $x$ direction:

$$v_x \frac{\partial C}{\partial x} \gg D_b \frac{\partial^2 C}{\partial x^2} \tag{5.2.5}$$

We now want to solve

$$v_x \frac{\partial C}{\partial x} = D_b \frac{\partial^2 C}{\partial y^2} \tag{5.2.6}$$

with the following boundary conditions:

$$\text{at } x = 0, \quad C = 0 \tag{5.2.7}$$

$$\text{at } y = 0, \quad C = C_s \tag{5.2.8}$$

The first condition expresses the assumption that the liquid is free of the soluble species upon entering the region $x = 0$. The second condition gives the solubility of the solid in the liquid, which we will take to be constant along the length $x = L$ of soluble surface. Further, we assume that the extent of dissolution is so small that the solid surface remains in the plane $y = 0$.

For a third boundary condition we will assume that the penetration of solute into the flow is small in comparison to some length scale of the flow field in the $y$ direction. This length scale might be the diameter of the pipe or the wall-to-wall separation of a channel through which the liquid flows. Really, what we require is a penetration length that is small compared to the distance over which the velocity profile changes, because we are going to assume that throughout the penetration region (i.e., the distance in the $y$ direction over which $C$ falls from $C_s$ to zero), the velocity profile is linear. This assumption will put Eq. 5.2.6 in a form that yields an analytical solution useful for computational purposes. That form is

$$\alpha y \frac{\partial C}{\partial x} = D_b \frac{\partial^2 C}{\partial y^2} \tag{5.2.9}$$

where $\alpha$ is simply the velocity gradient at $y = 0$. For different flow fields (e.g., tube flow, channel flow, a falling film), $\alpha$ would have different forms.

The third boundary condition is then written as

$$C = 0 \text{ for } y = \infty \qquad (5.2.10)$$

This "boundary value problem" may be solved by the method of combination of variables, and the result (see Problem 5.4) is

$$\frac{C}{C_s} = \frac{\int_\eta^\infty \exp(-\eta^3)\, d\eta}{\int_0^\infty \exp(-\eta^3)\, d\eta} \qquad (5.2.11)$$

where the function $\eta$ is defined as a combination of the space variables $x$ and $y$:

$$\eta = y \left( \frac{\alpha}{9 D_b x} \right)^{1/3} \qquad (5.2.12)$$

The denominator of Eq. 5.2.11 is a pure number: it is a so-called gamma function $\Gamma$, and in this case it is $\Gamma(4/3)$, and has the value $\Gamma(4/3) = 0.893$.

The local solute flux along the surface $y = 0$ is

$$N_y(x) = - D_b \frac{\partial C}{\partial y} \bigg|_{y=0} = D_b C_s \frac{(\alpha/9 D_b x)^{1/3}}{\Gamma(4/3)} \qquad (5.2.13)$$

(See the comment following Eq. 5.1.25.)

Per unit width $W$, the total rate at which moles are dissolved from a length $L$ of a surface is

$$M_L = - D_b \int_0^L \frac{\partial C}{\partial y} \bigg|_{y=0} dx = \frac{2 D_b C_s}{\Gamma(7/3)} \left( \frac{\alpha L^2}{9 D_b} \right)^{1/3} \qquad (5.2.14)$$

[The gamma function has the property $\Gamma(n + 1) = n\Gamma(n)$.]

We might be interested in the average molar concentration of the solute in the liquid after the position $x = L$ is passed. If $Q_w$ is the volumetric flowrate of liquid past the surface (*per unit width*), then the average concentration $\langle C \rangle$ is found from a molar balance in the form

$$Q_w \langle C \rangle_L = M_L \qquad (5.2.15)$$

(i.e., we are equating the rate of flow of moles past the position $x = L$ to the rate at which moles dissolve into the flowing liquid, all *per unit width* of the surface.)

It is important to note that if we had calculated the average concentration at the position $x = L$ from Eq. 5.2.11, by setting $x = L$ in Eq. 5.2.12 and then integrating $C(\eta)$ over the coordinate axis $y$, we would not have obtained the result yielded by Eq. 5.2.15. Accepting that statement (your alternative is to do the calculus!), you must wonder how we can have two different averages. The answer is that there are in fact two averages in this system, this is common to convective transport problems, and they have different physical meanings. Figure 5.2.2 serves to aid our discussion of this.

Imagine three tubes parallel to the flow field, labeled 1–3 in Fig. 5.2.2. These tubes do not disturb the flow or retard the diffusion; they are "control volumes" for bookkeeping on the flow of moles through the system. Coming out of tube 1 is a flow that has a high velocity but zero concentration of solute. In fact, if the penetration layer is thin with respect to some dimension of the tube or channel, or to the film thickness in the case of the falling film, a lot of the fluid that exits the system will be carrying no, or very little, solute. Tube 2 samples from a region that has a lower velocity but a higher

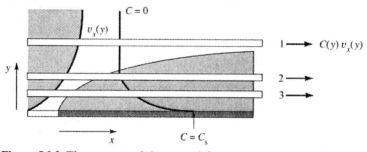

**Figure 5.2.2** The concept of the cup-mixing average concentration.

concentration of the soluble species. Tube 3 yields liquid with a very high concentration of solute, but the liquid exits that tube very slowly because it is near the solid boundary where the velocity goes to zero. If we took the liquid coming from each "pipeline" and mixed all these flows in a cup, we would obtain a concentration that is called the cup-mixing average. Analytically, the cup-mixing average is calculated from

$$C_{cm} = \frac{\int v_x(y) C(x, y)\, dA}{\int v_x(y)\, dA} \qquad (5.2.16)$$

We are to understand that these integrals are over the cross section of the pipe or channel or film through which there is flow. The denominator is just the volumetric flowrate $Q$ (per unit width). The numerator integrates the concentration profile, but it weights each concentration by the local velocity. Hence this integral accounts for the dependence of what comes out of the end of the pipe or channel not only on the concentration profile but also on the velocity profile. By contrast, the usual average concentration is just

$$\bar{C} = \frac{\int C(x, y)\, dA}{\int dA} \qquad (5.2.17)$$

Once the mathematical details have been settled, we can turn to this model and use it to calculate the rate at which a surface dissolves into a flowing solvent.

**EXAMPLE 5.2.1** *Cleaning a Solid Residue From Inside a Tube*

A long tube of inside diameter 2 cm is coated with solid benzoic acid over a region of length $L = 10$ cm centered about the midpoint of the tube. Water is flowing through the tube at a Reynolds number of 1200. How long will it take to dissolve all of the solid film?

Assume a temperature of 25°C, a diffusivity of benzoic acid in water of $1.2 \times 10^{-5}$ cm²/s, and a solubility in water of 0.0034 g/cm³. The initial film thickness $B_o$, which is uniform along the axis of the tube, is 10 $\mu$m.

We begin with the assumption that removal is limited by diffusion and convection into the solvent. We need an expression for the rate of dissolution. Equation 5.2.13 gives the *local* rate (actually, the flux) while Eq. 5.2.14 gives the *total* rate (per unit width) over the entire length $L$. We use the latter result. Note that the equation is written in terms of *molar* units, while our solubility data are in *mass* units. It does not take too much thought to realize that the same form of the equation holds if $C_s$ is in mass units. Take the density of solid benzoic acid as $\rho_s = 1.32$ g/cm$^3$. The initial mass of benzoic acid is

$$\pi DB_oL\rho_s = 3.14 \times 2 \times 0.001 \times 10 \times 1.32 = 0.083 \text{ g} \qquad (5.2.18)$$

Hence we must find the time required to dissolve 0.083 g.

The rate of mass dissolution is $\pi DM_L$ (Remember: M is the rate *per unit width* W. In this case the appropriate width is the circumference of the tube.) Thus the rate is

$$\pi DM_L = \frac{2\pi DD_bC_s}{\Gamma(7/3)}\left(\frac{\alpha L^2}{9D_b}\right)^{1/3} \qquad (5.2.19)$$

On the right-hand side, we know everything but $\alpha$, which was defined as the velocity gradient at the solid surface. For laminar Newtonian tube flow (review your fluid dynamics),

$$\alpha = \frac{8V}{D} \qquad (5.2.20)$$

where $V$ is the average velocity of flow. Since the Reynolds number is given as Re = 1200 and the density and viscosity of water are 1 g/cm$^3$ and 0.01 poise, respectively, we find

$$V = \frac{\mu \, \text{Re}}{\rho D} = \frac{0.01 \times 1200}{1 \times 2} = 6 \text{ cm/s} \qquad (5.2.21)$$

and so

$$\alpha = \frac{8 \times 6}{2} = 24 \text{ s}^{-1} \qquad (5.2.22)$$

We also need to look up the value of the gamma function (Appendix B: Eq. B.3.2 and Fig. B.3.1) and we find that $\Gamma(7/3) = 1.19$.

Now all the necessary numbers are available and we find, from Eq. 5.2.19, that that rate is

$$\pi DM_L = 1.2 \times 10^{-4} \text{ g/s} \qquad (5.2.23)$$

The required time, then, is

$$\frac{0.083 \text{ g}}{1.2 \times 10^{-4} \text{ g/s}} = 692 \text{ s} = 11.5 \text{ min} \qquad (5.2.24)$$

Wait a minute! We forgot something. The local dissolution rate *varies* along the surface of the solid film. It is possible that the region near $x = 0$ will be completely cleaned and free of solid benzoic acid film while there is still some undissolved film in the region toward $x = L$. Equation 5.2.19 assumes that the total length of the dissolving region remains constant, while in fact toward the end of the process the coated length $L$ will be decreasing with time. This will slow down the *total* dissolution rate (note that $\pi DM_L$ depends on $L$ to a positive power), but increase the *local* dissolution rate. Hence we have to carry out a more complicated analysis that includes the possibility that a growing portion of the length $L$ may be uncoated if we examine sufficiently long times.

We begin with Eq. 5.2.13:

$$N_y(x) = \frac{D_bC_s}{\Gamma(4/3)}\left(\frac{\alpha}{9D_bx}\right)^{1/3} \qquad (5.2.25)$$

From this result we find that the time required to dissolve a film of thickness $B_o$, at some point $x$, is just

$$t_d = \frac{B_o C_d}{N_y(x)} = \frac{\Gamma(4/3) B_o C_d}{D_b C_s} \left(\frac{9 D_b x}{\alpha}\right)^{1/3} \tag{5.2.26}$$

where $C_d$ is the molar density of the coating in the solid phase. This result is valid as long as no portion of the region $0 < x < L$ becomes completely depleted of the coating. For sufficiently long times, however, a bare region will appear and its length $x_d$ will grow in time, as suggested in Fig. 5.2.3.

We will develop an approximate model of the dissolution process along the following lines. We will use Eq. 5.2.26 as an approximation for the time required to deplete a length $x_d$ of surface, when $B_o$ is the *initial* coating thickness. We write the result in the form

$$t_d = \frac{\Gamma(4/3) B_o C_d}{D_b C_s} \left(\frac{9 D_b}{\alpha}\right)^{1/3} x_d^{1/3} \tag{5.2.27}$$

Next, we write a "dynamic" equation for the coating thickness as a function of time. As long as the entire length $L$ of the surface is coated, the rate of dissolution may be written as

$$-C_d \frac{\partial B}{\partial t} = \frac{D_b C_s}{\Gamma(4/3)} \left(\frac{\alpha}{9 D_b}\right)^{1/3} x^{-1/3} \tag{5.2.28}$$

To account for a portion of bared surface we replace $x$ with $x - x_d$:

$$-C_d \frac{\partial B}{\partial t} = \frac{D_b C_s}{\Gamma(4/3)} \left(\frac{\alpha}{9 D_b}\right)^{1/3} (x - x_d)^{-1/3} \tag{5.2.29}$$

But $x_d$ is a function of time, given by Eq. 5.2.27:

$$x_d = \left[\frac{D_b C_s}{\Gamma(4/3) B_o C_d} \left(\frac{\alpha}{9 D_b}\right)^{1/3}\right]^3 t^3 \tag{5.2.30}$$

When this result is substituted into Eq. 5.2.29, we obtain a model that accounts, approximately, for the effect of the growing bare region on the local flux. With the following definitions of dimensionless variables as

$$t^* = \left[\frac{D_b C_s}{\Gamma(4/3) B_o^{4/3} C_d} \left(\frac{\alpha}{9 D_b}\right)^{1/3}\right] t = \left(\frac{x_d}{B_o}\right)^{1/3}$$

$$x^* = \frac{x}{B_o} \qquad B^* = \frac{B(x^*, t^*)}{B_o} \tag{5.2.31}$$

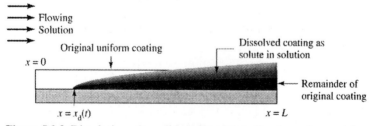

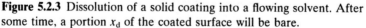

**Figure 5.2.3** Dissolution of a solid coating into a flowing solvent. After some time, a portion $x_d$ of the coated surface will be bare.

Eq. 5.2.29 becomes

$$-\frac{\partial B^*}{\partial t^*} = (x^* - t^{*3})^{-1/3} \tag{5.2.32}$$

with the initial condition

$$B^* = 1 \quad \text{at} \quad t^* = 0 \tag{5.2.33}$$

The solution may be written in the form

$$B^*(t^*, x^*) = 1 - \int_0^{t^*} \frac{ds^*}{(x^* - s^{*3})^{1/3}} \tag{5.2.34}$$

With this expression we could find the thickness profile $B(x^*, t^*)$, of a dissolving coating at any time $t^*$, by fixing $t^*$ in the upper limit of the integral above, and integrating numerically for a series of $x^*$ values. The integration, of course, would be over the "dummy" time variable $s^*$.

From Eq. 5.2.34 we can find the time required to remove all the soluble film. This corresponds to the condition

$$B^*(t^*, L^*) = 0 \tag{5.2.35}$$

where

$$L^* = \frac{L}{B_o} \tag{5.2.36}$$

Hence we need to find the time $t^*$ at which

$$\int_0^{t^*} \frac{ds^*}{(L^* - s^{*3})^{1/3}} = 1 \tag{5.2.37}$$

where $s^*$ is the "dummy variable" of integration. If we now define

$$\tau^* = \frac{t^*}{L^{*1/3}} \tag{5.2.38}$$

this integral may be written as

$$\int_0^{\tau^*} \frac{du^*}{(1 - u^{*3})^{1/3}} = 1 \tag{5.2.39}$$

where $u^* = s^*/L^{*1/3}$ is the new dummy variable of integration. It is easy to integrate this numerically to find

$$\tau^* = \frac{t^*}{L^{*1/3}} = 0.91 \tag{5.2.40}$$

This result takes into account the effect of the receding film from the leading edge of the surface. If we were to neglect this effect, Eq. 5.2.25 would provide the model for the local flux, and the amount of coating remaining after time $t$ would follow from the solution to

$$-\frac{dm}{dt} = M_w \int_0^L N_y(x)\, dx \tag{5.2.41}$$

Here we introduce the *mass* of coating, $m$, by using the molecular weight of the coating $M_w$.

It is instructive to follow this line of analysis and get some idea of the error associated with neglecting the effect of the receding film. The mass of coating remaining is $m(t)$. Upon introducing Eq. 5.2.25 into Eq. 5.2.41, we find

$$-\frac{dm}{dt} = \frac{M_w D_b C_s}{\Gamma(4/3)}\left(\frac{\alpha}{9D_b}\right)^{1/3}\int_0^L x^{-1/3}\,dx = \frac{3M_w D_b C_s}{2\Gamma(4/3)}\left(\frac{\alpha}{9D_b}\right)^{1/3} L^{2/3} \quad (5.2.42)$$

for the surface coated over the full length $L$. The initial mass of film is (per unit width)

$$m_o = M_w C_d B_o L \quad (5.2.43)$$

so

$$-\frac{d(m/m_o)}{dt} = \frac{3D_b C_s}{2\Gamma(4/3)C_d B_o L^{1/3}}\left(\frac{\alpha}{9D_b}\right)^{1/3} \quad (5.2.44)$$

or

$$-\frac{d(m/m_o)}{dt*} = \frac{3}{2L*^{1/3}} \quad (5.2.45)$$

Then it follows that

$$\frac{m(t)}{m_o} = 1 - \frac{3t*}{2L*^{1/3}} \quad (5.2.46)$$

From this, and setting $m(t) = 0$, we find that the time to completely dissolve the coating would be

$$\tau* = \frac{t*}{L*^{1/3}} = \frac{2}{3} \quad (5.2.47)$$

if we neglect the receding film, compared to the value $\tau* = 0.91$ when the effect of the receding film is accounted for.

Hence the time to dissolve the entire coating, arrived at in Ex. 5.2.1, is not 692 s, but

$$692\left(\frac{0.91}{2/3}\right) = 944 \text{ s} \quad (5.2.48)$$

## 5.3 THE ARTIFICIAL KIDNEY (DIALYSIS ANALYSIS)

An artificial kidney can be designed based on the principles of convective transport to a membrane in contact with (and separating) blood and a second diluting or "dialysis" fluid. Figure 5.3.1 shows the details of a single unit of the system, relevant to the model

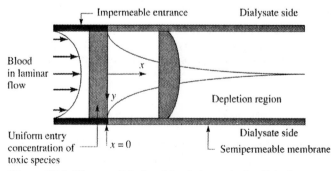

**Figure 5.3.1** The parallel plate blood channel of a dialysis system.

we will develop here. Because the red cells in the blood are fragile and easily damaged by high shear stresses, the blood flow is maintained in the laminar regime at a relatively low flowrate. Hence the blood side of the system provides the major resistance to mass transfer of species from the blood, across the membrane, and into the dialysis fluid. The dialysis fluid is pumped rapidly enough to ensure that the concentration of any species transferred into it across the membrane is effectively zero. While any fluid could serve that purpose, the dialysis fluid used in medical applications contains species that are in balance with various electrolytes in the blood that are *not* to be removed. Of course, in the medical application the dialysis fluid must be sterile, and free of any species that is not to be transported into the blood.

Our concern here is to develop a model of dialysis that indicates how specific design parameters, such as the flowrate of the blood, the channel separation, and the length of the channel, affect the performance of the system. The primary measure of performance is the fraction of toxic material removed in a single pass of blood through the membrane unit.

Blood enters in laminar flow with a uniform concentration of the toxic species. At $x = 0$ and beyond, the channel is permeable to that species, but not to the fluid (blood) itself. Hence the flow does not change, but the concentration of the permeable species decreases as the blood passes down the channel.

The toxic species is convected with the blood at a velocity $u(y)$. The flow is taken to be laminar and Newtonian, and it is easily shown that the velocity profile is

$$u(y) = \frac{3}{2} U \left[ 1 - \left( \frac{y}{H} \right)^2 \right]$$  (5.3.1)

where $U$ is the average velocity across the channel. (Note that $y = 0$ is the midplane.) Hence the flow per unit width of channel is

$$Q_W = 2UH$$  (5.3.2)

We assume, in this analysis, that the channel is very wide compared to $H$ and that the flow and the mass transport are two-dimensional (no variations in the $z$ direction).

For the steady state operation, the convective diffusion equation on the blood side of the system (i.e., in the region $[-H, H]$) takes the form

$$u(y) \frac{\partial C}{\partial x} = D_b \frac{\partial^2 C}{\partial y^2}$$  (5.3.3)

with $u(y)$ given by Eq. 5.3.1.

As is always the case, the specification of boundary conditions determines much of the character of the solution to this equation. Hence we discuss these conditions now.

We must supply an entrance condition regarding the concentration of the toxic species at the plane $x = 0$. We take the concentration to be uniform there, and write

$$C = C_o \quad \text{at} \quad x = 0 \quad \text{for all } y$$  (5.3.4)

We assume symmetry along the midplane of the channel, and so we write

$$\frac{\partial C}{\partial y} = 0 \quad \text{at} \quad y = 0 \quad \text{for all } x$$  (5.3.5)

At the blood/membrane interface we must express the continuous character of the transport of the toxic species across that plane. Hence we equate the fluxes at $y = H$. For the membrane we write a permeability relationship in the form

$$\text{molar flux} = PC_w \quad \text{at} \quad y = H$$  (5.3.6)

We define the flux to be positive if it is from the blood to the dialysate side. The membrane permeability P includes a partition coefficient for the species between the blood phase and the membrane, as well as a diffusivity of the toxic species through the membrane. $C_w$ is the concentration at the wall $(y = H)$, but on the blood side. This form of the flux through the membrane assumes that the species is highly diluted on the dialysate side. With this model for the membrane, we may write the expression of continuity of flux at the boundary $y = H$ in the form

$$-D_b \frac{\partial C}{\partial y} = P C_w \qquad \text{at} \quad y = H \quad \text{for all } x \qquad (5.3.7)$$

(The minus sign in front of the diffusive flux reflects our definition of a positive flux and our choice of coordinate plane $y = 0$.)

We now have enough boundary conditions to permit a solution to the convective diffusion equation. Note that the equation (Eq. 5.3.3) is linear, but because of the velocity term it has nonconstant coefficients. The equation, with this set of linear boundary conditions, can be solved analytically by a series method. Mathematically, we call the equation and boundary conditions a "Sturm–Liouville problem," for which a solution procedure can be found in a number of applied mathematics texts.[1] We simply present the solution here, without derivation. We first nondimensionalize the equations in the following manner:

*For the dependent variable:*

$$C^* = \frac{C}{C_o} \qquad (5.3.8)$$

*For the independent variables:*

$$y^* = \frac{y}{H} \qquad x^* = x \frac{D_b}{UH^2} \qquad (5.3.9)$$

In the course of the nondimensionalization, the boundary condition at the membrane surface will produce a nondimensional permeability, which we identify and write as

$$P^* = \frac{PH}{D_b} \qquad (5.3.10)$$

We refer to $P^*$ as a membrane Biot modulus.

Now the differential equation to be solved takes the form

$$\frac{3}{2}(1 - y^{*2}) \frac{\partial C^*}{\partial x^*} = \frac{\partial^2 C^*}{\partial y^{*2}} \qquad (5.3.11)$$

and the boundary conditions are

$$C^* = 1 \qquad \text{at} \quad x^* = 0 \quad \text{for all } y^* \qquad (5.3.12)$$

$$\frac{\partial C^*}{\partial y^*} = 0 \qquad \text{at} \quad y^* = 0 \quad \text{for all } x^* \qquad (5.3.13)$$

---

[1] For example, Rice and Do, *Applied Mathematics and Modeling for Chemical Engineers,* Wiley, New York, 1995.

$$-\frac{\partial C^*}{\partial y^*} = \mathsf{P}^\star C^* \qquad \text{at} \quad y^* = 1 \quad \text{for all } x^* \tag{5.3.14}$$

The analytical solution for $C^*$ takes the form

$$C^* = \sum_{m=1}^{\infty} A_m \exp\left(\frac{-2\lambda_m^2 x^*}{3}\right) \sum_{n=0}^{\infty} a_{nm} y^{*n} \tag{5.3.15}$$

To evaluate this infinite series solution, we must have values for the coefficients $a_{nm}$ and $A_m$, and for the eigenvalues $\lambda_m$. These numbers depend on the membrane Biot modulus $\mathsf{P}^*$.

We are not really interested in the $y^*$ dependence of $C^*$. We do want to know how much of the toxic material has passed through the membrane after any position $x^*$ downstream from the inlet has been reached. Alternatively, we want to know the change in concentration of the toxic species accomplished by passage of the blood down a length of the channel. A simple material balance tells us that the rate of removal of the toxic species, averaged over the length of the channel, is (per unit width of channel),

$$M = 2UH(C_o - C_{cm}) \tag{5.3.16}$$

where $C_{cm}$ is the cup-mixing averaged defined earlier (see Eq. 5.2.16). Hence we need to obtain the cup-mixing average from the concentration profile given in Eq. 5.3.15, so we must evaluate the integral

$$C_{cm} = \frac{\int_0^H \frac{3}{2} U \left(1 - \left[\frac{y}{H}\right]^2\right) C(x,y)\, dy}{\int_0^H \frac{3}{2} U \left(1 - \left[\frac{y}{H}\right]^2\right) dy} \tag{5.3.17}$$

Equation 5.3.17 can be written in terms of $C^*$ as

$$C_{cm}^* = \frac{3}{2} \int_0^1 (1 - y^{*2}) \left[\sum_{m=1}^{\infty} A_m \exp\left(\frac{-2\lambda_m^2 x^*}{3}\right) \sum_{n=0}^{\infty} a_{nm} y^{*n}\right] dy^* \tag{5.3.18}$$

This is not difficult to show. (You guessed it! It's a homework problem.)

After performing the indicated integration, Eq. 5.3.18 takes the form

$$C_{cm}^* = 3 \left[\sum_{m=1}^{\infty} A_m \exp\left(\frac{-2\lambda_m^2 x^*}{3}\right) \sum_{n=0}^{\infty} \frac{a_{nm}}{(n+1)(n+3)}\right] \tag{5.3.19}$$

This is getting to be a mess! We don't want to let the arithmetic obscure the physics. Let's review where we are at this point.

The dimensionless cup-mixing concentration of the toxic species is given by the double infinite series expression of Eq. 5.3.19. This is the primary measure of the performance of the dialysis system. To evaluate this expression, we need the eigenvalues $\lambda_m$, the coefficients $a_{nm}$, which depend on the specific $\lambda_m$ values, and the constants $A_m$, which depend on the coefficients $a_{nm}$. Hence everything revolves about the eigenvalues of this boundary value problem. The eigenvalues, in turn, depend on the membrane Biot modulus $\mathsf{P}^*$. If the mathematical details are examined we find the following results:

The $\lambda_m$ are the roots of a polynomial of the form

$$0 = \mathsf{P}^* - \left(\frac{2}{3} + \frac{5}{12}\mathsf{P}^*\right)\lambda_m^2 + \left(\frac{1}{20} + \frac{1}{45}\mathsf{P}^*\right)\lambda_m^4 - \cdots \tag{5.3.20}$$

The $a_{nm}$ depend on the eigenvalues as follows:

$$a_{0m} = 1 \qquad\qquad a_{1m} = 0$$

$$a_{2m} = \frac{-\lambda_m^2}{2} \qquad\qquad a_{3m} = 0$$

$$a_{4m} = \frac{\lambda_m^2 (2 + \lambda_m^2)}{24} \qquad\qquad a_{5m} = 0 \qquad\qquad (5.3.21)$$

$$a_{6m} = \frac{-\lambda_m^4 (14 + \lambda_m^2)}{720}, \ldots$$

The $A_m$ coefficients are more complicated and must be calculated from the following expression:

$$A_m = \frac{\displaystyle\sum_{n=0}^{\infty} \frac{a_{nm}}{(n+1)(n+3)}}{\displaystyle\sum_{n=0}^{\infty} \sum_{p=0}^{p=n} \frac{a_{pm} a_{(n-p)m}}{(n+1)(n+3)}} \qquad\qquad (5.3.22)$$

We can simplify our model to some degree. We find, for example, that except very near the entrance to the channel, the exponential terms in the summation in Eq. 5.3.19 decay very rapidly, and the first term of the series provides a useful approximation over most of the channel. Thus we need just the first eigenvalue. This is the smallest root of Eq. 5.3.20. To a good approximation we may write this as

$$\lambda_1^2 = \frac{P^*}{\dfrac{2}{3} + \dfrac{5}{12} P^*} \qquad\qquad (5.3.23)$$

We show $\lambda_1$ as a function of $P^*$ in Fig. 5.3.2. This approximation (Eq. 5.3.23) is quite accurate for small values of $P^*$ and is only about 10% low for $P^* \gg 1$. (The exact value of $\lambda_1$ in the limit of very large values of $P^*$ is 1.68.)

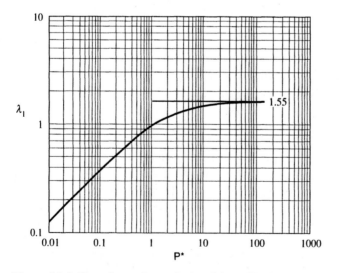

**Figure 5.3.2** First eigenvalue, calculated from Eq. 5.3.23.

The first term of the series solution for $C_{cm}^*$ then takes the form

$$C_{cm}^* = 3A_1 \left( \sum_{n=0}^{\infty} \frac{a_{n1}}{(n+1)(n+3)} \right) \exp \left( \frac{-2\lambda_1^2 x^*}{3} \right) \tag{5.3.24}$$

The coefficient $A_1$ is given by Eq. 5.3.22, with $m = 1$ in the sums, and we find the following expression:

$$A_1 = \frac{a_{01}/3 + a_{21}/15 + a_{41}/35 + \cdots}{a_{01}^2/3 + 2a_{01}a_{21}/15 + (2a_{01}a_{41} + a_{21}^2)/35 + \cdots} \tag{5.3.25}$$

With Eqs. 5.3.21 and 5.3.23 we may write this in terms of the membrane parameter $P^*$. In the same manner we may write the first parenthetical expression in Eq. 5.3.24 as a function of $P^*$. Then for a given value of $P^*$ we may write the cup-mixing concentration as a function of the distance down the length of the membrane channel. Figure 5.3.3 plots $C_{cm}(x^*)$ for several values of the membrane parameter $P^*$. Note that the $x$ axis is actually the product $P^*x^*$. We will see shortly that as $P^*$ gets small, all the lines converge to the lowest line, labeled $P^* \approx 0$.

Transport of the toxic species out of the blood channel requires diffusion through the liquid and permeation through the membrane. In a sense the toxic species diffuses through two resistances in series. When the membrane resistance to transport is high with respect to that in the blood, which corresponds to the case of small values of the parameter $P^*$, a simpler mathematical model can be derived. One route to the derivation relies on the complicated analytical solution we have just struggled through. We take the limit of small $P^*$ for the eigenvalues and the coefficients based on them and derive the desired result. Alternatively, we can begin at the beginning, a species balance along the channel, and derive the simpler model directly. We will illustrate the latter approach, replacing Fig. 5.3.1 above with Fig. 5.3.4. The difference is that we now assume that diffusion on the blood side is so efficient in the $y$ direction, compared to permeation across the membrane, that the blood side is essentially well mixed in the $y$ direction. Hence the concentration profiles are flat, as shown, and the concentration simply de-

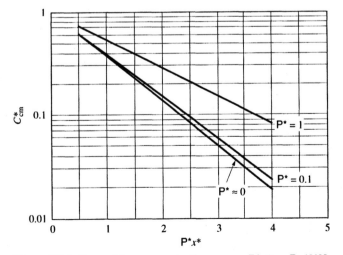

**Figure 5.3.3** Cup-mixing concentration versus $P^*x^* = Px/UH$. The parameter $P^*$ is the membrane Biot modulus.

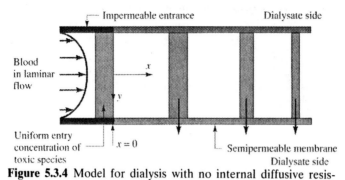

**Figure 5.3.4** Model for dialysis with no internal diffusive resistance.

creases as the blood flows down the channel. A simple differential species balance (Fig. 5.3.5) written over a region $dx$ would take the form

$$-2HU \, dC = 2PC(x) \, dx \qquad (5.3.26)$$

Both sides of Eq. 5.3.26 represent differential flowrates of the toxic species (per unit width of channel). The left-hand side is the net flow of the species across the planes at some position $x$, and one just downstream at $x + dx$. The right-hand side is the "flow" of the species by permeation across the membrane. The factor of 2 on the right-hand side reflects exposure of the blood to membrane surface on both sides of the channel ($y = H$ and $y = -H$). The model for permeation introduced earlier is reused. The flux is proportional to the concentration difference across the membrane, and we are assuming zero concentration of the toxic species on the dialysate sides. The assumption that the toxic species is uniformly distributed in the $y$ direction implies that the concentration we are writing in this balance is the same as the cup-mixing concentration.

We may solve Eq. 5.3.26 easily, using the entry condition given above as Eq. 5.3.5, and the result is

$$\frac{C(x)}{C_0} = \exp\left[-\left(\frac{P}{HU}\right)x\right]$$

or

$$C_{cm}^* = \exp(-P^* x^*) \qquad \text{for} \quad P^* \ll 1 \qquad (5.3.27)$$

This is obviously a much simpler model to work with than Eq. 5.2.24. It is not difficult to show that Eq. 5.3.27 follows from Eq. 5.3.24 if we introduce Eq. 5.3.23 for the first eigenvalue, and take just the first term of the series solution. In Fig. 5.3.3 the lowest

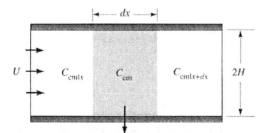

**Figure 5.3.5** Differential volume of the dialysis system.

curve is Eq. 5.3.27. The label $P^* \approx 0$ implies that Eq. 5.3.24 is used in the limit of very small $P^*$. We see from these results that when $P^*$ is below 0.1, the "well mixed" assumption is a good one, and the resistance to transfer out of the blood channel is dominated by the membrane resistance.

Everything we need to know about the concentration of the toxic species in the blood side of the system is given in the solution for $C_{cm}^*$. (For design purposes it is this cup-mixing average, rather than the detailed concentration *distribution* $C^*(x^*,y^*)$, that has the essential information.) It is useful at this point to calculate a measure of the rate of transfer of the toxic species across the membrane, and we usually do this by defining a mass transfer coefficient. The local convective mass transfer coefficient $k_{x,c}$ is defined in the following way. As shown in Fig. 5.3.5, the rate of transfer of the toxic species across the membrane, over any small length $dx$ of membrane, must just balance the change in the cup-mixing concentration in the fluid within that small volume. Hence a species balance takes the form (cf. Eq. 5.3.26)

$$-UH\, dC_{cm} = k_{x,c} C_{cm}\, dx \qquad (5.3.28)$$

In this expression the mass transfer coefficient is based on the difference between the average concentration on the blood side (and it is the cup-mixing average) and the concentration on the dialysis side, *which has been taken to be zero in this analysis*. Note also that Eq. 5.3.28 is for the transfer across the lower half of the channel; for both sides of the channel we would simply double each side of this expression.

Then we may calculate the mass transfer coefficient from

$$k_{x,c} = -\left(\frac{UH}{C_{cm}^*}\right)\left(\frac{D_b}{UH^2}\right)\left(\frac{dC_{cm}^*}{dx^*}\right) \qquad (5.3.29)$$

(We have changed from $dx$ to $dx^*$ in this expression, as well as from $C$ to $C^*$.)

Now we must go back to the solution for $C_{cm}^*$ (Eq. 5.3.19) and differentiate $C_{cm}^*$ with respect to $x^*$. The result may be written in the format

$$\frac{k_{x,c}H}{D_b} = \frac{-3\left[\sum\limits_{m=1}^{\infty}\left(-\frac{2}{3}\lambda_m^2\right)A_m \exp\left(\frac{-2\lambda_m^2 x^*}{3}\right)\sum\limits_{n=0}^{\infty}\frac{a_{nm}}{(n+1)(n+3)}\right]}{3\left[\sum\limits_{m=1}^{\infty}A_m \exp\left(\frac{-2\lambda_m^2 x^*}{3}\right)\sum\limits_{n=0}^{\infty}\frac{a_{nm}}{(n+1)(n+3)}\right]} \qquad (5.3.30)$$

where we recognize the left-hand side to be a Sherwood number based on the length scale $H$. This is a *local* Sherwood number, however, since the function on the right-hand side is dependent on $x^*$. For large enough values of $x^*$ we may use a one-term approximation to the two infinite series that appear in the numerator and denominator, with the result that

$$\frac{k_{x,c}H}{D_b} = \mathrm{Sh} = \frac{2}{3}\lambda_m^2 \qquad (5.3.31)$$

If we introduce Eq. 5.3.23 we may write this as

$$\mathrm{Sh} = \frac{\frac{2}{3}P^*}{\frac{2}{3} + \frac{5}{12}P^*} \qquad (5.3.32)$$

Again, with all the algebra flying around, we need to remind ourselves of what we have derived. Equations 5.3.31 (and 5.3.32) are valid only for distances $x^*$ that are in some sense far downstream from the entrance to the dialysis system. By carrying out a more detailed analysis, we could show that these equations are good approximations as long as $x^*$ exceeds a value of order 0.1. Keep in mind that $x^*$ is a dimensionless length. Whether this simple model is useful depends on whether most of the dialysis unit lies in the region $x^* > 0.1$, and of course this depends on the parameters $U$ and $D_b$, especially.

Let's turn away from the development of the theory for a short time, and look at some applications of the models we have in hand.

---

**EXAMPLE 5.3.1** *Determination of the Permeability of a Dialysis Membrane*

A membrane test cell is set up as suggested in the sketch shown as Fig. 5.3.6. A dialysis fluid is pumped rapidly through the upper section of the cell in a way that ensures maintenance of the concentration of the permeable solute at a value that is very low compared to its concentration in the lower region. The lower region contains the solution that is to be dialyzed. It is in a fixed volume $V$, and the exposed area of the membrane across which the solute will pass is $A$. We write a simple mass balance on the solute, equating the loss of solute from the lower section to the transport across the membrane. This balance takes the form

$$-V\frac{dC_b}{dt} = \frac{D_m \alpha C_b A}{H_m} \tag{5.3.33}$$

This expression is essentially the steady state solution for diffusion across a membrane of thickness $H_m$, where the solubility of the solute in the membrane is the parameter $\alpha$. Note that the model uses the diffusivity of the solute in the *membrane*, $D_m$, which may be different from that in the solution, $D_b$. These three membrane parameters are combined into a single parameter that is, in fact, what we have already introduced as the membrane permeability:

$$\mathsf{P} = \frac{D_m \alpha}{H_m} \tag{5.3.34}$$

(Note that $\mathsf{P}$ has units of centimeters per second in the cgs system.) On the assumption that the time scale for establishment of the steady (linear) solute profile in the membrane is very short in comparison to the time of the experiment, we may write the solution of this equation in the form

$$\frac{C_b}{C_{b,o}} = \exp\left[-\left(\frac{\mathsf{P}A}{V}\right)t\right] \tag{5.3.35}$$

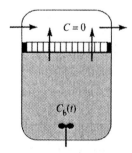

**Figure 5.3.6** A membrane test cell for determination of the permeability of the membrane.

where $C_{b,o}$ is the initial solute concentration. Now suppose we have the following experimental data:

$$V = 1000 \text{ cm}^3 \qquad A = 200 \text{ cm}^2$$

Concentration in the lower volume falls to 90% of its initial value in 22 min.

Then we find the permeability from Eq. 5.3.35 as

$$\ln\left(\frac{C_b}{C_{b,o}}\right) = \ln(0.9) = -\left(\frac{PA}{V}\right)t = -\frac{200(22 \times 60)P}{1000} \qquad \text{(5.3.36)}$$

or $P = 4 \times 10^{-4}$ cm/s.

Now let's use this value of permeability to assess the performance of a dialysis unit.

**EXAMPLE 5.3.2**  *Required Size of a Dialysis Unit for Removal of a Solute*

A membrane dialysis unit consists of a set of parallel channels and is designed with the following parameters:

Channel width $W = 8$ cm
Channel height $2H = 0.01$ cm
Membrane permeability for the solute of interest $P = 4 \times 10^{-4}$ cm/s
Diffusivity of the solute in the blood side $D_b = 2 \times 10^{-5}$ cm$^2$/s
Volumetric flowrate through *each* channel $Q = 1$ cm$^3$/min

If we desire to remove 95% of the solute on each pass through the unit, what is the required length?

We begin with the assumption that Fig. 5.3.3 is relevant to this system (we will have to check on this). We see that for $C_{cm}^* = 0.05$ we need $P^*x^* = 3.2$, since the value of $P^*$ is

$$P^* = P\frac{H}{D_b} = (4 \times 10^{-4})\frac{0.005}{2 \times 10^{-5}} = 0.1 \qquad \text{(5.3.37)}$$

Hence we require a dimensionless length

$$x^* = 32 \qquad \text{(5.3.38)}$$

Returning to the definition of $x^*$ (Eq. 5.3.9) we find

$$x = \frac{x^*}{D_b/UH^2} \qquad \text{(5.3.39)}$$

We need the average velocity $U$, and from the volumetric flowrate we find

$$U = \frac{Q}{2HW} = \frac{1}{60 \times 0.01 \times 8} = 0.208 \text{ cm/s} \qquad \text{(5.3.40)}$$

Finally, we find the required length $L$ to be

$$x = L = \frac{32}{\dfrac{2 \times 10^{-5}}{0.208 \times 0.005^2}} = 8.32 \text{ cm} \qquad \text{(5.3.41)}$$

This seems a reasonable length for a membrane system, and indeed it would appear that the system is quite compact.

Now we must assess one of the key assumptions in this analysis. Our calculation is based on Fig. 5.3.3, which in turn is based on Eq. 5.3.24. Recall that we assumed that the series converges so fast that we may use only the first term ($n = 0$) of the summation. A more detailed analysis tells us that this is a good assumption as long as $x^*$ is greater than 0.1. But in fact, for the calculation of the required length in this example to be accurate, *most* of the dialysis would have to take place beyond the region of $x^* > 0.1$. In other words, it is not enough that the length exceed 0.1; it should be *very large* in comparison to 0.1. In the example here, we have $x^* = 32$, which is more than two orders of magnitude larger than this criterion. Hence we expect that the estimate of the required length given here is accurate.

There are several issues to discuss before we can arrive at a suitable design for a dialysis unit based on the design and operating conditions given. One has to do with the flowrate through a channel, specified as $Q = 1$ cm³/min. This is a very low flowrate, and we would compensate for it by using a system with multiple channels in parallel, each operating according to the parameters above. For example, we could have a system that consists of 500 channels in parallel (and of course an equal number of channels on the dialysate side). With the spacing given ($2H = 0.01$ cm), we would have a unit with a thickness of the order of 10 cm—still fairly compact. We could then process $500Q = 500$ cm³/min. Although we would be reducing the solute concentration in the blood by a factor of 20 (a 95% reduction) on each pass, we would be returning the process blood to the body for recirculation. Let's examine some of the consequences of this feature of a dialysis system.

**EXAMPLE 5.3.3**   *The Effect of Dialysis on the Concentration of a Toxin in the Body*

Now we have to consider how a dialysis system interacts with the volume of fluid being dialyzed. Figure 5.3.7 suggests how we might examine this feature of the process. The patient has a blood volume $V_B$ which is being circulated through the dialysis unit at the rate $Q_{tot}$. This total flow is divided among a number of membrane channels in parallel, such that the flowrate per channel, the quantity that appears in our model, is $Q$. We make the following assumptions to develop a simple model of the change in

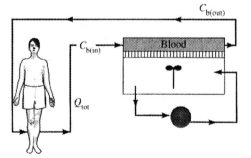

**Figure 5.3.7** Interaction of the dialysis unit with the patient.

body blood level of the toxic species:

1. The membrane permeability is small enough to allow us to use 5.3.27 as an approximation to the behavior of the dialysis unit itself.
2. The blood volume within the body of the patient, $V_B$, is well mixed and has the concentration $C_{b(in)}$. (In the following, all concentrations are understood to be cup-mixing averages.)

With these assumptions, a species balance on the body takes the form

$$-V_B \frac{dC_{b(in)}}{dt} = Q_{tot}(C_{b(in)} - C_{b(out)}) \qquad (5.3.42)$$

With Eq. 5.3.27 we may describe the performance of the dialysis unit by the relationship

$$\frac{C_{b(out)}}{C_{b(in)}} = \exp(-P^*x^*) \qquad (5.3.43)$$

Hence we may eliminate $C_{b(out)}$ and write Eq. 5.3.42 in the form

$$-V_B \frac{dC_{b(in)}}{dt} = Q_{tot} C_{b(in)} E \qquad (5.3.44)$$

where E, which is called the "extraction efficiency," is defined as

$$E = 1 - \frac{C_{b(out)}}{C_{b(in)}} = 1 - \exp(-P^*x^*) \qquad (5.3.45)$$

(E is just $1 - C_{cm}^*$ from Fig. 5.3.3.) Equation 5.3.44 has the solution

$$\frac{C_{b(in)}(t)}{C_{b(in)}(t=0)} = \exp\left(-\frac{Q_{tot}E}{V_B}t\right) \qquad (5.3.46)$$

which we will use in the following manner. We want to reduce the level of a specific toxic species in the blood by 95% in 3 h of continuous dialysis. Hence we require

$$\frac{C_{b(in)}(t)}{C_{b(in)}(t=0)} = 0.05 \qquad (5.3.47)$$

or

$$\frac{Q_{tot}E}{V_B}t = 3 \qquad (5.3.48)$$

The volume of blood in the body of an adult human male is approximately $V_B = 5$ L. A normal circulation rate from the heart is 5.5 L/min. Clearly we cannot take a high proportion of the cardiac output and put it through an external device. We will take 5% of the cardiac output, or

$$Q_{tot} = 0.275 \text{ L/min} \qquad (5.3.49)$$

The prescribed dialysis time is $t = 3$ h $= 180$ min. This lets us calculate a value required for E of

$$E = 3\frac{V_B}{Q_{tot}t} = \frac{3 \times 5}{0.275 \times 180} = 0.3 \qquad (5.3.50)$$

or, since E is just $1 - C_{cm}^*$,

$$C_{cm}^* = 0.7 \qquad (5.3.51)$$

From Fig. 5.3.3 (at $x = L$) we find

$$\frac{PL}{UH} = 0.36 \tag{5.3.52}$$

The design variable that is unspecified at this point is the average velocity $U$ in the blood channel. We find this by noting that

$$Q_{\text{tot}} = 2n(HUW) \tag{5.3.53}$$

where $n$ is the number of parallel blood channels required. If we use the parameters of Example 5.3.2 (in particular, $Q_{\text{per channel}} = 1$ cm$^3$/min) we find

$$n = 275 \text{ channels required} \tag{5.3.54}$$

This completes the design of this system with respect to the specific features we chose to consider. Of course there are other issues involved in the design of a dialysis system. For example, the shear stress imposed on the red cells by this flow field should not exceed a critical level for cell damage, and we would have to verify system performance in this respect.

## 5.4 DIFFUSION-CONTROLLED FILM GROWTH BY CHEMICAL VAPOR DEPOSITION

Compound semiconductors such as gallium arsenide (GaAs) are important materials in solar cells, lasers, and transistors. Such devices are fabricated by growing thin films (of a few hundred angstroms) of controlled purity and thickness uniformity over large areas. In the process most commonly used for this task, organometallic chemical vapor deposition (OMCVD), a carrier gas such as hydrogen conveys reactive vapors such as arsine (AsH$_3$) and an organometallic species such as trimethylgallium (TMG) to a heated substrate. Thermal decomposition occurs on the substrate, leading to the deposition of GaAs. The reaction is often carried out under conditions controlled by mass transfer; that is, the rate of reaction itself is so high that the transport of reactants to the surface controls the observed rate of growth of film. If, as is often the case, the arsine is in excess, the rate of reaction is determined by transport of the TMG to the surface. Hence the analysis of this reactive process is reduced to analysis of the mass transfer of TMG.

In the following we illustrate the analysis of OMCVD under some idealized conditions, and explore the effect of process changes on growth rate. We imagine a reactor designed along the lines suggested in Fig. 5.4.1. The reactor has a *rectangular* cross section.

Gases are conveyed to the reactor and mixed in a section just before the surface on which deposition is to occur. The susceptor is heated, and reaction occurs on this surface only. The susceptor is tilted, typically at a small angle of perhaps 10°, with the result that the mean velocity *increases with* $x$ as the gas moves along the susceptor. We shall see shortly why this effect is wanted.

OMCVD reactors are often operated at atmospheric pressure, and at flowrates of the order of 10 standard liters per minute (slm: i.e., 10 L/min, at one atmosphere and 273 K). For a reactor cross section of 5 cm height by 10 cm width, with the gas heated to 600 K, the average linear velocity $U$ is

$$U = \frac{(10 \times 1000/60) \text{ cm}^3/\text{s} \times 600/273}{50 \text{ cm}^2} = 7.33 \text{ cm/s} \tag{5.4.1}$$

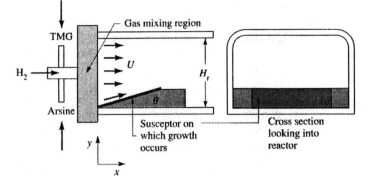

**Figure 5.4.1** A horizontal OMCVD reactor.

The kinematic viscosity of hydrogen at one atmosphere and 600 K is 3.5 cm²/s. This yields a Reynolds number (based on the reactor height) of

$$\text{Re} = \frac{7.33\,(5)}{3.5} = 10.5 \qquad (5.4.2)$$

Hence the flow is laminar through the reactor.

We need a model for the mass transfer coefficient in laminar flow through a duct of rectangular cross section. While this problem has a number of features in common with our earlier analysis of a dialysis system, there are enough differences to justify an independent model of this system. Figure 5.4.2 shows the essential features of the model.

We simplify the analysis by assuming that the reactor has such a large aspect ratio (ratio of width to height) that we may consider it to be a parallel plate reactor. Then we begin with the convective diffusion equation in the form

$$u(y)\frac{\partial C}{\partial x} = D\frac{\partial^2 C}{\partial y^2} \qquad (5.4.3)$$

and we again write the laminar velocity profile as

$$u(y) = \frac{3}{2}U\left[1 - \left(\frac{y}{H}\right)^2\right] \qquad (5.4.4)$$

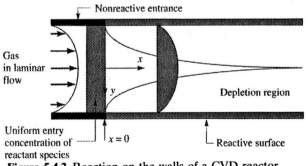

**Figure 5.4.2** Reaction on the walls of a CVD reactor.

One boundary condition is at the entrance, where it is assumed that

$$C = C_o \quad \text{at} \quad x = 0 \quad \text{for all } y \tag{5.4.5}$$

At the reactive surface we assume that the reaction is so fast, relative to diffusion, that the concentration of the reactive species is essentially zero along $y = H$:

$$C = 0 \quad \text{at} \quad y = H \quad \text{for all } x > 0 \tag{5.4.6}$$

We will find that under conditions typical of CVD, the region over which the concentration drops from its free stream value to the surface value (which is $C = 0$) is very thin in comparison to the height of the reactor. Thus we will introduce the assumption that a suitable boundary condition is

$$C = C_o \quad \text{at} \quad y = 0 \text{ (the midplane)} \tag{5.4.7}$$

To proceed, we now nondimensionalize these equations and find a familiar differential equation. We use the definitions already given in Eqs. 5.3.8 and 5.3.9:

$$C^* = \frac{C}{C_o} \qquad y^* = \frac{y}{H} \qquad x^* = \frac{xD_b}{UH^2} \tag{5.4.8}$$

and thus the differential equation becomes

$$\frac{3(1 - y^{*2})}{2} \frac{\partial C^*}{\partial x^*} = \frac{\partial^2 C^*}{\partial y^{*2}} \tag{5.4.9}$$

Next, we shift the coordinate system to make the reactive surface the basis for the $y$ coordinate by defining

$$s = 1 - y^* \tag{5.4.10}$$

and we consider only the region near the wall (i.e., near $s = 0$). Then Eq. 5.4.9 becomes

$$3s \frac{\partial C^*}{\partial x^*} = \frac{\partial^2 C^*}{\partial s^2} \tag{5.4.11}$$

Note in particular that we have approximated the velocity profile by its slope at the wall. This is similar to our treatment of the convective diffusion equation (Eq. 5.2.9) in the problem of the dissolving tube surface. The boundary conditions now take the forms

$$C^* = 1 \quad \text{at} \quad x^* = 0 \tag{5.4.12}$$
$$C^* = 0 \quad \text{at} \quad s = 0 \tag{5.4.13}$$

and

$$C^* = 1 \quad \text{at} \quad s = \infty \tag{5.4.14}$$

If we examine the boundary value problem of Section 5.2, we see that the equation and boundary conditions for $C/C_s$ are identical to the equations and boundary conditions for $1 - C^*$. The coefficient $\alpha$ of that problem (which is the velocity gradient at the wall) has the value $\alpha = 3U/H$ in this problem. Hence we may take over the solution of the problem in Section 5.2 and write the following results by inspection. (Well—almost! There is a little shifting of the coefficients to move the gamma function into the denominator. See Problem 5.14.)

$$1 - C^* = \frac{\int_\eta^\infty \exp(-\eta^3)d\eta}{\int_0^\infty \exp(-\eta^3)d\eta}$$

or

$$C^* = \frac{\int_0^\eta \exp(-\eta^3)d\eta}{0.893} \qquad (5.4.15)$$

where the function $\eta$ is now defined *almost* as in Section 5.2 (see Eq. 5.2.12)

$$\eta = s\left(\frac{1}{3x^*}\right)^{1/3} \qquad (5.4.16)$$

(Note the shift in the limits of the integral in the two parts of Eq. 5.4.15, from $\eta$ to $\infty$ in the first part and from 0 to $\eta$ in the second. This lets us get rid of the subtraction of unity from $C^*$.)

The local flux of reactant along the surface $s = 0$, which is the rate of reaction per unit of surface area for a diffusion-limited reaction, as we are assuming here, is found to be

$$N_s(x) = R(x) = -D\left.\frac{\partial C}{\partial s}\right|_{s=0} = \frac{DC_0}{H}\frac{(UH^2/3Dx)^{1/3}}{0.893} \qquad (5.4.17)$$

The reaction on the surface deposits a solid film of gallium arsenide. The film thickness increases according to

$$\dot{R} = N_s\left(\frac{M_w}{\rho}\right) = \left(\frac{M_w}{\rho}\right)C_0k_c(x) \qquad (5.4.18)$$

where $M_w$ is the molecular weight of the deposited species (for GaAs, $M_w = 144.6$) and $\rho$ (= 5.3 g/cm$^3$) is the mass density of the *solid* GaAs. In Eq. 5.4.18 we introduce a mass transfer coefficient $k_c$, defined so that

$$N_s = k_c\,\Delta C = kC_0 \qquad (5.4.19)$$

where we have assumed that $\Delta C = C_0$ (the feed concentration of GaAs). Using Eq. 5.4.17, we can find $k_c$ and write it as a Sherwood number:

$$\frac{k_cH}{D} = \text{Sh} = \frac{(UH^2/3Dx)^{1/3}}{0.893} \qquad (5.4.20)$$

Note that the length scale $H$ is used in the Sherwood number, and note especially that the Sherwood number decreases along the length of the reactor. Hence we would expect the film thickness to decrease along the length of the reactor. This is not a desirable feature for a reactor, since we normally require films of a high degree of uniformity.

Recall, now, that according to Fig. 5.4.1, the susceptor (the reactive surface) of the reactor is tilted at a small angle. Hence the velocity $U$ is actually *increasing* linearly with $x$, and as a consequence Eq. 5.4.20 leads to a more complex prediction of the profile of deposition than meets the eye.

We will use Eq. 5.4.20 for a tilted susceptor, even though the derivation is for the case of parallel surfaces. (This is similar to a quasi-steady analysis in which we derive a steady state solution and then let some parameter vary with time. Here we have derived a model for parallel surfaces, and then we apply it locally along the varying gap between the surfaces.) We write $H(x)$ in the form

$$H(x) = H_0 - x\tan\theta = 5 - 0.176x \qquad (5.4.21)$$

for the case of a 10° tilt. Note that while $U$ varies with $x$, the $UH$ product will remain constant if we neglect any compressibility effects. We are also implying that the reactor behaves isothermally. We can now proceed with the calculation of the growth rate.

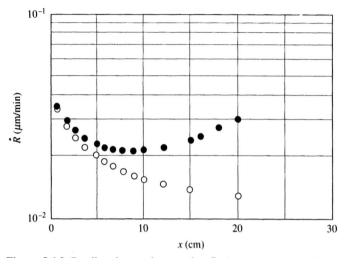

**Figure 5.4.3** Predicted growth rates for GaAs: upper curve, from
Eq. 5.4.18 with Eq. 5.4.17 and a tilt angle of 10°; lower curve, for
a zero tilt.

For the diffusivity of TMG in $H_2$ we find the following form recommended (which
follows approximately from the Chapman–Enskog theory):

$$D \, [\text{cm}^2/\text{s}] = \frac{0.052 T^{1.5}}{P_{\text{torr}}} \tag{5.4.22}$$

At one atmosphere and 600 K we find $D = 1$ cm²/s. For $C_o$ we use a partial pressure
of $7 \times 10^{-5}$ atm for TMG. This corresponds to a $C_o$ of $3.1 \times 10^{-9}$ mol/cm³. We take
$\theta = 10°$. We now have sufficient information to predict the growth rate of the GaAs
film. The results are shown in Fig. 5.4.3. Also shown is the growth rate predicted for a
susceptor angle of 0° (no tilt at all). We see that there is a region beyond the beginning
of the susceptor in which the growth rate is fairly uniform. In the case studied, this
occurs in the region from about 5 cm to 15 cm. By contrast, there is *no* region of nearly
uniform growth without tilt. The tilt of the susceptor leads to an increase in the velocity
$U$, which offsets the reduction in mass transfer coefficient associated with the growth
of the boundary layer thickness in this flow field.

## 5.5 BURNING OF A COAL PARTICLE

We turn next to the development of a model for an extremely complex physical chemical
process: the burning of a solid combustible material. We will outline a methodology of
modeling that puts our fundamental knowledge of mass transfer together with our
experience of reducing problems to the essential physics, in the hope of producing a
mathematical model that represents a good compromise between simplicity and fidelity
to observation. This is not an easy task, and we will not approach it in a timid manner.
Instead, we will work our way toward a testable model by boldly applying a series of
simplifications to the physics until we have what we can regard as the simplest model
that still has in it the essential physics of the process.

We begin by considering a small spherical carbon particle within a gas stream of
reactants. The particle is so small that it is effectively entrained by the flow. That is,
about each particle there is a spherical volume of gas (assumed large compared to the
particle volume) that is quiescent over a time scale appropriate to the burnout of the

particle. ("Burnout" is the time required for the particle mass to be consumed by the combustion reactions.)

A realistic model of combusion must account for multiple reactions, both homogeneous and heterogeneous. For example, on the carbon surface we expect several reactions, such as

$$C + \tfrac{1}{2}O_2 \rightarrow CO \tag{5.5.1}$$

In the gas phase adjacent to the surface we expect homogeneous reactions such as

$$CO + \tfrac{1}{2}O_2 \rightarrow CO_2 \tag{5.5.2}$$

Water vapor and nitrogen (when burning is in air) also participate in reactions on and about the particle.

Strong nonisothermal effects are characteristic of a combustion problem. Thus, the energy equation must be solved simultaneously with the equations for conservation of chemical species. This is a complex problem, but we will ignore thermal effects in this first attempt at an analysis, with the hope that the resultant model might hold in a regime where mass transfer limits the rate of combustion. We will find that this is in fact the case in many combustion problems.

We begin with the following physical ideas. The coal particle has an instantaneous radius $R(t)$, and the surface of the particle is impermable. All reactions occur either on the surface (Eq. 5.5.1) or in the region $r > R$ (Eq. 5.5.2). There is no convection, in the sense that the particle is entrained by the surrounding gas flow. *However, the diffusion of gaseous products away from the surface creates a net flow that must be accounted for in the diffusion equations.* **This is the novel feature of this example.**

We examine only the isothermal analysis here. We have three gaseous fluxes to account for:

$$
\begin{array}{lll}
O_2 & N_{O_2} = N_1 & \\
CO & N_{CO} = N_2 & \text{The } N_i \text{ are } \textit{molar} \text{ fluxes relative to the particle.} \\
CO_2 & N_{CO_2} = N_3 &
\end{array}
$$

We assume a steady state. We take the fluxes to be positive if directed outward. We assume spherical symmetry so that these fluxes are strictly radial.

We write Fick's law of diffusion in the form

$$N_i = -CD_{i,m}\frac{\partial x_i}{\partial r} + x_i\left(\sum_{j=1}^{3} N_j\right) \tag{5.5.3}$$

for each species. Further, we assume that all the diffusion coefficients are equal, since the species are of similar molecular weights. The $x_i$ are mole fractions.

The molar average velocity relative to stationary coordinates is

$$v^* = \frac{1}{C}\left(\sum_{j=1}^{3} N_j\right) \tag{5.5.4}$$

The mean molar density of the mixture, $C$, is assumed to be spatially uniform. For a constant pressure process, and for an ideal gas,

$$C = \frac{p}{R_G T} \tag{5.5.5}$$

We may now write the species balance equations individually—remember, we are in spherical coordinates, beginning with species 1, which is $O_2$:

$$0 = -\frac{1}{r^2}\frac{\partial(N_1 r^2)}{\partial r} + r_{11} + r_{12} \tag{5.5.6}$$

Here we use $r_{11}$ to refer to the rate of reaction of $O_2$ via reaction 1 (Eq. 5.5.1), and $r_{12}$ is the rate via reaction 2 (Eq. 5.5.2). Since reaction 1 does not occur in the gas phase, but is instead a surface reaction, it should not appear in the diffusion equation. Hence we delete $r_{11}$ from Eq. 5.5.6. It will come into the model through a boundary condition on the diffusion equation.

Using Eq. 5.5.4 in Eq. 5.5.3 we find (after differentiating $N_1 r^2$ in Eq. 5.5.6) the following diffusion equation:

$$Cv^* \frac{\partial x_1}{\partial r} + x_1 C \left( \frac{\partial v^*}{\partial r} + \frac{2}{r} v^* \right) = CD \frac{1}{r^2} \frac{\partial}{\partial r} \left( r^2 \frac{\partial x_1}{\partial r} \right) + r_{12} \qquad (5.5.7)$$

The coefficient of $x_1 C$ in the second term on the left-hand side is just $\nabla \cdot \mathbf{v}^*$ in spherical coordinates, assuming spherical symmetry. One might be tempted to invoke the continuity equation for a constant density fluid at this point, and write

$$\nabla \cdot \mathbf{v}^* = 0 \qquad (5.5.8)$$

but we must recall that $\mathbf{v}^*$ and $\mathbf{v}$ are different velocities. Whereas $\mathbf{v}$ is defined by

$$\mathbf{v} = \frac{1}{\rho} \sum_i \rho_i v_i \qquad (5.5.9)$$

the *molar*-average velocity $\mathbf{v}^*$ is

$$\mathbf{v}^* = \frac{1}{\rho} \sum_i \frac{\rho_i v_i}{M_{w,i}/M_w} \qquad (5.5.10)$$

We see from Eq. 5.5.9 that $\mathbf{v} = \mathbf{v}^*$ only if the molecular weights of all the species are the same. Among the species $N_2$, $O_2$, CO, and $CO_2$ we find molecular weights of 28, 32, 28, and 44, respectively. Since we have already assumed that the mean molar density $C$ is spatially uniform, the neglect of the molecular weight differences among the gas species implies that the *mass* density is also spatially uniform.

With these ideas in mind, we will set $\mathbf{v} = \mathbf{v}^*$ and $\nabla \cdot \mathbf{v}^* = 0$ in order to simplify Eq. 5.5.7 to the form

$$v \frac{\partial x_1}{\partial r} = D \frac{1}{r^2} \frac{\partial}{\partial r} \left( r^2 \frac{\partial x_1}{\partial r} \right) + \frac{r_{12}}{C} \qquad (5.5.11)$$

For CO (species 2), by similar arguments, we write

$$v \frac{\partial x_2}{\partial r} = D \frac{1}{r^2} \frac{\partial}{\partial r} \left( r^2 \frac{\partial x_2}{\partial r} \right) + \frac{r_{22}}{C} \qquad (5.5.12)$$

where $r_{22}$ is the rate of the *homogeneous reaction* of CO (see Eq. 5.5.2). From the stoichiometry

$$r_{12} = \tfrac{1}{2} r_{22} \qquad (5.5.13)$$

For $CO_2$ (species 3) we write

$$v \frac{\partial x_3}{\partial r} = D \frac{1}{r^2} \frac{\partial}{\partial r} \left( r^2 \frac{\partial x_3}{\partial r} \right) + \frac{r_{32}}{C} \qquad (5.5.14)$$

where $r_{32}$ is the rate of formation of $CO_2$ via the reaction given in Eq. 5.5.2. Again, stoichiometry gives

$$r_{32} = -r_{22} = -2r_{12} \qquad (5.5.15)$$

Although we have neglected convection by assuming that the particle is entrained in a quiescent volume of gas, all the diffusion equations written here have convective terms appearing on the left-hand sides. *These terms arise from the diffusion of the gases to and from the reactive surface. Thus, formally, convection does play a role in this problem.*

Before we can solve these species balances (the diffusion equations 5.5.11, 5.5.12, and 5.5.14) we must introduce kinetic models. To simplify the mathematics for this example we take

$$r_{32} = k_{12}x_1 \tag{5.5.16}$$

We could try to argue the basis for this assumption about the kinetics, but we will find that the point is irrelevant. We will later make the assumptions that the gas phase reactions are very fast and that the surface reactions control the rate of burning. Hence, while in principle we need a gas phase kinetic model here, we will find that the choice is not of importance.

Equation 5.5.11 is a second-order ordinary differential equation for $x_1$ ($O_2$), which we may solve to find $x_1(r)$. We need a boundary condition at the surface $R$, and this will be given in terms of the oxygen flux at $R$ due to the heterogeneous reaction (Eq. 5.5.1). If $r_{11}$ is the molar rate of *consumption* of $O_2$ per unit of surface area, then

$$-N_1(r = R) = r_{11} \tag{5.5.17}$$

We assume that the particle surface is impervious to diffusion. (This would not be the case if the fuel particle were porous, or if combustion led to a partial liquiefaction of the fuel particle and evolution of gas bubbles inside the particle.)

The second boundary condition will be imposed at $r = \infty$, in the form

$$x_1 = x_{1\infty} \quad \text{at} \quad r = \infty \tag{5.5.18}$$

The oxygen concentration in the free stream is taken to be known, and constant at the value $x_{1\infty}$.

The velocity $v$ must satisfy conservation of mass. Instead of writing $\nabla \cdot \mathbf{v} = 0$, as we did above in simplifying the diffusion equation, we now write a less restrictive constraint on velocity:

$$\frac{\partial}{\partial r}(\rho r^2 v) = 0 \tag{5.5.19}$$

or $\rho v r^2 = $ constant. At the surface, $r = R$, the mass flow is just the flow of carbon converted to carbon monoxide through the reaction of Eq. 5.5.1. Hence we write[1]

$$4\pi \rho v r^2 = 2r_{11}4\pi R^2 M_{w2} \tag{5.5.20}$$

where $M_{w2}$ is the molecular weight of CO. Since we earlier assumed equal molecular weights for all gaseous species we find, using $M_w = \rho/C$,

$$v = \frac{2r_{11}}{C}\frac{R^2}{r^2} \tag{5.5.21}$$

---

[1] Recall that $r_{11}$ is the rate of reaction of oxygen via reaction 1 (Eq. 5.5.1). Since two moles of CO are produced per mole of oxygen consumed, the surface rate $r_{11}$ is multiplied by a factor of 2 to yield the CO rate.

Now we introduce a simple first-order reaction model (assumption) for this *surface* reaction:

$$r_{11} = k_{11}x_{1,R} \tag{5.5.22}$$

where $x_{1,R} = x_1$ at $r = R$. The diffusion equation takes the form

$$\frac{dx_1}{d\xi} = \frac{K_D}{x_{1,R}}\frac{d}{d\xi}\left(\xi^2\frac{dx_1}{d\xi}\right) - \frac{K_r}{x_{1,R}}\xi^2 x_1 \tag{5.5.23}$$

where we have defined

$$K_r = \frac{Rk_{12}}{4k_{11}} \quad \text{and} \quad K_D = \frac{CD}{2Rk_{11}} \tag{5.5.24}$$

and $\xi = r/R$ is a dimensionless radial coordinate. The boundary condition given in Eq. 5.5.17 takes the form

$$N_1 = -CD\frac{dx_1}{dr} = -k_{11}x_{1,R} \quad \text{at} \quad r = R$$

or

$$2K_D\frac{dx_1}{d\xi} = x_{1,R} \quad \text{at} \quad \xi = 1 \tag{5.5.25}$$

To proceed, we now introduce the assumption mentioned earlier that gas phase reactions are unimportant (we set $K_r = 0$) and the diffusion equation for $x_1$ becomes

$$\frac{dx_1}{d\xi} = \frac{K_D}{x_{1,R}}\frac{d}{d\xi}\left(\xi^2\frac{dx_1}{d\xi}\right) \tag{5.5.26}$$

with boundary conditions

$$x_{1,R} = 2K_D\frac{dx_1}{d\xi} \quad \text{at } \xi = 1$$

$$x_1 = x_{1\infty} \quad \text{at } \xi \to \infty \tag{5.5.27}$$

The solution is easily found to be

$$x_1 = \frac{(x_{1\infty} - x_{1,R})e^{-\kappa/\xi} + (x_{1,R} - x_{1\infty}e^{-\kappa})}{1 - e^{-\kappa}} \tag{5.5.28}$$

where

$$\kappa = \frac{x_{1,R}}{K_D} \tag{5.5.29}$$

We must keep in mind that $x_{1,R}$ is not known at this stage of the analysis. It is obtained by application of the boundary condition at $\xi = 1$, with the result (after some algebra)

$$\frac{2}{1 - e^{\kappa}}(\kappa - \kappa_\infty) = \frac{1}{K_D} \tag{5.5.30}$$

where

$$\kappa_\infty = \frac{x_{1\infty}}{K_D} \tag{5.5.31}$$

The detailed concentration profile is not of major interest to us. We do want to know the degree to which diffusion affects the rate of burning of the carbon particle. We shall see that the parameter $\kappa$ is central to our understanding of this aspect of the physics. We may solve Eq. 5.5.30 for $\kappa$ as a function of the diffusion parameter $K_D$, defined earlier (Eq. 5.5.24) and as a function of the ambient oxygen mole fraction $x_{1\infty}$. Equation 5.5.30 is not an explicit equation for $\kappa$, and it must be solved numerically. In many cases, however, $\kappa$ is small compared to unity, and we may obtain an analytical solution as follows:

We write $e^{\kappa} \approx 1 + \kappa$ in Eq. 5.5.30 and we find

$$\kappa = \frac{2\kappa_{\infty}}{1/K_D + 2} \tag{5.5.32}$$

The physical significance of $\kappa$ follows from inspection of the boundary condition at the particle surface (Eq. 5.5.25), which takes the form

$$\kappa = 2\frac{dx_1}{d\xi} = \frac{2R}{CD}N_1 \tag{5.5.33}$$

From Eq. 5.5.25 we note that $N_1$ is proportional to the rate of reaction of oxygen at the surface of the particle. Thus we conclude that $\kappa$ is a dimensionless rate of reaction at the surface of the particle. It is defined in terms of the molar rate of consumption of oxygen via the reaction of oxygen at the surface (Eq. 5.5.1). If we define an effectiveness factor[2] as

$$\eta_{\text{eff}} = \frac{N_1}{k_{11}x_{1\infty}} \tag{5.5.34}$$

then we have a measure of the role of diffusion resistance in reducing the rate of burning of the particle, since $k_{11}x_{1\infty}$ would be the reaction rate (per unit of surface) at the free stream condition $x_1 = x_{1\infty}$. From these last two equations, then, we may write

$$\eta_{\text{eff}} = \frac{\kappa}{\kappa_{\infty}} \tag{5.5.35}$$

or, noting Eq. 5.5.32,

$$\eta_{\text{eff}} = \frac{2}{1/K_D + 2} \tag{5.5.36}$$

For $K_D \ll 1$,

$$\eta_{\text{eff}} = 2K_D \tag{5.5.37}$$

If we examine experimental data on the combustion of carbon particles, we find that $K_D$ is often very small, typically $O(10^{-2})$. From Eq. 5.5.36 we find an effectiveness factor of the order of 0.01, and we conclude that in such cases the effect of diffusion resistance is to reduce the rate of burning by nearly two orders of magnitude.

More generally, $\kappa$ must be found from Eq. 5.5.30, and $\eta_{\text{eff}}$ follows from Eq. 5.5.35. Figure 5.5.1 shows these results for two values of $x_{1\infty}$.

We turn now to a test of the fidelity of our simple combustion model to observations.

---

[2] So-called because it is the ratio of the actual flux to the surface to the flux that would exist if there were no mass transfer resistance.

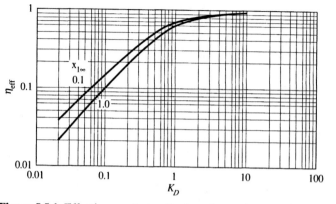

**Figure 5.5.1** Effectiveness factor for burning carbon particles.

**EXAMPLE 5.5.1**   *A Model for Fuel Particle Burnout*

Particulate combustion eventually leads to burnout—the disappearance of the particle through consumption. The analysis above is "quasi-static," since we have everywhere assumed that the particle radius $R$ is not changing. Now we turn to the problem of particle dynamics.

The loss of particle mass (assumed to be carbon) is written as

$$\frac{d}{dt}\rho_s\frac{4}{3}\pi R^3 = -4\pi R^2 N_1 (2M_{w,C}) \tag{5.5.38}$$

where we have balanced the loss of mass of the solid by the surface flux (times the particle area) of carbon monoxide. Equation 5.5.1 gives the relevant reaction and stoichiometry. Since $N_1$ is the flux of oxygen, we introduce the factor of 2 to convert the oxygen flux to the CO flux. (There is also a change of sign, since the fluxes are in opposite directions.) The mass density and the molecular weight of solid *carbon* appear in Eq. 5.5.38. Thus this relation is a *mass* balance, which we may write in a nondimensional form as

$$\frac{ds}{d\tau} = -\frac{\kappa}{s} \tag{5.5.39}$$

where $\kappa$ has been defined earlier, $s$ is a dimensionless radius

$$s = \frac{R(\tau)}{R_o} \tag{5.5.40}$$

and

$$\tau = \frac{CDM_{w,C}}{\rho_s R_o^2}t \tag{5.5.41}$$

Equation 5.5.39 has the solution (assuming $\kappa$ = constant)

$$s^2 = 1 - 2\kappa\tau \tag{5.5.42}$$

We now define the burnout time as the time at which $s = 0$. From Eq. 5.5.42 we write this as

$$\tau_\infty = \frac{1}{2\kappa} \tag{5.5.43}$$

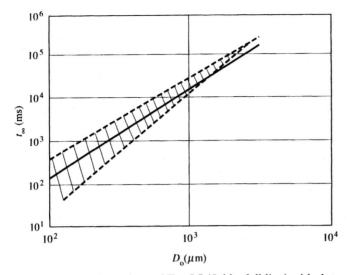

**Figure 5.5.2** A comparison of Eq. 5.5.48 (the full line) with data for burnout of carbon particles in air.

and for $\kappa \ll 1$ and $K_D \ll 1$ we find[3]

$$\tau_\infty = \frac{1}{4x_{1\infty}} \tag{5.5.44}$$

Inserting the definition of $\tau$, we find

$$t_\infty = \frac{\rho_s R_o^2}{CDM_{w,C}} \frac{1}{4x_{1\infty}} \tag{5.5.45}$$

We must recall at this point that the case being examined here, ($\kappa \ll 1$ and $K_D \ll 1$), corresponds to diffusion control of burning. This is seen in the absence of any chemical reaction rate constants in the final model for burnout time.

An extensive series of experiments is reviewed in a paper by Mulcahy and Smith [*Rev. Pure Appl. Chem.*, **19,** 81 (1969)], and we will compare our simple model to their data. We will take the case corresponding to the following parameters:

$$x_{1\infty} = 0.2 \text{ (air)} \qquad \rho_s = 1.5 \text{ g/cm}^3$$
$$p = 1 \text{ atm} \qquad T = 1600 \text{ K}$$

We use the ideal gas law for the molar density $C$, and the Chapman–Enskog equation for the diffusivity of CO in air. Hence we find

$$C = \frac{p}{R_G T} = \frac{1}{82 \times 1600} = 7.6 \times 10^{-6} \text{ mol/cm}^3 \tag{5.5.46}$$

and

$$D = 3.5 \text{ cm}^2/\text{s} \tag{5.5.47}$$

---

[3] Under these conditions we see, from examination of Eqs. 5.5.43 and 5.5.44, that $\kappa = 2x_{1\infty}$, thus confirming the assumption made in obtaining Eq. 5.5.42.

Using a molecular weight of 12 for carbon we find, from Eq. 5.5.45,

$$t_\infty = 1470 D_o^2 \tag{5.5.48}$$

where we have replaced the initial particle radius with the diameter $D_o$. The large number of data points examined in the original reference exhibit a wide range of burnout times, especially at small values of the particle diameter. The range of the data is indicated in Fig. 5.5.2 by the hatched lines. From this comparison we can conclude that a model of carbon burnout based on the assumption that diffusion of reactants to the burning surface controls the rate of burning is in gross agreement with data. The slope of the dependence of burnout time on initial particle size is predicted with reasonable accuracy, and the burnout times are predicted to within a factor of 3, except for the smallest particles. The wide range of the data is itself an indication that no simple model will adequately predict the behavior over a wide range of parameters. Among the factors that must be considered in a more realistic model are thermal effects, chemical kinetics, and the formation of an ash layer on the burning surface. Despite these reservations, we see that the application of our basic knowledge of mass transfer has permitted us to develop a model of a very complex process that mimics some features of the data with reasonable accuracy.

## SUMMARY

When the flow field is simple enough, it is possible to solve the steady state convective diffusion equation (Eq. 5.1.4) analytically. Often this simplification is brought about by assuming that the diffusion is confined to a region wherein the flow field is simple—a constant velocity, as in the case of evaporation from the free surface of a falling laminar film (Eq. 5.1.15)—or a linear velocity profile, as in the case of dissolution at the interface of a soluble solid and a parallel laminar flow (Eq. 5.2.9). The models that result from these simplifications permit the examination of a number of design issues of importance.

When a rapid chemical reaction occurs at a surface, diffusive mass transfer rather than the intrinsic kinetics of the reaction(s) may control the observed rates of reaction. In the model of the burning coal particle (Section 5.5), we find that the net flow of reactants to and from the surface introduces a convective term into the diffusion equations (Eqs. 5.5.11, 5.5.12, and 5.5.14), even though there is no externally induced flow in the neighborhood of the surface.

## PROBLEMS

**5.1** Show that Eq. 5.1.25 follows from Eq. 5.1.24.

**5.2** Write the convective diffusion equation and boundary conditions for the problem of a falling film *absorber*. Give the solution for the total moles absorbed over a length $L$ and compare your result to Eq. 5.1.26. Verify Eq. 5.1.45.

**5.3** Give the form of the coefficient $\alpha$ in Eq. 5.2.9 for

**a.** Laminar flow in a tube of circular cross section.

**b.** Laminar flow between infinite parallel plates.

**c.** Laminar gravity flow of a film of thickness $\delta$.

**5.4** Derive Eq. 5.2.11. The key step is to show that when we define a new variable $\eta$ that combines the $x$ and $y$ variables in the form

$$\eta = y \left( \frac{\alpha}{9 D_b x} \right)^{1/3} \tag{P5.4.1}$$

the *partial* differential equation collapses to an *ordinary* differential equation of the form

$$\frac{d^2 f}{d\eta^2} + 3\eta^2 \frac{df}{d\eta} = 0 \tag{P5.4.2}$$

(where $f = C/C_s$) and the three boundary conditions (Eqs. 5.2.7, 5.2.8, and 5.2.10) collapse to two boundary conditions on $\eta = 0$ and $\eta = \infty$ for $C/C_s$.

**5.5** Provide the mathematical details that yield Eqs. 5.2.13 and 5.2.14.

**5.6** Calculate $\langle C \rangle$ from Eqs. 5.2.14 and 5.2.15, and compare the result to $C_{cm}$ from Eq. 5.2.16, and $\overline{C}$ from Eq. 5.2.17.

**5.7** A fluid in laminar flow within a long tube of inside diameter $2R$ has a dissolved species with a concentration profile of the form

$$\frac{C}{C_o} = \frac{r}{R}\exp[-\lambda(R - r)] \qquad \text{(P5.7)}$$

where $\lambda$ is a positive constant. Calculate the cup-mixing average concentration $C_{cm}$.

**5.8** The maximum shear stress that can be imposed on red cells without causing irreversible damage is denoted by $\tau^*$. Present a plot of volume flowrate per unit width $Q_W$ in a parallel plate channel of spacing height $2H$, which satisfies this constraint. Choose values of $\tau^*$ in the range 50 to 500 dyn/cm$^2$ and $H$ in the range 50 to 3000 $\mu$m. Use a viscosity of blood of 3 mPa $\cdot$ s.

**5.9** Show the details that lead from Eq. 5.3.17 to Eq. 5.3.19.

**5.10** Demonstrate the derivation implied after Eq. 5.3.27.

**5.11** Find the Sherwood number that corresponds to the solution given in Eq. 5.3.27 for dialysis through a membrane with very low permeability, and show that your result is compatible with Eq. 5.3.31.

**5.12** Experiments are performed in a dialysis test unit for which $H = 0.0125$ cm. The test solution is sodium chloride in water at 37°C, for which the diffusivity is $1.2 \times 10^{-5}$ cm$^2$/s. Data were obtained over a range of flowrates with a Cuprophan membrane, and the Sherwood number was calculated from the mass transfer data to be Sh = 0.6, independent of flowrate. What is the permeability of the membrane?

**5.13** A membrane test cell was set up along the lines of the system shown in Fig. 5.3.6. The concentration fell by a factor of 2 in 110 min. If the volume of the test cell was 800 cm$^3$, and the membrane area was 90 cm$^2$, what was the permeability of the membrane?

**5.14** Show how the solution to the problem in Section 5.2 may be converted to the form given

as the solution to the problem in Section 5.4. In particular, verify Eqs. 5.4.15 and 5.4.16. This illustrates one of the easiest ways to solve mathematical problems: recognize that the problem has already been solved in some other form, and convert that solution to fit your problem.

**5.15** Using the parameters given in Section 5.4, define the position and length of a region on the susceptor over which the growth rate of solid film varies within 3% of the minimum value.

**5.16** Using the parameters given in Section 5.4, but with the tilt angle $\theta$ as a variable, determine the length of the region in which the growth rate of solid film varies within 3% of the minimum value, as a function of $\theta$.

**5.17** Using the parameters given in Section 5.4, plot the film growth rate as a function of temperature, for $T$ in the range [600, 900] K. Using an Arrhenius model, determine the apparent activation energy that corresponds to this result.

**5.18** In the example of a design for an oxygen absorber, we chose a design route that permitted us to use a model for absorption into a falling film that was limited to a small degree of penetration of the diffusing species into the film. We would like to have a model for an absorber that does not have this limitation. Formulate the boundary value problem that is free of this limitation. Do not attempt to solve it, but state clearly the boundary conditions that are appropriate to the absorber problem.

**5.19** With reference to the model of the sequence of absorbers in series, it should be apparent that a key parameter is the length of the liquid film $L_e$, in each unit. The performance of the system, as measured by the total length of required surface (the number of units times this length), will depend on this effective length in a complex way. We want to compare the performance of many short units versus fewer long units. Suppose we require that $\kappa = 0.95$, where the final concentration of the absorbed species is given by $C_N = \kappa C_{eq}$. For lengths of $L_e = 50$, 10, 5, and 2.5 cm, find the number of units $n$ and the total length of surface $nL_e$. Assume that the parameters are those that lead to Eq. 5.1.48.

**5.20** Using our simple model of a wiped film absorber, plot the required overall length as a function of the rotational speed $\Omega$ (in the range 0.1 to 1 rev/s), for values of $\kappa$ of 0.5, 0.9, and 0.99. Take all other parameters as in Example 5.1.1.

**5.21** Suppose we had neglected the convective term in Eq. 5.5.26. Find the solution for $x_1(\xi)$ in this case, and compare it to Eq. 5.5.28. Calculate $\eta_{eff}$ for this case, and compare it to Eq. 5.5.36. Discuss your result.

**5.22** In developing our model of particle burning we assumed that all the relevant diffusivities were identical. Calculate and plot the diffusivities of CO, $CO_2$, and $O_2$ in air at $p = 1$ atm for temperatures in the range of 800 to 1800 K.

**5.23** Is Eq. 5.5.35 dependent on the assumption that leads to Eq. 5.5.32? Why does Fig. 5.5.1 not show the behavior predicted by Eq. 5.5.37 at small values of $K_D$?

**5.24** What is the sensitivity of Eq. 5.5.48 to our choice of temperature? Plot values of the coefficient of $D_o^2$ against absolute temperature in the range of 800 to 1800 K.

**5.25** A stainless steel tube of inside diameter $D = 2$ mm and length $L = 0.2$ m has a uniform thin coating ($100 \ \mu$m) of oleic acid on its inner surface. Supercritical $CO_2$ flows through the inside of the tube at a flowrate of 1.5 cm³/min, measured at the pressure within the tube of $P = 16$ MPa and at the temperature of 313.2 K. Find the mass rate of removal of oleic acid from the tube, assuming that the removal mechanism is dissolution of the oleic acid into the supercritical $CO_2$ flow. At what time will 25% of the initial mass be removed? What is the Reynolds number for this flow? Verify that there is no significant pressure drop along the length of the tube.

The following physical property data are available, at the conditions of the experiment.

Density of oleic acid = 900 kg/m³
Density of supercritical $CO_2$ = 793 kg/m³
Solubility of oleic acid in supercritical $CO_2$ = 10 kg/m³
Diffusivity of oleic acid in supercritical $CO_2$ = $6.5 \times 10^{-9}$ m²/s
Viscosity of supercritical $CO_2$ = $7.1 \times 10^{-5}$ Pa · s

Oleic acid ($C_{18}H_{34}O_2$) has a molecular weight of 282.5.

**5.26** Using Eq. 5.2.34, prepare plots of film thickness $B^*$ as a function of time $t^*$, for $x^* = 100$ and for $x^* = 1000$. Evaluate the approximation that neglects the effect of the growing bare region on the time required to completely dissolve a coating up to a specific position $x^*$.

**5.27** Demonstrate the validity of Eq. 5.2.40. In addition, find the dimensionless time $\tau^*$ at which $B^* = 0.5$ at $x^* = L^*/2$.

**5.28** A tube of circular cross section is coated with a thin layer of a viscous soluble film and is cleaned by pumping a solvent in laminar flow through the tube. A second tube, twice the axial length of the first, with a coating of twice the thickness as that in the shorter tube, is to be cleaned in the same total time as the first tube. Assume that all the chemical components (film and solvent) are identical in both cases. How much more solvent is required in the second case? Assume that the solvent is not recycled.

**5.29** Find the time (as the dimensionless time $\tau^*$) at which half of the initial mass of coating has dissolved into a laminar tube flow. Plot the thickness profile $B^*(x^*)$ at that time.

**5.30** Return to Eq. 5.2.41, but account for the effect of the receding film. Do so by replacing $x$ in Eq. 5.2.25 by $x - x_d$, with $x_d$ given by Eq. 5.2.30.
**a.** Show that the rate of dissolution of mass at any time $t$ may be written in a dimensionless format in the form

$$-\frac{d(m/m_o)}{d\tau^*} = \int_{\tau^{*3}}^{1} (u^* - \tau^{*3})^{-1/3} \, du^* \quad \textbf{(P5.30)}$$

where $m/m_o$ is the fraction of *remaining* film; $\tau^*$ is defined as in Eq. 5.2.38, and $u^*$ is just a dummy variable of integration.
**b.** Prepare a plot of $1 - m/m_o$ as a function of $\tau^*$.
**c.** Plot the time $\tau^*$ required to yield a cleared length $x_d/L = \xi^*$, for $[0 < \xi^* < 1]$.

**5.31** Repeat Problem 5.25, but find the time at which 50% of the initial mass of oil will be removed. At that point in time, what fraction of the tube surface will be bare of oil?

**5.32** A stainless steel tube of inside diameter $D = 2$ mm and length $L = 0.2$ m has a uniform thin coating ($100 \ \mu$m) of naphthalene on its inner surface. Supercritical $CO_2$ flows through the inside of the tube at a flowrate of 1.5 cm³/min, measured at the pressure within the tube of $P = 14$ MPa and at the temperature of 308 K. Find the mass rate of removal of naphthalene from the tube, assuming that the removal mechanism is dissolution of the naphthalene into the supercritical $CO_2$ flow. At what time will all the naphthalene be removed?

The following physical property data are available, at the conditions of the experiment.

Density of naphthalene = 1100 kg/m³
Density of supercritical $CO_2$ = 803 kg/m³

Solubility of naphthalene in supercritical $CO_2$ = 0.012 mole fraction

Diffusivity of naphthalene in supercritical $CO_2$ (see Fig. P2.30)

Viscosity of supercritical $CO_2$ = $7.2 \times 10^{-5}$ Pa $\cdot$ s

The molecular weight of naphthalene ($C_{10}H_8$) is 128.

**5.33** A section of a chemical process system consists of a bank of 17,420 parallel tubes, all connecting two common headers. Each tube is 1 cm inside diameter, and 5 m in length. Periodically during the operation of this system the tube bank must be decontaminated to remove an interior oil film that degrades the performance of the system. Decontamination must be carried out when the oil film grows to a thickness of 5 $\mu$m. The oil has the properties of oleic acid. The procedure involves the dissolution of the contaminant film by flowing a solvent through the tubes.

Cleaning will be performed using supercritical carbon dioxide ($SCCO_2$) at 313.2 K and a pressure of 16 MPa. At these conditions the $CO_2$ has a viscosity of $7.1 \times 10^{-5}$ Pa $\cdot$ s, and a density of 793 kg/m$^3$. At these conditions the equilibrium solubility of the oil in $CO_2$ is $C_s$ = 10 kg/m$^3$. The density of the oil film is 900 kg/m$^3$. The molecular weight of oleic acid is 282.5. The diffusivity of oleic acid in supercritical $CO_2$ is $6.5 \times 10^{-9}$ m$^2$/s at the conditions of cleaning.

**a.** Find the cleaning time $t_c$ as a function of flowrate $m$ of $SCCO_2$ (in kg/h). Do this for Reynolds numbers in the range $10 < Re < 1000$. Plot $t_c$ (h) versus mass flowrate.

**b.** Plot the cost of cleaning versus mass flowrate for the following economic model. The total cost $C_T$ is the sum of the cost of the $SCCO_2$ plus the cost of lost production time:

$$C_T = amt_c + bt_c \qquad \text{(P5.33)}$$

Assume that $a$ = \$5/kg $SCCO_2$ and $b$ = \$2000/h of lost production time.

**c.** What is the optimum cleaning procedure, and what is the minimum cost, based on this model?

**d.** How does this optimum vary if the cost factor $a$ is doubled?

**5.34** A piece of process equipment is to be cleaned by a method in which the cleaning time depends on the flowrate of the cleaning solution according to

$$t_c = c_1 Q^{-n} \qquad \text{(P5.34)}$$

Suppose that the cost of cleaning depends on two additive factors: the cost of the solvent used and the cost of production time lost while cleaning is proceeding. (See Eq. P5.33 for such a cleaning cost model.) Develop a general algebraic model for the optimum flowrate $Q^*$ that minimizes the total cleaning cost.

**5.35** A synthetic fluorocarbon has been recommended as a blood substitute for use during open heart surgery. This liquid has a viscosity of 2 cP at 37°C and a density of 1.1 g/cm$^3$. Its virtue lies in its compatibility with human tissue and fluids, and in its high capacity for dissolved oxygen. The designer of a falling film (wetted-wall) oxygenator is unsure of whether increasing the volumetric flow to the top of the system will increase or decrease the oxygen content of the blood leaving the system. Assume the wetted-wall system is a vertical 4-inch-diameter tube, smooth walled, and operating with no wiping of the film.

Present the designer with an algebraic expression for the effect of flowrate on the concentration of oxygen in the fluid leaving the system. Film thickness should not appear in the desired expression.

**5.36** A tube, of inside diameter 1 cm, is coated with a solid that has a molecular weight of 80 and a vapor pressure of 100 mmHg at 25°C. Air at 25°C is pumped through the tube at 360 L/min. The tube is 20 cm long. The tube exit is at atmospheric pressure.

**a.** What is the air pressure at the tube entrance (mmHg)?

**b.** What is the rate of mass loss of the solid, into the airstream (mol/s)?

**5.37** Water at 300 K overflows the inside of a vertical open pipe of diameter 1 m and forms a smooth film of thickness 0.05 cm. Pure carbon dioxide, at a pressure of 1 atm, flows slowly through the inside of the pipe. If the pipe is 2.5 m long, and the water enters free of carbon dioxide, at what rate (mol/h) can carbon dioxide be removed from the system in the exiting water stream? The Henry's law constant for $CO_2$/water is $H$ = 500 atm, where $p$ (atm) = $Hx_{CO_2}$ when $x_{CO_2}$ is the mole fraction of $CO_2$ in the solution. The diffusion coefficient of $CO_2$ in water at 300 K is $2 \times 10^{-5}$ cm$^2$/s.

**5.38** An aqueous solution of salt is to be dewatered from 99% water (by weight) to 95% in a single-pass falling film evaporator. Design a system based on the following constraints.

Feed water will be conveyed to a distributor from which it will fall as a thin film adhering to the inside of a tube, or set of like tubes, through the inside of which dry air is conveyed. The liquid film thickness should not exceed 2 mm. Tube length should not exceed 20 ft. Tube diameters available are 2, 4, and 6 inches. Total feed rate is 500 gal/h, at 25°C.

In your design, give the tube length and diameter, and number of parallel identical tubes, which will meet these constraints.

**5.39** An existing falling film absorber operates in a manner that is well described by the model presented in Section 5.1. Under the original design and operating conditions, the absorber is capable of reaching $\phi = 8\%$ of the absorbtion capacity of an infinitely long absorber.
**a.** If the volume flowrate of absorbing liquid (per unit width of film) is doubled, what will be the new value of $\phi$?
**b.** If the volume flowrate of absorbing liquid (per unit width of film) is doubled, by what factor must the vertical length of the absorbing film be changed if we wish to maintain $\phi = 8\%$?
**5.40** A falling film evaporator is being used to dewater a viscous solution. The film "falls" over the outer surface of a 0.3 m diameter pipe. The film is exposed to dry air. The liquid is collected at the end of the pipe (which is 1 m below the inlet), and it is found that 8% (by mass) of the initial amount of water has been evaporated. If the pipe length (i.e., the length of the falling liquid film) is doubled, what percentage of water evaporation will result in the longer system?
**5.41** When a submerged disk coated with a soluble film rotates about a vertical axis, as suggested in Fig. P5.41, a flow field is established—at least at low Re—which leads to a uniform mass transfer coefficient independent of the radius of the disk. The theoretical value for the mass transfer coefficient is given by the expression

$$k_{rd} = 0.625\nu^{-1/6}D_{AB}^{2/3}\omega^{1/2} \qquad \textbf{(P5.41)}$$

where $\nu$ is the kinematic viscosity of the surrounding fluid (the solvent), $D_{AB}$ is the diffusion coefficient of the soluble species in the solvent, and $\omega$ is the rotational speed of the disk in radians per second.
**a.** Suppose a disk of radius $R$ is coated with a thin uniform solid film of mass $M$ of pure solute.

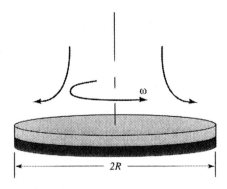

**Figure P5.41**

Derive an expression for the fraction of solute removed as a function of time, using the mass transfer model implied by Eq. P5.41.
**b.** Data on the rate of dissolution of behenic acid from stainless steel disks, using ethanol as the solvent, are tabulated below. Test the model derived in part **a** against these data. You will need the following physical property data: solubility of behenic acid in ethanol (mole fraction) = 0.001 at 304 K and 0.01 at 320 K; density of ethanol = 0.79 g/cm³; molecular weight of ethanol = 46; molecular weight of behenic acid = 341; and $R$ = 9 mm. For the estimation of the diffusion coefficient of behenic acid in ethanol, see Chapter 2, Problem 2.9.

### RATE OF DISSOLUTION OF BEHENIC ACID

| $\omega$ (rpm) | Rate (%/min) | Initial Mass of Solid Film (mg) |
|---|---|---|
| **30°C** | | |
| 30 | 3.2 | 39.3 |
| 30 | 2.3 | 21.1 |
| 80 | 3.5 | 39.2 |
| 80 | 3.6 | 41.2 |
| **40°C** | | |
| 30 | 5.7 | 33.6 |
| 30 | 5.8 | 33.9 |
| 40 | 8.96 | 30 |
| 40 | 6.41 | 38.4 |
| 80 | 8.2 | 39.2 |
| 80 | 9.9 | 41.2 |

*Source:* Grant, Perka, Thomas, and Caton, *AIChE J.*, **42**, 1465 (1996).

**5.42** In Problem 5.30 we corrected Eq. 5.2.25 for the presence of a growing bare region of length

$x_d$ by replacing $x$ in Eq. 5.2.25 by $x - x_d$. The resulting equation for $dm/dt$ (Eq. P5.30) is not exact because the result for $x_d$ (Eq. 5.2.26) is not exact. Equation P5.30 must be integrated numerically to yield a relationship between $m/m_o$ and $\tau^*$.

An alternative approximate method of obtaining a model for $dm/dt$ is to write

$$-\frac{dm}{dt} = M\left[\int_0^L N_y(x, t)\, dx\right.$$

$$\left.- \int_0^{x_d} N_y(x, t)\, dx\right] \quad \textbf{(P5.42.1)}$$

$$= M\int_{x_d}^L N_y(x, t)\, dx$$

and use Eq. 5.2.13 for the flux $N_y(x, t)$. Show that this approach leads to a differential equation of the form

$$-\frac{dm/m_o}{dt^*} = \frac{3}{2}(L^{*-1/3} - L^{*-1}t^{*2}) \quad \textbf{(P5.42.2)}$$

and the solution may be written as

$$\frac{m}{m_o} = 1 - \frac{3}{2}\left(\tau^* - \frac{1}{3}\tau^{*3}\right) \quad \textbf{(P5.42.3)}$$

Now we find the time to completely clear the surface (by setting $m = 0$) to be $\tau^* = 1$, instead of $\tau^* = 0.91$. Plot $1 - m/m_o$ versus $\tau^*$, using Eq. P5.42.3, and if possible, compare your result to that in Problem 5.30b.

**5.43** Using the results from Prob. 5.30, plot the $\tau^*$ required to yield a cleared length $x_d/L = \xi^*$. On the same coordinates, plot the fraction remaining $(m/m_o)$ versus $x_d/L$.

**5.44** With reference to Prob. P5.41, suppose such a disk were coated with a viscous liquid. From Eq. P5.41 we can find the rate of film thinning, $dh/dt$, due to evaporation. At the same time, the liquid flows centrifugally out radially off the disk, and the fluid dynamics of that "spin-off" flow leads to a model for the film thickness $h(t)$ of the form

$$\frac{h(t)}{h_o} = \left(1 + \frac{4\rho\omega^2 h_o^2}{3\mu}t\right)^{-1/2} \quad \textbf{(P5.44)}$$

where $h_o$ is the initial film thickness. Write expressions for $dh/dt$ by evaporation, and by flow, on the assumption that each occurs independently of the other.

Find an expression for the ratio of $dh/dt$ by spin-off to $dh/dt$ by evaporation, in an appropriate nondimensional format. Which of the two mechanisms of removal controls at large times? Which at small times?

**5.45** With reference to Eq. 5.2.34, show that a good approximation that avoids the need for numerical integration is

$$B^* = 1 - \frac{t^*}{(x^*)^{1/3}} \quad \textbf{(P5.45)}$$

for $x^* > x_d^*$, and $x_d/L \ll 1$. Plot and compare $B^*(t^*)$ using Eqs. 5.2.34 and P5.45, for $x^* = 1000$.

# Chapter 6

# Convective Mass Transfer Coefficients

In this chapter we deal with the issue of obtaining convective mass transfer coefficients in terms of the design and operating parameters of a system of interest. Some of these coefficients are found from analytical models such as those developed in Chapter 5. Others are based on experimental observations, and on correlations based on dimensional analysis. A series of examples illustrates the design and analysis of mass transfer systems in which convective processes dominate the rate of transfer across a phase boundary.

## 6.1 SOME EXAMPLES OF CONVECTIVE MASS TRANSFER

Thus far, when we wrote transport models that in some cases included a convective mass transfer coefficient, we have treated the coefficient as a known parameter, or at least as a parameter to which we have access. Now we must deal with a twofold question:

*How do we estimate or predict a value for a mass transfer coefficient, and how does the coefficient depend on characteristic parameters of the convective process?*

We are most often interested in mass transfer between a fluid and a solid (or between two fluids, across a mobile interface, as in a gas/liquid or liquid/liquid dispersion). Some typical examples, drawn from various areas of technology, follow. In subsequent sections of this chapter we will learn how to develop models for such complex mass transfer systems.

---

**EXAMPLE 6.1.1**   *Spinning Fibers from Solution*

Certain types of synthetic fiber are spun from a polymer solution into air, as shown schematically in Fig. 6.1.1. Solvent evaporation occurs at the fiber/air interface, and the solid fiber is formed. Mass transfer from the fiber surface to the surrounding air affects the rate at which the fiber forms, and so does internal mass transfer within the viscous polymer solution. For very fine fibers, the mass transfer could be controlled by the

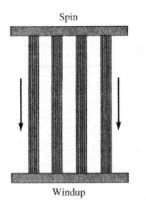

**Figure 6.1.1** Fiber spinning.

*external* convective process. To design a fiber spinning system, we need to be able to estimate mass transfer coefficients.

---

**EXAMPLE 6.1.2**   *Reaction in a Catalytic Converter*

Gas flows through a tube packed with catalytically active particles. The flow field is very tortuous (see Fig. 6.1.2), and convection of reactants to the catalyst surfaces is very nonuniform. How do we describe convective transport in this situation? How does the convective coefficient depend on the characteristics of the particulate phase?

---

**EXAMPLE 6.1.3**   *Burning of a Coal Particle in a Fluidized Bed Furnace*

A solid fuel combustion system is designed in which the fuel (powdered coal particles) is circulated by the action of a hot turbulent reactive gas flow. Reactants must be transported *to the surface* of the fuel particle, where a chemical reaction occurs, and reaction products must be transported *away from the surface*, as indicated in Fig. 6.1.3. We expect the rate of combustion (hence the heat release rate) to depend on how effectively the reactants can be brought to and from the particle surface. We want to know how this might depend on the external gas velocity and the particle size.

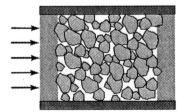

**Figure 6.1.2** Flow through a bed of catalyst particles.

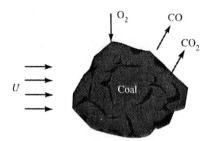

**Figure 6.1.3** Flow around a burning particle.

---

**EXAMPLE 6.1.4**    *Oxygen Transfer from Bubbles to Liquid*

Fermentors are often designed as stirred aerated tanks. The rate of oxygen transfer to the liquid is dictated in part by the convective transfer of the dissolved gas from the bubble surfaces into the bulk of the liquid. The total area depends on the bubble size distribution. (There will be a range of bubble sizes about some mean value.) In this turbulent liquid flow the mass transfer coefficient characteristic of the system is difficult to predict from first principles. Such systems, as represented by Fig. 6.1.4, provide examples of extremely complex interactions between the fluid dynamics and the mass transfer.

Convective mass transfer problems of real technological significance usually are so complex (especially because of the complex boundaries that are typical of such problems) that it is not possible to solve the equations of motion, written here (for an isothermal Newtonian flow) in vector format as

$$\rho \frac{D\mathbf{u}}{Dt} = -\nabla p + \mu \nabla^2 \mathbf{u} \tag{6.1.1}$$

$$\nabla \cdot \rho\mathbf{u} = 0 \tag{6.1.2}$$

The equations of motion must be coupled with the convective diffusion equation

$$\frac{DC_A}{Dt} = D_{AB}\nabla^2 C_A + R_A \tag{6.1.3}$$

We have recourse to several approaches, *each* of which can provide some guidance:

   **1. Dimensional Analysis**   We nondimensionalize the complete set of transport equations and discover which dimensionless groups control the behavior of the solutions.

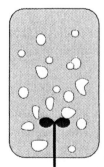

**Figure 6.1.4** Bubbles in a stirred tank.

This approach provides guidelines for the design of an experimental investigation of the system of interest to us.

2. *Approximate Fluid Dynamic Analysis*  We invoke boundary layer theory, or "film theory," which gives "exact" solutions to the flow field right next to the interface, and solve the convective diffusion equation analytically for that simplified flow field.

3. *Empirical Correlations of Data*  Guided by dimensional and approximate fluid dynamic analysis, we plot data in a dimensionless format and present empirical correlations (curve-fits) of the data.

We begin a discussion of methods of developing models of convective transfer by considering an approximate fluid dynamic analysis. We then extend the analysis to a more complex, but still tractable model of the flow. In the course of these analyses we will discover the key dimensionless groups that control the behavior of systems of these types.

## 6.2 FILM THEORY

The film model is the simplest approach to the prediction of mass transfer coefficients. It is not as elegant as boundary layer theory, which we take up and develop in Section 6.3. It is much easier to understand, however, and it gives surprisingly useful results.

We transform a complex convective transfer problem into a simpler soluble problem. The goal is to retain enough of the essential physics to ensure that the simple model will predict some of the observations of the complex system.

Schematically, we go through a series of "views" of some particular problem. For example, consider flow around a rigid object (Fig. 6.2.1). The geometry is not simple, and so the flow field is complex. Numerical solutions are possible at considerable expense and time.

We may take a closer look (Fig. 6.2.2) at the phase boundary across which transfer is occurring. We introduce the assumption that the flow field consists of two regions: an outer region of nearly uniform flow and a *thin* inner region, in which the velocity field is dominated by viscous effects as the "no-slip" surface condition is approached. For simple shapes, "exact" and integral boundary layer analysis can yield useful, and relatively simple, models.

The lowest order boundary layer model is the *film model*, in which the velocity field near the surface is linearized in some average sense over the whole surface (i.e., along the $x$ axis). Consider first the fluid dynamics, as suggested in Fig. 6.2.3. If the velocity field were really linear, the shear stress exerted on the surface by the fluid would be given by

$$\tau = \mu \frac{\partial v_x}{\partial y} = \mu \frac{U}{\delta} = \frac{F_s}{A} \qquad (6.2.1)$$

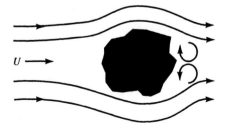

**Figure 6.2.1** Flow around a complex body.

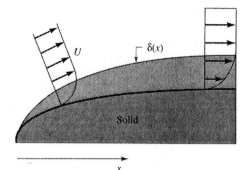

**Figure 6.2.2** A complex flow simplified near the surface of a boundary layer flow.

where $F_s/A$ is the shear force per unit of surface area acting in the $x$ direction. Solving for the film thickness, we find

$$\delta = \frac{\mu U}{\tau} \tag{6.2.2}$$

We usually cannot measure $\delta$ directly, but the shear force is measurable, and we have friction factor charts available for some systems. We may calculate $\delta$ from a knowledge of the friction factor $f$ (or the drag coefficient $C_f$ as it is presented in some texts), since $f$ is a dimensionless shear stress:

$$f \equiv \frac{\tau}{\frac{1}{2}\rho U^2} = C_f \tag{6.2.3}$$

This leads to

$$\delta = \frac{\mu U}{\frac{1}{2}f\rho U^2} = \frac{2\mu}{\rho U f} \tag{6.2.4}$$

or

$$\frac{\delta}{L} = \frac{2}{f}\frac{\mu}{\rho U L} = (\tfrac{1}{2}f\mathrm{Re})^{-1} \tag{6.2.5}$$

where $L$ is *some* characteristic length scale for the flow field.

The film model assumes that a *definable boundary layer* exists. It is meaningful only in the turbulent regime. For turbulent pipe flow, for example, we know that a good approximation is

$$f \sim \mathrm{Re}^{-0.25} \tag{6.2.6}$$

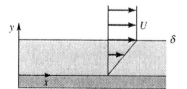

**Figure 6.2.3** A uniform film model of the boundary layer.

so, using Eq. 6.2.5, we find

$$\frac{\delta}{L} \sim \text{Re}^{-0.75} \tag{6.2.7}$$

Now we will apply this fluid model to convective mass transfer. We suppose that the concentration profile is uniform except in a thin layer near the interface, as suggested in Fig. 6.2.4. It is not clear that $\delta' = \delta$, so we keep a separate notation. (They are *not* the same.) We now write the molar flux of some species A in the $y$ direction by diffusion as

$$N_A = -D_{AB}\frac{\partial C_A}{\partial y} = D_{AB}\frac{\Delta C_A}{\delta'} = k_c \Delta C_A \tag{6.2.8}$$

We assume that Fick's law for a dilute binary system holds. In writing the far right-hand side of Eq. 6.2.8, we are *defining* the convective mass transfer coefficient $k_c$. The film model gives

$$k_c = \frac{D_{AB}}{\delta'} \text{ (cm/s)} \tag{6.2.9}$$

The friction factor is a useful measure of the *momentum transport* to a surface. It is a dimensionless parameter that in turn depends only on dimensionless characteristics of the flow field, such as the Reynolds number and the geometry of the boundaries of the flow. If $f$ is a dimensionless shear stress, what is a dimensionless mass flux?

We will define a dimensionless mass flux as

$$\frac{N_A}{D_{AB}(\Delta C_A/L)} \tag{6.2.10}$$

where $L$ is some appropriate length scale for the flow field. Then, according to this simple film model,

$$\frac{N_A}{D_{AB}(\Delta C_A/L)} = \frac{k_c L}{D_{AB}} \equiv \text{Sh}_L \tag{6.2.11}$$

where we have now introduced the Sherwood number as the dimensionless mass transfer coefficient. We subscript Sh as a reminder of its dependence on the choice of length scale. But $k_c = D_{AB}/\delta$, so it follows that

$$\text{Sh}_L = \frac{L}{\delta'} = \frac{L}{\delta}\left(\frac{\delta}{\delta'}\right) = \tfrac{1}{2}f\text{Re}_L\left(\frac{\delta}{\delta'}\right) \tag{6.2.12}$$

We can expect that $\delta/\delta'$ will turn out to be a function of the "diffusivities" for momentum and mass, or

$$\frac{\delta}{\delta'} = \text{fn}\left(\frac{\nu}{D_{AB}}\right) = \text{fn}(\text{Sc}) \tag{6.2.13}$$

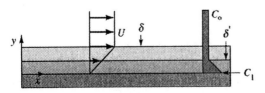

**Figure 6.2.4** Representation of the concentration field confined to a thin uniform film.

[We use the notation fn( ) here to denote "a function of ( )." The more usual $f(\ )$ notation would be confusing, since $f$ is a friction factor here.] It turns out that, using a more exact analytical method (we will develop that in the next section),

$$\mathrm{fn(Sc)} = \mathrm{Sc}^{0.33} \tag{6.2.14}$$

Hence

$$\mathrm{Sh}_L = \tfrac{1}{2} f\, \mathrm{Re}_L\, \mathrm{Sc}^{0.33} \tag{6.2.15}$$

A so-called $j$-factor for mass transfer is often (and at first, mysteriously) defined as

$$j_D \equiv \frac{\mathrm{Sh}}{\mathrm{Re}\ \mathrm{Sc}^{0.33}} \tag{6.2.16}$$

With this definition, Eq. 6.2.15 takes the form

$$j_D = \frac{f}{2} \tag{6.2.17}$$

which is sometimes called *the Chilton–Colburn analogy*. We see then that film theory leads to the prediction that there is a simple relationship between a characteristic dimensionless measure of convective mass transfer, the $j$-factor, and a well-known characteristic measure of the nature of a turbulent flow field, the friction factor.

The so-called *Reynolds analogy* is simply the Chilton–Colburn analogy for Sc = 1 (which is a good approximation for many common gas mixtures). In that case we find

$$\frac{k_c}{U} = \frac{f}{2} \tag{6.2.18}$$

We can make an immediate prediction here. For example, for turbulent pipe flow a good approximation to the friction factor curve at high Reynolds numbers is

$$f \approx 0.079\, \mathrm{Re}_D^{-0.25} \tag{6.2.19}$$

where the Reynolds number is based on the tube diameter $D$ as the length scale—the usual choice in tube flow. From Eq. 6.2.15 it follows that

$$\mathrm{Sh}_D = 0.04\, \mathrm{Re}_D^{0.75} \mathrm{Sc}^{0.33} \tag{6.2.20}$$

with Sh based on $D$ as the length scale, in place of $L$. This is a *testable* prediction, and we will examine its validity later.

Film theory, as we have presented it here, is very simple to understand. It is not a very good representation of the physics of interest, however, largely because it assumes that the boundary layers are *uniform along the direction of flow*. This is not the case in most real systems of interest to us, and the growth of the film thickness along the flow direction has a number of important consequences. To pursue this train of thought, we must turn to a boundary layer analysis that accounts for changes in the flow direction.

## 6.3 BOUNDARY LAYER THEORY FOR MASS TRANSFER FROM A FLAT PLATE (THE INTEGRAL METHOD)

We illustrate here the so-called integral boundary layer method. Instead of solving the Navier–Stokes equations, along with the continuity equation and a convective diffusion equation, we can *assume* a form for the velocity and concentration profiles, force those forms to satisfy integral conservation relationships, and obtain models for the shear stress (momentum transfer) and diffusive flux (mass transfer).

### 6.3.1  Momentum Boundary Layer Analysis

Since we are going to *assume* a functional relationship for the velocity field, we pick a form that is easy to manipulate. We choose a four-constant polynomial for $v_x$:

$$v_x = a_0 + a_1 y + a_2 y^2 + a_3 y^3 \tag{6.3.1}$$

and force $v_x$ to satisfy four boundary conditions:

$$\text{at } y = 0: \quad v_x = 0 \text{ (no slip)} \quad \text{and} \quad \frac{\partial^2 v_x}{\partial y^2} = 0 \left( = \frac{\partial p}{\partial x} \text{ [no pressure variation]} \right) \tag{6.3.2}$$

$$\text{at } y = \delta: \quad v_x = v_0 \quad \text{and} \quad \frac{\partial v_x}{\partial y} = 0 \tag{6.3.3}$$

Note that $y = \delta$ is the point beyond which the velocity takes on its free stream value. This is how we define the boundary layer thickness $\delta$. The second condition in Eq. 6.3.3 (the vanishing of the gradient at $y = \delta$) is simply an expression of our attempt to choose a velocity profile that has a continuous gradient as the velocity attains its free stream value.

These constraints on the form of the velocity field are satisfied if the constants have the values:

$$a_0 = a_2 = 0 \qquad a_1 = \frac{3}{2} \frac{v_0}{\delta} \qquad a_3 = -\frac{1}{2} \frac{v_0}{\delta^3} \tag{6.3.4}$$

Keep in mind that $\delta$ is an unknown function of $x$ at this point in the analysis. We have carried out what I like to call a *transformation of ignorance*. We have traded our lack of knowledge of the velocity function $v_x(x, y)$ for a different unknown function, $\delta(x)$. The point of this so-called integral method is that it will be easier to solve for $\delta(x)$ than it is to solve the Navier–Stokes equations for $v_x(x, y)$. The price we pay is having to deal with an approximate velocity field, and the issue is the accuracy of the result. The test of this method, *and the test of any modeling exercise,* comes in comparison of the predictions of the model to reality.

Conservation of mass is now imposed over a region of width $\Delta x$, and height L, as in Fig. 6.3.1. Note that per unit width in the $z$ direction (the $z$ axis comes out of the page), the $x$ component of the flux of mass is just $\rho v_x$. The mass balance just equates the net rate of flow of mass across the two vertical boundaries to the rate of flow across the horizontal boundaries. The lower horizontal boundary is the impermeable surface of

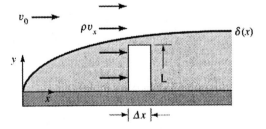

**Figure 6.3.1** Definition sketch for the integral momentum analysis.

the flat plate, and so there is no flow across it. Hence we find

$$\int_0^L \rho v_x \, dy \bigg|_x - \int_0^L \rho v_x \, dy \bigg|_{x+\Delta x} = \rho v_{y,L} \, \Delta x \tag{6.3.5}$$

Dividing by $\Delta x$ and letting $\Delta x$ vanish, we may solve for $v_{y,L}$ and find

$$v_{y,L} = -\frac{d}{dx} \int_0^L v_x \, dy \qquad \left(\begin{array}{c} \text{Compare to} \\ \dfrac{\partial v_y}{\partial y} = -\dfrac{\partial v_x}{\partial x} \end{array}\right) \tag{6.3.6}$$

So far we have imposed the principle of conservation of mass. Conservation of $x$ *momentum* is now imposed. Note that the *density* of $x$-directed momentum is $\rho v_x$, and the $x$-directed component of the *flux* of $x$-directed momentum is the momentum density times the velocity with which it crosses the boundary, or $[(\rho v_x) \, v_x]$. In the plane $y = L$ the $x$-directed momentum density is $\rho v_{x,L}$. Momentum crosses the plane $y = L$ at a rate $v_{y,L}$, so $[(\rho v_{x,L}) \, v_{y,L}]$ is the flux of $x$-directed momentum crossing that boundary. The momentum balance states that any difference in $x$-directed momentum flow must be equated to the net $x$-directed force acting on the solid surface. This arises from the shear stress[1] $-\tau_0$. The integral form of this balance then is

$$\int_0^L \rho v_x v_x \, dy \bigg|_x - \int_0^L \rho v_x v_x \, dy \bigg|_{x+\Delta x} - \rho v_{x,L} v_{y,L} \, \Delta x = -\tau_0 \, \Delta x \tag{6.3.7}$$

so the shear stress is

$$-\tau_0 = -\frac{d}{dx} \int_0^L \rho v_x^2 \, dy - \rho v_{x,L} v_{y,L} \tag{6.3.8}$$

We eliminate $v_{y,L}$ using Eq. 6.3.6 above and find:

$$-\tau_0 = \frac{d}{dx} \int_0^L (v_{x,L} - v_x) \rho v_x \, dy \tag{6.3.9}$$

or

$$\mu \frac{\partial v_x}{\partial y} \bigg|_{y=0} = \frac{d}{dx} \int_0^L (v_{x,L} - v_x) \rho v_x \, dy \tag{6.3.10}$$

Now we introduce the assumed polynomial expressions for the velocity profile, using Eqs. 6.3.1 and 6.3.4. We replace L by $\delta$ in the upper limit of the integral, since the integrand vanishes for $L > \delta$, where $v_x = v_{x,L} = v_0$. After some algebra, we find a differential equation for $\delta$.

$$\delta \frac{d\delta}{dx} = \frac{140}{13} \frac{\nu}{v_0} \tag{6.3.11}$$

or

$$\frac{\delta}{x} = 4.64 \left(\frac{v_0 x}{\nu}\right)^{-1/2} \tag{6.3.12}$$

Notice what we have achieved to this point. We have a simple model of the dependence of the boundary layer thickness on distance downstream, $x$. The price we have paid is

---

[1] $+\tau_0$ is the shear stress exerted *by* the solid surface *on* the fluid.

an uncertain degree of approximation in the assumed velocity field. Our next task is to apply this method to the concentration field.

### 6.3.2 Concentration Boundary Layer Analysis

We again assume a polynomial of third order, with four undetermined constants:

$$c = a_0' + a_1'y + a_2'y^2 + a_3'y^3 \tag{6.3.13}$$

and impose the following boundary conditions on the concentration field:

$$c = c_0 \quad \text{at} \quad y = 0 \tag{6.3.14}$$

$$\frac{\partial^2 c}{\partial y^2} = 0 \quad \text{at} \quad y = 0 \tag{6.3.15}$$

$$c = 0 \quad \text{at} \quad y = \delta_c \tag{6.3.16}$$

$$\frac{\partial c}{\partial y} = 0 \quad \text{at} \quad y = \delta_c \tag{6.3.17}$$

We have defined a concentration variable (and labeled it simply $c$ with no subscript) that vanishes beyond the boundary layer. Physically, this might correspond to the situation of a species dissolving from the surface of the plate and being transported by convection and diffusion into a fluid that is free of this species except very near the plate. (We could assume that there is a concentration $c_a$ in the fluid approaching the plate, in which case all concentrations in the analysis that follows would simply be converted to $c - c_a$.) The concentration at the solid boundary is taken as constant, with the value $c = c_0$.

Boundary conditions (6.3.15) and (6.3.17) are "smoothness" conditions. Physically we expect the diffusive flux to vary continuously across the boundary $y = \delta_c$. Since there is no diffusive flux outside the boundary layer, we force the flux to vanish right at the boundary layer. It follows that to satisfy these conditions, the concentration field has to take the form

$$\frac{c}{c_0} = 1 - \frac{3}{2}\frac{y}{\delta_c} + \frac{1}{2}\left(\frac{y}{\delta_c}\right)^3 \tag{6.3.18}$$

Now, a balance on the chemical species is made over the control volume. The net rate of convection of the species across the vertical planes of the control volume (we use the same control volume shown in Fig. 6.3.1) must be balanced by the rate corresponding to the flux $N_0$ from the solid surface:

$$\left.\int_0^L cv_x\,dy\right|_x - \left.\int_0^L cv_x\,dy\right|_{x+\Delta x} + N_0\,\Delta x - cv_{y,L}\,\Delta x = 0 \tag{6.3.19}$$

We solve for the species flux and find[2]

$$-N_0 = D_{AB}\left.\frac{\partial c}{\partial y}\right|_{y=0} = -\frac{d}{dx}\int_0^{\delta_c} cv_x\,dy = -k_c c_0 \tag{6.3.20}$$

(Note that we have introduced a mass transfer coefficient $k_c$ at this point. As usual, the mass transfer coefficient is defined as the ratio of the species flux to the concentration

---

[2] We integrate only up to $y = \delta_c$. Beyond that point, $c = 0$ and $v_{y,L} = 0$.

*difference* between the surface from which the convection is occurring and the concentration in the bulk of the fluid beyond the surface.)

Since $v_x$ is a known function of $x$ and $y/\delta$, and $c$ is given in terms of $y/\delta_c$, Eq. 6.3.20 yields a relationship between $\delta_c$ and $\delta$. If we assume for the moment (we will justify this later) that $\delta_c$ is much less than $\delta$, we find

$$\frac{4}{3} x \left(\frac{d}{dx}\right) \left(\frac{\delta_c}{\delta}\right)^3 + \left(\frac{\delta_c}{\delta}\right)^3 = \frac{13}{14} \frac{D_{AB}}{\nu} = \frac{13}{14} \text{Sc}^{-1} \tag{6.3.21}$$

(We introduce the Schmidt number here. It is the ratio of the diffusivities of momentum and species.) A trivial solution of this differential equation is (approximately)

$$\frac{\delta_c}{\delta} = \text{Sc}^{-1/3} \tag{6.3.22}$$

Finally we may find the flux from Eq. 6.3.20, the concentration boundary layer thickness $\delta_c$, and the dimensionless mass transfer coefficient Sh. The local (i.e., $x$-dependent) Sherwood number is defined with $x$ as the length scale:

$$\frac{k_c x}{D_{AB}} = \text{Sh}_x = 0.33 \left(\frac{x v_0}{\nu}\right)^{1/2} \text{Sc}^{1/3} \tag{6.3.23}$$

For a plate of finite length $L$ (note that this $L$ is not the L of the control volume in Fig. 6.3.1), the *average* Sherwood number is found from $\overline{k}_c$, which is found by integration down the plate:

$$\overline{k}_c = \frac{1}{L} \int_0^L k_c(x) \, dx \tag{6.3.24}$$

and the average Sherwood number over the distance $L$ is

$$\overline{\text{Sh}} = \frac{\overline{k}_c L}{D_{AB}} = 0.66 \left(\frac{L v_0}{\nu}\right)^{1/2} \text{Sc}^{1/3} = 0.66 \, \text{Re}_L^{1/2} \, \text{Sc}^{1/3} \tag{6.3.25}$$

We need to summarize what we have been doing here, and recall our original goal. We wish to have a means of prediction of the rate of mass transfer from surfaces in situations of a surface geometry and/or flow field so complex that we cannot solve analytically the exact forms of the conservation equations for momentum and species. The problem we have just examined, the convective transfer across a boundary layer on a flat plate subject to an external flow that is parallel to the plate, is in a sense an intermediate problem. We can solve for the velocity and concentration fields analytically, but only through the use of certain approximations that lead to the boundary layer analysis.

Many mass transfer problems are so complex that a simple boundary layer theory is not possible. An example is flow, either laminar or turbulent, normal to the axis of a long cylinder. No simple analytical solution exists for laminar flow, and turbulent flow creates a vortex structure behind the cylinder that does not permit the development of a simple boundary layer model. But one can certainly obtain experimental data on such a system. Figure 6.3.2 shows the results of the experiments of several investigators, over different ranges of Reynolds numbers, recast in the form of a plot of *j*-factor versus Reynolds number. It is not surprising that curves $A$ and $B$ do not overlap. The definition of the *j*-factor implies that the Schmidt number dependence of Sherwood number is $\text{Sc}^{1/3}$. In many cases this dependence is observed to be otherwise. Hence two sets of data obtained at very different Schmidt numbers, which is the case in comparing lines $A$ and $B$ in the Reynolds number range of 10–100, may not fall on a single curve of *j*-

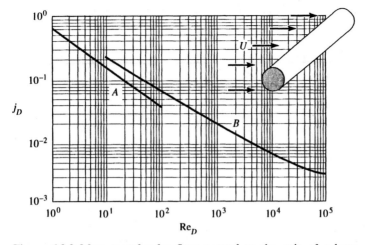

**Figure 6.3.2** Mass transfer for flow normal to the axis of a long cylinder. Curve $A$ is based on data of Vogtlander and Bakker [*Chem. Eng. Sci.*, **18**, 583 (1963)]. Curve $B$ is from a collection of data presented in Sherwood et al., *Mass Transfer*, (McGraw-Hill, 1975).

factor versus Reynolds number if the Schmidt number dependencies of the two sets of data are not the assumed $Sc^{1/3}$ dependence.

What good, then, is the boundary layer analysis we have just carried out? First of all, many problems do indeed have this boundary layer character, and we now have a model for such flows. But this model also teaches us how the mass transfer coefficient may be nondimensionalized to a Sherwood number, and how this Sherwood number depends on the Reynolds number and the Schmidt number. Indeed, we learn that Sh does depend on these groups, although we would have reached the same conclusions by dimensional arguments (dimensional analysis).

We assert now that the most convenient way to discuss convective mass transfer is through the Sherwood number, or the $j$-factor. Before we proceed to examine a series of complex mass transfer situations, let's go back to some simpler problems—those we examined in Chapter 5—and recast our solutions in terms of the Sherwood number.

## 6.4 PREDICTIONS OF SHERWOOD NUMBER FOR TRANSFER TO LAMINAR FILMS

For various flow situations we find different mass transfer models. In the case of a turbulent flow parallel to a flat plate, the boundary layer theory leads to predictions that we can express in the form of the relationship of the Sherwood number to the Reynolds number and the Schmidt number. In the case of mass transfer to a laminar film of liquid falling down a vertical plane, we obtain a very different model of mass transfer. (See Chapter 5, and the analysis of the falling film evaporator.) In any event, once we have a model, we need to assess its degree of correspondence to observation. *This is the ultimate test of a mathematical model of a physical system.*

For mass transfer across the *gas/liquid interface* of a laminar falling film on a solid surface, with mass transfer resistance *only* on the liquid side of the interface, we derived a model, valid for short "time" in the form (we are using a slightly different notation here, compared to that in Chapter 5)

$$\text{flux } N = c_0 \left(\frac{D_{AB}}{\pi t_E}\right)^{1/2} \qquad \text{(see Eq. 5.1.25)} \tag{6.4.1}$$

or

$$\frac{N}{c_0} \equiv k_c = \left(\frac{D_{AB}}{\pi t_E}\right)^{1/2} \qquad \text{and} \qquad t_E = \frac{z}{v_m} \tag{6.4.2}$$

Keep in mind that $v_m$ is the maximum, or surface, velocity, *not* the mean velocity of the liquid film. If we average over a time $T$, corresponding to some distance along the film,

$$\bar{k}_c = \frac{1}{T}\int_0^T k_c \, dt = 2\left(\frac{D_{AB}}{\pi T}\right)^{1/2} \tag{6.4.3}$$

The model that leads to this result is based on the assumption that the velocity profile is fully developed right from the beginning of the mass transfer region. Only the concentration field is changing along the length of the surface, in the direction of flow. A different model is obtained if the velocity field is also undergoing development. This latter case would occur if we considered a uniform flow field that suddenly encountered a flat plate parallel to the flow, as in consideration of the developing laminar boundary layers of velocity and concentration on a flat plate.

Note that we use a time, or "age" here, instead of the distance downstream, as in Chapter 5. Note also that the average coefficient at some point in "time" (or at some distance along the surface), in this case is exactly twice the local value. The relationship between time and distance is

$$T = \frac{L}{v_m} = \frac{2L}{3\langle v \rangle} \tag{6.4.4}$$

where $\langle v \rangle$ is the average film velocity.

Now we define a Sherwood number averaged along the length of the surface to which the film is attached, and a Reynolds number for this flow, both based on $\delta$ as the length scale:

$$\overline{\text{Sh}_\delta} = \frac{\bar{k}_c \delta}{D_{AB}} \qquad \text{and} \qquad \text{Re}_\delta = \frac{\delta \langle v \rangle}{\nu} \tag{6.4.5}$$

Keep in mind that we are talking about flow of a laminar film of thickness $\delta$, not a boundary layer flow. In this discussion $\delta$ is a constant, and it denotes a laminar film thickness—not a boundary layer thickness.

The result is a predictive model for the Sherwood number for transfer to a falling laminar film. Convective mass transfer occurs across the liquid/gas interface, and the mass transfer resistance is entirely in the liquid phase. The result for the averaged Sherwood number is

$$\overline{\text{Sh}_\delta} = \left(\frac{6}{\pi} \text{Re}_\delta \, \text{Sc} \, \frac{\delta}{L}\right)^{1/2} \tag{6.4.6}$$

Note that the Reynolds and Schmidt numbers appear here as a product, called the Peclet number (Pe):

$$\text{Re}_\delta \, \text{Sc} = \frac{\delta \langle v \rangle}{D_{AB}} = \text{Pe} \tag{6.4.7}$$

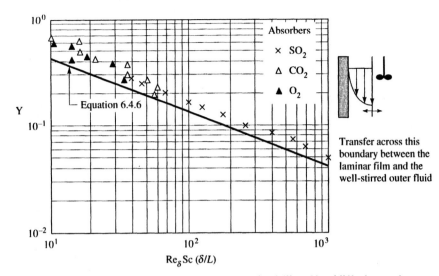

**Figure 6.4.1** Mass transfer to the free surface of a falling film (diffusion resistance in the liquid).

Keep in mind that in this problem we encounter diffusion across a nearly flat velocity profile. An experimental test of Eq. 6.4.6 is given in Fig. 6.4.1. Data for "short" contact times agree quite well with the theory. The deviations of the data from the theory at smaller values of the ordinate can be shown to be in good agreement with a model that is more appropriate to longer contact times.

The vertical axis, labeled $Y$ in Fig. 6.4.1, is not the Sherwood number, but instead the grouping

$$Y = \frac{\overline{Sh_\delta}}{Re_\delta \, Sc(\delta/L)} \tag{6.4.8}$$

where the Reynolds and Sherwood numbers use $\delta$ as in the length scale. This group $Y$ seems an odd choice of a parameter, and we will clarify this point shortly.

A similar problem arises in consideration of mass transfer at the *solid/liquid* boundary, again for the laminar falling film. (This is a case that would be relevant to the dissolution of a soluble solid into a flowing liquid.) We carried out an analytical model for this problem in Chapter 5 (see Eq. 5.2.13) and we found the species flux from the plate, in the region of parameter space where the extent of dissolution is confined to a thin layer near the surface of the plate. This permitted us to approximate the velocity profile near the surface as a linear profile. It can be shown that Eq. 5.2.13 leads directly to the following expression for the average Sherwood number:

$$\overline{Sh_\delta} = A \, (Re_\delta Sc)^{1/3} \left( \frac{\delta}{L} \right)^{1/3} \qquad \text{in a planar flow} \tag{6.4.9}$$

where $A = 1.165$. The Reynolds and Sherwood numbers are defined as in Eqs. 6.4.5 above, with the liquid film thickness and the *average* velocity as the length and velocity scales.

Although we have spoken of mass transfer to a falling liquid film across the solid/liquid boundary (not across the liquid/gas boundary, as in the earlier discussion) these results are unchanged if we wish to consider mass transfer from the wall of a parallel plate channel into a laminar flow that fills the channel. This is because the boundary

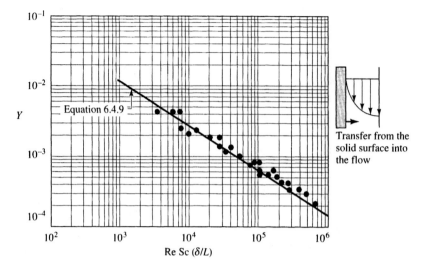

**Figure 6.4.2** Data for dissolution from a solid surface in contact with a laminar film flow ($Y$ defined in Eq. 6.4.8.)

condition used at the free surface of the film (Eq. 5.2.10) is identical to what we would use as the second boundary condition along the midplane of symmetry for flow in a channel. Hence the boundary value problems are identical and the solutions take the same functional forms.

Note, in Eq. 6.4.9, the change in exponents compared to Eq. 6.4.6. This difference arises because in the case of dissolution near the wall, we encounter diffusion across a nearly *linear* velocity profile, while near the gas/liquid interface (in the falling film absorber or evaporator), the profile is nearly *flat*. Data agree very well with this theory, for short lengths of surface (or short contact times $t_c = L/\langle v \rangle$), such that

$$\frac{D_{AB}t_c}{\delta^2} \leq 10^{-1} \tag{6.4.10}$$

We can show that with the definition of contact time $t_c = L/\langle v \rangle$ just used, based on the *average* film velocity, *not the maximum* velocity, the following is true:

$$\frac{D_{AB}t_c}{\delta^2} = \left( \mathrm{Re\ Sc} \frac{\delta}{L} \right)^{-1} \tag{6.4.11}$$

Data for dissolution of benzoic acid from flat plates are shown in Fig. 6.4.2. Again, $Y$ is given by Eq. 6.4.8, and Re and Sh use $\delta$ as the length scale, rather than $L$.

The foregoing model for mass transfer from the solid boundary of a laminar planar flow (Eq. 6.4.9) is limited to "short contact times." When the extent of penetration of the diffusing species into the flow is comparable to the film thickness, the assumption of a linear velocity field no longer holds. The actual velocity profile would be parabolic, and the form of the solution changes.[1] The average Sherwood number becomes a

---

[1] For the corresponding problem of dissolution from the wall of a circular tube or pipe, with uniform wall concentration, we find Eq. 6.4.9, but with $A = 1.29$, and Eq. 6.4.12, but with the Sherwood number equal to 1.83. For the circular tube case, the *radius of the tube* is used as the length scale in both Sh and Re, in place of the film thickness $\delta$, and the average velocity is used in Re. Many authors use the diameter instead of the radius in presenting these mass transfer results, and of course this changes the coefficients. See Problem 6.2.

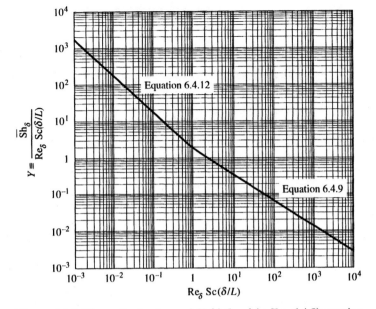

**Figure 6.4.3** Theoretical values of $Y$ (defined in Eq. 6.4.8) ranging from short to long "contact times" for transfer from the wall of a planar surface to a laminar flow parallel to the wall.

constant, in fact, and we find (see Problem 6.2)

$$\overline{Sh}_\delta = \frac{\bar{k}\delta}{D_{AB}} = 1.6 \tag{6.4.12}$$

It is appropriate to use this model when

$$\frac{D_{AB}t_c}{\delta^2} = \left( Re_\delta \, Sc \, \frac{\delta}{L} \right)^{-1} \gg 0.1 \tag{6.4.13}$$

Figure 6.4.3 shows the theoretical values of the Sherwood number (plotted as $Y$ defined in Eq. 6.4.8) over the whole range of contact times [plotted as the product $X = Re_\delta \, Sc(\delta/L)$]. The range between $X = 1$ and $X = 10$ is simply interpolated, and $Y$ so obtained is close to the actual theoretical values in this range.

Many other simple laminar flow problems yield analytical solutions for the flux of some species at a phase boundary that we may put into the form of a Sherwood number dependent on a Reynolds number and a Schmidt number. Let's pause here and review what we have learned by using some of these models for the prediction of mass transfer rates at a phase boundary. We will give special attention to distinguishing between local and average mass transfer coefficients, and we will come to a definition of a concentration driving force to use in mass balances written in terms of the average coefficient.

**EXAMPLE 6.4.2** *Mass Transfer in Flow Through a Parallel Plate Channel: Definition of a Concentration Driving Force*

The flow field and the developing concentration profile are depicted in Fig. 6.4.4. We assume that the concentration at the entrance to the channel is uniform, at the level $C_1$, as shown. The surface concentration of some dissolvable species is $C_s$, and dissolution

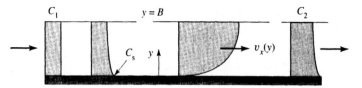

**Figure 6.4.4** Dissolution from the wall of a channel into a laminar fully developed flow.

and diffusion into the flow occurs from the solid boundary. At the end of the channel there is a concentration profile $C_2$, which will not be uniform unless the channel is so long that the concentration reaches the equilibrium value $C_s$. We assume that the laminar velocity profile is fully developed right from the entrance to the channel.

A differential species balance gives

$$Q \, d\overline{C} = 2k(C_s - \overline{C}) \, W \, dx \qquad (6.4.14)$$

where $\overline{C}$ is understood to be the *cup-mixing average* molar concentration across the channel. In Eq. 6.4.14, $k$ is defined as the local mass transfer coefficient and $Q$ is the volumetric flowrate through the channel. The factor of 2 accounts for transfer from both sides of the channel. If we take $C_s$ to be constant along the channel length, we may write Eq. 6.4.14 in the form

$$2 \langle v \rangle \, Bd(C_s - \overline{C})W = -2k(C_s - \overline{C}) \, W \, dx \qquad (6.4.15)$$

Equation 6.4.14 is actually a definition of the local mass transfer coefficient $k$, based on the difference in concentration between the wall value and the cup-mixing average of the fluid. We may integrate Eq. 6.4.15 between the entrance $x = 0$ and any arbitrary position $L$ downstream, with the result

$$\frac{C_s - \overline{C_L}}{C_s - \overline{C_1}} = \exp\left( - \int_0^L \frac{k}{\langle v \rangle \, B} \, dx \right) = \exp\left( - \frac{\overline{k}L}{\langle v \rangle \, B} \right) \qquad (6.4.16)$$

where we define the average mass transfer coefficient as

$$\overline{k} = \frac{1}{L}\left( \int_0^L k \, dx \right) \qquad (6.4.17)$$

Note that we have *not* assumed that $k$ is constant along the length of the channel.

If we examine the argument of the exponential function in Eq. 6.4.16, we find that

$$\frac{\overline{k}L}{\langle v \rangle B} = \frac{\overline{\text{Sh}}}{\text{Re Sc } (B/L)} \qquad (6.4.18)$$

This is the dimensionless group that we have labeled $Y$ in plotting Figs. 6.4.1 through 6.4.3, *except that* in this case the length scale used in the Sherwood and Reynolds numbers is the half-width $B$ of the channel. We conclude that this particular grouping arises naturally in the analysis of mass transfer into a flowing system. The definition of $Y$ introduced earlier was not arbitrary. It was made in anticipation that this particular grouping would appear.

There is another way of writing Eq. 6.4.16 that connects this result to older, classical, studies of mass transfer processes. We define a length denoted by $H_{\text{TU}}$ such that

$$\frac{L}{H_{\text{TU}}} = \frac{\overline{k}L}{\langle v \rangle B} = \frac{\overline{\text{Sh}}}{\text{Re Sc } (B/L)} \qquad (6.4.19)$$

Then we may write Eq. 6.4.16 in the form

$$\frac{C_s - \overline{C_L}}{C_s - \overline{C_1}} = \exp\left(-\frac{L}{H_{TU}}\right) = \exp(-Y) \tag{6.4.20}$$

with $Y$ defined as in Eq. 6.4.8. In this format we may identify the length $H_{TU}$ as a measure of the efficiency of mass transfer, in the sense that the length of channel, $L$, needed to reduce the initial departure of the fluid from concentration equilibrium by a factor of $1/e = 0.368$, is just $L = H_{TU}$. (A channel of length $L = 3H_{TU}$ brings the fluid within 95% of equilibrium with respect to the wall value.) Obviously, a small value of $H_{TU}$ corresponds to very efficient mass transfer. In the classical literature of chemical engineering, $H_{TU}$ is called the height of a theoretical unit.

From Eq. 6.4.19 we see that

$$\frac{H_{TU}}{B} = \frac{\text{Re Sc}}{\overline{\text{Sh}}} \tag{6.4.21}$$

(Again, keep in mind that Re and Sh use $B$ as the length scale, not $L$.) We sometimes find the inverse of this grouping appearing in dimensionless correlations of data. It is called the Stanton number for mass transfer:

$$\frac{1}{\text{St}} = \frac{H_{TU}}{B} = \frac{\text{Re Sc}}{\overline{\text{Sh}}} \tag{6.4.22}$$

Returning now to our initial task, we write the total transfer (not the flux) of the soluble species into the flow, over the whole length $L$ of the channel, in the form

$$B\langle v \rangle(\overline{C}_L - \overline{C}_1) = \overline{k}L\,\Delta C \tag{6.4.23}$$

Equation 6.4.23 does not follow from anything preceding! It is a definition of the average mass transfer coefficient $\overline{k}$. But we have already defined $\overline{k}$ in Eq. 6.4.17. What we have *not* defined is the concentration difference $\Delta C$ that is appropriate to use with this definition of $\overline{k}$. We solve for $\Delta C$ from Eq. 6.4.23 and we find

$$\Delta C = \frac{B\langle v \rangle(\overline{C}_L - \overline{C}_1)}{\overline{k}L} \tag{6.4.24}$$

If we use Eqs. 6.4.19 and 6.4.20 to eliminate $\overline{k}$, we may write $\Delta C$ in the form

$$\Delta C = \frac{(C_s - \overline{C}_1) - (C_s - \overline{C}_L)}{\ln\left[(C_s - \overline{C}_1)/(C_s - \overline{C}_L)\right]} = \Delta C_{lm} \tag{6.4.25}$$

This is the definition of the "logarithmic mean" concentration difference. From this analysis we conclude that the average transfer coefficient $\overline{k}$ that appears in the overall mass balance (Eq. 6.4.23) is identical to the average transfer coefficient defined directly by Eq. 6.4.17 only if we understand that the concentration "driving force" that appears in the mass balance, $\Delta C$, is the log–mean difference. The log–mean difference is not an arbitrary definition of a mean value that has some mystical advantage over the arithmetic average—the log–mean is *the proper difference* to use in the definition of the mean mass transfer coefficient that is introduced through the mass balance (Eq. 6.4.23).

We note finally that the argument presented in Example 6.4.1 would hold for other flow geometries, such as flow through a circular tube. Some of the details of the derivation would change slightly (such as the length scale in the Sh and Re numbers), but we would still conclude that Eq. 6.4.25 defines the appropriate concentration difference.

Now let's turn to an example of application of these ideas.

**EXAMPLE 6.4.3** *Evaporation of Naphthalene from Planar Boundaries of a Channel When the Flow is Laminar*

We suppose that air at 16°C flows through a thin channel bounded by large parallel planes. The planes are coated with solid naphthalene. We wish to prepare a plot that permits us to predict the concentration of naphthalene in the gas phase at the outlet of the channel, as a function of the gas flowrate. The following data are available:

$$p_{vap} = 0.037 \text{ mmHg} \qquad D_{AB} = 0.071 \text{ cm}^2/\text{s}$$

$$B = 0.1 \text{ cm} \qquad Q = 10\text{--}1000 \text{ cm}^3/\text{s}$$

$$Sc = 2.25 \qquad W = 10 \text{ cm} \qquad L = 20 \text{ cm}$$

We begin with a calculation of the Reynolds number for this flow. For the kinematic viscosity of air we use the given value of the Schmidt number to find

$$\nu_{air} = D_{AB} \, Sc = 0.071 \times 2.25 = 0.16 \text{ cm}^2/\text{s} \tag{6.4.26}$$

At the lowest flowrate ($Q = 10 \text{ cm}^3/\text{s}$), the Reynolds number is found from

$$Re = \frac{\langle v \rangle B}{\nu_{air}} = \frac{Q}{2 W \nu_{air}} = \frac{10}{2 \times 10 \times 0.16} = 3.1 \tag{6.4.27}$$

Thus the flow is laminar over the whole range of anticipated flowrates.

Next we calculate the parameter

$$Re \, Sc \, \frac{B}{L} = 3.1 \times 2.25 \times \frac{0.1}{20} = 0.035 \tag{6.4.28}$$

We will use Fig. 6.4.3 for the calculation of the Sherwood number. (The film thickness $\delta$ is replaced by the half-width of the channel, $B$.) We see that at the lowest flowrate we are in the regime of constant average Sherwood number (Eq. 6.4.12), and we find

$$Y = 46 = \frac{L}{H_{TU}} \qquad \text{or} \qquad H_{TU} = \frac{20}{46} = 0.44 \text{ cm} \tag{6.4.29}$$

From Eq. 6.4.20 we find immediately that the air exits this long channel in equilibrium with the naphthalene, since $\exp(-46) \approx 0$.

At the highest flowrate of the anticipated range we find, from Fig. 6.4.3 (or more accurately from Eqs. 6.4.8 and 6.4.9), $Y = 0.505$, and so

$$H_{TU} = \frac{20}{0.505} = 39.6 \text{ cm} \tag{6.4.30}$$

Since the channel is only 10 cm long, we expect that the air will not be near equilibrium in this case. From Eq. 6.4.20 we find

$$\frac{C_s - \overline{C_L}}{C_s - \overline{C_1}} = \exp(-Y) = \exp(-0.505) = 0.60 \tag{6.4.31}$$

With the assumption that the air is free of naphthalene at the entrance to the channel ($C_1 = 0$), we may write the outlet concentration as

$$\overline{C_L} = 0.4 C_s \tag{6.4.32}$$

For units on concentration we choose, in this case, to write the equilibrium concentration in terms of the vapor pressure of naphthalene and convert that to a concentration in units of parts per million by volume (ppmv). We find (noting that the pressure ratio is just the volume fraction in an ideal gas)

$$C_s = 0.037 \text{ mmHg} = \frac{0.037}{760} = 49 \times 10^{-6} = 49 \text{ ppmv} \qquad (6.4.33)$$

and so the outlet concentration of naphthalene in the air is

$$\overline{C_L} = 0.4 C_s = 0.4 \times 49 \approx 20 \text{ ppmv} \qquad (6.4.34)$$

This procedure is followed for other flowrates in the anticipated range, and the results are tabulated as follows:

| $Q$ (cm³/s) | $Y$ | $\overline{C_L}$ (ppmv) |
|---|---|---|
| 10 | 46 | 49 |
| 30 | 15.2 | 49 |
| 100 | 4.57 | 48.5 |
| 300 | 1.52 | 38 |
| 1000 | 0.5 | 20 |

Over this range of flowrates the flow is laminar. In Example 6.4.3 we consider the *turbulent* flow regime.

**EXAMPLE 6.5.1** *Evaporation of Naphthalene from Planar Boundaries of a Channel When the Flow is Turbulent*

At flowrates higher than those of Example 6.4.2, the flow becomes turbulent, and Fig. 6.4.3 is no longer applicable, since the theories that yield the models we have used are valid only for laminar fully developed flow in the channel. When the flow is turbulent we may hope to get a reasonable estimate of behavior through the use of film theory, and in particular through application of the Chilton–Colburn analogy. Thus we begin with

$$j_D \equiv \frac{\text{Sh}}{\text{Re Sc}^{1/3}} = \frac{f}{2} \qquad (6.4.35)$$

For turbulent flow in a circular tube, this result leads us to predict that

$$\text{Sh}_D = 0.04 \text{ Re}_D^{0.75} \text{Sc}^{0.33} \qquad \textit{for tube flow} \qquad (6.4.36)$$

Now it is important that we recall that in *this* model the Sherwood and Reynolds numbers use the *tube diameter* as the length scale. (That is why we have subscripted them as shown in Eq. 6.4.36.) For turbulent flow in a channel bounded by large parallel plates, we may use the hydraulic radius concept and replace the diameter everywhere by $4R_h$, where the hydraulic radius for a wide channel is given by

$$R_h = \frac{\text{cross-sectional area}}{\text{wetted perimeter}} = \frac{2BW}{2W + 4B} \approx B \qquad (6.4.37)$$

(We have assumed that $W$ is much greater than $B$.) Equation 6.4.36 is transformed to

$$\text{Sh}_B = 0.028 \text{ Re}_B^{0.75} \text{Sc}^{0.33} \qquad \textit{for channel flow} \qquad (6.4.38)$$

where the Reynolds and Sherwood numbers now use the half-width of the channel, $B$, as the length scale. This permits us to use Eq. 6.4.20 in the form

$$\frac{C_s - \overline{C_L}}{C_s - \overline{C}_1} = \exp(-Y) \qquad (6.4.39)$$

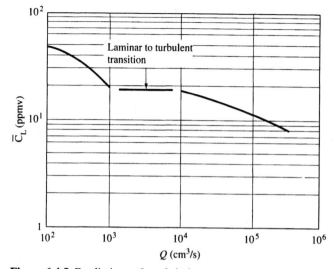

**Figure 6.4.5** Predictions of naphthalene evaporation into the turbulent regime.

with $Y$ defined as in Eq. 6.4.8 (and with $\delta$ replaced by $B$). A series of computations yields the predictions of Fig. 6.4.5, which also shows the laminar flow results from Example 6.4.2. In the region of transition from laminar to turbulent flow—somewhere in the interval $[10^3 < Q < 10^4]$—neither model is expected to be very accurate. One might estimate the concentration to be about 20 ppmv over that region, but if accuracy is required, an experiment in this transition regime is advisable.

## 6.5   TRANSFER TO A LAMINAR FALLING FILM WITH GAS SIDE RESISTANCE

In this section we look at the gas/liquid interface again, but this time at the *gas side mass transfer coefficient*. Normally, the gas flow relative to a film absorber or evaporator is turbulent. We will examine the Chilton–Colburn analogy for the purpose of developing a model of this system. First, though, we should determine the circumstances under which the gas side mass transfer resistance would be significant. Aren't diffusivities in gases so high that we might expect the resistance to reside almost entirely in the liquid? In general, the answer to this question is yes. In at least one important case, however, there would be no resistance on the liquid side of an evaporator, and we would need an estimate of the gas side coefficient to predict evaporation rates. This is the case of an evaporating liquid that is pure. As the species evaporates, the concentration does not change within the liquid, since the liquid remains pure. Hence there is *no* liquid side resistance *in evaporation from a pure liquid,* and only the gas side resistance matters, and therefore must be known.

We will estimate the gas side mass transfer coefficient using the Chilton–Colburn analogy:

$$j_D = \frac{f}{2} \tag{6.5.1}$$

We assume that the liquid film falls down the inside of a tube through which there is a gas flow with a high Reynolds number $\mathrm{Re}_g$. For the friction factor $f$, we use

$$f = 0.079\,\mathrm{Re}_g^{-0.25} \tag{6.5.2}$$

Then if the gas side flow is turbulent, the Chilton–Colburn analogy holds and we predict

$$j_D = \frac{\overline{Sh}}{Re_g \, Sc^{1/3}} = 0.04 \, Re_g^{-0.25} \tag{6.5.3}$$

or

$$\overline{Sh} = 0.04 \, Re_g^{0.75} Sc^{1/3} \tag{6.5.4}$$

A test of the Chilton–Colburn analogy is shown in Fig. 6.5.1. These data are for evaporation of various pure liquids "falling" inside a wetted-wall column. The data are consistently higher than predicted by Eq. 6.5.4.

Gilliland and Sherwood correlated these data for evaporation of pure liquids into air by

$$\overline{Sh} \equiv \frac{\overline{k}D}{D_{AB}} = 0.023 \, Re_g^{0.83} Sc^{0.44} \tag{6.5.5}$$

with $Re_g$ based on the gas velocity relative to the tube. Others have suggested using a Reynolds number based on the gas velocity relative to the liquid surface. In any case the Reynolds number uses the tube diameter as the length scale, as does the Sherwood number. When Eq. 6.5.5 is converted to the $j_D$ form, we see that the equation fits the data quite well.

If the *liquid film* is not laminar, some dependence on a liquid Reynolds number is observed. For example, one recommended correlation [Kafesjian, Plank, and Gerhard, *AIChE J.*, **7**, 463 (1961)] for the gas phase mass transfer coefficient is

$$\overline{Sh} \equiv \frac{\overline{k}D}{D_{AB}} = 0.00814 \, Re_g^{0.83} Sc^{0.44} \, (4 \, Re_L)^{0.15} \tag{6.5.6}$$

where

$$Re_L = \frac{\langle v \rangle \, \delta}{\nu_L} \tag{6.5.7}$$

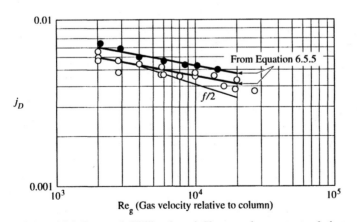

**Figure 6.5.1** Data of Gilliland and Sherwood as a test of the Chilton–Colburn analogy (the "$f/2$" line) and the empirical correlation of Eq. 6.5.5: solid circles, chlorobenzene (Sc = 2.2); open circles, water (Sc = 0.6). Gilliland and Sherwood, *Ind. Eng. Chem.*, **26**, 516 (1934).

Note that

$$\mathrm{Re}_g = \frac{u_{gas}D}{\nu_{gas}} \tag{6.5.8}$$

When the gas phase flow is turbulent, the most common correlation recommended is that of Johnstone and Pigford [*Trans. AIChE,* **38,** 25 (1942)]:

$$j_d = 0.0328 \, (\mathrm{Re}_g')^{-0.23} \tag{6.5.9}$$

where both $j_D$ and Re' use the velocity of the gas relative to the *surface* velocity of the liquid film. (Note the lack of a Sc dependence here.)

We must recall that the use of the Chilton–Colburn analogy is based on film theory, and film theory in turn is based on the supposition that the mass transfer resistance is confined to a thin film near the boundary across which transfer occurs. The film concept breaks down in the case that the convective flow (in this case, the gas flow through the pipe) is in *laminar* flow. Some data for evaporation of pure liquids in a falling film evaporator through which a gas is in *laminar* flow are shown in Fig. 6.5.2.

If we compare these data to the theoretical prediction for short contact times (which is Eq. 5.2.19) based on the assumption of a laminar flow in the pipe, with transfer across an interface that is fixed with respect to the flow, we find that the data are higher than predicted, which means that equilibrium is approached more efficiently than predicted. The reason for this discrepancy is not clear, but a potential factor that would enhance transfer across the interface is the mobility of the interface, particularly if the falling film is rippled, as is often the case even at low flow velocities.

In examining experimental data we have focused on comparing the data to simple theoretical models, and we have examined empirical correlations of the data. In the course of this exercise we have neglected to emphasize an important issue, which is the methodology by which we convert raw data on mass transfer rates into the Sherwood number. We pick this point up here.

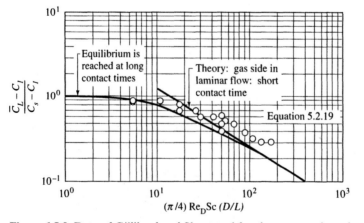

**Figure 6.5.2** Data of Gilliland and Sherwood for the evaporation of pure liquids from a falling film into a laminar gas flow in a pipe.

**EXAMPLE 6.5.1**     *Conversion of Evaporation Rate Data to Sherwood Number Data*

The defining equation for the mass transfer coefficient comes out of a species balance, such as Eq. 6.4.23 for the parallel plate channel. It would take a different form for transfer from the wall of a circular pipe, for example:

$$\frac{\pi D^2}{4} \langle v \rangle (\bar{C}_L - \bar{C}_1) = \bar{k} (\pi DL) \Delta C_{lm} = N \pi DL \tag{6.5.10}$$

where $N$ is the total molar flux of the evaporating species over the length of the evaporator. Thus we would write the Sherwood number as

$$\overline{Sh} \equiv \frac{\bar{k}D}{D_{AB}} = \frac{ND}{D_{AB}\Delta C_{lm}} \tag{6.5.11}$$

Now we consider the following experimental observation:

$n$-Butyl alcohol flows at a rate of 790 cm$^3$/min as a film inside a tube of diameter $D = 2.67$ cm and length $L = 117$ cm. Air flows at a rate of 100 g/min countercurrent to the liquid flow. The inlet air temperature is 52.4°C and the exiting air temperature is 47.6°C. (Evaporative cooling accounts for this change in temperature.) The liquid film is at the air temperature at any point along the tube. The absolute pressure in the tube is 820 mmHg and is tentatively assumed to be constant along the axis of the tube. (We assume that the pressure drop due to friction losses is small at these flow rates.) The vaporization rate is observed to be $q_{vap} = 6.9$ cm$^3$/min (based on the liquid, which has a density of 0.81 g/cm$^3$). The vapor pressure of $n$-butyl alcohol is 38 mmHg at the inlet temperature, and 28.7 mmHg at the outlet. The diffusivity of $n$-butyl alcohol is 0.09 cm$^2$/s at the mean temperature (50°C) and pressure (820 mmHg) of the gas. The molecular weight of $n$-butyl alcohol is 74.

From Eq. 6.5.11 we see that we need to calculate two quantities: the evaporative flux, and the concentration driving force.

## Evaporative Flux

The wetted area is

$$A = \pi DL = \pi \times 2.67 \times 117 = 981 \text{ cm}^2 \tag{6.5.12}$$

The molar vaporization rate (flux) is

$$N = \frac{\rho q_{vap}}{M_w A} = \frac{0.81 \times 6.9/60}{74 \times 981} = 1.28 \times 10^{-6} \, (\text{mol/cm}^2 \cdot \text{s}) \tag{6.5.13}$$

## Concentration Driving Force

The log–mean driving force is defined in Eq. 6.4.25. Since data are available for the vapor pressure, we will use partial pressures as the concentration units. Thus we write Eq. 6.4.25 in the form

$$\Delta p_{lm} = \frac{(p_{vap} - \bar{p}_1) - (p_{vap} - \bar{p}_L)}{\ln \left[ (p_{vap} - \bar{p}_1)/(p_{vap} - \bar{p}_L) \right]} \tag{6.5.14}$$

We take the inlet value of the partial pressure to be $p_1 = 0$. The vapor pressure of the liquid is different at the two ends, and so, in calculating the $\Delta p$ terms at each end, we use the inlet and outlet values of vapor pressure with the inlet (zero) and outlet values of the partial pressure of the alcohol in the airstream.

At the outlet we have a flowrate of alcohol in the gas of $0.81 \times 6.9 = 5.56$ g/min. The total pressure (820 mmHg) is the sum of the air and alcohol *partial* pressures. The partial pressure of *n*-butyl alcohol in the outlet gas is then found from

$$\overline{p_L} = \frac{5.56/74}{5.56/74 + 100/29} \times 820 = 17.5 \text{ mmHg} \qquad (6.5.15)$$

(We have used the fact that the air flowrate is 100 g/min.) Equation 6.5.15 is the statement that the partial pressure is the mole fraction times the total pressure.

Now we can calculate the driving force for mass transfer, *in partial pressure units*, as

$$\Delta p_{lm} = \frac{(38 - 0) - (28.7 - 17.5)}{\ln(38/11.2)} = 21.9 \text{ mmHg} \qquad (6.5.16)$$

Since $N$ is in *molar* units, we must convert the concentration difference $\Delta p_{lm}$ to molar units. This requires the use of the gas law, and we find, using

$$\Delta p_{lm} = R_G T \Delta C_{lm} \qquad (6.5.17)$$

that

$$\Delta C_{lm} = \frac{\Delta p_{lm}}{R_G T} = \frac{21.9/760}{82 \times 323} = 1.08 \times 10^{-6} \text{ mol/cm}^3 \qquad (6.5.18)$$

(Note that we converted pressure to atmosphere units and used $R_G = 82$. In Eq. 6.5.18 the factor 760 is the conversion of millimeters of mercury to atmospheres, and not the pressure in the tube, which is 820.)

From Eq. 6.5.11 we find

$$\overline{Sh} = \frac{ND}{D_{AB}\Delta C_{lm}} = \frac{1.28 \times 10^{-6} \times 2.67}{0.09 \times 1.08 \times 10^{-6}} = 34.8 \qquad (6.5.19)$$

We will want the Reynolds number for this set of data. The cross-sectional area of the tube is

$$\frac{\pi D^2}{4} = 0.7854 \times 2.67^2 = 5.61 \text{ cm}^2 \qquad (6.5.20)$$

For an airflow of 100 g/min we find

$$\rho \langle v \rangle = \frac{100}{60 \times 5.61} = 0.3 \text{ g/cm}^2 \cdot \text{s} \qquad (6.5.21)$$

The Schmidt number for *n*-butyl alcohol in air is Sc = 1.9, so the kinematic viscosity of the air is

$$\nu = \text{Sc } D_{AB} = 1.9 \times 0.09 = 0.17 \text{ cm}^2/\text{s} \qquad (6.5.22)$$

The density of air may be found from the gas law (assuming the alcohol contributes little to the density, since its mole fraction is small):

$$\rho = \frac{M_w p}{R_G T} = \frac{29 (820/760)}{82 \times 323} = 1.2 \times 10^{-3} \text{ g/cm}^3 \qquad (6.5.23)$$

The viscosity of the air then is found to be

$$\mu = \rho \nu = 1.2 \times 10^{-3} \times 0.17 = 2 \times 10^{-4} \text{ poise} \qquad (6.5.24)$$

Finally, the Reynolds number is

$$\text{Re} = \frac{(\rho \langle v \rangle) D}{\mu} = \frac{0.3 \times 2.67}{2 \times 10^{-4}} = 4000 \qquad \textbf{(6.5.25)}$$

The $j$-factor for mass transfer is

$$j_D = \frac{\overline{\text{Sh}}}{\text{Re Sc}^{1/3}} = \frac{34.8}{4000 \times 1.9^{1/3}} = 7.0 \times 10^{-3} \qquad \textbf{(6.5.26)}$$

The friction factor for tube flow at this Reynolds number is (using Eq. 6.5.2)

$$f = 0.079 \, \text{Re}^{-0.25} = 0.079 \times 4000^{-0.25} = 0.01 \qquad \textbf{(6.5.27)}$$

and so

$$\frac{f}{2} = 5 \times 10^{-3} \qquad \textbf{(6.5.28)}$$

which is only 30% below the $j$-factor. Hence the Chilton–Colburn analogy provides a reasonable prediction of these results if no data are available.

## 6.6 MASS TRANSFER INTO AERATED VESSELS

In real engineering systems interfacial transfer often occurs in such complex geometries that no simple theory is available for prediction of the mass transfer coefficient. A good example arises in consideration of mass transfer from bubbles. A number of technologically important systems involve the transfer of mass across the interface separating a continuous liquid from a swarm of gas bubbles. Examples include aerobic *fermentors,* which transfer oxygen into a broth of cells for biochemical transformations, *strippers,* which remove volatile species from liquids, and *bubblers,* which are used to deliver volatile liquids to chemical reactors as a gas phase. Special problems arise when the gas bubble size exceeds more than a few millimeters because the individual bubbles are no longer spherical, there is a distribution of bubble sizes, and the interfacial area itself is unknown and unpredictable *a priori.*

### 6.6.1 Bubble Dynamics

Let us consider first the issue of the bubble shape. We expect and observe the small bubbles rising slowly through a liquid, or entrained in an agitated liquid, to be spherically shaped. At the other extreme, we know from observation that very large bubbles, say several centimeters in diameter, released at the bottom of a column of water, will break up into several smaller bubbles of a range of diameters, and the larger of these "pieces" will rise as nonspherical bubbles. Further, the rise of these large bubbles may not be steady; instead, the bubbles may wobble or oscillate as they rise. Quite a bit of work has been done in the study of this class of problem, and certain features of this system emerge. (See, e.g., Chapter 2 in Clift, Grace, and Weber, *Bubbles, Drops and Particles,* Academic Press, 1978.) We restrict our attention here to bubbles rising in a quiescent liquid.

Bubble shape depends largely on three dimensionless groups. These are the Reynolds number based on the rise velocity $U$:

$$\text{Re} = \frac{\rho d_e U}{\mu} \qquad \textbf{(6.6.1)}$$

the Morton number (sometimes called the "property" number)

$$M = \frac{g\mu^4 \Delta\rho}{\rho^2 \sigma^3} \tag{6.6.2}$$

and the Bond number

$$Bo = \frac{g \Delta\rho \, d_e^2}{\sigma} \tag{6.6.3}$$

In these definitions $d_e$ is the "equivalent" diameter of the bubble, defined as

$$d_e = \left(\frac{6V}{\pi}\right)^{1/3} \tag{6.6.4}$$

where $V$ is the bubble volume.

Figure 6.6.1 shows the shape regimes for bubbles rising unhindered through liquids. In these dimensionless groups the *liquid* properties are used ($\rho$ and $\mu$), while $\Delta\rho$ is the density difference and $\sigma$ is the interfacial tension.

We can anticipate that the bubble shape will have some impact on the rise velocity of the bubble. We know that Stokes' law breaks down when Re exceeds a value of order unity. Data for rise velocity are also found to be strongly dependent on the cleanliness of the liquid, especially in the case of aqueous systems, which could contain surfactants as impurities. Figure 6.6.2 shows the range of observations for air bubbles rising through water.

We should expect bubble shape to affect the rate of mass transfer at the bubble interface with the surrounding liquid. Theoretical treatments of the convective mass transfer across the gas/liquid interface are possible only in the simplest cases, such as the *spherical bubble* rising slowly and steadily through a liquid. Several theoretical treatments exist which differ in specific assumptions about the flow field external to the bubble [e.g., creeping flow *vs.* inviscid (potential) flow], and one may find any of the following models recommended for prediction of the mass transfer coefficient to (or from) a bubble:

$$Sh = \frac{k_c d_e}{D_{AB}} = 0.991 \, Pe^{1/3} \tag{6.6.5}$$

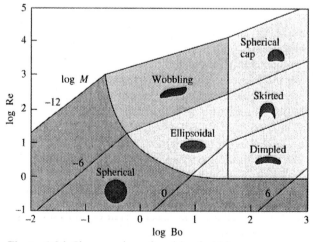

**Figure 6.6.1** Shape regimes for rising bubbles.

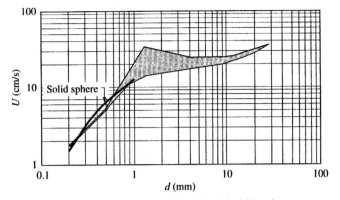

**Figure 6.6.2** Terminal rise velocity of air bubbles in water at 20°C. The line labeled "Solid sphere" follows from the friction factor chart for a sphere.

$$\text{Sh} = (4.0 + 1.21\,\text{Pe}^{2/3})^{1/2} \tag{6.6.6}$$

$$\text{Sh} = 0.991\,\text{Pe}^{1/3} + 0.92 \tag{6.6.7}$$

In these equations the Peclet number is defined (using the *liquid phase* diffusivity $D_{AB}$) as

$$\text{Pe} = \frac{U d_e}{D_{AB}} \tag{6.6.8}$$

It would appear that Eqs. 6.6.5 through 6.6.8 are in reasonable agreement with data for relatively small bubbles (consistent with the restriction that the bubbles be nearly spherical). In the case of fairly large bubbles, the most reliable set of data seems to be that of Johnson, Besik, and Hamielec [*Can. J. Chem. Eng.*, **47,** 559 (1969)], correlated as

$$\text{Sh} = 1.13\,\text{Pe}^{1/2}\left(\frac{d_e}{0.45 + 0.2d_e}\right)^{1/2} \tag{6.6.9}$$

*where $d_e$ is in centimeters.* This correlation is valid for $d_e$ in the range of 0.3 to 3 cm.

One problem with this correlation is that it requires *a priori* knowledge of the bubble size. Thus it may be necessary to have available a subsidiary correlation for $d_e$ as a function of the design of the system by which gas is introduced to the liquid, and the flowrate of the gas. More to the point, however, these correlations and models are for single bubbles rising in a quiescent liquid—a dynamic situation very different from that observed in an aerated tank, especially if external agitation is provided.

### 6.6.2   Capacity Coefficients in Aerated Systems

Often, aerated vessels are agitated by a stirring device (such as a high speed turbine impeller), to promote mass transfer through two mechanisms. Agitation increases convection by inducing additional turbulence. In addition, the energy imparted to the liquid by the power input through the agitator is available to disrupt the larger bubbles and reduce their diameters. For a given volume of gas held up in the vessel at any instant of time, smaller bubbles correspond to a greater area of contact, hence to enhanced mass transfer. Unfortunately, the interfacial area is not known a priori and itself must be obtained through experiment.

It is not difficult to devise a simple transient aeration of an agitated vessel to generate valuable data for correlation and prediction. Suppose a vessel is initially unsaturated with respect to some soluble species, and at time $t = 0$ a gas is suddenly and continuously bubbled through the liquid. The concentration of dissolved species in the *liquid* phase is $C(t)$, and $C_{sat}$ is the concentration *in the liquid* that would be in equilibrium with the gas. We assume that the gas passes continuously through the vessel and that only a small fraction of the soluble species dissolves from each bubble into the liquid as the individual gas bubbles rise through the liquid volume $V_L$ and leave the vessel. Then a species balance takes the form

$$\frac{d(V_L C)}{dt} = A k_c (C_{sat} - C) \tag{6.6.10}$$

where $A$ is the *total* interfacial area, and the solution is simply

$$\frac{C(t)}{C_{sat}} = 1 - \exp\left[-\left(\frac{k_c A}{V_L}\right)t\right] = 1 - \exp[-(k_c a_v)t] \tag{6.6.11}$$

where $a_v$ is the interfacial area per unit of total *liquid* volume $V_L$. In terms of the fractional approach to equilibrium, defined as

$$1 - \bar{Y} = \frac{C(t)}{C_{sat}} \tag{6.6.12}$$

this may be written as

$$k_c a_v = -\frac{1}{t} \ln \bar{Y} \tag{6.6.13}$$

Data on the degree of saturation as a function of time would yield the *product* $(k_c a_v)$, but not the individual values of either parameter. This product is called a "capacity coefficient."

Such experiments have been carried out on agitated, aerated fermentors [Fukuda et al., *J. Ferment. Technol. (Japan)*, **46**, 829 (1968)]. Figure 6.6.3 shows an example of results

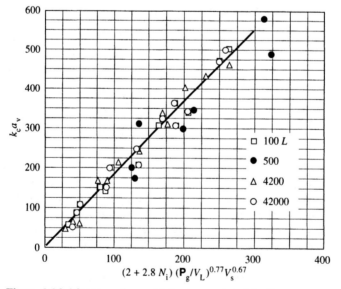

**Figure 6.6.3** Mass transfer coefficients correlated for fermentors of different sizes ($k_c a_v$ is in units of millimoles per liter-hour-atmosphere).

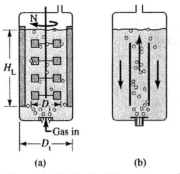

**Figure 6.6.4** Typical fermentors: (a) with multiple impellers and (b) gas-sparged type.

for the product $k_c a_v$. This correlation involves several parameters that characterize the operation of an agitated aerated vessel. In large vessels multiple impellers are aligned along the axis of the shaft (see Fig. 6.6.4) and $N_i$ is the number of impellers. The power input to the vessel is given as $\mathbf{P}_g/V_L$, in units of horsepower per 1000 L of liquid. The so-called *superficial gas velocity*, $V_s$, is the volumetric flowrate of gas (measured at the inlet pressure of the vessel) divided by the cross-sectional area of the vessel normal to the axis of the vessel. In this correlation $V_s$ has the units of centimeters per minute.

Note in particular first that the data correlate quite well over an extremely wide range in vessel size; this is very important because it gives us confidence in using the correlation for *scale-up* from laboratory data to large-scale systems. Second, note the *units* on the $k_c a_v$ product. These units arise in the reference cited because the measured gas concentration ($C$ on the left-hand side of the species balance above) was in units of millimoles per liter, while the "driving force" for mass transfer ($C_{sat} - C$) was given in units of partial pressure of dissolved gas. "Bastard" units like these are commonly found in the literature, and one must be prepared for them.

We have enough background now to be able to illustrate applications of these ideas to the design of some mass transfer systems. We look at a problem that arises in biochemical engineering: the need to provide sufficient oxygenation to sustain the growth of a cell culture in a fermentation vessel.

**EXAMPLE 6.6.1**   *Gas Flow to Sustain Growth in a Fermentor*

A cell culture is being grown in a stirred tank fermentor. The cell growth is sustained by continual supply of oxygen to the broth. Cell metabolism removes some of the dissolved oxygen from the broth. Mass transfer of oxygen from the gas bubbles to the broth is assumed to control the rate of metabolism of the cells. That is, we are assuming that the *resistance* to transfer of oxygen from the liquid to the cells is small in comparison to the resistance from the bubbles to the broth. We are also assuming that the metabolic rate is fast, and effectively limited by the rate of external mass transfer.

Then a species balance on dissolved oxygen in the broth takes the form

$$\frac{d(V_L C)}{dt} = A k_c (C_{sat} - C) - \mathfrak{R}_M V_L \tag{6.6.14}$$

where the notation is the same as in Eq. 6.6.10, except that we have now added a term to account for depletion of oxygen due to respiration of the cells. We treat this balance as if a *homogeneous* chemical reaction were occurring throughout the broth. For a well-mixed fermentor with a high number density of microscopic cells, we would expect the system to behave as if the reaction were indeed homogeneously distributed in the liquid phase.

The respiration term can be written once we have acquired some information from the microbiologists concerning this particular cell system. We define the "loading" of the reactor as the mass density of cells $\rho_c$ (*dry* mass of cells per liquid volume of broth). We suppose that the respiration rate is linear in dissolved oxygen concentration and is given by

$$\Re_M = \rho_c k_{resp} C \tag{6.6.15}$$

Note that by writing the reaction term in this form we introduce the coefficient $k_{resp}$ as a *specific* rate constant, in the sense that it is on a *per mass density of cells* basis. We usually use the *dry weight* of the cells in expressing the cell density.

From the microbiologists we have the additional information that the oxygen concentration in the broth must be maintained at or above some specified level $C^*$ if the fermentation is to produce the desired product over an extended period of time. Hence we are interested in the steady state solution to Eq. 6.6.14, and we specify $C = C^*$ as the critical operating condition. Thus we set the time derivative to zero in Eq. 6.6.14, and from this we find that the capacity coefficient must be at a level defined by the following relationship among the variables:

$$a_v k_c = \frac{\rho_c k_{resp} C^*}{C_{sat} - C^*} \tag{6.6.16}$$

where, as earlier, $a_v$ is the interfacial area per unit of total *liquid* volume $V_L$. Since everything on the right-hand side of Eq. 6.6.16 is available to us, we have a design constraint in terms of the required capacity coefficient.

The following parameters are known:

$$\rho_c = 0.05 \text{ g cells/cm}^3 \quad C^* = 1.2 \times 10^{-6} \text{ mol/cm}^3 \quad C_{sat} = 1.4 \times 10^{-6} \text{ mol/cm}^3$$
$$k_{resp} = 5 \text{ h}^{-1}/(\text{g cells/cm}^3) \quad N_i = 1 \quad P_g/V_L = 5 \text{ hp/1000 L}$$

The fermentor holds 64 L of broth, which may be considered to be equivalent to water in terms of any physical properties, and has a cross-sectional area (normal to the impeller axis) of 1600 cm$^2$. We want to specify the required gas rate. We use Fig. 6.6.3 on the assumption that this correlation works for small vessels. Note that the data are all for vessels of large volume relative to our case at hand.

From Eq. 6.6.16 we find

$$a_v k_c = 1.5 \text{ h}^{-1} \tag{6.6.17}$$

Note especially the units here, because Fig. 6.6.3 has mixed units (as noted in the caption), and we must convert to the units of the figure. This is done by using the gas law

$$C = \frac{P}{R_G T} \tag{6.6.18}$$

or simply recalling that 1 mole of gas occupies 22.4 L at standard conditions, so 1 atm of concentration difference corresponds to $1/22.4 = 0.045$ mol/L $= 45$ mmol/L. From this we find that the conversion is

$$45(a_v k_c)(\text{h}^{-1}) = (a_v k_c) \text{ mmol/L} \cdot \text{h} \cdot \text{atm} \tag{6.6.19}$$

Hence we find

$$a_v k_c = 1.5 \times 45 = 67 \text{ mmol/L} \cdot \text{h} \cdot \text{atm} \tag{6.6.20}$$

From Fig. 6.6.3 we now find a value of the $x$ axis of

$$30 = (2 + 2.8)(5)^{0.77} V_s^{0.67} \tag{6.6.21}$$

from which it follows that the gas velocity is

$$V_s = 2.44 \text{ cm/min} \tag{6.6.22}$$

From its definition, the superficial velocity is the volumetric flowrate divided by the cross-sectional area of the vessel (1600 cm$^2$). Hence we find the required volumetric flowrate to be

$$Q = 1600 \times 2.44 = 3900 \text{ cm}^3/\text{min} = 3.9 \text{ slm} \tag{6.6.23}$$

(Recall that slm means standard liters per minute; i.e., the volume is measured at 273 K and 1 atm pressure.)

One problem with the design thus far is that we are using Fig. 6.6.3 at the extreme low end of the correlation, and we are designing for a vessel that is outside the range of the data (which covered only volumes greater than 100 L). It would be wise to generate some data for the mass transfer capacity coefficient under conditions corresponding to the 64 L vessel we plan to use, and at gas flows and power inputs in the neighborhood of the design values used or found here. Then we can design with greater confidence.

The measurement of the capacity coefficient $k_c a_v$ in an agitated system is relatively straightforward, as Example 6.6.2 demonstrates.

**EXAMPLE 6.6.2**   *Experimental Determination of $k_c a_v$*

The data below were obtained by following the increase in dissolved oxygen concentration upon sparging *pure* oxygen into water that is initially saturated with respect to air. An oxygen electrode was used to measure the dissolved oxygen concentration, and the readings were in units labeled "parts per million." However, the instrument was not calibrated against pure oxygen, so we take the readings to be proportional to the actual concentration.[1]

| $t$ (s) | 0 | 10 | 20 | 30 | 40 | 50 | 60 | 70 | 80 | 90 | 120 |
|---|---|---|---|---|---|---|---|---|---|---|---|
| $x$ (ppm) | 40 | 94 | 107 | 128 | 146 | 149 | 153 | 156 | 157 | 158 | 160 |

Since no reaction takes place in this experiment (there are no viable cells in the water), we may expect to find that Eq. 6.6.10 holds. The solution to this equation (Eq. 6.6.11) must be modified, since in the present example the water has an initial concentration of dissolved oxygen, while in solving Eq. 6.6.10 we used the initial condition that the water was unsaturated. It is not difficult to show that the solution now takes the form

$$\frac{C(t) - C_{sat}}{C_o - C_{sat}} = f = \exp[-(k_c a_v)t] \tag{6.6.24}$$

---

[1] There seems to be a possible discrepancy here. We would expect the ratio of saturation concentrations to be 5:1 between pure oxygen and oxygen in air. The observed ratio was 160:40 = 4:1.

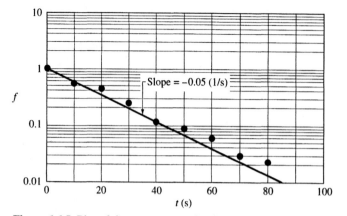

**Figure 6.6.5** Plot of the oxygen transfer data of Example 6.6.2.
The slope is $-k_c a_v$.

The data are plotted in Fig. 6.6.5 in accordance with this model. Concentrations are used directly in the "ppm" units read from the instrument, and $x_o = 40$ ppm and $x_{sat} = 160$ ppm. The slope yields a value for the capacity coefficient, which in this case is found to be

$$k_c a_v = 0.053 \text{ s}^{-1} \tag{6.6.25}$$

## 6.7  MASS TRANSFER IN A PARALLEL PLATE REVERSE OSMOSIS SYSTEM

Membrane desalination is a process by which relatively pure water may be obtained from seawater, or other waters containing high concentrations of salt(s). Mass transfer near the membrane plays a very important role in the efficiency of performance of such desalination systems. Figure 6.7.1 shows some important features of this mass transfer process.

We will consider here a laminar flow, parallel plate membrane system. The membrane rejects most of the salt, as a consequence of which the permeate water is relatively salt free. Membranes with rejection coefficients of 96 to 99% are available commercially. Water passes through the membrane under the influence of the pressure difference $P_H - P_L$. However, the osmotic pressure of the saline solution can be significant. (Seawater has a salt concentration of about 35,000 ppmw and an osmotic pressure of

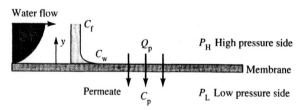

**Figure 6.7.1** Schematic of a membrane system for reverse osmosis.

about 400 psi.) We usually write the permeate flux (as a volume flowrate per unit of membrane area) in the form

$$Q_P = K_m[\{P_H - \Pi(C_w)\} - \{P_L - \Pi(C_P)\}] \tag{6.7.1}$$

where $K_m$ is a membrane permeability coefficient and $\Pi(C_i)$ is the osmotic pressure at the salt concentration $C_i$. Note that the osmotic pressure on the high pressure side is evaluated at the concentration $C_w$, not at the bulk or average concentration in the water on the high pressure side. Because the salt is almost perfectly rejected at $y = 0$, the osmotic pressure at the membrane can be many times higher than that in the feed water. This necessitates the use of very high pressures in membrane desalination, a requirement that in turn contributes to the high costs of this process.

From the foregoing remarks it should be apparent that it is important to have some understanding of the mass transfer process at the membrane surface, so that the salt concentration buildup, or *polarization* as it is usually called, can be estimated, and methods for its reduction instituted. As Fig. 6.7.1 suggests, there is a high concentration of salt near the membrane surface. Hence we expect transport of salt away from the membrane to occur at a rate that reflects the convection at that boundary. We use a film model here, with some assumed film thickness $\delta$ introduced into the model, but we choose shortly to work in terms of a mass transfer coefficient $k_c$ characteristic of convective diffusion for a laminar flow membrane system. We have already considered such a problem—the laminar flow dialysis system presented in Chapter 5—and we will use results from that analysis.

We introduce the mass transfer model by writing the flux of salt back into the mainstream flow in the form

$$N_{salt} = k_c(C_w - C_f) \tag{6.7.2}$$

Since we are using a film model as an approximation, we may write the concentration profile within the film in the form of a salt flux balance. If the membrane completely rejects salt, there is no net flow of salt across any plane parallel to the membrane, and we find

$$Q_P C(y) + D_{sw}\frac{\partial C}{\partial y} = 0 \tag{6.7.3}$$

where $D_{sw}$ is the diffusion coefficient of salt in water. (Note that we define $Q_P$ to be positive if it corresponds to flow toward the membrane.) Equation 6.7.3 may be integrated between the limits $y = 0$ and $y = \delta$ (using boundary conditions $C = C_w$ at $y = 0$ and $C = C_f$ at $y = \delta$), with the result

$$\frac{C_w}{C_f} = \exp\left(\frac{Q_P\delta}{D_{sw}}\right) \tag{6.7.4}$$

Now we introduce the film model in the form of the relationship between $k_c$ and $\delta$:

$$k_c = \frac{D_{sw}}{\delta} \tag{6.7.5}$$

This permits us to eliminate $\delta$. We also introduce the $j$-factor for mass transfer ($j_D = $ Sh/Re Sc$^{1/3}$) and the result is that Eq. 6.7.4 takes the form

$$\frac{C_w}{C_f} = \exp\left(\frac{Q_P \, Sc^{2/3}}{j_D U}\right) \tag{6.7.6}$$

where $U$ is the average velocity of the feed stream. (Keep in mind that $Q_P$ is the *permeate* flow, not the *feed* flow.)

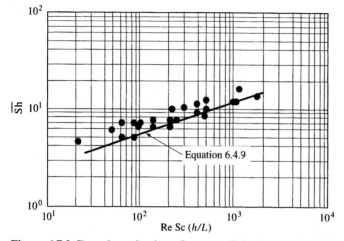

**Figure 6.7.2** Data for a laminar flow, parallel plate membrane system.

Equations 6.7.1 and 6.7.6 represent a pair of equations in the two unknowns, $Q_P$ and $C_w$, assuming that the mass transfer coefficient $j_D$ is known. Alternatively, we may measure $Q_P$, and with independent data for the osmotic pressure and the membrane coefficient $K_m$ we may find $C_w$. Then, with Eq. 6.7.6 we may determine the mass transfer coefficient $j_D$.

Experimental data generated by this latter approach—that is, data used to calculate the mass transfer coefficient from measured permeate flow—are available (Rautenbach and Albrecht, *Membrane Processes,* Wiley, 1989, p. 102). The results for laminar flow through a parallel plate reverse osmosis (RO) system are shown in Fig. 6.7.2. The data are tested against the predicted values of the average Sherwood number, as given by Eq. 6.4.9. (The half-height of the channel, $h$, is used as the length scale in Sh and Re.) We see that the data generally lie above this simple theory by about 15%.

From these results we may estimate the effect of concentration polarization on the production of permeate in a RO system of this type.

---

**EXAMPLE 6.7.1**  *Effect of Concentration Polarization on Permeate Flow*

We consider the performance of a parallel plate RO system under the following conditions:

Tested with pure water, the permeate flow is 0.05 gal/h at a pressure difference of 100 psi, with a membrane area of 0.25 ft².
The RO system is fed water that has 1% (by weight) sodium chloride.
The channel height is $2h = 2$ mm, the width is $W = 20$ cm, and the length is $L = 30$ cm. Feed rates will be in the range of 2 to 20 L/min.
The downstream pressure on the high pressure side is controlled at a constant value of 500 psi. (Is it valid to assume that the pressure is uniform along the channel length?)
The rejection coefficient of the membrane, defined as $r = 1 - C_p/C_w$, is $r = 0.98$ or 98%.

We want to find, as a function of feed rate, the fraction of the feed that appears as permeate, the quality of the permeate (i.e., the salt concentration), and the polarization (which we define as the ratio of salt concentration at the membrane surface to its value in the feed).

First we need the membrane permeability coefficient $K_m$ of Eq. 6.7.1. For pure water flow we find, using the data given,

$$K_m = \frac{Q_P}{P_H - P_L}$$

$$= \frac{0.05 \text{ gal/h} \times 3.8 \text{ L/gal}}{0.25 \text{ ft}^2 \times 100 \text{ psi}} = 7.6 \times 10^{-3} \text{ L/h} \cdot \text{ft}^2 \cdot \text{psi}$$

(6.7.7)

Let's pick a feed rate of 2 L/min. The Reynolds number (using the half-width $h$ as the length scale) is

$$\text{Re}_h = \frac{hU}{\nu} = \frac{Q}{2W\nu} = \frac{2000/60 \text{ cm}^3/\text{s}}{2 \times 20 \text{ cm} \times 0.01 \text{ cm}^2/\text{s}} = 83.3$$

(6.7.8)

(We use the volumetric flowrate $Q = 2hWU$ here.) For saltwater solutions, the Schmidt number is approximately Sc = 620. We may calculate a value of

$$\text{Re}_h \text{ Sc } (h/L) = 83.3 \times 620 \times \frac{0.1}{30} = 172$$

(6.7.9)

From Fig. 6.7.2 we find Sh $\approx$ 6.5, so the $j$-factor is

$$j_D = \frac{\text{Sh}}{\text{Re Sc}^{1/3}} = \frac{6.5}{83.3 \times 620^{1/3}} = 0.0092$$

(6.7.10)

We now have two unknowns that characterize the performance of this system. One is $C_w$, the salt concentration at the membrane surface, and the other is the permeate rate $Q_P$. These two variables are found by simultaneous solution of Eqs. 6.7.1 and 6.7.6. It is necessary, of course, to convert all parameters to a consistent set of units. We write Eq. 6.7.1 as

$$Q_P = K_m(500 - \Pi[C_w])$$

(6.7.11)

where we neglect the osmotic pressure of the permeate solution on the assumption that nearly pure permeate will have a negligible osmotic pressure. The osmotic coefficient for saltwater may be written as

$$\Pi = BC_w = 115 \text{ (psi/wt\%)} C_w$$

(6.7.12)

when the salt concentration $C_w$ is measured in units of weight percent (wt%). We convert $K_m$ given in Eq. 6.7.7 to

$$K_m = \frac{7.6 \text{ cm}^3}{3600 \text{ s/h} \times 929 \text{ cm}^2/\text{ft}^2} = 2.3 \times 10^{-6} (\text{cm}^3/\text{cm}^2 \cdot \text{s} \cdot \text{psi})$$

(6.7.13)

Then

$$Q_P = 2.3 \times 10^{-6} (500 - 115 C_w)$$

(6.7.14)

when concentration is in units of weight percent. In Eq. 6.7.6 we need the average linear velocity of the feed, which is easily found to be $U = 8.33$ cm/s at a feed rate of 2 L/min. Then we may write Eq. 6.7.6 as

$$\frac{C_w}{C_f} = \exp\left[\frac{620^{2/3} \times 2.3 \times 10^{-6}}{0.0092 \times 8.33} (500 - 115 C_w)\right]$$

$$= \exp[0.0022 (500 - 115 C_w)]$$

(6.7.15)

We may solve Eq. 6.7.15 by trial and error and find

$$C_w = 1.9 \text{ wt\%}$$

(6.7.16)

Then the permeate *flux* is (from Eq. 6.7.14)

$$Q_P = 6.6 \times 10^{-4} \, \text{cm}^3/\text{cm}^2 \cdot \text{s} \qquad (6.7.17)$$

The permeate *flow* is $Q_P$ times the membrane area, or (assuming both sides of the channel are membrane covered)

$$A_m Q_P = 30 \times 20 \times 2 \times 6.6 \times 10^{-4} = 0.8 \, \text{cm}^3/\text{s} \qquad (6.7.18)$$

This represents 2.4% of the feed flow. The water quality now follows from the stated rejection coefficient of 98%:

$$C_P = 0.02 \times 0.019 = 3.8 \times 10^{-4} = 380 \, \text{ppmw} \qquad (6.7.19)$$

Similar calculations may be carried out at other flow rates. At higher flow rates the mass transfer coefficient increases, the polarization decreases, and the water quality will improve.

## 6.8 MASS TRANSFER TO PARTICLES IN A TURBULENT STIRRED TANK

This problem was modeled in Chapter 4, Example 4.4.2. The analysis was carried out on the assumption that a laminar boundary layer of known thickness controlled the mass transfer process. The convective transport was encoded in the parameter $\delta^* = \delta/R_0$. Hence we must now address the question: How do we obtain estimates of $\delta$ under various conditions of agitation of a slurry of solid dissolving particles?

We begin by noting that the film thickness $\delta$ is equivalent to a convective mass transfer coefficient $k_c$:

$$k_c = \frac{D}{\delta} \qquad (6.8.1)$$

where $D$ is the diffusion coefficient of the solute.

Since solid particle dissolution typically is carried out in agitated vessels, we search for data on Sherwood numbers from particles in stirred tanks, correlated in terms of design and operating variables of the system.

The Sherwood number for mass transfer from a solid particle in an agitated tank could be defined as

$$\text{Sh} = \frac{2k_c R_0}{D} \qquad (6.8.2)$$

where $D$ is the diffusivity in the convecting (i.e., the fluid) phase. The factor of 2 reflects the usual definition of Sherwood number with the particle diameter used as the length scale, rather than the radius. One would expect to observe a correlation of the form

$$\text{Sh} = A \, \text{Re}^m \text{Sc}^n \qquad (6.8.3)$$

consistent with similar data and theory for single particles in well-defined flow fields. An immediate problem arises with respect to the definition of Re in this equation, and indeed whether Re is the only characteristic of the design and operation of the vessel that would appear in a correlation. The use of a Reynolds number implies a definition of a characteristic velocity $v_0$ for the agitated system. A logical choice is to take $v_0$ proportional to the tip speed of the impeller that is creating the turbulence in the system:

$$v_0 = \pi d_i N \qquad (6.8.4)$$

where $d_i$ is the diameter of the impeller, and $N$ is the rotational speed, in revolutions per second, for example.

An examination of experimental data indicates that Eq. 6.8.3 provides a basis for correlation of data in a specific agitated system. The coefficient $A$ depends on how efficiently the power is transmitted to the fluid through the impeller. This depends on the Reynolds number, and on geometric parameters such as the ratio of impeller to tank diameter, the specific geometry of the impeller, and the geometry of baffling, if any, used to inhibit vortex formation in the vessel. Another observation is that the convective coefficient is essentially independent of the particle size. This leads us to suggest a correlation in the form of Eq. 6.8.3, but with the Sherwood number defined with the vessel diameter $d_T$ as the length scale. In fact, a specific correlation of this form is suggested in the work of Kulov et al. [*Chem. Eng. Commun.*, **21**, 259 (1983)], which we write as

$$\text{Sh} = \frac{k_c d_T}{D} = 0.267 \, \text{Sc}^{1/4} \text{Re}^{3/4} N_P^{1/4} \left(\frac{d_T^4}{V d_i}\right)^{1/4} \tag{6.8.5}$$

where the Reynolds number Re is defined as

$$\text{Re} = \frac{N d_i^2}{\nu_L} \tag{6.8.6}$$

and $\nu_L$ is the kinematic viscosity of the liquid in the tank. We note that it is sometimes more common to define a Reynolds number for a stirred tank as

$$\text{Re}' = \frac{\pi N d_i^2}{\nu_L} = \pi \, \text{Re} \tag{6.8.7}$$

*When reading the literature on this topic, it is important to keep track of the definitions used for the various dimensionless groups that characterize the performance of the system.*

A new dimensionless group, the power number, appears here, and is defined as

$$N_P = \frac{\mathbf{P}}{(\rho_L N^3 d_i^5)} \tag{6.8.8}$$

where $\rho_L$ is the liquid density and $\mathbf{P}$ is the mixing power, in watts, input to the vessel filled to volume $V$ with the liquid. Correlations of power number as a function of Reynolds number are available for various geometries of mixing vessels, and an example of such a correlation is shown in Fig. 6.8.1. For the so-called standard turbine configuration (a six-bladed turbine in a tank filled to a height equal to the inside diameter of

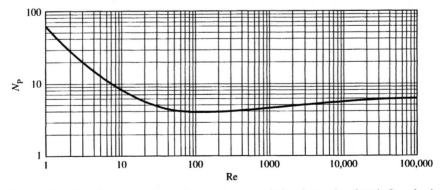

**Figure 6.8.1** Power number/Reynolds number correlation for a stirred tank. Standard turbine configuration.

the tank, with four full-length baffles at 90°), the power number $N_P$ is nearly independent of Reynolds number for Re above 5000 and has a value of $N_P = 6$. An opportunity to compare this model (Eq. 6.8.5) to experimental data is given in Problem 6.9.

**EXAMPLE 6.8.1** *Particle Dissolution in an Agitated Tank: Mass Transfer Coefficient*

A viscous aqueous solution ($\mu_L = 0.01$ Pa·s and $\rho_L = 1500$ kg/m³) is agitated in a tank of diameter $d_T = 2$ m filled to a height $H_T = 2$ m. A six-bladed turbine with $d_i = 0.5$ m rotates at $N = 90$ rpm. The tank is baffled, and the system corresponds to the so-called standard turbine configuration, for which Fig. 6.8.1 applies. What is the expected convective mass transfer coefficient between the agitated liquid and solid particles immersed in the liquid? The diffusion coefficient of the dissolved solute in the liquid is $5 \times 10^{-10}$ m²/s.

We begin with a calculation of the Reynolds number, using

$$\text{Re} = \frac{Nd_i^2\rho_L}{\mu_L} = \frac{(90/60)(0.5)^2(1500)}{0.01} = 56{,}000 \qquad (6.8.9)$$

From Fig. 6.8.1 we find $N_P = 6$, so

$$\text{P} = 6\rho_L N^3 d_i^5 = 6(1500)(1.5)^3(0.5)^5 = 950 \text{ W} \qquad (6.8.10)$$

The agitated volume is

$$V = \pi d_T^2 H_T/4 = \pi(2)^2(2)/4 = 6.3 \text{ m}^3 \qquad (6.8.11)$$

From Eq. 6.8.5 we find

$$\text{Sh} = 0.267 \, (1.33 \times 10^4)^{1/4}(5.6 \times 10^4)^{3/4}(6)^{1/4}\left(\frac{2^4}{6.3(0.5)}\right)^{1/4} = 2.5 \times 10^4 \quad (6.8.12)$$

and

$$k_c = \frac{2.5 \times 10^4 \, (5 \times 10^{-10})}{2} = 6.3 \times 10^{-6} \text{ m/s} \qquad (6.8.13)$$

Once the mass transfer coefficient has been established we can estimate the time required for dissolution of the particles. We assume in the example that follows that all the particles are of the same size.

**EXAMPLE 6.8.2** *Particle Dissolution in an Agitated Tank: Dissolution Time*

A highly soluble solid (sucrose) is to be dissolved in an agitated tank. Spherical particles of initial radius $R_0 = 6$ mm are available. Find the time to dissolve 30% of the initial mass, when $k_c$ has the value found in Example 6.8.1. For sucrose at 25°C, $\rho_s = 1600$ kg/m³, and $C_s = 1800$ kg/m³.

If we assume that convection of the dissolved sucrose from the solid/liquid interface is the rate-determining step in dissolution, then a simple mass balance takes the form

$$-\frac{dM}{dt} = k_c A C_s \qquad (6.8.14)$$

We are assuming that the dissolved sucrose is quickly dispersed throughout the agitated tank and that the tank volume is so large that the sucrose concentration in the bulk of the liquid is very small compared to the equilibrium solubility. Equation 6.8.14 is easily rearranged to the form

$$-\frac{dM/M_o}{dt} = \frac{3k_cC_s}{\rho_sR_o} \tag{6.8.15}$$

where $M_o$ is the initial mass of solid and the initial radius is $R_o$ of each particle. This equation is easily solved to yield

$$\frac{M}{M_o} = 1 - \frac{3k_cC_s}{\rho_sR_o}t \tag{6.8.16}$$

For this example we want the time at which $M/M_o = 0.7$. Hence we find

$$t = \frac{0.3\rho_sR_o}{3k_cC_s} = \frac{0.3\,(1600)\,(6\times10^{-3})}{3(6.3\times10^{-6})\,1800} = 85\text{ s} \tag{6.8.17}$$

We have assumed that the particle radius remains constant even though the particle is dissolving. It is not difficult to confirm that for 30% dissolution, the change in radius is small.

## 6.9   MASS TRANSFER IN A FLUIDIZED BED FURNACE

Another example of a very complex convective transfer system is a fluidized bed. Not only is the flow field turbulent, but in addition there is a particulate phase that itself undergoes a very chaotic motion. Mass transfer between the gas and particulate phases depends on the instantaneous relative velocity of the gas with respect to the particle. This is such a complex system that one would not expect to be able to predict mass transfer coefficients from solutions of the convective diffusion equation. Instead, experimental data are correlated in a nondimensional format that can be motivated and guided by analogy to simpler problems. That is the case in the example at hand, which is a consideration of a fluidized bed furnace for the burning of a particulate fuel such as fine coal particles. We assume, as in the example of the analysis of burning carbon particles in Chapter 5, that the rate of burning is controlled by mass transfer. Figure 6.9.1 shows the details of the system of interest to us.

Diffusion of oxygen to the fuel particles will be affected by the presence of the (usually smaller) inert particles. The void fraction of the bed, $\varepsilon$, is typically about 0.4, which means that 60% of the volume surrounding the fuel is occupied by the particulate phase. The inert particles affect the transport to the burning surface in two ways. First,

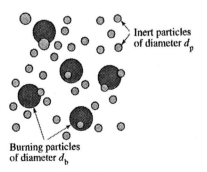

Inert particles
of diameter $d_p$

Burning particles
of diameter $d_b$

**Figure 6.9.1** Burning particles immersed in a fluidized bed of inert particles.

they promote convection by creating a stirring action in the neighborhood of the burning particles; in a sense, each particle is a tiny mixing element driven by the turbulence of the gas flow. In addition, the presence of the inert particles creates a barrier to diffusion, since oxygen must diffuse around the inert particles. That is, the diffusion path is not directly through the gas phase but, rather, exhibits a certain degree of tortuosity. This effect is usually accounted for by modifying the gas phase diffusivity $D$ by dividing it by a "tortuosity factor," often given as $\sqrt{2}$.

As a model for convective mass transfer to particles, we might take as a starting point one of the models used for convection to a single spherical particle held fixed in a flow of mean velocity $U$. For example, we might recall from our discussion of mass transfer to rising bubbles that a commonly recommended model is

$$Sh = (4.0 + 1.21\ Pe^{2/3})^{1/2} \tag{6.9.1}$$

In fact this does provide a starting point for an empirical correlation of mass transfer data to or from spheres in a fluidized bed, and we find the recommendation of Guedes de Carvalho et al. [*Trans. Inst. Chem. Eng.*, **69**, 63 (1991)] in the form

$$Sh' = [4 + 0.576\ Pe'^{0.78} + 1.28\ Pe' + 0.141\ (d_p/d_b)\ Pe'^2]^{1/2} \tag{6.9.2}$$

The following modified definitions of the Sherwood and Peclet numbers are used:

$$Sh' = \frac{\sqrt{2}\ k\ d_b}{\varepsilon D} \quad \text{and} \quad Pe' = \frac{\sqrt{2}\ U_{mf}\ d_b}{\varepsilon D} \tag{6.9.3}$$

Note that the fuel particle diameter is the length scale in these definitions. The inert particle diameter appears only in the fourth term on the right-hand side of Eq. 6.9.2, and is often negligible. The tortuosity factor of $\sqrt{2}$ has been included, and $D$ is the diffusivity of oxygen in the gas (e.g., air), in the case of carbon burning. The velocity that appears here is the minimum fluidization velocity $U_{mf}$, which is the linear velocity just capable of transforming the bed from a fixed to a fluidized state. We usually estimate this velocity from the application of the Ergun equation (review your fluid dynamics), which we write in the dimensionless form

$$150\frac{1-\varepsilon}{\varepsilon^3}\ Re_{mf} + 1.75\frac{1}{\varepsilon^3}(Re_{mf})^2 = Ga \tag{6.9.4}$$

where we find a Reynolds number and a new dimensionless group called the Galileo number.

$$Re_{mf} = \frac{\rho U_{mf}\ d_p}{\mu} \quad \text{and} \quad Ga = \frac{\rho\ d_p^3\ \rho_s g}{\mu^2} \tag{6.9.5}$$

The parameters $\rho$ and $\mu$ are based on gas properties, and $\rho_s$ is the density of the inert particulates.

A test of this proposed correlation is given in Fig. 6.9.2. Coke particles with diameters in the range of 3 to 4 mm were burned in a sand bed fluidized with air at 1220 K and at pressures from 1 to 4 atm. The sand particles were an order of magnitude smaller than the fuel (coke) particles and had a diameter of 0.46 mm. Flow rates corresponded to velocities that were anywhere from two to five times the minimum fluidization velocity.

The mass transfer coefficient was calculated in the following manner. It was assumed that burning was controlled by the diffusion of oxygen to the particle surface, where the following reaction (see Chapter 5, Eq. 5.5.1) occurred so rapidly that the oxygen concentration at the surface vanished:

$$C + \tfrac{1}{2}O_2 \rightarrow CO \tag{6.9.6}$$

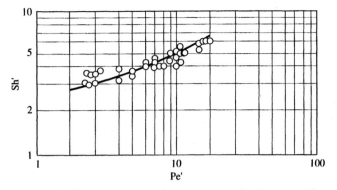

**Figure 6.9.2** Data on the rate of consumption of carbon particles converted to Sh'.

From Chapter 5 (Eq. 5.5.38) we have the relationship between the rate of consumption of the particle and the oxygen flux $N_1$:

$$\frac{d}{dt} \rho_s \frac{4}{3} \pi R^3 = -4\pi R^2 N_1 (2M_{w,C}) \tag{6.9.7}$$

The mass transfer coefficient is then defined as

$$N_1 = kC_{f,O_2} \tag{6.9.8}$$

where $C_{f,O_2}$ is the molar concentration of oxygen in the feed to the combustor. The Sherwood number then follows from Eq. 6.9.3.

On the whole, we must regard the correlation as simple to use, and adequate in its ability to describe the data in the range of parameters studied. As always, caution is necessary in applying this correlation to conditions much different from those on which it is based. Nevertheless, the correlation provides what we need—a starting point for design of a fluidized bed combustor.

## SUMMARY

When we develop models of *convective* mass transfer, the convective mass transfer coefficient appears in a boundary condition on the convective diffusion equation. In this chapter we track down the source of these coefficients. For turbulent flows it is often possible to introduce a film or boundary layer model that connects the mass transfer coefficient to the flow field. Nondimensionalization of the species flux leads naturally to a model relating the Sherwood number ($k_c L/D_{AB}$) to the Reynolds number and the Schmidt number, as in Eq. 6.2.15. In some cases we choose to define a dimensionless mass transfer coefficient as a $j$-factor (Eq. 6.2.16), and we then find the Chilton–Colburn analogy for the relationship of the dimensionless transport coefficients (the $j$-factor and the friction factor) for mass and momentum.

The film model is a poor approximation for many real systems, for which the film thickness is not observed to be uniform along the surface across which transfer is occurring. Boundary layer theory then provides the next level of modeling and leads to very simple predictions (such as Eq. 6.3.23) for the relationship of Sherwood number to Reynolds and Schmidt numbers.

For relatively simple *laminar* flows of the kind encountered in Chapter 5, we can solve the convective diffusion equation analytically (as we did in that chapter) and obtain the convective mass transfer coefficient by manipulating the solution. For example, for mass transfer across the free surface of a falling laminar liquid film, we find Eq. 6.4.6; data in support of this model are presented in Fig. 6.4.1. For mass transfer at the boundary between a solid surface and a fluid in laminar flow, we find Eq. 6.4.9 (for planar flow), which is supported by data such as shown in Fig. 6.4.2. Restrictions on these models include the assumption that the transfer is confined to a thin region near and parallel to the boundary across which transfer occurs. This set of limitations introduces the idea of short and long "contact times," and criteria for choosing the appropriate models.

The use of these analytical models is illustrated first through a series of examples of mass transfer in which the flow field is quite simple. But for more complex flows, such as those that occur in many real engineering situations (packed towers, falling film gas absorbers, aerated stirred tanks), we resort to empirical correlations of data. However, the *form* of these correlations is guided by our observations of the models of simpler flows. The models and correlations presented in this chapter are summarized briefly as follows.

### Planar Flows

Turbulent flow parallel to a planar surface, laminar boundary layer:

$$\mathrm{Sh}_x = 0.33 \mathrm{Re}_x^{1/2} \mathrm{Sc}^{1/3} \tag{6.3.23}$$

$$\overline{\mathrm{Sh}}_L = 0.66 \mathrm{Re}_L^{1/2} \mathrm{Sc}^{1/3} \tag{6.3.25}$$

Laminar film flow, transfer across the liquid/gas interface with resistance on the liquid side:

$$\overline{\mathrm{Sh}}_\delta = \left( \frac{6}{\pi} \mathrm{Re}_\delta \mathrm{Sc} \frac{\delta}{L} \right)^{1/2} \tag{6.4.6}$$

Laminar flow, film of thickness $\delta$, or fluid confined to a planar channel of height $2\delta$, transfer across the solid/fluid boundary with resistance on the fluid side:

$$\overline{\mathrm{Sh}}_\delta = 1.165 \, (\mathrm{Re}_\delta \mathrm{Sc})^{1/3} \left( \frac{\delta}{L} \right)^{1/3} \qquad \text{for} \qquad \mathrm{Re}_\delta \mathrm{Sc} \frac{\delta}{L} > 10$$

$$\overline{\mathrm{Sh}}_\delta = 1.6 \qquad \text{for} \qquad \mathrm{Re}_\delta \mathrm{Sc} \frac{\delta}{L} < 10 \tag{6.4.9}$$

### Flow in Tubes and Pipes

Laminar flow

$$\overline{\mathrm{Sh}}_R = 1.29 (\mathrm{Re}_R \mathrm{Sc})^{1/3} \left( \frac{R}{L} \right)^{1/3} \qquad \text{for} \qquad \mathrm{Re}_R \mathrm{Sc} \frac{R}{L} > 10$$

$$\overline{\mathrm{Sh}}_R = 1.83 \qquad \text{for} \qquad \mathrm{Re}_R \mathrm{Sc} \frac{R}{L} < 10$$

Turbulent flow

$$\mathrm{Sh}_D = 0.023 \, \mathrm{Re}_D^{0.83} \, \mathrm{Sc}^{0.44} \tag{6.5.5}$$

*Flow normal to the axis of a long cylinder*

See Fig. 6.3.2.

*Transfer to rising bubbles*

$$Sh_{d_e} = (4.0 + 1.21\ Pe^{2/3})^{1/2}$$

(6.6.6)

where $\qquad\qquad Pe = Ud_e/D_{AB}.$

*Transfer to particles in a stirred tank*

$$Sh_{d_T} = 0.267\ Sc^{1/4}\ Re_{d_i}^{3/4}\ N_P^{1/4}\left(\frac{d_T^4}{Vd_i}\right)^{1/4}$$

(6.8.5)

In all cases of the use of a mathematical model, or an algebraic or graphical correlation of data, it is essential to know *what length scale* is being used in the definitions of the Sherwood number and the Reynolds number.

## PROBLEMS

**6.1** Derive Eq. 6.4.12 by the following route. Go back to the analysis of the parallel plate dialyzer in Chapter 5. Argue that if we take the case of $P \rightarrow \infty$ in the boundary condition at the wall (Eq. 5.3.14), then the boundary value problem becomes (almost) identical to the equations that describe dissolution at the wall of a parallel plate channel into a laminar flow. Show that Eq. 5.3.32 then leads to Eq. 6.4.12.

**6.2** Read footnote 1 in Section 6.4 and give the equivalent theoretical results for mass transfer coefficients from the wall of a circular tube, using the tube diameter instead of the radius as the length scale.

**6.3** We have derived the following models of mass transfer coefficients across a planar interface. State clearly how the physics differs in each case that leads to one of these models.

$$\overline{Sh} = \frac{\bar{k_c}L}{D_{AB}} = 0.66\left(\frac{Lv_0}{\nu}\right)^{1/2} Sc^{1/3} = 0.66\ Re_L^{1/2}\ Sc^{1/3}$$

(6.3.25)

$$\overline{Sh} = \left(\frac{6}{\pi} Re\ Sc\ \frac{\delta}{L}\right)^{1/2}$$

(6.4.6)

$$\overline{Sh} = A\ (Re\ Sc)^{1/3}\left(\frac{\delta}{L}\right)^{1/3}$$

(6.4.9)

**6.4** A particular fluorinated lubricating oil (Krytox 143 AD) has a vapor pressure of $2 \times 10^{-6}$ torr at 386 K. Suppose a steel tube of length $L = 10$ cm and inside diameter 5 mm is coated with a uniform layer of this oil, with an initial film thickness of 1 $\mu$m. How long is required for 90% of the oil to evaporate into an airstream at

386 K, at flow rates in the range of 10 to 1000 cm$^3$/s? The molecular weight of the oil is 8250. The diffusion coefficient of the oil vapor in air is 0.04 cm$^2$/s. The density of the oil is 1900 kg/m$^3$.

**6.5** A thin film of water flows down the inner surface of a tube of diameter 1 cm. The film thickness is 0.01 cm. Dry air enters this tube at a temperature of 25°C, and at flow rates in the range of 10 to 1000 cm$^3$/s. Find the relative humidity of the exiting air. The tube length is 1 m.

**6.6** In Example 6.5.1, calculate the pressure drop across the ends of the tube due to friction losses for this flow. What is the implication of your result?

**6.7** Verify Eq. 6.6.24.

**6.8** Complete Example 6.7.1 and plot $C_w$, $C_P$, and $Q_P$ as functions of flow rate.

**6.9** Barker and Treybal [*AIChE J.*, **6**, 289 (1960)] present the data shown in Fig. P6.9 for mass transfer coefficients for dissolution of solids suspended in a stirred tank. Fit an equation of the form of Eq. 6.8.3 to these data. Does Eq. 6.8.5 fit these data?

**6.10** In Eq. 6.3.6 the upper limit on the integral is $y = L$. Find an expression for the $y$ component of velocity along a line of constant $\eta = L/\delta$. Is the velocity component $v_y$ positive or negative? Find the volume flowrate through the region $0 < y < L$. Does this value increase or decrease with increasing $x$? Are these two results compatible with each other?

Show that along the line $y = \delta(x)$ we find

$$v_\delta = -\frac{5}{8}\frac{d\delta}{dx}$$

(P6.10)

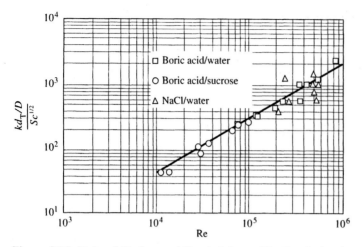

**Figure P6.9** Data of Barker and Treybal for solids dissolution in baffled vessels.

**6.11** For steady flow in a long straight pipe, a simple force balance yields the result that

$$\frac{\pi}{4} D^2 \Delta P = \pi D L \tau_0 \qquad \text{(P6.11)}$$

where $\tau_0$ is the shear stress along the pipe surface $r = R$. Suppose we use the laminar boundary layer theory to evaluate the shear stress. What do we find for the friction factor/Reynolds number relationship, and how does this compare to the observed relationship for turbulent pipe flow?

**6.12** Derive Eq. 6.3.11. Repeat the derivation for an assumed linear velocity profile:

$$v_x = \frac{v_0}{\delta} y \qquad \text{(P6.12)}$$

Derive Eq. 6.3.21, but use the result above for the velocity profile, and use a linear concentration profile, as well.

**6.13** Sherwood numbers for gas phase transfer usually exhibit a $Sc^{1/3}$ dependence, while Sherwood numbers for liquid phase transfer often exhibit a $Sc^{1/2}$ dependence. Show that the coefficient for mass transfer between a well-mixed gas and a laminar film flowing down a vertical surface satisfies an equation of the form $Sh = A\,(Re\;Sc)^{1/2}$.

**6.14** A highly toxic gas is produced as a by-product of a chemical process. It can be converted through a catalytic chemical reaction to harmless gaseous products. A proposal is made to pass the product gas through a pipe, the inside surface of which is coated with a suitable catalyst. Derive an expression for the pipe length $L^*$ sufficient to reduce the concentration of the toxic species by a factor of $10^{-4}$ of its inlet value. The surface reaction is first-order, with a surface rate constant defined by the expression

$$R_B = k_s C_B \qquad \text{(P6.14)}$$

Your final expression for $L^*$ should be in a nondimensional format. Answer for the limiting case of a very slow surface reaction. Suggest a suitable definition of small $k_s$.

Give a numerical value for $L^*$ under the conditions that the gas has the properties of air at 400 K, the pipe diameter is $D = 2$ cm, and the Reynolds number of the entering gas flow is $Re = 4\rho Q/\pi D \mu = 1000$. For $k_s$ use a value of $k_s = 0.05$ cm/s.

**6.15** We derived Eq. 6.4.16 for convective transport into a flow bounded by infinite parallel plates. Carry out the corresponding derivation for flow in a tube of circular cross section. Do we simply replace $B$ in Eq. 6.4.16 by $R$ for the tube flow case?

Show that for a duct of arbitrary but axially uniform cross section, the general solution analogous to Eq. 6.4.16 is

$$\frac{C_s - \overline{C_L}}{C_s - \overline{C_1}} = \exp\left(-\frac{\bar{k}L}{\langle v \rangle R_h}\right) \qquad \text{(P6.15.1)}$$

where the "hydraulic radius" $R_h$ is defined as

$$R_h = \frac{\text{cross-sectional area}}{\text{wetted perimeter}} \quad \textbf{(P6.15.2)}$$

**6.16** Show that for flow through a tube of circular cross section, the height of a transfer unit, defined as in Eq. 6.4.20, may be expressed as

$$\frac{H_{TU}}{L} = \frac{\text{Re Sc}(R/L)}{2\,\overline{\text{Sh}}} \quad \textbf{(P6.16)}$$

**6.17** We suppose that air at 16°C flows through a circular tube of diameter $D = 2$ mm and length $L = 20$ cm. The inner surface of the tube is coated with solid naphthalene. We wish to find the concentration of naphthalene in the gas phase at the outlet of the tube, as a function of the gas flow rate. The following data are available:

$$p_{vap} = 0.037 \text{ mmHg}$$

$$D_{AB} = 0.071 \text{ cm}^2/\text{s}$$

$$\text{Sc} = 2.25$$

Plot $\overline{C_L}$ as a function of flowrate in the range $6 < Q < 600$ sccm. The downstream end of the tube is at atmospheric pressure. Find the inlet pressure required to maintain the flow, as a function of the gas flowrate.

**6.18** At high enough flow rates, frictional losses will create a pressure gradient along the axis of the tube in Problem 6.17. Where, in the analysis of this problem, must the convective mass transfer model be modified to account for this pressure variation?

Rework Problem 6.17, but for the case that $L = 200$ cm and $D = 1$ mm. Show how to calculate $\overline{k_c}$ for this case. Assume that $D_{AB}$ varies inversely with absolute pressure.

**6.19** $n$-Butyl alcohol flows as a smooth laminar liquid film down the inside of a circular pipe of diameter 3 cm. Air flows through the pipe at a rate of 1 g/min. The air inlet temperature is 50°C. What length of pipe is required if we want to saturate the air with the alcohol? Physical property data are given in Example 6.5.1.

**6.20** Air at 16°C flows through a pipe of inside diameter 3 mm. The pipe has a length of 10 cm. The inside of the pipe is coated with naphthalene with an initial uniform thickness of 100 $\mu$m.

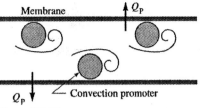

**Figure P6.22**

**a.** What air flowrate will just yield a naphthalene-saturated flow at the pipe exit?
**b.** How long can this system operate before the saturation drops below 90%?
Relevant data for naphthalene are given in Problem 6.17.

**6.21** Using Eq. P6.16, find the value of Re Sc $R/(L\overline{\text{Sh}})$ that is required to yield a 99% approach to equilibrium. Is this result consistent with the data of Fig. 6.5.2?

**6.22** The performance of the RO system described in Example 6.7.1 has been enhanced by the introduction of a "laminar convection promoter," as suggested in the sketch in Fig. P6.22. The particular promoter used leads to a doubling of the average convective coefficient, relative to that in the undisturbed laminar flow case, at a feed rate of 20 L/min. What are the percentage increases to the two key performance variables, the permeate flux $Q_P$ and the water quality $C_P$? How does you answer change if the feed flow is only 0.2 L/min? What conclusion do you draw regarding the value of convection promotion?

**6.23** Rework Example 6.7.1, but for the case that the feedwater salt concentration is 2%. What are the percentage decreases to the two key performance variables, the permeate flux $Q_P$ and the water quality $C_P$?

**6.24** Two membrane units, both identical to the one described in Example 6.7.1, are available. Examine the performance of each of the configurations sketched in Fig. P6.24. For each, give the total permeate flow (cm³/s), and the water quality (ppmw). Is the series arrangement better than the parallel arrangement, or vice versa?

**6.25** Repeat Example 6.7.1, but for the case of a membrane with twice the pure water permeability given in the example. What are the percentage changes in the two key performance variables, the permeate flux $Q_P$ and the water quality $C_P$?

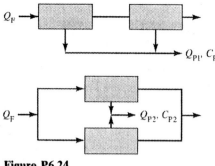

**Figure P6.24**

fraction. Use $M_w = 119.4$ for the molecular weight of chloroform.

| $V_G$ (cm/s) | $V_L$ (cm/s) | Mass Fraction Removed |
|---|---|---|
| 2.89 | 0.31 | 0.57 |
| 2.89 | 0.62 | 0.39 |
| 2.89 | 0.70 | 0.40 |
| 3.61 | 0.31 | 0.60 |
| 3.61 | 0.70 | 0.48 |
| 3.61 | 0.62 | 0.43 |
| 4.39 | 0.31 | 0.70 |
| 4.39 | 0.62 | 0.51 |
| 4.39 | 0.70 | 0.49 |

**6.26** Repeat Problem 6.22, but for the case of the membrane described in Problem 6.25.

**6.27** Chloroform is sometimes produced in water as an undesired consequence of the chlorination process. Because this gas has a fairly high vapor pressure, it can be removed from water by bubbling air through a column of water containing chloroform in solution. The EPA requires that chloroform be at a concentration below 100 ppbw (parts per *billion* by weight). An important characteristic of the solute (chloroform, in this case, but more generally as well) is its Henry's law constant $H$, defined as

$$p_{vap} \text{ [Pa]} = Hx \qquad \text{(P6.27)}$$

where $x$ is mole fraction in the liquid.

One way to measure the Henry's constant of a solute is to bubble air through the solution, under conditions that lead to saturation of the gas phase, and determine the mole fraction of the solute in the solute-saturated air. A somewhat different method is to mix air and solute-containing water at the inlet to a continuously flowing system, and measure the solute concentration in the liquid phase leaving the system, under conditions such that equilibrium is achieved at the outlet. Such experiments are described by Velazquez and Estevez [*AIChE J.*, **38**, 211 (1992)].

From the equilibrium data tabulated below, calculate a Henry's law constant for chloroform in water at 298 K. The velocities $V_G$ and $V_L$ are "superficial" velocities for the gas and liquid flows, defined in such a way that their ratio $V_G/V_L$ is identical to the ratio of the volume flowrates of the two phases through the system. Give a value for $H$ in units of megapascals per mole

**6.28** Several models have been presented for prediction of the convective mass transfer coefficient external to a sphere falling through a quiescent fluid. A commonly recommended expression is

$$Sh = 2 + 0.6 \, Re^{1/2} \, Sc^{1/3} \qquad \text{(P6.28.1)}$$

and another is

$$Sh = (4 + 1.21 \, Pe^{2/3})^{1/2} \qquad \text{(P6.28.2)}$$

where $Pe = Re \, Sc$, and both the Sherwood number and the Reynolds number are based on the drop diameter. The following data are available [Beard and Prupaccher, *J. Atmos. Sci.*, **28**, 1455 (1971)] for the Sherwood number for the evaporative mass transfer of water drops (Sc = 0.71) with initial diameters in the range 140 to 750 $\mu$m, falling through still air. The Reynolds number is based on the terminal velocity of the drops and uses the drop diameter as the length scale. Compare the data to Eqs. P6.28.1 and P6.28.2. What do you conclude?

| $Re^{1/2} \, Sc^{1/3}$ | Sh |
|---|---|
| 0.4 | 1.92 |
| 0.76 | 2.08 |
| 1.1 | 2.2 |
| 1.39 | 2.38 |
| 1.7 | 2.6 |

**6.29** Water drops with initial diameters in the range of 50 to 200 $\mu$m fall through still air at 0°C, 765 mbar, and 90% relative humidity. Plot the diameter of the drop after a fall of 300 m, as a function of its initial diameter. Assume no evaporative cooling of the drop.

**6.30** Equation 6.7.3 is a salt balance for the case that the membrane completely rejects salt ($r \equiv 1 - C_p/C_w = 1$). Show that for the general case $0 < r < 1$, for which Eq. 6.7.3 is replaced by

$$Q_P C(y) + D_{sw} \frac{\partial C}{\partial y} = Q_P C_w (1 - r)$$

**(P6.30.1)**

Eq. 6.7.6 takes the form

$$\frac{C_w}{C_f} = \frac{\exp\left(\dfrac{Q_P \, Sc^{2/3}}{j_D U}\right)}{r + (1 - r)\exp\left(\dfrac{Q_P \, Sc^{2/3}}{j_D U}\right)}$$

**(P6.30.2)**

**6.31** Derive Eq. 6.4.9. Be careful! To get the result given, you must be precise with definitions of several dimensionless groups and especially with the definition of the mass transfer coefficient. Read Section 6.2 again. Define Re and Sh based on the half-width of the parallel plate channel:

$$Re_\delta = \frac{\overline{U}\delta}{\nu} \qquad Sh = \frac{\overline{k}_c \delta}{D_{AB}} \qquad \textbf{(P6.31.1)}$$

Define the local mass transfer coefficient as

$$k_c = \frac{N_y(x)}{C_s} \qquad \textbf{(P6.31.2)}$$

and use Eq. 5.2.13 for the local flux. Why don't we define the local mass transfer coefficient as

$$k_c = \frac{N_y(x)}{C_s - \overline{C}(x)} \qquad \textbf{(P6.31.3)}$$

**6.32** The following data are available [Sensel and Myers, *Chem. Eng. Educ.* **26**, 156 (1992)] for the dissolution of sucrose spheres into an agitated tank of liquid. Assume that the liquid volume is much greater than the initial solids volume. Given $\rho_s = 1600$ kg/m³, and $C_s = 1800$ kg/m³, and an initial radius $R_o = 12$ mm, calculate the convective mass transfer coefficient under these conditions.

| $t$ (s) | $M/M_o$ |
|---|---|
| 120 | 0.8 |
| 180 | 0.71 |
| 240 | 0.63 |
| 300 | 0.55 |
| 360 | 0.47 |

**6.33** Experimental data have been presented by Nassif et al. [*Int. J. Heat Mass Transfer*, **38**, 691 (1995)] from which the mass transfer coefficient for the turbulent boundary layer on a flat plate can be calculated. The data can be correlated in the form

$$Sh_L = 0.035 \, Re_L^{0.8} \, Sc^{0.6} \qquad \textbf{(P6.33.1)}$$

for $5 \times 10^4 < Re_L < 2 \times 10^6$. Both Sh and Re use the plate length $L$ as the length scale.

**a.** The criterion usually given for transition from the laminar boundary layer to the turbulent boundary layer is $Re_L = 5 \times 10^5$. Hence some of the data of this study appear to be in the range of the laminar boundary layer. For the laminar boundary layer we find

$$Sh_L = 0.66 \, Re_L^{0.5} \, Sc^{0.33} \qquad \textbf{(P6.33.2)}$$

Using both these correlations in the appropriate regimes, prepare a plot of $Sh_L$ versus $Re_L$ for $Sc = 2.25$, over the range $2 \times 10^4 < Re_L < 2 \times 10^6$.

**b.** Suppose a planar naphthalene-coated surface is exposed to a parallel turbulent flow of air at 16°C with a mean velocity 40 m/s. The length of the surface in the flow direction is 0.6 m. How long is required for the average naphthalene thickness to decrease by 1 mm? Use the property data of Prob. 6.17.

**6.34** Suppose we wish to verify the data that led to the correlation shown in Fig. 6.3.2, in the range $10 < Re < 100$. Evaporation of naphthalene cylinders will be measured by periodically weighing the cylinders while evaporation is taking place. Air will be blown normal to the cylinder axis. Choose a cylinder diameter such that about 10% of the mass of the cylinder will be lost in 20 min.

**6.35** Vogtlander and Bakker [Chem. Eng. Sci., **18**, 583 (1963)] present data for convective mass transfer coefficients under conditions of liquid flow normal to the axis of cylinders (cross-flow). Their data were obtained in a Reynolds number range of 5 to 75, and their Schmidt numbers ranged from 1300 to 2000. They recommended the following correlation:

$$Sh_D = 0.38 \, Sc^{0.2} + (0.56 \, Re_D^{0.5} + 0.001 Re_D) \, Sc^{0.33}$$

**(P6.35)**

Calculate the *j*-factor that corresponds to this correlation, and plot the result on Fig. 6.3.2.

**6.36** Figure 6.5.2 is based on the classical paper of Gilliland and Sherwood [*Ind. Eng. Chem.,* **26,** 516 (1934)]. These authors comment on the finding that much of the data at large values of the *x* axis is higher than predicted by the laminar flow theory. In their experiments, the wetted section was 117 cm long, and the inside diameter of the tube was 2.67 cm. Schmidt numbers for the evaporative species were approximately 2, with the exception of water (Sc = 0.8). Approximately, what Reynolds number range was covered by their data?

**6.37** Show that Eq. 5.2.19 yields the line indicated on Fig. 6.5.2. Give the equation of the line, in the form $y = Ax^n$.

**6.38** Suppose, instead of the assumed laminar parabolic flow that leads to Eq. 6.4.18, we had a laminar flow but with a velocity profile exhibiting uniform "plug flow" across the tube radius. Derive an expression from which a line may be placed on Fig. 6.5.2 corresponding to this flow, in the regime where $ReSc(D/L)$ is large.

**6.39** Begin with the equation for the flux at the interface of a laminar "falling film," and derive an expression for the average Sherwood number over a height $L$ of film, as a function of Peclet number Pe. Define $Sh = kL/D_{AB}$, and $Pe = \bar{v}L/D_{AB}$. Note that we choose to define Pe based on the *average* velocity $\bar{v}$.

**6.40** A long cylindrical rod of naphthalene, 3 mm in diameter, is subjected to a cross-flow of air at 34°C, with a velocity of 3 ft/s. How much time will pass before 10% of the mass of the rod has been lost to the air?

At 34°C, the diffusivity of naphthalene in air is $8 \times 10^{-6}$ m²/s, and its vapor pressure is 26 Pa. Its molecular weight is 128, and the solid density is 1.15 g/cm³.

State clearly the method you use for estimation of the mass transfer coefficient.

**6.41** A liquid droplet (density $\rho = 1.2$ g/cm³) of diameter 50 $\mu$m falls through still air at 25°C and atmospheric pressure. Initially, the drop has 1% (by weight) of volatile solute dissolved uniformly throughout its volume. The diffusivity of solute through the liquid is $10^{-5}$ cm²/s, and through the air is $10^{-1}$ cm²/s. How far does the drop fall before its center concentration is reduced to 1% of its initial value?

**6.42** Find the rate of transfer of oxygen into a fermentation broth under the following conditions. Air bubbles at 1 atm pressure and having diameters of 100 $\mu$m are injected into water at 37°C. The water has no dissolved oxygen initially. The solubility of $O_2$ from air is $2.26 \times 10^{-7}$ g · mol $O_2$ per cubic centimeter of water, in water at 37°C. The diffusivity of $O_2$ in water at 37°C is $3.25 \times 10^{-9}$ m²/s.

**a.** Find the flux of oxygen, in units of kilogram-moles of oxygen per second-meter squared.

**b.** What is the Reynolds number characteristic of the bubble motion? Define and give its value. For a convective transfer coefficient, use

$$Sh = Pe^{1/3} \qquad \textbf{(P6.42)}$$

**6.43** Several models have been presented for prediction of the convective mass transfer coefficient external to a sphere moving through a fluid. Examples include

$$Sh = (4 + 1.21 \, Pe^{2/3})^{1/2} \qquad \textbf{(P6.43.1)}$$

and

$$Sh = Pe^{1/3} \qquad \textbf{(P6.43.2)}$$

both of which have a theoretical basis. Calderbank and Jones [*Trans. Inst. Chem. Eng. (London),* **39,** 363 (1961)] correlate data for mass transfer to bubbles rising in a liquid, using

$$k_L = 0.38 \, Sc^{-2/3} \left( \frac{g\mu\Delta\rho}{\rho^2} \right)^{1/3} \qquad \textbf{(P6.43.3)}$$

**a.** What limitation of Eq. P6.43.2 is accounted for by Eq. P6.43.1?

**b.** Show how an equation nearly identical to Eq. P6.43.3 may be derived from Eq. P6.43.2.

**6.44** An "artificial gill" is proposed to support human respiration underwater. A hollow fiber design is contemplated, as sketched in Fig. P6.44. Upon exhalation, the fibers fill with $CO_2$, which is transported into the water by diffusion/convection. Upon inhalation, $O_2$ is transported into the fiber interior. The fibers are microporous polypropylene, a hydrophobic polymer (Celgard, a Celanese product). These fibers have an o.d. of 0.01 cm and a wall thickness of 0.003 cm. The fiber bundle is packaged so that 10% of the volume is occupied by fibers. This low volume fraction promotes more efficient external mass transport to the water, which flows normal to the long axis of the fiber bundle.

For these fibers, mass transfer is controlled by

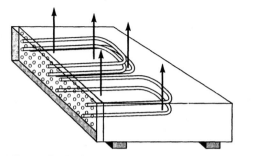

**Figure P6.44** Water flows transverse to a bundle of fibers.

the resistance on the water side. Data of Yang and Cussler [*AIChE J.*, **32,** 1910 (1986)], taken in a 750-fiber module with cross-flow of water, were correlated by

$$\text{Sh} = 1.38 \, \text{Re}^{0.34} \, \text{Sc}^{0.33} \qquad \textbf{(P6.44)}$$

where

$$\text{Re} = \frac{d_f V_L}{\nu_L}$$

$$\text{Sh} = k_L d_f / D$$

and $d_f$ is the outside diameter of the fiber. The oxygen requirement to support a man is 2500 cm³ $O_2$ per minute. The water is two-thirds saturated with $O_2$. What is the required volume of this hollow fiber module?

**6.45** One of the most important features of a mathematical model is the parametric dependence of the key variables on the various parameters of the system. As an example, Levich (*Physicochemical Hydrodynamics*, Prentice Hall, 1962) obtained a theoretical model for the convective mass transfer coefficient at the surface of a sphere, when the sphere is in creeping flow (Re < 0.1) with respect to the surrounding fluid. The model takes the form

$$\text{Sh} \equiv \frac{k_c d_p}{D^\infty} = \text{Pe}^{1/3} \qquad \textbf{(P6.45.1)}$$

where $d_p$ is the sphere diameter and the Peclet number is

$$\text{Pe} = \text{Re Sc} \qquad \textbf{(P6.45.2)}$$

Show that if Stokes' law applies, then $k_c$ is independent of $d_p$.

**6.46** In most technologically important interphase transfer problems, there are convective and/or diffusive resistances on both sides of the interface. Sometimes one or the other side of the interface controls the transfer. In some cases we have one chemical species being transferred in one direction across the interface and a different species being transferred in the other direction. If both species are dilute, the transfer of one may be independent of the transport of the other.

A gas bubble is formed at a submerged orifice. When it is released from the orifice it has a diameter of $d_p = 0.05$ cm and an internal pressure of 15 psia. The gas is pure oxygen initially, and the bubble rises through water at 20°C. The height of the water above the orifice is $H = 10$ cm.
**a.** What fraction of the oxygen initially in the bubble is transferred into the water?
**b.** What is the mole fraction of water vapor in the bubble after 10 cm of rise?

Vapor pressure of water at 20°C = 17.5 mmHg
Diffusivity of $O_2$ in liquid water at 20°C, $D^\infty = 2 \times 10^{-5}$ cm²/s
Diffusivity of $H_2O$ vapor in $O_2$ at 20°C, $D = 0.3$ cm²/s
Solubility of $O_2$ in $H_2O$ at 20°C and 1 atm = 1.4 $\times 10^{-6}$ mol $O_2$/cm³ $H_2O$, which corresponds to

$$\alpha = \frac{\text{mol } O_2/\text{cm}^3 O_2}{\text{mol } O_2/\text{cm}^3 H_2O} = 29.7 \qquad \textbf{(P6.46.1)}$$

If Re for bubble rise is not small, a recommended correlation is

$$\text{Sh}_D = 2 + 0.6 \, \text{Re}_D^{1/2} \text{Sc}^{1/3} \qquad \textbf{(P6.46.2)}$$

**6.47** A stream of a pure viscous oil enters and falls down the inner surface of a 2 inch i.d. pipe. The liquid film thickness is 1 mm. The oil absorbs a gas species from an airstream that is flowing at a linear velocity of 3 m/s upward through the pipe. We also know the following:

Liquid viscosity = 50 cP
Liquid phase diffusivity of the dissolved gas species, $D = 10^{-6}$ cm²/s
Pipe length, $L = 5$ m
Gas phase diffusivity of the gas species, $D^\infty = 0.25$ cm²/s
**a.** Estimate a convective mass transfer coefficient on the gas side of the interface, at 20°C.
**b.** Do the same for the liquid side coefficient.

**6.48** A silicon film is grown on a heated silicon surface as shown in Fig. P6.48. The feed gas

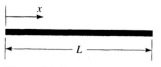

**Figure P6.48**

is silane (SiH$_4$) diluted in nitrogen. The mole fraction is 0.05 = [SiH$_4$/(SiH$_4$ + N$_2$)]. The pressure is 1 atm and the temperature is 500°C. The feed gas is at a linear velocity of 1 m/s just upstream of the surface, and the gas flow is parallel to the surface. The surface is planar with $L$ = 10 cm and has a width of 5 cm. Assume that the gas flow is turbulent. No reactions occur in the gas phase. The surface reaction is

$$SiH_4 \rightarrow Si + 2H_2 \qquad \textbf{(P6.48)}$$

and is so fast that the film deposition is diffusion controlled. Calculate the solid film thickness profile $h(x)$ after 10 min of growth. The density of solid silicon is $\rho_s$ = 2.3 g/cm$^3$. The molecular weight of silicon is $M_{w,Si}$ = 28.

**6.49** Tetraethoxysilane (TEOS) is a liquid source used in the semiconductor industries to grow silicon dioxide. One proposed method of delivery of the TEOS to a reactor involves creation of a falling film evaporator, as shown in Fig. P6.49. The TEOS evaporates at the gas/liquid interface and diffuses into the inert gas stream to be carried to the reactor. We need a quick approximate estimate of the mole fraction of TEOS in the carrier gas for a

system operating under the following conditions:

| | |
|---|---|
| Inner diameter of the tube | 2 cm |
| Tube length (wetted) | 25 cm |
| Pressure at gas inlet | 1 atm |
| Carrier gas flow | $2 \times 10^5$ sccm |
| TEOS flow | 100 mL/min |
| Temperature | 60°C |

The carrier gas is helium. For the viscosity of helium, use the elementary kinetic theory result in the form

$$\mu = 26.69 \frac{(M_w T)^{1/2}}{\sigma^2} \, micro\text{poise} \qquad \textbf{(P6.49)}$$

where $M_w$ is the molecular weight of the gas species. Use $\sigma$ = 2.6 Å for helium. We have the following data for TEOS, all at 60°C.

Vapor pressure, $P_{vap}$ = 16 torr
$\sigma$ = 7.6 Å
$\varepsilon/k_B$ = 522 K
Viscosity of the *liquid* TEOS, $\mu$ = 0.03 poise
Liquid density = 0.95 g/cm$^3$
The molecular weight of TEOS [Si(OC$_2$H$_5$)$_4$] is 208.

**6.50** A gas bubble rises through water. The bubble is saturated with water vapor. The bubble initially has 20 mol% oxygen in addition to the water vapor. The bubble rises at a steady velocity of 20 cm/s. The bubble has an initial diameter of 3 mm, and its diameter does not change significantly as it rises. Use the data listed to find the height of liquid above the point of introduction of the bubble that must be provided if the goal is to permit sufficient time for the bubble to reach equilibrium, with respect to oxygen, with the surrounding water.

Vapor pressure of water at 20°C = 17.5 mmHg
Diffusivity of O$_2$ in liquid water at 20°C, $D^*$ = $2 \times 10^{-5}$ cm$^2$/s
Diffusivity of H$_2$O vapor in O$_2$ at 20°C, $D$ = 0.3 cm$^2$/s
Solubility of O$_2$ in H$_2$O at 20°C and 1 atm = 1.4 $\times 10^{-6}$ mol O$_2$/cm$^3$ H$_2$O, which corresponds to

$$\alpha = \frac{mol\ O_2/cm^3\ H_2O}{mol\ O_2/cm^3 O_2} = 0.034$$

**6.51** Benzoic acid is a solid at 300 K and is soluble in water. It has an equilibrium solubility in

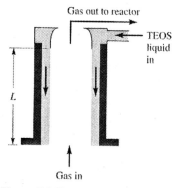

Gas out to reactor

TEOS liquid in

$L$

Gas in

**Figure P6.49**

water of $0.003$ g/cm$^3$. The diffusion coefficient of benzoic acid in water is $3 \times 10^{-5}$ cm$^2$/s. The mass density of solid benzoic acid is $1.3$ g/cm$^3$. One hundred grams of benzoic acid particles of initial radius $R = 0.1$ cm is introduced to a highly agitated stirred tank containing 10 L of water. The water is initially free of dissolved benzoic acid. After 10 min the benzoic acid concentration in the water is found to be $0.0003$ g/cm$^3$. What is the value of the mass transfer coefficient under these conditions?

What is the value of the Sherwood number?

**6.52** A drop of water with an initial diameter of 1 mm falls through still air that has a relative humidity of 15%. Assume that the air and the drop are at 25°C, at which the vapor pressure is 23.8 mmHg. How far does the drop fall before its diameter is reduced by 10%?

**6.53** A nitrogen/water vapor mixture at 25°C and 100% relative humidity flows through a pipe of inside diameter 10 cm at a mean velocity of 200 cm/s. The inside surface of the pipe is covered with a laminar falling water film of thickness $H = 0.15$ cm. The water is saturated with oxygen by bubbling air through the feed supply prior to delivery of the water to the top of the pipe. Oxygen will be stripped from the water as the liquid film falls. The air is at atmospheric pressure.

Which phase, gas or liquid, controls the rate of stripping?

**6.54** Naphthalene (molecular weight = 128) coats the inside of a 2 cm diameter pipe through which air at 16°C flows at a mean velocity of 100 cm/s. The pipe has a length of 1 m. Find the mass loss of the naphthalene, in grams per hour, under these conditions. Assume that the inside of the pipe is everywhere coated with the naphthalene film throughout the history of the evaporation.

Before working through the details of this problem, you must modify Eq. 6.4.16 to be valid for flow through a circular pipe. Present that modification clearly as part of your solution to this problem.

**6.55** Naphthalene (molecular weight = 128) coats the inside of a 2 cm diameter pipe through which air at 16°C flows at a mean velocity of 1000 cm/s. The pipe has a length of 1 m. The air entering the pipe has 500 ppmv naphthalene

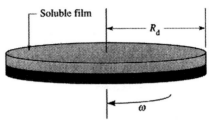

**Figure P6.56** Spinning disk test device.

vapor. The pipe has an initial uniform coating of solid naphthalene of thickness equal to 30 $\mu$m. What is the coating thickness after 20 min?

**6.56** A "spinning disk" mass transfer device is designed as suggested in Fig. P6.56. A solid disk rotates about an axis normal to its circular face and is immersed in a large volume of fluid. Except for the motion induced by the rotation of the disk, there is no imposed convective flow in the system. The surface of the disk is coated with a solid film of a material that is soluble in the surrounding fluid. A significant motion is induced by the rotation of the disk, and this motion gives rise to significant convective mass transfer from the disk surface.

Studies have shown that the convective mass transfer coefficient is given by

$$k_c = 0.62 D_{AB}^{2/3} \nu^{-1/6} \omega^{1/2} \qquad \text{(P6.56.1)}$$

where $k_c$ = convective mass transfer coefficient (cm/s)

$\qquad D_{AB}$ = diffusion coefficient of the solute in the fluid (cm$^2$/s)

$\qquad \nu$ = kinematic viscosity of the fluid (cm$^2$/s)

$\qquad \omega$ = rotational speed of the disk (rad/s)

Convert Eq. P6.56.1 into a nondimensional format of the form

$$\text{Sh} = A \, \text{Re}^n \, \text{Sc}^m \qquad \text{(P6.56.2)}$$

Define each of the dimensionless groups in terms of the parameters that appear in Eq. P6.56.1 and/ or the figure. Give the values for the coefficients $A$, $n$, and $m$.

**6.57** The spinning disk device described in Problem 6.56 is coated with a uniform film of naphthalene of thickness 2 mm. Plot the fractional re-

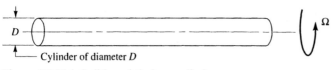

Cylinder of diameter $D$

**Figure P6.58.1** Solid naphthalene cylinder.

moval of naphthalene as a function of time for rotational speeds in the range of 30 to 90 rpm. The disk radius is $R_d = 1$ cm.

**6.58** A long cylinder of solid naphthalene rotates in air about its axis, as shown in Fig. P6.58.1. The rotation induces motion in the surrounding air, but no external flow other than this is imposed on the system. Fig. P6.58.2 shows a friction factor curve, obtained by measuring the torque exerted by the air on the rotating cylinder. (See Sherwood, Pigford, and Wilke, *Mass Transfer*, McGraw-Hill, 1975, Chapter 6.)

The cylinder diameter is initially $D = 1$ cm and the rotational speed is $\Omega = 600$ rpm. How long is required for the diameter of the cylinder to change by 10%? The density of solid naphthalene is 1.15 g/cm³. The vapor pressure of naphthalene at the temperature of the experiment (290 K) is 0.037 mmHg. $D_{AB}$ for naphthalene/air is 0.07 cm²/s. The molecular weight of naphthalene is 128.

**6.59** The spinning disk device described in Problem 6.56 is coated with an initially uniform film of lubricant oil of thickness 10 $\mu$m. The molecular weight of the oil is 7000. At a temperature of

300 K the oil has a vapor pressure of $10^{-6}$ torr, a diffusion coefficient of the oil vapor in air of 0.05 cm²/s, a density of 1800 kg/m³, and a viscosity of 1.5 Pa·s. The disk rotates at 120 rpm. Oil is lost by two mechanisms: spin-off of the liquid by centrifugal flow, and evaporation. Plot film thickness versus time associated with each mechanism, as if it occurred separately without the other, down to a film thickness of 0.1 $\mu$m. Use log–log coordinates for the plotting. A study of the fluid dynamics of the "spin-off" flow leads to a model for the film thickness $h(t)$ of the form

$$\frac{h(t)}{h_o} = \left(1 + \frac{4\rho\omega^2 h_o^2}{3\mu}t\right)^{-1/2} \quad \textbf{(P6.59)}$$

where $h_o$ is the initial film thickness.

**6.60** A highly agitated bath of an aqueous solution is enclosed in a chamber of volume $V_c$. The bath volume is $V_b$. A solute A evaporates at the bath surface and is transferred into the surrounding chamber. The rate of mass transfer is controlled by the convective coefficient on the gas side of the interface, $k_c$. The interfacial area

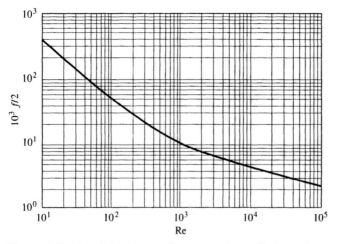

**Figure P6.58.2** Friction factor for a rotating cylinder, where $\text{Re} = \pi D^2 \Omega / \nu_{\text{air}}$ and $\Omega$ is in revolutions per second.

is $A_b$. The mole fraction of solute in the bath is initially $x_o$, and the surrounding gas is initially free of the solute. The vapor/liquid equilibrium is described by Henry's law, in the form

$$p_{vap}[\text{atm}] = Hx \qquad \textbf{(P6.60)}$$

**a.** Develop a mathematical model for the mole fraction $y$ of the solute in the surrounding chamber, as a function of time. Define a dimensionless parameter that controls the rate of buildup of $y$ in the chamber.

**b.** Find the time for this system to reach 99% of equilibrium, and give the mole fractions $x$ and $y$ at equilibrium, for the following conditions:

$$A_b = 100 \text{ cm}^2 \qquad V_b = 1 \text{ L}$$

$$V_c = 10 \text{ L} \qquad k_c = 5 \text{ cm/s}$$

$$T = 35°C \qquad x_o = 0.001$$

for $SO_2$ ($H = 50$ atm/mol fraction) and CO ($H = 8 \times 10^4$).

# Chapter 7

# Continuous Gas/Liquid Contactors

**C**ommercial mass transfer systems are often so complex that it is not possible to develop a mathematical model of the detailed concentration profiles throughout the system. As a consequence, it is not possible to predict the overall mass transfer rate from a knowledge of the geometry, operating conditions, and physical properties. The source of this difficulty usually lies in the complexity of the flow field throughout the system. In this chapter we examine one class of mass transfer device—a packed column—in which gas and liquid flow continuously through a tortuous geometry designed to enhance the amount of interfacial area, and the convective coefficients across that area. Design equations are derived, in terms of mass transfer coefficients that account for the resistances to mass transfer on both the gas and liquid sides of the complex boundary separating the two phases. We learn that thermodynamics has an important effect on the rate of mass transfer that can be achieved.

In earlier chapters we examined convective mass transfer across interfaces under conditions such that the flow on either side of the interface was, in some sense, simple. In the falling film evaporator of Chapter 5, for example, the liquid flow is laminar, unidirectional, and fully developed, and the gas side is well mixed. Commercial mass transfer equipment often involves two-phase flows under turbulent conditions of such geometrical complexity that there is no hope of modeling the details of the flow field in either phase. An example would be a "packed column," in which liquid flows (down) by gravity through a large diameter pipe, and gas is introduced into the bottom of the pipe. The pipe is filled with a particulate packing, the shape of which promotes intimate contact between the gas and liquid streams. (See Fig. 7.1.1.) Fluid dynamics principles seldom produce predictions of such important system characteristics as the mean interfacial area per unit of volume in the column. We resort instead to empirical correlations in which the interfacial area is related to the gas and liquid flowrates.

In such systems there may be significant resistance to mass transfer on the liquid side, or on the gas side, of the interface, and often the resistances on *both* sides are significant. We begin this chapter by developing a method of accounting for the overall effect of resistances in series. Then we derive a mass balance for a continuous contactor in which the area across which mass transfer occurs is ill defined and unmeasurable,

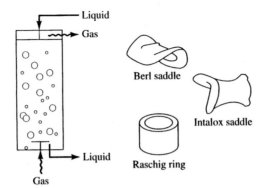

**Figure 7.1.1** A packed column, and some typical packings.

except through measurement of the rate of transfer itself. This mass balance yields a "design equation" with which we can relate the performance of a gas/liquid contactor to the design and operating variables of the system. Several examples will illustrate the use of this design equation. This chapter does not provide a comprehensive treatment of the design and analysis of mass transfer equipment. Recommended references are *Mass Transfer* (Sherwood, Pigford, and Wilke, McGraw-Hill, 1975) and *Mass-Transfer Operations*, 3rd ed. (Treybal, McGraw-Hill, 1987).

## 7.1  TWO-RESISTANCE FILM THEORY

We focus our attention on the interface between a gas and liquid, and on the flux of some species across that interface. The solid packing is irrelevant in the sense that while it aids in the creation of the gas/liquid interface, there is no transfer across any *solid*/fluid boundary. We will begin by using partial pressure units for the gas phase concentration of the species of interest, and molar concentration for the species in the liquid. Later we will see that other units for concentrations are often used.

We assume steady state transfer and thermodynamic equilibrium at the interface. Then the molar flux across the interface is continuous, and for any species A we may write

$$N_A^G = N_A^L \tag{7.1.1}$$

At the interface

$$p_{A,i} = f_e(c_{A,i}) \tag{7.1.2}$$

is given by an *equilibrium relationship*. In this case, Eq. 7.1.2 is the equilibrium partial pressure of species A above a solution of molar concentration $c_{A,i}$.

The "driving force" for mass transfer, on the gas side, is $p_{A,G} - p_{A,i}$. (On the liquid side, it is $c_{A,i} - c_{A,L}$.) Figure 7.1.2 shows these compositions. It may be easier to think about compositions in this system if we look first at a simple case. In Fig. 7.1.3, we suppose that the liquid side is "well stirred," in the sense that turbulence in the liquid reduces the boundary layer thickness on the liquid side to a very small value. Then we would find that the concentration on the liquid side is uniform, at the value $c_{A,L}$. Hence

$$c_{A,i} = c_{A,L} \tag{7.1.3}$$

and

$$p_{A,i} = f_e(c_{A,L}) = p_A^* \tag{7.1.4}$$

and the gas side driving force is $p_{A,G} - p_A^*$.

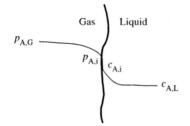

**Figure 7.1.2** Transfer at a gas/liquid interface.

By virtue of the assumption that the liquid side is well mixed, the only resistance to mass transfer is on the gas side. On a $p/c$ diagram (see Fig. 7.1.4) we may draw the equilibrium curve defined by the function $f_e$ (Eq. 7.1.2). We define the partial pressure $p_A^*$ as the partial pressure in equilibrium with the bulk liquid composition $c_{A,L}$. This definition holds regardless of whether the liquid side is well mixed, but the first equality in Eq. 7.1.4 holds only in the well-stirred liquid case. We could draw a similar picture to Fig. 7.1.3 for the case of a well-stirred gas side.

In the general case, where neither fluid is well stirred, we have finite mass transfer resistances on *both* sides of the interface. We may define gas and liquid side mass transfer coefficients as

$$N_A^G = k_G(p_{A,G} - p_{A,i}) = N_A^L = k_L(c_{A,i} - c_{A,L}) \tag{7.1.5}$$

It follows that

$$\frac{k_L}{k_G} = \frac{p_{A,G} - p_{A,i}}{c_{A,i} - c_{A,L}} \tag{7.1.6}$$

If, in Fig. 7.1.4, $O$ represents compositions in each of the bulk phases, and if steady transfer is occurring with no interfacial accumulation, then in keeping with Eq. 7.1.6 $I$ represents the interface compositions, for a given $k_L/k_G$.

The problem with using these individual gas side and liquid side coefficients is that *the interface compositions themselves are not measurable.* We want to express fluxes in terms of *bulk* composition differences, which *are* measurable. To achieve this, we carry out the following analysis, using Fig. 7.1.5 as a guide. We define *overall* coefficients as

$$N_A = K_G(p_{A,G} - p_A^*) = K_L(c_A^* - c_{A,L}) \tag{7.1.7}$$

Since $p_{A,G} - p_A^*$ is not (except when the liquid is well stirred), a true driving force in the gas phase (the same holds for $c_A^* - c_{A,L}$ in the liquid phase), then $K_G$ is not a gas side coefficient. $K_G$ accounts, in some way, for resistance on *both* sides of the interface, but it is written in terms of a gas side (pseudo) driving force.

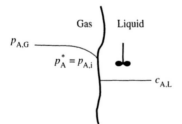

**Figure 7.1.3** Well-stirred liquid side.

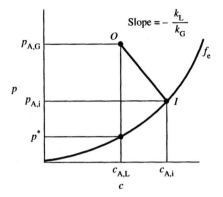

**Figure 7.1.4** Compositions on a $p/c$ diagram.

The first thing we want to do is show the relationship between the *overall* coefficients $K_G$ and $K_L$, and the *individual* coefficients $k_G$ and $k_L$, defined in Eq. 7.1.5. If a linear equilibrium relationship exists for Eq. 7.1.4, say

$$p_{A,i} = mc_{A,i} \qquad \text{(a form of Henry's law)} \tag{7.1.8}$$

then

$$p_{A,G} = mc_A^* \tag{7.1.9}$$

and

$$p_A^* = mc_{A,L} \tag{7.1.10}$$

From the $K_G$ definition given in Eq. 7.1.7,

$$N_A = K_G(p_{A,G} - p_A^*) \tag{7.1.11}$$

and we find

$$\frac{1}{K_G} = \frac{p_{A,G} - p_A^*}{N_A} = \frac{p_{A,G} - p_{A,i} + p_{A,i} - p_A^*}{N_A} = \frac{p_{A,G} - p_{A,i}}{N_A} + m\frac{c_{A,i} - c_{A,L}}{N_A} \tag{7.1.12}$$

or

$$\frac{1}{K_G} = \frac{1}{k_G} + \frac{m}{k_L} \tag{7.1.13}$$

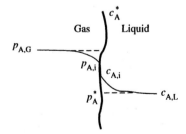

**Figure 7.1.5** Overall driving forces.

By a similar method we can show that

$$\frac{1}{K_L} = \frac{1}{mk_G} + \frac{1}{k_L} \tag{7.1.14}$$

These are the desired relationships between the overall ($K$) coefficients and the individual ($k$) coefficients. We will later have to address the issue of how we obtain the individual coefficients.

Two limiting cases follow easily from the expressions above. For a *highly soluble gas* (e.g., $NH_3$ in $H_2O$), $m$ is small and

$$\frac{1}{K_G} \approx \frac{1}{k_G} \tag{7.1.15}$$

and this indicates that there is no significant liquid phase resistance.

By contrast, for a *sparingly soluble gas* (e.g., $CO_2$ in $H_2O$), $m$ is large and

$$\frac{1}{K_L} \approx \frac{1}{k_L} \tag{7.1.16}$$

and the liquid phase resistance controls the flux.

The choice of $p$ and $c$ as composition variables in the two phases is not unique. We often use the *mole fractions* $x$ and $y$ as composition variables. Then we may write the flux expression as (we use $p = Py$, Dalton's law, in Eq. 7.1.5, where $P$ is the total pressure)

$$N_A = k_G P(y_G - y_i) \tag{7.1.17}$$

for an ideal gas, and

$$N_A = k_L C(x_i - x_L) \tag{7.1.18}$$

where $c = Cx$. Figure 7.1.6 shows a composition diagram in terms of mole fractions. Note that as $k_L/k_G$ gets either large or small, $x_i \to x_L$ or $y_i \to y_L$, as expected.

Now we can address the matter of extracting the individual ($k$) coefficients for a particular system. Suppose we establish $k_G$ for a specific gas/liquid contactor by using the transfer of $NH_3$ between the two phases (note Eq. 7.1.15). We would attempt to correlate the data in the form, typically, of a Sherwood number/Reynolds number/ Schmidt number relationship:

$$Sh_G = \frac{k_G h}{D_{NH_3,G}} = A_G \, Re_G^{m_G} \, Sc_{NH_3,G}^{1/3} \tag{7.1.19}$$

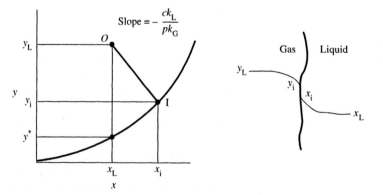

**Figure 7.1.6** Compositions on a mole fraction diagram.

where $h$ is some appropriate length scale for the system. (The subscript G is a reminder that the parameters are for the gas phase.) We could also establish $k_L$, for the same contactor, by using transfer of (sparingly soluble) $CO_2$, and find

$$Sh_L = \frac{k_L h}{D_{CO_2,L}} = A_L \, Re_L^{m_L} \, Sc_{CO_2,L}^{1/2} \qquad (7.1.20)$$

(Note that the exponents on the Schmidt number Sc are often different for gas and liquid forms of convection.)

For the transfer of some *arbitrary* species X in the *same* transfer system, we need either $K_G$ or $K_L$

$$\frac{1}{K_{G,X}} = \frac{1}{k_{G,X}} + \frac{m_X}{k_{L,X}} \qquad (7.1.21)$$

where

$$m_x \equiv \frac{p_x}{c_x} \qquad (7.1.22)$$

and $m_x$ is the $m$ value for the species X.

We assume

$$\frac{k_{G,X} h}{D_{X,G}} = A_G \, Re_G^{m_G} \, Sc_{X,G}^{1/3} \qquad (7.1.23)$$

$$\frac{k_{L,X} h}{D_{X,L}} = A_L \, Re_L^{m_L} \, Sc_{X,L}^{1/2} \qquad (7.1.24)$$

From these assumptions it follows that

$$\frac{k_{G,X}}{k_{G,NH_3}} = \frac{D_{X,G} \, Sc_{X,G}^{1/3}}{D_{NH_3,G} \, Sc_{NH_3,G}^{1/3}} = \left( \frac{D_{X,G}}{D_{NH_3,G}} \right)^{2/3} \qquad (7.1.25)$$

assuming X and $NH_3$ are dilute in the carrier gas (usually air) and $\nu_{air/X} = \nu_{air/NH_3}$.

A similar result follows for $k_{L,X}$ in terms of $k_{L,CO_2}$, but with a different exponent on the ratio of diffusion coefficients:

$$\frac{k_{L,X}}{k_{L,CO_2}} = \frac{D_{X,L} \, Sc_{X,L}^{1/2}}{D_{CO_2,L} \, Sc_{CO_2,L}^{1/2}} = \left( \frac{D_{X,L}}{D_{CO_2,L}} \right)^{1/2} \qquad (7.1.26)$$

These relationships provide a method of estimating $k_G$ and $k_L$ for some particular species X, if we first obtain *data* on $k_G$ for a highly soluble gas like $NH_3$, and for $k_L$ for a sparingly soluble gas like $CO_2$. This idea is important, since we often design transfer units for gases and liquids for which no mass transfer coefficient data are available.

Once the *individual* coefficients are established for each side of the interface, Eq. 7.1.13 or 7.1.14 yields the *overall* coefficients, assuming that the thermodynamic data (a value for $m$ at the operating temperature) are available. It is the *overall* coefficients that we must use in the mass balances because only the overall compositions—not the interfacial compositions—are known and measurable.

To establish the individual coefficients using highly and/or sparingly soluble species such as $NH_3$ and $CO_2$, we must develop a performance relationship between the mass transfer coefficients and the overall (i.e., measurable) changes in composition of the two streams that flow through the column. We turn now to a derivation of that relationship.

## 7.2   ANALYSIS OF CONTINUOUS CONTACT TRANSFER

Now we carry out a mass balance and a species balance on a gas/liquid contactor. We suppose that the gas and liquid mix so intimately that the two-phase fluid may be thought of as a continuum in the sense that at any axial position $z$ there is a well-defined composition for each phase. We also assume that there are no radial gradients in the contactor, with the result that each composition variable is a function only of axial position along the column. The gas flow can be in the same direction as the liquid flow (concurrent) or opposite the liquid flow (countercurrent). The latter case is the more common, and we illustrate the analysis for that case here.

Consider *countercurrent* flow in a mass transfer unit, as sketched in Fig. 7.2.1.

We will now use mole fractions $x$ (liquid) and $y$ (gas). We define the molar fluxes (sometimes called "velocities," although they do not have units of velocity) denoted by $G$ and $L$, with units of moles/time–cross-sectional area. The area used in this definition is the *empty* cross-sectional area of the column. Note that the molar flowrate is $GA$ for the gas and $LA$ for the liquid, with units of moles/time.

We make two mass balances across the ends of the contactor.

**1. *Overall molar balance***

$$L_2 + G_1 = L_1 + G_2 \tag{7.2.1}$$

This assumes no reactions.

**2. *Overall component balance for each species***

$$L_2 x_2 + G_1 y_1 = L_1 x_1 + G_2 y_2 \tag{7.2.2}$$

Next we make a pair of balances around one end of the column and an arbitrary interior plane at some $z$ (see Fig. 7.2.2).

***Overall***

$$L_2 + G = L + G_2 \tag{7.2.3}$$

***Component***

$$L_2 x_2 + Gy = Lx + G_2 y_2 \tag{7.2.4}$$

We solve for $y$:

$$y = \frac{L}{G}x + \frac{G_2 y_2 - L_2 x_2}{G} \tag{7.2.5}$$

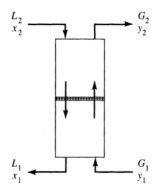

**Figure 7.2.1**  Definition sketch for material balances on a countercurrent contactor.

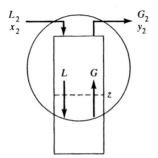

**Figure 7.2.2** Balance around one end of a column.

This $y(x)$ equation is called the "*operating line*" because it gives the actual composi-tions $x$, $y$ at any axial position through the column. The operating line is shown on Fig. 7.2.3, along with the equilibrium curve for the gas/liquid pair. The equilibrium curve could be nonlinear by virtue of the thermodynamics of the system. From Eq. 7.2.5 we see that the operating line could be nonlinear if $L/G$ is not constant. This could be the case, for example, if a large fraction of the volatile species is transferred across the phase boundary.

Figure 7.2.3 is drawn for the case of an *absorber,* where the mole fraction in the gas phase is *above* the equilibrium value in the liquid. For a "stripper," in which the volatile component is transferred *from* the liquid *to* the gas, the operating curve is *below* the equilibrium curve. The analyses carried out here are valid for both absorbers and strippers.

Usually, $L$ and $G$ may be regarded as streams of a solution consisting of an inert, or nontransferable, material along with components that can transfer between phases. In the simple case of a single transferred solute, we may write

$$L(1 - x) = L' = L_s \quad \text{and} \quad G(1 - y) = G' = G_s \tag{7.2.6}$$

where $L'$, $G'$ are the "*inert*" molar flows. $L'$ and $G'$ do not vary through the column. (We designate the inert flows with either a prime or subscript "s" for $L$ and for $G$, since both notations are common in the engineering literature.) Then we may write

$$(1 - x_2)\frac{L_2 x_2}{1 - x_2} + (1 - y)\frac{Gy}{1 - y} = (1 - x)\frac{Lx}{1 - x} + (1 - y_2)\frac{G_2 y_2}{1 - y_2} \tag{7.2.7}$$

or

$$L_2' X_2 + G' Y = L' X + G_2' Y_2 \tag{7.2.8}$$

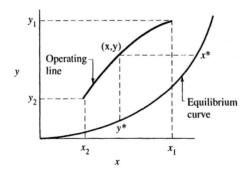

**Figure 7.2.3** Operating line on the $x$-$y$ diagram.

where $X \equiv x/(1 - x)$, $Y \equiv y/(1 - y)$ are mole fraction *ratios*. Note that

$$G' = G'_2 = G_s \tag{7.2.9}$$

and

$$L' = L'_2 = L_s \tag{7.2.10}$$

Hence we find

$$L'(X_2 - X) = G'(Y_2 - Y) \tag{7.2.11}$$

or

$$Y = \frac{L'}{G'}X + \frac{G'Y_2 - L'X_2}{G'} \tag{7.2.12}$$

Since $L_s = L'$ and $G' = G_s$ are constants, the $Y(X)$ relationship is linear. However, as suggested in the sketch of Fig. 7.2.4, a linear equilibrium relationship might become nonlinear in $X$, $Y$ space. Sometimes the advantage of working with constant inert flows in the material balances outweighs any disadvantage of nonlinear equilibrium lines, especially if the equilibrium lines are already nonlinear.

Our next task is to take a closer look at the interfacial area. This area is, in general, not measurable; hence it is usually unknown. But the interfacial area must somehow appear in the final form of the design equations, since we certainly expect that the performance of a contactor will depend on the efficiency with which we can create interfacial area for transfer between phases. Consider the flux $N_A$ in either of the forms

$$N_A = k_G(p_{A,G} - p_{A,i}) = K_G(p_{A,G} - p_A^*) \tag{7.2.13}$$

and consider (see Fig. 7.2.5) a *differential* volume of the exchanger, of cross-sectional area $A$ and axial length $dz$. The *interfacial area* ($dA_i$) within $dz$ is related to the geometry of distribution of the two phases in a packed column. We describe this area by an area per (column) volume factor $a_v$. Then we find

$$dA_i = a_v dV = a_v A \, dz \tag{7.2.14}$$

and the rate of transfer is given by

$$(N_A a_v) \, dV = N_A a_v A \, dz = (K_G a_v)(p_{A,G} - p_A^*)A \, dz \tag{7.2.15}$$

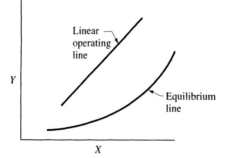

**Figure 7.2.4** The operating line is linear in $X$, $Y$ space.

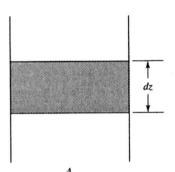

$A$

**Figure 7.2.5** A differential section of the exchanger.

The product $(K_G a_v)$ is called a *capacity coefficient*. It includes two unknown quantities: the mass transfer coefficient and the interfacial area. [A similar relationship may be written for the liquid side, and we could define $(K_L a_v)$ capacity coefficients.]

Our goal now is to derive an expression for the relationship of the total height of a countercurrent column to the change in compositions of the gas and liquid streams. To do so we make a *differential* balance over an *interior* section of a column. For some specific component, a differential balance gives

$$-d(Gy)A = d(Lx)A = N_A a_v A \, dz \qquad (7.2.16)$$

where $x$ and $y$ are mole fractions of the transferring species. We may write the molar flux $N_A$ in a number of forms. For example:

$$d(Lx)A = (K_x a_v)(x^* - x)A \, dz = (k_x a_v)(x_i - x)A \, dz \qquad (7.2.17)$$

or, alternatively,

$$-d(Gy)A = (K_y a_v)(y - y^*)A \, dz = (k_y a_v)(y - y_i)A \, dz \qquad (7.2.18)$$

(Note that we have now defined some new $k$ or $K$ coefficients. For example, on comparing Eqs. 7.2.15 and 7.2.18 we see that $K_G P = K_y$, since $p_A = P y_A$.)

*For equimolar diffusion*, or *dilute unidirectional* transfer, $L$ and $G$ do not vary through the column. Then, integrating Eq. 7.2.17, we find the height of the column to be

$$Z_T = \int_0^{Z_T} dz = \int_{x_2}^{x_1} \frac{d(Lx)A}{(K_x a_v)A(x^* - x)} \approx \frac{L}{(K_x a_v)} \int_{x_2}^{x_1} \frac{dx}{x^* - x} \qquad (7.2.19)$$

or, alternatively, integrating Eq. 7.2.18, we find

$$Z_T = \int_0^{Z_T} dz = \int_{y_2}^{y_1} \frac{d(Gy)A}{(K_y a_v)A(y - y^*)} \approx \frac{G}{(K_y a_v)} \int_{y_2}^{y_1} \frac{dy}{y - y^*} \qquad (7.2.20)$$

If we used, instead of Eqs. 7.2.17, and 7.2.18

$$d(Lx)A = (K_X a_v)(X^* - X)A \, dz \qquad (7.2.21)$$

or

$$-d(Gy)A = (K_Y a_v)(Y - Y^*)A \, dz \qquad (7.2.22)$$

we would get different forms of the same result. For example,

$$-d(Gy) = -d\left(G(1-y)\frac{y}{1-y}\right) = -d(G_s Y) = -G_s dY \qquad (7.2.23)$$

and

$$Z_T = \int_0^{Z_T} dz = \frac{G_s}{(K_Y a_v)} \int_{Y_2}^{Y_1} \frac{dY}{Y - Y^*} \tag{7.2.24}$$

Note that these results *assume equimolal counterdiffusion* or relatively little gas transfer (we assumed $G$ constant) and also that the *capacity coefficient* is *constant*, in taking $(G/K_y a_v)$ across the integral sign.

The specific form of the result for the height of a column depends on what form we choose for the mass transfer coefficient. That choice is usually dictated by the available data. Previous investigators may have left us with mass transfer information in one of several formats, and it is then convenient to use an expression for $Z_T$ that uses the particular form of the capacity coefficient that is available to us.

More complex results are possible, for example, beginning with

$$d(GAy) = (k_G a_v)(p - p_i)A \, dz \tag{7.2.25}$$

or

$$dz = \frac{d(Gy)}{(k_G a_v)(p - p_i)} \tag{7.2.26}$$

But if $G$ is not a constant through the column, as assumed above, then we write

$$G = \frac{G_s}{1 - y} \tag{7.2.27}$$

and $G_s$ is constant regardless of whether we have equimolar counterdiffusion. Now we can write

$$d(Gy) = d\left(\frac{G_s}{1 - y} y\right) = G_s d\left(\frac{y}{1 - y}\right)$$

$$= G_s \left[\frac{1}{1 - y} + \frac{y}{(1 - y)^2}\right] dy = G_s \frac{1}{(1 - y)^2} dy \tag{7.2.28}$$

Hence

$$d(Gy) = \frac{G_s}{1 - y} \frac{dy}{1 - y} = G \frac{dy}{1 - y} \tag{7.2.29}$$

Then

$$dz = \frac{G}{k_G a} \frac{dy}{(1 - y)(p - p_i)} = \frac{G_s dy}{(k_G a_v)(1 - y^2)(p - p_i)} \tag{7.2.30}$$

If $p = Py$, then

$$Z_T = \int dz = \int_{y_2}^{y_1} \frac{G_s dy}{(k_G a_v)P(1 - y)^2(y - y_i)} = \left(\frac{G_s}{k_G a_v P}\right) \int_{y_2}^{y_1} \frac{dy}{(1 - y)^2(y - y_i)} \tag{7.2.31}$$

These various forms (Eqs. 7.2.19, 7.2.20, 7.2.24, or 7.2.31), each of them a relationship of $Z_T$ to end compositions, and each of them in terms of a different capacity coefficient, are identical except in form. Our choice of a working equation depends on available data, as noted above.

## 7.3  THE PERFORMANCE EQUATION FOR DILUTE SYSTEMS

Let's select Eq. 7.2.20 as one of the choices for relating the performance of the system (the overall change in gas phase composition $y_1 - y_2$) to design and operating parameters such as the column height $Z_T$ and the gas flowrate $G$. If only a small amount of transfer takes place, a good approximation is the constancy of the gas flowrate $G$, and we may write

$$Z_T = \int_0^{Z_T} dz = \int_{y_2}^{y_1} \frac{d(Gy)A}{(K_y a_v)A(y - y^*)} = \frac{G}{K_y a_v} \int_{y_2}^{y_1} \frac{dy}{y - y^*} \qquad (7.3.1)$$

Generally, unless both the equilibrium and operating lines are *linear,* numerical or graphical solution of Eq. 7.3.1 is necessary. Linearity of the equilibrium relationship is often a good approximation in very dilute systems. In that case analytical solutions are possible, since the integration of Eq. 7.3.1 can be performed analytically.

Obviously (see Fig. 7.3.1) linearity implies that $y - y^*$ is a linear function of $y$, say

$$y - y^* = ay + b \qquad (7.3.2)$$

We define a number $N_G$, characteristic of the difficulty of transfer, since it is related to how far the system is from equilibrium, as

$$\int_{y_2}^{y_1} \frac{dy}{y - y^*} \equiv N_G \qquad (7.3.3)$$

and $N_G$ can be obtained analytically, now.

The coefficients $a$ and $b$ of Eq. 7.3.2 take the forms

$$a = \frac{(y_1 - y_1^*) - (y_2 - y_2^*)}{y_1 - y_2}$$

$$b = (y_1 - y_1^*) - \frac{y_1}{y_1 - y_2} [(y_1 - y_1^*) - (y_2 - y_2^*)] \qquad (7.3.4)$$

This can be verified by substitution of $y = y_1$, and $y = y_2$ into Eq. 7.3.2 above. (Remember that at $y = y_1$, $y^* = y_1^*$, and the same holds for $y = y_2$.)

After integration, we find

$$N_G = \frac{y_1 - y_2}{(y - y^*)_1 - (y - y^*)_2} \ln \frac{(y - y^*)_1}{(y - y^*)_2} \qquad (7.3.5)$$

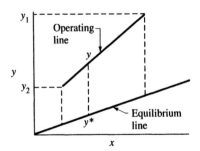

**Figure 7.3.1** The case of linear equilibrium and operating lines.

$N_G$, the integral defined by Eq. 7.3.3, is called the *number of transfer units*. Since we are talking about a continuous contactor, this "number" does not refer to any discrete set of actual units or elements of the column. Roughly speaking, if the end compositions $y_1$ and $y_2$ are specified, $N_G$ is the composition change that we propose to achieve in the column, relative to the departure of the absorbing liquid from equilibrium. (See Eq. 7.3.7 and the comment that follows it.) Hence we would expect large columns to be required for large separations (large $N_G$), and vice versa.

If we define the so-called *log–mean driving force* (in gas phase mole fraction units here) by the following expression:

$$\Delta y_{lm} \equiv \frac{(y - y^*)_1 - (y - y^*)_2}{\ln[(y - y^*)_1/(y - y^*)_2]} \qquad (7.3.6)$$

then we may write Eq. 7.3.5 as

$$N_G = \frac{y_1 - y_2}{\Delta y_{lm}} \qquad (7.3.7)$$

(In view of the preceding comment on the interpretation of $N_G$, this indicates that $\Delta y_{lm}$ is a measure of the average departure of the absorbing liquid from equilibrium.)

Then from Eq. 7.3.1 we find

$$Z_T = \frac{G}{K_y a_v} \int_{y_2}^{y_1} \frac{dy}{y - y^*} = \left(\frac{G}{K_y a_v}\right) N_G \qquad (7.3.8)$$

and hence

$$Z_T = \frac{G}{K_y a_v} \frac{y_1 - y_2}{\Delta y_{lm}} = \frac{G}{K_G a_v P} \frac{y_1 - y_2}{\Delta y_{lm}} \qquad (7.3.9)$$

The definition of $N_G$ given by Eq. 7.3.3 is general, but Eq. 7.3.5 is valid *only* for linear equilibrium and operating lines.

It is common practice to define something called the "height of a transfer unit," by inspection of Eq. 7.3.8, as

$$H_G \equiv \frac{G}{K_y a_v} \qquad (7.3.10)$$

(More precisely, $H_G$ as defined here is the height of a *gas phase* transfer unit based on the *overall* mole fraction driving force $y - y^*$.) As a consequence, we may write Eq. 7.3.8 in the form

$$Z_T = N_G H_G \qquad (7.3.11)$$

We can illustrate applications of these ideas with several examples now.

**EXAMPLE 7.3.1**   *How to Find $K_y a_v$ from Performance Data on a Column*

An $SO_2$ "scrubber" operates as shown in Fig. 7.3.2. Pure water enters the top of the column and is used to absorb or "scrub" $SO_2$ from an airstream that enters the bottom of the column. From data on the performance of this column, find $K_y a_v$. Subscripts T and B refer to the top and bottom of the column.

Assume that $G$ and $L$ are constant with values given in Fig. 7.3.2. An overall balance on $SO_2$ yields

$$GA(y_B - y_T) = LA(x_B - x_T) \qquad (7.3.12)$$

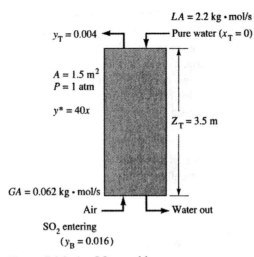

**Figure 7.3.2** An $SO_2$ scrubber.

If three compositions and both flows are given, the fourth composition is fixed by an overall balance. Hence with the values given on the figure we find $x_B$:

$$0.062(0.016 - 0.004) = 2.2(x_B - 0) \qquad \text{or} \qquad x_B = 0.00034 \qquad \textbf{(7.3.13)}$$

This gives the $SO_2$ content in the exiting water stream.

Since we now have all four compositions, we can find $N_G$. From its definition (Eq. 7.3.5), we see that values for $y^*$ (equilibrium data for the $SO_2$/air/water system) are needed. Henry's law holds under dilute conditions, and at the temperature of operation we can find the equilibrium relationship in the form $y^* = 40x$. Then

$$N_G \equiv \int_{y_T}^{y_B} \frac{dy}{y - y^*} = \frac{y_B - y_T}{(y - y^*)_{lm}} = \frac{y_B - y_T}{(y - y^*)_B - (y - y^*)_T} \ln \frac{(y - y^*)_B}{(y - y^*)_T}$$

$$= \frac{0.016 - 0.004}{[0.016 - 40(0.00034)] - [0.004 - 40(0)]} \ln \frac{0.016 - 40(0.00034)}{0.004}$$

$$\textbf{(7.3.14)}$$

Thus we find

$$N_G = 3.8 \qquad \textbf{(7.3.15)}$$

Since $Z_T$, $G$, and $A$ are given, then from the relationships

$$Z_T = 3.5m = N_G \frac{G}{K_y a_v} \qquad \textbf{(7.3.16)}$$

and

$$G = \frac{AG}{A} = \frac{0.062 \text{ kg} \cdot \text{mols/s}}{1.5 \text{ m}^2} \qquad \textbf{(7.3.17)}$$

we find

$$K_y a_v = \frac{N_G G}{Z_T} = \frac{3.8(0.062/1.5)}{3.5} = 0.045 \text{ kg} \cdot \text{mol/m}^3 \cdot \text{s} \cdot \Delta y \qquad \textbf{(7.3.18)}$$

(Note $\Delta y$ in the units for $K_y a_v$. While $\Delta y$ is dimensionless, we display it in the units as a reminder that the mole fraction of the gas is used as the driving force. The subscript $y$ in $K_y a_v$ also serves as a reminder in this respect.)

If we operated this system over a range of gas and liquid flowrates, and measured two flowrates and three compositions, we could determine values for the capacity coefficient $K_y a_v$ as a function of the two flowrates and attempt a correlation of the data that would be useful for design and scale-up.

Often we specify the desired performance of a system, and require an estimate of the required height. Then we must have some data on the capacity coefficients. The next example illustrates this.

**EXAMPLE 7.3.2** *Find Tower Height for a Specified Performance*

Find the tower height to remove 90% of the cyclohexane ($C_6$) in an airstream. Conditions of operation are shown on Fig. 7.3.3. The pure component vapor pressure of $C_6$ is $p^o_{vap} = 121$ mmHg at 30°C. Assume that the solutions obey Raoult's law:

$$p = p^o_{vap} x \qquad (7.3.19)$$

We are given the following data on the mass transfer coefficients:

$$k_x a_v = 0.32 \text{ mol/h} \cdot \text{ft}^3 \qquad (7.3.20)$$

$$k_y a_v = 14.2 \text{ mol/h} \cdot \text{ft}^3 \qquad (7.3.21)$$

presumably at the operating conditions of this contactor. Note that since the coefficients $k_x$ and $k_y$ are given, we will choose to work with $x$ and $y$ as the composition variables. The equilibrium line will be linear, since we have assumed Raoult's law. Because the $C_6$ is so dilute in the entering stream, the operating line will be nearly linear. Hence we can solve this problem analytically, and we do not need to use a graphical integration scheme.

First we must find the molar fluxes $G$ and $L$:

$$G = \left(\frac{363 \text{ ft}^3/\text{min}}{2 \text{ ft}^2}\right)\left(\frac{1}{399} \frac{\text{lb} \cdot \text{mol}}{\text{ft}^3}\right) \text{ (at 30°C = 545 R)} \qquad (7.3.22)$$

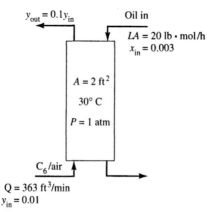

**Figure 7.3.3** An oil absorber for cyclohexane.

and

$$G = 0.46 \text{ lb} \cdot \text{mol/min} \cdot \text{ft}^2 \tag{7.3.23}$$

Likewise we find

$$L = \frac{20}{2}\frac{1}{60} = 0.167 \text{ mol/min} \cdot \text{ft}^2 \text{ (note unit change from h to min)} \tag{7.3.24}$$

A molar balance on $C_6$ gives us

$$G(y_{in} - y_{out}) = L(x_{out} - x_{in}) \tag{7.3.25}$$

or

$$0.46(0.01 - 0.001) = 0.167(x_{out} - 0.003) \tag{7.3.26}$$

Hence

$$x_{out} = 0.028 \tag{7.3.27}$$

We assumed Raoult's law:

$$p = p^o_{vap}x \tag{7.3.28}$$

Hence the equilibrium data may be expressed as

$$y = \frac{p}{P_T} = \frac{p^o_{vap}}{P_T}x = \frac{121}{760}x = 0.16x = m'x \tag{7.3.29}$$

Then we find

$$y^*_B = 0.16(0.028) = 0.00446 \tag{7.3.30}$$

$$y^*_T = 0.16(0.003) = 0.00048 \tag{7.3.31}$$

We now have enough information to find the number of transfer units that correspond to this operation.

$$\begin{aligned}
N_G &= \frac{y_B - y_T}{(y - y^*)_B - (y - y^*)_T} \ln \frac{(y - y^*)_B}{(y - y^*)_T} \\
&= \frac{0.01 - 0.001}{(0.01 - 0.00446) - (0.001 - 0.00048)} \ln \frac{5.54 \times 10^{-3}}{5.2 \times 10^{-4}}
\end{aligned} \tag{7.3.32}$$

or

$$N_G = 4.25 \tag{7.3.33}$$

From Eq. 7.3.29,

$$y = m'x = 0.16x \tag{7.3.34}$$

Then we can find the overall capacity coefficient from

$$\frac{1}{K_y a_v} = \frac{1}{k_y a_v} + \frac{m'}{k_x a_v} = \frac{1}{14.2} + \frac{0.16}{0.32} = 0.57 \text{ h} \cdot \text{ft}^3/\text{mol} \tag{7.3.35}$$

Since

$$G = 0.46(60) = 27.5 \text{ lb} \cdot \text{mol/h} \cdot \text{ft}^2 \tag{7.3.36}$$

it follows that the required tower height is

$$Z_T = 4.25(27.5)(0.57) = 66.7 \text{ ft} \tag{7.3.37}$$

Note our use of foot-pound-hour units here. Much of the classic literature, and the corresponding data, for these contactors are in these units.

## 7.4 MINIMUM FLOWRATE RATIO $(L/G)_{min}$

Usually, in a mass transfer design problem, the flowrate of one phase and three end compositions are specified by process requirements. There is a *minimum L/G* ratio such that operation at $(L/G)_{min}$ requires a column of infinite height. Obviously we would not operate at the minimum flowrate. However, knowledge of the minimum flowrate for a particular set of design constraints is useful. In fact, experience shows that it is often efficient to operate a process at a fixed factor times the minimum flowrate in order to approach optimal economic operation. This factor is usually such that we operate at a flowrate that is 30 to 50% above the minimum. This is a rule of thumb, and as such it provides a quick means of estimating the performance of a system. Often a quick estimate provides enough information to produce an economic decision about the feasibility of a proposed process.

We can understand the concept of minimum flowrate, as well as the means by which it is calculated, by examination of Fig. 7.4.1. Suppose we specify values for $(X_2, Y_2)$ and $Y_1$ for a process. Depending on the $L/G$ ratio, any of the operating lines of Fig. 7.4.1. is possible. As we reduce the $L/G$ ratio we see a series of lines leaning over toward the equilibrium line. The line that just touches the equilibrium line at $Y = Y_1$ defines the minimum $L/G$ ratio, since the required value $Y_1$ cannot be achieved by an operating line with a smaller $L/G$.

From Eq. 7.2.12, the slope of the operating line is seen to be $L'/G'$ $(= L_s/G_s)$. Note that the operating line is exactly linear only in the $(X, Y)$ coordinates, where we use the solute-free flowrates $L_s$ and $G_s$. From the geometry of the figure we see that

$$\left(\frac{L_s}{G_s}\right)_{min} = \frac{\Delta Y}{\Delta X} \tag{7.4.1}$$

where

$$Y_1 - Y_2 = \Delta Y \tag{7.4.2}$$

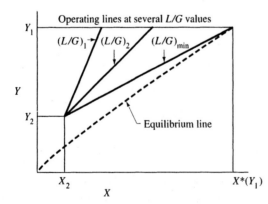

**Figure 7.4.1** Definition of the minimum flowrate ratio $(L/G)_{min}$.

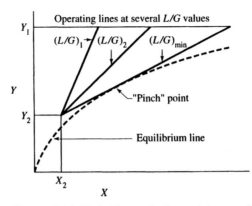

**Figure 7.4.2** Definition of the minimum flowrate ratio $(L/G)_{min}$, for a very nonlinear equilibrium line.

is known from the process specification. To find $\Delta X$ we need the equilibrium line, since (see Fig. 7.4.1)

$$\Delta X = X^*(Y_1) - X_2 \tag{7.4.3}$$

Hence

$$\left(\frac{L_s}{G_s}\right)_{min} = \frac{Y_1 - Y_2}{X^*(Y_1) - X_2} \tag{7.4.4}$$

This analysis is for *counter*current flow, but a similar analysis is possible for the case of *co*current flow.

In deriving Eq. 7.4.4 we did not require the assumption of a linear equilibrium line. In some cases of very nonlinear equilibrium lines, however, the minimum $L_s/G_s$ does not follow from Eq. 7.4.4 and is obtained graphically, as suggested in Fig. 7.4.2. A "pinch point" exists, rendering the operating line at the minimum $L/G$ value just tangent to the equilibrium line. To find the slope of this minimum line, it is necessary to find the $X, Y$ coordinates of the tangent, at the pinch point.

## 7.5  MAXIMUM VAPOR FLOWRATE (FLOODING)

Suppose that a packed tower is operating at a set of conditions corresponding to specified liquid and vapor flowrates. Imagine, now, that the gravity-driven downward liquid flow is held constant, and the upward (countercurrent) vapor flow is increased. If we think about the hydrodynamics of flow through a packed column, it becomes apparent that at some vapor flowrate the vapor will retard the flow of the liquid phase so significantly that the liquid can no longer move uniformly downward through the void space of the packing. At this point the tower is said to be "flooded," and the vapor leaving the top of the tower carries liquid with it.

The flooding velocity is the result of a mechanical force balance and occurs independent of the extent or rate of mass transfer in the system. We can expect the onset of flooding to depend on the vapor and liquid flowrates, the viscosity of the liquid, the densities of each phase, and geometrical parameters that characterize the packing in the tower. Because of the complexity of two-phase flow through a packed tower, no simple mathematical model of the onset of flooding is available, and critical conditions for the onset of flooding are determined from an empirical and approximate graphical

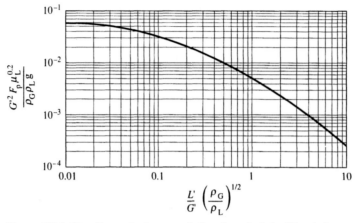

**Figure 7.5.1** Flooding velocity correlation ($y$ axis is in SI units).

presentation of the form shown in Fig. 7.5.1. Predicted flooding velocities based on this figure can deviate from observations by a factor of 2 in some cases. A common rule of thumb is to operate towers at gas phase flows that are half the predicted value, in view of the uncertainty of the correlation.

Note that the $y$ axis of Fig. 7.5.1 is not dimensionless. One must use SI units with this figure, and these are as follows:

$$G' \; [=] \; (\text{kg/s} \cdot \text{m}^2) \qquad L' \; [=] \; (\text{kg/s} \cdot \text{m}^2)$$
$$\mu \; [=] \; (\text{Pa} \cdot \text{s}) \qquad \rho_L, \rho_G \; [=] \; (\text{kg/m}^3)$$
$$g = 9.8 \; \text{m}^2/\text{s}$$

Note that the flowrates used here are *mass*, not molar flowrates. The factor $F_p$ is a packing (shape) factor that depends on the type of packing. It has units of area/volume, or $F_p \; [=] \; (\text{m}^{-1})$. Perry's *Chemical Engineers' Handbook* (Perry and Chilton, Eds., 5th ed., McGraw-Hill, 1973; Table 18.5) gives packing factors for various shapes and sizes of packing.

**EXAMPLE 7.5.1**    *Choosing a Tower Diameter*

A tower packed with 1-inch ceramic Raschig rings is to be used in the treatment of a gas that enters the tower at a rate of 1000 m³/h. The mass flow of the liquid matches that of the gas, so $L'/G' = 1$. The temperature is 20°C and the tower is at atmospheric pressure. Choose a tower diameter $D_T$ such that the gas flow is half the flooding limit.

Since Fig. 7.5.1 is so inexact, we will use rough estimates for the physical properties of the streams. Hence we take the gas to have a nominal density of $\rho_G = 1 \text{ kg/m}^3$ and for the liquid $\rho_L = 1000 \text{ kg/m}^3$. We will take the liquid to have the properties of water, so $\mu_L = 0.001 \text{ Pa} \cdot \text{s}$.

From Fig. 7.5.1 we find, for

$$\frac{L'}{G'} \left( \frac{\rho_G}{\rho_L} \right)^{1/2} = 0.0316 \tag{7.5.1}$$

a value of

$$\frac{G'^2 F_p \mu_L^{0.2}}{\rho_G \rho_L g} = 0.05 \tag{7.5.2}$$

From Table 18.5 of Perry we find

$$F_p = 155 \text{ ft}^{-1} = 509 \text{ m}^{-1} \tag{7.5.3}$$

We solve Eq. 7.5.2 for $G'$ and find

$$G'_{\text{flood}} \approx 2 \text{ kg/s} \cdot \text{m}^2 \tag{7.5.4}$$

We "size" the tower by operating at half the flooding velocity, so we use $G' = 1$ kg/s · m². But the gas flowrate is specified as $Q = 1000$ m³/h, and this is related to $G'$ through

$$Q = \frac{1000}{3600} \text{ m}^3/\text{s} = \frac{\pi D_T^2}{4} \frac{G'}{\rho_G} \tag{7.5.5}$$

From this we find

$$D_T^2 = \frac{4 \times 1000 \times 1}{3600\pi} = 0.35 \text{ m}^2 \tag{7.5.6}$$

or

$$D_T = 0.6 \text{ m} \tag{7.5.7}$$

## 7.6 A DESIGN PROCEDURE FOR LINEAR SYSTEMS

In a typical design problem for an absorber, the two end gas compositions are specified, since we require the gas composition to change from its entering value by a prescribed amount. The liquid stream inlet composition is also specified. The outlet composition of the liquid stream is unknown. Usually the gas stream flowrate is specified, since we need to treat an existing gas stream. A typical design task is to "size" the tower (specify its height and diameter) and specify the required liquid flowrate that will permit the required extent of absorption to occur. We need thermodynamic information (primarily, the coefficient $m$ of Eq. 7.1.9, or a Henry's law constant), and we need mass transfer (capacity) coefficients, such as $k_G$ and $k_L$. It is these latter coefficients that are so hard to come by, since they depend on the nature of the transferable species, the type and size of packing, and the gas and liquid flowrates. If the liquid flowrate is unknown, it may be necessary to use an iterative procedure in which a guess of a liquid flowrate permits an estimate of $k_L$.

While the liquid flowrate $L$ may be unspecified, we know that there is a minimum flowrate dictated by the thermodynamics (see Fig. 7.4.1). A common rule of thumb is to operate at some fixed factor larger than the minimum $L/G$, say $L/G = 1.2(L/G)_{\min}$. Experience shows that this arbitrary choice nevertheless produces a design that is near an economic optimum. We will see that there are other rules of thumb on this issue, as well.

In any event, a major issue in design is the estimation of the transport coefficients leading to an estimate for $H_G$, as defined in Eq. 7.3.10. When $H_G$ is available, Eq. 7.3.11 yields the required height of the tower, given the value of $N_G$. However, $N_G$ is not known from Eq. 7.3.5, for example, because we do not know the outlet liquid composition $x_2$, hence $y_2^*$. But we can manipulate Eq. 7.3.5 by introducing the operating line (Eq

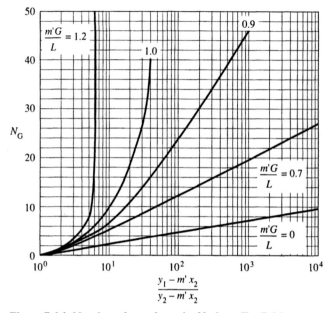

**Figure 7.6.1** Number of transfer units $N_G$ from Eq. 7.6.1, assuming dilute transfer.

7.2.11), with the result that

$$N_G = \frac{1}{1 - m'G/L} \ln\left[\left(\frac{y_1 - m'x_2}{y_2 - m'x_2}\right)\left(1 - \frac{m'G}{L}\right) + \frac{m'G}{L}\right] \tag{7.6.1}$$

where $m'$ is the coefficient in the equilibrium relationship

$$y^* = m'x \tag{7.6.2}$$

We plot Eq. 7.6.1 in Fig. 7.6.1. We can illustrate the use of this figure in the simple example that follows.

---

**EXAMPLE 7.6.1**   *Find the Required Tower Height*

In Example 7.3.2 we specified the flowrates of the liquid and gas phases. Let's rework this example, but regard the liquid flowrate as an unknown. We will pick $L$ based on some rule of thumb, and then determine the tower height.

The minimum flowrate follows from the analog to Eq. 7.4.4 in terms of $L$, $G$, $y$, and $x$:

$$\left(\frac{L}{G}\right)_{min} = \frac{y_1 - y_2}{x^*(y_1) - x_2} \tag{7.6.3}$$

In this example we have

$$y_1 = 0.01 \qquad y_2 = 0.001 \qquad x_2 = 0.003 \qquad m' = 0.16$$

Hence

$$x^*(y_1) = \frac{y_1}{m'} = \frac{0.01}{0.16} = 0.0625 \tag{7.6.4}$$

and

$$\left(\frac{L}{G}\right)_{min} = \frac{0.01 - 0.001}{0.0625 - 0.003} = 0.15 \qquad (7.6.5)$$

and so

$$L_{min} = 0.15(0.46) = 0.069 \text{ lb} \cdot \text{mol/ft}^2 \cdot \text{min} \qquad (7.6.6)$$

Note that this is less than half of the value assumed in Example 7.3.2.

One rule of thumb for design is to use a liquid flow that is 20% above the minimum. Then we would choose

$$L = 1.2 L_{min} = 0.084 \text{ lb} \cdot \text{mol/ft}^2 \cdot \text{min} \qquad (7.6.7)$$

Another rule of thumb is based on the parameter $m'G/L$, and the recommendation is to choose $L$ such that

$$\frac{m'G}{L} = 0.7 \qquad (7.6.8)$$

This would yield

$$L = 0.105 \text{ lb} \cdot \text{mol/ft}^2 \cdot \text{min} \qquad (7.6.9)$$

These two values (from Eqs. 7.6.7 and 7.6.9) are not so different, and we select

$$L = 0.1 \text{ lb} \cdot \text{mol/ft}^2 \cdot \text{min} = 6 \text{ lb} \cdot \text{mol/ft}^2 \cdot \text{h} \qquad (7.6.10)$$

This is only 60% of the value used in Example 7.3.2.

Now we need $N_G$. We calculate the parameter

$$\frac{y_1 - m'x_2}{y_2 - m'x_2} = \frac{0.01 - 0.16 \times 0.003}{0.001 - 0.16 \times 0.003} = 18.3 \qquad (7.6.11)$$

and from Fig. 7.6.1 we find

$$N_G = 6.5 \qquad (7.6.12)$$

Note that this is about 50% larger than the $N_G$ of Example 7.3.2.

Since the liquid flowrate is smaller than that of Example 7.3.2, we expect that $k_x a_v$ will be smaller. Observations suggest that a rough approximation is that $k_x a_v$ is proportional to $\sqrt{L}$, so

$$k_x a_v = 0.32 \left(\frac{4.25}{6.5}\right)^{1/2} = 0.26 \qquad (7.6.13)$$

and

$$\frac{1}{K_y a_v} = \frac{1}{k_y a_v} + \frac{m'}{k_x a_v} = \frac{1}{14.2} + \frac{0.16}{0.26} = 0.69 \text{ h} \cdot \text{ft}^3/\text{lb} \cdot \text{mol} \qquad (7.6.14)$$

Finally,

$$Z_T = \frac{N_G G}{K_y a_v} = 6.5 \times 27.5 \times 0.69 = 124 \text{ ft} \qquad (7.6.15)$$

We see that when a lower liquid flowrate is used, the tower must be considerably larger in height—in this case, nearly 86% larger. This requirement is compensated to some extent by the possibility of using a smaller tower diameter (compared to that of Example 7.3.2) when the flooding criterion is applied (see Problem 7.1).

## 7.7 CORRELATIONS FOR PERFORMANCE CHARACTERISTICS OF PACKED TOWERS

Various empirical correlations are available for use in the estimation of mass transfer coefficients for packed towers. Typically, they are presented in terms of the height of a transfer unit. For example, many references recommend that the height of a gas phase transfer unit in a tower packed with Raschig rings be estimated from the following equation [see Cornell, Knapp, and Fair, *Chem. Eng. Prog.*, **56**(8), 68 (1960)]

$$H_G = \frac{\psi \, Sc_G^{0.5}}{(f_1 f_2 f_3)^{0.6}} (D_T)^{1.24} \left(\frac{Z}{10}\right)^{0.33} L'^{-0.6} \qquad (7.7.1)$$

This is not a dimensionless equation: $H_G$ has units of feet when the other parameters are defined in the following way:

$$L' = \text{superficial } \textit{mass} \text{ flowrate (lb/h} \cdot \text{ft}^2)$$
$$D_T = \text{tower diameter (ft)}$$
$$Z = \text{tower (packed) height (ft)}$$

The parameters $f_i$ account for viscosity, density, and surface tension values relative to those of water:

$$f_1 = \left(\frac{\mu_L}{2.42}\right)^{0.16} \qquad f_2 = \left(\frac{62.4}{\rho_L}\right)^{1.25} \qquad f_3 = \left(\frac{72.8}{\sigma}\right)^{0.8} \qquad (7.7.2)$$

where

$$\mu_L = \text{liquid phase viscosity (lb/ft} \cdot \text{h)}$$
$$\rho_L = \text{liquid phase density (lb/ft}^3)$$
$$\sigma = \text{interfacial tension (dyn/cm)}$$

In Eq. 7.7.1 the dependence on the liquid phase flowrate appears explicitly. The function $\psi$ depends on the type and size of packing, and the liquid and gas phase flowrates, through the extent to which the flooding condition is approached. Figure 7.7.1 shows $\psi$ for Raschig rings. The $\psi$ values obtained from this figure must be used with the units noted in connection with Eq. 7.7.1. (See Problem 7.6.)

In another form of correlation, the gas and liquid flowrate dependencies are explicitly displayed:

$$H_G = \alpha G'^\gamma L'^\lambda Sc_G^{0.5} \qquad (7.7.3)$$

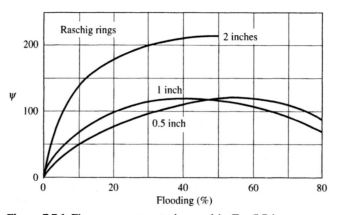

**Figure 7.7.1** Flow parameters to be used in Eq. 7.7.1.

where $G'$ and $L'$ are superficial *mass* flowrates. For 25.4 mm Raschig rings one finds

$$\alpha = 0.56 \qquad \gamma = 0.32 \qquad \lambda = -0.51$$

as the recommended values when the flowrates are in the following ranges:

$$0.27 < G' < 0.81 \text{ kg/s} \cdot \text{m}^2 \qquad 0.68 < L' < 6.1 \text{ kg/s} \cdot \text{m}^2$$

With the stated coefficients, SI units must be used throughout Eq. 7.7.3. An extensive discussion of this and other empirical correlations, as well as a display of experimental data, can be found in Chapter 18 of the 5th edition of *The Chemical Engineers' Handbook.*

## EXAMPLE 7.7.1  *Use of Empirical Correlations*

Water at 293 K and one atmosphere pressure enters a tower packed with 25.4 mm Raschig rings. A gas flow consisting of an air/$SO_2$ mixture enters with 10% (by volume) $SO_2$. The design specifies that 90% of the $SO_2$ is to be absorbed. The gas flow is $G = 0.007 \text{ kg} \cdot \text{mol/s} \cdot \text{m}^2$, and the liquid flow is $L = 0.5 \text{ kg} \cdot \text{mol/s} \cdot \text{m}^2$. The tower has a cross-sectional area of 0.1 $\text{m}^2$. What height of packed section is required?

We will calculate the $H_G$ value using Eq. 7.7.1, and then Eq. 7.7.3, and compare the results.

We have the liquid and gas flowrates specified on a molar basis. We will also need the superficial *mass* flowrates:

$$G' = 0.007 \times 29 = 0.2 \text{ kg/s} \cdot \text{m}^2 \tag{7.7.4}$$

and

$$L' = 0.5 \times 18 = 9 \text{ kg/s} \cdot \text{m}^2 \tag{7.7.5}$$

Note that we do not bother to use the average molecular weight for the gas mixture, which contains a substantial amount of high molecular weight $SO_2$. Instead, we use the average molecular weight of air. The level of uncertainty in the mass transfer correlations that we will use does not warrant a high level of precision in the subsidiary calculations.

To use Eq. 7.7.1 we must examine the flooding limit. First, we find

$$\frac{L'}{G'}\left(\frac{\rho_G}{\rho_L}\right)^{1/2} = \frac{9}{0.2}\left(\frac{1.2}{1000}\right)^{1/2} = 1.56 \tag{7.7.6}$$

(Recall that the flooding correlation is in terms of the mass flowrates—not the molar flowrates.) From Fig. 7.5.1,

$$\frac{G'^2 F_p \mu_L^{0.2}}{\rho_G \rho_L g} = 0.003 \tag{7.7.7}$$

From Table 18.5 of Perry we find

$$F_p = 155 \text{ ft}^{-1} = 509 \text{ m}^{-1} \tag{7.7.8}$$

and it follows from Eq. 7.7.7 that

$$G'_{\text{flood}} = \left(\frac{0.003 \, \rho_G \rho_L g}{F_p \mu_L^{0.2}}\right)^{1/2} = \left(\frac{0.003 \times 1.2 \times 1000 \times 9.8}{509(0.001)^{0.2}}\right)^{1/2} = 0.53 \text{ kg/s} \cdot \text{m}^2 \tag{7.7.9}$$

Hence the operating gas flow ($0.2$ kg/s $\cdot$ m$^2$) is 40% of the flooding velocity. From Fig. 7.7.1 we find

$$\psi = 120 \tag{7.7.10}$$

We are going to use Eq. 7.7.1. We will approximate the liquid stream as pure water, as a consequence of which all the $f_i$ factors are unity. The Schmidt number for air/SO$_2$ is of the order of unity, so we will approximate

$$Sc^{1/2} = 1 \tag{7.7.11}$$

in Eq. 7.7.1. We must use British units in Eq. 7.7.1. (Or see Problem 7.7.) Hence

$$L' = 9 \text{ kg/s} \cdot \text{m}^2 = 9 \times 2.2 \times 3600 \times (0.3048)^2 = 6620 \text{ lb/h} \cdot \text{ft}^2 \tag{7.7.12}$$

With a cross-sectional area of $0.1$ m$^2$, the tower has a diameter of

$$D_T = 1.2 \text{ ft} \tag{7.7.13}$$

Inspection of Eq. 7.7.1 reveals that we need to know the tower height to estimate $H_G$. But the tower height is unknown. Hence we write Eq. 7.7.1 in the form

$$H_G = A Z_T^{1/3} \tag{7.7.14}$$

But we also have Eq. 7.3.11, so we may eliminate the unknown $H_G$ and solve for the tower height from

$$Z_T = N_G H_G = N_G A Z_T^{1/3} \tag{7.7.15}$$

or

$$Z_T = (N_G A)^{3/2} \tag{7.7.16}$$

The coefficient $A$ is calculated as

$$A = \psi Sc_G^{0.5} (D_T)^{1.24} (10)^{-0.33} L'^{-0.6} = 120 \times 1 \times (1.2)^{1.24} \times (10)^{-0.33} \times (6620)^{-0.6} = 0.36 \tag{7.7.17}$$

For SO$_2$ in water, the Henry's law constant is found (Fig. C2.1 in Appendix C) at 293 K as

$$H = 25 \text{ atm} \tag{7.7.18}$$

At a pressure of one atmosphere the coefficient $m'$ has the same value (but it is dimensionless). With $m' = 25$ we find

$$\frac{m' G}{L} = \frac{25 \times 0.007}{0.5} = 0.35 \tag{7.7.19}$$

From Eq. 7.6.1, noting that $y_1 = 0.1$, $y_2 = 0.01$, and $x_2 = 0$ (pure water feed) we find

$$N_G = 3 \tag{7.7.20}$$

The packed tower height follows from Eq. 7.7.16 as

$$Z_T = (3 \times 0.36)^{3/2} = 1.1 \text{ ft} \tag{7.7.21}$$

This is a very short tower.

We can also calculate the tower height using, as an alternative, the correlation of Eq. 7.7.3 and the values of the coefficients listed there:

$$H_G = \alpha G'^{\gamma} L'^{\lambda} Sc_G^{0.5} = 0.56 \times (0.2)^{0.32} \times 9^{-0.51} \times 1^{0.5} = 0.11 \text{ m} \tag{7.7.22}$$

Keep in mind that SI units must be used with this value of $\alpha$. Equation 7.7.20 still holds, so we find

$$Z_T = N_G H_G = 3 \times 0.11 = 0.33 \text{ m} = 1.1 \text{ ft} \qquad \textbf{(7.7.23)}$$

the same as we found using Eq. 7.7.1.

**EXAMPLE 7.7.2**   *Effect of Flowrates on the Height of a Gas Phase Transfer Unit* ($H_G$)

Equation 7.7.1 takes no account of any effect of the gas flowrate on $H_G$, while Eq. 7.7.3 includes both flowrates. The dependence on gas flowrate can be relatively weak under some conditions, as the data of Fig. 7.7.2 suggest. These data were obtained at 86°F (= 303 K) for the absorption of ammonia by water in a tower packed with 1-inch Raschig rings. Over a fivefold range of gas flowrates, the efficiency of mass transfer does not vary strongly. Let's examine the degree to which Eq. 7.7.3 mimics these data.

For $L' = 500 \text{ lb/h} \cdot \text{ft}^2 = 0.68 \text{ kg/s} \cdot \text{m}^2$ we may write Eq. 7.7.3 in the form (using SI units, as required)

$$H_G = \alpha G'^{\gamma} L'^{\lambda} \text{Sc}_G^{0.5} = 0.56(0.68)^{-0.51}(0.67)^{0.5} G'^{0.32} = 0.56 G'^{0.32} \qquad \textbf{(7.7.24)}$$

For the Schmidt number we have used a kinematic viscosity of air of 0.16 cm²/s and a diffusion coefficient of ammonia in air of 0.24 cm²/s, calculated from the Chapman–Enskog theory (Chapter 2). The square-root dependence on the Schmidt number reduces the necessity for a high degree of precision for this value, especially since Eq. 7.7.3 is not regarded as a precise correlation based on extensive data.

At a gas flowrate of $G' = 0.3 \text{ kg/s} \cdot \text{m}^2$ (220 lb/h · ft²), we calculate

$$H_G = 0.56 G'^{0.32} = 0.56(0.3)^{0.32} = 0.38 \text{ m} = 1.25 \text{ ft} \qquad \textbf{(7.7.25)}$$

and similar calculations can be made for other $G'$ and $L'$ values. The lines corresponding to Eq. 7.7.3 for the three liquid flowrates studied by Fellinger are shown on Fig. 7.7.2. The predicted behavior is within about 20% of the observations, and the dependence of $H_G$ on $G'$ is roughly approximated by the empirical model. The maximum value of $H_G$ is observed in systems in which the flooding point is being approached.

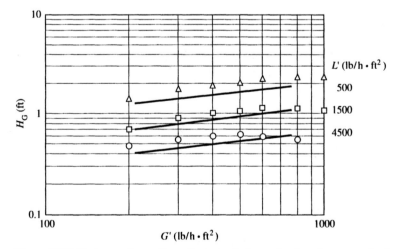

**Figure 7.7.2** Data on absorption of ammonia by water in a tower packed with 1-inch Raschig rings. Fellinger, as reported by Sherwood, Pigford, and Wilke in *Mass Transfer*.

## SUMMARY

In earlier chapters we encountered problems featuring convective resistance to transfer on one side of a phase boundary and diffusive resistance on the other. In many examples of commercial mass transfer equipment we encounter convective resistance on *both* sides of the interface. The complexity of the flow fields on both sides is usually so great that we have no analytical models for the individual convective coefficients. In this chapter we begin by developing a two-resistance film model. The presentation is difficult because we have many choices of driving forces (molar concentrations, mass fractions, partial pressures), and as a result we have specific mass transfer coefficients connected to each of these choices. Thermodynamics plays an important role, and as a consequence we find that the overall coefficients that sum the resistances on each side of the interface include a solubility parameter, often through Henry's law in the common case of gas/liquid transfer (see, e.g., Eq. 7.1.13).

In Section 7.2 we write mass and species balances that lead us to design or performance equations (e.g., Eq. 7.2.20) that relate the height $Z_T$ required to achieve a specified composition change $y_1 - y_2$ to the "*capacity coefficient*" $(K_y a_v)$ and the equilibrium data, given, for example, as the $y^*(x)$ curve. Important characteristics of the system include the number of transfer units $N_G$ (e.g., Eq. 7.3.3) and the height of a transfer unit $H_G$ (Eq. 7.3.10). The number of transfer units is found from the departure of the system from equilibrium, and the required transfer. The height of a transfer unit depends on the resistances to mass transfer on either side of the phase boundary.

To a great extent the performance of a specific system is described by an empirical correlation of the dependence of the height of a transfer unit on the flowrates of the two phases. The relationship includes parameters characteristic of the type of packing and its size. Examples of such correlations are presented: we can use Eq. 7.7.1, which requires Fig. 7.7.1, or we can use Eq. 7.7.3, which requires knowledge of the parameters $\alpha$, $\gamma$, and $\lambda$. Arguments in support of various empirical correlations can be found elsewhere (e.g., Chapter 18 of the 5th edition of *The Chemical Engineers' Handbook*).

## PROBLEMS

**7.1** Derive Eq. 7.1.14.

**7.2** Linear equilibrium relationships may be written in a number of forms, such as in Eq. 7.1.8. The more common form of Henry's law is

$$p_A = Hx_A \qquad \text{(P7.2)}$$

where $H$ is the Henry's law constant, in units of partial pressure per mole fraction in the liquid. Figure C2.1 in Appendix C gives Henry's law constants for several common gases over their aqueous solutions. What forms do Eqs. 7.1.13 and 7.1.14 take if we use Eq. P7.2 in place of Eq. 7.1.8?

**7.3** Show that the coefficient $m'$ defined in Eq. 7.6.2 has the same numerical value as the Henry's law constant defined in Eq. P7.2 when the total pressure on the system is one atmosphere.

**7.4** Based on a flooding criterion of 70% of $G_{flood}$, find the tower diameter $D_T$ in Examples 7.3.2 and 7.6.1.

**7.5** An experimental study of the performance of a packed bed absorber for $CO_2$ gives the following result for the capacity coefficients (units of mol/ft$^3 \cdot$ h $\cdot$ atm) as a function of the liquid flowrate (lb/ft$^2 \cdot$ h), at 20°C:

$$K_{G,CO_2} a_v = 0.003 \, L^{0.3} \qquad \text{(P7.5)}$$

for $L$ in the range [1000, 10000] lb/ft$^2 \cdot$ h. Estimate $K_G$ for arsine in water as a function of temperature, for $15 < T < 45$°C. In both cases the bed is packed with 1.5-inch ceramic Raschig rings.

**7.6** By analogy to the development that leads to the definition of $H_G$ (Eq. 7.3.10), present a definition of a height of a liquid phase transfer unit $H_L$ based on the overall liquid phase driving force $x^* - x$.

**7.7** Rewrite Eq. 7.7.1 using SI units for all parameters that appear. Replot Fig. 7.7.1 so that it may be used with SI units in Eq. 7.7.1.

**7.8** Repeat Example 7.7.1 and plot $Z_T$ versus liquid flowrate for flowrates in the range of $L = 0.05$ to $0.5$ kg · mol/s · m$^2$.

**7.9** Repeat Example 7.7.1 and plot $Z_T$ versus tower diameter for diameters in the range of 0.1 to 0.3 m.

**7.10** Compare the $H_G$ values predicted by Eqs. 7.7.1 and 7.7.3, for 1-inch Raschig rings, for the system described in Example 7.7.1. Use the parameters of that example, but find $H_G$ as a function of liquid flowrate, for

$$0.1 < L < 1 \text{ kg} \cdot \text{mol/s} \cdot \text{m}^2 \quad \textbf{(P7.10)}$$

**7.11** Benzene is stripped from water with air in a countercurrent packed column. The Henry's law constant for benzene at 25°C is 240 atm. The following set of results is available:

Column height (packed) = 10 ft
Inside diameter of column = 23 inches
Liquid flowrate = 27 gal min$^{-1}$ ft$^{-2}$
Air flowrate = 1620 gal min$^{-1}$ ft$^{-2}$
Benzene concentration entering = 40,000 $\mu$g/L
Benzene concentration leaving = 3700 $\mu$g/L

Find the mass transfer coefficient $K_L a (\text{s}^{-1})$.

**7.12** For the column described in Problem 7.11, find the height required to remove 99.9% of the benzene.

**7.13** An air/ammonia mixture ($y_{NH_3} = 0.009$) enters a packed bed countercurrent scrubber at 450 ft$^3$/min (68°F and 1 atm). The bed has a diameter of 6 inches. Pure water enters the top of the scrubber at 29 lb$_m$/min. Assume $K_y a = 4.5$ lb · mol/min · ft$^3$. Estimate the tower height required to reduce the ammonia mole fraction to $y_{NH_3} = 0.00036$. Take the equilibrium relationship to be $y = 1.25x$ for the NH$_3$/H$_2$O system.

**7.14** A gas containing 1.3 mol % CO$_2$ enters the bottom of a packed tower of cross-sectional area 0.90 m$^2$, at a flowrate of 2.5 g · mol/s. The gas is to leave containing only 0.04% CO$_2$. Absorption is accomplished by countercurrent flow of an organic liquid at 15°C. The liquid flow is 0.5 g · mol/s. The entering gas (1.25% CO$_2$) would

be in equilibrium with the organic solution if the liquid contained 7.5 mol % CO$_2$. The capacity coefficient $K_G a_v$ has been found to be $10^{-6}$ mol/s · cm$^3$ · atm.

**a.** What is the required height of the tower?

**b.** What is the minimum liquid flowrate for this operation?

**7.15** Suppose $k_L = D_L/(0.01 \text{ cm})$ and $k_c = D_G/(0.01 \text{ cm})$, where

$$k_L \equiv N/(c_{i,L} - c) \quad \text{and} \quad k_c \equiv N/(c - c_{i,G}) \quad \textbf{(P7.15.1)}$$

$k_L$ is a liquid side coefficient using liquid concentrations, and $k_c$ is a gas side coefficient, using gas side concentrations, in molar units ($c$ has units of moles per cubic centimeter).

An overall coefficient $K_L$ is defined by

$$N = K_L(c^* - c) \quad \textbf{(P7.15.2)}$$

where

$$c^* = \frac{p}{H} \quad \textbf{(P7.15.3)}$$

Given a Henry's law constant of $4.4 \times 10^4$ atm, $D_G = 0.23$ cm$^2$/s, and $D_L = 2.1 \times 10^{-5}$ cm$^2$/s, find the value of $K_L$ in centimeters per second.

**7.16** Sherwood and Holloway (see Chapter 18 of *The Chemical Engineer's Handbook*, 5th ed) recommend the following expression for the calculation of $H_G$ values in aqueous systems:

$$H_G = 1.01 G'^{0.31} L'^{-0.33} \quad \textbf{(P7.16)}$$

where both flowrates are in units of pounds per hour–foot squared and $H_G$ has units of feet. Use this equation to predict the data shown in Fig. 7.7.2.

**7.17** A set of data for the absorption of ammonia in water shows the dependence of $H_G$ on the liquid flowrate in a tower packed with 1-inch Raschig rings. (Wen, as reported in Chapter 18 of *The Chemical Engineer's Handbook*, 5th ed., Fig. 18.80). These data are replotted in Fig. P7.17. Predict these data, using Eqs. 7.7.3 and P7.16.

**7.18** Sherwood and Holloway [*Trans. AIChE.,* **36**, 39 (1940)] correlated data for the desorption of oxygen from water in packed towers and found

$$\frac{k_L a_v}{D_{AB}} = \alpha \left(\frac{L'}{\mu}\right)^{1-n} \text{Sc}^{0.5} \quad \textbf{(P7.18.1)}$$

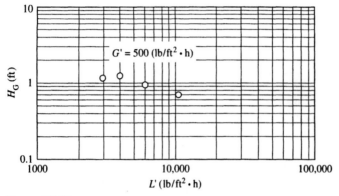

**Figure P7.17**

when the units are

$k_L a_v$ [=] lb mol/h·ft$^3$·(lb mole/ft$^3$)
$L'$ [=] lb/h·ft$^2$
$\mu$ [=] lb/h·ft

and $D_{AB}$ [=] ft$^2$/h is the diffusion coefficient of oxygen in water.

For 1-inch ceramic Raschig rings, these authors give $\alpha = 100$ and $n = 0.22$.

**a.** Convert Eq. P7.18.1 to SI units. What is the value of $\alpha$ if SI units are used? Use the *gram-mole* instead of the kilogram-mole.)
**b.** Assuming no gas phase resistance, write Eq. P7.18.1 in terms of the height $H_L$:

$$H_L \equiv \frac{L'}{k_L a_v \rho_L} \qquad \text{(P7.18.2)}$$

# Chapter **8**

# Membrane Transfer and Membrane Separation Systems

**In this chapter we examine membrane devices as systems for achieving separations of components from a mixture. After developing some relatively simple models of mass transfer rates and separation efficiency, based on fundamental principles, we derive some design equations for membrane systems of several different types and illustrate their applications in a series of design problems.**

## 8.1  INTRODUCTION

We have already treated several examples of membrane-mediated mass transfer. In a very general sense, a membrane is a semipermeable barrier that selectively permits the transport of certain species while blocking or retarding the transport of others. As a consequence, a membrane provides a means of selectively separating components from a fluid. For the most part, the analysis of membrane systems is based on ideas already introduced in the text. Nevertheless, the growing importance of membrane technology justifies the allocation of a separate chapter to a discussion of the principles of membrane separations, and to the presentation of examples of the design of membrane separation systems.

A generic view of a membrane separation system is shown in Fig. 8.1.1. A feed solution is in contact with a membrane across which the solvent may pass preferentially with respect to some solute. Thus the solute is rejected to a certain degree, and the solution that permeates the membrane is "purified" in the sense that it contains less solute than the feed. The development of membrane separation technology has required the cooperative efforts of the polymer scientist (for the synthesis of new polymeric membrane materials with special physical and chemical properties), the polymer process engineer (for the development of techniques for the large-scale production of membrane materials), and the chemical engineer (for the design and analysis of the membrane separation systems that utilize these membranes in compact efficient units).

Membrane systems are available in a variety of forms, and an excellent reference for

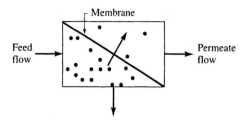

**Figure 8.1.1** Schematic of a membrane separation system: solution passes across the membrane freely with respect to the solute, resulting in a reduced solute concentration for the permeate. Solid dots represent solute molecules.

further study of this topic is Rautenbach and Albrecht (*Membrane Processes,* Wiley, 1989). We will consider just a few examples of membrane systems in this chapter.

## 8.2 MASS TRANSFER ANALYSES OF SOME MEMBRANE PHENOMENA

Let's begin by considering a simple batch, well-stirred tank containing an aqueous binary solution. We denote the concentration of the *dilute* component as $C_1$ (moles/volume). One portion of the surface of the tank is a membrane, as shown in Fig. 8.2.1. A positive pressure difference $\Delta P$ is imposed across the membrane. If the membrane is permeable to water (the solvent) and the solute, there will be a flux of solution across the membrane. In a sense, the membrane leaks. If the membrane has some *selectivity,* the solution passing through (called the "permeate") will have a composition different from that of the solution on the high pressure side of the membrane.

The most common observation is that the permeate flux is given by

$$N_{\text{permeate}} = N_{\text{B}} = \mathscr{P}\Delta P_{\text{eff}} \qquad (8.2.1)$$

where $\mathscr{P}$ is called the membrane permeability coefficient. We will look at its units in a moment.

Many of the solutions that are treated by a membrane process have solutes with a significant osmotic pressure. Hence there is a tendency for the solute to move across the membrane under the action of the osmotic pressure, and the "effective" pressure difference that appears in Eq. 8.2.1 is defined as

$$\Delta P_{\text{eff}} = \Delta P - [\Pi(x_1 C_1) - \Pi(x_3 C_3)] \qquad (8.2.2)$$

where $C_1$ is the molar concentration of the solution and $x_1$ is the mole fraction of the solute (and similarly for $x_3 C_3$). The osmotic pressure, denoted by $\Pi$, is a function

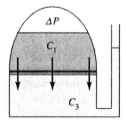

**Figure 8.2.1** Schematic of a batch membrane system.

of the concentration of the solute. For the common case of dilute solutions, $C_1 = C_3 = C$, which is the molar concentration of the solvent. For water, for example, $C \approx 1/18 = 0.055$ mol/cm$^3$.

Note that in the absence of an externally imposed pressure difference, there would be a flow of solution arising from the osmotic pressure difference between the solutions on the two sides of the membrane. This is the usual osmotic flow discussed in most standard physical chemistry texts. To achieve a permeate flow from the high solute concentration side to the low solute concentration side, it is necessary to impose an external pressure greater than the net osmotic pressure. Since this reverses the direction of the naturally occurring osmotic flow, we speak of this situation as "reverse osmosis."

If the membrane is semipermeable, and if the solute is retarded relative to the solvent, we say that the membrane is characterized by a "rejection coefficient" $r$, defined in terms of the compositions of the solute on each side of the membrane:

$$x_3 = (1 - r)x_1 \tag{8.2.3}$$

As $r$, the rejection coefficient of the membrane for the particular solvent/solute pair, approaches unity, we get nearly pure solvent as the permeate (since $x_3$ gets very small).

---

**EXAMPLE 8.2.1** *Calculation of the Pure Water Permeability*

In a membrane test cell containing pure water, water permeates the membrane at a rate of 0.05 gal/h at a pressure difference of 100 psi, with a membrane area of 0.25 ft$^2$. Find the permeability coefficient $\mathscr{P}$. For this case, with no solute,

$$N_B = \mathscr{P}\Delta P_{eff} = \mathscr{P}\Delta P \, (\text{mol/h} \cdot \text{ft}^2) \tag{8.2.4}$$

(Note that the permeate flux is written here in *molar* units, although the data are often presented in volumetric units, as in this example.) We first must convert gallons of water to moles.

Pure water and dilute aqueous solutions have the molar concentration (taking the density of water as 1 g/cm$^3$):

$$C = 1/18 = 0.055 \text{ g–mol/cm}^3 \tag{8.2.5}$$

Then

$$1 \text{ gallon} = 3.8 \text{ liters} = 3800 \text{ grams} = \frac{3800}{18} = 211 \text{ gram-moles} \tag{8.2.6}$$

The observed flow of 0.05 gal/h across 0.25 ft$^2$ of membrane, substituted into Eq. 8.2.4, leads us to find

$$\mathscr{P} = \frac{N_B}{\Delta P} = \frac{0.05 \times 211 \text{ g–mol}}{0.25 \text{ ft}^2 \times 100 \text{ psi}} = 0.42 \text{ (g–mol/h} \cdot \text{ft}^2 \cdot \text{psi)} \tag{8.2.7}$$

The units used here are hybrid (a mixture of British and cgs). It is actually commonplace to express the permeability as the number of gallons of water per square foot *per day* at some arbitrary pressure. For example, this membrane produces

$$\frac{0.05 \text{ gal/h}}{0.25 \text{ ft}^2} \times 24 \text{ h/day} = 4.8 \text{ GFD (@ 100 psi)} \tag{8.2.8}$$

The usual nomenclature is GFD for "gallons per square foot per day," and the pressure at which that flux is achieved is given in parentheses, as above.

A common assumption, based on observation, is that the pure water flux is propertional to the applied pressure. Hence, at a higher pressure, say 250 psi, we would "rate" this membrane by giving the permeability parameter as

$$\frac{\mathscr{P}}{C} = 4.8 \text{ GFD (@ 100 psi)} \times \frac{250}{100} = 12 \text{ GFD (@ 250 psi)} \qquad \textbf{(8.2.9)}$$

### EXAMPLE 8.2.2   Rejection Coefficient of a Membrane

The membrane of Example 8.2.1 is used to desalt a 1 wt% NaCl solution, and the following observations are made.

At 200 psi the permeate flow is 0.04 gal/h for a 0.25 ft$^2$ membrane, and the permeate has a salt concentration of $x_3 = 0.02$ wt%. What is the rejection coefficient of the membrane?

Using Eq. 8.2.3 we find

$$1 - r = \frac{x_3}{x_1} = \frac{0.02}{1} = 0.02 \qquad \text{or} \qquad r = 0.98 \qquad \textbf{(8.2.10)}$$

(Note that we have used "wt%" as the concentration unit here, instead of molar concentration. Since Eq. 8.2.3 gives a *ratio* of concentrations, we do not have to convert any units to another set.)

It is interesting to see whether we achieved the expected permeate flow with this example. With pure water we found 0.05 gal/h at a pressure difference of 100 psi, for the same membrane. At 200 psi we would expect twice this flow, since the flow is found to be linear with the imposed pressure in this range of pressures. However, we must correct for the osmotic pressure effect, since we have a "feed" solution with 1% salt.

We can find data on osmotic pressure as a function of salt concentration, and Fig. 8.2.2 shows such data. At a concentration of 1 wt%, NaCl has an osmotic pressure of 115 psi. From the linearity of the data, we can see that the permeate has an osmotic pressure of $(0.02/1) \times 115 = 2.3$ psi. Hence, from Eq. 8.2.2, the *expected* permeate flow is actually not twice 0.05 gal/h but considerably less:

$$Q = 0.1 \times \frac{200 - 115 + 2.3}{200} = 0.0437 \text{ gal/h} \qquad \textbf{(8.2.11)}$$

The observed flow of 0.04 gal/h is about 10% less than this prediction. We will see (in an analysis presented in the next section) that this lower rate is due to the rejection of salt on the high pressure side of the membrane, as a consequence of which the salt

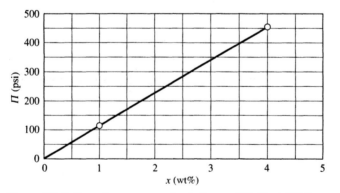

**Figure 8.2.2** Osmotic pressure of solutions of NaCl at 25°C.

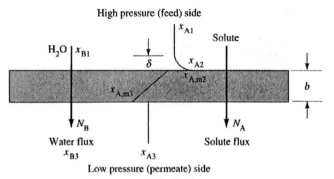

High pressure (feed) side

Water flux
Solute flux

Low pressure (permeate) side

**Figure 8.3.1** Transport of a binary solution across a membrane.

concentration on that side, *right at the membrane surface,* is higher than that of the bulk of the fluid. This phenomenon of salt buildup is called "concentration polarization," and we must now turn to a mass transfer analysis to predict the extent to which it occurs, and its effect on membrane transport. We note at this point, however, that the rejection coefficient calculated from Eq. 8.2.10 is only an "apparent" rejection coefficient, because in the presence of polarization, the concentration right at the membrane surface, on the high pressure side, is not $x_1$ but a larger value $x_{A2}$ (see Fig. 8.3.1).

## 8.3  A SOLUTION–DIFFUSION MODEL OF STEADY MEMBRANE TRANSPORT

Figure 8.3.1 depicts the situation we are going to study. We assume that the membrane is permeable to the solvent (species B). We assume that the solute (species A) is soluble in the membrane and that it diffuses through the membrane with a diffusion coefficient $D_{A,m}$.

Using the notation of Fig. 8.3.1 we write Eq. 8.2.2 in the form

$$\Delta P_{\text{eff}} = \Delta P - [\Pi(x_{A2}) - \Pi(x_{A3})] \tag{8.3.1}$$

We assume a linear relationship between osmotic pressure and solute concentration:

$$\Pi(x_A) = \beta x_A \tag{8.3.2}$$

The solute is assumed to diffuse across (i.e., within) the membrane according to Fick's law for dilute diffusion through a solid:

$$N_A = \frac{D_{A,m}}{b}(C_{m2}x_{A,m2} - C_{m3}x_{A,m3}) \tag{8.3.3}$$

where $C_m$ is the molar density of species A in the solid (membrane) phase, and $x_{A,m}$ is the mole fraction of species A in the membrane phase. We assume that species A partitions itself between the aqueous and the membrane phases according to a linear solubility relationship, which we write in the form

$$Cx_A = KC_m x_{A,m} \tag{8.3.4}$$

where $C$ is the molar density of the liquid phase.

Now Eq. 8.3.3 takes the form

$$N_A = \frac{D_{A,m}}{Kb}(C_2 x_{A2} - C_3 x_{A3}) \tag{8.3.5}$$

(For dilute aqueous phases on both sides of the membrane, $C_2 = C_3 = C = 0.055$ g–mol/cm$^3$).

The mole fraction of species A appearing on the low pressure (permeate) side is $x_{A3}$. A simple molar flux balance gives

$$N_A = x_{A3}(N_A + N_B) \tag{8.3.6}$$

or

$$N_A = \frac{x_{A3}N_B}{1 - x_{A3}} \approx x_{A3}N_B \tag{8.3.7}$$

(The indicated approximation, if we choose to use it, follows for the case that $x_{A3} \ll 1$.)

Then (using Eqs. 8.3.5 and 8.3.7 without the indicated approximation), the permeate flux of the solvent becomes

$$N_B = \frac{D_{A,m}}{Kb} \frac{1 - x_{A3}}{x_{A3}} (C_2 x_{A2} - C_3 x_{A3}) \tag{8.3.8}$$

Now we want to examine the concentration of solute in the near neighborhood of the membrane, on the high pressure side. (See Fig. 8.3.1.)

We imagine that the solution on the high pressure side is well mixed to within a boundary layer of thickness $\delta$. The magnitude of $\delta$ will be estimated from a knowledge of the hydrodynamics on the high pressure side of the membrane. The flux of solute (species A) in the region $0 < z < \delta$ may be written as the sum of the convective and diffusive fluxes:

$$
\begin{aligned}
N_A &= x_A(N_A + N_B) - D_{A,s}C_1 \frac{dx_A}{dz} \\
&= N_A \text{ on the low pressure side} \\
&= x_{A3}(N_A + N_B)
\end{aligned}
\tag{8.3.9}
$$

Note that this is a flux through the boundary layer, *in the liquid,* and so the diffusion coefficient that appears here ($D_{A,s}$) is that for the solute through the solution, not through the membrane!

We write this differential equation for $x_A$ in the form

$$\frac{dx_A}{dz} - x_A \frac{N_A + N_B}{D_{A,s}C_1} = -\frac{x_{A3}(N_A + N_B)}{D_{A,s}C_1} \tag{8.3.10}$$

subject to boundary conditions

$$
\begin{aligned}
x_A &= x_{A1} && \text{at } z = 0 \\
x_A &= x_{A2} && \text{at } z = \delta
\end{aligned}
\tag{8.3.11}
$$

Note that $x_{A1}$ is the concentration of solute in the bulk fluid on the high pressure side, and $x_{A2}$ is the unknown concentration of solute right at the membrane surface. In this analysis, we are after $x_{A2}$ because this concentration, through its osmotic pressure, can strongly affect the performance of the membrane. Physically, we are examining the extent to which the solute "piles up" at the membrane surface because of its rejection by the membrane. This phenomenon is called "concentration polarization."

Returning to Eq. 8.3.10, we may show that the solution to the differential equation takes the form

$$x_A = a_1 + a_2 \exp\left(\frac{N_A + N_B}{D_{A,s}C_1} z\right) \tag{8.3.12}$$

After applying the boundary conditions, we find the concentration at $z = \delta$ to be

$$x_{A2} = x_{A3} + (x_{A1} - x_{A3}) \exp\left(\frac{N_A + N_B}{D_{A,s}C_1}\delta\right)$$  (8.3.13)

We may rearrange this solution into a format that will be useful shortly:

$$\frac{N_A + N_B}{kC_1} = \ln\frac{x_{A2} - x_{A3}}{x_{A1} - x_{A3}}$$  (8.3.14)

where

$$k = \frac{D_{A,s}}{\delta}$$  (8.3.15)

is a convective mass transfer coefficient at the high pressure side of the membrane surface.

The flux of solute (the retarded species) is generally much less than the flux of solvent:

$$N_A \ll N_B$$  (8.3.16)

In what follows, we will take $C = C_1 = C_2 = C_3$. From Eq. 8.3.7, making the approximation that $x_{A3} \ll 1$, we find

$$x_{A3} = \frac{N_A}{N_B}$$  (8.3.17)

The same approximations let us write Eq. 8.3.14 in the form

$$N_B = kC \ln\frac{x_{A2} - x_{A3}}{x_{A1} - x_{A3}}$$  (8.3.18)

Using Eq. 8.3.5 for $N_A$, and Eq. 8.3.18 for $N_B$, we may convert Eq. 8.3.17 into an expression for the relationship of $x_{A2}$ and $x_{A3}$:

$$x_{A3} = \frac{x_{A2}}{1 + \dfrac{k}{D_{A,m}/Kb}\ln\left(\dfrac{x_{A2} - x_{A3}}{x_{A1} - x_{A3}}\right)}$$  (8.3.19)

At the same time, Eqs. 8.3.3 and 8.3.4, with Eq. 8.2.3 (and 8.2.2 and 8.3.2) permit us to write Eq. 8.3.17 as a second relationship among these composition variables:

$$x_{A3} = \frac{D_{A,m}C}{Kb\mathscr{P}\Delta P}\frac{x_{A2} - x_{A3}}{1 - (\beta/\Delta P)(x_{A2} - x_{A3})}$$  (8.3.20)

(In writing Eq. 8.2.2, we replace $x_1$ by $x_2$, since this is the concentration on the high-pressure side of the membrane, at the membrane surface.)

We may define a permeate velocity as

$$v_p = \frac{N_B}{C} = \frac{\mathscr{P}\Delta P}{C}\left[1 - \frac{\beta}{\Delta P}(x_{A2} - x_{A3})\right]$$  (8.3.21)

Before we move too far forward, we need to review what we have achieved. Equations 8.3.19, 8.3.20, and 8.3.21 are three independent equations in the unknown compositions $x_{A2}$ and $x_{A3}$, and the permeate flow $v_p$. We assume that $x_{A1}$ is known independently, from feed water conditions.

A limiting case of interest is that in which there is no osmotic effect on the permeate side ($x_{A3} = 0$) and no buildup of rejected solute on the high pressure side of the membrane surface ($x_{A2} = x_{A1}$). Then the maximum permeate water velocity would be

$$v_p^0 = \frac{\mathcal{P}\Delta P}{C}\left(1 - \frac{\beta x_{A1}^0}{\Delta P}\right) \tag{8.3.22}$$

In a typical design problem $x_{A1}^0$ and $\Delta P$ are specified, and the physical properties ($C$ and $\beta$) are known. We seek the following performance variables:

1. The permeate flux, or $v_p$.
2. The permeate quality, or $x_{A3}$.

The degree of polarization is unknown and comes out of the analysis.

To facilitate the solution of a design problem, we will now nondimensionalize Eqs. 8.3.18 to 8.3.21. We normalize all mole fractions using the input mole fraction $x_{A1}^0$:

$$c_1 \equiv \frac{x_{A1}}{x_{A1}^0} \qquad c_2 \equiv \frac{x_{A2}}{x_{A1}^0} \qquad c_3 \equiv \frac{x_{A3}}{x_{A1}^0} \tag{8.3.23}$$

We define three dimensionless characteristic parameters:

$$\gamma \equiv \frac{\beta x_{A1}^0}{\Delta P} = \frac{\Pi(x_{A1}^0)}{\Delta P} \qquad \theta \equiv \frac{C}{\mathcal{P}\Delta P}\frac{D_{A,m}}{Kb} \tag{8.3.24}$$

$$\lambda \equiv \frac{k}{D_{A,m}/Kb}$$

Now we may write Eqs. 8.3.18 to 8.3.21 in the following forms:

$$c_3 = \frac{c_2}{1 + \lambda \ln \dfrac{c_2 - c_3}{c_1 - c_3}} \tag{8.3.25}$$

$$c_2 = \frac{c_3(1 + \theta + \gamma c_3)}{\theta + \gamma c_3} \tag{8.3.26}$$

$$\frac{v_p}{v_p^0} = \frac{1 - \gamma(c_2 - c_3)}{1 - \gamma} \tag{8.3.27}$$

Given the physical property data, we may solve these three equations for the performance variables. For small permeate flow, relative to the feed flow, we may take $c_1 = 1$. Equations 8.3.25 to 8.3.27 are three coupled nonlinear algebraic equations that can be solved numerically by a trial-and-error procedure. An alternative graphical procedure will be described later.

Before we proceed further, we should comment on the procedure used for determining the coefficients that appear in Eqs. 8.3.25 to 8.3.27. For a given binary solution, the osmotic coefficient $\beta$ is assumed to be known. Hence, under a given set of operating conditions ($\Delta P$ and $x_{A1}^0$), the coefficient $\gamma$ is known.

If the osmotic coefficient is not known (we may have a solution of unknown chemistry), we may still find $\gamma$. First, the membrane is operated with pure water (or another solvent for a nonaqueous system), and from the measured permeate rate, Eq. 8.3.22 (with $\beta = 0$ for a solute-free liquid) yields the coefficient $\mathcal{P}\Delta P/C$. Then if we test the membrane in a well-stirred test cell, so that there is no polarization, Eq. 8.3.22 yields $\gamma$ from a measurement of $v_p^0$.

Next, we may perform an experiment in which polarization is eliminated by stirring, so that $c_2 = c_1$, and further we may hold $c_1 = 1$. Equation 8.3.26 takes the form

$$c_3 = \frac{\theta(1 - c_3)}{1 - \gamma(1 - c_3)} \tag{8.3.28}$$

By measuring $c_3$, and knowing $\gamma$, we may find $\theta$.

To find $\lambda$ from experimental data, we rearrange Eq. 8.3.25 into the form

$$\lambda = \frac{c_2 - c_3}{c_3 \ln \dfrac{c_2 - c_3}{c_1 - c_3}} \tag{8.3.29}$$

Measurements of all three concentrations, under a given set of experimental conditions, yields the coefficient $\lambda$. Since $\lambda$ includes the convective mass transfer coefficient, we would expect to find that $\lambda$ is a function of the flowrate and the detailed hydrodynamics of the solution on the high pressure side of the membrane.

**EXAMPLE 8.3.1**   *Coefficients for the Membrane of Examples 8.2.1 and 8.2.2*

Our goal is to find values for the coefficients that characterize the membrane described in Examples 8.2.1 and 8.2.2.

We find a value for $\gamma$ immediately as

$$\gamma = \frac{\Pi(c_1)}{\Delta P} = \frac{115}{200} = 0.575 \tag{8.3.30}$$

First we must find the polarization, $c_2$. From the data given earlier we know that $c_3 = 0.02$.

In the statement of Example 8.2.1 we find that we get 0.05 gal/h at 100 psi with pure water. Hence, at 200 psi (the pressure of Example 8.2.2) we expect 0.1 gal/h of pure water. With the feed solution of 1% salt, the observed flow is 0.04 gal/h. Hence

$$\frac{v_p}{v_p^o} = \frac{0.04}{0.10} = 0.4 \tag{8.3.31}$$

From Eq. 8.3.27 we find

$$\frac{v_p}{v_p^o} = 0.4 = \frac{1 - 0.575\,(c_2 - 0.02)}{1 - 0.575} \tag{8.3.32}$$

The solution of this equation for $c_2$ gives the polarization as

$$c_2 = 1.46 \tag{8.3.33}$$

From Eq. 8.3.29 we now find $\lambda$ as

$$\lambda = \frac{1.46 - 0.02}{0.02 \ln \dfrac{1.46 - 0.02}{1 - 0.02}} = 187 \tag{8.3.34}$$

We find $\theta$ by rearranging Eq. 8.3.28 to the form

$$\theta = \frac{c_3[1 - \gamma(c_2 - c_3)]}{c_2 - c_3} = \frac{0.02[1 - 0.575(1.46 - 0.02)]}{1.46 - 0.02} = 2.39 \times 10^{-3} \tag{8.3.35}$$

**EXAMPLE 8.3.2**   *Mass Transfer Coefficient for the Membrane*

From the data given in Examples 8.2.1 and 8.2.2, we may also find the convective mass transfer coefficient on the high pressure side of the membrane. We already have a value for $\lambda$ from Eq. 8.3.34. From the definition of $\lambda$ the convective mass transfer coefficient is given by

$$k = \lambda \frac{D_{A,m}}{Kb} \tag{8.3.36}$$

while

$$\frac{D_{A,m}}{Kb} = \frac{\mathscr{P}\Delta P}{C} \theta \tag{8.3.37}$$

The pure water (hence $\gamma = 0$) flow at 200 psi is given as

$$Q_o = a_m v_p^o = \frac{\mathscr{P}\Delta P}{C} a_m = 0.1 \text{ gal/h} \tag{8.3.38}$$

where $a_m$ is the membrane area ($= 0.25 \text{ ft}^2$). Hence we find

$$\frac{\mathscr{P}}{C} = \frac{Q_o}{a_m \Delta P} = \frac{0.1}{200 \times 0.25} = 0.002 \frac{\text{gal/h}}{\text{psi/ft}^2} \tag{8.3.39}$$
$$= 2.66 \times 10^{-4} \text{ ft/h} \cdot \text{psi}$$

Using the $\theta$ value found earlier, we calculate the desired parameter from Eq. 8.3.37, with the result

$$\frac{D_{A,m}}{Kb} = \frac{\mathscr{P}\Delta P}{C} \theta = (2.66 \times 10^{-4})200 \times 0.00239 = 1.27 \times 10^{-4} \text{ ft/h} \tag{8.3.40}$$

and, from Eq. 8.3.36,

$$k = \lambda \frac{D_{A,m}}{Kb} = 187 \times 1.27 \times 10^{-4} = 0.0238 \text{ ft/h} \tag{8.3.41}$$

Note that in Example 8.3.2 there is no information about the hydrodynamics on the high pressure side of the membrane. We simply calculate the mass transfer coefficient from the observed performance of the system. In Example 8.3.3 we will specify the hydrodynamics, estimate the convective coefficient from a correlation discussed earlier, and predict the performance of the system.

### EXAMPLE 8.3.3  *Performance of a Tubular Membrane*

The membrane of the preceding examples is used in a tubular configuration to create a 1-inch inside diameter, 10-foot-long membrane unit. Water containing 1 wt% NaCl is pumped through the tube at a rate of 6 gal/min.

The operating pressure is 600 psig. If we assume that there is a negligible pressure loss along the axis of the tube, due to friction losses, we may assume that the *transmembrane* pressure difference is $\Delta P = 600$ psi.

a. What fraction of the entering flow is produced as permeate? (This is called the "fractional recovery.")
b. What is the "quality" of the permeate (i.e., what is $c_3$)?

We will continue our use of British units here, since these units are used so commonly in the existing literature of the membrane industry. From Example 8.3.2 we have

$$\frac{\mathscr{P}}{C} = 0.002 \frac{\text{gal/hr}}{\text{psi} \cdot \text{ft}^2} \tag{8.3.42}$$

and

$$\frac{D_{A,m}}{Kb} = 1.27 \times 10^{-4} \, \text{ft/h} = \frac{1.27 \times 10^{-4}}{0.133} = 9.6 \times 10^{-4} \, \text{gal/h} \cdot \text{ft}^2 \tag{8.3.43}$$

This yields

$$\theta = \frac{C}{\mathscr{P}\Delta P} \frac{D_{A,m}}{Kb} = \frac{9.6 \times 10^{-4}}{0.002 \times 600} = 8 \times 10^{-4} \tag{8.3.44}$$

The osmotic pressure of 1% NaCl is 115 psi, so

$$\gamma = \frac{\Pi}{\Delta P} = \frac{115}{600} = 0.19 \tag{8.3.45}$$

We now need a value for

$$\lambda = \frac{k}{D_{A,m}/Kb} \tag{8.3.46}$$

Hence we need an estimate of the convective coefficient $k$.

Assume that the flow through the tubular membrane unit is turbulent. The Reynolds number is

$$\text{Re} = \frac{4Q\rho}{\pi D \mu} \tag{8.3.47}$$

The flowrate of 6 gpm (gal/min) is converted to

$$Q = 6 \, \text{gpm} \times 3800 \, \text{cm}^3/\text{gal} \times \frac{1}{60} \, (\text{min/s}) = 380 \, \text{cm}^3/\text{s} \tag{8.3.48}$$

and with $D = 2.54$ cm and $\mu = 0.01$ poise, we find

$$\text{Re} = 19,000 \tag{8.3.49}$$

To estimate the mass transfer coefficient for turbulent flow inside a circular pipe, we will use the empirical Sieder–Tate equation [*Ind. Eng. Chem.*, **28**, 1429 (1936)]. (See Problem 8.20.)

$$\text{Sh} = 0.023 \, \text{Re}^{0.8} \, \text{Sc}^{1/3} \tag{8.3.50}$$

The Schmidt number for NaCl/water at 25°C is 620. Hence the Sherwood number is found to be

$$\text{Sh} = 0.023 \, (19,000)^{0.8} \, 620^{1/3} = 518 \tag{8.3.51}$$

The Sherwood number is defined as

$$\text{Sh} = \frac{kD}{D_{A,s}} \tag{8.3.52}$$

where $D$ is the inner diameter of the tube and $D_{A,s}$ is the diffusion coefficient of the solute (NaCl) in the *solvent* (water). Using a value of

$$D_{A,s} = 6 \times 10^{-5} \, \text{ft}^2/\text{h} \tag{8.3.53}$$

we find

$$k = 518 \frac{6 \times 10^{-5} \, \text{ft}^2/\text{h}}{0.083 \, \text{ft}} = 0.375 \, \text{ft/h} \tag{8.3.54}$$

and

$$\lambda = \frac{0.375 \text{ ft/h}}{1.27 \times 10^{-4} \text{ ft/h}} = 2950 \tag{8.3.55}$$

Now we may find the concentrations, and the permeate flow, from Eqs. 8.3.25 to 8.3.27. Equations 8.3.25 and 8.3.26 are easily solved by trial and error for $c_2$ and $c_3$. In most cases, the fractional recovery is so small that it is a good approximation to say that $c_1 = 1$ everywhere along the axis of the membrane tube. In that case, $c_2$ and $c_3$ are also constant along the axis. If this is not the case (i.e., if the fractional recovery is not small), the balance equations we have written must be modified. We will examine that problem later.

For the set of parameters in this problem,

$$\lambda = 2950 \qquad \gamma = 0.19 \qquad \theta = 8 \times 10^{-4}$$

we find the solutions to be

$$c_2 = 1.35 \quad \text{and} \quad c_3 = 1.4 \times 10^{-3}$$

Since the feed water is 1% = 10,000 ppm salt, the permeate quality is $1.4 \times 10^{-3} \times$ 1% or

$$x_3 = 14 \text{ ppm}$$

The permeate water flux is

$$v_p = \frac{\mathscr{P}\Delta P}{C}[1 - \gamma(c_2 - c_3)] \tag{8.3.56}$$

$$= (2 \times 10^{-3}) \times 600[1 - 0.19(1.35 - 0.0014)] = 0.89 \text{ gal/h} \cdot \text{ft}^2$$

For a single 1-inch, 10-foot-long tubular membrane unit, the membrane area is

$$a_m = \pi DL = \frac{10 \, \pi}{12} = 2.62 \text{ ft}^2 \tag{8.3.57}$$

Hence the volumetric permeate flow is

$$q_p = v_p a_m = 0.89 \times 2.62 = 2.33 \text{ gal/h} \tag{8.3.58}$$

Since the feed flowrate is 6 gal/min (= 360 gal/h), the fractional recovery $\psi$ is found to be

$$\psi = \frac{2.33}{360} = 6.5 \times 10^{-3} \tag{8.3.59}$$

In this example we find (from Eq. 8.3.59) a fractional recovery that, from a commercial point of view, is very small. Under these conditions, however, the assumption that $c_1 = 1$ throughout the membrane tube is good, since such a small amount of water is removed from the solution flowing down the tube axis.

The rejection coefficient of the membrane was defined in Eq. 8.2.3 (in terms of the mole fractions $x_2$ and $x_3$). For the conditions of this example, the apparent rejection coefficient is found to be

$$r = 1 - \frac{x_3}{x_1} = 1 - \frac{c_3}{c_1} = 1 - \frac{0.0014}{1} = 0.9986 \tag{8.3.60}$$

We call this an "apparent" rejection coefficient because the actual rejection coefficient, when there is polarization, is

$$r = 1 - \frac{x_3}{x_2} = 1 - \frac{c_3}{c_2} = 1 - \frac{0.0014}{1.35} = 0.99896 \tag{8.3.61}$$

The primary performance variables of this reverse osmosis system are the permeate flow $v_p$ and the water quality $c_3$. Trial-and-error or graphical methods are required for their solution from the governing equations. It is useful to have a simple graphical representation of the solutions that can be used for quick estimates of performance. We examine the means to produce such a representation now.

For the permeate flow we may combine Eqs. 8.3.18 and 8.3.21, and use the (usually good) approximation that $c_3 \ll c_2$. It is not difficult to rearrange these two equations into a single implicit equation for $v_p$, in the form

$$\vartheta \equiv \frac{v_p}{v_p^{pw}} = 1 - \gamma \exp\left(\frac{1}{\lambda\theta}\frac{v_p}{v_p^{pw}}\right) = 1 - \gamma \exp\left(\frac{\vartheta}{\lambda\theta}\right) \tag{8.3.62}$$

where the pure water permeation rate is

$$v_p^{pw} = \frac{\mathscr{P}\Delta P}{C} \tag{8.3.63}$$

Thus we may plot $\vartheta$ as a function of $\lambda\theta$, with $\gamma$ as a parameter. Figure 8.3.2 shows the results, for a few values of $\gamma$.

We observe that for large values of the product $\lambda\theta$, the flux function $\vartheta$ approaches

$$\vartheta = 1 - \gamma \quad \text{for} \quad \lambda\theta \gg 1 \tag{8.3.64}$$

This is essentially a thermodynamic limitation to the performance of the system, under conditions of no convective mass transfer limitation.

It is also useful to plot the permeate quality as a function of parameters. By manipulating Eqs. 8.3.25 and 8.3.26 we may find an implicit equation for $c_3$ in the form

$$\frac{1}{\lambda} = (\theta + \gamma c_3) \ln\left[c_3\left(1 + \frac{1}{\theta + \gamma c_3}\right)\right] \tag{8.3.65}$$

(Again, this is subject to the approximations $c_3 \ll c_2$ and $c_1 = 1$.) This result is plotted in Fig. 8.3.3 for the parameters shown. In the limit of large $\lambda\theta$, which implies no convective mass transfer limitation, the permeate quality approaches the value

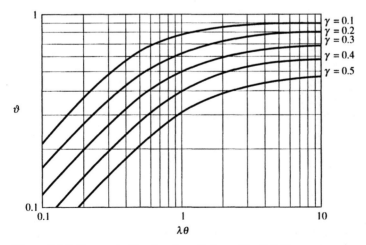

**Figure 8.3.2** Permeate flux function $\vartheta$, from Eq. 8.3.62.

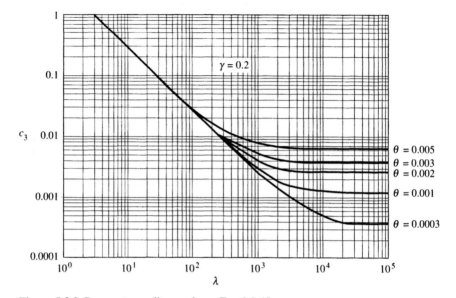

**Figure 8.3.3** Permeate quality $c_3$, from Eq. 8.3.65.

$$c_3 = \frac{\theta}{1 - \gamma} \quad \text{for} \quad \lambda\theta \gg 1 \tag{8.3.66}$$

Equations 8.3.64 and 8.3.66 provide a pair of analytical approximations that are valid under conditions commonly achieved in commercial reverse osmosis operations. Hence they are useful for performance estimates or a quick design at an initial stage.

## 8.4 DESIGN EQUATIONS FOR A TUBULAR MEMBRANE SYSTEM WITH LARGE FRACTIONAL RECOVERY

In the preceding section we carried out an analysis that assumes that $c_1$ remains constant, and in fact that the fraction of solvent (water) removed is so small that $c_1$ remains at the value $c_1 = 1$. In commercial systems the goal is to remove a significant fraction of the feed water as "pure" permeate, and so $c_1$ will change in the direction of flow. Equations 8.3.25 to 8.3.27 now hold "locally," and we must add a pair of equations to account for the loss of solute and solvent along the direction of flow.

For the solute we have

$$-\frac{d}{dz} c_1 Q = v_p a'_m c_3 \tag{8.4.1}$$

where $c_1$, $c_3$, $v_p$, and $Q$ are all functions of position in the direction of flow, and $a'_m$ is the membrane area per unit of axial length. For a tubular membrane, $a'_m = \pi D$.

For the solvent

$$-\frac{d}{dz} Q = v_p a'_m \tag{8.4.2}$$

Since we have added a new unknown variable, $Q$, now we must solve for $c_1$, $c_2$, $c_3$, $v_p$, and $Q$ from the five equations available: Eqs. 8.3.25 to 8.3.27, and 8.4.1, and 8.4.2. There is another "hidden" variable, and that is the transmembrane pressure drop $\Delta P$, which also changes along the flow direction because of friction losses. Thus the parameters $\gamma$ and $\theta$ will vary along the flow direction.

Equations 8.4.1 and 8.4.2 are subject to the initial conditions

$$Q = Q_o \quad \text{and} \quad c_1 = 1 \quad \text{at} \quad z = 0 \qquad (8.4.3)$$

Clearly this set of equations, involving so many interacting variables and parameters, is best solved numerically through an appropriate computer code. The problem is far too complex to permit an illustration of a solution procedure here. We can get some of the flavor of the procedure for evaluating the performance of such a system, however, by considering a simple, though special, case.

We will consider the behavior of a tubular RO system; its operating conditions permit the parameter $\lambda$ to be so large that polarization effects may be neglected. Then $c_1 = c_2$, and Eq. 8.4.1 may be written in the form

$$-\frac{d}{dz} c_1 Q = v_p a'_m c_1 (1 - r) \qquad (8.4.4)$$

where $r$ is the true (not the "apparent") rejection coefficient of the membrane. Now let us regard the permeate flow as a constant, with the average value $v_p$, over some length $Z_1$. With this approximation, Eq. 8.4.2 may be integrated directly to yield, on the range $0 < z < Z_1$,

$$Q = Q_o - v_p a'_m z \qquad (8.4.5)$$

When this result for $Q(z)$ is substituted into Eq. 8.4.4, we obtain a first-order differential equation with separable variables that may be integrated directly to yield the solution

$$c_1 = \left(1 - \frac{v_p a'_m}{Q_o} z\right)^{-r} \qquad (8.4.6)$$

on the range $0 < z < Z_1$.

A simple computational algorithm can be expressed in the following form. We break the membrane unit into a series of identical units of length $Z_1$. At the outlet of the unit that we label $(i + 1)$ we write Eq. 8.4.6 in the form

$$\frac{c_1^{i+1}}{c_1^i} = \left(1 - \frac{v_p^i a'_m}{Q_i} Z_1\right)^{-r} \qquad (8.4.7)$$

The permeate flow $v_p$ is given by Eq. 8.3.27 (with Eq. 8.3.22), which we write with the assumptions that $c_2 = c_1$ (negligible polarization) and $c_3 \ll c_1$ (rejection coefficient near unity). Since the solute concentration is increasing along the direction of flow, the permeate flow will be decreasing, because of the increased osmotic pressure. Note that we use $v_p$ and $Q$ from the $i$th "tube" in calculating $c_1$ exiting the $(i + 1)$th tube. We could improve the precision of the algorithm, which takes $v_p$ and $Q$ as constant in each tube section, by calculating an average concentration over the length $Z_1$ as

$$\overline{c_1^{i+1}} = \frac{c_1^{i+1} + c_1^i}{2} \qquad (8.4.8)$$

and then calculating an average permeate flow from

$$\overline{v_p^{i+1}} = \frac{\mathcal{P}\Delta P}{C}(1 - \gamma \overline{c_1^{i+1}}) \tag{8.4.9}$$

Equations 8.4.7 to 8.4.9 may be manipulated iteratively to yield a value for $c_1^{i+1}$ at the position $Z_1$. Equation 8.4.5 gives a value for $Q = Q_{i+1}$ at the position $Z_1$, but in the form

$$Q_{i+1} = Q_{i+1} - \overline{v_p^{i+1}} a_m' Z_1 \tag{8.4.10}$$

We then continue the procedure into the $(i + 2)$th section to calculate the behavior of the system over another length $Z_1$ down the flow direction. In this way we "march" down the flow direction toward the exit of the membrane system.

In Eq. 8.4.8 an average transmembrane pressure $\Delta P$ is used. The pressure drop $\Delta P_f$ over the length $Z_1$ must be found from a friction factor relationship. If $Q_o$ is specified, we must estimate the total $\Delta P_f$ along the flow direction. This yields a value (an estimate) for the $\Delta P$ to be used in Eq. 8.4.9. This procedure is continued step by step until the end of the system is reached. The transmembrane pressure must match the pressure at the end of the system. If it does not, a new estimate of the total $\Delta P_f$ must be made, and the entire procedure repeated. Under most practical conditions, the pressure drop due to the friction loss is a small fraction of the operating pressure, and we may take $\Delta P$ as constant along the flow direction.

Let's illustrate this calculational algorithm with a simple example.

**EXAMPLE 8.4.1**   *Performance Analysis of a Tubular Membrane System with No Polarization Effects*

A tubular membrane module is used to desalt a 1 wt% NaCl solution. The water is fed at a rate of 250 cm$^3$/s to the system, which is equivalent to a single tube of length 30 m, with an inside diameter of $5/\pi$ cm. The rejection coefficient of the membrane is $r = 0.99$. The membrane is rated at 25 GFD (@ 100 psi). A restricting valve downstream of the module permits control of the inlet pressure at 600 psig. The low pressure (external) side of the membrane module is at atmospheric pressure.

The pure water permeability coefficient is found to be

$$\frac{\mathcal{P}}{C} = 25 \text{ gal/ft}^2 \cdot \text{day} = \frac{25 \times 3800}{(30.48)^2 \, 24 \times 3600} \tag{8.4.11}$$

$$= 1.2 \times 10^{-3} \text{ cm}^3/\text{cm}^2 \cdot \text{s} \text{ (@ 100 psi)} = 1.2 \times 10^{-5} \text{ cm}^3/\text{cm}^2 \cdot \text{s} \cdot \text{psi}$$

We will assume that polarization is minimal under these flow conditions. The parameter $\gamma$ was found earlier, under the same relevant conditions, to be $\gamma = 0.19$.

To minimize the computations we will break the 30 m of tubing into three 10-meter sections. To start the calculational procedure, the permeate flux from the first computational section of the module is

$$v_p^1 = \frac{\mathcal{P}\Delta P}{C}(1 - \gamma) = 1.2 \times 10^{-5} \times 600 \,(1 - 0.19) = 5.83 \times 10^{-3} \text{ cm/s} \tag{8.4.12}$$

For a first approximation we can ignore the small increase of osmotic pressure over the first 10 meters of the unit and consider the permeate flux given in Eq. 8.4.12 to be constant over the whole length. Then at the position $Z_1 = 10$ m the volumetric flowrate is reduced to

$$Q_1 = Q_0 - v_p a_m' Z = 250 - 5.83 \times 10^{-3} \left[ \pi \left( \frac{5}{\pi} \right) 1000 \right] = 221 \text{ cm}^3/\text{s} \tag{8.4.13}$$

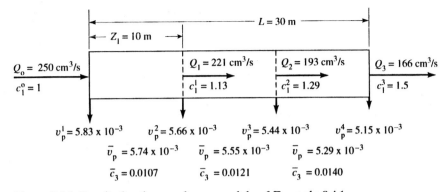

**Figure 8.4.1** Results for the membrane module of Example 8.4.1.

Consistent with this approximation, we use $Z_1 = 10$ m in Eq. 8.4.7. This yields an outlet concentration, on the high pressure side, of

$$c_1^1 = \left(1 - \frac{5.83 \times 10^{-3} \times 5 \times 1000}{250}\right)^{-0.99} = 1.13 \qquad \textbf{(8.4.14)}$$

We could improve the accuracy by using the average concentration (see Eq. 8.4.8) in a second iteration to get new values of $v_p$ and $Q$, as indicated in Eqs. 8.4.9 and 8.4.10. We will not do that in this example. Figure 8.4.1 shows the results for the first third of the module.

It is not difficult to show that the frictional loss of pressure is negligible compared to the transmembrane pressure $\Delta P$. Hence $\Delta P$ remains at 600 psi. In the second section, then, we find[1]

$$v_p^2 = \frac{\mathscr{P}\Delta P}{C}(1 - \gamma c_1^1) = 1.2 \times 10^{-5} \times 600\,(1 - 0.19 \times 1.13) = 5.66 \times 10^{-3}\ \text{cm/s}$$

$$\textbf{(8.4.15)}$$

and

$$\frac{c_1^2}{c_1^1} = \left(1 - \frac{5.66 \times 10^{-3} \times 5 \times 1000}{221}\right)^{-0.99} = 1.15 \qquad \textbf{(8.4.16)}$$

or

$$c_1^2 = 1.13 \times 1.15 = 1.29 \qquad \textbf{(8.4.17)}$$

The exiting flowrate is

$$Q_2 = 221 - (5.66 \times 10^{-3} \times 5 \times 1000) = 193\ \text{cm}^3/\text{s} \qquad \textbf{(8.4.18)}$$

In the third section we find

$$v_p^3 = \frac{\mathscr{P}\Delta P}{C}(1 - \gamma c_1^2) = 1.2 \times 10^{-5} \times 600\,(1 - 0.19 \times 1.29) = 5.44 \times 10^{-3}\ \text{cm/s}$$

$$\textbf{(8.4.19)}$$

---

[1] Remember! The superscripts refer to the axial section over which the calculation is being made. We are not squaring $v_p$.

and

$$Q_3 = 193 - (5.44 \times 10^{-3} \times 5 \times 1000) = 166 \text{ cm}^3/\text{s} \qquad \text{(8.4.20)}$$

Then

$$\frac{c_1^3}{c_1^2} = \left(1 - \frac{5.44 \times 10^{-3} \times 5 \times 1000}{193}\right)^{-0.99} = 1.16 \qquad \text{(8.4.21)}$$

and

$$c_1^3 = 1.16 \times 1.29 = 1.50 \qquad \text{(8.4.22)}$$

The permeate flows in each computational section follow simply from

$$q_p = v_p a_m \qquad \text{(8.4.23)}$$

The results of this simple approximation scheme are given in Figs. 8.4.1 and 8.4.2.

To find the quality of the permeate, we simply use the rejection coefficient ($r = 0.99$) and the calculated values of the concentrations $c_1^i$. Hence we find

$$\overline{c_3^1} = (1 - r)\,\overline{c_1^1} = 0.01\,\frac{1.13 + 1.00}{2}\,10^4 = 0.0107 \qquad \text{(8.4.24)}$$

which corresponds to 107 ppmw. Notice that we have averaged $c_1^i$ between the inlet and the end of the first 10-meter section. Similar calculations over the second and third computational section of the module yield the results plotted in Fig. 8.4.2. The permeate flows yielded by each computational section are easily found as

$$Q_p^i = Q_{i-1} - Q_i \qquad \text{(8.4.25)}$$

with the results

$$\begin{aligned}
Q_p^1 &= 250 - 221 = 29 \text{ cm}^3/\text{s}\\
Q_p^2 &= 221 - 193 = 28 \text{ cm}^3/\text{s} \qquad \text{(8.4.26)}\\
Q_p^3 &= 193 - 166 = 27 \text{ cm}^3/\text{s}
\end{aligned}$$

For the entire module, then, the yield is

$$\phi = \frac{29 + 28 + 27}{250} = 0.34 \qquad \text{(8.4.27)}$$

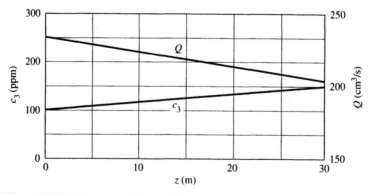

**Figure 8.4.2** Permeate quality and flowrate along the membrane module of Example 8.4.1.

and the total permeate quality follows from

$$\overline{c_3} = \frac{Q_p^1 c_3^1 + Q_p^2 c_3^2 + Q_p^3 c_3^3}{Q_p^1 + Q_p^2 + Q_p^3} = \frac{29\,(107) + 28\,(121) + 27\,(140)}{84} = 122\ \text{ppmw}$$

(8.4.28)

## 8.5  HOLLOW FIBER MEMBRANE SYSTEMS

Another common commercial configuration for a membrane separation system utilizes a bundle of hollow fibers contained in a "shell" through which the feed solution circulates. The fibers have inside diameters of a few tens of micrometers, with wall thicknesses of the same order. Because of the microscopic size of these fibers, it is possible to pack an enormous surface area of membrane into a small volume. Ratios of surface area to volume of the order of $10^4\ \text{m}^2/\text{m}^3$ are typical.

Figure 8.5.1 shows a schematic of the geometry of a single loop of hollow fiber—one of perhaps 10,000 that would make up the fiber bed of a module. The feed flow, under pressure, passes across and along the fiber axes, and the permeate is driven across the fiber (membrane) wall under the positive pressure difference between the shell and fiber sides of the membrane. A densely packed shell might show a significant pressure loss on the shell side, between the inlet and outlet. Under some circumstances, a very slow shell side flow and a lower packing density of fibers could produce a negligible pressure loss in comparison to the shell side inlet pressure. Under these conditions the inlet pressure would serve as a good approximation to the shell side (operating) pressure.

By analyzing the flow into and along the axis of a hollow fiber, we will be able to develop a performance model of this type of membrane separation system. As is usual, in the interest of obtaining an analytical model, we introduce a number of simplifying assumptions. The primary assumptions are the following:

1. The shell side fluid is well mixed with respect to concentration of solute.
2. There is no significant loss of pressure on the shell side.
3. Polarization effects are negligible. At any point along the fiber axis, the permeate and shell side solute concentrations are connected by the rejection coefficient, $r$.
4. Flow in the fibers is laminar and uniaxial. We neglect the radial flow associated with the transmembrane flow.

Figure 8.5.2 shows a schematic of the flow field we will analyze. The midpoint of the loop is taken as a plane of symmetry across which there is no internal flow. This is the position $z = 0$. The length of active (permeable) fiber is $L_a$, and there is a length $L_s$ where the fiber is "potted" into the outlet header. No permeation occurs over this length.

The "pure" water permeability of the fiber is given by

$$N_{\text{permeate}} = N_B = \mathscr{P}\Delta P_{\text{eff}}$$

(8.5.1)

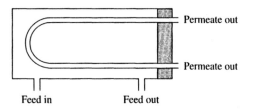

Figure 8.5.1  Schematic of a single loop of a hollow fiber in a pressurized shell.

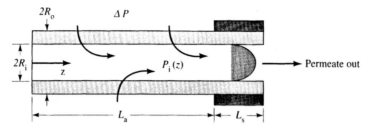

**Figure 8.5.2** Flow into a hollow fiber.

where the effective pressure is

$$\Delta P_{\text{eff}} = \Delta P - P_i(z) - \Pi(x_{1,\text{shell}}C) \qquad \textbf{(8.5.2)}$$

Here we use $\Delta P$ to denote the shell side pressure relative to the ambient, and $P_i(z)$ is the internal fiber pressure (also relative to ambient) that is developed inside the fiber by the viscous friction losses along the fiber axis. We will assume that the rejection coefficient of the membrane is close enough to unity to permit us to neglect the osmotic effect of the permeate side.

We will assume that the flow within the fiber is given locally by Poiseuille's law, and in terms of the volume flowrate $Q(z)$ we write this as

$$Q(z) = -\frac{\pi R_i^4}{8\mu}\frac{dP_i(z)}{dz} \qquad \textbf{(8.5.3)}$$

The internal flowrate increases in the direction of increasing $z$ by virtue of the permeate flow that enters the fiber:

$$\frac{dQ(z)}{dz} = 2\pi R_o v_p \qquad \textbf{(8.5.4)}$$

Note that the permeate velocity $v_p$ is based on the outer surface of the fiber, of radius $R_o$. We continue to define the permeate velocity as in Eq. 8.3.21, so that

$$v_p = \frac{N_B}{C} \qquad \textbf{(8.5.5)}$$

Now we may differentiate Eq. 8.5.3 with respect to $z$ and use the other equations above to yield a second-order ordinary differential equation for the pressure profile along the fiber axis. The result may be written in the form

$$-\frac{\pi R_i^4}{8\mu}\frac{d^2 P_i(z)}{dz^2} = \frac{2\pi R_o \mathscr{P}}{C}[\Delta P - P_i(z) - \Pi(x_{1,\text{shell}}C)] \qquad \text{on } 0 \le z \le L_a \quad \textbf{(8.5.6)}$$

For one boundary condition we have the physical statement that there is no flow at $z = 0$. This, with Eq. 8.5.3, implies

$$\frac{dP_i(z)}{dz} = 0 \qquad \text{at } z = 0 \qquad \textbf{(8.5.7)}$$

At $z = L_a$ the internal pressure is unknown. We denote it as

$$P_i(L_a) \equiv P_a \qquad \text{at } z = L_a \qquad \textbf{(8.5.8)}$$

This is the second boundary condition on Eq. 8.5.6. We now write Eq. 8.5.6 in the form

$$\frac{d^2 P_i(z)}{dz^{*2}} - \kappa^2 [P_i(z) - (\Delta P - \Pi)] = 0 \qquad \textbf{(8.5.9)}$$

by defining

$$\kappa \equiv \frac{4 L_a}{R_i^2} \left( \frac{\mu R_o \mathscr{P}}{C} \right)^{1/2} \quad \text{and} \quad z^* \equiv \frac{z}{L_a} \qquad (8.5.10)$$

[We now use the notation $\Pi(x_{1,\text{shell}} C) = \Pi_{\text{shell}} = \Pi$.]

The general solution to this differential equation may be written as

$$P_i(z) - (\Delta P - \Pi) = a \cosh \kappa z^* + b \sinh \kappa z^* \qquad (8.5.11)$$

From Eq. 8.5.7 we find

$$\frac{dP_i(z)}{dz} = 0 = a \sinh(0) + b \cosh(0) = 0 + b \quad \text{or} \quad b = 0 \qquad (8.5.12)$$

Using Eq. 8.5.8 we write

$$P_a - (\Delta P - \Pi) = a \cosh \kappa \qquad (8.5.13)$$

and so

$$a = \frac{P_a - (\Delta P - \Pi)}{\cosh \kappa} \qquad (8.5.14)$$

Then the solution for the pressure may be written as

$$\frac{P_i(z) - (\Delta P - \Pi)}{P_a - (\Delta P - \Pi)} = \frac{\cosh \kappa z^*}{\cosh \kappa} \qquad (8.5.15)$$

Now we have to evaluate $P_a$. Since no further permeation occurs across the region $z > L_a$, the volume flowrate through that region is constant and identical to $Q$ at $z^* = 1$. Hence Poiseuille's law gives

$$Q_a = \frac{\pi R_i^4}{8 \mu L_s} P_a \qquad (8.5.16)$$

From Eqs. 8.5.3 and 8.5.15 we also have

$$Q_a = -\frac{\pi R_i^4}{8 \mu} \left( \frac{dP_i}{dz} \right)_{z = L_a} = -\frac{\pi R_i^4}{8 \mu L_a} [P_a - (\Delta P - \Pi)] \kappa \tanh \kappa \qquad (8.5.17)$$

We can equate the two $Q_a$ expressions to eliminate the pressure $P_a$, with the result

$$P_a = \frac{(L_s/L_a)(\Delta P - \Pi) \kappa \tanh \kappa}{1 + (L_s/L_a) \kappa \tanh \kappa} \qquad (8.5.18)$$

Then we may introduce this result for $P_a$ into Eq. 8.5.15 and write it in the form

$$\frac{P_i(z^*)}{\Delta P - \Pi} = 1 - \frac{\cosh \kappa z^*}{\cosh \kappa + (L_s/L_a) \kappa \sinh \kappa} \qquad (8.5.19)$$

The permeate flow is simply $Q_p = Q_a$, and from Eqs. 8.5.16 and 8.5.18 we find

$$Q_p = \frac{\pi R_i^4}{8 \mu L_a} (\Delta P - \Pi) \frac{\kappa \tanh \kappa}{1 + (L_s/L_a) \kappa \tanh \kappa} \qquad (8.5.20)$$

If there were no frictional buildup of pressure inside the hollow fiber, the transmembrane pressure $(\Delta P - \Pi_{\text{feed}})$ would be fully available along the permeable length $L_a$ of the fiber, and the product flow would be given by

$$Q_p^o = 2\pi R_o L_a \left(\frac{\mathscr{P}}{C}\right)[\Delta P - \Pi] \qquad (8.5.21)$$

The frictional losses reduce the effectiveness of the system, and we may define an effectiveness factor and calculate it as

$$\eta \equiv \frac{Q_p}{Q_p^o} = \frac{(\tanh \kappa)/\kappa}{1 + (L_s/L_a)\,\kappa \tanh \kappa} \qquad (8.5.22)$$

In the development just completed, we have treated the shell side fluid as if it were well mixed. This permitted us to use $\Pi_{shell}$ as a constant in the integration of Eq. 8.5.6. To know $\Pi_{shell}$, however, we must make an additional solute balance on the system.

The rate at which solute leaves the system in the permeate must balance the loss of solute from the feed solution. We express this statement in the form

$$-Q_{out}Cx_{out} + Q_{feed}Cx_{feed} = n_f Q_p C x_{shell}(1 - r) \qquad (8.5.23)$$

where $n_f$ is the number of fibers in the shell. (Each loop counts as two fibers.) A solution balance may be written as

$$Q_{out} = Q_{feed} - n_f Q_p \qquad (8.5.24)$$

The fractional recovery is defined as

$$\Psi \equiv \frac{n_f Q_p}{Q_{feed}} \qquad (8.5.25)$$

The well-mixed-shell assumption implies that

$$x_{out} = x_{shell} \qquad (8.5.26)$$

As a consequence, we find

$$x_{shell} = \frac{x_{feed}}{1 - \Psi r} \qquad (8.5.27)$$

Hence Eqs. 8.5.20 and 8.5.27 must be solved by trial and error for the two unknowns, $Q_p$ and $\Pi_{shell}$. When $\Psi$ is small, as is often the case, we may take $x_{shell} = x_{feed}$, and Eq. 8.5.20 gives $Q_p$ directly. (In either case, we require Eq. 8.3.2 to relate $\Pi$ to the mole fraction $x$.)

---

**EXAMPLE 8.5.1**   *Performance Characteristics of a Hollow Fiber Membrane*

A solution containing 10 kg of NaCl per cubic meter of solution, is pumped through a hollow fiber module at a flowrate of 100 cm³/s. The shell side is pressurized to 40 bar ($\approx 4 \times 10^6$ Pa). The hollow fiber module contains 10 m² of permeable (external) surface area. The rejection coefficient of the membrane is $r = 0.96$. The fiber parameters are

$$L = 0.2\,\text{m} \qquad L_a = 0.18\,\text{m} \qquad R_i = 10\,\mu\text{m} \qquad R_o = 20\,\mu\text{m} \qquad \frac{\mathscr{P}}{C} = 4 \times 10^{-13}\,\text{m/Pa·s}$$

Using mass concentration units, we may write the osmotic pressure/concentration relationship (Eq. 8.3.2) as

$$\Pi = \beta' \rho_{salt} \qquad (8.5.28)$$

and with density in units of kilograms per cubic meter, $\beta'$ has the value, for NaCl at 25°C, of

$$\beta' = 0.78 \text{ bar} \cdot \text{m}^3/\text{kg} \tag{8.5.29}$$

We will take the viscosity of the permeate flow (which should be nearly pure water) to be

$$\mu = 0.001 \text{ Pa} \cdot \text{s} \tag{8.5.30}$$

We begin by calculating the parameter $\kappa$ defined in Eq. 8.5.10:

$$\kappa \equiv \left(\frac{4L_a}{R_i^2}\right)\left(\frac{\mu R_o \mathscr{P}}{C}\right)^{1/2} = \frac{4(0.18)}{10^{-10}}[10^{-3}(20 \times 10^{-6})(4 \times 10^{-13})]^{1/2} = 0.64 \tag{8.5.31}$$

We may use Eq. 8.5.20, with the tentative assumption that the shell side is at the feed concentration (small recovery), and find the permeate flow for the unit. The fiber area is given as 10 m$^2$ of permeable (external) surface area. Then the number of fibers is found to be

$$n_f = \frac{10}{2\pi R_o L_a} = \frac{10}{2\pi(20 \times 10^{-6})(0.18)} = 4.42 \times 10^5 \tag{8.5.32}$$

and Eq. 8.5.20 (multiplied by $n_f$) yields

$$Q_p = (4.42 \times 10^5)\frac{\pi(10 \times 10^{-6})^4[(40 - 7.8)10^5]}{8(0.001)(0.18)} \times \frac{0.64 \tanh(0.64)}{1 + (0.02/0.18) \times 0.64 \tanh 0.646}$$
$$= 1.7 \times 10^{-5} \text{ m}^3/\text{s} \tag{8.5.33}$$

This corresponds to a recovery of

$$\Psi = \frac{17 \text{ cm}^3/\text{s}}{100 \text{ cm}^3/\text{s}} = 0.17 \tag{8.5.34}$$

Hence a corrected value of the osmotic pressure must be used in a second iteration, since

$$x_{\text{shell}} = \frac{x_{\text{feed}}}{1 - 0.17 \times 0.96} = 1.2x_{\text{feed}} \tag{8.5.35}$$

and the osmotic pressure will be increased by the same factor. We will not pursue the second iteration here (but see Problem 8.12).

An analytical model permits one to explore easily the dependence of the performance of the system on design parameters. For example, we should suspect the existence of an optimum fiber diameter for the design of a hollow fiber module. However, we must be careful here, since we must state clearly what is being held constant when the optimization procedure is carried out. We will address this point in a moment, but first let's see why there might be an optimum fiber diameter.

The major advantage of the hollow fiber system is that the very small fiber diameter yields an enormous area per volume ratio. That is, with very small fibers one can create a compact module with an enormous active area for separation. Thus one is tempted to make hollow fiber units with very small diameter fibers. However, we know from our analysis that the effectiveness of the unit is reduced by the viscous friction that retards the permeate flow in the fiber. These comments suggest that an optimum fiber diameter may exist.

We will explore this optimum under the following constraints of design:

1. The ratio $R_o/R_i$ will be held constant.
2. The "porosity" of the module, that is, the volume fraction $\varepsilon$ of open space through which the feed flow circulates, will be held constant.
3. The ratio $L_a/L_s$ will be held constant, but $L = L_a + L_s$ itself may vary.
4. Membrane physical properties will be held constant.
5. We make the approximation that the fractional recovery is small.

Now we must address the question: What function do we seek an optimum for? A simplistic answer would be that we should optimize (i.e., maximize) the permeate flow. We could do this. However, we note that since the ratio $R_o/R_i$ is being held constant, fibers of different diameters will have different wall thicknesses, and presumably will have different pure water permeability coefficients. Also, the module volume will change, since we are keeping the ratio $R_o/R_i$ and $\varepsilon$ constant while changing the size of the fibers. We will therefore maximize productivity (permeate flow) while keeping module volume and transmembrane pressure constant. Hence the "objective function" that we maximize is

$$\Gamma = \frac{n_f Q_p}{(\Delta P/\mu) V_M} \tag{8.5.36}$$

(The permeate viscosity is used in the definition of $\Gamma$ to make it dimensionless.) We may think of $\Gamma$ as the productivity of the membrane unit, per shell side volume $V_M$, per unit of operating pressure.

Using Eq. 8.5.20 we may write this as

$$\Gamma = \frac{n_f \pi R_i^4 (1 - \Pi_{shell}/\Delta P)}{8 L_a V_M} \frac{\kappa \tanh \kappa}{1 + (L_s/L_a) \kappa \tanh \kappa} \tag{8.5.37}$$

The internal (shell side) module volume, the porosity, and the fiber geometry are all connected by the relationship

$$\varepsilon = 1 - \frac{n_f \pi R_o^2 L_a}{V_M} \tag{8.5.38}$$

which may be used to eliminate $n_f$ from Eq. 8.5.37.

We now write $\Gamma$ in the form

$$\Gamma = \frac{1 - \varepsilon}{2} \left[ 1 - \frac{\Pi_{shell}}{\Delta P} \right] \left( \frac{R_i}{R_o} \right)^{3/2} \left( \frac{\mu \mathscr{P}}{C} \right)^{1/2} \frac{R_i^{1/2}}{L_a} \frac{\tanh \kappa}{1 + (L_s/L_a) \kappa \tanh \kappa}$$

where

$$\kappa = \frac{4 L_a}{L} \left( \frac{\mu \mathscr{P} R_o}{C R_i} \right)^{1/2} \frac{L}{R_i^{3/2}} \tag{8.5.39}$$

In this form we display $\Gamma$ as a function of parameters that will be held constant in the optimization ($\varepsilon$, $\Pi/\Delta P$, $R_i/R_o$, $L/L_a$, $\mu \mathscr{P}/C$), and as a function of $R_i$ and $L$. We can find the optimum fiber radius by examining

$$\left( \frac{\partial \Gamma}{\partial R_i} \right)_{\varepsilon, R_o/R_i, \dots} = 0 \tag{8.5.40}$$

The result will yield a value of $R_i$ that maximizes $\Gamma$, as a function of the module length $L$. As an alternative (which we choose here) we may simply plot $\Gamma$ as a function of $R_i$ with $L$ as a parameter. The results for two choices of $L$ are shown in Fig. 8.5.3. Other parameters are listed in the caption, in SI units for $\mu$ and $\mathscr{P}/C$.) It would appear that shorter modules yield the more efficient operation, where $\Gamma$ is the measure of the efficiency.

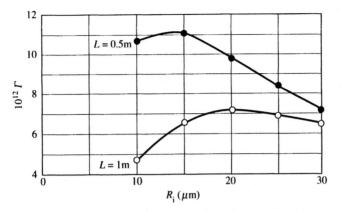

**Figure 8.5.3** The effect of fiber radius on module productivity: $\varepsilon = 0.25$, $\mu = 0.001$, $\Pi/\Delta P = 0.25$, $R_i/R_o = 0.5$, $L/L_a = 1.2$, $L_s/L_a = 0.2$, $\mathcal{P}/C = 4 \times 10^{-13}$.

## SUMMARY

The ideas developed in earlier chapters are applied to the development of design and performance equations for relatively simple membrane separation systems. We restrict attention to systems for which the permeate flux is linearly related to the effective pressure difference. Account is taken of osmotic effects, again through a simple linear model relating osmotic pressure to solute concentration. For the membrane itself, a solution–diffusion model is adopted, and parameters characteristic of such a membrane are defined in Section 8.3. With the assumptions that solute transport to the membrane is controlled by a convective diffusion process and that transport across the membrane itself is diffusive, a set of performance equations is derived. In nondimensional format these are Eqs. 8.3.25 to 8.3.27. A series of examples shows how to calculate some membrane characteristics from performance data. Another example, in Section 8.4, illustrates the use of these performance equations for a quick but approximate analysis and design of a tubular reverse osmosis system.

Viscous losses occurring as the permeate flows in the microscopic tubes of the membrane module impose an additional complication on the design of systems incorporating hollow fiber membranes. A model for the performance of a simple hollow fiber module is presented. With such a model we can consider issues such as optimum design.

## PROBLEMS

**8.1** Reverse osmosis is carried out in a module that consists of a 0.75-inch-i.d. tube, 10 ft long. The feed water is 2% NaCl (by weight). The system is designed to operate at a Reynolds number of 20,000. The inlet pressure is 800 psig. We have the following data:

$$D_{A,m}/Kb = 0.01 \text{ ft/h; Sc} = 620; \text{ membrane}$$
$$\text{"rated" at 10 GFD (@ 250 psi).}$$

**a.** Find the weight fraction of salt in the perme-

ate, the permeate production rate (gal/day), the fractional recovery, and the polarization.

**b.** Suppose the Reynolds number based on the inlet flow varies from $10^4$ to $10^5$. Plot, as a function of Reynolds number, the performance variables listed in part a.

**8.2** The membrane described in Problem 8.1 is available in tube sizes from 0.5 to 2 inches (i.d.). We wish to treat 10,000 gal/day of water containing 1% NaCl (by weight). Inlet pressure will be maintained at 500 psig. Plot recovery, recovery per square foot of membrane surface, and permeate quality versus tube size. Assume all tubes are 10 ft long.

**8.3** A cellulose acetate membrane is to be used for desalination. It has a pure water permeability of $4 \times 10^{-7}$ g–mol/s·cm²·atm and a salt transport parameter of $D_{A,m}/Kb = 6 \times 10^{-6}$ cm/s. We wish to recover 5000 gal/day of potable water containing 400 ppm (by weight) of salt (or less) from 16,700 gal/day of seawater (35,000 ppm salt). The feed pressure is 1500 psig and the temperature is 40°F. Membrane tubes, with a 1 inch "nominal" diameter, 10 ft long, are used in a series arrangement. (These "nominal" 1-inch i.d. tubes have a measured i.d. of 2.174 cm.) The pressure drop due to friction in each 10 ft section is 0.023 atm. The tubes are connected by a short "U-bend" over which there is an additional pressure drop of 0.014 atm.

How many such tubes in series are required? Assume that the salt transport parameter is an average value for the mixture of solutes in seawater. Use the data shown in Fig. P8.3 for osmotic pressure.

**8.4** The system examined in Example 8.3.3 gave a very small recovery. Repeat the analysis with the following changes:

**a.** Increase the flowrate to 60 gal/min, and use 10 tubes in series, each tube identical to that used in the example.

**b.** Repeat part **a** but use 20 tubes in series; each tube is 10 ft long and 0.5 inch in inner diameter. Is the recovery improved in proportion to the increased membrane area available? In both cases, examine the frictional pressure loss and account for it if necessary.

**8.5** Derive Eq. 8.3.66.

**8.6** Polarization effects become small as $\lambda$ becomes large. If we take

$$\frac{v_p}{v_p^{pw}} = 0.95(1 - \gamma) \qquad \textbf{(P8.6.1)}$$

as a criterion, show that a rule of thumb for negligible polarization effects is

$$\frac{\gamma}{\lambda \theta} \le 0.05 \qquad \textbf{(P8.6.2)}$$

**8.7** In Example 8.3.3, what flowrate is required to reduce the osmotic pressure at the inner surface of the membrane to within 10% of the osmotic pressure of the feed water?

**8.8** A $CaCl_2$ solution (4 wt%) must be treated in a reverse osmosis system. The goals are to produce 1000 gal/day of water with less than 50 ppmw solute. Design a simple tubular membrane system to meet these goals. Specify a tube diameter, total length of tubing, and the feed flowrate to meet these goals. At 25°C the solute diffusivity is $1.3 \times 10^{-5}$ cm²/s. Other properties are essentially those of water. Osmotic pressure data are shown in Fig. P8.8.

**8.9** Hollow fiber membranes are packed in a square-pitch array, as shown in Fig. P8.9. Find the ratio of surface area to volume, based on the external area of the fiber.

**8.10** Show that under conditions such that $c_3 \ll c_2$ and $c_1 \approx 1$, the polarization may be calcu-

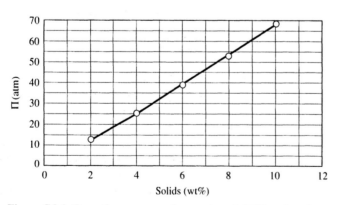

**Figure P8.3** Osmotic pressure of seawater at 40°F and various concentrations.

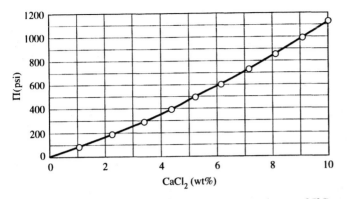

**Figure P8.8** Osmotic pressure of $CaCl_2/H_2O$ solutions at 25°C.

lated from

$$(1 + \lambda \ln c_2)^{-1} = \theta + \frac{\gamma c_2}{1 + \lambda \ln c_2}$$

$$\text{(P8.10)}$$

Plot polarization against wt% NaCl in water (in the range 0.5–3.4%) under the following conditions: $\Delta P = 600$ psig, $\lambda = 3000$, and $\theta = 0.001$. Explain the result.

**8.11** Repeat the analysis of Section 8.5, for the hollow fiber system, but for the case of fibers open at both ends, instead of looped (see Fig. P8.11). Compare $\eta$ (Eq. 8.5.22) for the two systems. Compare $\Gamma$ for the two systems. What conclusions do you draw?

**8.12** Complete Example 8.5.1 by correcting the osmotic pressure and performing a second iteration for the permeate flowrate.

**8.13** A polyimide membrane is being considered as a candidate for separation of $CO_2$ from air. The diffusion coefficients for $CO_2$ and $N_2$

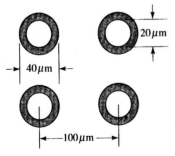

**Figure P8.9**

through the polymer are $0.2 \times 10^{-9}$ and $12 \times 10^{-9}$ cm²/s, respectively, at 25°C. The solubility coefficients (Henry's law constants) are 0.63 and 0.025, respectively, in units of cubic centimeters, at STP, of gas per cubic centimeter of polymer·cmHg partial pressure. Find the relative rates at which $CO_2$ and $N_2$ pass through the polymer, if the partial pressures of $CO_2$ and $N_2$ are 0.1 and 0.79 atm, respectively.

**8.14** Water containing 3.5 wt% sodium chloride flows at a rate of 0.2 L/s through a tubular reverse osmosis membrane. The tube diameter (inside) is 2 cm. The membrane permeability corresponds to a pure water permeate flow of 5 gal/day per square foot of membrane at a pressure differential of 100 psi. Operating temperature is 25°C. The tube is 1 m long. The downstream pressure in the tube is maintained at 650 psia. The membrane has a rejection coefficient of 97%. Find the production rate of permeate water and the water quality.

**8.15** A set of hollow fibers connects two tubes, as shown in Fig. P8.15. Pure nitrogen enters the fibers at a pressure of 20 psia. The pressure in the downstream tube is 15 psia. The set of fibers is surrounded by a large high pressure chamber containing pure oxygen maintained at 200 psia. The oxygen concentration is measured in the flow exiting the set of fibers and is found to be 1 mol %. The temperature is 273 K. The *permeance* of the fibers is defined in terms of the rate of transmission of oxygen through the fiber wall. Specifically, the permeance is the oxygen flux, in cubic centimeters of $O_2$, at STP, per day per square centimeter of *external* fiber surface

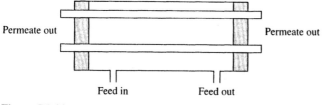

**Figure P8.11**

area, per atmosphere of $O_2$ pressure difference across the fiber wall. All the fibers are 30 cm long, and the inside and outside diameters are 100 and 150 mm, respectively. There are 10,000 fibers in the fiber bundle connecting the two tubes. Find the permeance of the fiber wall.

**8.16** A parallel plate RO system operates under the following conditions. It is fed with brackish water containing 3000 ppm (by weight) sodium chloride. The channel height is $2h = 4$ mm, the width is $W = 20$ cm, and the length is $L = 50$ cm. The feed rate is 20 L/min. The downstream pressure on the high pressure side is controlled at a constant value of 500 psi. The rejection coefficient of the membrane is $r = 0.98$ or 98%. In a test of this membrane with pure water, the permeate flow is observed to be 0.1 gal/h at a pressure difference of 100 psi, with a membrane area of 0.1 ft². Find the permeate flow $F_P$ and the water quality $C_P$.

**8.17** Derive Eq. 8.3.62.

**8.18** Following the method described in Example 8.5.1, prepare a plot of the optimum fiber radius as a function of length $L$ (in meters), for $[0.1 < L < 1]$.

**8.19** A system is to be designed using the RO membrane tubes and the feed conditions described in Problem 8.3. The total number of tubes

available for use is 202. Examine the performance of the three configurations shown in Fig. P8.19.

**a.** With 202 tubes in series.

**b.** With four modules in series. Each module consists of 50 tubes arranged as ten parallel units of 5 tubes in series.

**c.** With three modules in series. The first module consists of 110 tubes arranged as ten parallel units of 11 tubes in series. The second module consists of 56 tubes arranged as eight parallel units of 7 tubes in series. The third and final module consists of 36 tubes arranged as four parallel units of 9 tubes in series.

For a 10-foot length of tube, plus one U-bend, pressure losses are described by the relationship

$$\Delta P \text{ (psig)} = 0.004 \, Q^2 \qquad \textbf{(P8.19)}$$

where $Q$ is in units of gallons per minute. Two-thirds of that pressure drop is over the straight section of tube and one-third is associated with the U-bend. Neglect any pressure losses in the headers of each parallel module and in the connections between modules.

Compare each design on the basis of productivity and permeate quality. Interpret the results.

**8.20** In Section 8.3 we used the Sieder–Tate equation to calculate a Sherwood number. Com-

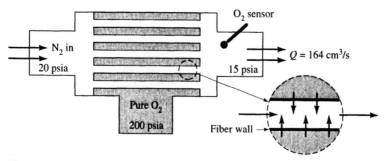

**Figure P8.15**

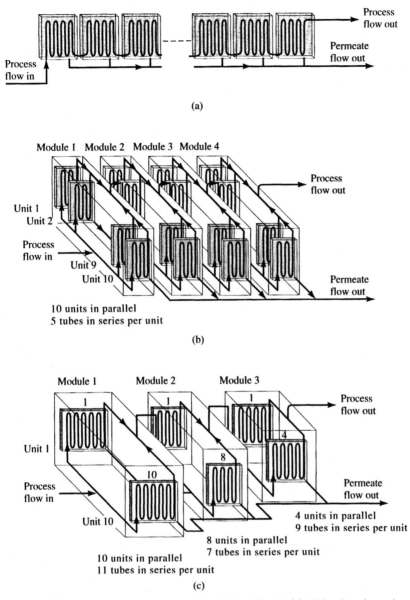

**Figure P8.19** Three membrane system configurations: (a) 202 tubes in series, (b) 200 tubes in four modules in series, and (c) 202 tubes in three modules in series.

pare this value to the ones we find using alternative suggestions presented in Chapter 6:

Gilliland and Sherwood correlated data for evaporation of pure liquids into air in turbulent pipe flow by

$$Sh_D = 0.023\, Re_D^{0.83}\, Sc^{0.44} \qquad \textbf{(6.5.5)}$$

The Chilton–Colburn analogy for turbulent flow in a circular pipe leads to

$$Sh_D = 0.04\, Re_D^{0.75}\, Sc^{0.33} \qquad \textbf{(6.4.36)}$$

# Part Two

# HEAT TRANSFER

# Chapter **9**

# Introduction

**I**n this final part of the text we examine problems in which heat transfer is the central physical phenomenon. We begin with a very brief introduction that should give you an idea of the types of thermal systems you will learn to model.

## 9.1  SOME HEAT TRANSFER PROBLEMS

Heat transfer analysis often plays a central role in the design of chemical processes and in the development of an understanding of the performance of process systems. A few "simple" problems illustrate the kinds of ideas we will be dealing with.

### 9.1.1  Production of a Polymeric Film

A polymer melt, which is a very high viscosity liquid, is extruded as a thin liquid film onto a "chill roll," a cold surface that solidifies the liquid into a *solid* but flexible film. What operating conditions permit us to remove the solidified film from the roll? Obviously, it is necessary for the polymer to solidify sufficiently to have mechanical integrity, yet it must retain some flexibility so that it can be stripped from the roll, somewhere downstream, as a continuous film, as indicated in Fig. 9.1.1.

We might expect the following parameters to be important:

- Roll speed, $U_R$
- Thermal conductivity of the polymer, $k_p$
- Rate of supply of cold air, $Q$
- Film thickness, $H$
- Temperatures of the roll surface $T_R$ and of the air $T_A$

The goal of a heat transfer analysis of this process is the estimation of where, downstream, the polymer has cooled enough to permit it to be removed from the chill roll, and the parametric dependence of this position on the operating and design parameters of the process.

### 9.1.2  Desalination by Vaporization/Condensation

Salt water (brine) is pumped into a pipe through which a hot oil is flowing. We wish to vaporize the water and later condense the vapor as salt-free water. The cold brine

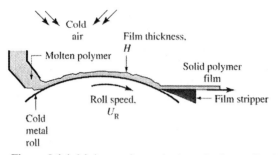

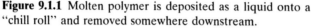

**Figure 9.1.1** Molten polymer is deposited as a liquid onto a "chill roll" and removed somewhere downstream.

is immiscible in the surrounding oil, and shortly past the point of injection of the brine, the brine stream will be broken up into droplets. These droplets will be heated by the oil, and eventually water vapor bubbles will evolve from the droplets. The flow field consists of three phases: oil, water, and gas. This is a very complex flow. Such a system is illustrated schematically in Fig. 9.1.2. What parameters affect its design and performance?

We might expect fluid dynamics to play an important role here, in addition to heat transfer, particularly in terms of the drop size distribution created by the flow. Some important parameters would be:

- Reynolds number of the oil flow, Re
- Viscosities of water and oil, $\mu_w$, $\mu_o$
- Interfacial tension, $\sigma$
- Mean water drop size, $D_p$
- Thermal conductivities, $k_w$, $k_o$

### 9.1.3 A Solar Hot Water Heater

Water is pumped through a system of hollow tubes made of a black polymer. The system is exposed to solar radiation. Absorption of this radiation heats the tube surfaces, and this heat is transferred into the water flowing within the tubes, as suggested in Fig. 9.1.3. Heat will also be lost by convection from the tube surfaces to the surrounding air, which is normally cooler than the tube surfaces. What factors affect the temperature rise that we can expect for water flowing through this system? How might we design a

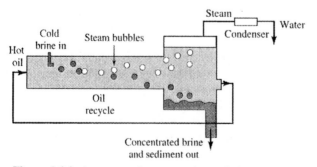

**Figure 9.1.2** A process for recovering salt-free water from brine.

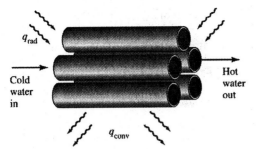

**Figure 9.1.3** Radiant heating of water flowing inside small tubes.

system that maximizes the heat transfer into the water while minimizing the losses to the atmosphere?

We might expect the following characteristics of the system to be important:

- Water flowrate, $Q$
- Tube diameter, $D$
- Tube length, $L$
- Air temperature, $T_a$
- Solar radiation flux, $q$
- Absorptivity of the polymer, $a$

A mathematical model of the system shows how system performance depends on these parameters.

### 9.1.4 Contaminant Entrainment by Buoyancy

A vertical reactor containing silicon wafers has been used to carry out a specific stage in the fabrication of a set of semiconductor devices. Upon completion of the reaction, the wafer load must be cooled, but the wafers cannot be exposed to contaminants in the air external to the reactor. An inert gas flow is used to "flush" the reactor (i.e., to envelop the wafer load and prevent diffusion or convection of external gas to the load). However, the reactor wall is at 1000 K, and a buoyancy-driven flow (Fig. 9.1.4) is generated by the rise of the hot, low density gas near the reactor wall. As a consequence, if the buoyancy flow is large enough relative to the inert flow, external gas may be

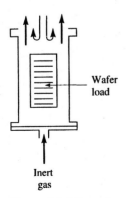

**Figure 9.1.4** Intrusion of air through buoyancy-driven flow.

sucked into the reactor. If this external gas contains particulates, or a chemically active species, the wafer load may be ruined.

We would expect the key factors that determine the dynamics of this system to be:

- Temperature dependence of the gas density, $\partial \rho / \partial T$
- Height of the reactor, $H_R$
- Difference in temperature between the wall and the flushing gas, $\Delta T$
- Flowrate of the flushing gas, $Q_f$

## 9.2   THE ROLES OF MODES OF HEAT TRANSFER

In each of the four problem areas just introduced, different modes of heat transfer play differing roles. Analysis will show us that in the problem illustrated in Section 9.1.1 the cooling of the polymer film is dominated by *conduction* within the polymer. In Section 9.1.2 conduction and *convection* interact. In Section 9.1.3 *radiation* delivers energy to the pipe wall which must be picked up by the water, presumably by convection. In Section 9.1.4 *buoyancy* establishes the convective flow field that competes with the flushing flow.

We will study these various modes of heat transfer—conduction, convection, and radiation—and learn how each is characterized. We will begin by examining simple systems with only one mechanism of heat transfer. Eventually, by putting together what we know about the individual mechanisms, we will be able to look at complex heat transfer problems and carry out analyses and/or designs of thermal systems of technological importance.

# Chapter 10

# Heat Transfer by Conduction

**W**e begin with a study of conductive heat transfer, confining our attention in this chapter to steady state systems. Although the examples selected are very simple, we will find that some technologically important heat transfer problems can be modeled with the information developed in this chapter.

## 10.1  FOURIER'S LAW OF CONDUCTIVE HEAT TRANSFER

The term "heat transfer" implies that the thermal energy of the matter that occupies some region of space is somehow transferred to the matter that occupies another region of space. This transfer can occur by three mechanisms:

*Conduction*  Collision of molecules causes the thermal energy to be transferred from one molecule to another. Very energetic molecules will lose energy in the transfer process, and the lower energy molecules will receive energy. Motion, as we understand the term in fluid dynamics, is not necessary. Heat can be conducted through a *stationary solid,* for example. The only motion is on a molecular level.

*Convection*  When macroscopic flow does occur, the energy associated with a "parcel" of fluid is carried—convected—to another region of space. We call this "convective heat transfer." Clearly we will have to use our knowledge of fluid dynamics to understand convective heat transfer.

*Radiation*  Molecular vibrations give rise to electromagnetic radiation, the amount of which is related in some way to the temperature of the matter. This radiation transmits energy through space, *including through a vacuum containing no matter,* and when it impinges on other molecules some of this radiant energy is absorbed by the receiving molecules.

We begin our discussion of heat transfer by considering conduction. With reference to Fig. 10.1.1, we could do the following experiment. A well-stirred liquid is partitioned into two regions separated by a solid planar sheet of material, of thickness $L$ and surface area $A$ (per side). In one region liquid is pumped in at temperature $T_o$ and mass flowrate $w$. Heat is transferred into the second region by conduction across the partition. Both regions are well stirred, and the temperature of the upper (closed) region is $T_1(t)$. The

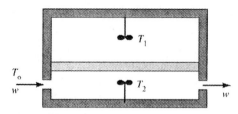

**Figure 10.1.1** Heat is transferred across a sheet separating two regions of fluid.

lower region is at the temperature $T_2(t)$. The term

$$Q = wC_p(T_o - T_2)$$   (10.1.1)

gives the rate of heat transfer from the lower fluid to the upper fluid, in units of joules/s or watts, when

$$w = \text{mass flowrate of fluid (kg/s)}$$
$$C_p = \text{heat capacity of the fluid (J/kg} \cdot \text{K)}$$

We actually have assumed that $C_p$ is independent of temperature in the range $[T_2, T_o]$. We also assume that the entire system is insulated, so that the heat loss from the lower fluid *all* goes to the upper fluid.

When we perform this experiment, we observe the following:

The rate of heat transfer is proportional to the area $A$, proportional to the temperature difference $(T_2 - T_1)$, and inversely proportional to the thickness of the sheet that separates the two regions, $L$. We may express this observation as

$$Q = k \frac{A(T_2 - T_1)}{L}$$   (10.1.2)

The proportionality coefficient $k$ is called the *thermal conductivity* of the solid sheet. Its units are abbreviated as J/s $\cdot$ m $\cdot$ K = W/m $\cdot$ K in the SI system of units (and as cal/s $\cdot$ cm $\cdot$ C° in the cgs system). (The British system of units for conductivity is Btu/h $\cdot$ ft $\cdot$ °F. Despite the motivation to move toward consistent use of SI units, you are very likely to encounter references that employ either cgs or British units, rather than SI units. Hence we will employ a variety of units in this text.)

We define the heat flow, *per unit area normal to the heat flow*, as the *heat flux, q.* Hence we may write Eq. 10.1.2 in the form

$$q = k \frac{\Delta T}{L}$$   (10.1.3)

If we had the ability to measure the heat flux within the solid, across any plane *parallel* to the surface of the sheet, we would find that Eq. 10.1.3 holds on a microscopic or differential scale, and we would conclude that

$$q_x = -k \frac{dT}{dx}$$   (10.1.4)

at every point within the solid. If we could perform a highly refined set of experiments in which we measured the heat flux across any surface within the sheet, regardless of its orientation relative to the normal vector to the surface of the sheet, we would find heat flux proportional to temperature gradient in the direction normal to any chosen surface. Hence we would express this observation in the *vector* format:

$$\mathbf{q} = -k \, \nabla T \qquad\qquad (10.1.5)$$

We refer to this observation, written in the vector form, as Fourier's law of heat conduction. Note the minus sign, which simply reflects the observation that heat is transferred from hot regions to cold regions.

We will begin our study of heat conduction with *one-dimensional, steady state* heat flow problems. First we examine a design problem.

### 10.1.1   A Design Problem, Measurement of $k$

We plan to measure thermal conductivity of a metal rod in the system shown in Fig. 10.1.2. The rod, of axial length $L$ and diameter $D$, is insulated except at the upper and lower faces, each of which is exposed to water in a well-stirred reservoir. The upper reservoir will be maintained at a fixed temperature by contact with a stirred ice bath. There is a constant flow of water through the lower reservoir at a *volumetric* flowrate $F$. The entering temperature of this flow will be measured, and fixed at $T_{H1}$; the outlet temperature is measured *after the system has achieved a steady state* and found to be at $T_{H2}$. We assume that the lower reservoir is well mixed, with the result that the outlet temperature is the same as the uniform temperature within the lower reservoir. The upper reservoir has an excess of ice, so it remains at constant temperature.

We want to find an expression (a mathematical model) for the thermal conductivity $k$ of the metal rod in terms of measured variables for this system. We do this in the context of designing the system for carrying out this experiment.

*The goal is to recommend a flowrate F (cm³/min) and an entering temperature $T_{H1}$(°F). For the recommended conditions, give the outlet temperature $T_{H2}$.*

We will assume that Fourier's law holds, and we write a simple heat balance. Heat lost by the fluid in the lower reservoir is transferred by conduction to the upper reservoir. The heat balance is

$$Q = \rho_w F_w C_{p,w}(T_{H1} - T_{H2}) = \frac{k}{L}\frac{\pi D^2}{4}(T_{H2} - T_c) \qquad\qquad (10.1.6)$$

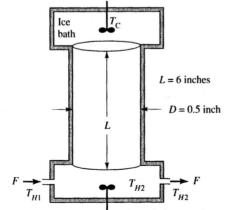

**Figure 10.1.2** Experimental system for measurement of $k$.

We use the fact that the heat capacity $C_p$ is defined as the heat content per unit mass of fluid. We have assumed that the heat capacity is independent of temperature, a reasonable assumption over a small temperature range, and we have neglected any change in density of the water as its temperature changes. If we solve for the conductivity $k$, we obtain a "design equation" for $k$ which takes the form

$$k = \frac{4L}{\pi D^2} (\rho C_p)_w F_w \left( \frac{T_{H1} - T_{H2}}{T_{H2} - T_c} \right) \tag{10.1.7}$$

We are free to set the values of several operating parameters. Let's pick

$$T_c = 32°F \text{ (ice bath)}$$
$$T_{H1} = 80°F \text{ (easily attained)}$$
$$F_w = 100 \text{ mL/min} = 13.2 \text{ lb}_m/h = 0.21 \text{ ft}^3/h$$

These operating parameters are easy to achieve in the laboratory. In addition, water is a convenient "working fluid," and the flowrate $F_w$ is easy to produce by gravity flow or with a simple inexpensive pump.

If we examine the design equation, we see that it is important that the temperature differences $(T_{H1} - T_{H2})$ and $(T_{H2} - T_c)$ be large enough to allow us to measure them with some accuracy. We should design so that these differences are *tens* of degrees, if we can.

The conductivity $k$ is unknown, of course. We should examine typical values for metals. Consider the following:

| | $k$ (cal/s · cm · K) | $k$ (Btu/h · ft · °F) |
|---|---|---|
| Aluminum | 0.45 | 108 |
| Copper | 0.9 | 216 |
| Steel | 0.13 | 32 |

Copper has an unusually high value. We will take the value of $k$ for steel for the sake of a design estimate. Then the system will be well designed for the study of metals whose conductivities are of the order of that of steel.

Assume $k = 32$ Btu/ft · h · °F for the metal. For water, $C_{pw} = 1$ Btu/lb$_m$ · °F, and $\rho = 62$ lb$_m$/ft$^3$. We find, upon rearranging Eq. 10.1.7, that

$$\frac{T_{H1} - T_{H2}}{T_{H1} - T_c} = \frac{\theta}{1 + \theta} = 6.3 \times 10^{-3} \tag{10.1.8}$$

where

$$\theta \equiv \frac{\pi D^2 k}{4L(\rho C_p)_w F_w}$$

Thus

$$T_{H1} - T_{H2} = 0.3°F \tag{10.1.9}$$

which is too small for accurate measurement. This small temperature change is predicted because the liquid flowrate originally specified is much too large. Let's try a smaller value of $F_w$.

Pick $F_w = 1$ mL/min $= 0.132$ lb/h $= 0.0021$ ft$^3$/h. We still use $T_c = 32°F$, but we increase the water inlet to $T_{H1} = 150°F$. Now we find

$$\frac{T_{H1} - T_{H2}}{T_{H1} - T_c} = \frac{\theta}{1 + \theta} = \frac{0.63}{1.63} = 0.387 \qquad \textbf{(10.1.10)}$$

and, after performing the required algebra,

$$T_{H2} = 104°F \qquad \textbf{(10.1.11)}$$

This is a "good" outlet temperature, although $F_w$ is on the low side. (It is sometimes difficult to control a very low flowrate.) An increase of $F_w$ to 5 mL/min will give $T_{H2}$ = 137°F.

We have completed the design problem, and we have an idea of properties and operating conditions that will permit us to measure the thermal conductivity of metals with some precision.

### 10.1.2   Conductive Heat Loss Across a Window Pane

Much of the heat loss from residences in cold climates is through the windows. We carry out an analysis here that provides some guidelines to reduction of this loss. We begin by considering the situation illustrated in Fig. 10.1.3, in which a plane slab of solid material with a conductivity $k$ is exposed to two different temperatures. We assume that the air on either side of the window pane is well mixed by convection, so that a good approximation is $T_h = T_1$ and $T_c = T_2$. We also assume steady state conditions ($q$ is constant) and that Fourier's law is valid:

$$q_x = -k\frac{dT}{dx} = \frac{k(T_1 - T_2)}{L} \qquad \textbf{(10.1.12)}$$

If we know $k$, $T_1$, $T_2$, and $L$, we can find $q_x$.

Suppose $T_1$ = 60°F, $T_2$ = 20°F, and $L$ = 1/8 inch. For glass, we find a thermal conductivity value of

$$k = 0.0017 \text{ cal/s} \cdot \text{cm} \cdot \text{K} = 2.4175 \times 10^2(0.0017) \text{ Btu/h} \cdot \text{ft} \cdot °F$$
$$= 0.41 \text{ Btu/h} \cdot \text{ft} \cdot °F \qquad \textbf{(10.1.13)}$$

and it follows that the $x$ component of the heat *flux,* which is the rate of conduction per unit of area, in the $x$ direction, is

$$q_x = \frac{0.41(60 - 20)}{(8 \times 12)^{-1}} = 1574 \text{ Btu/h} \cdot \text{ft}^2 \qquad \textbf{(10.1.14)}$$

Is this a large or a small rate of heat loss?

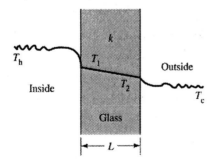

**Figure 10.1.3** Sketch for analysis of heat loss across a window pane.

Suppose we have a room, insulated except for a window having an area of 6 ft². The room has dimensions $9 \times 12 \times 8 = 864$ ft³. How long will it take for the room to cool from 60°F to 45°F? Assume that the air in the room is well mixed and that the outside temperature remains constant at 20°F. The rate of heat loss is

$$Q = \frac{k(T_1 - T_2)A}{L} \tag{10.1.15}$$

The rate of change of the heat content of the room is

$$\frac{d}{dt}[V\rho C_p(T_1 - T_{ref})] = -Q \tag{10.1.16}$$

where $V$ is the volume of air in the room, and $\rho$ and $C_p$ are the mass density and heat capacity (per mass) of the *room air*. The heat content is always measured relative to some arbitrary reference value. We define $T_{ref}$ as the temperature at which the heat content is arbitrarily set equal to zero. Note that since $T_{ref}$ appears in the derivative term, we do not have to specify its value. The energy balance gives

$$\frac{dT_1}{dt} = -\frac{kA}{\rho C_p VL}(T_1 - T_2) \tag{10.1.17}$$

and the initial condition is written in the form

$$T_1 = T_1^o \quad \text{at} \quad t = 0 \tag{10.1.18}$$

The solution to Eq. 10.1.17 is

$$\frac{T_1 - T_2}{T_1^o - T_2} = \exp\left(-\frac{kA}{\rho C_p VL}t\right) \tag{10.1.19}$$

For the problem at hand we have the following data:

$$
\begin{array}{ll}
T_2 = 20°F & k = 0.41 \text{ Btu/h} \cdot \text{ft} \cdot °R \\
T_1^o = 60°F & A = 6 \text{ ft}^2 \\
T_1 = 45°F & V = 864 \text{ ft}^3 \\
& L = 1/96 \text{ ft}
\end{array}
$$

and the physical properties of air are

$$\rho_{air} = 0.07 \text{ lb}_m/\text{ft}^3 \qquad C_{air}^p = 0.24 \text{ Btu/lb}_m \cdot R$$

We find

$$\tau \equiv \frac{\rho C_p VL}{kA} = 6.2 \times 10^{-2} \text{ h} \tag{10.1.20}$$

and

$$\frac{T_1 - T_2}{T_1^o - T_2} = \frac{45 - 20}{60 - 20} = \frac{25}{40} = e^{-t/\tau} \tag{10.1.21}$$

Then

$$\frac{t}{\tau} = -\ln\frac{25}{40} = -\ln 0.625 = 0.47 \tag{10.1.22}$$

and the time for the room to cool down is

$$t = 0.47\,\tau = 0.47 \times 6.2 \times 10^{-2} = 2.9 \times 10^{-2} \text{ h} = 1.75 \text{ min} \tag{10.1.23}$$

This is a very short time, and this result is contrary to our intuition, based on experience. Is something wrong with the model?

One of the *assumptions* of the model is contrary to our experience. A room is not normally well mixed. We might expect there to be a boundary layer of stagnant air adjacent to the window glass, on the inside. Can we modify the model to account for a second resistance to conduction? The answer, of course, is yes. Before we produce the appropriate model that accounts for a stagnant film of air, we develop a general model for conduction across multilayer, or composite, materials.

## 10.2 THE COMPOSITE SOLID

### 10.2.1 The Planar Solid

Consider a "composite" solid made up of two planar solid sheets in contact, as shown in Fig. 10.2.1. A temperature difference $T_1 - T_2$ is maintained across the outer surfaces, and the system has come to a steady state temperature distribution within the pair of solids. We want to find that temperature distribution, as well as the heat flux normal to the surface of the composite.

For each sheet we write a flux expression in the form of Fourier's law:

$$q_1 = \frac{k_1(T_1 - T_{12})}{L_1} \qquad q_2 = \frac{k_2(T_{12} - T_2)}{L_2} \tag{10.2.1}$$

What can we say about the heat fluxes $q_1$ and $q_2$ at the interface between the two solids? While we don't know the magnitude of either flux at this point, we can say that these two fluxes must be equal. If they were not, any difference between the rate at which heat was conducted *to* the boundary and the rate at which it is conducted *away* from the boundary would give rise to an accumulation of heat in a region of vanishingly small volume. This is not a physically appealing notion, so we conclude that the heat flux must be continuous. (An exceptional case exists if there is a phase change, e.g., melting, at the boundary.)

We can use this assertion to eliminate the unknown temperature at the boundary, $T_{12}$:

$$\frac{k_1(T_1 - T_{12})}{L_1} = \frac{k_2(T_{12} - T_2)}{L_2} \tag{10.2.2}$$

It follows that

$$T_{12} = \frac{\dfrac{k_1 T_1}{L_1} + \dfrac{k_2 T_2}{L_2}}{\dfrac{k_2}{L_2} + \dfrac{k_1}{L_1}} \tag{10.2.3}$$

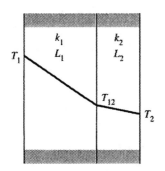

**Figure 10.2.1** Heat conduction across a planar composite solid.

and we find

$$q = q_1 = q_2 = \frac{k_1}{L_1}(T_1 - T_{12}) \qquad \text{(10.2.4)}$$

We manipulate this result algebraically:

$$
\begin{aligned}
q &= \frac{k_1}{L_1}\left(T_1 - \frac{k_1/L_1 T_1 + k_2/L_2 T_2}{k_2/L_2 + k_1/L_1}\right) \\
&= \frac{k_1/L_1}{k_2/L_2 + k_1/L_1}\left(\frac{T_1 k_2}{L_2} + \frac{T_1 k_1}{L_1} - \frac{k_1 T_1}{L_1} - \frac{k_2 T_2}{L_2}\right) \\
&= \frac{k_1/L_1}{k_2/L_2 + k_1/L_1}\left(\frac{T_1 k_2}{L_2} - \frac{k_2 T_2}{L_2}\right) \\
&= \frac{(k_1/L_1)(k_2/L_2)(T_1 - T_2)}{k_2/L_2 + k_1/L_1} = \frac{k_1 k_2}{k_2 L_1 + k_1 L_2}(T_1 - T_2)
\end{aligned}
\qquad \text{(10.2.5)}
$$

Finally, we may write the flux in terms of the external—measurable—temperatures as

$$q = \frac{T_1 - T_2}{L_1/k_1 + L_2/k_2} \qquad \text{(10.2.6)}$$

If we introduce the notion of a single effective conductivity of the composite, defined so that

$$q_{\text{comp}} = \frac{k_{\text{comp}}}{L}(T_1 - T_2) \qquad \text{(10.2.7)}$$

where the total thickness of the composite is

$$L = L_1 + L_2 \qquad \text{(10.2.8)}$$

we easily find that

$$\frac{1}{k_{\text{comp}}} = \frac{1}{(1 + L_2/L_1)}\frac{1}{k_1} + \frac{1}{(L_1/L_2 + 1)}\frac{1}{k_2} = \frac{L_1}{Lk_1} + \frac{L_2}{Lk_2} \qquad \text{(10.2.9)}$$

or

$$\frac{L}{k_{\text{comp}}} = \frac{L_1}{k_1} + \frac{L_2}{k_2} \qquad \text{(10.2.10)}$$

Let us use this result immediately by returning to the problem of heat loss through a window pane.

Suppose we put a plastic film in front of the glass, as in Fig. 10.2.2, in such a way that it traps a *stagnant* air space of thickness $L_2 = 0.5$ inch. If the air film is really stagnant, it acts like a solid film, with a thermal conductivity that we can find to be $k_{\text{air}} = 0.014$ Btu/ft·h·°F. From Eq. 10.2.6 we now find the flux as

$$q = \frac{60 - 20}{\left[\dfrac{1}{96(0.41)}\right] + \left[\dfrac{1}{24(0.014)}\right]} = \frac{40}{0.025 + 2.98} = 13 \text{ Btu/h·ft}^2 \qquad \text{(10.2.11)}$$

which is only 0.8% of the case given by Eq. 10.1.14. This result indicates that a half-inch layer of *stagnant* air would reduce the heat loss rate by a factor of 125. If we compare Eq. 10.2.6 to Eq. 10.1.12 we see that the *composite* solid, made up of a layer of glass and a "layer" of air, behaves like a homogeneous solid with a greatly reduced thermal conductivity.

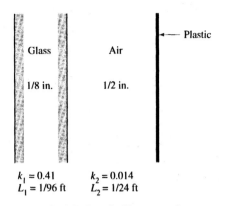

**Figure 10.2.2** An air film on a glass pane.

If we return to Eq. 10.1.20 and replace $k$ by $k_{comp}$, we will find

$$\tau = 7.8 \text{ h} \tag{10.2.12}$$

and the time to reach 45°F is now extended to

$$t = 3.6 \text{ h} \tag{10.2.13}$$

This seems much more in line with our expectations.

### 10.2.2 A Convective Boundary Condition

Consider the case suggested in Figure 10.2.3. Heat is transferred by conduction across a sheet of glass, as before. The outside air is very well mixed, and the outside surface of the glass is at the ambient temperature $T_a$. The air in the room is mixed, but not perfectly. The inside surface of the glass will be below the room temperature $T_r$, to a degree that depends on how well mixed the room air is. We are describing convective heat transfer at the inner surface. The usual observation is embodied in a *definition* of a heat transfer coefficient $h$, such that the heat flux to the inner surface is given by

$$q = h(T_r - T_i) \tag{10.2.14}$$

Convection

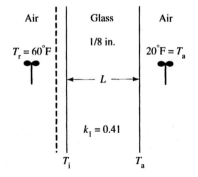

**Figure 10.2.3** Sketch for discussion of a convective boundary condition.

The inside glass surface temperature $T_i$ is unknown at this point. We can find it by writing a second heat flux equation—this one for conduction through the glass pane:

$$q = \frac{k(T_i - T_a)}{L} \tag{10.2.15}$$

Conduction

Note that by this notation we have introduced the assertion that the two fluxes, the flux to the window surface and the flux through the window pane, are equal.

We have two unknowns in this pair of equations: $T_i$ and $q$. We may solve for each, using the two equations above, and find

$$T_i = \frac{1}{1 + m} T_r + \frac{m}{1 + m} T_a \tag{10.2.16}$$

where a dimensionless group $m$ occurs:

$$m = \frac{k}{hL} \tag{10.2.17}$$

The inverse of $m$ is called the Biot (pronounced "B.O.") modulus for heat transfer:

$$\text{Bi} = \frac{hL}{k} \tag{10.2.18}$$

Note that $T_i$ is close to $T_r$ or $T_a$ for very small, or very large, values of $m$, respectively.

Equation 10.2.14, the *definition* of $h$, is usually referred to as *Newton's law of cooling*. It is simply a statement that heat transfer by convection is linear in the temperature difference between the surface and the ambient fluid.

It is not difficult to show that a *convective* heat transfer coefficient can be thought of as arising from *conduction* across a stagnant film of fluid, and that

$$h = \frac{k_{\text{fluid}}}{\delta_H} \tag{10.2.19}$$

where $\delta_H$ is the fluid film thickness. Hence we may think of convection in terms of a convective coefficient, or in terms of an equivalent stagnant film that separates well-mixed fluid from the solid surface. We will return to this idea later when we apply boundary layer theory to heat transfer.

## 10.2.3  Heat Transfer Across a Composite Cylindrical Solid

The case of cylindrical resistances in series is illustrated in Fig. 10.2.4. We assume that Fourier's law in the vector form (Eq. 10.1.5) holds. If there is symmetry about the cylindrical axis, the only heat flux is in the radial direction, and in the radial direction Fourier's law takes the form

$$q_r = -k \frac{dT}{dr} \tag{10.2.20}$$

But at what value of $r$ is this equation written, and why did we not raise this question in the earlier example of the planar solid?

The complication arises because the area normal to the heat flow is a function of radial position, and it is not the *flux*, but the *rate* $Q$ that must be constant, at steady

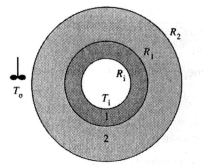

**Figure 10.2.4** Conduction through a composite cylinder.

state. At any cylindrical surface $r$, of axial length $L$, the rate is the product of the area times the flux:

$$Q = -k_1(2\pi rL)\frac{dT_1}{dr} = -k_2(2\pi rL)\frac{dT_2}{dr} = \text{constant} = C \qquad \textbf{(10.2.21)}$$

Note that we have defined *two* unknown temperature *functions* $T_1(r)$ and $T_2(r)$, each in the domains $[R_i, R_1]$ and $[R_1, R_2]$. For $T_1(r)$ we find

$$\frac{dT_1}{dr} = -\frac{C}{2\pi Lk_1 r} \qquad \textbf{(10.2.22)}$$

Separating the variables we find

$$dT_1 = -\frac{C}{2\pi Lk_1}\frac{dr}{r} \qquad \textbf{(10.2.23)}$$

Integration yields

$$T_1(r) = -\frac{C}{2\pi Lk_1}\ln r + a_1 \qquad \textbf{(10.2.24)}$$

where $a_1$ is an integration constant. Similarly, for $T_2(r)$ we find

$$T_2(r) = -\frac{C}{2\pi Lk_2}\ln r + a_2 \qquad \textbf{(10.2.25)}$$

We suppose that the physics is such that we know the temperatures $T_i$ and $T_o$. Further, for this example, we suppose that heat is transferred to the exterior by convection, and the convective coefficient $h_c$ is known. We impose the following boundary conditions:

at $r = R_i$     $T_1 = T_i$ (the inner surface is at a known temperature)    **(10.2.26)**

at $r = R_2$     $q_2 = h_c(T_2 - T_o)$ (convection at the outer surface)    **(10.2.27)**

at $r = R_1$     $\left\{\begin{matrix} T_1 = T_2 \\ q_1 = q_2 \end{matrix}\right\}$              **(10.2.28)**
                                                      **(10.2.29)**

You should ask yourself the following questions: Why is $q_1$ equal to $q_2$ here, but not before (see the comment preceding Eq. 10.2.21), and why do we appear to have four boundary conditions, but only three unknown constants, $a_1$, $a_2$, and $C$?

After doing the algebra to establish the integration constants, we arrive at the final result: the functions $T_1(r)$, $T_2(r)$. We are also interested in the heat loss, which is given by Eq. 10.2.21, once we know the functions $T_1(r)$ and $T_2(r)$. We find the rate of heat transfer in the form

$$Q = C = 2\pi L \frac{T_i - T_o}{\dfrac{\ln(R_1/R_i)}{k_1} + \dfrac{\ln(R_2/R_1)}{k_2} + \dfrac{1}{R_2 h_c}} \tag{10.2.30}$$

A common format for heat transfer through composite bodies is

$$Q = UA\,\Delta T \tag{10.2.31}$$

which really defines the $(UA)$ product. If we arbitrarily specify an area to use in this equation, then $U$ has a specific form. For example, if

$$A_2 = 2\pi R_2 L \text{ (the outside area)} \tag{10.2.32}$$

and we denote the denominator of Eq. 10.2.30 as $D$,

$$\frac{2\pi L(T_i - T_o)}{D} = 2\pi R_2 L U_o(T_i - T_o) \tag{10.2.33}$$

and we find

$$U_2 = \frac{1}{DR_2} = \left[ \frac{R_2}{k_1} \ln\left(\frac{R_1}{R_i}\right) + \frac{R_2}{k_2} \ln\left(\frac{R_2}{R_1}\right) + \frac{1}{h_c} \right]^{-1} \tag{10.2.34}$$

where $U_2$ is called the overall heat transfer coefficient. Note that it is based on the area $A_2$, and it is based on the overall temperature difference, $\Delta T = T_i - T_o$.

Under some conditions it might be the case that

$$h_c \ll \begin{cases} \dfrac{k_1}{R_2} \dfrac{1}{\ln(R_1/R_i)} \\[2mm] \dfrac{k_2}{R_2} \dfrac{1}{\ln(R_2/R_1)} \end{cases} \tag{10.2.35}$$

Then we find, from Eq. 10.2.34,

$$U_2 \approx h_c \tag{10.2.36}$$

This is expected, since if $k_1/R_2$ and $k_2/R_2$ are large, or $R_1/R_i$ and $R_2/R_1 \approx 1$, either of which would lead to the inequality in Eq. 10.2.35, conduction is very rapid compared to the external convective loss.

## 10.2.4  The Use of Additional Material to Reduce Heat Loss

From the model of the composite cylinder, we may examine how $Q$ depends on $R_2$. Specifically, we ask the following question: If we add insulation to the outer surface of a pipe, does $Q$ increase or decrease? Consider the region $[R_1, R_2]$ to be an insulating (low $k$) material. Consider $\partial Q/\partial R_2$. A negative value means that increasing $R_2$ insulates the pipe against heat loss. Write Eq. 10.2.30 in a form that displays only the dependence on $R_2$:

$$Q = \frac{a}{b + c\ln(R_2) + d/R_2} = \frac{a}{f(R_2)} \tag{10.2.37}$$

where we write the denominator as the function $f(R_2)$. To find the dependence of $Q$ on $R_2$, we differentiate $Q$ and find

$$\frac{\partial Q}{\partial R_2} = -\frac{a}{f^2}\frac{\partial f}{\partial R_2} = \frac{-a}{[b + c\ln(R_2) + d/R_2]^2}\left(\frac{c}{R_2} - \frac{d}{R_2^2}\right) \tag{10.2.38}$$

We conclude that $\partial Q/\partial R_2$ is $< 0$ or $> 0$ depending on the sign of $(c/R_2 - d/R_2^2)$. Hence we define a "critical radius" by

$$c - \frac{d}{R_2} = 0 \tag{10.2.39}$$

In writing Eq. 10.2.37 we used $d = 1/h_c$ and $c = 1/k_2$. Hence we find, from Eq. 10.2.39, that the critical radius is

$$R_2^* = \frac{k_2}{h_c} \tag{10.2.40}$$

Only if $R_2$ exceeds $k_2/h_c$ will the "insulation" *begin to decrease* the heat loss, and so actually behave as insulation. However, $R_2$ will have to exceed $R_2^*$ in order to bring the heat loss *below* the value in the absence of any insulation. (See Problem 10.30.)

## 10.3 FINNED HEAT EXCHANGERS

Sometimes our goal is to *promote* heat loss from a hot surface. An automobile radiator is one example. A common approach is to add highly conductive surfaces that extend the area of the system out into the convecting fluid. This is most commonly done by adding "fins" to the surface, and we refer to such a system as a finned heat exchanger. Figure 10.3.1 shows a wide rectangular fin on a planar surface. By "wide" we imply that the width $w$ in the $y$ direction is much greater than the fin thickness $2B$. The wide rectangular fin is our first example because it gives the simplest mathematical model. This makes it easier to focus on the physics.

### 10.3.1 Analysis of a Simple Finned Surface

We begin (see Fig. 10.3.1) with a steady state heat balance across a *differential* slice of the fin:

*Conductive flux in, across the face at z:*

$$-k\frac{dT}{dz}\bigg|_z \tag{10.3.1}$$

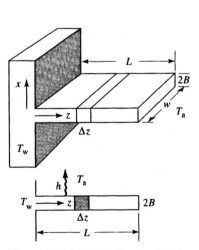

**Figure 10.3.1** Sketch for model of a rectangular fin.

*Conductive flux out, across the face at z + Δz:*

$$-k\frac{dT}{dz}\bigg|_{z+\Delta z} \tag{10.3.2}$$

*Convective flux out, across the surface normal to z:*

$$h[T(z) - T_a] \tag{10.3.3}$$

Multiplying these fluxes by the appropriate areas and adding, we find

$$\left[-k\frac{dT}{dz}\bigg|_z - \left(-k\frac{dT}{dz}\right)_{z+\Delta z}\right]2Bw - h[T(z) - T_a]2w\,\Delta z = 0 \tag{10.3.4}$$

We have set this sum to zero because at steady state the sum of the *rates* of heat transfer in and out of the differential volume element must vanish—there is no net increase or decrease of heat content within the volume. We now divide these rates by Δz and take the limit as Δz → 0. The result is

$$\frac{d}{dz}\left(k\frac{dT}{dz}\right) = \frac{h}{B}(T - T_a) \tag{10.3.5}$$

This procedure yields a differential equation for the temperature distribution $T(z)$ along the fin. We have assumed that in the thin $(x)$ direction the temperature is uniform, and hence there is no conduction of heat in the $x$ direction. This should be a good assumption if $B \ll L$. In addition, we have assumed that the fin is long in the direction of the width $w$ (i.e., $w \gg B$).

We need two boundary conditions to solve this equation.

We assume that the base of the fin is at the wall temperature $T_w$. We also assume that the tip of the fin $(z = L)$ has a negligible heat loss compared to that from the larger surfaces. Then the boundary conditions read

$$\text{at } z = 0: \qquad T = T_w \tag{10.3.6}$$

$$\text{at } z = L: \qquad q_L = -k\left[\frac{dT}{dz}\right]_L = 0 \tag{10.3.7}$$

We now define dimensionless variables:

$$\theta = \frac{T - T_a}{T_w - T_a} \qquad \zeta = \frac{z}{L} \qquad N = \left(\frac{hL^2}{kB}\right)^{1/2} \tag{10.3.8}$$

Equation 10.3.5 becomes

$$\frac{d^2\theta}{d\zeta^2} - N^2\theta = 0 \tag{10.3.9}$$

subject to boundary conditions

$$\theta = 1 \quad \text{at} \quad \zeta = 0 \tag{10.3.10}$$

$$\frac{d\theta}{d\zeta} = 0 \quad \text{at} \quad \zeta = 1 \tag{10.3.11}$$

The solution can be written either in terms of hyperbolic functions or exponential functions:

$$\theta = A \sinh N\zeta + B \cosh N\zeta = A'e^{N\zeta} + B'e^{-N\zeta} \tag{10.3.12}$$

We will use the hyperbolic functions. After imposing the boundary conditions, we find

$$\theta = \cosh N\zeta - (\tanh N) \sinh N\zeta \qquad (10.3.13)$$

We can calculate the heat loss from the fin in either of two ways:

**Rate into fin through the base**

$$-k\frac{dT}{dz}\bigg|_{z=0} (2Bw) = Q \qquad (10.3.14)$$

**Rate of convection from surfaces**

$$\int_0^L 2wh(T - T_a)\, dz = Q \qquad (10.3.15)$$

We will make a choice momentarily. First, let's calculate lower and upper bounds on the heat loss. If there were no fin, heat loss across the base area $2Bw$ would be

$$2Bwh(T_w - T_a) = Q_0 \qquad (10.3.16)$$

assuming that $h$ is unaffected by the presence of the fin. This is the minimum $Q$.

If the fin were uniformly at the wall temperature $T_w$, its heat loss would be the maximum possible value:

$$2wLh(T_w - T_a) = Q_{\max} \qquad (10.3.17)$$

It is usually easier to calculate $Q$ by differentiation of the solution than by integration. From Eq. 10.3.13, we can write

$$\frac{T - T_a}{T_w - T_a} = \cosh N\frac{z}{L} - (\tanh N) \sinh N\frac{z}{L} \qquad (10.3.18)$$

and we find, using Eq. 10.3.14,

$$\frac{1}{T_w - T_a}\frac{dT}{dz} = \frac{N}{L}\sinh N\frac{z}{L} - \left(\frac{N}{L}\tanh N\right)\cosh N\frac{z}{L} \qquad (10.3.19)$$

Since

$$\frac{1}{T_w - T_a}\frac{dT}{dz}\bigg|_{z=0} = -\frac{N}{L}\tanh N \qquad \text{at} \quad z = 0 \qquad (10.3.20)$$

then

$$Q = (2Bw)\frac{k(T_w - T_a)N}{L}\tanh N \qquad (10.3.21)$$

We define the "effectiveness" of the fin as $Q/Q_{\max}$:

$$\eta = \frac{Q}{Q_{\max}} = \frac{k(T_w - T_a)N \tanh N(2Bw)}{L\, 2wL\, h(T_w - T_a)} = \frac{\tanh N}{N} \qquad (10.3.22)$$

Figure 10.3.2 plots $\eta$ versus $N$. Extrapolation for $N \gg 1$ is made easier by noting that

$$\lim_{N \gg 1} \eta = \frac{1}{N} \qquad \text{since } \tanh N \to 1 \quad \text{as } N \gg 1$$

Obviously $N$ plays an important role in describing the efficiency of heat loss of the extended surface. Note that $N$ (actually, $N^2$) is a kind of Biot number, since

$$N^2 = \frac{h}{k}\left(\frac{L^2}{B}\right) \qquad (10.3.23)$$

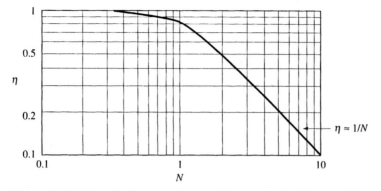

**Figure 10.3.2** Fin effectiveness for a wide rectangular fin on a flat surface.

One can find fins of various cross-sectional shapes. Triangular fins are common, for example. Hence graphs are available to describe the efficiency of various fin shapes. Two such graphs are presented in Fig. 10.3.3. (Note the use of arithmetic coordinates here, vs. logarithmic coordinates in Fig. 10.3.2.)

For fin shapes other than planar with rectangular cross section, one must solve differential equations other than Eq. 10.3.9 above. Qualitatively, the results for $\eta$ are similar in shape to Fig. 10.3.2.

The dependent variable (the abscissa) is often written as

$$N' = L^{3/2}\left(\frac{h}{kA_m}\right)^{1/2} \qquad (10.3.24)$$

but we can show that this is similar to $N$. If $A_m$ is the cross-sectional area, in a section normal to the direction of the width $w$, then for a rectangular cross section,

$$A_m = 2BL \qquad (10.3.25)$$

It is not hard to see that Eq. 10.3.23 is equivalent to

$$\frac{N}{\sqrt{2}} = L^{3/2}\left(\frac{h}{kA_m}\right)^{1/2} \qquad (10.3.26)$$

A corrected length $L_c$ is often substituted for the geometrical length $L$ in Eq. 10.3.26, as in Fig. 10.3.3a, where

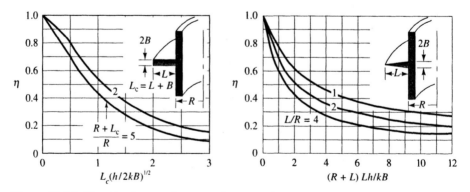

**Figure 10.3.3** Fin effectiveness for several geometries.

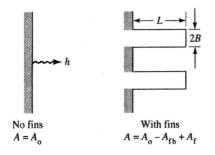

**Figure 10.3.4** Effect on surface area for convection when fins are added.

$$L_c = L + B \tag{10.3.27}$$

This accounts approximately for the fact that the boundary condition (Eq. 10.3.7) ignores some small loss across the end of the fin.

For finned heat exchangers, the total rate of heat loss is given by (defining $\Delta T = T_w - T_a$)

$$Q = h\,\Delta T[A_o' + \eta A_f] \tag{10.3.28}$$

where $A_o$ is the original surface area without fins that is suffering convective loss. With fins we add a surface area $A_f$, but since the base of the fins (with area $A_{fb}$) covers some of the original area $A_o$, we correct $A_o$ to $A_o' = A_o - A_{fb}$. Figure 10.3.4 illustrates this.

Recall our assumptions that there was no heat loss from the edges of the fin (the area we would see in cross section) and that the same was true of the tip of the fin, which has the area $2Bw$ for a rectangular fin. We usually compensate for the failure of these assumptions by making slight corrections to the fin length $L$, as noted above.

---

**EXAMPLE 10.3.1**   *Finding $\eta$ and the Heat Loss from a Finned Surface*

Aluminum circumferential fins of rectangular cross section are arrayed as shown in Fig. 10.3.5 on a tube of diameter 2.5 cm. Values for $k$ and $h$ are given (note the units).

We must calculate the following geometrical parameters (see Fig. 10.3.3):

$$L_c = L + B = 1.5 + 0.05 = 1.55 \text{ cm} \tag{10.3.29}$$

$$R = \frac{2.5}{2} = 1.25 \text{ cm} \tag{10.3.30}$$

$$R + L_c = 1.25 + 1.55 = 2.8 \text{ cm} \tag{10.3.31}$$

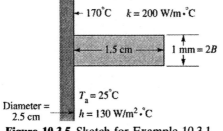

**Figure 10.3.5** Sketch for Example 10.3.1.

Then we find

$$\frac{R + L_c}{R} = \frac{2.8}{1.25} = 2.24 \qquad (10.3.32)$$

$$A_m = 2BL_c = 1.55 \times 10^{-5} \text{ m}^2 \qquad (10.3.33)$$

$$L_c^{3/2} \left( \frac{h}{kA_m} \right)^{1/2} = 0.4 \qquad (10.3.34)$$

From Fig. 10.3.3 we find

$$\eta_f = 82\% \qquad (10.3.35)$$

The maximum rate of heat loss would occur if the entire surface were at the temperature of $T_w = 170°C$:

$$Q_{max} = 2\pi[(R + L)^2 - R^2]h(T_w - T_a) = 71 \text{ W} = 244 \text{ Btu/h} \qquad (10.3.36)$$

Hence we find

$$Q_f = \eta_f Q_{max} = 58 \text{ W} = 200 \text{ Btu/h} \qquad (10.3.37)$$

This gives the loss from a single fin.

---

**EXAMPLE 10.3.2**   *Calculation of Improved Heat Transfer with Fins*

Water and air are separated by a 0.125-inch-thick steel sheet. It is proposed to increase the heat transfer rate between these fluids by adding straight triangular fins of 0.05-inch base thickness ($t$) and 1-inch length ($L$), in the direction normal to the sheet (see Fig. 10.3.6). The fins will be spaced with a distance of 0.5 inch between successive centerlines of the bases. The fins are very wide in the direction normal to the triangular cross section. The heat transfer coefficients on the air and water sides are 2 and 45 Btu/ h · ft² · °F, respectively.

What percentage increase in heat transfer can be realized if the fins are on the airside? Suppose instead we put them on the waterside. Does this make a difference?

The first issue we should examine is whether the conductive resistance of the steel sheet is significant in comparison to the convective resistances on either of its surfaces. To do so, return to Eq. 10.2.6, which gives the heat flux for conduction normal to two planar resistances in series. We may write this expression in terms of the *rate* of heat transfer as

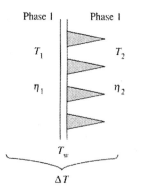

**Figure 10.3.6** Sketch for Example 10.3.2.

$$Q = \frac{\Delta T}{\Sigma R} \qquad (10.3.38)$$

where $\Sigma R$ represents the sum of the resistances. In the development of that equation we learned that a purely conductive resistance may be expressed as

$$R = \frac{L_i}{k_i} \qquad (10.3.39)$$

where $L$ and $k$ are the thickness and thermal conductivity of the resistance, respectively. Later, consideration of convective heat flow revealed that we may write a convective resistance in terms of a heat transfer coefficient $h$ and that convection is, in some sense, equivalent to conduction across a stagnant film of thickness

$$\delta = \frac{k_{fluid}}{h} \qquad (10.3.40)$$

It follows then that if we write a convective resistance in the form of Eq. 10.3.39 we find

$$R_{conv} = \frac{\delta}{k_{fluid}} = \frac{1}{h} \qquad (10.3.41)$$

For the case considered here—a steel sheet with convective resistances on each side—the total resistance may be written as

$$\Sigma R = \frac{L_{steel}}{k_{steel}} + \frac{1}{h_{air}} + \frac{1}{h_{water}} \qquad (10.3.42)$$

A typical conductivity value for steel is $k_{steel} = 26$ Btu/ft $\cdot$ h $\cdot$ °F. Hence we find

$$\Sigma R = \frac{\frac{1}{8} \times \frac{1}{12}}{26} + \frac{1}{2} + \frac{1}{45} = 0.0004 + 0.5 + 0.022 = 0.522 \qquad (10.3.43)$$

Clearly, the conductive resistance of the steel wall is unimportant in this analysis; the resistance is dominated by convection.

Let's turn then to the analysis of heat transfer enhancement through the use of fins. As a basis for the calculation, take 1 ft$^2$ of area (a 1 ft $\times$ 1 ft square) in which there are 24 fins. By definition, $A_o = 1$ ft$^2$.

The unfinned area available for convective transfer is

$$A = 1 \text{ ft}^2 - 24 \text{ fins} \left( \frac{0.05 \text{ in.} \times 1 \text{ ft}}{12 \text{ in./ft}} \frac{1}{\text{fin}} \right) = 0.9 \text{ ft}^2 \qquad (10.3.44)$$

The finned area is

$$A_f = (24 \text{ fins})(1 \text{ ft}) \left( 2 \times \frac{1}{12} \text{ ft} \right) = 4 \text{ ft}^2 \qquad (10.3.45)$$

(We neglect the cosine of the fin angle in calculating $A_f$ because the angle is so small.)
We will use the corrected length in calculating $N$ from Eq. 10.3.26:

$$L_c = L + \frac{t}{2} = 1.025 \text{ in.} = 0.085 \text{ ft} \qquad (10.3.46)$$

$$A_m = \frac{t L_c}{2} = 0.025(1.025) = 0.0256 \text{ in.}^2 = 1.78 \times 10^{-4} \text{ ft}^2 \qquad (10.3.47)$$

Assume the fins are mild steel ($k = 26$ Btu/ft $\cdot$ h $\cdot$ °F).

**Air Side $h = 2$ Btu/h $\cdot$ ft$^2$ $\cdot$ °F**     **Water Side $h = 45$ Btu/h $\cdot$ ft$^2$ $\cdot$ °F**

$$N = L_c^{3/2} \left( \frac{h}{kA_m} \right)^{1/2} = 0.515 \qquad N = L_c^{3/2} \left( \frac{h}{kA_m} \right)^{1/2} = 2.44 \qquad \text{(10.3.48)}$$

From an efficiency relationship for straight triangular fins (see Problem 10.2), we have

$$\eta_f = 0.89 \qquad \text{and} \qquad \eta_f = 0.38 \qquad \text{(10.3.49)}$$

The wall temperature $T_w$ is unknown, and its value depends on which side (air or water) is finned. In the absence of fins, Eq. 10.3.38 gives

$$Q = \frac{A_o \, \Delta T}{R} \qquad \text{(10.3.50)}$$

with

$$R = \frac{1}{h_1} + \frac{1}{h_2} \qquad \text{(10.3.51)}$$

$$Q_0 = \frac{A_o \, \Delta T}{\dfrac{1}{h_1} + \dfrac{1}{h_2}} = \frac{A_o \, \Delta T}{\dfrac{1}{2} + \dfrac{1}{45}} = 1.91 \, A_o \, \Delta T \qquad \text{(10.3.52)}$$

We have neglected the conductive resistance of the wall in writing the fin base temperature $T$ as $T_1$.

In the following we label the finned side as side 2.

$$\begin{aligned} Q_f &= h_1 A_o (T_1 - T_w) \\ &= h_2 A_f (T_w - T_2) \eta_f + h_2 A (T_w - T_2) \end{aligned} \qquad \text{(10.3.53)}$$

with fins

where $A = A_o - A_{fb}$, and $A_{fb}$ is the area of the fin base(s).

Eliminating the unknown $T_w$ is the next task. This requires a few lines of algebra:

$$h_1 A_o (T_1 - T_w) = h_2 (T_w - T_2)(A_f \eta_f + A) \qquad \text{(10.3.54)}$$

$$T_w [-h_1 A_o - h_2 (A_f \eta_f + A)] = -h_2 T_2 (A_f \eta_f + A) - h_1 T_1 A_o \qquad \text{(10.3.55)}$$

$$T_w = \frac{h_1 T_1 A_o + h_2 T_2 (A_f \eta_f + A)}{h_1 A_o + h_2 (A_f \eta_f + A)} \qquad \text{(10.3.56)}$$

$$Q_f = h_1 A_o (T_1 - T_w) = h_1 A_o \left[ T_1 - \frac{h_1 T_1 A_o + h_2 T_2 (A_f \eta_f + A)}{h_1 A_o + h_2 (A_f \eta_f + A)} \right] \qquad \text{(10.3.57)}$$

$$Q_f = h_1 A_o \left[ \frac{T_1 h_2 (A_f \eta_f + A) - T_2 h_2 (A_f \eta_f + A)}{h_1 A_o + h_2 (A_f \eta_f + A)} \right] \qquad \text{(10.3.58)}$$

$$Q_f = \frac{A_o (T_1 - T_2)}{\dfrac{h_1 A_o + h_2 (A_f \eta_f + A)}{h_1 h_2 (A_f \eta_f + A)}} = \frac{A_o (T_1 - T_2)}{\dfrac{1}{h_2} \dfrac{A_o}{A + A_f \eta_f} + \dfrac{1}{h_1}} \qquad \text{(10.3.59)}$$

In the absence of fins, the heat loss from the bare surface was found (Eq. 10.3.52) to be $(Q_0/A_o \Delta T) = 1.91$ Btu/h $\cdot$ ft$^2$ $\cdot$ °F. Let's calculate this quantity for the finned surface.

**Air Side Fin ($h$ in Btu/h·ft²·°F)** | **Water Side Fin**

$h_2 = 2$, $h_1 = 45$, $\eta_f = 0.89$      $h_2 = 45$, $h_1 = 2$, $\eta_f = 0.38$

$$\frac{Q_f}{A_o\,\Delta T} = \frac{1}{\dfrac{1}{2[0.9+4(0.89)]}+\dfrac{1}{45}} = 7.4$$

$$\frac{Q_f}{A_o\,\Delta T} = \frac{1}{\dfrac{1}{45}\dfrac{1}{[0.9+4(0.38)]}+\dfrac{1}{2}} = 1.96 \qquad \textbf{(10.3.60)}$$

The relative increase in heat loss is      The relative increase in heat loss is

$$\frac{Q_f-Q_o}{Q_o}=\frac{7.4-1.91}{1.91}=2.87$$

$$\frac{Q_f-Q_o}{Q_o}=\frac{1.96-1.91}{1.91}=0.026$$

or $Q_f/Q_o = 3.87$      or $Q_f/Q_o = 1.026$ $\qquad$ **(10.3.61)**

We conclude that to improve the efficiency of heat loss, fins should be placed on the side with the smaller $h$, the air side in this example.

### EXAMPLE 10.3.3   Optimum Fin Length for Fixed Weight

We can use the model for the rectangular fin to illustrate an optimization problem. We wish to find the length of a rectangular fin that maximizes the heat loss from the fin, while holding the *weight* of the fin constant. That is, we seek the optimum fin geometry (the $L/B$ ratio) under the constraint of fixed fin weight. We start with Eq. 10.3.21, which we write in the form

$$Q = \frac{mk(T_w - T_a)}{\rho L^2} N \tanh N \qquad \textbf{(10.3.62)}$$

where the mass of the fin is

$$m = 2\rho BLw \qquad \textbf{(10.3.63)}$$

where $\rho$ is the density of the fin material, and we write $N$ as

$$N = \frac{m}{2w\rho}\left(\frac{h}{k}\right)^{1/2} B^{-3/2} \qquad \textbf{(10.3.64)}$$

We choose this format because we have written $L$ in terms of the fixed mass $m$ and the optimization variable $B$, and we have done the same with $N$. As a consequence of these substitutions, we may write the heat transfer rate in the form

$$Q = cN^{-1/3} \tanh N \qquad \textbf{(10.3.65)}$$

To find the optimum shape we examine

$$\frac{\partial Q}{\partial N} = 0 = -\frac{1}{3}N^{-4/3}\tanh N + \frac{1}{N^{1/3}\cosh^2 N} \qquad \textbf{(10.3.66)}$$

The solution is found by a simple trial-and-error procedure to be

$$N_{opt} = 1.419 \qquad \textbf{(10.3.67)}$$

from which we find, using the definition of $N$ in Eq. 10.3.8,

$$\left(\frac{L}{B}\right)_{opt} = 1.419\left(\frac{k}{hB}\right)^{1/2} \qquad \textbf{(10.3.68)}$$

This result holds specifically for the rectangular fin. The optimum shape will depend on the geometry of the fin.

## 10.4 CONDUCTION WITH INTERNAL HEAT GENERATION

The passage of electrical current through conductors generates inside the conductor heat that must be dissipated to prevent damage to the circuits and components of the electronic or electrical device. We can develop a simple model of the dissipation of this heat by conduction through a cylindrical wire and its surrounding insulation. With reference to Fig. 10.4.1, we consider first a heat balance on an element of volume within the electrical wire. We let $k_e$ be the *electrical conductivity* (the inverse of the electrical resistance $R$) of the wire (in units of $\Omega^{-1} \cdot cm^{-1}$), and we describe the amount of current being carried by the wire by $I$ (A/cm$^2$).

Electrical or resistive heating is often described as "$I^2R$ heating" because the rate of heat production per unit of conductor volume is given by

$$S_e = I^2R = \frac{I^2}{k_e} \tag{10.4.1}$$

If we assume that we have cylindrical symmetry about the axis, then the radial heat flux within the wire is just

$$q_r = -k\frac{dT}{dr} \tag{10.4.2}$$

across any surface of radius $r$.

A heat balance on a differential annular volume within the wire, of inner and outer radii $r$ and $r + dr$ and axial length $L$, then takes the form

$$([rq]_r - [rq]_{r+dr})\, 2\pi L + S_e\, 2\pi r\, dr\, L = 0 \tag{10.4.3}$$

We divide both sides by the differential volume $2\pi r\, dr\, L$ and then let $dr$ vanish, with the result that we obtain a differential equation in the form

$$\frac{d}{dr}[rq_r] = S_e r \tag{10.4.4}$$

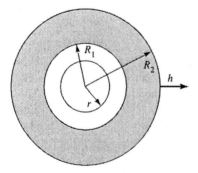

**Figure 10.4.1** Sketch for analysis of heat transfer through a current-carrying wire.

With Fourier's law we then find a differential equation for temperature:

$$-\frac{d}{dr}\left[r\frac{dT}{dr}\right] = \frac{S_e}{k}r \qquad (10.4.5)$$

Two integrations lead to a general solution for $T(r)$ in the form

$$T = -\frac{S_e}{4k}r^2 - a\ln(r) + b \qquad (10.4.6)$$

where $a$ and $b$ are integration constants. Since the temperature must be finite along the wire axis, the integration constant $(a)$ multiplying $\ln r$ must be zero.

This solution is valid only within the wire (i.e., within $0 < r < R_1$). Now we must find the temperature distribution within the insulation *surrounding* the wire, in the region $R_1 < r < R_2$. Actually, we may go right back to Eq. 10.4.5, but set $S_e = 0$ (there is no generation within the insulation) and change the notation to show clearly that we are finding the temperature $T_i$ in the insulation. Hence we solve

$$-\frac{d}{dr}\left[r\frac{dT_i}{dr}\right] = 0 \qquad (10.4.7)$$

and find

$$T_i = c\ln(r) + d \qquad (10.4.8)$$

Now we must impose three boundary conditions to find the constant $b$ in Eq. 10.4.6 and the constants $c$ and $d$ in Eq. 10.4.8. The physical statements that we make are the following:

$$T = T_i \qquad \text{at the interface of the two materials (i.e., at } r = R_1) \qquad (10.4.9)$$

$$k\frac{dT}{dr} = k_i\frac{dT_i}{dr} \qquad \text{at } r = R_1 \qquad (10.4.10)$$

These two statements reflect our physical intuition (as well as our experience) that the temperature, and the heat flux, must be continuous across the boundary of the two materials.

The third boundary condition is based on the assumption that the heat loss from the exterior surface of the insulation obeys Newton's law of cooling, with a heat transfer coefficient $h$:

$$-k_i\frac{dT_i}{dr} = h(T_i - T_a) \qquad \text{at } r = R_2 \qquad (10.4.11)$$

When we apply these three boundary conditions to Eqs. 10.4.6 and 10.4.8 (noting $a = 0$ is already established), we find the following solutions for the two temperature distributions:

**T(r)**

$$T - T_a = \frac{S_e R_1^2}{4k}\left[1 - \left(\frac{r}{R_1}\right)^2 - \frac{2k}{k_i}\ln\frac{R_1}{R_2} + \frac{2k}{hR_2}\right] \qquad (10.4.12)$$

**T_i(r)**

$$T_i - T_a = \frac{S_e R_1^2}{2k_i}\left(\ln\frac{R_2}{r} + \frac{k_i}{hR_2}\right) \qquad (10.4.13)$$

With this model we may explore the dependence of the temperature inside the conductor on the thickness of the surrounding insulation. Of particular importance is the determination of the amount of insulation that minimizes the temperature rise due to resistive heating. The addition of a small amount of insulation decreases the temperature in the conductor because the additional surface area performs the function of an extended surface—a kind of "fin." Ultimately, however, additional insulation begins to perform as a true insulator, and the conductor temperature rises with additional insulation beyond that point. If our goal is the minimization of the temperature at the conductor/insulator interface, we can achieve it by selecting a radius $R_2$ that satisfies the condition

$$\left[\frac{\partial T}{\partial R_2}\right]_{r=R_1} = 0 \tag{10.4.14}$$

It is not difficult to show (from differentiation of Eq. 10.4.12) that the solution is

$$R_2 = \frac{k_i}{h} \tag{10.4.15}$$

---

**EXAMPLE 10.4.1**    *Convective Cooling of an Insulated Wire*

A copper wire of diameter 2 mm carries a current "density" (really, a current *per area*) of 1000 A/cm². It is insulated with a material that has a thermal conductivity of $10^{-3}$ W/cm · K. The insulated wire is exposed to air moving in a manner that yields a convective heat transfer coefficient of 0.002 W/cm² · K. The copper has an electrical resistance $R$ of $2 \times 10^{-6}$ Ω · cm. Plot the temperature at the wire/insulator interface as a function of insulation radius.

From Eq. 10.4.15 the optimum radius is

$$R_2 = \frac{k_i}{h} = \frac{0.1}{20} = 0.005 \text{ m} \tag{10.4.16}$$

We will use Eq. 10.4.13:

$$T_i = T_a + \frac{S_e R_i^2}{2k_i}\left[\ln\left(\frac{R_2}{R_1}\right) + \frac{k_i}{hR_2}\right] \tag{10.4.17}$$

We must first calculate $S_e$. From the data given we find

$$S_e = I^2 R = (10^3 \text{ A/cm}^2)^2\, 2 \times 10^{-6}\, \Omega \cdot \text{cm} = 2 \text{ W/cm}^3 \tag{10.4.18}$$

At the optimum radius, the temperature of the copper/insulator interface *relative to the ambient* is

$$T_i - T_a = \frac{2(0.1)^2}{2(0.001)}\left[\ln\left(\frac{0.5}{0.1}\right) + \frac{0.001}{0.002(0.5)}\right] = 26 \text{ K} \tag{10.4.19}$$

At some other value of $R_2$, say $R_2 = 0.002$ m, we find

$$T_i - T_a = \frac{2(0.1)^2}{2(0.001)}\left[\ln\left(\frac{0.2}{0.1}\right) + \frac{0.001}{0.002(0.2)}\right] = 32 \text{ K} \tag{10.4.20}$$

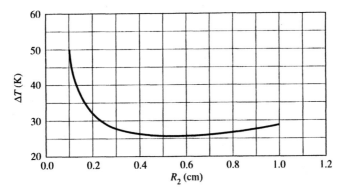

**Figure 10.4.2** Dependence of interface temperature rise on insulation radius.

For a larger value than optimum, say $R_2 = 0.01$ m, we find

$$T_i - T_a = \frac{2(0.1)^2}{2(0.001)} \left[ \ln\left(\frac{1}{0.1}\right) + \frac{0.001}{0.002(1)} \right] = 28 \text{ K} \qquad \textbf{(10.4.21)}$$

Figure 10.4.2 shows the results of similar computations.

## SUMMARY

Heat conduction in most materials is well described by Fourier's law, which really defines the thermal conductivity of the material. With Fourier's law it is possible to write very simple models for the one-dimensional *steady state* heat flow through regions of simple geometry, such as slabs, cylinders, and spheres. One of our first applications of these ideas is to the analysis of extended surface or finned heat exchangers. The resulting models can be cast in the form of effectiveness factors specific to each geometry.

When heat conduction occurs under *unsteady-state* conditions, the mathematical analysis is more complicated, but Fourier's law is still at the heart of the analysis. Unsteady behavior is treated in the next chapter.

Methods for obtaining values of the thermal properties of materials are summarized in Appendix C.

## PROBLEMS

**10.1** A well-insulated box loses heat to the ambient only through a single-pane glass window of area 1 m$^2$ and thickness 0.5 cm. The ambient medium is at 10°C, and the external heat transfer coefficient is 10 W/m$^2$·K. The interior of the box is maintained at 25°C and is well mixed, giving the window an inside surface temperature of 25°C. At \$0.15/kW·h (kilowatt-hour), what is the annual cost of maintaining this system? Use a glass thermal conductivity of 0.75 W/m·K.

Someone suggests a double-glazed window, which consists of two panes of glass, each of thickness 0.25 cm, separated by an air space of thickness 1 mm. What would be the annual savings in energy cost with this window? Assume that heat is transferred across the air space between the panes only by conduction.

**10.2** Derive a differential equation (do not attempt to solve it) for the steady temperature distribution in a straight triangular fin as shown

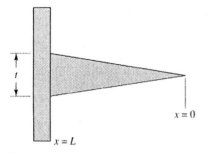

**Figure P10.2**

in Fig. P10.2. Assume the temperature ($T = T_L$) at $x = L$ is known and that convective losses occur to the ambient medium at temperature $T_a$. Give the necessary boundary conditions for solution of the equation.

**10.3** In a graphite-moderated nuclear reactor, heat is generated uniformly in a set of uranium rods of 0.05 m o.d. at the rate of $8 \times 10^7$ W/m³. These rods are jacketed by an annulus through which water at an average temperature of 120°C is circulated. The water cools the rods, and the average heat transfer coefficient is estimated to be 30,000 W/m² · K. The thermal conductivity of uranium is 30 W/m · K. Determine the center temperature, and the surface temperature, of the uranium fuel rods.

**10.4** To increase the heat dissipation from a 3 cm o.d. tube, circumferential fins made of aluminum ($k = 200$ W/m · K) are soldered to the outer surface. The fins are 0.1 cm thick and have an outer diameter of 6 cm. If the tube surface temperature is 100°C, the environmental temperature is 25°C, and the heat transfer coefficient between the fin and the environment is 65 W/m² · K, calculate the rate of heat loss from a fin.

**10.5** To determine the thermal conductivity of a 1 m long, solid 2.5 cm diameter rod, one half was inserted into a furnace at 300°C while the other half was left projecting into air at 30°C. After steady state had been reached, the temperature at the midpoint of the exposed length was measured and found to be 125°C. The heat transfer coefficient over the surface of the rod exposed to the air was estimated to be 30 W/m² · K. What is the thermal conductivity of the rod?

**10.6** For the situation described in Problem 10.5, find the rate of heat loss from the rod. Plot the

rate of heat loss against the rod diameter $D$, for the furnace and ambient conditions stated in Problem 10.5, but for rod diameters in the range of 1 to 3 cm. Make one of the following assumptions:

**a.** The heat transfer coefficient over the surface of the rod is independent of rod diameter $D$.

**b.** The heat transfer coefficient over the surface of the rod varies as $D^{-1/2}$.

**10.7** In Eq. 10.2.34 we find a model for the overall heat transfer coefficient for conduction through a composite cylindrical solid, with convection at the exterior surface. Develop the corresponding expression for a composite *spherical* solid, with the inner spherical surface at a known temperature $T_i$.

**10.8** A composite planar "solid" consists of two panes of glass, each of thickness 0.25 cm, separated by an air space of thickness 1 mm. Assume that heat is transferred across the air space between the panes by conduction alone. What is the overall heat transfer coefficient $U$ for this system? Suppose that heat is transferred across the air space by convection. For what magnitude of convective heat transfer coefficient would the double-pane system be thermally equivalent (within 1%) to a single pane of thickness 0.5 cm?

**10.9** A spherical copper shell, of inside radius 5 cm and wall thickness 1 cm, loses heat to the surrounding air, which is maintained at 25°C. The inside surface of the shell is maintained at $T_i = 100$°C. The external heat transfer coefficient is $h = 10$ W/m² · K. What is the rate of heat loss to the ambient?

Suppose that an additional layer of material is coated onto the exterior of the shell. Is the rate of heat loss increased or decreased by this additional layer? Specifically, answer the following questions.

**a.** If the thermal conductivity of the additional material is one-tenth that of copper, plot the heat loss against the thickness of the additional material for thicknesses in the 0.01–1 cm range.

**b.** If the additional material has a thickness of 1 mm, plot the heat loss against the thermal conductivity of the material, for thermal conductivities in the range from one-tenth that of copper to 10 times that of copper.

**10.10** Develop a mathematical model for the problem of current-induced temperature rise in an *uninsulated* electrical wire. Use Newton's law of cooling for the boundary condition at the surface.

**a.** Solve for the temperature difference between the wire center and the ambient.

**b.** For a 1/16-inch-diameter copper wire, exposed to air at 20°C, find the center temperature when the voltage drop is 40 V over a 5 m length, assuming (i) infinite $h$ and (ii) finite $h = 25$ W/ $m^2 \cdot K$. Take the rate of heat production as

$$S_e = \frac{I^2}{k_e} \text{ per unit volume} \quad \textbf{(P10.10.1)}$$

For the electrical conductivity $k_e$, use the Wiedemann–Franz–Lorenz equation:

$$\frac{k}{k_e T} = L = \text{constant} \quad \textbf{(P10.10.2)}$$

($T$ in kelvins)

For copper, the Lorenz number L is $2.2 \times 10^{-8}$ $V^2/K^2$.

**10.11** The walls of an insulated lunch box are made of inch-thick styrofoam with a thermal conductivity of 0.05 W/m · K. If an ice cube, one inch on a side, is placed in the box and the box placed in an ambient medium at 90°F with an external heat transfer coefficient of $h = 10$ W/ $m^2 \cdot K$, how long is required for half the ice cube to melt? State clearly all the assumptions you make in formulating your model of this problem.

**10.12** Return to Section 10.1.1, the design problem for measurement of $k$. Since there is a temperature gradient across the length $L$, and $k$ is a function of temperature, we must address the question: What temperature do we assign to the value of $k$ that results from a particular choice of operating and design conditions? A simple, but vague, answer is that the measured $k$ represents the value at the average of the temperatures $T_c$ and $T_{H2}$. Is this an approximation, or is it exact? Answer by considering two models for $k(T)$:

$$k = a + bT \quad \textbf{(P10.12.1)}$$

and

$$k = a + bT + cT^2 \quad \textbf{(P10.12.2)}$$

Derive an expression analogous to Eq. 10.1.7 for each case, in the form

$$\bar{k} = \frac{4L}{\pi D^2} (\rho C_p)_w F_w \left( \frac{T_{H1} - T_{H2}}{T_{H2} - T_c} \right) f \quad \textbf{(P10.12.3)}$$

where $k$ is evaluated at the arithmetic average temperature over the length $L$. The question is, how different is the "correction factor" $f$ from unity?

**10.13** In deriving Eq. 10.3.13 we assumed that there was no heat loss from the tip of the fin (the boundary condition given by Eq. 10.3.11). Using Eq. 10.3.13, write an expression for the temperature at the tip of the fin, and from this calculate the heat loss from the tip. Give an expression for the ratio of the heat loss from the tip to the loss from the two large fin surfaces of area $Lw$. For what value of the parameter $N$ is the tip loss less than 10% of the fin surface loss?

**10.14** Suppose a rectangular fin of length $L$ has a tip temperature $T_L$. It loses heat by convection if $T_L$ exceeds the ambient temperature $T_a$. Suppose we add another length of identical material to the end of the fin, so that the total length is now $L_c$. This additional length is actually insulated at its tip (i.e., at $z = L_c$), but it does lose heat by convection from its upper and lower surfaces. Find the length $L_c$ such that the additional lateral heat loss of the extended fin is identical to the loss from the tip of the fin of original length $L$. In other words, for what "corrected length" $L_c$ does a fin having an insulated tip give the same heat loss as calculated by assuming no heat loss at the tip for a fin having an uninsulated tip?

**10.15** Solve Eq. 10.3.9, with Eq. 10.3.10, but replace Eq. 10.3.11 with a nondimensional form of

$$-k \left[ \frac{dT}{dx} \right]_{x=L} = h(T_L - T_a) \quad \textbf{(P10.15)}$$

(i.e., allow for convective loss at the tip).

Compare $\eta$ for this case to $\eta$ of Eq. 10.3.22. What parameter determines how close Eq. 10.3.22 is to this more realistic model? For what value of this parameter are the two $\eta$s within 10% of each other?

**10.16** The differential equation derived in Problem 10.2 can be put in the form

$$x \frac{d^2 y}{dx^2} + \frac{dy}{dx} - \beta y = 0 \quad \textbf{(P10.16.1)}$$

Show that a solution $y(x)$, satisfying

$$\begin{array}{ll} y = y_o & \text{at } x = L \\ y' = \frac{dy}{dx} = 0 & \text{at } x = 0 \end{array} \quad \textbf{(P10.16.2)}$$

may be found by assuming

$$y = \sum_{n=0}^{\infty} a_n x^n \qquad \textbf{(P10.16.3)}$$

and show further that the coefficients $a_n$ are such that the solution is

$$\frac{y}{y_0} = \frac{\sum_{n=0}^{\infty} \dfrac{(\beta x)^n}{(n!)^2}}{\sum_{n=0}^{\infty} \dfrac{(\beta L)^n}{(n!)^2}} \qquad \textbf{(P10.16.4)}$$

The series solution actually is one of the Bessel functions, that is,

$$\sum_{n=0}^{\infty} \frac{(\beta x)^n}{(n!)^2} = I_0(2\sqrt{\beta x}) \qquad \textbf{(P10.16.5)}$$

The function $I_0$ can be found tabulated or plotted in various sources. See Appendix B.

**10.17** A two-dimensional fin has a curved profile, as shown in Fig. P10.17. The temperature profile $T(x)$ is observed to be exactly linear, that is,

$$T(x) = ax + T_a \qquad \textbf{(P10.17)}$$

where $a$ is some constant. Show that the thickness profile $B(x)$ is parabolic.

**10.18** In Section 10.3, where the model for a rectangular fin is developed, the boundary condition of Eq. 10.3.11 implies that there is no heat loss from the tip of the fin. Does this imply that the tip temperature is the same as the ambient temperature? One way to answer this question is to calculate $\theta$ at $\zeta = 1$ from Eq. 10.3.13. Is it zero? Plot $\theta$ at $\zeta = 1$ as a function of $N$. What conclusion do you reach?

**10.19** Use the result given in Problem 10.16 to show that the fin efficiency for a long triangular

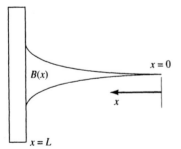

**Figure P10.17**

fin is given by

$$\eta = \frac{I_1(2N)}{N I_0(2N)} \qquad \textbf{(P10.19.1)}$$

where

$$N = L^{3/2} \left( \frac{h}{k A_m} \right)^{1/2} \qquad \textbf{(P10.19.2)}$$

**10.20** Use the result given in Problem 10.19 to find the optimum length of a *triangular* fin of fixed weight. Compare the optimum $L/B$ to that for a rectangular fin (Eq. 10.3.68).

**10.21** A furnace wall consists of 200 mm of firebrick ($k = 1.52$ W/m · °C) on the flame side, clad with 6 mm of steel ($k = 45$ W/m · °C) on the outside. The hot side temperature is 1150°C and the cold side surface is at 30°C. A very accurate measurement of heat flow shows that the steady heat flux through the wall is 826 W/m$^2$.

It is suspected that the steel cladding has separated from the firebrick, leaving a thin layer of air to separate these two materials. If this is true, calculate the air film thickness. Use $k = 0.016$ Btu/ft · h · °F for air.

**10.22** A thin, high conductivity metal wall separates two fluids. On each side (1 or 2) of the wall, the fluid has a bulk temperature $T_1$ (or $T_2$) and a convective heat transfer coefficient $h_1$ (or $h_2$). Fins are placed on side 2 of the wall, in an attempt to displace the wall temperature away from the value

$$T_w = \frac{T_1 + (h_2/h_1)T_2}{1 + h_2/h_1} \qquad \textbf{(P10.22.1)}$$

that would exist in the absence of fins.

Show that if the fin efficiency is 100%, the wall temperature will be

$$T_w = \frac{T_1 + T_2}{2} F_1(H_{21}) + T_2 F_2(H_{21}) \qquad \textbf{(P10.22.2)}$$

where

$$F_1(H_{21}) = \frac{2}{1 + H_{21}} \qquad F_2(H_{21}) = \frac{H_{21} - 1}{H_{21} + 1} \qquad \textbf{(P10.22.3)}$$

and

$$H_{21} = \frac{h_2}{h_1} \qquad \textbf{(P10.22.4)}$$

**10.23** An alloy metal fin ($k = 64$ Btu/h·ft·°F) is to be used on a wall, in a situation where the surface heat transfer coefficient is 50 Btu/ h·ft·°F. The fin has a rectangular cross section, a base 0.05 inch thick, and a length of 1 inch. Assume that the fin is 4 inches wide.

**a.** Find the fin efficiency.

**b.** What is the rate of heat loss from a single fin if the wall temperature is 200°F and the air is at 80°F?

**10.24** Liquid nitrogen is stored in a thin-walled metallic sphere of radius $r_1 = 0.25$ m. The sphere is covered with an insulating material ($k = 0.0017$ W/m·K) of thickness 25 mm. The exposed surface of the insulation is surrounded by air at 300 K. The convective heat transfer coefficient to the air is $h = 20$ W/m²·K. The latent heat of vaporization and density of liquid nitrogen are $2 \times 10^5$ J/kg and 804 kg/m³, respectively. Liquid nitrogen boils at 77 K at one atmosphere. What is the rate of liquid boil-off, in kilograms per day?

**10.25** A hollow sphere of aluminum, with an inside radius of 0.01 m and an outside radius of 0.02 m has its outer surface maintained at 100°C. Cold water at 0°C is available to cool the inside surface of this spherical shell, and its flow is controlled at a rate that serves to maintain the inside surface at 10°C. Assuming that the inside heat transfer coefficient between the water and the inner surface is very high, and that the water within the inner region is very well mixed, what is the required flow rate of the cooling water? The thermal conductivity of aluminum is 229 W/m·K. The heat capacity of water is 4180 W·s/kg·K.

**10.26** Derive Eq. 10.4.14.

**10.27** Electrical current flows through a bare electrical wire, of radius $R_w = 0.5$ mm. The current and electrical resistance are such that heat is generated at a rate of 1.5 W per meter of axial length. The wire loses heat to the surrounding air. It is observed that the air temperature is 25°C, and the wire surface temperature is 125°C, when the system is at steady state.

**a.** What is the convective heat transfer coefficient under these conditions?

**b.** An annular plastic insulation will be put on the wire. The insulation has a thermal conductivity of $k = 0.25$ W/m·K. At steady state, what is the rate of heat loss (per meter of wire)?

**c.** What will the exposed surface temperature of the insulation be if the insulation thickness is 1 mm? Assume the convective coefficient is 5 W/m²·K.

**d.** What will the inner surface temperature of the insulation be if the insulation thickness is 1mm?

**10.28** A hollow tube serves as a handle for a metal pot that holds boiling water. The tube is stainless steel ($k = 16$ W/m·K), and has the dimensions shown in Fig. P10.28. If the base temperature of the handle is 95°C and the external convective coefficient is $h = 10$ W/m²·K, estimate the distance from the base at which the handle temperature falls to 40°C. State clearly what assumption you make about heat transfer from the *interior* surface of the handle.

**10.29** Derive a differential equation (do not attempt to solve it) for the steady temperature distribution in a circumferential rectangular fin as shown in Fig. P10.29. Assume that the temperature ($T = T_1$) at $r = r_1$ is known and that convective losses occur to the ambient medium at temperature $T_a$. Give a sufficient set of physically meaningful boundary conditions for solution of the equation.

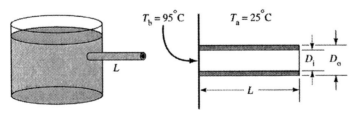

$$T_b = 95\,°C \qquad T_a = 25\,°C$$

$L = 20$ cm   $D_i = 1.5$ cm   $D_o = 1.9$ cm

**Figure P10.28**

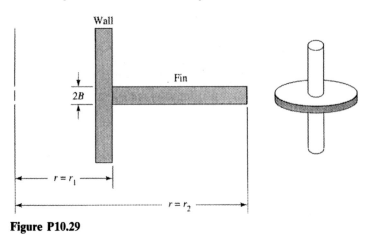

Wall

Fin

$2B$

$r = r_1$

$r = r_2$

**Figure P10.29**

**10.30** In Section 10.2.4 we find a critical insulation radius $R_2^*$ such that in the range $R_1 < R_2 < R_2^*$ the addition of "insulation" actually increases the heat loss. For $R_2 > R_2^*$ we find $\partial Q / \partial R_2 < 0$, but the heat loss may still lie *above* the value for the uninsulated pipe ($R_2 = R_1$). Develop a relationship from which we may find the insulation radius $R_2^{**}$ such that for $R_2 > R_2^{**}$ the heat loss lies *below* that of the uninsulated pipe. Show that the result may be plotted as $R_2^{**}/R_2^*$ as a function of $m_2 = k_2/R_1 h_c$. What does your analysis imply if $m_2 < 1$?

# Chapter 11

# Transient Heat Transfer by Conduction

In this chapter we introduce a simple form of the transient heat transfer equation, valid for conduction in incompressible materials. We look primarily at one-dimensional heat conduction problems, for which analytical solutions of the transient equation are available in many cases of interest. Through a series of examples, we demonstrate the use of these models in the analysis and design of engineering systems.

## 11.1 UNSTEADY HEAT TRANSFER ACROSS THE BOUNDARIES OF SOLIDS

### 11.1.1 The Solid with Uniform Internal Temperature

In general, the temperature in a conducting material may be a function of position and time. To describe such complex systems, it is necessary to derive, and eventually to solve, the partial differential equation that expresses the principle of transport of thermal energy. Before going to that general case, we present a simpler transient heat transfer problem—one in which the temperature distribution *in* the solid is uniform *spatially* but varies with time as the solid exchanges heat by convection with its surroundings. With reference to Fig. 11.1.1, we consider a solid of volume $V$, density $\rho$, heat capacity (per unit mass)[1] $C_p$, and surface area $A$. We *assume* that the body has a spatially uniform but time-dependent temperature $T(t)$ and loses heat to the surroundings by convection from its surface. A heat balance on the solid states that the time rate of change of the thermal energy within the solid body is due to the convective heat exchange with the surroundings. If the rate of convection is taken to follow the usual assumption of a linear function of the temperature difference between the surface and the surroundings, the heat balance takes the form

$$\frac{d}{dt}\rho C_p V T = -hA(T - T_a) \tag{11.1.1}$$

[1] For a solid the two heat capacities $C_p$ and $C_v$ are essentially equal, and we arbitrarily write the equation in terms of $C_p$.

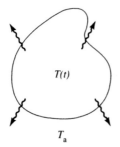

**Figure 11.1.1** A solid body exchanging heat with its surroundings.

with the initial condition

$$T = T_o \quad \text{at} \quad t = 0 \tag{11.1.2}$$

where $h$ is the convective heat transfer coefficient.

Equation 11.1.1 is easily solved by separating the variables:

$$\frac{d(T - T_a)}{(T - T_a)} = -\frac{hA}{\rho C_p V} dt \tag{11.1.3}$$

We integrate both sides from $t = 0$, at which $T = T_o$, to some arbitrary time and temperature $(t, T)$ and find

$$\ln\left(\frac{T - T_a}{T_o - T_a}\right) = -\frac{hA}{\rho C_p V} t \tag{11.1.4}$$

We write this in a form explicit in temperature as

$$\frac{T - T_a}{T_o - T_a} = \exp\left(-\frac{hA}{\rho C_p V} t\right) \tag{11.1.5}$$

Defining dimensionless variables by inspection of this equation, we can write this solution as

$$\Theta = e^{-\tau} \tag{11.1.6}$$

This is a good example of how the definition of dimensionless variables simplifies the *format* (the appearance) of the solution to a problem.

While this is a very simple problem, we may use it immediately to aid us in the design of an experiment.

---

**EXAMPLE 11.1.1** *A Design Problem: Measurement of the Convective Transfer Coefficient*

Our goal is to design an experiment to measure the convective heat transfer coefficient from a solid sphere to a moving airstream, as a function of air speed.

*Proposal* Measure the transient cooling of the sphere under conditions of a spatially uniform internal temperature for the sphere.

*Design Constraints* Measurement time should be convenient, say $\Theta$ (the dimensionless temperature change—Eq. 11.1.6) goes from 1.0 to 0.1 (from the initial state to within 10% of steady state) for times no shorter than $t = 20$ s and no longer than 2 min.

When $\Theta = 0.1$, we find the dimensionless time at which the body is nearly at thermal equilibrium from Eq. 11.1.6:

$$\exp(-\tau) = 0.1 \quad \text{or} \quad \tau = 2.3 \qquad (11.1.7)$$

Hence we should design the experiment so that an easily measured real time to reach the dimensionless time $\Theta = 0.1$ satisfies

$$\frac{hA}{\rho C_p V} t = 2.3 \qquad (11.1.8)$$

We need estimates for $h$ to determine the size of the sphere to be used in the experiment.

The lowest convection rate corresponds to heat transfer strictly by steady *conduction* from the surface of the sphere ($r = R$) through the air, radially out toward $r = \infty$. If we have only one-dimensional (radial) conduction in the surrounding air, then the radial flux is

$$q_r = -k \frac{dT}{dr} \qquad (11.1.9)$$

(here $k$ is the conductivity of the *exterior*—convective—medium).

At steady state the *rate* (not the flux) of heat conduction through the surrounding air must be constant with respect to radial position:

$$4\pi r^2 q_r = -4\pi k r^2 \frac{dT}{dr} = C \qquad (11.1.10)$$

The solution of this equation is

$$T = \frac{a}{r} + b \qquad (11.1.11)$$

in the air exterior to the sphere (i.e., in the region $r > R$).

Boundary conditions are stated in terms of the surface temperature and the ambient temperature:

$$T = T_R \quad \text{at} \quad r = R \qquad (11.1.12)$$
$$T = T_a \quad \text{at} \quad r \to \infty \qquad (11.1.13)$$

Hence

$$T - T_a = (T_R - T_a) \frac{R}{r} \qquad (11.1.14)$$

and we now can find the steady flux at the surface $r = R$ as

$$q_R = -k \left( \frac{\partial T}{\partial r} \right)_R = k \frac{T_R - T_a}{R} \quad \text{at} \quad r = R \qquad (11.1.15)$$

We define a *convective* heat transfer coefficient as

$$h = \frac{q_r}{T_R - T_a} = \frac{k}{R} \qquad (11.1.16)$$

(Pretending to be unaware that the heat transfer is due to conduction, we calculate the equivalent convective coefficient as the ratio of the heat flux to the temperature difference between the surface and the ambient.)

The minimum convective coefficient is given by this simple steady state analysis. Under any realistic conditions of external airflow, $h$ could be several orders of magnitude greater. We define a dimensionless heat transfer coefficient as

$$\frac{2hR}{k} = \text{Nu} = \text{Sh}_h = 2 \tag{11.1.17}$$

This dimensionless group is called a Nusselt number, and sometimes is referred to as the Sherwood number for heat transfer. The value of $\text{Nu} = 2$ follows from Eq. 11.1.16. We conclude that this is the minimum Nusselt number for convection from a sphere, and it corresponds physically to the case of an ambient fluid that is actually still, and heat transfer is strictly by steady conduction through the surrounding fluid.

Now we can return to the design problem. We know that the minimum Nusselt number is

$$\text{Nu} = 2 = \frac{hD}{k_{\text{air}}} \tag{11.1.18}$$

(Note that we now introduce $D = 2R$.) Guess that a more realistic value, under conditions featuring some degree of induced airflow, is $\text{Nu} = 10\text{--}100$, so $h$ is in the range

$$h = 10\frac{k_{\text{air}}}{D} \quad \text{to} \quad 100\frac{k_{\text{air}}}{D} \tag{11.1.19}$$

For air we may find $k_{\text{air}} = 0.014$ Btu/ft $\cdot$ h $\cdot$ °F.

Take $D = 1$ cm as a first try at specifying the size of the sphere to be used in the experiment. Then for the lower value of $h$

$$h = \frac{10(0.014)}{(1/30)\,\text{ft}} \,\text{Btu/ft} \cdot \text{h} \cdot \text{°F} = 4.2 \,\text{Btu/h} \cdot \text{ft}^2 \cdot \text{°F} \tag{11.1.20}$$

(or possibly 10 times higher for the higher value of $h$).

Convert to units of seconds for time:

$$h = 4.2 \,\text{Btu/h} \cdot \text{ft}^2 \cdot \text{°F} = \frac{4.2}{3600} = 1.17 \times 10^{-3} \,\text{Btu/s} \cdot \text{ft}^2 \cdot \text{°F} \tag{11.1.21}$$

To estimate thermal properties $\rho$ and $C_p$, take the solid material to be iron, for which $\rho = 436$ lb/ft$^3$ and $C_p = 0.12$ Btu/lb $\cdot$ °F. We found earlier that (near) steady state occurs at

$$\frac{hA}{\rho C_p V}t = 2.3 \tag{11.1.22}$$

and we can now solve for $t$. For a sphere we use

$$\frac{V}{A} = \frac{R}{3} = \frac{D}{6} \tag{11.1.23}$$

From

$$t = \frac{2.3\,\rho C_p V}{hA} = \frac{2.3(436)(0.12)}{1.17 \times 10^{-3}}\frac{(1/30)}{6} \,\text{s} \tag{11.1.24}$$

we find

$$t = 571 \,\text{s} \quad \text{for} \quad \text{Nu} = 10 \tag{11.1.25}$$

or

$$t = 57 \,\text{s} \quad \text{for} \quad \text{Nu} = 100 \tag{11.1.26}$$

It appears that an iron sphere, about a half-inch in diameter, is a good size because it yields a transient that takes place over a time scale that is neither too short for precision nor too long from a practical point of view. Of course this conclusion is based on the assumption that the Nusselt number lies in the range of 10 to 100.

There are also some hidden assumptions here. One is that the sphere is at a spatially uniform temperature throughout its volume. How do we evaluate this assumption? We have to look more carefully at the transient cooling process, *within the solid.*

## 11.1.2   Unsteady Heat Conduction Within Bounded Solids

For a solid sphere with symmetry, we may derive a thermal energy balance by equating the sum of the conductive flows of heat in and out of a spherical volume element (a region between radii $r$ and $r + dr$) to the change of thermal energy in the volume element. For a spherical surface the area is $4\pi r^2$ at any radius $r$, so the balance takes the form[2]

$$4\pi (r^2 q_r)_r - 4\pi (r^2 q_r)_{r+dr} = \frac{\partial}{\partial t} \rho C_p 4\pi r^2 dr\, T \tag{11.1.27}$$

Dividing both sides by $4\pi r^2 dr$ and letting $dr$ go to zero, we find,[3] using $q_r = -k\partial T/\partial r$,

$$-\frac{1}{r^2}\frac{\partial}{\partial r}(r^2 q_r) = \frac{k}{r^2}\frac{\partial}{\partial r}\left(r^2\frac{\partial T}{\partial r}\right) = \rho C_p\frac{\partial T}{\partial t} \tag{11.1.28}$$

Note that we have assumed that the $\rho C_p$ product is independent of temperature, hence of time. Equation 11.1.28 is often called Fourier's second law, or simply the transient heat conduction equation. In this specific format it is valid only for *spherically symmetric conductive heat transfer.* A transient heat balance for the more general three-dimensional case is derived in Section 11.2. Equation 11.1.28 is usually written in the form

$$\frac{\partial T}{\partial t} = \frac{k}{\rho C_p}\frac{1}{r^2}\frac{\partial}{\partial r}\left(r^2\frac{\partial T}{\partial r}\right) = \alpha\frac{1}{r^2}\frac{\partial}{\partial r}\left(r^2\frac{\partial T}{\partial r}\right) \tag{11.1.29}$$

We have defined $\alpha = k/\rho C_p$, which is called the *thermal diffusivity.* As an initial condition we take

$$T = T_o(r) \quad \text{at} \quad t = 0 \tag{11.1.30}$$

In the following we will assume that the initial temperature distribution is uniform across $r$, so $T_o(r) = T_o = $ constant. Boundary conditions are

$$\frac{\partial T}{\partial r} = 0 \quad \text{at} \quad r = 0 \text{ (symmetry)} \tag{11.1.31}$$

and Newton's law of cooling on the surface:

$$-k\frac{\partial T}{\partial r} = h(T(R) - T_a) \quad \text{at} \quad r = R \tag{11.1.32}$$

---

[2] Strictly speaking, this is the internal energy balance for an *incompressible* material, such as a solid or liquid. In this book we will not deal with cases where the energy balance is altered by compressibility effects. Thus, our application of this equation will not be appropriate to gases undergoing any significant pressure changes.

[3] For a solid undergoing modest temperature changes, the simplifying approximation used here—namely, that the thermal properties $C_p$ and $k$, and the density $\rho$ are constant with respect to temperature—is reasonable.

For this simple case, the solution to Eq. 11.1.29 is given by an infinite series, and we may write the temperature in the form $T = T(r, t; \alpha, T_o, R, k, h, T_a)$.

We can reduce the number of independent parameters if we first make the problem dimensionless with the following choices:

$$\Theta = \frac{T - T_a}{T_o - T_a} \qquad \xi = \frac{r}{R} \qquad X_{Fo} = \frac{\alpha t}{R^2} \qquad Bi_R = \frac{hR}{k_{solid}} \qquad \text{(11.1.33)}$$

The dimensionless time variable $X_{Fo}$ is called the Fourier number. Keep in mind that it is not a characteristic dimensionless *constant* for the system—it is an *independent variable*. It is a dimensionless time. Note the appearance of Bi—the Biot modulus for heat transfer—which is based on the $k$ of *the solid*.

Then with these changes Eq. 11.1.29 may be written as

$$\frac{\partial \Theta}{\partial X_{Fo}} = \frac{1}{\xi^2} \frac{\partial}{\partial \xi}\left(\xi^2 \frac{\partial \Theta}{\partial \xi}\right) \qquad \text{(11.1.34)}$$

with initial and boundary conditions

$$\Theta = 1 \qquad \text{at} \quad X_{Fo} = 0 \qquad \text{(11.1.35)}$$

$$\frac{\partial \Theta}{\partial \xi} = 0 \qquad \text{at} \quad \xi = 0 \qquad \text{(11.1.36)}$$

$$-\left.\frac{\partial \Theta}{\partial \xi}\right|_1 = Bi\,\Theta(1) \qquad \text{at} \quad \xi = 1 \qquad \text{(11.1.37)}$$

This change in format that nondimensionalization brings about does not make it any easier to solve the differential equation for the temperature field. However, it *does* aid our decision about how we will present the solutions in a form that is most convenient for computation.

The solution must be such that

$$\Theta = \Theta(\xi, X_{Fo}, Bi) \qquad \text{(11.1.38)}$$

The simplest case is $Bi \to \infty$. This implies that the external convection is so efficient that the surface of the sphere is maintained at the ambient temperature $T_a$. This corresponds to the boundary condition

$$\Theta = 0 \qquad \text{at} \quad \xi = 1 \qquad \text{(11.1.39)}$$

Then

$$\Theta = \Theta(\xi, X_{Fo}) \text{ only} \qquad \text{(11.1.40)}$$

A compact representation of this large Biot modulus case is given in Fig. 11.1.2, which shows the temperature profiles across the radius at several selected values of time (as the Fourier number). The *average* temperature throughout the sphere can also be obtained, and this is given in Fig. 11.1.3, again for the large Biot modulus case. (This graph also shows the corresponding solutions to the transient average temperature changes in a long circular cylinder with axial symmetry, and in an infinite planar slab with symmetry about the midplane.)

For the case of a finite Biot modulus, graphical solutions are available as well. Because of the additional parameter, the Biot modulus, it is more difficult to present these results in a compact format. One method of doing this is based on some properties of the analytical solutions to Eq. 11.1.34 and the boundary conditions that follow. In the cases

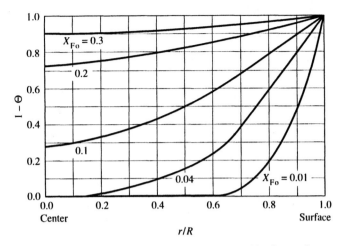

**Figure 11.1.2** Transient radial temperature profiles in a sphere. Surface maintained at $T = T_a$ (i.e., Bi $\gg$ 1).

of other simple geometries, such as the long cylinder and the thin rectangular slab, we must examine the solutions to the corresponding transient conduction equations that replace Eq. 11.1.34 for those geometries.

For the simple geometries of a thin rectangular slab (with the same boundary conditions on both exposed surfaces, i.e., equal convective losses from both surfaces), the long cylinder (for which the heat flow is strictly radially directed) and the sphere, the analytical solutions are available to us, and they are expressible as rapidly converging infinite (Fourier or Bessel) series. Thus, except for very short times (defined as $X_{Fo} <$ 0.1), a good approximation to the exact solution for the slab is found to be of the form

***Thin slab***

$$\Theta_1 = A_1 \exp(-\lambda_1^2 X_{Fo})\cos(\lambda_1 \xi) \tag{11.1.40a}$$

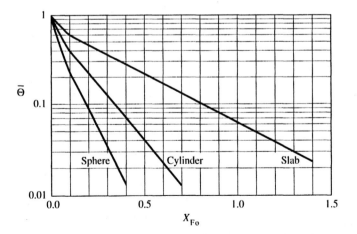

**Figure 11.1.3** Spatially averaged temperature at various times. Surface maintained at $T = T_a$ and Bi $\gg$ 1. In $X_{Fo} = \alpha t/R^2$, the length scale $R$ is the radius for the sphere or cylinder, and the half-thickness for a slab.

The subscript 1 is our reminder that we are using the first term of the series solution. For $\xi = 0$, which is along the midplane of a slab, the solution may be written in the form

$$\Theta_1^o = A_1 \exp(-\lambda_1^2 X_{Fo}) \tag{11.1.40b}$$

This permits us to write Eq. 11.1.40a in the format

$$\Theta_1(\xi, X_{Fo}) = \Theta_1^o(X_{Fo}) \cos(\lambda_1 \xi) \tag{11.1.40c}$$

One common method of presenting these results graphically is in the form of a plot of the midplane temperature (Eq. 11.1.40b) as a function of time ($X_{Fo}$) along with a plot of the cosine function $\cos(\lambda_1 \xi)$ by which we "correct" the midplane temperature to obtain the temperature at any position $\xi$ between the midplane and the surface. Essential to this graphical technique is a set of values of the coefficients $A_1$ and $\lambda_1$ as functions of the Biot number. That presentation is simplified by the recognition that for Bi $< 1$ a good approximation is $\lambda_1 \approx (n\text{Bi})^{0.5}$, where $n$ is 1, 2, and 3 for the slab, cylinder, and sphere, respectively. The plots in Fig. 11.1.4 present the results of these ideas in a form suitable for computation of transient behavior. This method of graphical presentation (pairs of figures such as Figs. 11.1.4b and c for a slab) is called a "Heisler" chart. It presents information similar to that in the Gurney–Lurie charts for the corresponding mass diffusion problems described earlier in this book, but in a different format.

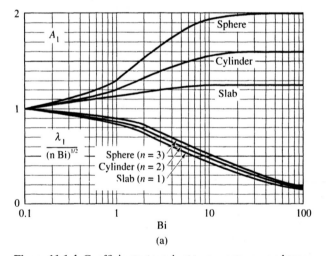

**Figure 11.1.4** Coefficients, transient temperatures, and transient temperature functions for a slab, a long cylinder, and a sphere. (a) Coefficients for Eqs. 11.1.40a, 11.1.40d, and 11.1.40e. (b) Transient midplane temperature for a slab for several Biot numbers. (c) Transient surface ($\xi = 1$) temperature function for a slab (Eq. 11.1.40c). (d) Transient axial temperature for a long cylinder for several Biot numbers. (e) Transient surface temperature function for a long cylinder for several Biot numbers. (f) Transient center temperature for a sphere for several Biot numbers. (g) Transient surface temperature function for a sphere for several Biot numbers.

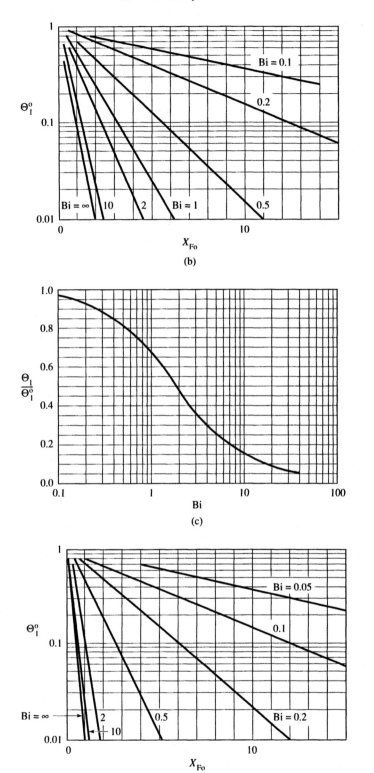

**Figure 11.1.4** (*Continued*)

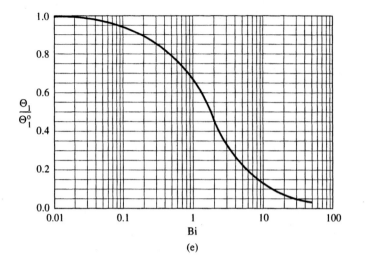

(e)

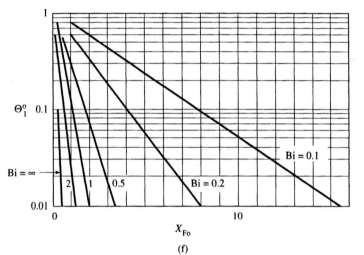

(f)

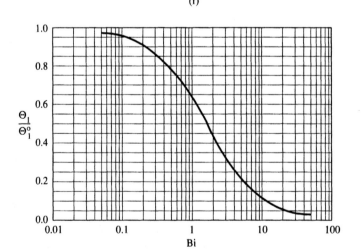

(g)

**Figure 11.1.4** (*Continued*)

For a long cylinder the analytical solution to the heat conduction equation in cylindrical coordinates may be approximated by the first term of an infinite series of Bessel functions:

**Long cylinder**

$$\Theta_1 = A_1 \exp(-\lambda_1^2 X_{Fo}) J_o(\lambda_1 \xi) = \Theta_1^o J_o(\lambda_1 \xi) \qquad (11.1.40d)$$

Figure 11.1.4d gives the function $\Theta_1^o$. Figure 11.1.4e gives the Bessel function $J_o(\lambda_1) = \Theta_1/\Theta_1^o$ for use in finding the surface temperature (at $\xi = 1$) with this expression.

For a sphere the one-term approximation takes the form

**Sphere**

$$\Theta_1 = A_1 \exp(-\lambda_1^2 X_{Fo}) \frac{\sin(\lambda_1 \xi)}{\lambda_1 \xi} = \Theta_1^o \frac{\sin(\lambda_1 \xi)}{\lambda_1 \xi} \qquad (11.1.40e)$$

Figure 11.1.4f shows the function $\Theta_1^o$ for this geometry, while Fig. 11.1.4g gives the required surface function $[\sin(\lambda_1 \xi)/\lambda_1 \xi]$ for use with this equation at $\xi = 1$. If there was a requirement for the temperature "history" at some value of $\xi$ between 0 and 1, we would plot, on Fig. 11.1.4g, curves of $\sin(\lambda_1 \xi)/\lambda_1 \xi$ for other $\xi$ values. (Of course we could provide similar plots for the slab and cylinder, as well.)

Often we are more interested in the *average* temperature history than with the detailed temperature profile history in a body. For the *average* temperature in a sphere, for example, we can solve Eq. 11.1.34 for $\Theta$, and find the average of $\Theta$ over the radius. The result is an infinite series, but for values of the Fourier number of order one and larger, a one-term approximation is given by

**Sphere (averaged) for Bi ≫ 1**

$$\overline{\Theta}_1 \equiv \frac{T_a - \overline{T}}{T_a - T_o} = 0.608 \exp(-9.87 X_{Fo}) \qquad (11.1.41)$$

This particular result is for Bi $= \infty$, and it is for a *sphere* with constant surface temperature $T_a$, with the sphere initially at $T_o$.

For a *planar slab* initially at $T_o$ and with constant surface temperature $T_a$ (this is also for the case Bi $= \infty$) the corresponding result for the *average* temperature is

**Thin planar slab (averaged) for Bi ≫ 1**

$$\overline{\Theta}_1 \equiv \frac{T_a - \overline{T}}{T_a - T_o} = 0.81 \exp(-2.47 X_{Fo}) \qquad (11.1.42)$$

For a *long circular cylinder* the corresponding result (again, for Bi $= \infty$) is

**Long cylinder (averaged) for Bi ≫ 1**

$$\overline{\Theta}_1 \equiv \frac{T_a - \overline{T}}{T_a - T_o} = 0.692 \exp(-5.78 X_{Fo}) \qquad (11.1.43)$$

In each of these cases the Fourier number is defined using as the length scale the radius of the sphere or cylinder, or the half-thickness of the slab.

> *Keep in mind that each of Eqs. 11.1.41 to 11.1.43 is valid only for the case of very large Biot numbers, and thus the assumption that the surface is maintained at ambient temperature $T_a$ is valid.*

Similar results are, of course, possible for finite values of Bi, and we could develop them from Eqs. 11.1.40a, 11.1.40d, and 11.1.40e, and Fig. 11.1.4a.

**EXAMPLE 11.1.2** *Cooling of a Long Copper Cylinder*

A long copper cylinder of diameter 0.25 inch was held in an airstream at a temperature $T_a = 100°F$. After 30 s the average cylinder temperature increased from its initial value of 50°F to 80°F. Estimate the heat transfer coefficient between the cylinder and the airstream (in units of Btu/h · ft² · °F).

Let us *assume* that because copper is such a good conductor of heat there is no internal conductive resistance to heat loss. *If* this is true, then Eq. 11.1.5 describes the transient behavior:

$$\frac{\overline{T} - T_a}{T_o - T_a} = \exp\left(\frac{-hAt}{\rho C_p V}\right) \tag{11.1.44}$$

(Note that we have used the notation for *average* temperature here.) For copper, $k = 220$ Btu/h · ft · °F and the thermal diffusivity is found to be $\alpha = 1.1$ cm²/s. (Note that we often find physical property values in different sources with different systems of units.) Then from

$$\frac{k}{\rho C_p} = \alpha = 1.1 \text{ cm}^2/\text{s} = 4.26 \text{ ft}^2/\text{h} \tag{11.1.45}$$

we find

$$\rho C_p = \frac{k}{\alpha} = \frac{220}{4.26} = 51.6 \text{ Btu/ft}^3 \cdot °F \tag{11.1.46}$$

For this problem

$$\frac{\overline{T} - T_a}{T_o - T_a} = \frac{80 - 100}{50 - 100} = 0.4 \tag{11.1.47}$$

so

$$e^{-\tau} = 0.4 \tag{11.1.48}$$

or

$$\tau = 0.92 = \frac{hAt}{\rho C_p V} \tag{11.1.49}$$

For $D = 0.25$ inch $= 0.0208$ ft we find

$$\frac{A}{V} = \frac{\pi DL}{\pi D^2 L/4} = \frac{4}{D} = 192 \text{ ft}^{-1} \tag{11.1.50}$$

From Eq. 11.1.49 we find (at $t = 30$ s $= 8.33 \times 10^{-3}$ h)

$$h = \frac{\rho C_p V}{At}(0.92) = \frac{51.6}{192}\frac{0.92}{8.33 \times 10^{-3}} = 29.7 \text{ Btu/h} \cdot \text{ft}^2 \cdot °F \tag{11.1.51}$$

We need to assess our original assumption (that convection controls the heat loss). To do so we examine the Biot number:

$$\text{Bi} = \frac{hR}{k} = \frac{29.7(0.0104)}{220} = 1.4 \times 10^{-3} \tag{11.1.52}$$

Since this value is much less than unity, the assumption that the heat loss is controlled by external convection, rather than by internal conduction, is a good one. This follows from the interpretation of the Biot number as a ratio of thermal resistances, since we can express the Biot number in the form

$$\text{Bi} = \frac{hR}{k} = \frac{R/k}{1/h} = \frac{\text{conductive resistance}}{\text{convective resistance}} \qquad (11.1.53)$$

A very small Biot number implies that the convective *resistance* to heat transfer is much greater than the conductive resistance, so heat transfer is limited by the rate of convection from the surface. In a loose sense, this means that conduction is so efficient (relative to convection) that the conductive flux can be achieved with a very small temperature gradient. Hence the internal temperature profile is relatively uniform when convection limits the heat exchange, which corresponds to Bi ≪ 1.

Another way to draw the same conclusion is through observation of Fig. 11.1.4e. For Bi < 0.01, the surface temperature function is essentially unity. From inspection of Eq. 11.1.40d this means that the radial dependence of temperature vanishes for Bi < 0.01.

## EXAMPLE 11.1.3   *Time to Cool a Solid Sphere, at any Biot Number*

Our goal is to prepare a graph of the time required for a solid sphere, initially at uniform temperature $T_0$, to cool to the ambient temperature $T_a$, for large and small values of the Biot number.

Our first task is to interpret the question, since it is ambiguous at this point. What do we mean, for example, when we ask for the time to cool to the ambient temperature? The answer is immediate: it's an infinite time! Hence what we require is a more practical definition of the *approach* to equilibrium. We will do that in terms of the dimensionless temperature $\Theta$, as defined in Eq. 11.1.33. We make the arbitrary choice that $\Theta = 0.01$ is close enough to equilibrium. This value of $\Theta$ means that the departure of the temperature from the steady state (ambient) value is within 1% of the initial temperature difference $T_0 - T_a$.

The second issue is to determine the point in the sphere at which we wish to have the solid reach (nearly) equilibrium. Because there is a temperature *distribution*, a value of $\Theta = 0.01$ will be reached at different times at different points. We will choose the *average* temperature, but we could have selected the center temperature as well.

Now we can use the models that we have at our disposal. For very small values of the Biot number we have Eq. 11.1.5, which we now rewrite in the form

$$\Theta = \exp\left(-\frac{hA}{\rho C_p V}t\right) = \exp(-3\text{Bi }X_{\text{Fo}}) \qquad (11.1.54)$$

To obtain this form, we have introduced the definitions of the Fourier and Biot numbers, and for a sphere we note that $A/V = 3/R$. Now, setting $\Theta = 0.01$ we find

$$X_{\text{Fo}} = \frac{1.535}{\text{Bi}} \qquad (11.1.55)$$

At the other extreme, when Bi is large and internal conduction is the dominant resistance, compared to external convection, we may use Eq. 11.1.41 in the form

$$\overline{\Theta} = 0.01 = 0.608 \exp(-9.87 X_{\text{Fo}}) \qquad (11.1.56)$$

(We have dropped the subscript 1 on $\Theta$—the reminder that we are using a one-term

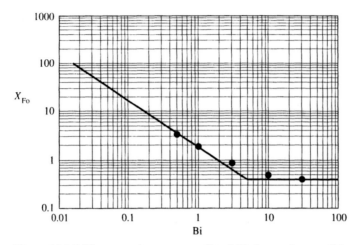

**Figure 11.1.5**  Time to reduce average $\Theta$ to 0.01, for a sphere; solid points indicate exact values.

approximation to the full infinite series solution.) It follows that

$$X_{\mathrm{Fo}} = 0.415 \qquad \text{(11.1.57)}$$

(We may also get this number, approximately, from Fig. 11.1.3.)

The next task is to find a model for the averaged $\Theta$ for arbitrary values of Bi. However, if we prepare a plot of the results for high and low values of Bi we see that we can interpolate for at least an approximate solution in the intermediate range. Figure 11.1.5 shows this, along with the exact values for this case.

**EXAMPLE 11.1.4**  *Dependence of Cooling Rate on Sphere Radius*

Two investigators are studying cooling of solid polyethylene spheres by placing different sized spheres in a temperature bath. One finds that if he doubles the radius of the sphere, its cooling rate is halved. The other finds that if she doubles the radius of the sphere, its cooling rate is reduced by a factor of nearly 4. All other conditions, such as physical properties, are identical, but sphere size may have differed in the respective experiments. Without assuming that either investigator is in error, explain these results.

The basic question here is: What is the effect of radius on cooling rate? We have two cooling models, for the extremes in the Biot number, to aid our search for an answer.

For Bi $\ll$ 1 we examine Eq. 11.1.53 and we find that the coefficient of time in the exponential (this coefficient is the rate of cooling) is inversely proportional to $R$. Hence, at very small values of Biot number, where external convection controls the rate of heat loss, the rate of cooling depends linearly (inversely) on the radius. If we double the radius, the cooling time is doubled, or the *rate* is halved.

For Bi $\gg$ 1 we see from Eq. 11.1.56 that the Fourier number is constant, independent of Biot number. Hence $t/R^2$ is constant. Thus if we double the radius we require *four* times the time to reach equilibrium. We conclude that as a result of having chosen different sphere sizes, the two investigators were operating at opposite extremes of heat transfer. In one case, internal conduction controlled the rate of cooling, and in the other it was the external convection that controlled.

---

**EXAMPLE 11.1.5**   *Temperature Across the Radius of a Sphere*

A solid sphere is cooling under conditions that $Bi = 0.2$. We want the (dimensionless) temperature at the center and surface of the sphere, and at a position $r = R/2$, all at the (dimensionless) time $X_{Fo} = 5$.

For the center temperature we use Fig. 11.1.4f, from which we find $\Theta_1^o = 0.06$ at $X_{Fo} = 5$, $Bi = 0.2$.

For the surface temperature we find, using Fig. 11.1.4g, $\Theta_1/\Theta_1^o = 0.9$ at $Bi = 0.2$. Hence

$$\Theta_1 = 0.9\Theta_1^o = 0.9(0.06) = 0.054 \tag{11.1.58}$$

at the surface $r = R$.

For $r = R/2$ ($\xi = 0.5$) Eq. 11.1.40e yields the result

$$\Theta_1 = \Theta_1^o \frac{\sin(\lambda_1 \xi)}{\lambda_1 \xi} = \Theta_1^o \frac{\sin(0.5\lambda_1)}{0.5\lambda_1} \cdot \tag{11.1.59}$$

Now we need the value of $\lambda_1$ for $Bi = 0.2$ for a sphere. From Fig. 11.1.4a, we find

$$\frac{\lambda_1}{\sqrt{3Bi}} \approx 1 \tag{11.1.60}$$

so

$$\lambda_1 = \sqrt{3(0.2)} = 0.78 \tag{11.1.61}$$

Then

$$\Theta_1 = \Theta_1^o \frac{\sin(0.39)}{0.39} = 0.6 \left(\frac{0.38}{0.39}\right) = 0.059 \tag{11.1.62}$$

Note that the temperature halfway between the center and the surface is much closer to the center temperature than it is to the surface temperature.

## 11.2   THE HEAT CONDUCTION EQUATION

In this section we derive the general partial differential equation that describes the temperature in a material at some point in space, as a function of position and time. We often refer to this equation as the *thermal energy equation,* or the *heat transfer equation.* It is an *internal energy* balance. It is derived in a manner very similar to that used to obtain the continuity equation and the species diffusion equation. We derive a special case of a more general energy equation here, because all the transient heat transfer applications considered in this introductory text are for solids or incompressible fluids.

As in similar derivations before, we select a control volume fixed in space and "keep book" on the transport (in this case) of energy across the boundaries of the volume element. *With the restriction to constant density materials,* we consider the internal energy $U$ per unit volume, defined by

$$U = \rho C_p T \tag{11.2.1}$$

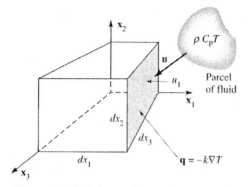

**Figure 11.2.1** Volume element for energy balance.

where $\rho$ is the mass density and $C_p$ is the heat capacity at constant pressure,[1] per unit of mass.

We pick a volume element bounded by the surfaces of a rectangular parallelepiped of dimensions $dx_1 dx_2 dx_3$, as depicted in Fig. 11.2.1. The procedure is very much like that used in deriving the continuity equation, which should be reviewed. Internal energy crosses the boundaries of the volume element by two mechanisms—conduction and convection. The convective flux $\mathbf{C}$ is simply the volumetric flux times the density of internal energy in an element of fluid passing across the boundary:

$$\mathbf{C} = \mathbf{u}U \tag{11.2.2}$$

The conductive flux will be given by Fourier's law in a vector format:

$$\mathbf{q} = -k\nabla T \tag{11.2.3}$$

Then internal energy crosses the face normal to the $x_1$ axis, at the plane $dx_1$, at a rate given by

$$R_{dx_1} = \left[ -u_1 \rho C_p T + k \frac{\partial T}{\partial x_1} \right]_{dx_1} dx_2 dx_3 \tag{11.2.4}$$

Note the signs that appear here. If $u_1$ were positive (pointing toward the right in Fig. 11.2.1), the parcel would be *leaving* the volume. If the temperature gradient were positive (temperature increasing toward the right), heat would be conducted *into* the volume.

Similar terms can be written for transport across the other five surfaces of the volume element, with appropriate changes in notation. When all six terms are added, with proper consideration of the signs of each, we find the net rate of transport of thermal energy across the surfaces in the form

$$\begin{aligned}
\Re = & \left[ -u_1 \rho C_p T + k \frac{\partial T}{\partial x_1} \right]_{dx_1} dx_2 dx_3 - \left[ -u_1 \rho C_p T + k \frac{\partial T}{\partial x_1} \right]_{x_1=0} dx_2\, dx_3 \\
& + \left[ -u_2 \rho C_p T + k \frac{\partial T}{\partial x_2} \right]_{dx_2} dx_1 dx_3 - \left[ -u_2 \rho C_p T + k \frac{\partial T}{\partial x_2} \right]_{x_2=0} dx_1\, dx_3 \\
& + \left[ -u_3 \rho C_p T + k \frac{\partial T}{\partial x_3} \right]_{dx_3} dx_1 dx_2 - \left[ -u_3 \rho C_p T + k \frac{\partial T}{\partial x_3} \right]_{x_3=0} dx_1\, dx_2
\end{aligned} \tag{11.2.5}$$

---

[1] Because of the assumption of constant density, we use $C_p$ instead of $C_v$.

The rate of accumulation of thermal energy (i.e., an increase or decrease of the amount of internal energy in the element over time) is then expressed by

$$\left[\frac{\partial}{\partial t}\rho C_p T\right] dx_1 dx_2 dx_3 = \Re + \mathscr{P} dx_1\, dx_2\, dx_3 \qquad (11.2.6)$$

where we allow for the possibility that energy may be transformed within the volume element at a rate (per unit volume) denoted $\mathscr{P}$. Examples of energy transformation include chemical or nuclear reactions, viscous friction, and electrical resistance heating.

When the bookkeeping details are finished and we examine the limit of $(dx_1 dx_2 dx_3) \to 0$, we obtain a partial differential equation, which we may write in a compact vector notation as

$$\frac{\partial}{\partial t}(\rho C_p T) + \nabla \cdot (\mathbf{u}\rho C_p T) = \nabla \cdot (k\nabla T) + \mathscr{P} \qquad (11.2.7)$$

For an incompressible medium, the convective term may be further simplified through the use of the continuity equation, with the result

$$\rho C_p \left[\frac{\partial T}{\partial t} + \mathbf{u} \cdot \nabla T\right] = k\, \nabla^2 T + \mathscr{P} \qquad (11.2.8)$$

(We have also assumed that the physical properties are constant in simplifying the terms.)

For the case of heat transfer in solids, or in fluids which are not in motion, the convective term vanishes, and we have

### The heat conduction equation

$$\frac{\partial T}{\partial t} = \frac{k}{\rho C_p}\nabla^2 T + \frac{\mathscr{P}}{\rho C_p} \qquad (11.2.9)$$

As noted earlier, the term in front of the Laplacian operator is called the thermal diffusivity, denoted

$$\alpha \equiv \frac{k}{\rho C_p} \qquad (11.2.10)$$

In Table 11.2.1 we present the component forms of Eq. 11.2.8 in the "top three" coordinate systems.

**EXAMPLE 11.2.1** *Cooling of a Long Plastic Cylinder*

A cylindrical rod of polyethylene, with a diameter of $D = 0.01$ m, at an initial uniform temperature of 200°C, is suddenly exposed to an airstream at 25°C for which the convective heat transfer coefficient has been found to be 23.2 Btu/h·ft²·°F. (Note Example 11.1.2, which indicates how we might determine a heat transfer coefficient under a given set of external cooling conditions.) How long is required for the surface temperature to drop from 200°C to 42.5°C? What is the *average* temperature at that time? What is the *centerline* temperature at that time?

Polyethylene has a thermal diffusivity of $\alpha = 1.3 \times 10^{-7}$ m²/s, and a conductivity of 0.33 kg·m/s²·K, in the temperature range of interest.

**Table 11.2.1** Component Forms of Eq. 11.2.8 (But with No Transformation Term)

Cartesian Coordinates

$$\rho C_p \left( \frac{\partial T}{\partial t} + u_x \frac{\partial T}{\partial x} + u_y \frac{\partial T}{\partial y} + u_z \frac{\partial T}{\partial z} \right) = k \left( \frac{\partial^2 T}{\partial x^2} + \frac{\partial^2 T}{\partial y^2} + \frac{\partial^2 T}{\partial z^2} \right) \qquad \textbf{(11.2.8a)}$$

Cylindrical Coordinates

$$\rho C_p \left( \frac{\partial T}{\partial t} + u_r \frac{\partial T}{\partial r} + \frac{u_\theta}{r} \frac{\partial T}{\partial \theta} + u_z \frac{\partial T}{\partial z} \right) = k \left[ \frac{1}{r} \frac{\partial}{\partial r} r \left( \frac{\partial T}{\partial r} \right) + \frac{1}{r^2} \frac{\partial^2 T}{\partial \theta^2} + \frac{\partial^2 T}{\partial z^2} \right] \qquad \textbf{(11.2.8b)}$$

Spherical Coordinates

$$\rho C_p \left( \frac{\partial T}{\partial t} + u_r \frac{\partial T}{\partial r} + \frac{u_\theta}{r} \frac{\partial T}{\partial \theta} + u_z \frac{\partial T}{\partial z} \right) = k \left[ \frac{1}{r^2} \frac{\partial}{\partial r} \left( r^2 \frac{\partial T}{\partial r} \right) + \frac{1}{r^2 \sin \theta} \frac{\partial}{\partial \theta} \left( \sin \theta \frac{\partial T}{\partial \theta} \right) + \frac{1}{r^2 \sin^2 \theta} \frac{\partial^2 T}{\partial \phi^2} \right]$$

$$\textbf{(11.2.8c)}$$

We begin by calculating the Biot modulus. We first convert the given value of the heat transfer coefficient to SI units and find $h = 132 \ \text{kg/s}^2 \cdot \text{K}$. Then

$$\text{Bi} = \frac{hR}{k} = \frac{132 \times 0.005}{0.33} = 2 \qquad \textbf{(11.2.11)}$$

The desired surface temperature corresponds to a value of $\Theta$ of

$$\Theta = \frac{T - T_a}{T_0 - T_a} = \frac{42.5 - 25}{200 - 25} = 0.1 \qquad \textbf{(11.2.12)}$$

From Fig. 11.1.4e for a cylinder we find, at $\text{Bi} = 2$, that the surface temperature function is

$$\frac{\Theta_1}{\Theta_1^o} = 0.45 \qquad \textbf{(11.2.13)}$$

At the same time, the axial temperature is the ratio of the surface temperature (in terms of $\Theta$) to this surface temperature function, or $\Theta^o = \Theta/0.45 = 0.1/0.45 = 0.22$. From Fig. 11.1.4d we find $X_{\text{Fo}} = 0.7$ or

$$t = \frac{R^2}{\alpha} X_{\text{Fo}} = \frac{0.005^2}{1.3 \times 10^{-7}} \times 0.7 = 135 \ \text{s} \qquad \textbf{(11.2.14)}$$

To find the average temperature we may begin (since the time is long) with Eq. 11.1.40d. We can find an expression of the form[2]

---

[2] To average $\Theta_1$ in Eq. 11.1.40d we must calculate the integral

$$\overline{\Theta}_1 \equiv \frac{1}{\pi R^2} \int_0^R \Theta_1(r) \, 2\pi r \, dr = 2 \int_0^1 \Theta_1(\xi) \xi \, d\xi$$

We also need to know that

$$\int_0^1 \xi J_0(\lambda \xi) \, d\xi = \frac{J_1(\lambda)}{\lambda}$$

$$\overline{\Theta} = \frac{2A_1 J_1(\lambda_1)}{\lambda_1} \exp(-\lambda_1^2 X_{\text{Fo}}) \tag{11.2.15}$$

With Fig. 11.1.4a we find the required values of $A_1 = 1.35$ and $\lambda_1 = 1.6$ for Bi $= 2$.

At the time $X_{\text{Fo}} = 0.7$, the average temperature is found from Eq. 11.2.15 to be $\Theta_1 = 0.14$ or $\overline{T} = 49.5°C$.

At the centerline we find, from $\Theta_1^o = 0.22$, that $T(0) = 63.5°C$. (Note that if we had used Fig. 11.1.3, which is valid only for Bi $\gg 1$, we would have found $\Theta_1^o = 0.012$, and our result would have been $\overline{T} = 27.1°C$, a value that is low by 22°C.)

At this point we should be wondering what to do if we are interested in the short-time ($X_{\text{Fo}} < 1$) behavior, where the one-term approximations (such as Eq. 11.1.40a) are not accurate. The next section explains how to proceed.

## 11.3   HEAT TRANSFER AT THE SURFACE OF A SEMI-INFINITE MEDIUM

With reference to Fig. 11.3.1, suppose the plane $y = 0$ is the boundary that separates a fluid at temperature $T_a$ from a solid at some initial uniform temperature $T_o$. The solid is large, in the sense that the entire region $0 < y < \infty$ is occupied by the solid. The boundary plane ($y = 0$) is infinite in the $x$ and $z$ directions. We call the region $y > 0$ a *semi-infinite medium*. We want to find the temperature history $T(y, t)$ that follows when the surface $y = 0$ is suddenly (at $t = 0$) brought in contact with the fluid. The transient heat conduction equation is simply

$$\frac{\partial T}{\partial t} = \alpha \frac{\partial^2 T}{\partial y^2} \tag{11.3.1}$$

and the initial and boundary conditions can now be written.

The initial condition is simply

$$T = T_o \quad \text{for} \quad t \leq 0 \tag{11.3.2}$$

Far from the plane $y = 0$ we have

$$T = T_o \quad \text{as} \quad y \to \infty \tag{11.3.3}$$

For the boundary condition at the exposed surface, let's first assume that the surface heat transfer coefficient is so large that the surface is brought to the ambient temperature instantly for $t > 0$. Then

$$T = T_a \quad \text{on} \quad y = 0 \quad \text{for} \quad t > 0 \tag{11.3.4}$$

Now we define a dimensionless temperature as

$$\Theta = \frac{T - T_a}{T_o - T_a} \tag{11.3.5}$$

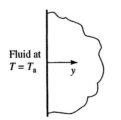

**Figure 11.3.1** The semi-infinite medium.

and write the transient heat conduction equation in the format

$$\frac{\partial \Theta}{\partial t} = \alpha \frac{\partial^2 \Theta}{\partial y^2} \tag{11.3.6}$$

$$\Theta = 1 \qquad \text{for} \quad t \leq 0 \tag{11.3.7}$$

$$\Theta = 0 \qquad \text{for} \quad y = 0 \tag{11.3.8}$$

$$\Theta \to 1 \qquad \text{as} \quad y \to \infty \tag{11.3.9}$$

There is no *natural* length scale in this problem, since the solid is unbounded. Hence we do not define a dimensionless space variable.

We can show that if we define a new (combined) variable as

$$\eta = \frac{y}{\sqrt{4\alpha t}} \tag{11.3.10}$$

then the *partial* differential equation becomes an *ordinary* differential equation:

$$\frac{d^2 \Theta}{d\eta^2} + 2\eta \frac{d\Theta}{d\eta} = 0 \tag{11.3.11}$$

with boundary conditions

$$\Theta = 0 \qquad \text{for} \quad \eta = 0 \tag{11.3.12}$$

$$\Theta = 1 \qquad \text{for} \quad \eta = \infty \tag{11.3.13}$$

(Note how two of the boundary conditions given earlier—the initial condition and the condition far from the surface—collapse to a single boundary condition for $\eta = \infty$.)

The solution of this problem is (see Problem 11.21)

$$\Theta = \text{erf } \eta = \text{erf} \left[ \frac{y}{(4\alpha t)^{1/2}} \right] \tag{11.3.14}$$

With this solution we can define a "thermal penetration length" $\delta_T$ such that the solid "senses" the presence of a heat flux at the surface $y = 0$ only within the region $0 \leq y \leq \delta_T$. A simple definition of $\delta_T$ would be the distance at which $\Theta$ goes from the surface temperature $\Theta = 0$ to (nearly) the initial temperature (which is $\Theta = 1$), say $\Theta = 0.99$. For $y > \delta_T$, the temperature will have hardly changed from its initial value. With this definition, that is,

$$\Theta = 0.99 = \text{erf} \left[ \frac{\delta_T}{(4\alpha t)^{1/2}} \right] \tag{11.3.15}$$

we find

$$\frac{\delta_T}{(4\alpha t)^{1/2}} = 2 \tag{11.3.16}$$

or

$$\delta_T = 4(\alpha t)^{1/2} \tag{11.3.17}$$

One conclusion we draw from this result is that for as long a time as a solid of finite thickness is thicker than $\delta_T$, it behaves like a semi-infinite solid. This suggests the use of Eq. 11.3.14 as an alternative to Eq. 11.1.40a when our interest is in the temperature history at "short" times. The advantage of this approach is that for short times the one-term approximation (Eq. 11.1.40a) is not accurate, and it is tedious to use the infinite

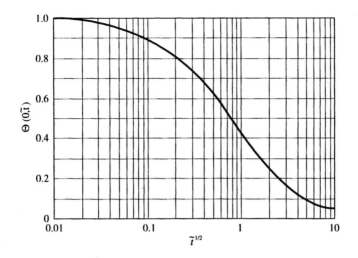

**Figure 11.3.2** The surface temperature function of Eq. 11.3.21.

series solution, which would require the summation of many terms to yield an accurate result. On the contrary, Eq. 11.3.14 is a simple analytical representation that requires only values of the error function.

In this model we have assumed that $T = T_a$ at $y = 0$. This is equivalent to the assumption that the surface is at the "bath" temperature. This is not so if there is a *finite* heat transfer coefficient at $y = 0$. In that case, the boundary condition at the surface becomes

$$-k\frac{\partial T}{\partial y}\bigg|_{y=0} = h(T_a - T) \tag{11.3.18}$$

Now the solution for $\Theta$ takes the form

$$\Theta = \text{erf}\frac{\bar{y}}{2\bar{t}^{1/2}} + \exp(\bar{y} + \bar{t})\,\text{erfc}\left(\frac{\bar{y}}{2\bar{t}^{1/2}} + \bar{t}^{1/2}\right) \tag{11.3.19}$$

where

$$\bar{y} \equiv \frac{hy}{k} \quad \text{and} \quad \bar{t} \equiv \frac{h^2 t}{\rho C_p k} = \left(\frac{h}{k}\right)^2 \alpha t \tag{11.3.20}$$

In the case of a finite heat transfer coefficient, we are often most interested in the history of the surface temperature:

$$\Theta(0, \bar{t}) = \exp \bar{t}\,\text{erfc}\,\bar{t}^{1/2} \tag{11.3.21}$$

This result is plotted in Fig. 11.3.2.

**EXAMPLE 11.3.1**   *Surface Temperature of a Cooling Sheet*

Polyethylene is coated onto an insulated substrate that is moving at 20 cm/s. The molten polymer is coated at a uniform melt temperature of 400°F, and cooling is achieved by blowing air, at 80°F, across the film. From earlier heat transfer studies it is known that the heat transfer coefficient is $h = 0.08$ cal/cm · s · °C. The coating thickness is 0.1 cm. At what point downstream does the *surface* temperature fall to 144°F?

In this example we want to find the surface temperature as a function of time. We have $B = 0.1$ cm and

$$h = 0.08 \text{ cal/cm}^2 \cdot \text{s} \cdot \text{C} = 3.35 \times 10^3 \text{ W/m}^2 \cdot \text{K} \qquad \text{(11.3.22)}$$

For polyethylene we take

$$k = 0.33 \text{ W/m} \cdot \text{K} \qquad \text{(11.3.23)}$$

and

$$\alpha = 1.3 \times 10^{-7} \text{ m}^2/\text{s} \qquad \text{(11.3.24)}$$

We find the Biot number to be

$$\text{Bi} = \frac{hB}{k} = \frac{3350 \times 10^{-3}}{0.33} \approx 10 \qquad \text{(11.3.25)}$$

On the surface, the dimensionless temperature of interest is

$$\Theta_s = \frac{T_s - T_a}{T_o - T_a} = \frac{144 - 80}{400 - 80} = \frac{64}{320} = 0.2 \qquad \text{(11.3.26)}$$

From Fig. 11.1.4c, at Bi = 10, the surface temperature function is

$$\frac{\Theta}{\Theta_1^o} = 0.15 \qquad \text{(11.3.27)}$$

From

$$\Theta_s = \Theta_1^o \frac{\Theta}{\Theta_1^o} = 0.2 \qquad \text{(11.3.28)}$$

we calculate the midplane temperature to be

$$\Theta_1^o = \frac{0.2}{0.15} > 1 \qquad \text{(11.3.29)}$$

Our plan was to find $\Theta_1^o$ and use Fig. 11.1.4b, for Bi = 10, to find $X_{\text{Fo}}$. But $\Theta_1^o$ cannot exceed unity! What's wrong?

A solution procedure based on Figs. 11.1.4b and 11.1.4c (and the corresponding pairs 11.1.4d/e and 11.1.4f/g) will be accurate only to the extent that Eq. 11.1.40a (or 11.1.40d and 11.1.40e) is a good approximation to $\Theta(\xi, X_{\text{Fo}})$. For short times we know that the one-term approximations are not accurate. Hence we will get into trouble if we use Figs. 11.1.4b/c under conditions that correspond to "short times." In general, this means that we must avoid cases of small $X_{\text{Fo}}$ (say $X_{\text{Fo}} < 1$) and large Bi (say Bi > 10), which would correspond to very rapid temperature changes.

For "short times" we use the semi-infinite slab solution (in this case, Eq. 11.3.21)

$$\Theta(0, \tilde{t}) = \exp \tilde{t} \text{ erfc } \tilde{t}^{1/2} = 0.2 \qquad \text{(11.3.30)}$$

and find, from Fig. 11.3.2,

$$\tilde{t}^{1/2} = 2.65 \qquad \text{(11.3.31)}$$

or

$$\tilde{t} = 7.02 = \frac{h^2 t}{\rho C_p k} \qquad \text{(11.3.32)}$$

For the $\alpha$ and $k$ values given above, and because

$$\rho C_{\mathrm{p}} = \frac{k}{\alpha} = 2.5 \times 10^6 \text{ J/m}^3 \cdot \text{K}$$

we find

$$7.02 = \frac{(3.35 \times 10^3)^2 t}{0.33(2.5 \times 10^6)} \tag{11.3.33}$$

and

$$t = 0.52 \text{ s} \tag{11.3.34}$$

Then the distance along the cooling line over which the sheet has traveled is

$$z = Ut = 20 \text{ cm/s } (0.52 \text{ s}) = 10.4 \text{ cm} \tag{11.3.35}$$

This completes the task set at the beginning of the example.

As an alternative to our use of Eq. 11.3.30, we may work with the more exact (i.e., not truncated) series solution for the slab of finite thickness. This is known to be:

$$\Theta = \sum_{n=1}^{\infty} \frac{4 \sin \lambda_n}{2\lambda_n + \sin 2\lambda_n} \exp(-\lambda_n^2 X_{\mathrm{Fo}}) \cos(\lambda_n \xi) \tag{11.3.36}$$

The eigenvalues $\lambda_n$ are the roots to

$$\lambda_n \tan \lambda_n = \text{Bi} \tag{11.3.37}$$

(For Bi = 10, the first three roots are $\lambda_1 = 1.429$, $\lambda_2 = 4.306$, $\lambda_3 = 7.228$.)

At $\xi = 1$ (the exposed surface), we write

$$\Theta(1) = \sum_{n=1}^{\infty} \frac{4 \sin \lambda_n}{2\lambda_n + \sin 2\lambda_n} \exp(-\lambda_n^2 X_{\mathrm{Fo}}) \cos \lambda_n \tag{11.3.38}$$

If we now set $\Theta(1) = 0.2$, and write the first two terms of the series above, we find

$$\Theta(1, X_{\mathrm{Fo}}) = 0.178 \exp(-2.04 \, X_{\mathrm{Fo}}) + 0.155 \exp(-18.5 \, X_{\mathrm{Fo}}) \tag{11.3.39}$$

By trial-and-error procedures, we find, for $\Theta(1, X_{\mathrm{Fo}}) = 0.2$,

$$X_{\mathrm{Fo}} = 0.068 = \frac{t\alpha}{B^2} \tag{11.3.40}$$

Then

$$t = \frac{0.068(10^{-3})^2}{1.3 \times 10^{-7}} = 0.52 \text{ s} \tag{11.3.41}$$

which is the same value we obtained using the short-time solution.

Although we speak of this solution procedure as one that corresponds to "short time" in the thermal history of the solid, we have not given a useful definition of "short time." Let's go back to the criterion of Eq. 11.3.17, and write it in the form

$$\frac{\delta_{\mathrm{T}}}{B} = 4 \left( \frac{\alpha t}{B^2} \right)^{1/2} \tag{11.3.42}$$

A definition of "short time" for the slab geometry would be that the penetration thickness be no more than the half-thickness $B$:

$$4 \left( \frac{\alpha t}{B^2} \right)^{1/2} \leq 1 \tag{11.3.43}$$

or

$$X_{Fo} = \frac{\alpha t}{B^2} \le \frac{1}{16} \le 0.06 \qquad \textbf{(11.3.44)}$$

Keep in mind that this is for the case of a large heat transfer coefficient: $Bi \gg 1$.

## 11.4 MODELING AND DIMENSIONAL ANALYSIS; GEOMETRICAL SIMILARITY

Suppose we wish to investigate heat transfer in a solid of complex geometry. We are not able to solve the heat conduction equation analytically because of the complexity of the geometry. The heat transfer is determined by the balance between conduction within the solid and convection at the surface to the external fluid. The magnitude of the Biot number will determine whether one or the other mechanism dominates. Here we want to learn how to design an experiment and interpret the data, with the goal of being able to extrapolate our results to other cases of a specific class of heat transfer problem. We often do this by building a *physical* model of the system of interest. Sometimes the model is small, as in Fig. 11.4.1, but it could be larger in some cases. The important point is that it must be a *scale model*.

The notion of a scale model is very simple. All linear dimensions are changed by the same factor in converting the system of interest to a scale model. We can also think of this in terms of a photograph. Two systems are scale models of each other if we cannot tell which is which from a photograph of either. When two bodies are scaled in this way we say that they are *geometrically similar*. Figure 11.4.2 gives another, simpler, picture of a system and a scale model, and introduces the idea of dimensionless geometrical parameters called "shape factors."

In this example, the ratios $L_1/L_2$ and $L_1/L_3$ define the shape of the body. They are the shape factors. Geometrical similarity requires that all corresponding shape factors to be identical between the solid and its scale model.

Returning to the bodies of Fig. 11.4.1, we note that heat transfer in either S or M is described by the same heat conduction equation:

$$\frac{\partial T_S}{\partial t} = \alpha_S \left( \frac{\partial^2 T_S}{\partial x^2} + \frac{\partial^2 T_S}{\partial y^2} + \frac{\partial^2 T_S}{\partial z^2} \right) = \alpha_S \nabla^2 T_S \qquad \textbf{(11.4.1)}$$

for S and similarly for M. The initial condition is taken to be

$$T_S = T_{o,S} \quad \text{at} \quad t = 0 \qquad \textbf{(11.4.2)}$$

One possible boundary condition is that of uniform and constant surface temperature:

$$T_S = T_a \quad \text{on the surface} \quad f(x, y, z) = 0 \qquad \textbf{(11.4.3)}$$

An alternative boundary condition could take the form of Newtonian cooling:

$$-k \nabla T = h(T - T_a) \qquad \textbf{(11.4.4)}$$

**Figure 11.4.1** An example of a complex body, S, and a scale model, M, of the body.

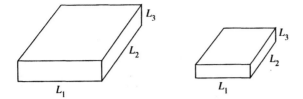

**Figure 11.4.2** A solid and its scale model.

If we make the equations dimensionless in the usual way:

$$\Theta = \frac{T - T_a}{T_o - T_a} \qquad X_{Fo} = \frac{\alpha t}{L_1^2} \qquad x' = \frac{x}{L_1} \qquad \textbf{(11.4.5)}$$

we find, for each solid,

$$\frac{\partial \Theta_S}{\partial X_{Fo,S}} = \nabla'^2 \Theta_S \qquad\qquad \frac{\partial \Theta_M}{\partial X_{Fo,M}} = \nabla'^2 \Theta_M \qquad \textbf{(11.4.6)}$$

$$\Theta_S = 1 \quad \text{at} \quad X_{Fo,S} = 0 \qquad \Theta_M = 1 \quad \text{at} \quad X_{Fo,M} = 0 \qquad \textbf{(11.4.7)}$$

The boundary conditions are one or the other of

$$\begin{cases} \Theta_S = 0 \text{ on the surface} & \Theta_M = 0 \text{ on the surface} \\ \nabla' \Theta_S = -Bi_S \Theta_s & \nabla' \Theta_M = -Bi_M \Theta_M \end{cases} \qquad \begin{matrix}\textbf{(11.4.8)}\\ \textbf{(11.4.9)}\end{matrix}$$

It follows that the solutions of either set of equations depend on the variables and parameters in the equations. Hence for either S or M we may say that

$$\Theta = \Theta(x', X_{Fo}; Bi) \qquad \textbf{(11.4.10)}$$

We conclude that if we use geometrically similar systems in which the Biot numbers are identical, the two systems have the same dependence of the dimensionless temperature on dimensionless position $x'$, and dimensionless time $X_{Fo}$.

---

**EXAMPLE 11.4.1**    *Experimental Design and Analysis of Data*

A solid object of odd shape is to be inserted as a temperature probe into a fluid. The object will heat up, and it is necessary to protect the interior, which contains a sensing device, from thermal damage. Design a model experiment to determine how long the probe can be exposed to the fluid.

We have the following data for the probe shown in Fig. 11.4.3.

$$L_1 = 0.10 \text{ cm} \qquad L_2 = 0.02 \text{ cm} \qquad L_3 = 0.005 \text{ cm}$$
$$\alpha = 10^{-7} \text{ m}^2/\text{s} = 10^{-3} \text{ cm}^2/\text{s}$$
$$Bi \gg 1 \qquad T_o = 70°F \qquad T_a = 800°F$$

Assume that damage occurs at a critical value $T^* = 455°F$ at a specific point within the probe. Thus

$$\Theta_S^* = \frac{T^* - T_a}{T_o - T_a} = \frac{455 - 800}{70 - 800} = \frac{345}{730} = 0.47 \qquad \textbf{(11.4.11)}$$

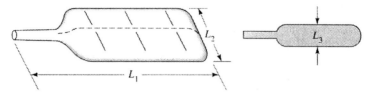

**Figure 11.4.3** A temperature probe.

We must design a model, measure its thermal response, find the time at which $\Theta_M = \Theta_S^* = 0.47$, and then calculate $X_{Fo,S}^*$ from $X_{Fo,M}^*$.

*Preliminary Design*   We choose

$$L_1 = 5 \text{ cm} \qquad \text{(linear scale-}up\text{ by } 50\times)$$
$$\alpha_M = 10^{-1} \text{ cm}^2/\text{s}$$

We cannot use the same $T_a$ because the model is going to be constructed of an inexpensive material that will melt. The maximum useful $T_a$ is 400°F. Choose $T_o = 70$°F, and find $T_M^*$. For the value of critical temperature $T_M^*$ (Eq. 11.4.11), we find

$$\Theta^* = 0.47 = \frac{T^* - T_a}{T_o - T_a} = \frac{T^* - 400}{70 - 400} = \frac{T^* - 400}{-330} \tag{11.4.12}$$

$$T_M^* = 245°F \tag{11.4.13}$$

If the systems are geometrically similar, and if the same physical phenomena are occurring, then

$$\Theta_M = \Theta_S \tag{11.4.14}$$

if

$$\left\{ \begin{array}{l} \mathbf{x}_M' = \mathbf{x}_S' \\ X_{Fo,M} = X_{Fo,S} \\ Bi_M = Bi_S \end{array} \right\}$$

because the equations and boundary conditions are identical.

Suppose we are interested in a specific solution, say a particular $T$ profile, or some average temperature, or a specified degree of approach to equilibrium. Then $\Theta$ is specified, so

$$\Theta_M^* = \Theta_S^* \tag{11.4.15}$$

This will occur at a particular value of time

$$X_{Fo,S} = X_{Fo,S}^* \tag{11.4.16}$$

If we do an experiment such that[1] $Bi_M = Bi_S$ and measure the time at which the *critical thermal situation* $\Theta_S^*$ occurs in the model, say $t_M = t_M^*$, then we can say that

$$\Theta_S^* = \Theta_M^* \qquad \text{when} \qquad X_{Fo,S}^* = X_{Fo,M}^* \tag{11.4.17}$$

---

[1] Actually, since $Bi_S \gg 1$, we only need to have $Bi_M \gg 1$, not $Bi_S = Bi_M$.

or

$$\left(\frac{T - T_a}{T_o - T_a}\right)_S^* = \left(\frac{T - T_a}{T_o - T_a}\right)_M^* \qquad (11.4.18)$$

when

$$\frac{\alpha_S t_S^*}{L_{1S}^2} = \frac{\alpha_M t_M^*}{L_{1M}^2} \qquad (11.4.19)$$

Hence

$$t_S^* = \left(\frac{L_{1S}}{L_{1M}}\right)^2 \frac{\alpha_M}{\alpha_S} t_M^* = \left(\frac{0.1}{5}\right)^2 \frac{10^{-1}}{10^{-3}} t_M^* \qquad (11.4.20)$$

and we find that

$$t_S^* = 0.04 t_M^* = \frac{t_M^*}{25} \qquad (11.4.21)$$

This leaves open the question of whether $t_M^*$ is easily measured (not too short or too long). To estimate $t_M^*$, we will use a simple model for a slab, of thickness $L_3 = 50(0.005)$ cm $= 0.25$ cm. (The factor of 50 is our scale-up factor.) The half-thickness is then 0.125 cm. At $\Theta^* = 0.47$, Fig. 11.1.4b gives (for Bi $\gg 1$)[2]

$$X_{Fo} = \frac{\alpha t^*}{R^2} \approx 0.4 \qquad (11.4.22)$$

at the center. (Note that $2R = L_3$ is used.) Then we find

$$t_M^* = \frac{0.4 R^2}{\alpha} = \frac{0.4(0.125)^2}{0.10} = 0.0625 \text{ s} \qquad (11.4.23)$$

This is a bad design! While our result is just an order of magnitude estimate, we find that $t_M^*$ is much too short to permit us to gather data accurately. We need to decrease $\alpha$ and increase $R$ in the model. Note that if this order of magnitude estimate is at all reasonable, the *system* response will be in the neighborhood of

$$t_S^* = \frac{t_M^*}{25} = 2.5 \times 10^{-3} \text{ s} \qquad (11.4.24)$$

## SUMMARY

The rate at which a "body" gains or loses heat with respect to its surroundings is a function of the rate of internal conduction and the rate of transfer across the interface of the body with the surroundings. Analytical solutions are available when the transfer occurs to (or from) bodies of very simple geometries, and especially when the initial and boundary conditions are symmetric with respect to the geometry. We develop or

---

[2] The graph is not easy to read with any precision in this region. It is more accurate to use Eq. 11.1.40a, in the limit of large Bi. This turns out to be

$$\Theta_1 = \frac{4}{\pi} \exp\left(-\frac{\pi^2}{4} X_{Fo}\right)$$

and we find $X_{Fo} = 0.404$ for $\Theta_1^* = 0.47$.

present models for one-dimensional conduction in the cases of the thin slab, the long cylinder, and the sphere, but with uniform initial and boundary conditions. The solutions can be written as infinite series, and graphical methods exist for facilitating computation of the local and average temperatures as functions of time. A key parameter is a measure of the efficiency of convective heat transfer across the interface of the body with the ambient medium. The Biot modulus is this dimensionless measure of convective heat transfer.

The general heat conduction equation is derived and presented in vector format (Eq. 11.2.8) and in component forms in Table 11.2.1.

The concept of conduction in a *semi-infinite medium* is introduced because of the mathematical simplification that follows when one boundary is put off to infinity. This leads to the concept of "short time" in the history of heat exchange.

When body shape or external conditions are too complex to allow analytical solutions of Eq. 11.2.8, we may gain valuable insight into the heat transfer process through dimensional analysis. This in turn can guide the design of an experimental study of the history of heat transfer in complex engineering situations.

## PROBLEMS

**11.1** The data presented below were obtained by Banan *et al.* [*J. Cryst. Growth*, **113**, 557 (1991)] in an attempt to measure heat transfer coefficients in a furnace used for growth of large single crystals. The experiment was performed using a cylinder of diameter 1.1 cm and length 7 cm, in the center of which was a thermocouple. The cylinder was brought to a uniform temperature of 875°C and quickly inserted into the furnace. The thermal diffusivity and thermal conductivity of the cylinder were $\alpha = 0.0628$ cm$^2$/s and $k = 0.146$ W/cm · K, respectively, at $T \approx 850$°C. From these data, find the heat transfer coefficient (in units of W/cm$^2$ · K).

| Time (s) | 0 | 25 | 50 | 75 | >200 |
|---|---|---|---|---|---|
| $T$(°C) | 871 | 831 | 810 | 800 | 792 |

**11.2** Give the definition of the dimensionless time $\tau$ that appears in Eq. 11.1.6, and relate $\tau$ to the Fourier number and the Biot number.

**11.3** Three solids of equal area per volume are made of the same material. Using Eqs. 11.1.41 to 11.1.43, determine the rank order in which they approach thermal equilibrium from an initially uniform temperature, for the case of large Biot number. Does the rank change if the solids have the same values of $R$, instead of the same area/volume ratio?

**11.4** Add curves for the slab and the long cylinder to Fig. 11.1.5.

**11.5** A long copper wire, 0.25 inch in diameter, was held in an airstream of temperature 100°F. After 30 s, the *average* wire temperature had increased from 50°F to 80°F. Estimate the heat transfer coefficient between the wire and the airstream (Btu/h · ft$^2$ · °F). Use Table P11.5 for thermal properties of copper.

**11.6** A stainless steel sphere of radius 1.5 inches, initially at 600°F, is exposed to an ambient medium at 100°F. After 10 s the *average* temperature is 230°F. Can the external heat transfer coefficient be estimated with these data? Comment on the precision of the method, if the temperature measurement is accurate to ±1°F. Use

**Table P11.5** Thermal Properties of a Few Selected Materials at 25°C

| | $\alpha_T$(m$^2$/s) | $k$(W/m · K) | | $\alpha_T$(m$^2$/s) | $k$(W/m · K) |
|---|---|---|---|---|---|
| Aluminum | $9.2 \times 10^{-5}$ | 228 | Brick | $6 \times 10^{-7}$ | 1.2 |
| Copper | $10^{-4}$ | 386 | Glass (Pyrex) | $6 \times 10^{-7}$ | 1.1 |
| Iron | $2 \times 10^{-5}$ | 73 | Air | $2.1 \times 10^{-5}$ | 0.027 |
| Stainless steel | $4.4 \times 10^{-6}$ | 16 | Water | $1.4 \times 10^{-7}$ | 0.59 |

Table P11.5 for thermal properties of stainless steel.

**11.7** A stainless steel ball, 5 cm in diameter, and initially at a uniform temperature of 450°C, is suddenly placed in a bath maintained at 100°C. The convective heat transfer coefficient is $h = 1.76$ Btu/h·ft²·°F. Calculate the number of hours required for the ball to attain a temperature of 150°C. Use Table P11.5 for thermal properties of stainless steel.

**11.8** A solid with initial temperature 60°C is placed in a well-stirred bath maintained at 70°C. The same solid, but with initial temperature 95°C, is placed in a well-stirred bath maintained at 35°C. Which material reaches steady state faster? Explain your answer.

**11.9** Two investigators are studying cooling of solid polyethylene spheres by placing spheres of different sizes in a low temperature bath. All experimental conditions are identical, except for the sphere size. One investigator uses spheres of radii 2 and 4 cm. The other uses radii of 0.025 and 0.05 cm. Find the time for the spheres to reach steady temperature in each case. By what factor does the cooling time change in going from 2 cm to 4 cm? From 0.025 cm to 0.05 cm?

Assume that $\alpha = 10^{-3}$ cm²/s, $k = 4 \times 10^{-4}$ cal/s·cm·K, and $h = 2 \times 10^{-3}$ cal/s·cm²·K.

**11.10** Prepare a graph from which the time for a long solid cylinder to reach thermal equilibrium may be found for any Biot modulus. Assume the cylinder is initially at uniform temperature and then is suddenly placed in a constant temperature bath having a constant heat transfer coefficient.

**11.11** Determine the temperature response [plot $T(t)$] of a 0.10 cm diameter copper wire originally at 150°C when suddenly immersed in (a) water at 40°C ($\bar{h}_c = 80$ W/m²·K) and (b) air at 40°C ($\bar{h}_c = 10$ W/m²·K). Use Table P11.5 for thermal properties of copper.

**11.12** A steel cylinder 2 m long and 0.2 m in diameter, with $k = 40$ W/m·K and $\alpha = 1.0 \times 10^{-5}$ m²/s, initially at 400°C, is suddenly immersed in water at 50°C. If the surface heat transfer coefficient is 200 W/m²·K, calculate the following after 20 min of immersion:

**a.** the center temperature

**b.** the surface temperature

**c.** the heat transferred to the water during the initial 20 min

**11.13** A stainless steel cylinder ($k = 14.4$ W/m·K, $\alpha = 3.9 \times 10^{-6}$ m²/s) is heated to 593°C preparatory to undergoing a forming process. If the minimum temperature permissible for forming is 482°C, how long may the cylinder be exposed to air at 38°C if the average external heat transfer coefficient is 85 W/m²·K? The axial length is 200 cm and the diameter is 10 cm.

**11.14** A long copper cylinder 0.6 m in diameter and initially at a uniform temperature of 38°C is placed in a water bath at 93°C. Assuming that the heat transfer coefficient between the copper and the water is 1248 W/m²·K, calculate the time required to heat the center of the cylinder to 66°C. As a first approximation, neglect the temperature gradient within the cylinder; then repeat your calculation without this simplifying assumption and compare your results.

**11.15** Polyethylene is coated onto an insulated substrate that is moving at 30 cm/s. The molten polymer is coated at 400°F, and cooling is achieved by blowing air, at 80°F, across the film. From earlier heat transfer studies it is known that the heat transfer coefficient is $h = 0.08$ cal/cm²·s·°C. The coating thickness is 0.15 cm. At what point downstream does the surface temperature fall to 125°F?

**11.16** Derive Eq. 11.2.15, and the corresponding results for a slab and a sphere. Plot average (dimensionless) temperature versus Fourier number, for several values of the Biot number, for the slab, the long cylinder, and the sphere.

**11.17** We can find the first eigenvalue $\lambda_1$ for the transient solution to heat transfer via conduction in a large slab from Fig. 11.1.4a. It would be useful to have the second and third eigenvalues to use if the truncated solution (Eq. 11.1.40a) is not accurate. The eigenvalues $\lambda_n$ are the roots to

$$\lambda_n \tan \lambda_n = \text{Bi} \qquad \textbf{(P11.17.1)}$$

Plot the first three eigenvalues as a function of Bi on the range $[0.1 < \text{Bi} < 100]$. Show that for very large Bi

$$\lambda_n = \frac{(2n-1)\pi}{2} \qquad \textbf{(P11.17.2)}$$

**11.18** A hot copper surface, initially at 120°C, is suddenly exposed to a convective air flow. The air temperature is 30°C and the heat transfer coefficient is $h = 100$ W/m²·K. Plot the surface temperature against time over the period required for the surface temperature to fall to 40°C. Use a logarithmic scale for time. Assume a thick copper medium.

**11.19** A large slab of copper, of thickness $2B = 1$ cm, initially at 120°C, is suddenly exposed to a convective air flow on both surfaces. The air temperature is 30°C and the heat transfer coefficient is $h = 100$ W/m² · K. Plot the surface temperature against time over the period required for the surface temperature to fall to 40°C. Compare your plot to the solution to Problem 11.18.

**11.20** A copper sphere of diameter 5 cm, initially at 120°C, is suddenly exposed to a convective airflow. The air temperature is 30°C and the heat transfer coefficient is $h = 100$ W/m² · K. Plot the surface temperature against time over the period required for the surface temperature to fall to 40°C. Solve first using Eq. 11.3.21, and then with Figs. 11.1.4f and 11.1.4g. Where the curves differ, which is a more accurate representation of the physics? Why?

**11.21** Derive Eqs. 11.3.11 to 11.3.14.

**11.22** In writing Eq. 11.3.39 we neglected the third term of the series. Show that the third term is negligible relative to the sum of the first two terms.

**11.23** A large sheet of glass 2 inches thick is initially at 300°F throughout. It is plunged into a stream of running water at 60°F. How long will it take to cool the glass to an average temperature of 100°F? For this glass, $k = 0.4$ Btu/ft · h · °F, $\rho = 155$ lb/ft³, and $C_p = 0.2$ Btu/lb · °F.

**11.24** An agitated tank, shown schematically in Fig. P11.24, contains 6200 kg of a dilute aqueous solution. The inside heat transfer coefficient due to agitation is 5861 W/m² · °C. The steel walls of the tank are 10 mm thick. Steam at 110°C condenses on the outside of the steel surface and gives a heat transfer coefficient of 10 kW/m² · °C. The heat transfer area is 14 m². How long does it take to heat the vessel contents from 20°C to 60°C?

**11.25** A material fabrication process involves the following steps:

**1.** A sphere of the material, radius $r_0 = 5$ mm, is removed from a 400°C furnace, where it had achieved thermal equilibrium. It is cooled in air at 20°C for a period of time $t_a$ until its center temperature has reached the value 335°C. The convective heat transfer coefficient is $h_a = 10$ W/m² · K.

**2.** After this first step the sphere is cooled further for a time $t_w$ in a well-stirred bath of water at 20°C wherein the convective coefficient is $h_w = 6000$ W/m² · K, until the center temperature falls to 50°C.

Material properties are $k = 20$ W/m · K, $C_p = 1000$ J/kg · K, and $\rho = 3000$ kg/m³.

Find the times $t_a$ and $t_w$, and the surface temperature at the end of the second step.

**11.26** A thermocouple junction (see Fig. P11.26) consists of two wires connected and embedded in a small ball of solder. The ball has a diameter $D = 0.71$ mm. The thermocouple is used to measure temperature in a gas stream. Under the conditions used, the convective heat transfer coefficient is $h = 400$ W/m² · K.

If the junction is initially at 25°C and is suddenly placed in the gas stream, how long a time is required for the thermocouple to reach 199°C if the gas is at 200°C? Material properties are $k = 20$ W/m · K, $C_p = 400$ J/kg · K, and $\rho = 8500$ kg/m³.

**11.27** Starting with Eq. 11.2.15, show that Eq. 11.1.5 is obtained in the limit as Bi gets very small.

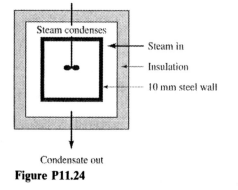

**Figure P11.24**

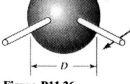

**Figure P11.26**

**11.28** In their retirement years, having purchased the *Enterprise* at auction, Kirk and Spock perform surveys for private parties. Their current assignment is to perform a survey of the planet Mongo. Kirk plans the following manuever. He will quickly bring the *Enterprise* to a position approximately 1000 ft above the surface of Mongo. He will then circumnavigate the planet several times, over a period of 10 h, gathering the required data.

The atmosphere of Mongo is a hot gas, of average molecular weight 35, and average temperature $T_a$ = 750°F. The hull of the *Enterprise* can withstand such a high temperature, but it is necessary to protect the interior of the ship, and its crew. The air-conditioning system of *Enterprise* has a capacity of $2 \times 10^4$ norguks per hour. (One norguk is the amount of energy required to raise the temperature of a standard Gorgon Slime Klak by one degree Fahrenheit.) Unfortunately, the air-conditioning system cannot be operated during this maneuver.

Spock has designed and executed the following experiment. A scale model of the *Enterprise* (linear scale factor of $10^{-2}$) was constructed of the same materials as the mighty starship itself. Spock exposed it to a stream of Mongo atmosphere, at 750°F, and found that the interior surface temperatures within the model were acceptable for a period of 5 s after exposure.

Spock recommends that the mission proceed according to Kirk's plan. Do you agree? Outline your reasoning.

*Hint 1:* Assume that heat flows through the hull of *Enterprise* only by conduction.

*Hint 2:* Spock is rarely wrong, unless it is mating season on Vulcan.

*Hint 3:* Assume that the external Biot numbers exceed 10 in both the model experiment as well as in the case of the *Enterprise*.

**11.29** Equation 11.1.42 gives the average dimensionless temperature of a slab for the case of large Bi. Starting with Eq. 11.1.40a, derive a more general expression for $\overline{\Theta}_1(X_{Fo}, Bi)$ in terms of $\lambda_1$ and $A_1$, and then give the values of these coefficients for the specific case of Bi = 8 so that you can write $\overline{\Theta}_1$ in the form

$$\overline{\Theta}_1 = ae^{-bX_{Fo}} \qquad \text{(P11.29)}$$

with specific numerical values of the coefficients $a$ and $b$ for the case Bi = 8.

**11.30** Find the time required to drop the center temperature of a cucumber from 80°F to 40°F by convective cooling with a fluid of temperature 35°F. The convective heat transfer coefficient is 180 W/m² · K. The cucumber has a radius of 1.9 cm and a length of 16 cm. The thermal conductivity is $k$ = 0.62 W/m · K, and the thermal diffusivity is $\alpha$ = 1.46 × 10⁻⁷ m²/s.

**11.31** The following data are available for the center temperature of grapes being cooled by convection with a fluid at 4°C. The grapes have a diameter of 11 mm and are initially at a uniform temperature of 21.5°C. The thermal conductivity of grapes is $k$ = 0.55 W/m · K, and the thermal diffusivity is $\alpha$ = 1.35 × 10⁻⁷ m²/s.

Estimate the convective heat transfer coefficient from these data.

| $t$ (s) | 112 | 224 | 448 | 672 | 896 | 1120 |
|---|---|---|---|---|---|---|
| $T_c$ (°C) | 16.8 | 15 | 10 | 7.4 | 5.8 | 5.0 |

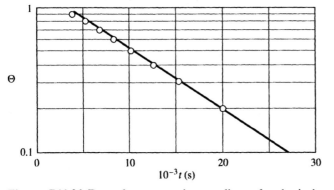

**Figure P11.34** Data for convective cooling of spherical watermelons.

**11.32** Show that $\tilde{t}$ defined in Eq. 11.3.20 may be written as

$$\tilde{t} = \text{Bi}^2 \, X_{\text{Fo}} \qquad \text{(P11.32)}$$

**11.33** A stainless steel pipe of diameter $D = 1$ m and wall thickness $b = 40$ mm is well insulated on its outer surface. The pipe wall is initially at a uniform temperature of $-20°C$. A hot fluid at a temperature of $60°C$ is then pumped through the inside of the pipe, under conditions that yield an internal heat transfer coefficient of $h = 500$ W/m$^2 \cdot$ K. Plot the exterior pipe surface temperature as a function of time. What is the difference between the inside and outside pipe wall temperatures at steady state? State clearly any approximations you make to solve this problem.

**11.34** Dincer and Dost [*Int. J. Energy Res.*, **20**, 351 (1996)] present the data shown in Fig. P11.34, for the convective cooling of watermelons. The center temperature was measured as a function of time and converted to the dimensionless values of $\Theta$. The watermelons are spherical, with a radius of 0.1 m. The thermal properties are $k = 0.6$ W/m $\cdot$ K and $\alpha = 1.4 \times 10^{-7}$ m$^2$/s. What is the value of the convective heat transfer coefficient?

**11.35** A spherical watermelon of radius $R = 0.1$ m is initially at a uniform temperature of $30°C$. It is cooled in a bath of water maintained at $1°C$. How long is required for the center temperature to fall to $5°C$? The convective heat transfer coefficient is known to be 30 W/m$^2 \cdot$ K, and thermal properties are given in Problem 11.38.

# Chapter 12

# Convective Heat Transfer

**In** this chapter we examine heat transfer by convection. After presenting boundary layer theory, which establishes the key dimensionless parameters and their interrelationships, we examine heat transfer in simple flows, for which mathematical models can be formulated and solved analytically. These theoretical results then provide a background for the presentation of empirical correlations of heat transfer coefficients determined in more complex flow situations.

## 12.1 SOME PRELIMINARY CONCEPTS REGARDING CONVECTIVE HEAT TRANSFER

In several problems presented in earlier chapters we have written boundary conditions that included a convective heat transfer coefficient. We have treated the coefficient as a known parameter, or at least as a parameter to which we have access. Now we must deal with a two-pronged question:

> *How do we estimate or predict a value for a heat transfer coefficient, and how does the coefficient depend on characteristic parameters of the convective process?*

We are interested in heat transfer between a fluid and a solid (or between two fluids, across a mobile interface, as in a gas/liquid or liquid/liquid dispersion). The range of problems of interest to us is quite broad. Consider the following examples.

**EXAMPLE 12.1.1** *Cooling and Solidification of Carbon Fibers*

One method of producing filaments for use in high strength composites is extrusion through tiny holes in a plate of a viscous tarlike material called "pitch" (see Fig. 12.1.1). To ensure a yield of fibers of maximum strength, the resultant strands of "spaghetti" must be rapidly cooled. This is achieved by blowing a stream of cold air over the fibers. Since the fibers are very thin, their internal temperature is nearly uniform, and heat transfer is dominated by the rate of external convection. (In the language of Chapter 11, the Biot number is very small.) We need a means of estimating the heat transfer coefficient as a function of air speed and filament diameter, and orientation of the airstream relative to the direction of extrusion.

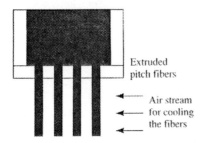

**Figure 12.1.1** Cooling of carbon fibers.

### EXAMPLE 12.1.2   *Temperature Control of an Evaporator*

Very pure gases are produced for semiconductor fabrication by evaporating highly purified liquids in relatively small batches. A typical system is shown in Fig. 12.1.2. Heat is transferred across the walls of the container by conduction, and then into the liquid, primarily by convection, where the internal motion of the liquid is driven by the buoyancy of the relatively hot liquid at the container walls. We need to know how to estimate convective heat transfer coefficients for this system as a function of liquid properties and the shape of the containing vessel. In particular, we want to know how to predict the rate of evaporation of the liquid from the design and operating parameters of the system.

### EXAMPLE 12.1.3   *Temperature Control in a Blood Oxygenator*

Blood passes through an oxygenator, and before being returned to the body it must be brought back to an appropriate temperature. We are to design a heat exchanger for blood that will minimize the contact area of the blood with foreign surfaces; we know that to minimize damage to the fragile red cells, the level of shear stress must be low. Both these constraints produce very poor convective heat transfer. We need to understand how heat transfer to a liquid flowing through a small heat exchanger is affected by the choice of tube size and flowrate.

In each of the cases described in Examples 12.1.1 to 12.1.3, an appropriate goal is the development of a mathematical model of a thermal system. Unfortunately, convective heat transfer problems of real technological significance usually are so complex that it is not possible to solve the equations of motion

$$\rho\left(\frac{\partial \mathbf{u}}{\partial t} + \mathbf{u}\cdot\nabla\mathbf{u}\right) = -\nabla p + \mu\nabla^2\mathbf{u} \qquad (12.1.1)$$

$$\nabla\cdot\rho\mathbf{u} = 0 \qquad (12.1.2)$$

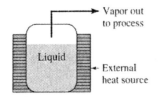

**Figure 12.1.2** An evaporator.

coupled with the convective energy equation

$$\rho C_p \left[ \frac{\partial T}{\partial t} + \mathbf{u} \cdot \nabla T \right] = k \, \nabla^2 T + \mathscr{P} \tag{12.1.3}$$

especially with the complex boundaries that are typical of such problems.

To learn something about the thermal behavior of a system, we have recourse to several approaches, *each* of which can provide some input to subsequent analysis and design:

1. **Dimensional Analysis**  We nondimensionalize the complete set of transport equations and discover which dimensionless groups control the behavior of the solutions to the equations. This approach provides guidelines for the design of an experimental investigation of any system of interest.
2. **Approximate Fluid Dynamic Analysis**  Under some conditions we can invoke laminar boundary layer theory, which offers solutions to the flow field right next to the interface; then we can solve the convective energy equation analytically for that flow field.
3. **Empirical Correlations of Data**  Guided by the dimensional and approximate fluid dynamic approaches above, we plot data in a dimensionless format and present empirical correlations (curve-fits) of the data. From these correlations, we interpolate or extrapolate to the desired conditions.

We turn first to the development of a simple fluid dynamic analysis, which is useful in a variety of convective heat transfer problems.

### 12.1.1   Film Theory

The film model is the simplest approach to prediction of heat transfer coefficients. Although not as elegant as the boundary layer theory that follows in Section 12.2, it is much easier to understand, and it gives surprisingly useful results.

We transform a complex convective transfer problem into a simpler, more easily solved, problem. The goal is to retain enough of the essential physics to ensure that the simple model will predict some of the important features of the complex system.

Schematically, we go through a series of "views" of a particular problem. For example, consider flow around a solid object, as in Fig. 12.1.3. The geometry is not simple, and so the flow field is complex. Numerical solutions are possible but expensive and time-consuming. We may take a closer look at Fig. 12.1.4, to focus on the phase boundary across which transfer is occurring.

We introduce the assumption that the flow field consists of two regions: an outer region of nearly uniform flow, and a *thin* inner region within which the velocity field is dominated by viscous effects as the "no-slip" surface condition is approached. For simple shapes, "exact" laminar boundary layer analysis can yield useful, and relatively simple, models.

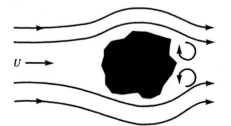

**Figure 12.1.3** Flow around a complex body.

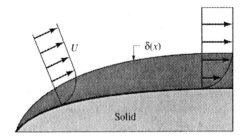

**Figure 12.1.4** A complex flow simplified near the surface to a boundary layer flow.

The lowest order boundary layer model is the *film model,* in which the velocity field near the surface is linearized in some average sense over the whole surface (i.e., along the *x* axis: see Fig. 12.1.5).

Consider first the fluid dynamics. If the velocity field were really linear, then the shear stress *on* the surface would be given by

$$\tau = \mu \frac{\partial u_x}{\partial y} = \mu \frac{U}{\delta} = \frac{F_s}{A} \tag{12.1.4}$$

where $F_s/A$ is the shear force acting in the *x* direction per unit of surface area. Solving for the film thickness, we find

$$\delta = \frac{\mu U}{\tau} \tag{12.1.5}$$

We usually cannot measure $\delta$ directly, but the shear force is measurable, and we have friction factor charts available for some systems. We may calculate $\delta$ from a knowledge of the friction factor $f$ (or the drag coefficient $C_D$ as it is presented in some texts) since $f$ is a dimensionless shear stress:

$$f \equiv \frac{\tau}{\frac{1}{2} \rho U^2} = C_D \tag{12.1.6}$$

This leads to

$$\delta = \frac{\mu U}{\frac{1}{2} f \rho U^2} = \frac{2\mu}{\rho U f} \tag{12.1.7}$$

or

$$\frac{\delta}{L} = \frac{2}{f} \frac{\mu}{\rho U L} = \left( \frac{f \, \text{Re}}{2} \right)^{-1} \tag{12.1.8}$$

where *L* is *some* characteristic length scale for the flow field.

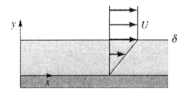

**Figure 12.1.5** A uniform film model of the boundary layer.

The film model assumes that a *definable boundary layer* exists. The concept is meaningful only in the turbulent regime. For pipe flow, for example, we know that a good approximation is

$$f \sim \mathrm{Re}^{-0.25} \qquad (12.1.9)$$

so, using Eq. 12.1.8, we find

$$\frac{\delta}{L} \sim \mathrm{Re}^{-0.75} \qquad (12.1.10)$$

Now we will apply this fluid model to convective heat transfer. We suppose (see Fig. 12.1.6) that the temperature profile is uniform except in a thin layer $\delta'$ near the surface. It is not clear that $\delta' = \delta$, so we keep a separate notation. (They are *not* the same.) We now write the flux of thermal energy by conduction across the film of thickness $\delta'$ as

$$q_y = -k\frac{\partial T}{\partial y} = k\frac{\Delta T}{\delta'} = h_c \, \Delta T \qquad (12.1.11)$$

In writing Eq. 12.1.11, we are *defining* the convective heat transfer coefficient $h_c$ as the ratio of the flux to the temperature difference. This film model gives

$$h_c = k/\delta' \qquad (12.1.12)$$

While conceptually this is a useful result, we still do not know the convective coefficient—we have simply transformed our ignorance of the coefficient $h_c$ to the unknown film thickness $\delta'$.

The friction factor is a useful measure of the *momentum transport* to a surface. It is a dimensionless parameter that in turn depends only on dimensionless characteristics of the flow field, such as the Reynolds number and the geometry of the boundaries of the flow. If $f$ is a dimensionless shear stress, what is a dimensionless *heat* flux?

We will define a dimensionless heat flux as

$$\frac{q_y}{k(\Delta T/L)} \qquad (12.1.13)$$

where $L$ is some appropriate length scale for the flow field. Then, according to this simple film model,

$$\frac{q_y}{k\,(\Delta T/L)} = \frac{h_c L}{k} \equiv \mathrm{Nu} \qquad (12.1.14)$$

where we have now introduced Nu, the so-called Nusselt number, as the dimensionless heat transfer coefficient. But $h_c = k/\delta'$, so it follows that

$$\mathrm{Nu} = \frac{L}{\delta'} = \frac{L}{\delta}\left(\frac{\delta}{\delta'}\right) = \frac{f\,\mathrm{Re}}{2}\left(\frac{\delta}{\delta'}\right) \qquad (12.1.15)$$

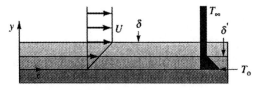

**Figure 12.1.6** Representation of the temperature field confined to a thin uniform film.

We have not specified $L$ in this case (keep in mind that we are no longer talking about a pipe flow, simply flow parallel to a plane), but whatever the choice, it must be the same choice in Nu and Re. We can expect to find that $\delta/\delta'$ is a function of the "diffusivities" for momentum and heat. The diffusivity for momentum is the kinematic viscosity $\nu = \mu/\rho$, with units of meters squared per second. The diffusivity for heat conduction is not the conductivity, but the thermal diffusivity $\alpha = k/\rho C_p$, with the same units. Hence we expect that

$$\frac{\delta}{\delta'} = \text{fn}\left(\frac{\nu}{\alpha}\right) = \text{fn (Pr)} \tag{12.1.16}$$

where $\text{Pr} = \nu/\alpha$ is a physical property called the Prandtl number for the fluid. [We use the notation fn( ) to denote "a function of ( )", so that there will be no confusion with the friction factor $f$.] It turns out that, using a more exact analytical method (we will develop that in the next section),

$$\text{fn(Pr)} = \text{Pr}^{1/3} \tag{12.1.17}$$

Hence

$$\text{Nu} = \frac{f}{2}\,\text{Re}\ \text{Pr}^{1/3} \tag{12.1.18}$$

A so-called $j$-factor for heat transfer is often (and at first, mysteriously) defined as

$$j_H \equiv \frac{\text{Nu}}{\text{Re}\ \text{Pr}^{1/3}} \tag{12.1.19}$$

We see then that film theory leads to the prediction that there is a simple relationship between a characteristic dimensionless measure of convective heat transfer, the $j$-factor, and a well-known characteristic measure of the nature of a turbulent flow field, the friction factor. This very simple looking result is

$$j_H = \frac{f}{2} \tag{12.1.20}$$

which is called *the Colburn analogy for heat transfer.*

We can make an immediate prediction here. For example, for turbulent pipe flow a good approximation to the friction factor curve at high Reynolds numbers is

$$f \approx 0.079\ \text{Re}^{-0.25} \tag{12.1.21}$$

From Eq. 12.1.18 it follows that

$$\text{Nu} = 0.04\ \text{Re}^{0.75}\ \text{Pr}^{0.33} \tag{12.1.22}$$

This is a *testable* prediction, and we will examine its validity later. (See Problem 12.1.) For pipe flow, since we use $L = D$ in Re, we also use $L = D$ in Nu.

Film theory, as we have presented it here, is very simple to understand, but it is not a very good representation of the physics of interest. The major criticism pinpoints the assumption that the boundary layers are *uniform* along the direction of flow. This is not the case in most real systems of interest to us, and the growth of the film thickness along the flow direction has a number of important consequences. In particular, the heat transfer coefficient will vary along the direction of flow. To pursue the implications of this point, we must turn to a boundary layer analysis that accounts for changes that occur in the flow direction.

## 12.2   AN EXACT LAMINAR BOUNDARY LAYER THEORY: HEAT TRANSFER FROM A FLAT PLATE

We do not have to introduce the artifice of a film of finite thickness into a boundary layer model. Instead, we will examine a solution of the appropriate equations for transport of momentum and heat, to see how the equivalent of a film thickness might be defined. With the assumptions of constant physical properties and no external forces, the steady state equations of motion and energy take the forms[1]

$$\frac{\partial u_x}{\partial x} + \frac{\partial u_y}{\partial y} = 0 \tag{12.2.1}$$

$$u_x \frac{\partial u_x}{\partial x} + u_y \frac{\partial u_x}{\partial y} = \nu \frac{\partial^2 u_x}{\partial y^2} \tag{12.2.2}$$

$$u_x \frac{\partial T}{\partial x} + u_y \frac{\partial T}{\partial y} = \alpha \frac{\partial^2 T}{\partial y^2} \tag{12.2.3}$$

Appropriate boundary conditions are (see Fig. 12.1.6 for the definitions of $U$ and $T_o$ and $T_\infty$, but note that we no longer assume that the velocity and temperature fields are linear)

$$u_x = U, T = T_\infty \quad \text{at} \quad x \le 0 \quad \text{or} \quad y = \infty \tag{12.2.4}$$

$$u_x = 0, T = T_o \quad \text{at} \quad y = 0 \tag{12.2.5}$$

We may integrate Eq. 12.2.1 to yield

$$u_y = -\int_0^y \frac{\partial u_x}{\partial x} dy \tag{12.2.6}$$

We note that Eqs. 12.2.2 and 12.2.3 are identical in form, except for the name of the independent variable, $u_x$ or $T$. Then we may represent Eq. 12.2.2 and 12.2.3 by a *single* equation of the form

$$u_x \frac{\partial \Pi}{\partial x} - \left( \int_0^y \frac{\partial u_x}{\partial x} dy \right) \frac{\partial \Pi}{\partial y} = \frac{\nu}{\Lambda} \frac{\partial^2 \Pi}{\partial y^2} \tag{12.2.7}$$

with the boundary conditions

$$\Pi = 1 \quad \text{at} \quad x \le 0 \quad \text{or} \quad y = \infty \tag{12.2.8}$$

$$\Pi = 0 \quad \text{at} \quad y = 0 \tag{12.2.9}$$

We have defined a dimensionless variable $\Pi$ that can stand for either the velocity or the temperature variable. Equation 12.2.7 stands for the velocity equation (12.2.2) by letting

$$\Pi = \Pi_u \equiv \frac{u_x}{U} \tag{12.2.10}$$

and

$$\Lambda = \Lambda_u \equiv 1 \tag{12.2.11}$$

---

[1] One of the key assumptions of boundary layer theory is that gradients of $u_x$ and $T$ are small in the main flow direction. Hence $\partial^2 T/\partial x^2$ does not appear on the right-hand side of Eq. 12.2.3, and the same approximation simplifies Eq. 12.2.2.

and for the energy equation (12.2.3) by letting

$$\Pi = \Pi_T \equiv \frac{T - T_o}{T_\infty - T_o} \tag{12.2.12}$$

and

$$\Lambda = \Lambda_T \equiv \frac{\nu}{\alpha} = \text{Pr} \qquad \text{(the Prandtl number)} \tag{12.2.13}$$

*We have not reduced two equations to one. We are just representing both equations by Eq. 12.2.7.*

If we now define[2] a new combined independent variable $\eta$ as

$$\eta = \frac{y}{2} \left( \frac{U}{\nu x} \right)^{1/2} \tag{12.2.14}$$

we may show that Eq. 12.2.7 becomes an *ordinary* differential equation, in which $\Pi$ is a function only of $\eta$:

$$\Lambda \left[ -\int_0^\eta 2\Pi_u \, d\eta \right] \Pi' = \Pi'' \tag{12.2.15}$$

where the prime denotes differentiation with respect to $\eta$. This would be a simple ordinary differential equation except that the bracketed term depends on the unknown velocity field. However, we may easily write a formal solution to Eq. 12.2.15 after defining the integral as

$$\phi(\eta) = \int_0^\eta 2\Pi_u \, d\eta \tag{12.2.16}$$

and noting that the new boundary conditions on Eqs. 12.2.15, with respect to $\eta$, become

$$\Pi = 0 \qquad \text{at} \quad \eta = 0 \tag{12.2.17}$$

$$\Pi = 1 \qquad \text{at} \quad \eta = \infty \tag{12.2.18}$$

Equation 12.2.15 becomes

$$-\Lambda \phi \Pi' = \Pi'' \tag{12.2.19}$$

and is solved directly by integration twice to give

$$\Pi(\eta, \Lambda) = \frac{\int_0^\eta \exp\left( -\Lambda \int_0^\eta \phi \, d\eta \right) d\eta}{\int_0^\infty \exp\left( -\Lambda \int_0^\eta \phi \, d\eta \right) d\eta} \tag{12.2.20}$$

Note that, since $\phi$ is a function of $\Pi_u$, Eq. 12.2.20 is not an *explicit* solution for $\Pi_u$. When we set $\Lambda = 1$ in Eq. 12.2.20, we obtain an *implicit* equation for $\Pi_u$, which must be solved numerically. Once a solution for $\Pi_u$ has been obtained, then $\phi$ is known, and Eq. 12.2.20 is an *explicit* solution for $\Pi_T$ for any value of $\Lambda_T$.

Figure 12.2.1 shows the velocity profile $\Pi_u$ and one temperature profile (for a Prandtl number of 2) as a function of $\eta$. We see that the velocity is very close to the free stream value (i.e., $\Pi_u \approx 1$) at $\eta = 2.5$. Hence one could *define* a film thickness $\delta$ as the distance from the solid surface over which the velocity field reaches nearly the free stream

---

[2] We will later need the fact that, with this definition,

$$\frac{\partial}{\partial y} = \frac{1}{2} \left( \frac{U}{x\nu} \right)^{1/2} \frac{d}{d\eta}$$

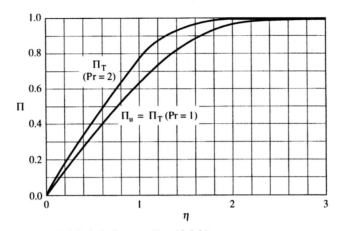

**Figure 12.2.1** Solutions to Eq. 12.2.20.

value. Thus

$$\frac{\delta}{2}\left(\frac{U}{\nu x}\right)^{1/2} = 2.5 \qquad (12.2.21)$$

or

$$\delta = 5\left(\frac{\nu x}{U}\right)^{1/2} \qquad (12.2.22)$$

(Compare this to Eq. 6.3.12.)

We anticipate that we will also want to define an equivalent film thickness $\delta_T$ *for heat transfer,* by the equation

$$-k\left.\frac{\partial T}{\partial y}\right|_{y=0} = k\frac{T_o - T_\infty}{\delta_T} \qquad (12.2.23)$$

Reverting to dimensionless variables, we find

$$\delta_T = 2\frac{\left(\frac{x\nu}{U}\right)^{1/2}}{\Pi'_T(0, \Lambda_T)} \qquad (12.2.24)$$

(Remember, the prime on $\Pi$ denotes a derivative with respect to $\eta$. Also, see the footnote 2, bearing on Eq. 12.2.14.) We may obtain an analytical solution for $\delta_T$ if we introduce an approximate velocity profile into Eq. 12.2.20. From Fig. 12.2.1 we find the slope of the velocity profile, at the surface, to be

$$\Pi'_u(0) = 0.664 \qquad (12.2.25)$$

If we use a linear approximation to the velocity, then, we find

$$\Pi_u = 0.664 \, \eta \qquad (12.2.26)$$

and

$$\phi = \int_0^\eta 2\Pi_u \, d\eta = 0.664 \, \eta^2 \qquad (12.2.27)$$

From Eq. 12.2.20 we find

$$\Pi_T'(0, \Lambda_T) = \left[ \int_0^\infty \exp\left( -\Lambda_T \int_0^\eta \phi \, d\eta \right) d\eta \right]^{-1} \tag{12.2.28}$$

$$= \left[ \int_0^\infty \exp\left( -\frac{0.664}{3} \Lambda_T \eta^3 \right) d\eta \right]^{-1} = 0.68 \Lambda_T^{1/3}$$

(Despite the linear approximation to $\Pi_u$, this turns out to be within 10% of the exact solution, for values of $\Lambda_T$ of order one.)

This permits us to write

$$\delta_T = 2.94 \left( \frac{x\nu}{U} \right)^{1/2} Pr^{-1/3} \tag{12.2.29}$$

where, now, we have replaced $\Lambda_T$ by the more standard notation for the Prandtl number. As expected, $\delta_T$ also depends on $x$ and increases in the downstream direction.

The analysis that led to Eq. 12.2.28 did not require that we *impose* a film model on the analysis of heat transfer to (or from) the surface. $\delta_T$ is really just a length scale defined (by Eq. 12.2.23) in terms of the temperature gradient at the surface. The system behaves *as if* conduction were occurring across a stagnant layer of thickness $\delta_T$, and Eq. 12.2.29 lets us estimate the magnitude of that thickness.

In Section 12.1 we defined a dimensionless heat transfer coefficient, the Nusselt number, as

$$\frac{q_y}{k(\Delta T/L)} = \frac{h_c L}{k} \equiv Nu \tag{12.2.30}$$

Because the heat flux varies along the $x$ direction according to the analysis we have just completed, we now define a *local* Nusselt number, in which the length scale is the variable position $x$ along the surface, measured from the origin. Thus we define

$$Nu_x = \frac{h_c x}{k} = \frac{q_y x}{k \, \Delta T} \tag{12.2.31}$$

If we now use Eqs. 12.2.29 and 12.2.23, we find

$$\frac{h_c x}{k} = Nu_x = 0.34 \left( \frac{xU}{\nu} \right)^{1/2} Pr^{1/3} \tag{12.2.32}$$

For a plate of finite length $L$, the *average* heat transfer coefficient is found by integration of $h_c$ down the plate:

$$\bar{h}_c = \frac{1}{L} \int_0^L h_c(x) \, dx \tag{12.2.33}$$

and the average Nusselt number over the distance $L$ is defined by

$$\overline{Nu} = \frac{\bar{h}_c L}{k} = 0.68 \left( \frac{LU}{\nu} \right)^{1/2} Pr^{1/3} = 0.68 \, Re_L^{1/2} \, Pr^{1/3} \tag{12.2.34}$$

(Note that $\overline{Nu}$ uses the length in the $x$ direction, $L$, and $Nu_x$ uses the variable $x$ as the length scale.)

We need to summarize what we have been doing here, and recall our original goal. We wish to have a means of predicting the rate of heat transfer from surfaces when we cannot solve the exact forms of the conservation equations for momentum and energy analytically because of the complexity of the surface geometry and/or the flow field. The problem we have just examined, the convective transfer across a boundary layer on a flat plate subject to an external flow that is parallel to the plate, is in a sense an intermediate problem. We can solve for the velocity and temperature fields analytically, but only through the use of certain approximations, which lead to the boundary layer analysis. Many heat transfer problems are so complex that a simple boundary layer theory is not possible.

What good, then, is the analysis that we have just carried out? First of all, there are many problems that do indeed have this boundary layer character, and so we now have a model for such flows. But we also learn from this model how the heat transfer coefficient may be nondimensionalized to a Nusselt number, and how this Nusselt number depends on the Reynolds number and the Prandtl number. We could have concluded by dimensional arguments (dimensional analysis) alone that the Nu does depend on these groups. But the form of the dependence, namely,

$$\overline{\text{Nu}} = A \, \text{Re}^m \, \text{Pr}^n \qquad (12.2.35)$$

is a consequence of the boundary layer analysis, not dimensional analysis. We find that in a number of convective heat transfer systems, including many of much more complex geometry than that of heat transfer from a planar surface to a (nearly) parallel flow, the behavior follows Eq. 12.2.35. The exponent $m$ on the Reynolds number is often in the range $0.5 < m < 0.8$, and the exponent on the Prandtl number is typically in the range $0.3 < n < 0.5$.

## 12.3 CORRELATIONS OF HEAT TRANSFER DATA

### 12.3.1 Turbulent Flow Inside Pipes

For complex systems for which it is not feasible to develop an analytical model, a useful strategy is to obtain data over a wide range of Reynolds numbers, using fluids with various Prandtl numbers, and then correlate the data according to Eq. 12.2.35. However, several alternative methods of correlation have been used which differ from Eq. 12.2.35 primarily in format, and since they appear so commonly in the heat transfer literature, we will examine them. In fact, we displayed earlier the prediction based on simple film theory, given in Eq. 12.1.20, that for turbulent flow inside pipes of circular cross section

$$j_H = \frac{f}{2} \qquad (12.3.1)$$

where the "$j$-factor" is defined by

$$j_H \equiv \frac{\text{Nu}}{\text{Re} \, \text{Pr}^{1/3}} \qquad (12.3.2)$$

As noted earlier, this is the so-called Colburn analogy between friction and heat transfer. Observations of data for heat transfer between a fluid and the wall of a pipe through which it flows confirm the utility of this expression. At high Reynolds numbers and Prandtl numbers, a better correlation of data was recommended by Friend and Metzner (see Problem 12.1) in the form

$$\text{St} \equiv \frac{\text{Nu}}{\text{Re} \, \text{Pr}} = \frac{h}{C_p G} = \frac{f/2}{1.2 + 11.8 \, (f/2)^{1/2} \, (\text{Pr} - 1) \, \text{Pr}^{-1/3}} \qquad (12.3.3)$$

where $G = \rho U$ is the "mass velocity," in units of mass flow rate per unit of cross-sectional area, with $U$ as the average velocity in the pipe. The so-called Stanton number

(St) is defined by this equation as the ratio of the Nusselt number to the product of the Reynolds and Prandtl numbers. We shall find shortly that the Stanton number is a natural choice of a dimensionless group for pipe heat transfer problems.

We could use Eq. 12.1.21 for the relationship of friction factor to Reynolds number. Another recommendation, for use with Eq. 12.3.3, is

$$f = 0.0014 + \frac{0.125}{\mathrm{Re}^{0.32}} \qquad (12.3.4)$$

The combination of Eqs. 12.3.3 and 12.3.4 provides a correlation of heat transfer data for flow in pipes that has been tested over a very broad range in Reynolds numbers ($3000 < \mathrm{Re} < 3 \times 10^6$) and Prandtl numbers ($0.46 < \mathrm{Pr} < 590$). Note specifically that this correlation is *not* appropriate for use under *laminar flow* conditions. The primary reason for this limitation is that the boundary layer concept makes sense *only* in turbulent flow. A different modeling approach is required for laminar flows, and a different form of prediction and correlation is obtained.

### 12.3.2 Flow Outside and Across Pipes and Cylinders

In many practical circumstances we require an estimate of a heat transfer coefficient for flow external to a pipe or cylinder. The typical case is that of flow *normal* to the cylindrical axis. Figure 12.3.1 gives a correlation of data (the data points are not shown) for the flow of air normal to single cylinders.

For fluids other than air, one expects a dependence of Nu on Pr. A recommended correlation is the following:

$$\frac{\mathrm{Nu}}{\mathrm{Pr}^{0.3}} = 0.35 + 0.56\,\mathrm{Re}^{0.52} \qquad (12.3.5)$$

Within the observed scatter in the experimental data, we find that Eq. 12.3.5 gives a plot of $\mathrm{Nu}/\mathrm{Pr}^{0.3}$ that is very close to the line on Fig. 12.3.1. Hence we can recommend the use of Fig. 12.3.1 for any Prandtl number by replacing Nu with $\mathrm{Nu}/\mathrm{Pr}^{0.3}$. With this change, Fig. 12.3.1 may be used for fluids other than air.

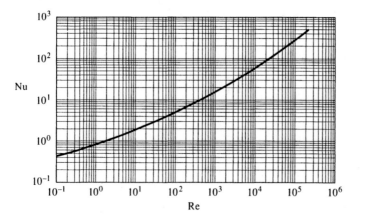

**Figure 12.3.1** Nusselt number correlation of data for the flow of air normal to pipes and cylinders. Both Nu and Re use the diameter $D$ as the length scale. For fluids other than air, replace Nu with $\mathrm{Nu}/\mathrm{Pr}^{0.3}$.

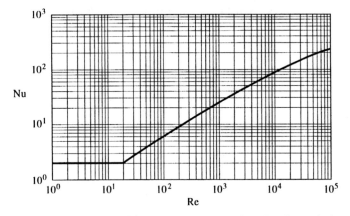

**Figure 12.3.2** Correlation of heat transfer data for flow of air past spheres. The diameter is used in both Re and Nu.

### 12.3.3 Flow Past Single Spheres

In another case of practical importance, heat transfer occurs between a fluid and a solid sphere. This is a very complex flow, especially at high Reynolds numbers. Figure 12.3.2 shows a typical correlation of data (again, without the data points) for this situation, and again this graph is limited to the flow of air. An examination of correlations for fluids other than air suggests that we may use Fig. 12.3.2 as a good approximation to the behavior of other fluids if we replace the Nusselt number by $Nu/Pr^{1/3}$, for $Nu > 2$.

For most gases, the Prandtl number lies in the range $0.65 < Pr < 0.8$ and is a very weak function of temperature and pressure. Since the fractional root of the Prandtl number appears in the correlations, we can just use a value for air ($Pr = 0.72$) and suffer an error that is probably smaller than the uncertainty in the correlation itself. For liquids the Prandtl number varies from one liquid to another, and is a strong function of temperature. Figure 12.3.3 shows Pr $(T)$ for water.

### 12.3.4 The Temperature Dependence of Physical Properties Used in the Correlations

In correlations of the types just examined, we must evaluate groups such as the Reynolds and Prandtl numbers, which contain physical properties—especially viscosity—that are

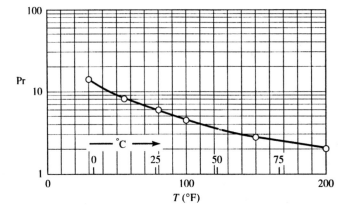

**Figure 12.3.3** Prandtl number as a function of temperature for water.

temperature dependent. Since the temperature varies spatially in the systems of interest to us, we do not have a single temperature, characteristic of the system, at which all physical properties are to be evaluated. In most cases the temperature dependence of physical properties is accounted for through a choice of a characteristic temperature that has some logic behind it and, based on extensive examination of data, permits us to fit available data with acceptable precision.

For flow inside tubes and channels, there are two temperatures often defined as characteristic of the system. One is the mean bulk temperature, which is simply the arithmetic average of the inlet and outlet average temperatures:

$$T_m \equiv \tfrac{1}{2}(\overline{T}_{in} + \overline{T}_{out}) \tag{12.3.6}$$

For internal flows

Another is called the "film" temperature. It represents an average of the bulk fluid temperature and the temperature of the heat transfer surface:

$$T_f \equiv \frac{\tfrac{1}{2}(\overline{T}_{in} + \overline{T}_{out}) + T_s}{2} \tag{12.3.7}$$

For internal flows

(We assume here that the surface is at uniform temperature $T_s$.)

For external flows in which transfer takes place between some ambient fluid at a uniform temperature $T_a$ and a surface at some other temperature $T_s$, the film temperature is defined as

$$T_f \equiv \frac{T_a + T_s}{2} \tag{12.3.8}$$

For external flows

For external flows, the physical properties are usually evaluated at the film temperature. For internal flows it is common practice to use a correlation such as Eq. 12.3.3 with all properties evaluated at $T_m$ defined above (Eq. 12.3.6).

In addition, for heat transfer to flows inside tubes and channels, one often finds the so-called Sieder–Tate correction applied to a correlation. The correction factor is most often written as

$$\chi \equiv \left(\frac{\mu_m}{\mu_s}\right)^{0.14} \tag{12.3.9}$$

This has the ratio of the viscosity at the mean bulk temperature $T_m$ to the viscosity at the surface temperature $T_s$. The 0.14 power is based on examination of data; other values are sometimes presented in the literature. Hence one would write Eq. 12.3.3 as

$$St \equiv \frac{Nu}{Re\,Pr} = \frac{h}{C_p G} = \frac{f/2}{1.2 + 11.8\,(f/2)^{1/2}\,(Pr - 1)\,Pr^{-1/3}}\left(\frac{\mu_m}{\mu_s}\right)^{0.14} \tag{12.3.10}$$

### 12.3.5   A Simple Model of the Effect of a Temperature-Dependent Viscosity

We can develop a simple model that indicates how a temperature-dependent viscosity enters into a heat transfer analysis at a surface whose temperature is very different from that of the ambient fluid. The starting point is the film model of turbulent flow near a surface (Fig. 12.3.1). We assume that the velocity profile satisfies the one-dimensional momentum balance, which we write in the form

$$\frac{d}{dy}\left[\mu(y)\frac{du_x}{dy}\right] = 0 \tag{12.3.11}$$

where we permit the viscosity to vary over the region $[0, \delta]$. We write the viscosity–temperature dependence in the form (see Problem 12.19)

$$\frac{\mu}{\mu_o} = \left(1 + \gamma\frac{T - T_o}{T_w - T_o}\right)^{-1} \tag{12.3.12}$$

where $\gamma$ is defined in terms of the ratio of viscosities at the wall temperature and in the bulk flow:

$$\gamma \equiv \frac{\mu_o}{\mu_w} - 1 \tag{12.3.13}$$

Equation 12.3.11 is solved subject to boundary conditions

$$u_x = 0 \quad \text{on} \quad y = 0 \tag{12.3.14}$$

and

$$u_x = U \quad \text{on} \quad y = \delta \tag{12.3.15}$$

Now we introduce a very simple temperature profile assumption. It takes the form

$$\frac{T - T_o}{T_w - T_o} = 1 - \frac{y}{\delta_T} \quad \text{for} \quad y \le \delta_T$$

$$\frac{T - T_o}{T_w - T_o} = 0 \quad \text{for} \quad y \ge \delta_T \tag{12.3.16}$$

When this equation is used with Eq. 12.3.12, we may integrate Eq. 12.3.11 for the velocity profile and find (see Problem 12.38) the following expression for the velocity near the surface:

$$\frac{u(y)}{U} = \left(\frac{\delta}{\delta_T} + \frac{\gamma}{2}\right)^{-1}\left[(1 + \gamma)\frac{y}{\delta_T} - \frac{\gamma}{2}\left(\frac{y}{\delta_T}\right)^2\right] \quad \text{for} \quad y \le \delta_T \tag{12.3.17}$$

and

$$\frac{u(y)}{U} = \left(\frac{\delta}{\delta_T} + \frac{\gamma}{2}\right)^{-1}\left(\frac{\gamma}{2} + \frac{y}{\delta_T}\right) \quad \text{for} \quad y \ge \delta_T \tag{12.3.18}$$

In Fig. 12.3.4 we show velocity profiles for two values of $\gamma$, one ($\gamma = -0.5$) corresponding to the case of $\mu_w = 2\mu_o$ (for a liquid, a cold wall), and the other ($\gamma = 1$) for the opposite case of a liquid with a hot wall, and specifically for $\mu_w = \mu_o/2$. We see that the velocity profiles near the wall are distorted relative to the "isothermal" profile.[1] In view of these results, we should expect the rate of heat transfer to or from the surface to depend on the temperature dependence of the viscosity, if the wall and the bulk liquid are at substantially different temperatures.

The next challenge is to use this model to estimate a correction factor to the isothermal heat transfer coefficient. We see that the effect of the temperature-dependent viscosity

---

[1] "Isothermal" refers to the case of the physical properties being independent of temperature. If heat is being transferred, of course, the system is not isothermal.

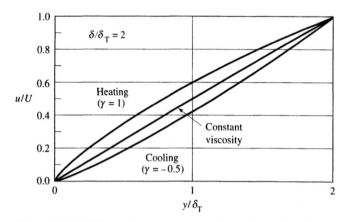

**Figure 12.3.4** Effect of a temperature-dependent viscosity on the velocity profile across a film of liquid.

is to alter the slope of the velocity profile near the heated (or cooled) surface. Let's go back to the "exact" boundary layer heat transfer model of Section 12.2. We will use this model to learn how the velocity gradient at the surface affects the rate of heat transfer at that surface. Equation 12.2.28 gives the relationship of the velocity [through the function $\phi(\eta)$] to the temperature gradient at the wall, hence to the heat flux at the wall. For the nonisothermal case we now have an approximate model for the velocity near the wall—Eq. 12.3.17—from which we find the velocity gradient (in the notation of Section 12.2)

$$\frac{d\Pi_u}{dy} = \frac{1+\gamma}{\delta_T \left( \dfrac{\delta}{\delta_T} + \dfrac{\gamma}{2} \right)} \qquad \text{at} \quad y = 0 \tag{12.3.19}$$

If we use the isothermal result (Eq. 12.2.29) to approximate $\delta_T$ on the right-hand side of this equation we may write a "corrected" velocity gradient as

$$\frac{d\Pi_u}{dy} = \frac{\mathrm{Pr}^{1/3}(1+\gamma)}{2.94 \left( \dfrac{\delta}{\delta_T} + \dfrac{\gamma}{2} \right)} \left( \frac{U}{x\nu} \right)^{1/2} \qquad \text{at} \quad y = 0 \tag{12.3.20}$$

or, very near the wall,

$$\Pi_u = \frac{\mathrm{Pr}^{1/3}(1+\gamma)}{2.94 \left( \dfrac{\delta}{\delta_T} + \dfrac{\gamma}{2} \right)} \left( \frac{U}{x\nu} \right)^{1/2} y = \frac{\mathrm{Pr}^{1/3}(1+\gamma)}{1.47 \left( \mathrm{Pr}^{1/3} + \dfrac{\gamma}{2} \right)} \eta \tag{12.3.21}$$

(We have replaced $\delta/\delta_T$ in the denominator of Eq. 12.3.21 with the "isothermal" result: $\mathrm{Pr}^{1/3}$.) Now we can pick up the algebra at Eq. 12.2.27, replacing the coefficient of $\eta$ in Eq. 12.2.26 by the result given in Eq. 12.3.21. After manipulating the resulting integral we find

$$\Pi'_T(0, \mathrm{Pr}) = 0.68\, \mathrm{Pr}^{1/3} \left[ \frac{\mathrm{Pr}^{1/3}(1+\gamma)}{\left( \mathrm{Pr}^{1/3} + \dfrac{\gamma}{2} \right)} \right]^{1/3} \tag{12.3.22}$$

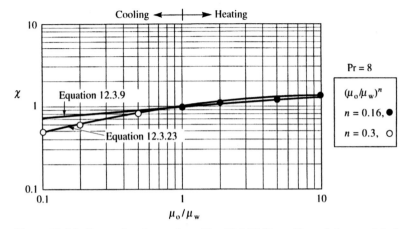

**Figure 12.3.5** Correction factor from Eq. 12.3.23 (Pr = 8), and the empirical Sieder–Tate correction.

Hence we find that the temperature gradient at the surface, the heat flux, and the Nusselt number are all "corrected" by a factor $\chi$ defined as

$$\chi \equiv \left[ \frac{Pr^{1/3}(1 + \gamma)}{\left( Pr^{1/3} + \dfrac{\gamma}{2} \right)} \right]^{1/3} \tag{12.3.23}$$

This correction factor is plotted in Fig. 12.3.5, where it is compared to the oft-recommended Eq. 12.3.9. This simple model agrees quite well with the empirical Sieder–Tate correction for values of $\mu_o/\mu_w > 1$, which is the regime in which the liquid is heated by the surface. An exponent of 0.16 does a slightly better job of matching Eq. 12.3.23 in this regime. In the range of $\mu_o/\mu_w < 1$, the regime in which the liquid is being *cooled* by the surface, the Sieder–Tate form of correction factor would agree with the model developed here, but with an exponent more like 0.3.

We must emphasize at this point that the simple model developed in this section is not necessarily more accurate than the empirical Sieder–Tate correction. In other words, we are not suggesting that Eq. 12.3.23 should replace Eq. 12.3.9 just because Eq. 12.3.23 has a theoretical basis. It is, after all, an *approximate* theoretical model in view of the many assumptions that entered into its development. The point is that this simple model does capture some of the observed features of the effect of a temperature-dependent viscosity on the convective heat transfer at a surface. The derivation of Eq. 12.3.23 does not provide a *better* model, but it does help us to understand better the physical factors that relate a temperature-dependent viscosity to its effect on heat transfer at a surface.

## 12.4   HEAT TRANSFER ANALYSIS IN PIPE FLOW

### 12.4.1   The Basic Heat Balance

Consider the problem of flow in a long pipe of circular cross section, with the pipe wall ($r = R$) maintained at a uniform temperature $T_R$. Fluid enters the pipe at the uniform temperature $T_1$. The goal of this analysis is to provide a model from which the average temperature of the fluid may be calculated at any downstream position as a function of flowrate, pipe length and diameter, and the (assumed known) heat transfer coefficient at the inside surface of the pipe. We initiate the model by writing the thermal energy equation in the form (see Table 11.2.1)

$$\rho C_p u_z \frac{\partial T}{\partial z} = k \left[ \frac{1}{r} \frac{\partial}{\partial r} \left( r \frac{\partial T}{\partial r} \right) + \frac{\partial^2 T}{\partial z^2} \right] \tag{12.4.1}$$

In simplifying the general equation for heat transfer in cylindrical coordinates, we have assumed that the flow field is one-dimensional ($\mathbf{u} = [u_z(r), 0, 0]$), the system has axial symmetry, steady state conditions have been achieved, and physical properties are independent of temperature. We now introduce an approximation, to the effect that the heat *conduction* along the pipe axis (the second term on the right-hand side of Eq. 12.4.1) is much slower than *convection* in that direction (the left-hand side of the equation). Then Eq. 12.4.1 simplifies to

$$\rho C_p u_z \frac{\partial T}{\partial z} = k \left[ \frac{1}{r} \frac{\partial}{\partial r} \left( r \frac{\partial T}{\partial r} \right) \right] \tag{12.4.2}$$

Boundary conditions are

$$T = T_1 \qquad \text{at } z = 0 \tag{12.4.3}$$

$$T = T_R \qquad \text{at } r = R \tag{12.4.4}$$

$$\frac{\partial T}{\partial r} = 0 \qquad \text{at } r = 0 \tag{12.4.5}$$

Instead of solving this equation for the detailed temperature profile, we seek a solution for the *average* temperature. We begin by multiplying Eq. 12.4.2 by $r\, dr$, and then we integrate the result from 0 to $R$:

$$\int_0^R \rho C_p u_z \frac{\partial T}{\partial z} r\, dr = \int_0^R k \frac{\partial}{\partial r} \left( r \frac{\partial T}{\partial r} \right) dr \tag{12.4.6}$$

This may be written in the form

$$\rho C_p \frac{\partial}{\partial z} \int_0^R u_z T r\, dr = k \left[ r \frac{\partial T}{\partial r} \right]_0^R = kR \frac{\partial T}{\partial r}\bigg|_R - 0 \tag{12.4.7}$$

We next recognize and define the integral on the left-hand side as a kind of average temperature by writing

$$\langle T \rangle = \frac{\int_0^R u_z T 2\pi r\, dr}{\int_0^R u_z 2\pi r\, dr} = \frac{\int_0^R u_z T 2\pi r\, dr}{Q} \tag{12.4.8}$$

Note that this is not $T$ averaged over the cross section. It is $T$ *times* the velocity at any radial position, averaged across the cross section. Hence it is an average temperature that is weighted by the flow. It is sometimes called the "flow-averaged" temperature, but more commonly we use the term "cup-mixing temperature," because it is in fact the temperature of the fluid that we would measure if we collected the efflux from the pipe in a cup and mixed the contents to yield a single uniform temperature. (This concept was introduced in the Mass Transfer part, in Chapter 5, Section 5.2.)

Then Eq. 12.4.8 is rearranged to the form

$$\int_0^R u_z T r\, dr = \frac{Q}{2\pi} \langle T \rangle \tag{12.4.9}$$

and we find

$$\rho C_p \frac{d}{dz}\left[\frac{Q}{2\pi}\langle T\rangle\right] = kR\frac{\partial T}{\partial r}\bigg|_R \tag{12.4.10}$$

Since the mass flowrate $w$ is constant:

$$\rho Q = w = \text{constant} \tag{12.4.11}$$

we may write this as

$$wC_p\frac{d\langle T\rangle}{dz} = k2\pi R\frac{\partial T}{\partial r}\bigg|_R \tag{12.4.12}$$

We now define a heat transfer coefficient by the usual general definition—the ratio of the heat flux to the temperature difference between the surface and the fluid:

$$-k\frac{\partial T}{\partial r}\bigg|_R = h[\langle T\rangle - T_R] \tag{12.4.13}$$

Then Eq. 12.4.12 takes the form

$$wC_p\frac{d\langle T\rangle}{dz} = \pi D h[T_R - \langle T\rangle] \tag{12.4.14}$$

where we have replaced twice the radius by the diameter $D$.

Let's pause for a moment and recall what we are going after. We want to find the average temperature of the fluid as a function of flowrate, pipe length and diameter, and the heat transfer coefficient at the inside surface of the pipe, which is assumed to be known. Since we are not interested in the radial profile of temperature, we have simplified the partial differential equation for $T(r, z)$ to an ordinary differential equation for $\langle T\rangle(z)$. We have achieved this, and we now examine solutions to Eq. 12.4.14.

We begin by rewriting Eq. 12.4.14 in the form

$$\frac{d(T_R - T)}{T_R - T} = -\frac{\pi D h}{wC_p}dz \tag{12.4.15}$$

where $T$ is now understood to be the average temperature $\langle T\rangle(z)$. *Assuming that $h$ is not a function of axial position $z$,* we may solve this equation and find[1]

$$\frac{T - T_R}{T_1 - T_R} = \exp\left(-\frac{\pi D h z}{wC_p}\right) = \exp\left(-4\text{St}\frac{z}{D}\right) \tag{12.4.16}$$

where we have defined the Stanton number as

$$\text{St} = \frac{h}{\rho C_p U} = \frac{\text{Nu}}{\text{Re Pr}} = \frac{\text{Nu}}{\text{Pe}} \tag{12.4.17}$$

We have also introduced the so-called Peclet number for heat transfer:

$$\text{Pe} = \text{Re Pr} \tag{12.4.18}$$

---

[1] If $h$ is a function of axial position we still obtain Eq. 12.4.16, but with $\overline{\text{St}} = \bar{h}/\rho C_p U$, where

$$\bar{h} = \frac{1}{L}\int_0^L h\, dz$$

We may now find the outlet temperature $T_2[= T(z = L)]$ of the fluid as

$$\frac{T_2 - T_R}{T_1 - T_R} = \exp\left(-4\mathrm{St}\,\frac{L}{D}\right) = \exp\left(-\frac{\pi D L h}{w C_\mathrm{p}}\right) \qquad \textbf{(12.4.19)}$$

We often define a heat transfer coefficient by

$$Q_\mathrm{H} = h A (\Delta T) \qquad \textbf{(12.4.20)}$$

where $\Delta T$ is the temperature difference across the pipe/fluid interface that drives the heat transfer. It is not clear what $\Delta T$ should be used, however, especially if we demand that the $h$ of Eq. 12.4.20 be compatible with the $h$ of Eq. 12.4.19. The term $\Delta T$ is *not* the temperature change in the fluid, $T_1 - T_2$. Let's find out what $\Delta T$ should be so that these definitions of $h$ are compatible.

We know that

$$Q_\mathrm{H} = +w C_\mathrm{p}(T_1 - T_2) \qquad \textbf{(12.4.21)}$$

We use the convention that $Q_\mathrm{H} > 0$ corresponds to heat flow *out* of the pipe, for which the temperature change $T_1 - T_2$ is positive, since the outlet temperature then lies below that of the inlet.

From Eqs. 12.4.20 and 12.4.21, we find

$$\Delta T = +\frac{w C_\mathrm{p}(T_1 - T_2)}{\pi D L h} \qquad \textbf{(12.4.22)}$$

But from Eq. 12.4.19 we see that

$$-\frac{\pi D L h}{w C_\mathrm{p}} = \ln \frac{T_2 - T_R}{T_1 - T_R} \qquad \textbf{(12.4.23)}$$

Hence, on comparison of Eqs. 12.4.22 and 12.4.23, we find

$$\Delta T = \frac{T_2 - T_1}{\ln\left(\dfrac{T_2 - T_R}{T_1 - T_R}\right)} = \frac{T_1 - T_2}{\ln\left(\dfrac{T_1 - T_R}{T_2 - T_R}\right)} = \frac{(T_1 - T_R) - (T_2 - T_R)}{\ln\left(\dfrac{T_1 - T_R}{T_2 - T_R}\right)} = \Delta T_{\mathrm{ln}} \qquad \textbf{(12.4.24)}$$

where $\Delta T_{\mathrm{ln}}$ is called the *log–mean temperature difference*. Equation 12.4.24 is the answer to the question: What $\Delta T$ do we use in Eq. 12.4.20? The $h$ defined in Eq. 12.4.20 is sometimes written as $h_{\mathrm{ln}}$ because when $h$ is used in Eq. 12.4.20, the $\Delta T$ that *must be* used is $\Delta T_{\mathrm{ln}}$. It is the proper definition of a heat transfer coefficient for flow in a pipe with a constant wall temperature, and Eq. 12.4.24 is a natural consequence of this definition.

---

**EXAMPLE 12.4.1**   *Flow Through a Chilled Tube*

Water is fed from a reservoir to a small capillary immersed in a well-stirred ice bath. Find $T_\mathrm{L}$ for the system shown in Fig. 12.4.1. Make and list any reasonable assumptions.

Our first task is to find the flowrate through the capillary. Then we may calculate a Reynolds number. This will lead us to a Nusselt number and a heat transfer coefficient. Then we calculate a Stanton number, and use Eq. 12.4.19 to find the temperature drop across the length $L$.

We assume that there are no significant pressure losses due to friction in the plumbing between the reservoir and the capillary entrance. Then we take the pressure at the

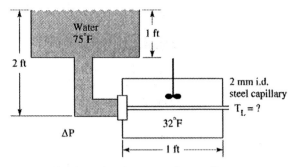

**Figure 12.4.1** A simple system for cooling water.

capillary entrance to be hydrostatic, and we find (using cgs units)

$$\Delta P = \rho g H = (1)\,980\,(2)\,(30.48) = 5.9 \times 10^4 \text{ dyn/cm}^2 \qquad (12.4.25)$$

Assume that the flow is turbulent, and estimate the friction factor from

$$f \approx 0.08 \text{ Re}^{-1/4} \qquad (12.4.26)$$

We introduce the definition of the friction factor,

$$f \equiv \frac{D \, \Delta P}{2 L \rho U^2} \qquad (12.4.27)$$

and we replace $U$ in terms of Re and find an equation for Re:

$$\frac{\rho D^3 \, \Delta P}{2 L \mu^2} \frac{1}{\text{Re}^2} = 0.08 \text{ Re}^{-1/4} \qquad (12.4.28)$$

Everything is known in this equation except Re, so we introduce the known parameters and find

$$\text{Re} = 2700 \quad \text{and} \quad f = 0.011 \qquad (12.4.29)$$

Now we need a correlation for the heat transfer coefficient. We have several choices, and we illustrate two. We may assume that the Colburn analogy holds, which allows us to write

$$j_H = \frac{f}{2} = 0.005 \qquad (12.4.30)$$

We need a Prandtl number for water, but at what temperature do we evaluate it? The water temperature will obviously be in the range of $32 < T < 75°$F. Guess an average temperature of 50°F, at which the Prandtl number (see Fig. 12.3.3) is Pr = 8.6. Then

$$\text{Pr}^{2/3} = 4.2 \qquad (12.4.31)$$

and

$$\text{St} = \frac{j_H}{\text{Pr}^{2/3}} = 0.0012 \qquad (12.4.32)$$

From Eq. 12.4.19

$$\frac{T_L - T_R}{T_1 - T_R} = \exp\left(-4\,\frac{30}{0.20}\,0.0012\right) = \exp(-0.72) = 0.49 \qquad (12.4.33)$$

and

$$T_L = T_R + (T_1 - T_R)(0.49) = 32 + 43(0.49) = 53°F \qquad (12.4.34)$$

We see that the average temperature along the pipe length is 64°F, not the value of 50°F assumed in calculating the Prandtl number. We correct the Prandtl number (Pr = 7.1 at 64°F) and do the calculation again. We find

$$T_L = 48°F \qquad (12.4.35)$$

We continue iterating, using a more accurate estimate of the average temperature each time, to obtain a more accurate estimate of Prandtl number. We converge on the value

$$T_L = 48°F \qquad (12.4.36)$$

on the next iteration. Considering that we are using an approximation for the friction factor (Eq. 12.4.26) and an approximation for the Stanton number (Eq. 12.4.32), there is no point in carrying out a lengthy iterative process.

A commonly recommended empirical correlation for heat transfer in turbulent pipe flow is the Dittus–Boelter equation (with the exponent on Pr of −0.7 for the case that the fluid is being *cooled*).

$$\frac{\overline{Nu}}{Re\,Pr} = \overline{St} = 0.023\,Re^{-0.2}\,Pr^{-0.7} \qquad (12.4.37)$$

(Compare this to Eq. 12.1.22.) This yields a Stanton number of

$$St = 0.023\,(2700)^{-0.2}\,(8.6)^{-0.7} = 0.0011 \qquad (12.4.38)$$

which gives us (see Eqs. 12.4.32 to 12.4.34)

$$T_L = 54°F \qquad (12.4.39)$$

which is just slightly different from what we find in Eq. 12.4.34. The iterative procedure, using a better estimate of Pr in Eq. 12.4.37, will lead us to essentially the same result found in Eq. 12.4.36. Hence these two correlations for h, Eq. 12.4.30 and Eq. 12.4.37, are quite similar under these flow conditions.

**EXAMPLE 12.4.2**  *A Tubular Heater for Air*

Find the length of 2-inch-i.d. tubing required to heat 70 lb/h of air at 1 atm from 70°F to 230°F. The tube wall is at 250°F along the entire length. We begin with Eq. 12.4.19 in the form

$$\ln\left(\frac{T_L - T_R}{T_1 - T_R}\right) = -\frac{4L}{D}St = \ln\left(\frac{230 - 250}{70 - 250}\right) = -2.2 \qquad (12.4.40)$$

Thus the required length is given by

$$L = \frac{2.2D}{4\,St} \qquad (12.4.41)$$

We need the Stanton number. Again, we examine two routes to finding St.

## Colburn Analogy

$$j_H \equiv \frac{\text{Nu}}{\text{Re Pr}^{1/3}} = \frac{f}{2} = \text{Pr}^{2/3}\,\text{St} \qquad (12.4.42)$$

We find the Reynolds number from

$$\text{Re} = \frac{4w}{\pi D \mu} \qquad (12.4.43)$$

We need an air temperature estimate so that we can find a value for the viscosity of air. In this problem the average air temperature along the pipe axis is known (150°F), and at this temperature

$$\mu_b = 2 \times 10^{-5}\ \text{Pa} \cdot \text{s} = 0.048\ \text{lb}_m/\text{h} \cdot \text{ft} \qquad (12.4.44)$$

Then it follows that

$$\text{Re} = \frac{4w}{\pi D \mu} = \frac{4\,(70)}{\pi\,(2/12)\,0.048} = 10^4 \qquad (12.4.45)$$

At this temperature the Prandtl number for air is

$$\text{Pr} = 0.7 \qquad (12.4.46)$$

Using Eq. 12.4.26 we find

$$f = 0.008 \qquad (12.4.47)$$

and it follows that

$$\text{Pr}^{2/3}\,\text{St} = \frac{f}{2} = 0.004 \qquad (12.4.48)$$

and

$$\text{St} = 0.0051 \qquad (12.4.49)$$

Hence we find the required length to be

$$L = 18\ \text{ft} \qquad (12.4.50)$$

## Dittus–Boelter Equation

We begin with

$$\text{St} = 0.023\ \text{Re}^{-0.2}\ \text{Pr}^{-0.6} \qquad (12.4.51)$$

(with the exponent on Pr of $-0.6$ for the case that the fluid is being *heated*) and find

$$\text{St} = 0.0045 \qquad (12.4.52)$$

Hence we find the required length to be

$$L = 20\ \text{ft} \qquad (12.4.53)$$

### 12.4.2  Constant Heat Flux at the Pipe Surface

We have restricted our attention to problems in which the pipe surface is maintained at a uniform temperature along its length. It is also possible to operate a heat exchanger in such a manner that the *flux* of heat across the pipe wall is constant along the length of the pipe, in which case the temperature of the heat transfer surface *varies* in the axial direction. If this is the case, the mathematical model we have been using is no longer valid (it assumed constant $T_R$), and we must derive a different model. Fortunately this is very simple to do.

We begin with a heat balance on a differential section of the pipe, in the form

$$wC_p \, dT = dQ_H = h\pi D \, dz \, (T_R - T) \tag{12.4.54}$$

(We are to understand here that $T$ denotes the cup-mixing average fluid temperature.) If the heat flux is constant then the *rate of heat transfer* $dQ_H$ is constant along the pipe axis. Hence, if $h$ is constant along the pipe length, the temperature difference $T_R - T$ remains constant, and we may write Eq. 12.4.54 in the form

$$\frac{dT}{dz} = \frac{dQ_H/dz}{wC_p} = \frac{h\pi D}{wC_p}(T_R - T) \tag{12.4.55}$$

We may solve this equation subject to the boundary condition

$$T = T_1 \qquad \text{at } z = 0 \tag{12.4.56}$$

with the result

$$T_L - T_1 = \frac{dQ_H/dz}{wC_p} L = \frac{\pi D L h}{wC_p}(T_R - T) \tag{12.4.57}$$

This very simple solution states that the temperature changes along the pipe axis linearly with the length of the pipe. Of course, to calculate the temperature rise, we must know the constant rate of heat input, or the constant temperature difference $(T_R - T)$.

### EXAMPLE 12.4.3   A Tubular Heater for Air: Constant Heat Input

Air entering a 1-inch-i.d. tube at 2 atm and 200°C is heated as it flows at a linear velocity of 10 m/s. A constant heat flux condition is maintained at the wall, and the wall temperature is found to be 20°C above the air temperature, *all along the length of the tube.* What is the temperature rise over a 3 m length of the tube?

We first calculate the Reynolds number. The properties of air at 200°C are (note the consistent set of SI units) as follows:

$$\rho = \frac{pM_w}{R_G T} = \frac{(2)(1.0132 \times 10^5) \, 29}{(8314)(473)} = 1.5 \text{ kg/m}^3$$

$$\text{Pr} = 0.68$$

$$\mu = 2.6 \times 10^{-5} \text{ kg/m} \cdot \text{s}$$

$$k = 0.0386 \text{ W/m} \cdot \text{°C}$$

$$C_p = 1025 \text{ J/kg} \cdot \text{°C}$$

$$\text{Re} = \frac{\rho U D}{\mu} = \frac{(1.5)(10)(0.0254)}{2.6 \times 10^{-5}} = 1.4 \times 10^4$$

The flow is turbulent. We will use the Dittus–Boelter equation:

$$\text{Nu} = \frac{hD}{k} = 0.023 \, \text{Re}^{0.8} \, \text{Pr}^{0.4} = (0.023)(14,000)^{0.8}(0.68)^{0.4} = 44 \tag{12.4.58}$$

$$h = \frac{k}{D} \text{Nu} = \frac{(0.0386)(44)}{0.0254} = 67 \text{ W/m}^2 \cdot \text{°C} \tag{12.4.59}$$

The mass flowrate is

$$w = \rho U \frac{\pi D^2}{4} = (1.5)(10)\pi \frac{0.0254^2}{4} = 7.6 \times 10^{-3} \text{ kg/s} \qquad \textbf{(12.4.60)}$$

so, from Eq. 12.4.57, we find

$$\Delta T = \frac{\pi (0.0254)(3)(67)(20)}{7.6 \times 10^{-3} (1025)} = 41°C \qquad \textbf{(12.4.61)}$$

## 12.5  MODELS OF CONVECTION IN LAMINAR FLOWS

In a number of practical and important situations, viscous fluids pass in laminar flow across surfaces with which they exchange heat. In this section we develop and examine some models for the convective heat transfer coefficient under laminar flow.

### 12.5.1   A Parallel Plate Heat Exchanger with Isothermal Surfaces

Figure 12.5.1 shows a laminar flow confined between parallel planes, which are maintained at a fixed temperature $T_H$. A fluid at a uniform inlet temperature $T_1$ enters the heated section and flows at a mean velocity $U$ down the channel. We neglect any effect of temperature variation on the viscosity of the fluid.

The flow is taken to be laminar and Newtonian, and it is easily shown that the velocity profile is

$$u(y) = \frac{3}{2} U \left[ 1 - \left( \frac{y}{H} \right)^2 \right] \qquad \textbf{(12.5.1)}$$

where $U$ is the average velocity across the channel. Hence the volumetric flowrate per unit width of channel is

$$q_w = 2UH \qquad \textbf{(12.5.2)}$$

We assume, in this analysis, that the channel is very wide compared to $H$ and that the flow and the heat transport are two-dimensional (i.e., no variations in the $z$ direction).

The convective energy equation in the region $[-H, H]$ for steady state operation takes the form

$$u(y) \frac{\partial T}{\partial x} = \alpha \left( \frac{\partial^2 T}{\partial y^2} + \frac{\partial^2 T}{\partial x^2} \right) \qquad \textbf{(12.5.3)}$$

The rate of heat conduction along the axis (the $\partial^2 T/\partial x^2$ term in Eq. 12.5.3) is generally much smaller than the rate at which heat is transferred by convection (the term on the

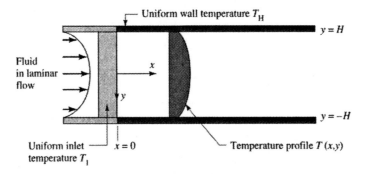

**Figure 12.5.1** Laminar flow between heated parallel planes.

left-hand side). Hence a good approximation (see Problem 12.65) is to neglect axial conduction and write Eq. 12.5.3 in the form

$$u(y)\frac{\partial T}{\partial x} = \alpha\frac{\partial^2 T}{\partial y^2} \tag{12.5.4}$$

As always, the specification of boundary conditions determines much of the character of the solution to this equation. Hence we discuss these conditions now.

We must supply an entrance condition regarding the temperature of the fluid at the plane $x = 0$. We take the temperature to be uniform there, and write

$$T = T_1 \qquad \text{at } x = 0 \qquad \text{for all } y \tag{12.5.5}$$

We assume symmetry along the midplane of the channel, and so we write

$$\frac{\partial T}{\partial y} = 0 \qquad \text{at } y = 0 \qquad \text{for all } x \tag{12.5.6}$$

At the fluid/solid interface we take the case of an isothermal wall, so we write

$$T = T_H \qquad \text{at } y = \pm H \qquad \text{for all } x \geq 0 \tag{12.5.7}$$

We now have enough boundary conditions to permit a solution to the convective energy equation. Note that the equation (Eq. 12.5.4) is linear, but because of the velocity term it has nonconstant coefficients. The equation, with this set of linear boundary conditions, can be solved analytically by a series method. Mathematically, we call the equation and boundary conditions a Sturm–Liouville problem, for which a solution procedure can be found in a number of applied mathematics texts. We simply present the solution here, without derivation. (We treated a very similar problem in Section 5.3 of Mass Transfer.)

We first nondimensionalize the equations in the following manner. We define a dimensionless temperature as

$$\Theta \equiv \frac{T - T_H}{T_1 - T_H} \tag{12.5.8}$$

and dimensionless space variables

$$y^* = \frac{y}{H} \qquad x^* = \frac{x\alpha}{UH^2} \tag{12.5.9}$$

Now the differential equation to be solved takes the form

$$\frac{3}{2}(1 - y^{*2})\frac{\partial\Theta}{\partial x^*} = \frac{\partial^2\Theta}{\partial y^{*2}} \tag{12.5.10}$$

and the boundary conditions are simplified to

$$\Theta = 1 \qquad \text{at } x^* = 0 \qquad \text{for all } y^* \tag{12.5.11}$$

$$\frac{\partial\Theta}{\partial y^*} = 0 \qquad \text{at } y^* = 0 \qquad \text{for all } x^* \tag{12.5.12}$$

$$\Theta = 0 \qquad \text{at } y^* = \pm 1 \qquad \text{for all } x^* \geq 0 \tag{12.5.13}$$

An analytical solution for $\Theta$ can be obtained, which takes the form of a product of an infinite series of polynomials in $y^*$ times exponentially decaying functions of $x^*$. The form of the solution is

$$\Theta = \sum_{m=0}^{\infty}\left[A_m \exp\left(\frac{-2\lambda_m^2 x^*}{3}\right)\sum_{n=0}^{\infty} a_{nm} y^{*n}\right] \tag{12.5.14}$$

We are not really interested in the $y^*$ dependence of $\Theta$. What we do want to know is the change in *mean* temperature of the fluid accomplished by passage of the fluid down a length $L$ of the channel. A simple energy balance tells us that the rate of heat transfer, averaged over the length of the channel, is (per unit width of channel)

$$Q_H = 2UH\rho C_p(T_1 - T_{cm}) \tag{12.5.15}$$

where $T_{cm}$ is (except for notation) the cup-mixing average defined earlier (see Eq. 12.4.8). Hence we need to obtain the cup-mixing average from the temperature profile given in Eq. 12.5.14, so we must evaluate the integral (in terms of $\Theta$ instead of $T$)

$$\Theta_{cm} \equiv \frac{\int_0^H \frac{3}{2} U \left(1 - \left[\frac{y}{H}\right]^2\right) \Theta(x, y)\, dy}{\int_0^H \frac{3}{2} U \left(1 - \left[\frac{y}{H}\right]^2\right) dy} \tag{12.5.16}$$

It is possible to show that this can be written in the form

$$\Theta_{cm} = \sum_{m=0}^{\infty} G_m \exp\left(\frac{-2\lambda_m^2 x^*}{3}\right) \tag{12.5.17}$$

where the first three coefficients are:

| $m$ | $G_m$ | $\lambda_m^2$ |
|---|---|---|
| 0 | 0.910 | 2.83 |
| 1 | 0.0533 | 32.1 |
| 2 | 0.0153 | 93.4 |

A one-term approximation to the infinite series solution, valid for values of $x^* > 0.1$, is

$$\Theta_{cm} = 0.91 \exp(-1.89x^*) \tag{12.5.18}$$

Everything we need to know about the average temperature of the fluid is given in the solution for $\Theta_{cm}$. (For purposes of design or analysis of an existing system, the essential information is found in this cup-mixing average, rather than in the detailed temperature *distribution* $\Theta(x^*, y^*)$.) It is useful at this point to calculate a measure of the rate of transfer of heat across the isothermal boundary, and we usually do this by first defining a heat transfer coefficient. The local convective heat transfer coefficient $h_{ln}$ is defined in terms of a heat balance on the fluid:

$$-\rho C_p UH\, dT_{cm} = h_{ln}(T_{cm} - T_H)\, dx \tag{12.5.19}$$

Note that in this expression the heat transfer coefficient is based on the difference between the average temperature in the fluid (and it is the *cup-mixing* average) and the wall temperature. Note also that Eq. 12.5.19 is for the transfer across the lower half of the channel. For both sides of the channel, we would simply double each side of this expression.

Then we may convert to the dimensionless format and calculate the heat transfer coefficient from

$$h_{ln} = -\left(\frac{UH}{\Theta_{cm}}\right)\left(\frac{k}{UH^2}\right)\left(\frac{d\Theta_{cm}}{dx^*}\right) \tag{12.5.20}$$

(We have changed from $dx$ to $dx^*$ in this expression, as well as from $T$ to $\Theta$.)

Now we must go back to the solution for $\Theta_{cm}$ (Eq. 12.5.17) and differentiate $\Theta_{cm}$ with respect to $x^*$. The result may be written in the format

$$\frac{4h_{ln}H}{k} = \mathrm{Nu}_{ln} = \frac{\sum\limits_{m=0}^{\infty} \tfrac{8}{3}\lambda_m^2 G_m \exp(-2\lambda_m^2 x^*/3)}{\sum\limits_{m=0}^{\infty} G_m \exp(-2\lambda_m^2 x^*/3)} \tag{12.5.21}$$

where the Nusselt number is based on the length scale $4H$.[1] However, this is a *local* Nusselt number, since the function on the right-hand side is $x^*$ dependent. For large enough values of $x^*$, we may use a one-term approximation to the infinite series that appear in the numerator and denominator, with the result that the Nusselt number approaches a constant value given by

$$\mathrm{Nu}_{ln} = \tfrac{8}{3}\lambda_0^2 = 7.55 \tag{12.5.22}$$

Again, with all the algebra flying around, we need to remind ourselves of what we have derived. Equation 12.5.22 is valid only for distances $x^*$ that are in some sense far downstream from the entrance to the heat exchanger. By carrying out a more detailed analysis, we could show that this is a good approximation as long as $x^*$ exceeds a value of order 0.1. Keep in mind that this $x^*$ is a dimensionless length. Whether this simple model is useful depends on whether most of the heat exchanger lies in the region $x^* > 0.1$, and of course this depends on the parameters $U$ and $\alpha$, especially.

We will find soon that we often need the *averaged* value of the Nusselt number more than the *local* value, in design problems. The average is defined as

$$\overline{\mathrm{Nu}_L} \equiv \frac{1}{x_L^*} \int_0^{x_L^*} \mathrm{Nu}_{ln}(x^*)\, dx^* \tag{12.5.23}$$

where $x_L^*$ is simply $x^*$ (Eq. 12.5.9) at $x = L$. Looking at Eq. 12.5.21 we see that we would have to evaluate the integral of a ratio of two infinite series. A much simpler algebraic form results if we use an energy balance approach, and this is left as a homework exercise (Problem 12.17). The result is that the mean Nusselt number can be calculated quite simply from

$$\overline{\mathrm{Nu}_L} \equiv \frac{4}{x_L^*} \ln\left(\frac{1}{\Theta_{cm}(x_L^*)}\right) \tag{12.5.24}$$

This is the mean that has been averaged over the length $x_L^*$.

---

[1] We adopt the common convention that uses the "hydraulic diameter" as the length scale in the Nusselt and Reynolds numbers. The hydraulic diameter is defined, for a tube of any cross-sectional shape, as

$$D_h \equiv \frac{4 \times \text{cross-sectional area}}{\text{wetted perimeter}}$$

For the circular cross section this gives the expected result

$$D_h \equiv \frac{4(\pi D^2/4)}{\pi D} = D$$

For parallel plates of width $W$ spaced a distance $2H$ apart, we find

$$D_h \equiv \frac{4(2HW)}{2W + 4H} = 4H \qquad \text{for} \quad W/H \gg 1$$

### 12.5.2  The Thermal Entry Region of a Parallel Plate Heat Exchanger with Isothermal Surfaces

Although Eq. 12.5.17 is applicable for small values of $x^*$, it is not a computationally efficient model because of the need to sum over a very large number of terms when $x^*$ is small. Since we would like to avoid the work of evaluating a large number of the $G_m$ and $\lambda_m$ coefficients, we shall examine a model that is useful for very short exchangers. It turns out that we can find a solution to Eq. 12.5.3 in a very simple form that is valid *only* for small values of $x^*$.

We begin, with reference to Fig. 12.5.2, by considering a region of small $x$ such that the temperature change in the fluid is confined to a region that is close to the wall, in comparison to the separation $(2H)$ between the plates. Beyond the point $x = 0$, the surface is at the temperature $T = T_H$, and a temperature profile develops as heat is conducted into the flowing stream. In developing a model for the entrance region of the exchanger, we assume that the change in the temperature field is confined to a region close enough to the solid surface to permit the approximation of the velocity profile as a linear function of $y$. We will see why we want to assume this in a moment.

We now want to solve

$$u(y)\frac{\partial T}{\partial x} = \alpha \frac{\partial^2 T}{\partial y^2} \tag{12.5.25}$$

with the following boundary conditions:

$$T = T_1 \quad \text{at } x = 0 \quad \text{for all } y \tag{12.5.26}$$

$$T = T_H \quad \text{at } y = 0 \quad \text{for all } x \geq 0 \tag{12.5.27}$$

(In comparison to the preceding analysis, we have shifted the $y$ axis so that the lower surface is at $y = 0$ and the central plane is at $y = H$.) Otherwise, these are the same boundary conditions applied earlier, and there is no reason to change them. It is the final boundary condition that we change. Although we still have symmetry about the central plane $y = H$, we no longer write the boundary condition in that form.

For a third boundary condition we will assume that the penetration of heat is small in comparison to some length scale of the flow field in the $y$ direction. Really, we require that the penetration length be small compared to the distance over which the velocity profile changes, because we are going to assume that the velocity profile is linear throughout the penetration region (i.e., the distance in the $y$ direction over which $T$ changes from $T_H$ to $T_1$). This assumption will put Eq. 12.5.25 in a form that yields an analytical solution useful for computational purposes. That form is

$$\beta y \frac{\partial T}{\partial x} = \alpha \frac{\partial^2 T}{\partial y^2} \tag{12.5.28}$$

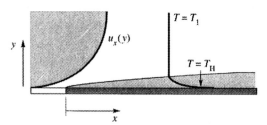

**Figure 12.5.2** Heat transfer in the entrance region of a parallel plate exchanger. The second plate, which is parallel to the one shown, is in the plane $y = 2H$.

where $\beta$ is simply the velocity gradient at $y = 0$ (the solid surface). This follows because we have linearized the velocity profile (Eq. 12.5.1) by writing

$$u_x = u_x(y = 0) + \left[\frac{\partial u_x}{\partial y}\bigg|_{y=0}\right] y + \cdots \approx \beta y \qquad (12.5.29)$$

where

$$\beta = \frac{3U}{H} \qquad (12.5.30)$$

The third boundary condition is then written as

$$T = T_1 \qquad \text{for} \quad y = \infty \qquad (12.5.31)$$

This reflects the notion that if the penetration region is very small, there will be no measurable change in temperature in the central core of the fluid, and the system will behave with respect to heat transfer as if it were infinite in the $y$ direction.

This "boundary value problem" may be solved by the method of combination of variables, and the result (see Section 5.4 of the *Mass Transfer* section for the details, where the same mathematical model arises) is

$$\Theta = 1 - \frac{\int_\eta^\infty \exp(-\eta^3)\, d\eta}{\int_0^\infty \exp(-\eta^3)\, d\eta} \qquad (12.5.32)$$

The dimensionless temperature $\Theta$ is defined as before (Eq. 12.5.8). The space variables are handled in a very different way now, because of our assumption that the fluid is infinite in the $y$ direction. In a sense, there is no longer a length scale $H$ in this formulation of the problem. The space variables have been combined into a single grouping, $\eta$, defined as

$$\eta = y \left(\frac{\beta}{9\alpha x}\right)^{1/3} \qquad (12.5.33)$$

The definite integral in the denominator of Eq. 12.5.32 is a pure number. (It is a so-called gamma function, in this case $\Gamma(4/3)$, and has the value $\Gamma(4/3) = 0.893$.

The local heat flux along the surface $y = 0$ is found from

$$q_y(x) = -k \frac{\partial T}{\partial y}\bigg|_{y=0} = k(T_H - T_1) \frac{(\beta/9\alpha x)^{1/3}}{\Gamma(4/3)} \qquad (12.5.34)$$

Per unit width $W$, the total rate at which heat is transferred across a length $L$ of a surface is

$$\frac{Q_H}{W} = -k \int_0^L \frac{\partial T}{\partial y}\bigg|_{y=0} dx = \frac{3k(T_H - T_1)}{2\Gamma(4/3)} \left(\frac{\beta L^2}{9\alpha}\right)^{1/3} \qquad (12.5.35)$$

We may use Eq. 12.5.34 to define and calculate a local heat transfer coefficient as

$$h(x) \equiv \frac{q_y(x)}{T_H - T_1} = k \frac{(\beta/9\alpha x)^{1/3}}{\Gamma(4/3)} \qquad (12.5.36)$$

With $\beta$ from Eq. 12.5.30, we may ultimately write this equation in the form

$$\text{Nu}(x) \equiv \frac{4h(x)H}{k} = \frac{4(UH^2/3\alpha x)^{1/3}}{\Gamma(4/3)} = 3.12(x^*)^{-1/3} \qquad (12.5.37)$$

where $x^*$, defined as in Eq. 12.5.9, is $x^* = (4x/H)/\text{Re Pr}$, and the Reynolds number is based on the length scale $4H$:

$$\text{Re} = \frac{4UH}{\nu} \qquad (12.5.38)$$

Equation 12.5.37 gives the local Nusselt number. For design purposes it is often useful to have a Nusselt number averaged over the length $L$. A useful definition of an averaged $h$ is

$$\bar{h}_L \equiv \frac{Q_H/W}{L(T_H - T_1)} = \frac{3k}{2\Gamma(4/3)} \left( \frac{\beta}{9\alpha L} \right)^{1/3} \qquad (12.5.39)$$

and the corresponding averaged Nusselt number is

$$\overline{\text{Nu}}_L \equiv \frac{4\bar{h}_L H}{k} = 2.95 \left( \frac{\text{Re Pr}}{L/H} \right)^{1/3} = 4.425 x_L^{*-1/3} \qquad (12.5.40)$$

where

$$x_L^* = \left( \text{Re Pr} \frac{H}{4L} \right)^{-1} \qquad (12.5.41)$$

To find the averaged temperature at some position down the length of the exchanger, we could use Eq. 12.5.32, but it is easier to work with the heat balance. We have $Q_H$ in Eq. 12.5.39, since

$$\frac{Q_H}{W} = \rho C_p (T_{cm} - T_1) q_w = \bar{h}_L L (T_H - T_1) = \frac{3k(T_H - T_1)}{2\Gamma(4/3)} \left( \frac{\beta L^2}{9\alpha} \right)^{1/3} \qquad (12.5.42)$$

Since Eq. 12.5.42 gives the heat transferred from half of the system (the lower plate), we use half the flowrate per unit width: $q_w/2 = UH$. When we rearrange Eq. 12.5.42 and identify and solve for $\Theta_{cm}$, we find

$$\Theta_{cm} \equiv 1 - \frac{T_{cm} - T_1}{T_H - T_1} = 1 - \frac{k}{UH\rho C_p} \frac{3}{2\Gamma(4/3)} \left( \frac{\beta L^2}{9\alpha} \right)^{1/3} \qquad (12.5.43)$$

which we may write, after some further rearrangement, in the form

$$\Theta_{cm} = 1 - 2.95 (\text{Re Pr } H/L)^{-2/3} = 1 - 1.17 (x_L^*)^{2/3} \qquad (12.5.44)$$

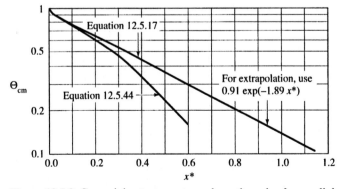

**Figure 12.5.3** Cup-mixing temperature along the axis of a parallel plate, laminar flow heat exchanger.

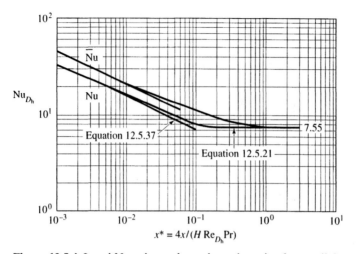

$$x^* = 4x/(H \, \text{Re}_{D_h} \text{Pr})$$

**Figure 12.5.4** Local Nusselt numbers along the axis of a parallel plate, laminar flow heat exchanger. Both surfaces at the same constant temperature. Also shown is the averaged Nusselt number plotted against $x_L^*$; $D_h = 4H$.

Figure 12.5.3 shows the variation of the cup-mixing (dimensionless) temperature along the axis of the channel. Equation 12.5.44 is valid only for *small* values of $x^*$, since the primary restriction on this model is that the degree of penetration of the wall temperature into the fluid must lie within a region so close to the surface that the linearization of the velocity profile is a good approximation. We also show the predicted temperature from Eq. 12.5.17, which does not have that restriction. However, Eq. 12.5.17 is not accurate unless we use a large number of terms of the infinite series solution, for $x^*$ below 0.1. Hence these two solutions cover the range of small and large $x^*$.

While Fig. 12.5.3 provides a graphical means of calculating the cup-mixing temperature, and this is normally sufficient information to solve a heat exchanger design problem, it is interesting to examine the *local* heat transfer coefficient for the parallel plate system. This is shown in Fig. 12.5.4. Again, we use the infinite series solution (but just with a few terms) for large $x^*$, and we use Eq. 12.5.35 for small values of $x^*$.

### 12.5.3 The Tubular Heat Exchanger

Similar analyses can be carried out for the case of flow through a heated tube of circular cross section, rather than between parallel surfaces. The cup-mixing temperature as a function of axial distance $z$ is found in a form similar (see Problem 12.16) to that given in Eq. 12.5.17, and an equation that provides a good approximation under many practical conditions is

$$\Theta_{cm} = 0.82 \exp(-3.66 \, z^*) \tag{12.5.45}$$

where

$$z^* = \frac{\alpha z}{UR^2} = \frac{4z/D}{\text{Pr} \, \text{Re}_D} \tag{12.5.46}$$

This equation is a good approximation to the infinite series solution (not Eq. 12.5.17, but the analogous result for the circular tube case) under the conditions

$$z^* > 0.3 \tag{12.5.47}$$

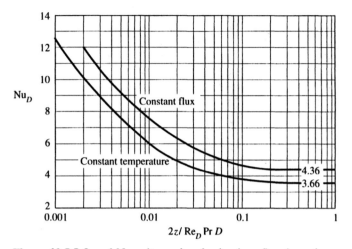

**Figure 12.5.5** Local Nusselt number for laminar flow in a circular pipe.

Equation 12.5.45 is plotted in Fig. P12.8 in the Problems for this chapter. In addition, we find it useful to have the local Nusselt number as a function of axial position for the case of laminar flow in a tube. This is shown in Fig. 12.5.5, where the dimensionless axial coordinate used is not $z^*$, as defined just above, but half that value.

When the approximation given in Eq. 12.5.47 above is valid, the Nusselt number is found to be a constant ($Nu_D = 3.66$), as seen in Fig. 12.5.5. Closer to the entrance to the tube, the Nusselt number is a function of axial position, and is given by (following a development similar to that in Section 12.5.2)

$$Nu_D(z^*) = 1.076 \left( \frac{z/D}{Pr\ Re_D} \right)^{-1/3} \qquad \text{for } \frac{z/D}{Pr\ Re_D} < 0.01 \qquad \textbf{(12.5.48)}$$

Note that in common with the usual conventions, the Nusselt number and the Reynolds number are defined with the *diameter* $D$ as the length scale. In reading the literature on this and related problems, one must be very careful to identify which length scale is used in correlations of Nu versus Re.

In Fig. 12.5.5 we have the *local* Nusselt number along the axis of the pipe or tube. We are also interested in the *average* value of the Nusselt number, as a function of the length $L$ of the pipe over which the heat transfer occurs. From Fig. 12.5.5 we see that for large values of $z^*$, the local Nusselt number becomes constant, and so the average value approaches this asymptotic value also. For small values of $z^*$ (i.e., near the entrance of the pipe), we must integrate Eq. 12.5.48 to find the average as a function of $L$. The result is

$$\overline{Nu}_D(L) = 1.614 \left( \frac{L/D}{Pr\ Re_D} \right)^{-1/3} \qquad \text{for } \frac{L/D}{Pr\ Re_D} < 0.03 \qquad \textbf{(12.5.49)}$$

and this is plotted in Fig. 12.5.6.

### 12.5.4 Heat Transfer to a Tube with Constant Surface Flux

While it is common practice to operate heat exchangers with surfaces that are maintained at a uniform temperature, it is possible and frequently convenient to operate under

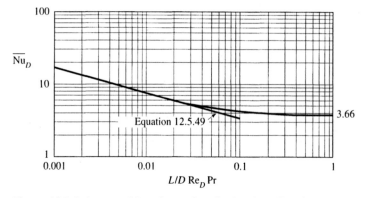

**Figure 12.5.6** Average Nusselt number for laminar flow in a circular pipe, with constant wall temperature.

conditions serving to keep the heat flux into the flowing stream uniform along the exchanger surface. Examples include electrical resistance heating and solar heating. If the heat flux is specified, the determination of the cup-mixing averaged temperature is a trivial application of a heat balance to the system.

Suppose we have a flow, *either laminar or turbulent*, through a tube or pipe of diameter $D$. At any axial position $z$, the change in the cup-mixing averaged temperature of the fluid is simply balanced by the total flow of heat across the tube wall up to that point:

$$\rho C_p Q(T_{cm} - T_1) = \int_0^L h(z)\,[T_R(z) - T_{cm}]\pi D\,dz = q_R \pi DL \qquad (12.5.50)$$

In Eq. 12.5.50, $T_R(z)$ is the unknown and varying surface temperature of the wall of the tube, and $q_R$ is the heat flux, assumed to be uniform and so independent of $z$. Hence we ignore the integral in the middle of this equation and work with the ends.

The cup-mixing temperature follows immediately in the form

$$T_{cm}(L) = T_1 + \frac{\pi DL q_R}{w C_p} \qquad (12.5.51)$$

where $w$ is the *mass* flowrate.

If we also have a model for the local heat transfer coefficient $h(z)$, such as the upper curve in Fig. 12.5.5, we may find the local surface temperature, since

$$h(z)[T_R(z) - T_{cm}] = q_R \qquad (12.5.52)$$

If $h(z)$ is constant, we see immediately that Eq. 12.5.51 leads to the conclusion that the temperature difference between the fluid and the wall is constant along the axis of the tube. We can illustrate this case with an immediate example.

**EXAMPLE 12.5.1**   *Design of a Solar Heater: Constant h*

Water flows at a rate of 0.1 kg/s through a 60 mm diameter pipe, which is receiving solar radiation at a *net* flux of 2000 W/m². What length of pipe is required to raise the water temperature from 20°C to 60°C? What is the surface temperature of the pipe at the downstream end?

We will calculate the required physical properties at the mean water temperature of 40°C. We will need the heat capacity ($C_p$ = 4200 J/kg·K), the thermal conductivity

($k = 0.61$ W/m·K), the viscosity ($\mu = 6.5 \times 10^{-4}$ Pa·s), and the Prandtl number ($\text{Pr} = 4.8$). First we must estimate a heat transfer coefficient.

The Reynolds number for this flow is

$$\text{Re} = \frac{4w}{\pi D \mu} = \frac{4 \times 0.1}{\pi (0.06)(0.00065)} = 3265 \tag{12.5.53}$$

The flow is therefore turbulent, and if the pipe is sufficiently long we should be able to use a constant value of a heat transfer coefficient for this design. We select the Dittus–Boelter equation (see Problem 12.37) for estimating the heat transfer coefficient.

$$\text{Nu} = 0.023\, \text{Re}^{0.8}\, \text{Pr}^{0.4} = 0.023 \times 3265^{0.8} \times 4.8^{0.4} = 28 \tag{12.5.54}$$

This yields a heat transfer coefficient of

$$h = \frac{k}{D}\, \text{Nu} = \frac{0.61}{0.06} \times 28 = 285\ \text{W/m}^2 \cdot \text{K} \tag{12.5.55}$$

The required pipe length follows from Eq. 12.5.50 in the form

$$L = \frac{wC_p}{\pi D q_R}(T_{cm} - T_1) = \frac{(0.1)(4200)}{\pi (0.06)(2000)}(60 - 20) = 45\ \text{m} \tag{12.5.56}$$

The surface temperature at the downstream end is

$$T_R = T_{cm} + \frac{q_R}{h} = 60 + \frac{2000}{285} = 67^\circ\text{C} \tag{12.5.57}$$

If the flow is laminar, and the heat flux is constant and specified, the temperature along the axis still follows from Eq. 12.5.51, which is valid for laminar *or* turbulent flow, and we must use Fig. 12.5.5 for the local heat transfer coefficient (using the constant flux curve) only if we need to estimate the temperature $T_R(L)$ along the surface of the tube.

**EXAMPLE 12.5.2**  *Design of a Parallel Tube Heat Exchanger*

A liquid with $\text{Pr} = 50$ (at 70°C) flows through a set of heated parallel tubes that are plumbed between common headers. The liquid has a kinematic viscosity $\nu$ of $10^{-5}$ m²/s at 70°C. We need to heat 0.002 m³/s (31.6 gals/min) of this liquid from 20°C to 60°C, using a uniform tube wall temperature of 100°C. How many parallel tubes are required? How do we select a combination of $L$ and $D$ for the tubes?

Since the number of tubes is unknown, the Reynolds number in each tube is not known. In anticipation that we will have to solve a number of similar problems, we will first prepare a "design chart" for this class of heat exchanger. The design chart will simply be a plot of the performance of this system, in terms of $\Theta_{cm}$, as a function of Reynolds number, with the tube $L/D$ as a parameter. The Prandtl number will be held constant at $\text{Pr} = 50$.

For laminar flow, say $\text{Re}_D < 2000$, we will use

$$\overline{\text{Nu}}_D = 3.66 \quad \text{for} \quad \frac{\text{Re}_D\, \text{Pr}\, D}{L} < 3 \tag{12.5.58}$$

and

$$\overline{\text{Nu}}_D = 1.6 \left(\frac{\text{Re}_D\, \text{Pr}\, D}{L}\right)^{1/3} \quad \text{for} \quad \frac{\text{Re}_D\, \text{Pr}\, D}{L} > 30 \tag{12.5.59}$$

Between these boundaries on $\text{Re}_D\, \text{Pr}\, D/L$ we will use Fig. 12.5.6.

**Table 12.5.1** Data for Design Chart: Pr = 50, $L/D$ = 100

| $Re_D$ | $Re_D$ Pr $D/L$ | $\overline{Nu}$ | St | $\Theta_{cm}$ |
|---|---|---|---|---|
| 1 | 0.5 | 3.66 | 0.073 | $\approx 0$ |
| 3 | 1.5 | 3.8 | 0.025 | $\approx 0$ |
| 6 | 3 | 3.9 | 0.013 | $5.5 \times 10^{-3}$ |
| 10 | 5 | 4 | $8 \times 10^{-3}$ | 0.041 |
| 20 | 10 | 4.2 | $4.2 \times 10^{-3}$ | 0.19 |
| 30 | 15 | 4.4 | $2.9 \times 10^{-3}$ | 0.31 |
| 100 | 50 | 6 | $1.2 \times 10^{-3}$ | 0.62 |
| 1,000 | 500 | 12.7 | $2.5 \times 10^{-3}$ | 0.9 |
| 5,000 | 2,500 | 152 | $6.1 \times 10^{-4}$ | 0.78 |
| 30,000 | 15,000 | 639 | $4.3 \times 10^{-4}$ | 0.84 |

For turbulent flow we will use

$$\overline{Nu}_D = 0.023Re_D^{0.8}\,Pr^{0.4} \qquad \text{for} \quad Re_D > 5000 \tag{12.5.60}$$

For $2000 < Re_D < 5000$, the region where the transition to turbulence occurs, Nusselt number correlations are not so well established.

The design chart is created in the following manner. We select some $L/D$, say $L/D = 100$. We then pick a series of Reynolds numbers, and for each one we calculate a Nusselt number and, using Pr = 50, a Stanton number. Equation 12.4.19 then yields a $\Theta_{cm}$ value, since the left-hand side is just $\Theta_{cm}$. The results are given in Table 12.5.1 and in Fig. 12.5.7.

We begin with a "base case design," choosing $D = 2$ cm and $L = 100D = 2$ m. For the specified temperatures we find

$$\Theta_{cm} = \frac{60 - 100}{20 - 100} = 0.5 \tag{12.5.61}$$

For any number $n_T$ of parallel tubes the flowrate in each tube is just

$$Q = \frac{Q_{tot}}{n_T} \tag{12.5.62}$$

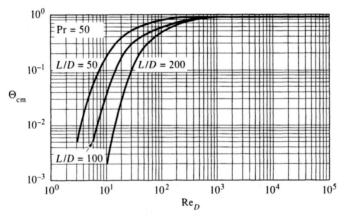

**Figure 12.5.7** Design chart: Pr = 50.

Hence the Reynolds number in each tube is

$$\mathrm{Re}_D = \frac{4Q_{tot}}{\pi D \nu n_T} \tag{12.5.63}$$

We have to find the $\mathrm{Re}_D$ at which $\Theta_{cm} = 0.5$, for the selected $L/D$ and Pr. From Fig. 12.5.7 we find, for $L/D = 100$, $\mathrm{Re}_D \approx 70$. With Eq. 12.5.63 this yields

$$n_T = \frac{4Q_{tot}}{\pi D \nu \, \mathrm{Re}_D} \tag{12.5.64}$$

which we can rearrange to the form

$$n_T L = \frac{4Q_{tot}(L/D)}{\pi \nu \, \mathrm{Re}_D} \tag{12.5.65}$$

Notice that for a specified value of $\Theta_{cm}$, this product $n_T L$ depends on the $L/D$ for the exchanger, but not on a specific choice of either $L$ or $D$. Hence we are free to select $D$ based on considerations outside the thermal demand. For example, we might choose $D$ to keep the required pressure drop within some limit. Or we might choose $L$, which then fixes $D$, to keep the exchanger length below some limitation on space. If we select $D = 2$ cm and $L/D = 100$, for example, Eq. 12.5.64 will yield the requirement

$$n_T = 182 \tag{12.5.66}$$

## 12.6   DESIGN AND ANALYSIS OF A LAMINAR FLOW HEAT EXCHANGER FOR A BLOOD SUBSTITUTE

We consider a design problem that illustrates how to put together several of the ideas and procedures presented up to this point. We select as a context a biomedical application.

**EXAMPLE 12.6.1**   *Required Length for a Compact Laminar Flow Heat Exchanger*

In a surgical procedure on a kidney it is necessary to take the kidney "off line" and nourish it with a blood substitute, using an extracorporeal (out of the body) system. The blood substitute is passed through an oxygenator before it enters the organ, and it is necessary to raise the temperature of the liquid from 32°C to 38°C before returning it to the organ. A preliminary design under consideration has the following features.

The blood substitute will flow through a parallel plate exchanger with a width $W = 10$ cm and a channel height $2H = 2$ mm. The plates are maintained at $T_H = 40°C$. The required capacity of the exchanger is a flowrate of 1 L/min.

The liquid has the following thermal properties:

$$\mathrm{Pr} = 4.5 \qquad k = 0.6 \ \mathrm{W/m \cdot K} \qquad \mu = 3 \ \mathrm{mPa \cdot s} \qquad \rho = 1250 \ \mathrm{kg/m^3}$$

Our first task is to estimate the required axial length of the exchanger.

Let's find the Reynolds number for this flow first, since this will determine the heat transfer correlation needed as part of the design procedure.

For a parallel plate channel, the Reynolds number is defined as

$$\mathrm{Re} = \frac{4UH\rho}{\mu} = \frac{4H\rho}{\mu} \frac{Q}{2HW} = \frac{2\rho Q}{\mu W} \tag{12.6.1}$$

For the specified $Q$ [remember to convert to consistent units (SI) here] and the physical properties given, we find

$$\text{Re} = \frac{2 \times 1250 \, (1.67 \times 10^{-5})}{0.003 \times 0.1} = 139 \tag{12.6.2}$$

We have a laminar flow exchanger. We could use Fig. 12.5.4 for calculating the Nusselt number, but this is not necessary, since our goal is the determination of the required exchanger length. We can simply go to Fig. 12.5.3 now that we know that the exchanger is in the laminar regime. We calculate $\Theta_{cm}$ from the given design requirements on temperatures:

$$\Theta_{cm} = \frac{T_H - T_L}{T_H - T_1} = \frac{40 - 38}{40 - 32} = 0.25 \tag{12.6.3}$$

From Fig. 12.5.3 we find

$$x_L^* = 0.68 \tag{12.6.4}$$

Actually, it is more accurate to use Eq. 12.5.18 than to read the graph. But we must confirm that we are in the asymptotic region of the graph, where Eq. 12.5.18 is accurate. It is clear from the graph that for this value of $\Theta$ we are in that regime.

From its definition (Eq. 12.5.41 is the most useful form for our calculation here), we find, with $H$ in meters,

$$L = \frac{H x_L^* \, \text{Re Pr}}{4} = \frac{0.001 \times 0.68 \times 139 \times 4.5}{4} = 0.106 \text{ m} \tag{12.6.5}$$

We conclude that the exchanger must be a little over 10 cm in axial length. This seems like a reasonably compact exchanger.

Other features of the design bear consideration. We have not said anything about the nature of the blood substitute used in this device. Some such liquids are emulsions of microscopic droplets. It could be important to "manage" the shear rate in this exchanger, since high shear rates might alter the drop size distribution of the emulsion. If, instead of a blood substitute, blood itself were used, high shear rates would be a potential problem because the corresponding shear stresses have the capacity to disrupt the integrity of the red cells and reduce the oxygen-carrying capacity of the blood. It is not difficult to show that the shear stress is small in this design.

One of the most important benefits of a mathematical model of a system is the perspective it gives on how design and operating parameters affect system performance. For example, if it were necessary to increase the flowrate of the blood substitute by a factor of 10, would we increase or decrease the plate separation in order to use an exchanger of the same length, while achieving the same temperature rise? Often one answers such questions from a combination of intuition and knowledge. In this case, if we increase the flowrate, the residence time of the liquid in the exchanger would be decreased, since the liquid would be moving through the system faster. Hence our intuition might favor increasing the channel separation ($2H$) since at a fixed flowrate, this will reduce the linear velocity and thereby increase the residence time. With this modification, there should be sufficient contact time to raise the temperature through the required change. But, an increase in $H$ also makes the exchanger less efficient, since heat must be conducted through a thicker layer. So our qualitative intuition and understanding are less clearly useful here. And this is exactly why mathematical models are so valuable in the design process.

Look at Fig. 12.5.3. If $\Theta$ is specified as a design constraint, then $x_L^*$ is fixed. Now look at Eq. 12.6.1. The Reynolds number is proportional to $Q$, but it is unaffected by $H$. Then to maintain the same $\Theta$ without changing $L$, we see from Eq. 12.6.5 that we must keep the $H$ Re product constant. We conclude that if we increase $Q$ by a factor of 10, we must decrease $H$ by the same factor. We must also be careful that in considering a change in operating conditions (here, an increase in $Q$) we do not find the physics of the system changed in a way that invalidates our design model. In the case in hand, for example, an increase in $Q$ by a factor of 10 yields a Reynolds number of 1390, which is still in the laminar range. If a Reynolds number outside the laminar range had been found, conclusions based on the laminar flow model might be in error.

**EXAMPLE 12.6.2** *Temperature Requirement for the External Heat Transfer Fluid*

As a final example of the use of this model, let's consider one more issue. In designing the heat exchanger, we might choose to make the channel from a material with a significant resistance to heat flow. Hence, if we require a surface temperature of 40°C, as specified in the design, how do we actually ensure that this temperature is achieved on the *inside* surface in contact with the blood substitute? Figure 12.6.1 will serve for the analysis.

To maintain the inner surface of the channel at the temperature $T_H$, there must be a higher temperature $T_o$ on the other side of the channel surface. The issue is whether the resistance of the channel wall is sufficient to require that the temperature $T_o$ be significantly greater than $T_H$. In the steady state, a simple heat balance gives us

$$\frac{k_{\text{wall}}(T_o - T_H)}{b} = h(T_H - T_{\text{cm}}) \qquad (12.6.6)$$

where $k_{\text{wall}}$ and $b$ are the channel wall conductivity and thickness, respectively. In general, $h$ and $T_{\text{cm}}$ vary along the length of the channel. Hence, if $T_o$ is held constant, we will find that $T_H$ varies along the direction of flow. Of course, the mathematical model we have used includes the assumption that $T_H$ is constant.

We will assume the following physical properties for the wall:

$$b = 2\,\text{mm} \qquad k_{\text{wall}} = 60\,\text{W/m} \cdot \text{K}$$

At the downstream end, where $T_{\text{cm}}$ has risen to 38°C (Example 12.6.1), we evaluate $h$ and find the following results:

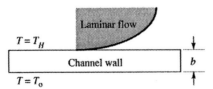

**Figure 12.6.1** Heat transfer across a channel wall.

At $x_L^* = 0.68$ (Eq. 12.6.4), from Fig. 12.5.4 (or with precision, from Eq. 12.5.21), the Nusselt number is Nu = 8.3. Then

$$h = \frac{k\,\text{Nu}}{4H} = \frac{0.6 \times 8.3}{0.004} = 1244 \text{ W/m}^2 \cdot \text{K} \tag{12.6.7}$$

We now solve Eq. 12.6.6 for $T_o$ required to yield $T_H = 40°C$:

$$T_o = T_H + \frac{bh(T_H - T_{cm})}{k_{wall}} = 40 + \frac{0.002 \times 1244\,(40 - 38)}{60} = 40.1°C \tag{12.6.8}$$

Obviously there is a very small temperature drop across the channel wall, at the downstream end. What about closer to the entrance, where $h$ is greater, and much more heat is being transferred? Suppose we consider a region that is only 10% of the downstream length. This corresponds to a distance $x_L^* = 0.068$, and from Fig. 12.5.4 we estimate Nu = 9. This yields an $h$ value of about 1350 W/m$^2 \cdot$K. We need an estimate of $T_{cm}$ at this point. From Fig. 12.5.3 we estimate $\Theta = 0.8$ at $x^* = 0.068$, which gives a temperature of

$$T_{cm}(x^*) = T_H - \Theta(T_H - T_1) = 40 - 0.8(40 - 32) = 33.6°C \tag{12.6.9}$$

Now, from Eq. 12.6.6 we find

$$T_o = T_H + \frac{bh(T_H - T_{cm})}{k_{wall}} = 40 + \frac{0.002 \times 1350(40 - 33.6)}{60} = 40.3°C \tag{12.6.10}$$

We conclude that an exterior temperature of 40°C maintained on the outer surface of the channel will yield an inner surface temperature very close to the design value. Hence this particular choice of channel material does not create a problem with the accuracy of this design procedure. For a channel made of a poorly conducting polymeric material, it could be a different story.

## 12.7  DESIGN OF A CONTINUOUS STERILIZER FOR A BIOPROCESS WASTE STREAM

Waste streams from commercial bioprocesses often contain viable organisms that must be inactivated before the stream can be discharged to the sewer or to a waste treatment plant. The process of sterilization may be thought of and analyzed as a chemical reaction in which the concentration of viable organisms $C_{vi}$ changes with time in accordance with some kinetic model. One of the simplest such models makes the assumption that the inactivation is a first-order process that follows a rate expression of the form

$$\frac{dC_{vi}}{dt} = -k_i C_{vi} \tag{12.7.1}$$

where the rate constant for inactivation is a strong function of temperature, typically assumed to obey

$$k_i = A \exp\left(\frac{-E}{R_G T}\right) \tag{12.7.2}$$

where $T$ is absolute temperature. The constants $A$ and $E$ must be obtained from experimental inactivation studies. With this model we expect that the reduction in concentration of the viable organism follows an expression of the form

$$\frac{C_{vi}}{C_{vi}^o} = \exp(-k_i t) \tag{12.7.3}$$

Conceptually, the simplest sterilization process would be one in which a batch of viable organisms is suddenly raised to the sterilization temperature and held there long enough for the viable concentration to fall, in accordance with Eqs. 12.7.2 and 12.7.3, below some specified level for safe disposal or subsequent processing. Reality is often quite distinct from concept, however. For example, the microscopic organism may be embedded in a macroscopic particulate phase. If this is the case, it is necessary to provide for the time required for heat to penetrate, by conduction, to the interior of the particles. Otherwise organisms near the center of the particles will not be at the sterilization temperature for as long as organisms near the surface. Thus to ensure that only a small fraction of viable organisms remain in the central portions of the particulate phase, it may be necessary to "overkill" a large fraction of the organisms.

In this section we simplify a number of issues associated with sterilization for the sake of focusing on a demonstration of the use of ideas presented in Chapters 10 and 11 to estimate the design parameters required for a continuous sterilizer. Figure 12.7.1 shows a schematic of the system. The biowaste stream of interest is an aqueous slurry of particulate material. Viable organisms are uniformly distributed throughout the particulate phase. The slurry passes to a holding tank, where it may accumulate until it is pumped at a steady flowrate through a steam-jacketed heat exchanger. This preheater raises the temperature of the aqueous phase to the sterilization temperature. The preheater is followed by a segment of piping (the sterilizer) long enough to provide the residence time needed to bring the viable organism count below the specified level for safe discharge. An ideal sterilizer would be a very well-insulated jacket surrounding the piping that comes from the preheater. In a real system, some additional heat input must be provided to make up for the losses through the insulation.

In examining a design problem such as this one, it is useful to focus on several issues that complicate the analysis and make the validity of the design uncertain.

The thermal properties of the waste stream need to be known. If the stream is very dilute, we might get away with assuming that the stream has the properties of water.

Heat transfer to the particulate phase will depend in part on the magnitude of the convective heat transfer coefficient across the interface between the moving stream and the particles. The particles are moving with the stream. How do we estimate heat transfer coefficients for such a system?

Even if we had all the thermophysical properties and convective coefficients, the thermal history of the interior of the particles would be very complex. For one thing, the

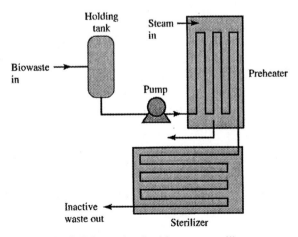

**Figure 12.7.1** Schematic of a biowaste sterilizer.

ambient temperature surrounding the particles is rising as the stream moves through the system. Hence our transient conduction models, which assume a constant ambient temperature, may not serve us well.

The most logical approach to this design problem would proceed by first operating a small-scale pilot plant to enable the measurement of some of the unknown performance characteristics. One could then scale up from the pilot plant to commercial scale operations. *But the pilot plant itself must be designed on some basis.* Hence we use our crude models to design a pilot plant, and then use the pilot plant as the laboratory for developing modifications to the crude model. This would permit us to design the commercial system with some confidence.

Let's proceed along this line then. We will use our simple models (that is all the background we have at this point) to design the pilot plant.

---

**EXAMPLE 12.7.1**   *Estimating Heat Transfer Characteristics in a Pilot Scale Sterilization System*

We will design a pilot scale sterilizer with the following characteristics and constraints:

All piping will be stainless steel with an inside diameter of 1.37 inches.
The feed rate will be 25 gal/min. The feed inlet temperature will be 20°C.
The pipe wall temperature can be maintained at 100°C in both the preheater and the sterilizer.
The thermophysical properties of the aqueous waste stream are those of water, but the viscosity is $\mu = 3$ mPa·s at 20°C. We will assume that the temperature dependence of the viscosity is the same as that of water. Looking at data for $\mu(T)$ for water, we conclude that the waste stream has a viscosity that is three times that of water, at any temperature.
The particles are spherical, with a diameter of 3 mm.
It is necessary to reduce the viable cell concentration by a factor of $10^{24}$.

We first want to determine the pipe length required to bring the stream from 20°C to 95°C. (Keep in mind that we would require an *infinite* pipe length to bring the stream to *exactly* the wall temperature of 100°C.)

A good starting point is a calculation of a Reynolds number. The mean temperature of the process stream between the entrance and the exit is

$$T_b = \frac{20 + 95}{2} = 57.5°C \qquad \text{(12.7.4)}$$

We will evaluate the viscosity of the stream at the mean between the wall temperature $T_R$ and the mean $T_b$. We call this the mean "film" temperature. It is a temperature characteristic of the boundary layer near the pipe surface, across which most of the convective transfer occurs to heat the slurry.

$$T_f = \frac{57.5 + 100}{2} = 79°C \qquad \text{(12.7.5)}$$

At this temperature we find the viscosity of water to be $\mu = 0.35$ mPa·s. Hence the waste stream has a vicosity three times this, or

$$\mu_b = 1.05 \text{ mPa·s} \qquad \text{(12.7.6)}$$

Now we can get a Reynolds number for the 25 gal/min flow. (We use SI units here.)

$$\text{Re} = \frac{4Q\rho}{\pi D\mu} = \frac{4\dfrac{25 \times (3.785 \times 10^{-3})}{60}\,1000}{\pi(1.37 \times 0.0254)\,(1.05 \times 10^{-3})} = 5.5 \times 10^4 \qquad (12.7.7)$$

The flow is highly turbulent, and we will use an empirical correlation for turbulent heat transfer in a pipe. We select Eq. 12.4.50, calculating the Stanton number because we will use Eq. 12.4.19 for the required pipe length:

$$4\,\text{St}\,L/D = -\ln\Theta = -\ln\frac{T(L) - T_R}{T_{in} - T_R} = -\ln\frac{95 - 100}{20 - 100} = 2.77 \qquad (12.7.8)$$

To calculate the Stanton number we need the Prandtl number. For water at $T_f = 79°C$ we find $\text{Pr} = 2.1$. We are assuming that all thermal properties of the slurry are those of water, but the viscosity is three times that of water. Since the Prandtl number has viscosity in the numerator, we estimate the Prandtl number for the slurry as three times that of water:

$$\text{Pr} = 6.3 \qquad (12.7.9)$$

Then the Stanton number follows as

$$\text{St} = 0.023\,\text{Re}^{-0.2}\,\text{Pr}^{-0.6} = 8.6 \times 10^{-4} \qquad (12.7.10)$$

(We choose the Dittus-Boelter equation (see Problem 12.37) with $n = 0.4$.) From Eq. 12.7.8 we find the required pipe length to be

$$L = \frac{2.77\,(1.37 \times 0.0254)}{4\,(8.6 \times 10^{-4})} = 32\text{ m} \qquad (12.7.11)$$

Up to this point along the pipe axis (i.e., at 32 m from the entrance to the heat exchanger), some of the organisms have been killed, but very little sterilization occurs at the lower temperatures, in accordance with Eqs. 12.7.1 and 12.7.2. As a simplification, we ignore any reduction in viable cells over the preheater section of the system.

In the next conceptual stage of this analysis we find the time required to bring the average temperature of the particles to the sterilization temperature. This requires some estimate of the Biot number. What basis shall we use? The particles are entrained by the liquid, and so on the average the particles are actually stationary with respect to the liquid. If that is the case, then heat transfer to the particles is strictly by conduction. However, the flow is turbulent, so there is some degree of agitation of the particles, with an enhancement of the heat transfer coefficient. We just don't have any information at this point. The best we can say is that the Nusselt number is no less than $\text{Nu} = 2$, since this is the lower limit due to conduction (i.e., no convection for a sphere that is stationary with respect to the surrounding liquid: review Section 11.1.1). The Biot number differs from the Nusselt number by a factor of the ratio of the thermal conductivities of the particle and the liquid. We will assume that the particle has the thermal properties of water. In addition, the Biot number is defined with the *radius* of the particle, while the Nusselt number uses the diameter. Hence we estimate that $\text{Bi} = 1$.

The ambient fluid is at 95°C in this section and remains at that temperature if good insulation is provided. We will assume that the center of the particles is at 20°C at the entrance to this section. It must in fact be somewhat higher, but we do not know how much without a more exact transient analysis. Hence we will find the time required to bring 20°C particles to an average temperature of 90°C, with the ambient fluid at 95°C, and the Biot number Bi = 1. Using

$$\Theta = \frac{90 - 95}{20 - 95} = 0.067 \tag{12.7.12}$$

we can construct a graph similar to Fig. 11.1.5, which was done for the case $\Theta = 0.01$. When this has been done (we don't show it here—you'll have to make up your own figure), it can be estimated that at Bi = 1,

$$X_{Fo} = 1 \tag{12.7.13}$$

and the required time is

$$t = 1 \frac{R^2}{\alpha} = \frac{R^2 \, Pr}{\nu} \tag{12.7.14}$$

Inspection of data for $\nu(T)$ and $Pr(T)$ for water (see Fig. C10.1, Appendix C) shows that the ratio is nearly independent of $T$ over the range of concern here. Hence we will evaluate $\nu$ and Pr at the ambient liquid temperature of 95°C:

$$\nu = 0.3 \times 10^{-6} \, m^2/s \qquad Pr = 1.8 \tag{12.7.15}$$

*We do not use the factor of 3 here.* Instead, we treat the liquid surrounding the particles as water. It is the presence of the particles that gives the *slurry* an elevated viscosity.

With a particle radius of 1.5 mm we find, from Eq. 12.7.14

$$t = \frac{(0.0015)^2 \times 1.8}{0.3 \times 10^{-6}} = 13.6 \, s \tag{12.7.16}$$

From the volume flowrate the average velocity in the pipe is

$$U = \frac{4Q}{\pi D^2} = 4 \frac{(25/60)(3.785 \times 10^{-3})}{\pi (1.37 \times 0.0254)^2} = 1.7 \, m/s \tag{12.7.17}$$

Hence the required pipe length for this section is

$$L = Ut = 1.7 \times 13.6 = 22.6 \, m \tag{12.7.18}$$

Let's review what we have at this point. We estimated that the preheater length must be approximately 32 m just to raise the slurry temperature to 95°C. We assumed that the particles remain cold through that section, which is clearly not true. We then calculated the pipe length required to bring cold particles from 20°C to an average temperature of 90°C when conveyed by the 95°C stream. We neglected any "death" of viable cells up to that point. We concluded that we should design a heating system with a total length of about 55 m of piping, just to bring the particles to the sterilization temperature.

The next step is to estimate the residence time required to produce the required degree of sterilization, at an average particle temperature of 90°C. This is a very conservative design, since some loss of viable cells will have occurred up to this point. Conservative designs are "safe," in the sense that we can certainly do the job required with the estimated length of piping. But conservative designs are also costly, since we have more heat exchanger than necessary. Let's come back to this point later.

We examine the following question: What residence time (pipe length) is required to sterilize the target organism if it is held at a uniform temperature of 90°C? Now, to be able to estimate the coefficient $k_i$ of Eq. 12.7.3, we need some kinetic data. We will assume that such data are available and that we know that

$$k_i = 5.5 \, s^{-1} \text{ at } 90°C \tag{12.7.19}$$

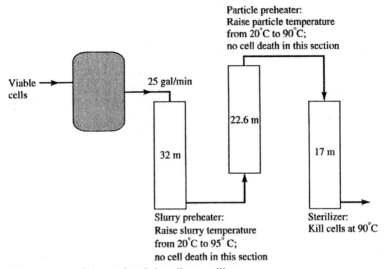

**Figure 12.7.2** Schematic of the pilot sterilizer.

In the problem specifications we stated that the viable cell population was to be reduced by a factor of $10^{24}$. Hence, using the $k_i$ value given here, we find that the required sterilization time at 90°C is

$$t = -\frac{\ln(C_{vi}/C_{vi}^0)}{k_i} = -\frac{2.3\log(C_{vi}/C_{vi}^0)}{k_i} = -\frac{2.3(-24)}{5.5} = 10 \text{ s} \qquad \textbf{(12.7.20)}$$

With the velocity given in Eq. 12.7.17, we find that the required length of this section is 17 m. Hence our analysis suggests that the pilot plant be designed as sketched in Fig. 12.7.2.

From what we have learned thus far, we could build the pilot plant and observe its performance. This would permit us to evaluate some of the simplifications and idealizations that went into the design, providing a basis for design of a larger (process scale) system. An alternative would be to carry out the pilot scale design using more sophisticated modeling techniques. An illustration of this approach is given in Example 12.7.2.

**EXAMPLE 12.7.2**  *A More Exact Analysis of the Performance of a Pilot Scale Sterilization System*

A key assumption of the prior analysis was that no cell death occurred in the preheating section of the sterilizer. This was based on the argument that since the rate of loss of viability is such a strong function of temperature, no significant loss occurs in the relatively cold preheater section. How would we assess the error of that assumption, and how would we develop an improved model of the process? We can gain some insight into this issue by solving the following problem.

A bioproduct is being produced which initially contains viable organisms. The feed solution has all the thermophysical properties of water. The initial concentration of viable cells is $10^{10}$ cells/mL. The cells are microscopic, and they are uniformly dispersed in an aqueous solution.

The cells in the final product must be nonviable, and the required safety level is specified in terms that correspond to less than one viable cell per billion ($10^9$) 100-liter batches of product. The kinetic model of Eq. 12.7.3 holds, and it is known that $k_i = 5.5$ s$^{-1}$ at 90°C and $k_i = 1.1$ s$^{-1}$ at 70°C.

The feed flowrate is 25 gal/h. Band heaters are available which control the wall temperature along the axis, providing a constant rate of heat input all along the axis of the pipeline, which is 50 m long and has an inside diameter of 1.25 inches. The heated pipeline must raise the temperature of the stream from 20°C to 90°C. This constraint will determine the power requirement for the heaters.

Our goal is to find the concentration of viable organisms at the end of the 50 m length of heated pipeline.

We can start by making two quick calculations. The first is the extent of "kill" for a process stream at 90°C over the entire length of pipeline. We need the residence time in the pipeline, and this follows from

$$t_{res} = \frac{L}{U} \tag{12.7.21}$$

with the average velocity given by

$$U = \frac{4Q}{\pi D^2} = \frac{4(25/60 \times 3.785 \times 10^{-3})}{\pi(1.25 \times 0.0254)^2} = 2 \text{ m/s} \tag{12.7.22}$$

Then the residence time is

$$t_{res} = \frac{50}{2} = 25 \text{ s} \tag{12.7.23}$$

If the cells were exposed to a temperature of 90°C for 25 s, the viable cell concentration would fall by a fraction given by

$$\log \frac{C_{vi}}{C_{vi}^o} = -\frac{k_i}{2.3} t_{res} = -\frac{5.5 \times 25}{2.3} = -60 \tag{12.7.24}$$

This fractional loss of viability ($10^{-60}$) seems like a huge decrement in viable cells. What decrement do we require? The initial cell concentration is $10^{10}$ cells/mL. A billion 100-liter batches corresponds to $10^{14}$ mL. Hence we must reduce the concentration from $10^{10}$ to $10^{-14}$, or

$$\log \frac{C_{vi}}{C_{vi}^o} = -24 \tag{12.7.25}$$

We don't need 25 s of residence time at 90°C to achieve this degree of kill—10 s would be sufficient. (This follows from $24 = 5.5 \, t_{res}/2.3$.)

We can make a second quick (but not very useful) calculation: the extent of "kill" for a process stream at an average temperature of 55°C over the entire length of pipeline. To do this calculation we need to evaluate the coefficients that appear in Eq. 12.7.2. Since we are given $k_i$ at two temperatures, we can evaluate the two coefficients, and we find that

$$k_i = 5.06 \times 10^{12} \exp\left(\frac{-10^4}{T}\right) \text{ s}^{-1} \tag{12.7.26}$$

with $T$ in kelvins. Now we find that at the average temperature of 55°C, $k_i = 0.29$ s$^{-1}$, and

$$\log \frac{C_{vi}}{C_{vi}^o} = -\frac{k_i}{2.3} t_{res} = -\frac{0.29 \times 25}{2.3} = -3.16 \tag{12.7.27}$$

This is far below the required decrement in viable cell concentration. In fact, because of the strongly nonlinear temperature dependence of $k_i$, this will turn out to be a very poor estimate of the reduction in viable cells for this process.

With these sets of numbers available for reference, we proceed to develop a model for estimation of the extent of viable cell reduction in the system of interest. Since the heat flux is assumed to be constant along the axis of the pipeline, the temperature rise will be a linear function of distance along the axis:

$$T(x) = T_{in} + \frac{T_{out} - T_{in}}{L} x \tag{12.7.28}$$

We rewrite Eq. 12.7.1 by replacing time with axial position through Eq. 12.7.21. The result is

$$\frac{dC_{vi}(x)}{dx} = -\frac{k_i}{U} C_{vi} \tag{12.7.29}$$

Since $k_i$ is a function of $T$, which in turn is a function of $x$, we must write the "solution" of this differential equation in the form of an undetermined integral:

$$\ln \frac{C_{vi}(x)}{C_{vi}^o} = 2.3 \log \frac{C_{vi}(x)}{C_{vi}^o} = -2.53 \times 10^{12} \int_0^x \exp \left( -\frac{10^4}{T_{in} + (\Delta T/L) x} \right) dx \tag{12.7.30}$$

where $\Delta T = T_{out} - T_{in}$. The integration must be performed numerically to yield the concentration of viable cells as a function of distance along the pipe axis. Of particular interest to us is the concentration at the end, $x = L = 50$ m, which we find to be

$$\log \frac{C_{vi}(x)}{C_{vi}^o} = -10.8 \tag{12.7.31}$$

This result is still far below the required decrement $(10^{-24})$ in viable cell concentration. Nevertheless, it is not a small reduction in viable cell concentration. It is small only relative to the required decrement. If we now provide an additional section of pipe that maintains the process stream at 90°C, we may find the required additional length from the required additional residence time:

$$t = -\frac{2.3 \log(C_{vi}/C_{vi}^l)}{k_i} = -\frac{2.3[-24 - (-10.8)]}{5.5} = -\frac{2.3(-13.2)}{5.5} = 5.5 \text{ s} \tag{12.7.32}$$

where $C_v^l$ is the outlet concentration from the preheater, which becomes the inlet concentration to the following section of pipe. This residence time corresponds to a pipe length of 11 m. If we completely ignore the death of viable cells in the preheater, we arrive at a requirement for 20 m beyond the preheater. Hence the more precise analysis makes a substantial difference in the design.

## 12.8   CONVECTIVE HEAT TRANSFER FROM A CYLINDRICAL FIBER IN STEADY AXIAL MOTION

In the manufacture of continuous filaments from polymers, ceramics, and glass, a key step is the cooling of the filament or fiber subsequent to extrusion. Under most conditions of interest, the flow external to the fiber is a boundary layer type of flow. It is possible

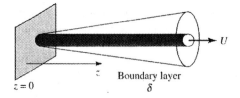

**Figure 12.8.1** A cylindrical fiber issues continuously from an orifice at $z = 0$. The fiber is surrounded by an axisymmetric boundary layer of radius $\delta(z)$.

to develop a model of the heat transfer coefficient along the fiber surface, and from this we may estimate the cooling rate of a fiber as a function of fiber diameter and extrusion speed. The method we outline here is along the lines of an *integral* analysis of the boundary layer equations. Figure 12.8.1 shows the geometry of the system.

The boundary layer equations for the steady flow induced by the axial motion of the fiber are

$$\frac{\partial u_z}{\partial z} + \frac{1}{r}\frac{\partial}{\partial r}(ru_r) = 0 \tag{12.8.1}$$

$$u_z\frac{\partial u_z}{\partial z} + u_r\frac{\partial u_z}{\partial r} = \frac{\nu}{r}\frac{\partial}{\partial r}\left(r\frac{\partial u_z}{\partial r}\right) \tag{12.8.2}$$

$$u_z\frac{\partial T}{\partial z} + u_r\frac{\partial T}{\partial r} = \frac{\alpha_T}{r}\frac{\partial}{\partial r}\left(r\frac{\partial T}{\partial r}\right) \tag{12.8.3}$$

Boundary conditions are

$$u_z = U, u_r = 0, T = T_R \qquad \text{at} \quad r = R \tag{12.8.4}$$

$$u_z = 0, u_r = 0, T = T_\infty \qquad \text{at} \quad r \to \infty \tag{12.8.5}$$

We assume that $R$, the radius of the fiber, is independent of axial position.[1] Of course, Eqs. 12.8.1 through 12.8.3 imply steady Newtonian flow in the fluid surrounding the fiber.

If we assume that all physical properties are independent of temperature, we may first solve for the velocity field induced by the motion of the fiber. When the velocity field is available, we may solve Eq. 12.8.3 for the temperature field. The local Nusselt number will then follow directly from the temperature gradient at the fiber surface. We will illustrate the use of an integral method of solving these boundary layer equations. It is similar to that used earlier in modeling the laminar boundary layer on a flat plate, but the axisymmetric geometry introduces some algebraic complications. As a result, we are not able to obtain a simple polynomial expression for the velocity field. This geometry also calls for some numerical evaluation of a complicated functional expression.

The procedure begins by integrating the momentum equation over the area $2\pi r\, dr$ in the region from the fiber surface, $r = R$, to infinity. (We drop the factor of $2\pi$ in all the integrals.) The first term on the left of Eq. 12.8.2 becomes

$$\int_R^\infty r u_z \frac{\partial u_z}{\partial z}\, dr = \frac{1}{2}\int_R^\infty r\frac{\partial u_z^2}{\partial z}\, dr \tag{12.8.6}$$

---

[1] In commercial fiber spinning the fiber is usually "drawn" in such a way that the diameter decreases along the spinline. This fact is ignored in the model developed here.

The second term requires some manipulation, including the use of the continuity equation (Eq. 12.8.1) to eliminate $u_r$ at one point. The steps and the result are the following:

$$\int_R^\infty r u_r \frac{\partial u_z}{\partial r}\, dr = \int_R^\infty \left( \frac{\partial r u_r u_z}{\partial r} - u_z \frac{\partial r u_r}{\partial r} \right) dr \tag{12.8.7}$$

$$= [r u_r u_z]_R^\infty + \int_R^\infty u_z r \frac{\partial u_z}{\partial z}\, dr = 0 + \frac{1}{2} \int_R^\infty r \frac{\partial u_z^2}{\partial z}\, dr$$

Then the sum of the two integrated terms on the left-hand side of Eq. 12.8.3 may be written in the form

$$\int_R^\infty r \frac{\partial u_z^2}{\partial z}\, dr = \frac{d}{dz} \int_R^\infty r u_z^2\, dr = \frac{d}{dz} \int_0^\infty (R + y) u_z^2\, dy \tag{12.8.8}$$

where $y = r - R$ is a radial coordinate measured from the *surface* of the fiber—not from the axis.

The right-hand side of Eq. 12.8.2 is easy to integrate:

$$\int_R^\infty \nu \frac{\partial}{\partial r} \left( r \frac{\partial u_z}{\partial r} \right) dr = -\nu R \left( \frac{\partial u_z}{\partial y} \right)_{y=0} \tag{12.8.9}$$

Then the integral method requires us to find a velocity profile that satisfies

$$\frac{d}{dz} \int_0^\infty (R + y) u_z^2\, dy = -\nu R \left( \frac{\partial u_z}{\partial y} \right)_{y=0} \tag{12.8.10}$$

In addition, our choice of velocity profile must satisfy some boundary conditions. We will impose two conditions here. One is the no-slip condition along the fiber surface:

$$u_z = U \quad \text{on} \quad y = 0 \tag{12.8.11}$$

The other is that the form of the velocity profile should satisfy the momentum equation far from the fiber surface (i.e., near the outer edge of the boundary layer). We expect that for distances such that $y \approx \delta$ the flow field is nearly axial, and the inertial terms can be neglected. Then Eq. 12.8.2 takes the form

$$0 = \frac{\nu}{r} \frac{\partial}{\partial r} \left( r \frac{\partial u_z}{\partial r} \right) \tag{12.8.12}$$

and the general solution is

$$u_z = a - b \ln r = a - b \ln(R + y) \tag{12.8.13}$$

We may select the constant $a$ so that we satisfy Eq. 12.8.11, and the result is[2]

$$\frac{u_z}{U} = 1 - \frac{1}{\alpha} \ln \left( \frac{R + y}{R} \right) \tag{12.8.14}$$

where $\alpha = 1/b$ is a redefined constant of integration. While Eq. 12.8.14 behaves properly right at the fiber surface, it does not give us a velocity profile that changes along the $z$ axis as the boundary layer grows out into the ambient fluid. The only way to account for this behavior is to permit the integration constant $\alpha$ to be a function of $z$. Hence our approximate velocity profile, which we will force to satisfy the integrated momentum

---

[2] In the rest of this section we use the notation $\alpha$ as defined in Eq. 12.8.14. Don't confuse this with $\alpha_T$, the thermal diffusivity.

equation (Eq. 12.8.10), may be written in the form

$$\frac{u_z}{U} = 1 - \frac{1}{\alpha(z)} \ln\left(\frac{R+y}{R}\right) \tag{12.8.15}$$

Before we proceed, let's review what we have done, and summarize what we are trying to do. We are carrying out an integral boundary layer analysis. This means that we integrate the fluid dynamic equations over the region normal to the fiber axis and then select an approximate velocity profile that is forced to satisfy conservation of momentum and mass. In the planar boundary layer case, we are able to satisfy the integral equations by selecting a simple polynomial that involves the boundary layer thickness explicitly. The axisymmetric case does not permit such a simple analysis. However, we may define a boundary layer thickness $\delta(z)$ as the radial position along which $u_z$ vanishes. Then, from Eq. 12.8.15 we find

$$\frac{\delta(z)}{R} = e^{\alpha(z)} - 1 \tag{12.8.16}$$

The computational task now is to substitute Eq. 12.8.15 into Eq. 12.8.10 and obtain an equation (hopefully, one we can solve!) for $\alpha(z)$. The integral equation takes the form

$$\frac{d}{dz} \int_0^\alpha \left(1 - \frac{\xi}{\alpha}\right)^2 e^{2\xi} d\xi = \frac{\nu}{\alpha U R^2} \tag{12.8.17}$$

after the transformation of variables

$$e^\xi = \frac{y}{R} + 1 \tag{12.8.18}$$

is introduced. When the indicated integrations (using a table of integrals) are performed, we find a differential equation for $\alpha(z)$:

$$\frac{2\nu}{UR^2} = \left(\frac{e^{2\alpha}}{\alpha} - \frac{e^{2\alpha}}{\alpha^2} + \frac{1}{\alpha} + \frac{1}{\alpha^2}\right) \frac{d\alpha}{dz} \tag{12.8.19}$$

The formal solution may be written as the integral

$$\frac{2\nu z}{UR^2} = \int_0^{\alpha(z)} \left(\frac{e^{2s}}{s} - \frac{e^{2s}}{s^2} + \frac{1}{s} + \frac{1}{s^2}\right) ds \tag{12.8.20}$$

In this form we use a dummy variable of integration, $s$, and the right-hand side is just some function of the upper limit $\alpha(z)$. Thus we could pick a value of the upper limit $\alpha(z)$, evaluate the integral numerically, and find thereby a value for the left-hand side of the equation. The result is shown in Fig. 12.8.2.

With the velocity profile known, we may solve the energy equation (Eq. 12.8.3), and again we use an integral technique. Equation 12.8.3, integrated across the region ($0 < y < \infty$), yields

$$\frac{d}{dz} \int_0^\infty (R+y) u_z (T - T_\infty) dy = -\frac{\nu R}{\text{Pr}} \left(\frac{\partial T}{\partial y}\right)_{y=0} \tag{12.8.21}$$

Once again we must assume a form for the temperature profile in the ambient medium, and again we select a function that satisfies a boundary condition on temperature at the fiber surface,

$$T = T_R \quad \text{at} \quad y = 0 \tag{12.8.22}$$

and satisfies Eq. 12.8.3 far from the surface (i.e., as the thermal boundary layer is

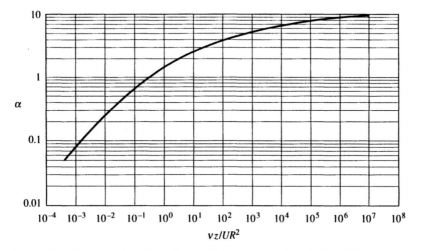

**Figure 12.8.2** $\alpha$ as a function of axial distance, according to Eq. 12.8.20.

approached). That function is

$$\frac{T - T_\infty}{T_R - T_\infty} = 1 - \frac{1}{\beta(z)} \ln\left(\frac{R + y}{R}\right) \tag{12.8.23}$$

When Eq. 12.8.23 for $T$ and Eq. 12.8.14 for $u_z$ are substituted into Eq. 12.8.21, we may, after a great deal of algebra, write a differential equation for the function $\beta(z)$ in terms of the known function $\alpha(z)$. We will not display the details here.[3] The key result is the local Nusselt number along the fiber axis. This is shown in Fig. 12.8.3 for the case Pr = 0.72, which is the Prandtl number of air at room temperature. We define the local Nusselt number as

$$\mathrm{Nu}_D = \frac{2hR}{k} \tag{12.8.24}$$

where $k$ is the thermal conductivity of the ambient fluid.[4]

We can turn now to some examples of application of this heat transfer model. Part of our goal in doing this is to assess the extent to which this model permits us to predict the temperature history of a cooling fiber.

**EXAMPLE 12.8.1**   *Temperature Uniformity across the Radius of a Fiber*

A glass fiber is being spun under the following conditions.

Fiber radius = 14.5 $\mu$m      Initial fiber temperature $T_R$ = 500°C
Fiber speed $U$ = 300 cm/s   Ambient medium is air at 20°C
Spinline length $L$ = 50 cm

In the fiber spinning process it is important to know whether substantial temperature gradients exist across the fiber radius during the cooling process, since these gradients

---

[3] For a detailed discussion of the mathematical analysis of this problem see the reference on which our presentation is based: Bourne and Elliston, *Int. J. Heat Mass Transfer,* **13**, 583 (1970).
[4] In the reference cited, Bourne and Elliston define a Nusselt number that is larger than the standard definition by a factor of $\pi$.

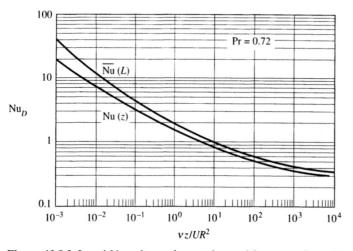

**Figure 12.8.3** Local Nusselt number at the position $z$ on the axis of a moving fiber, and the average Nusselt number over the length $L$. For the latter, replace $z$ with $L$ on the $x$ axis of the figure. Note that $Nu_D$ uses the fiber diameter $D$ as its length scale.

could lead to the development of internal strains that could have negative effects on the mechanical properties of the fiber. Our goal here is to evaluate the assumption that the fiber temperature is uniform across the radius of the fiber. Intuitively this seems quite reasonable, since the fiber is so thin. To support the assumption quantitatively, we should calculate the Biot number. Thus we start with a calculation of the Nusselt number from $z = 0$ to $z = 50$ cm. The worst case will correspond to the largest Nusselt number. Of course the Nusselt number is infinitely large right at the beginning of the spinline, $z = 0$, but this is not physically meaningful. Let's look at the Nusselt number near the plane $z = 0$ but not at it—say $z = 5$ cm, which is just 10% of the length of the cooling region. At this point we find

$$\frac{\nu_{air} z}{UR^2} = \frac{(0.43 \text{ cm}^2/\text{s}) \, 5 \text{ cm}}{(300 \text{ cm/s}) \, (14.5 \times 10^{-4} \text{ cm})^2} = 3400 \tag{12.8.25}$$

From Fig. 12.8.3 we find

$$Nu = 0.32 \tag{12.8.26}$$

In this calculation we evaluate the kinematic viscosity of the air at the mean temperature between the surface, taken as 500°C (although some cooling will have occurred), and the air at 20°C.

The Biot number is

$$Bi_R = \frac{Nu_D}{2} \frac{k_{air}}{k_{fiber}} \tag{12.8.27}$$

(Recall that the Biot number is based on $R$ as the length scale, while the Nusselt number uses $D$.) For the conductivities we will use the following values:

$$k_{air} = 0.043 \text{ W/m} \cdot \text{K at 260°C} \quad \text{and} \quad k_{glass} = 1.0 \text{ W/m} \cdot \text{K at 500°C}$$

Then the Biot number is

$$Bi = \frac{1}{2} \times 0.32 \times \frac{0.043}{1.0} = 0.007 \tag{12.8.28}$$

Such a small Biot number implies that internal conduction is much more efficient than external convection. Hence the rate of heat loss is controlled by the external convection process, and the temperature within the fiber is quite uniform.

**EXAMPLE 12.8.2**   *Evaluation of the Total Cooling of a Fiber*

In this example we suppose that a fiber is cooling according to our convective model, and we wish to determine the length of a cooling section required to bring the fiber temperature down to a specified level. Our first task is to develop a model for the cooling of the fiber. This follows easily from a heat balance—see Fig. 12.8.4—along the lines of those we have made in similar problems. With the assumptions that the fiber temperature is radially uniform and that the only heat loss is by convection, we derive a differential heat balance for the fiber of the form

$$\rho_{fiber} C_{p,fiber} \pi R^2 U \, dT_R = h 2\pi R (T_R - T_\infty) \, dz \qquad (12.8.29)$$

Hence the fiber temperature is the solution to the differential equation

$$\frac{dT}{dz} = \frac{2h(z)}{\rho C_p R U}(T - T_\infty) \qquad (12.8.30)$$

(For simplicity, we drop the subscripts on the fiber temperature and on the fiber properties.) We may write the solution for temperature as an integral involving the local heat transfer coefficient:

$$\Theta = \frac{T - T_\infty}{T^\circ - T_\infty} = \exp\left(\frac{-k_{air}}{\rho C_p R^2 U}\int_0^z \mathrm{Nu}(z)\, dz\right) = \exp\left(\frac{-k_{air}\,\overline{\mathrm{Nu}}}{\rho C_p R^2 U} z\right) \qquad (12.8.31)$$

(Note that since the Nusselt number uses the air conductivity, it is $k_{air}$ that appears in the coefficient. But $\rho C_p$ is based on the *fiber* properties.)

An interesting thing happens here. Even though the radius of the fiber might be changing along the spinline because of "drawing," the product $\rho R^2 U$ is constant (conservation of mass), and so this term can be brought outside the integral. As a consequence, an averaged Nusselt number appears here. At any point $z$ along the spinline, the averaged Nusselt number is the average over the distance $z$ from the origin. This averaging must be performed numerically, since we have no analytical expression for $\mathrm{Nu}(z)$ that can be integrated.

In this example our goal is to determine the cooling length $z = L$ required to bring the fiber temperature down to a specified level. Hence, in the model given by Eq. 12.8.31, $\Theta$ is given, but the average $\overline{\mathrm{Nu}}$ and $L$ are unknown. We begin a solution

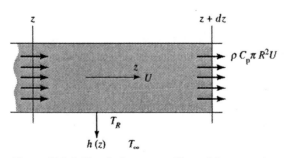

**Figure 12.8.4** Heat balance on a fiber with convective cooling.

procedure by inverting Eq. 12.8.31 to find

$$-\frac{\rho C_p R^2 U}{k_{air}} \ln \Theta = \overline{Nu}\, L \tag{12.8.32}$$

In a design problem involving total cooling of a fiber, the left-hand side of this equation will be known. There are two unknowns in this equation: $\overline{Nu}$ and $L$. The second "equation" is really the $\overline{Nu}$-versus-$L$ relationship that follows from numerical integration of the Nu($z$) relationship, for various values of $z = L$. That result is shown in Fig. 12.8.3 for Pr = 0.72.

We will illustrate the solution procedure for the following case:

*Polyethylene terephthalate (PET), a polyester fiber, is spun from the melt with an initial temperature on extrusion of $T^o = 290°C$. The ambient air is at $22°C$. The mass flow from the spinneret is w = $6.3 \times 10^{-3}$ g/s, and the initial fiber radius is 15 $\mu m$. The melt density is 1.2 $g/cm^3$, and $\rho C_p = 0.5\ cal/cm^3 \cdot K$. We want to find the cooling length L required to drop the fiber temperature to T = $100°C$.*

Physical properties of the ambient air will be evaluated at the mean of 290 and 22°C. These are

$$k_{air} = 0.036\ \text{W/m} \cdot \text{K} = 8.6 \times 10^{-5}\ \text{cal/cm} \cdot \text{s} \cdot \text{K at } 156°C$$
$$\nu_{air} = 0.30\ \text{cm}^2/\text{s at } 156°C$$

For the specified temperatures, we find

$$\Theta = \frac{T - T_\infty}{T^o - T_\infty} = \frac{100 - 22}{290 - 22} = \frac{78}{268} = 0.29 \tag{12.8.33}$$

We may write Eq. 12.8.32 in terms of the mass flowrate, since that is specified in the problem statement. The result is

$$\overline{Nu}\, L = -\frac{(w/\rho)\rho C_p}{\pi k_{air}} \ln \Theta = \frac{(5.25 \times 10^{-3})\, 0.5}{(8.6 \times 10^{-5})\pi} \times 1.23 = 12\ \text{cm} \tag{12.8.34}$$

This equation must be solved simultaneously with the $\overline{Nu}$-versus-$L$ relationship. We use Fig. 12.8.3, but we convert the dimensionless cooling length to $L$ using the data given here. Figure 12.8.5 shows the graphical solution to this pair of equations. The curves intersect at a point corresponding to a cooling length of

$$L = 32\ \text{cm} \tag{12.8.35}$$

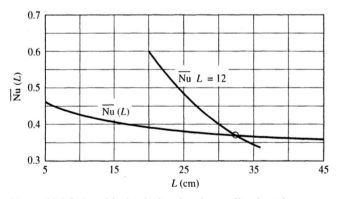

**Figure 12.8.5** Graphical solution for the cooling length.

Experimental data of Hill and Cuculo [*J. Appl. Polym. Sci.,* **18,** 2569 (1974)] indicate that the fiber temperature falls to 100°C at $L = 28$ cm, in very good agreement with this model.

Our purpose in illustrating the application of this model was to indicate that heat transfer coefficients for this complex flow can be calculated and used with some confidence in design. Real fiber spinning is usually more complex than suggested here, however. Mixed convection often occurs, with air being blown either normal to the fiber axis or parallel to it, to promote high heat transfer rates. The fibers themselves often do not move exactly along a stationary axis; instead, there may be some vibration in the spinline, creating motion of the filament normal to its axis, which of course enhances heat transfer. If the filament is extruded at very high temperatures, radiation losses may also add to the rate of heat transfer. If the fiber undergoes a phase change, as is usually the case, account must be taken of a heat of fusion or crystallization in the heat balance that yields Eq. 12.8.31. Despite these precautions, which are to be taken seriously, one will find that an understanding of the analysis presented here will make it possible to evaluate or develop analyses that go beyond this simple case.

## 12.9  OPTIMAL GEOMETRY FOR CONVECTIVE COOLING

Electronic devices often have heat-generating components stacked on parallel circuit boards in an arrangement such as that shown in Fig. 12.9.1. Some type of external cooling must be provided to keep system temperatures below a level that could degrade performance and stability. Often a means is provided for forcing a cooling fluid through the spaces between the boards. In other circumstances cooling is achieved only by buoyancy-driven natural convection (Chapter 14). A number of interesting heat transfer problems arise in this area of technology, and we can analyze them using the information developed up to this point in our study of convective heat transfer.

### 12.9.1  Optimal Design for Multiple Parallel Isothermal Surfaces

Suppose that the volume $LWH$ is fixed, and the available pressure drop $\Delta P$ is also fixed. The cooling flow that can be promoted is then governed by the separation $D$ between boards. As the number of boards increases, the flow will decrease and convective cooling will be less efficient. At issue is the effect on the rate of heat removal of the number of boards, and whether there might be an optimum number of boards.

In any optimization problem we must define the objective of the optimization. Suppose, for example, that we were considering the design of a heat exchanger made up

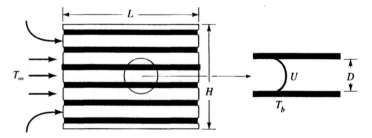

**Figure 12.9.1** Heat transfer from a set of parallel plates.

of parallel plates that were each electrically heated. The goal of this design would be to transfer the maximum amount of heat to the ambient fluid passing between the plates. While the addition of plates would decrease the rate of heat transfer per plate (by reducing the convecting flow), there would be a larger number of plates from which heat could be transferred. It is not obvious whether the *total* rate of heat transfer, the product of the heat transfer per plate times the number of plates, would increase or decrease. In fact, the *total* rate of heat transfer might go through a maximum—there may indeed be an optimum number of plates.

From the brief discussion above it should be clear that this convective heat transfer problem is intimately coupled with the fluid dynamics of flow parallel to a set of plates. It is important to arrive at an appropriate model for convective heat transfer in this system. When the plates are very close, we might assume laminar fully developed flow through a parallel plate channel, for which we have models for the flow and the heat transfer. When the plates are far apart, it is possible that an entry region (see Section 12.5.2) or boundary layer type of flow exists, but whether this is so will depend as well on the Reynolds number of the induced flow. Hence a second area of inquiry must be the dependence on the spacing $D$ (or number of plates) of the Reynolds number of the induced flow.

We will begin the analysis of this problem by assuming that when the number of plates is very large ($D$ very small), the induced flow will be laminar and fully developed. Correspondingly, we will assume that for $D$ large the flow is of the boundary layer or entry region type. We will carry out heat transfer analyses for these two asymptotic or extreme cases and see whether the results provide insight into the more general performance of the system.

*Large D/L (Boundary Layer Flow)*  We begin with a momentum balance of the form[1]

$$\Delta PDW = 2\bar{\tau}_w LW \qquad (12.9.1)$$

where $W$ is the width normal to $H$ and $L$. For the laminar boundary layer, the mean shear stress is found (using results in Section 6.3) from

$$\bar{\tau}_w = \frac{0.66\rho U^2}{\sqrt{\mathrm{Re}_L}} \qquad (12.9.2)$$

From these two equations we may solve for the Reynolds number of the flow to find

$$\mathrm{Re}_L^{bl} = \left(\frac{\Delta PDL}{1.32\,\rho\nu^2}\right)^{2/3} \qquad (12.9.3)$$

In the laminar boundary layer the heat transfer model begins with Eq. 12.2.34:

$$\overline{\mathrm{Nu}}_L = 0.68\,\mathrm{Pr}^{1/3}\,\mathrm{Re}_L^{1/2} \qquad (12.9.4)$$

The rate of heat transfer from both sides of a single plate is

$$Q_h = 2\bar{h}LW(T_b - T_\infty) = 1.36\,kW(T_b - T_\infty)\mathrm{Pr}^{1/3}\,\mathrm{Re}_L^{1/2} \qquad (12.9.5)$$

With Eq. 12.9.3 we may write this as

$$Q_h = 1.36\,kW(T_b - T_\infty)\,\mathrm{Pr}^{1/3}\left(\frac{\Delta PDL}{1.32\,\rho\nu^2}\right)^{1/3} \qquad (12.9.6)$$

---

[1] In writing the momentum balance in this form we are ignoring pressure losses due to entrance effects, considering only the shear stress along the plates bounding the flow.

Now let's introduce the number of plates, $n$, which is related to the spacing $D$ by

$$n = \frac{H}{D} - 1 \approx \frac{H}{D} \qquad (12.9.7)$$

(We assume that $n$ is large.)

Then the total rate of heat transfer from $n$ plates is

$$Q_{tot}^{bl} = nQ_{h}^{bl} = 1.36 \, kW(T_b - T_\infty) \, Pr^{1/3} H \left(\frac{\Delta PL}{1.32 \, \rho \nu^2}\right)^{1/3} D^{-2/3} \qquad (12.9.8)$$

**Small D/L (Laminar Fully Developed Flow)**  In the fully developed laminar flow regime, the mean velocity is

$$U = \frac{D^2 \Delta P}{12 \, \mu L} \qquad (12.9.9)$$

Hence the Reynolds number is

$$Re_D^{fd} = \frac{UD\rho}{\mu} = \frac{D^3 \rho \, \Delta P}{12 \, \mu^2 L} = \left(\frac{\Delta PL^2}{12\rho \nu^2}\right)\left(\frac{D}{L}\right)^3 \qquad (12.9.10)$$

For sufficiently large $L/D$, the fluid exits at the plate temperature $T_b$, and the rate of heat transfer per plate is simply[2]

$$Q_h^{fd} = UDW\rho C_p (T_b - T_\infty) \qquad (12.9.11)$$

For $n$ plates, the total rate of heat transfer in this regime is

$$Q_{tot}^{fd} = UHW\rho C_p (T_b - T_\infty) \qquad (12.9.12)$$

Using Eq. 12.9.9 we may write this in the form

$$Q_{tot}^{fd} = \frac{HW\rho C_p (T_b - T_\infty) \, \Delta PD^2}{12 \, \mu L} \qquad (12.9.13)$$

At this point we have derived two asymptotic models, one (Eq. 12.9.8) valid for very large $D/L$, where the flow and heat transfer are in the entry region or boundary layer regime, and Eq. 12.9.13 where, for very small $D/L$, the flow is fully developed laminar flow and the fluid exits in thermal equilibrium with the plate surfaces. We can nondimensionalize these two models and write them in the following forms:

$$Y = 1.2 \, X^{-2/3} \qquad \text{for } D/L \gg 1 \qquad (12.9.14)$$

and

$$Y = 0.0833 \, X^2 \qquad \text{for } D/L \ll 1 \qquad (12.9.15)$$

where

$$Y \equiv \frac{Q_{tot}}{WHC_p (T_b - T_\infty)(\rho \, \Delta P/Pr)^{1/2}} \qquad (12.9.16)$$

is a dimensionless rate of heat transfer, and

$$X \equiv \left(\frac{\Delta PL^2}{\mu\alpha}\right)^{1/4} \frac{D}{L} \qquad (12.9.17)$$

is a dimensionless pressure drop.

---

[2] Note that with this assumption we do not need a heat transfer (Nu) model because we *assume* that the outlet temperature is $T_b$.

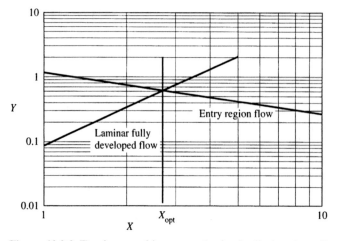

**Figure 12.9.2** Total rates of heat transfer in the limits of small and large $D/L$. $X$ and $Y$ defined in Eqs. 12.9.16 and 12.9.17.

Figure 12.9.2 shows these asymptotic models. The conclusion one draws from this plot is that there is some value of $X$ (hence of $D/L$) at which the total rate of heat transfer is maximized. The true value of $X_{opt}$, which lies in a region where neither asymptotic model is valid, could not be found without a model for this intermediate range of $D/L$. Crudely, but quickly, one can estimate $X_{opt}$ by taking the intersection of the two asymptotic models. This leads to the estimate that

$$X_{opt} = 2.7 \tag{12.9.18}$$

It turns out that this is a very good estimate of the value one gets with a more exact analysis. Bejan and Sciubba [*Int. J. Heat Mass Transfer,* **35,** 3259 (1992)] find that $X_{opt}$ is in the range of 3.03 to 3.08 for Pr in the range [0.72, 1000]. The value of $Y$ at the maximum, also a weak function of Pr, is found by Bejan and Sciubba to be $Y_{max} = 0.5 \pm 4\%$. From Fig. 12.9.2, which only approximates the behavior in the intermediate region, we see that $Y_{max} = 0.6$.

We may summarize the result of this optimization analysis with an approximate expression for the optimum geometry, using $X_{opt} = 3$:

$$\left(\frac{D}{L}\right)_{opt} = 3\left(\frac{\mu\alpha}{\Delta P L^2}\right)^{1/4} \tag{12.9.19}$$

According to this model, for a specified fluid ($\mu$ and $\alpha$) and for a given $L$ value, we can find an optimum spacing ($D/L$) for any choice of $\Delta P$. But to develop this model we must assume that the state of flow is either laminar and fully developed, or that we have an entry region flow with a laminar boundary layer. Neither assumption is guaranteed by *any* arbitrary choice of $\Delta P$. If we select or impose too large a value for $\Delta P$, we will find that we have a turbulent boundary layer, in which case the expression used for the Nusselt number, Eq. 12.9.4, must be changed to one appropriate to the higher flowrate.

We can write the Reynolds number at the optimum condition and develop a criterion for application of this model. From Eq. 12.9.3, in combination with Eq. 12.9.19, we find

$$Re_L^{opt} = 1.75\left(\frac{\Delta P L^2}{\mu\alpha}\right)^{1/2} Pr^{-2/3} \tag{12.9.20}$$

and from Eq. 12.9.10, combined with Eq. 12.9.19, we find

$$\text{Re}_D^{\text{opt}} = 2.25 \left(\frac{\Delta PL^2}{\mu\alpha}\right)^{1/4} \text{Pr} \qquad (12.9.21)$$

With these expressions, and a given set of design and operating conditions, we can assess the state of flow and evaluate the applicability of this model.

**EXAMPLE 12.9.1** *Conditions for Optimum Performance*

Consider the performance of a system of heated *isothermal* parallel plates with $L = 1$ m and $D = 5$ mm. Air at 20°C is drawn through the system by a small exhaust fan. What is the $\Delta P$ that yields maximum heat transfer, and what are the flow conditions at this $\Delta P$?

The system is designed with $D/L = 0.005$. From Eq. 12.9.19 the optimum $\Delta P$ would correspond to

$$\left(\frac{\Delta PL^2}{\mu\alpha}\right)^{1/4} = \frac{3}{D/L} = 600 \qquad (12.9.22)$$

For air at this temperature, we find

$$\mu = 1.8 \times 10^{-5} \text{ Pa·s} \qquad \alpha = 2 \times 10^{-5} \text{ m}^2/\text{s}$$

hence

$$\Delta P = \left(\frac{\mu\alpha}{L^2}\right) 600^4 = 47 \text{ N/m}^2 \qquad (12.9.23)$$

This is a very small pressure.

Under these conditions we find

$$\text{Re}_L^{\text{opt}} = 1.75 \times 360{,}000 \times 0.72^{-2/3} = 7.8 \times 10^5 \qquad (12.9.24)$$

and

$$\text{Re}_D^{\text{opt}} = 2.25 \times 600 \times 0.72 = 972 \qquad (12.9.25)$$

We conclude, from the value of $\text{Re}_D$, that the flow is laminar, but this does not imply that the flow is fully developed throughout the length $L$. A rough estimate of the entry length $L_e$, the axial distance required to achieve the parabolic velocity profile, is given by

$$\frac{L_e}{D} = 0.055 \, \text{Re}_D = 0.055 \times 972 = 53.5 \qquad (12.9.26)$$

Since the channel length is $200D$, we see that the flow is fully developed and laminar over 75% of the channel. Hence, while our model is not exact, the primary requirement that the flow field be somewhere between fully developed laminar flow and the laminar boundary layer flow is met. We should be able to use this model with some confidence in this case.

It is not difficult to show that design and operating conditions can be specified outside the range of applicability of this model.

**EXAMPLE 12.9.2** *Design of an Optimal Cooling System*

We again consider the design of a system for cooling a set of isothermal parallel plates. As before, cooling is provided by room air at 20°C. Air is drawn through the system by an exhaust fan that develops a pressure difference across the length $L$ of 0.2 atm ($2 \times 10^4$ N/m$^2$). $L$ is 20 cm.

From Eq. 12.9.19, we write

$$\left(\frac{D}{L}\right)_{\text{opt}} = 3 \left(\frac{1.8 \times 10^{-5} (2 \times 10^{-5})}{2 \times 10^4 (0.2)^2}\right)^{1/4} = 0.0025 \qquad (12.9.27)$$

and

$$D_{\text{opt}} = 0.0005 \text{ m} = 0.5 \text{ mm} \qquad (12.9.28)$$

This is a very narrow separation, perhaps too narrow for good mechanical design. Before we manipulate these numbers further, let's examine the state of the flow. From Eq. 12.9.27 it is easy to see that

$$\left(\frac{\Delta P L^2}{\mu \alpha}\right)^{1/4} = \frac{3}{0.0025} = 1200 \qquad (12.9.29)$$

The Reynolds numbers are

$$\text{Re}_L = 1.75 \, (1200)^2 (0.72)^{-2/3} = 31 \times 10^5 \qquad (12.9.30)$$

and

$$\text{Re}_D^{\text{opt}} = 2.25 \times 1200 \times 0.72 = 1944 \qquad (12.9.31)$$

From the second Reynolds number we conclude that the flow is not fully developed laminar flow over any significant portion of the length, $L$, *if it is laminar at all.* From the first Reynolds number we conclude that the boundary layer is turbulent over most of the length $L$.[3] Hence an analysis based on the model presented in this section is not warranted for the conditions of this example. It would be necessary to work out the heat transfer characteristics for this higher Reynolds number flow.

## 12.9.2 Optimal Design and Operation for a Single Heat-Generating Surface

A problem closely related to the one we have just discussed arises when there is a cooling flow in the region between a pair of plane parallel surfaces: an insulated surface and one that has heat-generating devices in its plane. Figure 12.9.3 suggests the geometry. We will take a simple case, in which it is assumed that the rate of heat generation is uniform along the length of the board, in the direction of flow. Hence we are considering a constant heat flux surface.

Two planar parallel surfaces are separated by a distance $D$. The lower surface has electronic components that generate heat uniformly at a rate per unit of area (i.e., a flux) $q$. Although the electronic components are positioned *discretely* along the surface, we assume that the heat flux $q$ is continuously uniform at any point along the surface. The upper surface is passive (i.e., it has no electronic components), and we assume that there is no heat transfer through that surface. A fluid is caused to move through the

---

[3] The criterion for the transition from the laminar to the turbulent boundary layer is $\text{Re}_x = 5 \times 10^5$, where $x$ is the distance downstream from the channel entrance.

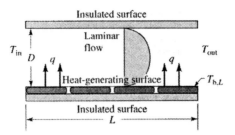

**Figure 12.9.3** Convective cooling of a heat-generating surface.

parallel planar region and provide convective cooling of the lower surface. Hence this fluid heats up as it moves downstream. If the heat flux is constant, then the lower surface temperature increases, and the maximum surface temperature occurs at the downstream end, $L$, and we denote this temperature as $T_{b,L}$. To protect the performance of the device, we would want to minimize the maximum temperature $T_{b,L}$.

As in any optimization problem, we must state clearly what parameters are held constant. In the case considered here, we assume that $L$ is fixed and that a flow is generated by imposition of a fixed pressure drop $\Delta P$ across the length $L$. We carry out an analysis of the response of the temperature $T_{b,L}$ to a reduction in $D$. The convective model for the system takes the form (since $q$ is constant)

$$q = q_L = h(T_{b,L} - T_{out}) \tag{12.9.32}$$

If we denote the mean velocity of the convective flow by $U$, the heat balance is simply

$$\rho C_p U D W (T_{out} - T_{in}) = qWL \tag{12.9.33}$$

where $W$ is the width normal to the page, and $T_{in}$ and $T_{out}$ are the mean fluid temperatures at the entrance and exit of the channel, respectively.

For small spacings $D$ it is reasonable to assume laminar fully developed flow, for which

$$U = \frac{\Delta P D^2}{12 \mu L} \tag{12.9.34}$$

With these equations we may solve for the maximum temperature in the dimensionless form

$$\frac{T_{b,L} - T_{in}}{qL/k} = \frac{12}{\Pi}\left(\frac{L}{D}\right)^3 + \frac{2}{Nu_{2D}}\frac{D}{L} \tag{12.9.35}$$

where a dimensionless pressure parameter $\Pi$ is defined by

$$\Pi \equiv \frac{\Delta P L^2}{\mu \alpha} \tag{12.9.36}$$

In the case being considered here, $\Pi$ is constant, $qL/k$ is constant, and $Nu_{2D}$ is assumed to be constant and is defined by

$$Nu_{2D} = \frac{2hD}{k} \tag{12.9.37}$$

The Nusselt number for this specific set of boundary conditions has not been discussed before. We have here the case of laminar convection between parallel planes, with one plane insulated and the other losing heat at a constant rate. It is possible to find the Nusselt number for this case analytically, and if $L/(D\,Re\,Pr)$ is large enough, the value is

$$\text{Nu}_{2D} = 5.385 \qquad (12.9.38)$$

The minimum value of $T_{b,L}$ follows by differentiating Eq. 12.9.35 with respect to $L/D$, and setting the derivative to zero, with the result

$$\left(\frac{T_{b,L} - T_{in}}{qL/k}\right)_{min} = 1.554 \, \Pi^{-1/4} \qquad (12.9.39)$$

which occurs at the optimum plate separation

$$\left(\frac{D}{L}\right)_{opt} = 3.14 \, \Pi^{-1/4} \qquad (12.9.40)$$

An example of the application of this model follows.

**EXAMPLE 12.9.3** *Sensitivity of Performance to Deviations from Optimum Design*

Air cools a surface on which heat is generated at a constant rate $q$. The geometry is as discussed in the analysis above, and $L = 0.2$ m and $\Delta P = 100$ Pa. Find the optimum spacing $D$, if the maximum allowable temperature is 50°C and the air intake is at 25°C. What is the maximum rate of heat dissipation under this constraint on temperature?

Use the property values for air of $k = 0.03$ W/m·K, $\alpha = 2 \times 10^{-5}$ m²/s, and $\mu = 1.8 \times 10^{-5}$ Pa·s.

We begin by calculating the optimum spacing from

$$\left(\frac{D}{L}\right)_{opt} = 3.14 \, \Pi^{-1/4} = 3.14 \left(\frac{\Delta PL^2}{\mu\alpha}\right)^{-1/4} = 3.14 \left[\frac{100 \, (0.2)^2}{(1.8 \times 10^{-5})(2 \times 10^{-5})}\right]^{-1/4} = 0.0096 \qquad (12.9.41)$$

or

$$D_{opt} = 0.0019 \text{ m} = 1.9 \text{ mm} \qquad (12.9.42)$$

From Eq. 12.9.34 we find the mean velocity of the cooling air as

$$U = \frac{\Delta PD^2}{12 \, \mu L} = \frac{100 \, (1.9 \times 10^{-3})^2}{12(1.8 \times 10^{-5}) \, 0.2} = 8.4 \text{ m/s} \qquad (12.9.43)$$

This corresponds to a Reynolds number of

$$\text{Re} = \frac{UD\rho}{\mu} = \frac{8.4 \, (1.9 \times 10^{-3})(1.2)}{1.8 \times 10^{-5}} = 1060 \qquad (12.9.44)$$

which is in the laminar regime. We should also check the value of the parameter

$$\frac{\text{Re Pr } D}{L} = \frac{1060 \, (0.7) \, 0.0019}{0.2} = 7 \qquad (12.9.45)$$

From inspection of Fig. 12.5.5, which is for the case of a circular pipe—not a parallel plate channel—we can infer that the local Nusselt number is probably *near* its asymptotic constant value but that under these conditions there is some error in assuming $h$ constant over the entire length $L$. The use of the smaller asymptotic value, compared to the actual varying value of $h$, will give us a conservative estimate for the maximum flux $q$. Hence we can dissipate heat at the rate (inverting Eq. 12.9.39)

$$q = 0.644 \, \frac{k(T_{b,L} - T_{in})}{L} \Pi^{1/4} = 0.644 \, \frac{0.03 \, (50 - 25)}{0.2} \, 325 = 784 \text{ W/m}^2 \qquad (12.9.46)$$

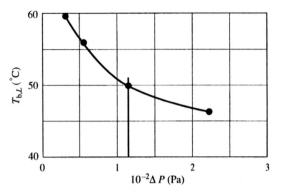

**Figure 12.9.4** Effect on maximum surface temperature when $\Delta P$ is varied.

At this point we have an estimate of the optimum performance of this system, and it is important to determine how sensitive the performance is to variations in parameters. For example, suppose we cannot control the pressure $\Delta P$ at the value of 100 Pa. To what extent does the system depart from its optimum performance? To examine this, we go back to Eq. 12.9.35. We suppose that we have designed the system with $D = 0.0019$ m and that the electronics are generating 784 W/m$^2$ uniformly along the length of the plate. We want to find the maximum temperature $T_{b,L}$ as a function of $\Delta P$, in the neighborhood of the value $\Delta P = 100$ Pa, for the case that $D = 1.9$ mm. Keep in mind that this $D$ is an optimum only when $\Delta P = 100$ Pa.

To do this calculation we simply return to Eq. 12.9.35 and calculate $T_{b,L}$ at various values of $\Pi$. The results are shown in Fig. 12.9.4. At lower pressure drops the air picks up more heat because it moves more slowly through the channel. (The Nusselt number remains unchanged in this flow regime.) As a consequence, the maximum surface temperature rises above the required limit. For larger pressure drops the air remains cooler, and this permits the surface temperature to drop, at constant heat flux. Keep in mind, however, that as $\Delta P$ increases we move into a turbulent flow regime. The Nusselt number will increase above the laminar flow value, and the surface temperature will be below the values calculated with the laminar Nusselt number.

The sensitivity of the performance of the system to changes in *geometry*, specifically to deviations of $D/L$ from the optimum value, can be seen in Fig. 12.9.5, which again

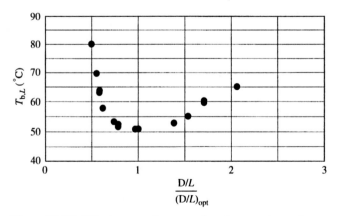

**Figure 12.9.5** Effect on maximum surface temperature when $D/L$ is varied.

follows from Eq. 12.9.35, for the parameters of this example. We see that the maximum temperature is always above that under optimum conditions, as $D$ varies either above or below the optimum value. The temperature increases very rapidly as $D/L$ falls below the optimum value. This is due to the cubic term $(L/D)^3$ in Eq. 12.9.35, which grows rapidly as $D/L$ gets smaller.

## SUMMARY

Convective coefficients reflect the fluid dynamics in the neighborhood of the interface across which heat is transferred. Film theory provides a crude model of convection and displays the appropriate dimensionless groups that describe convection. It also gives rise to the Colburn analogy (Eq. 12.1.20) relating heat transfer to friction. Boundary layer theory, developed here only for the *laminar* boundary layer on a flat plate, improves on film theory by showing how the convective coefficient varies along the direction of flow. For more complex flows and geometries, we resort to empirical correlations of data, often guided by the predictions of boundary layer theory. Examples are given in Figs. 12.3.1 and 12.3.2. Heat transfer analysis for pipe flow leads to the conclusion that the *log–mean temperature difference* is the appropriate measure of the temperature "driving force" for heat transfer. For *laminar flows in simple geometries* (e.g., pipe flow) and with simple boundary conditions (e.g., constant wall temperature or constant wall heat flux), analytical solutions are available from which theoretical predictions of the temperature change in the fluid and the heat transfer coefficient follow. Figures 12.5.3 and 12.5.4 give examples, in graphical form, for laminar flow in a parallel plate heat exchanger. Design examples which utilize this information are presented. Then, in Section 12.8 we illustrate an "integral method" of solving the convective diffusion equation for a fairly complex problem—heat transfer to a moving fiber. Finally, Section 12.9 illustrates the application of these ideas to optimal design of a convective cooling system.

## PROBLEMS

**12.1** Friend and Metzner [*AIChE J.*, **4**, 393 (1958)] presented the following data for heat transfer to corn syrup solutions flowing through a heated 0.75-inch (nominal inside diameter) brass pipe. Test the ability of Eqs. 12.1.22 and 12.3.3 to describe these data.

| **SOLUTION I:** Pr = 590 | | **SOLUTION II:** Pr = 93 | |
|---|---|---|---|
| **Re** | **Nu** | **Re** | **Nu** |
| 2578 | 80 | 3841 | 88.2 |
| 3746 | 135 | 5444 | 124.8 |
| 5000 | 200 | 7342 | 181 |
| 6000 | 235 | 9419 | 223.5 |
| 7200 | 280 | 11,210 | 266.5 |
| | | 17,130 | 367.7 |
| | | 21,120 | 440.7 |

**12.2** Return to Eq. 12.5.3. Nondimensionalize it using Eqs. 12.5.8 and 12.5.9. Define a dimen-

sionless group that determines the relative roles of axial conduction and axial convection of heat.

**12.3** Equation 12.2.29 gives a thermal boundary layer thickness for flow over a flat plate. Suppose fluid enters a parallel plate heat exchanger from a large reservoir, such that we may consider the inlet velocity and temperature profiles to be uniform. We might try to use Eq. 12.2.32 for the local Nusselt number, on the grounds that so long as $\delta_T < H$, the half-width of the channel, the system behaves in a manner similar to that of the boundary layer model.

**a.** Compare Eq. 12.2.32 to Eq. 12.5.37. To what physical phenomena do you ascribe the differences?

**b.** If we require $\delta_T/H < 0.1$, what value of $x/H$ does this correspond to?

**c.** Plot the two Nusselt number predictions versus Reynolds number in the range $[0.01 < \text{Re} < 10]$ where $\text{Re} = UH/\nu$, and comment on your observations.

**12.4** Give the derivation of Eq. 12.2.15 in detail.

**12.5** Carbon fibers are extruded from a spinneret as a viscous pitch which is then "frozen" to the solid carbon filament by air-cooling. Spinning conditions correspond to a linear velocity of extrusion of $U = 5$ cm/s, and the filament diameter is $D = 50$ μm. The Prandtl number of the pitch at the extrusion temperature ($T_o = 230°C$) is Pr $= 4 \times 10^4$. The thermal conductivity is 1.5 W/m·K. The thermal diffusivity is $10^{-4}$ m²/s. Air ($T_a = 25°C$) is blown normal to the filament at a velocity of 10 m/s. How much cooling length must be provided if the fiber is to be cooled to a center temperature of 150°C?

**12.6** In the solution of Example 12.4.3, are we dependent on an assumption that the pressure does not vary along the axis of the tube? Note, for example, that we calculated the density at the inlet pressure. Answer this question in general, and then for the specific data of this example, find the change in pressure across the ends of the tube, relative to the mean pressure.

**12.7** For flow through a pipe with an isothermal wall, we have a model for the change in temperature of the fluid which we write in the form (Eq. 12.4.19)

$$\frac{T_2 - T_R}{T_1 - T_R} = \exp\left(-\frac{4StL}{D}\right) \quad \textbf{(P12.7.1)}$$

Suppose the inlet temperature is 50°C and the pipe surface temperature is 90°C. Take

$$\frac{L}{D} = 60 \qquad Pr = 6 \qquad Re = \frac{40}{\pi}$$

Find $T_2$ (the outlet temperature).

**12.8** A viscous oil (Pr $= 200$, $k = 0.15$ W/m·K, $\mu = 1$ Pa·s, $\rho = 950$ kg/m³) is pumped through a circular pipe of diameter $D = 2$ cm, the surface of which is maintained at 400 K. What length of pipe is required to permit a flow of 10 m³/h to be raised from 300 K to 350 K? Figure P12.8 gives the theoretical solution for the cup-mixing averaged temperature for laminar flow through a circular pipe with an isothermal surface.

**12.9** Parallel the derivation of Eq. 12.4.16, but for the case of flow between parallel planes. Is Eq. 12.5.24 valid in this case?

**12.10** Derive Eq. 12.5.34.

**12.11** Prove that Nu $= 3.66$ follows from Eqs. 12.5.45 and 12.5.47.

**12.12** Wedekind [*Trans. ASME (J. Heat Transfer)* **111**, 1098 (1989)] presents data on the convective heat transfer coefficients from thin circular disks whose circular faces were parallel to the mean airflow used to provide convection. Reynolds numbers, based on the disk diameter, ranged from $10^3$ to $3 \times 10^4$, and disk diameters were in the range 5 to 16 mm. Wedekind correlates the data with the following expression:

$$\frac{Nu_d}{Pr^{1/3}} = 0.591\, Re_d^{0.564} \quad \textbf{(P12.12)}$$

If you took classical boundary layer theory for a wide flat plate, and averaged the heat transfer coefficient over a length equivalent to the disk diameter, how would the resulting expression for the averaged Nusselt number compare to Eq. P12.12? Would the comparison be improved by choosing another equivalent length, such as the ratio of surface area to perimeter? Comment on

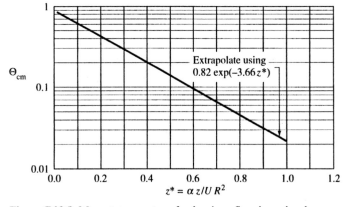

**Figure P12.8** Mean temperature for laminar flow in a circular tube with an isothermal surface.

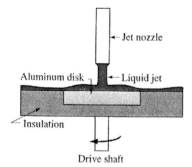

**Figure P12.13** Liquid jet cooling of a rotating disk.

the differences between the wide flat plate and a finite diameter disk.

**12.13** Carper et al. [*Trans. ASME (J. Heat Transfer)* **108**, 540 (1986)] present data on convective heat transfer from the circular face of a disk rotating about its axis. Convection is provided by an impinging liquid (oil) jet that strikes the disk normal to the disk face and is aligned with the axis of rotation. The resulting liquid flow is spun out by centrifugal force and by its own momentum. Figure P12.13 indicates the geometry. Data were correlated in the following form:

$$\overline{\mathrm{Nu}_D} = 0.097\,\mathrm{Re}_{\mathrm{rot}}^{0.384}\mathrm{Re}_{\mathrm{jet}}^{0.459}\mathrm{Pr}^{0.448}$$

$$(P12.13.1)$$

The Nusselt number and the rotational Reynolds number are based on the disk diameter $D$, while the jet Reynolds number is based on the jet nozzle diameter:

$$\overline{\mathrm{Nu}_D} = \frac{\overline{h}D}{k_{\mathrm{oil}}},\quad \mathrm{Re}_{\mathrm{rot}} = \frac{\omega D^2}{\nu_{\mathrm{oil}}},\quad \mathrm{Re}_{\mathrm{jet}} = \frac{dU_{\mathrm{jet}}}{\nu_{\mathrm{oil}}}$$

$$(P12.13.2)$$

The rotational speed of the disk is $\omega$ rad/s. The oil jet Reynolds numbers were in the laminar range (180–1300). The rotational Reynolds numbers were in the range of 16,000 to 545,000. Disk diameters varied from 10 to 28 cm. Oil Prandtl numbers varied with the temperature of the experiment, but were in the range of 87 to 400. Use this information to design a process to cool aluminum disks, 10 cm in diameter and 3 mm thick, that exit a furnace at a uniform temperature of 100°C and must be cooled to an *average* temperature of 35°C in 500 s. The disks exit the furnace rotating at 200 rpm. The lowest usable oil temperature is −20°C.

The oil properties (in SI units) are given as a function of absolute temperature:

$$k = 0.16 - 1.4 \times 10^{-4}T \qquad C_p = 686 + 3.8T$$
$$\rho = 1063 - 0.71T$$
$$\nu = 10^{-6}\{\log^{-1}[\log^{-1}(9.94 - 3.9\log T)] - 0.6\}$$

**12.14** Finn et al. study heat transfer coefficients for "wind convectors," large-scale outdoor heat exchangers for heat pumps. The original papers should be studied for background information on this interesting practical application of the material in this chapter. [Finn et al., *Trans. ASME, J. Solar Eng.,* **112**, 280, 287 (1990)]. The results of these authors, who present experimental data for the overall heat transfer coefficient $U$ as a function of average wind speed, are shown in Fig. P12.14.

The heat exchanger consists of a 14 m long copper tube with inside and outside diameters of 0.011 and 0.016 m, respectively. The test fluid is a "brine" made up of 84% water and 16%

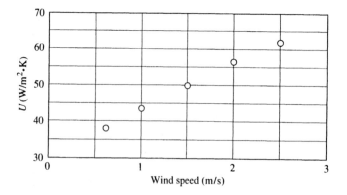

**Figure P12.14**

methanol. (This mixture provides a heat transfer fluid that can be used at temperatures below 0°C.) The test fluid has a density of 965 kg/m³ and a heat capacity of 3906 J/kg · K, at 20°C. The data shown were obtained at a brine flow of 125 L/h, which corresponds to a Reynolds number of 1400. No data were given for thermal conductivity of the brine. Pure methanol has a thermal conductivity of 0.2 W/m · K.

Using the correlations presented in this chapter, predict $U$ versus wind speed and compare your prediction to the data. How sensitive is $U$ to variations in brine flowrate? Show curves of $U$ versus wind speed for brine flows 50% above and below the one studied.

**12.15** A liquid is pumped through a pipe with a constant temperature maintained along the inner wall surface. Plot the product $w\Delta T = w(T_{cm} - T_{in})$ as a function of the mass flowrate $w$, over the full range of laminar flowrates, for tube diameters of 0.1 and 1 cm. The liquid has all the properties of water, except that its viscosity is 10 times that of water, at any temperature. Inlet temperature is 10°C, and the tube wall is maintained uniformly at 20°C. For both tubes, $L/D = 25$.

**12.16** The analysis of heat transfer to a laminar flow through a tube with a constant surface temperature can be carried out, and the details are similar to those illustrated in Section 12.5.1 for the parallel plate exchanger with isothermal surfaces. The solution for the cup-mixing temperature may be written in the form

$$\Theta_{cm} = \sum_{m=0}^{\infty} G_m \exp(-\lambda_m^2 z^*)$$

$$(P12.16.1)$$

where the dimensionless axial variable is defined as

$$z^* \equiv \frac{\alpha z}{UR^2} \quad (P12.16.2)$$

In the circular tube case the first three coefficients are:

| $m$ | $G_m$ | $\lambda_m^2$ |
|---|---|---|
| 0 | 0.82 | 3.66 |
| 1 | 0.098 | 22.3 |
| 2 | 0.0326 | 56.9 |

For $m > 2$, $\lambda_m = 2.828 m + 1.885$
$$G_m = 3.61 (\lambda_m^2)^{-7/6}$$

Using this solution, calculate the cup-mixing temperature profile along the tube axis and plot your results on double logarithmic coordinates over the range $0.001 < z^* < 1$. Are three terms of the series sufficiently accurate for $z^* < 0.01$?

**12.17** By analogy to Eq. 12.5.19 for the parallel plate heat exchanger, we may define a local heat transfer coefficient for tube flow, in the constant wall temperature case, by

$$-wC_P dT_{cm} = h_{ln}(T_{cm} - T_R)\pi D \, dz$$
$$(P12.17.1)$$

Show that this may be written in the form

$$Nu_{z*} \equiv \frac{h_{ln} D}{k} = -\frac{1}{\Theta_{cm}} \frac{d\Theta_{cm}}{dz^*} \quad (P12.17.2)$$

Now go back to Eq. 12.4.15, the heat balance for flow through a tube with a constant wall temperature. Show that this leads to an *average* Nusselt number (based on the diameter $D$ as the length scale) that can be calculated in terms of $\Theta_{cm}$ as

$$\overline{Nu}_D \equiv \frac{\overline{h}_{ln} D}{k} = \frac{1}{z_L^*} \ln\left(\frac{1}{\Theta_{cm}}\right)$$
$$(P12.17.3)$$

where

$$z_L^* \equiv \frac{4\alpha L}{UD^2} = \frac{4 L/D}{Re_D \, Pr} \quad (P12.17.4)$$

and the average heat transfer coefficient along the length of the tube is defined as

$$\overline{h}_{ln} \equiv \frac{1}{L} \int_0^L h_{ln}(z) \, dz \quad (P12.17.5)$$

Using the solution given in Eq. P12.16.1, calculate the local and average Nusselt numbers and plot your results on Figs. 12.5.5 and 12.5.6. Are three terms of the series sufficiently accurate for $z^* < 0.01$?

One often finds graphs such as Fig. 12.5.6 plotted against a dimensionless group called the Graetz number, which is defined as $Gz \equiv wC_p/kL$. Show that

$$z_L^* = \frac{\pi}{Gz} \quad (P12.17.6)$$

**12.18** Mineral oil is pumped through a concentric tube heat exchanger in order to raise its temperature from 20°C to 50°C over a length of

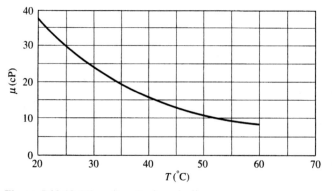

**Figure P12.18** Viscosity of mineral oil.

$L = 0.1$ m. The outer diameter of the inner tube is 5 mm, and the inner diameter of the outer tube is 7 mm. Both tubes are thin-walled, and they are maintained at a uniform temperature of 60°C. What mass flowrate will yield the desired outlet temperature?

The density of mineral oil is a linear function of the Celsius temperature scale in the range of interest, with

$$\rho_{20\,C} = 860 \text{ kg/m}^3 \quad \text{and} \quad \rho_{60\,C} = 833 \text{ kg/m}^3$$

The thermal conductivity is constant in this range, with

$$k = 0.147 \text{ W/m} \cdot \text{K}$$

The heat capacity $C_p$ is a linear function of temperature, with

$$C_{p,30\,C} = 1.93 \text{ kJ/kg} \cdot \text{K}$$

and

$$C_{p,60\,C} = 2.07 \text{ kJ/kg} \cdot \text{K}$$

The viscosity is a strong function of temperature, and data are shown in Fig. P12.18.

Solve this problem using the constant property solution appropriate to this geometry, with the physical properties evaluated at the mean film temperature defined as

$$T_{\text{film}} \equiv \frac{\frac{1}{2}(T_{\text{in}} + T_{\text{wall}}) + \frac{1}{2}(T_{\text{out}} + T_{\text{wall}})}{2}$$

$$\text{(P12.18)}$$

By what factor does the predicted mass flowrate differ if the wall temperature is used for evaluation of physical properties?

**12.19** The use of the constant physical properties solution for Problem 12.18 is certainly in error when the viscosity is such a strong function of temperature, as is the case with mineral oil. The heat transfer literature contains a number of solutions in which the temperature dependence of viscosity is accounted for.

One approach, described by Yang [*J. Heat Transfer*, **84**, 353 (1962)], uses an algebraic model to describe the viscosity temperature behavior. The model used is

$$\frac{\mu}{\mu_R} = \left(1 + \gamma' \frac{T - T_R}{T_1 - T_R}\right)^{-1} \quad \text{(P12.19)}$$

where $\gamma'$, $T_R$, and $T_1$ are constants used to fit the data, and $\mu_R$ is the viscosity at temperature $T_R$. Does this simple model do a good job of fitting real viscosity behavior? Test it by fitting the data of Fig. P12.18 with this model. Make the model fit the data exactly at $T = T_R = 60°$C and $T = T_1 = 20°$C.

**12.20** In the reference cited in Problem 12.19, Yang presents solutions for the *local* Nusselt number when the temperature dependence of viscosity is accounted for. He uses Eq. P12.19 for the $\mu(T)$ behavior. Figure P12.20 shows Yang's results for two values of $\gamma'$, corresponding to the case of a liquid entering a tube with the wall hot relative to the entering temperature of the liquid. Also shown is the $\gamma' = 0$ case, which corresponds to a constant viscosity solution. Go back to Problem 12.18. Provide a more accurate solution, or argue that there is no need to do so.

**12.21** Derive Eq. 12.5.24 from a thermal energy balance along the channel axis.

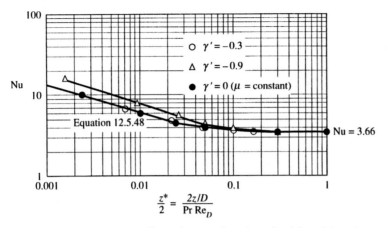

**Figure P12.20** Local Nusselt number as a function of axial position along the tube axis. Wall temperature is constant, and the viscosity depends on $T$ as in Eq. P12.19, with the $\gamma'$ values indicated.

**12.22** For the conditions of the heat exchanger design in Section 12.6, calculate the maximum shear rate and the maximum shear stress in the channel. Assume that the velocity profile is given by Eq. 12.5.1.

**12.23** Repeat Example 12.6.2, but take the channel walls to be made of a polymeric material with $k = 0.2$ W/m·K and $b = 2$ mm. What external bath temperature would you recommend to meet the design specifications?

**12.24** Repeat Example 12.6.2, but take the channel walls to be a composite solid made of the metal of the example, with $b = 2$ mm and $k_{wall} = 60$ W/m·K, which is coated on the inside surface with a polymeric material with the properties $k_{poly} = 0.2$ W/m·K and $b_{poly} = 50$ $\mu$m. What external bath temperature would you recommend to meet the design specifications?

**12.25** In Example 12.7.1 we find the pipe length required to bring the aqueous stream from 20°C to 95°C, when the wall temperature is 100°C. How much longer must the pipe be if we wish to bring the average temperature to 98°C?

**12.26** In Example 12.7.1 we ignored the "kill" that occurs in the preheater section, since most of that section is not at the sterilization temperature, and the rate of kill is such a strong function of temperature. Calculate the fraction of viable organisms killed in the preheater. Assume that $k_i = 5.5$ s$^{-1}$ at 95°C and $k_i = 1.1$ s$^{-1}$ at 75°C.

**12.27** A bioproduct is being produced which contains recombinant DNA (rDNA) organisms. The rDNA must be nonviable in this product,

and the required safety level is specified in terms that correspond to less than one viable organism per billion ($10^9$) 100-liter batches of product. The kinetic model of Eq. 12.7.3 holds, and it is known that $k_i = 4$ s$^{-1}$ at 90°C and $k_i = 0.8$ s$^{-1}$ at 70°C. Design the heat exchangers for a sterilization system with the following characteristics:

The feed solution has all the thermophysical properties of water. The initial concentration of viable rDNA is $10^{10}$ cells/mL. The cells are microscopic, and they are uniformly dispersed in an aqueous solution.

The feed flowrate is 20 gal/h. Band heaters are available that will provide a constant rate of heat input all along the axis of a pipeline which has an inside diameter of 1.25 inches. The first section of heated pipeline must raise the temperature of the stream from 20°C to 90°C over a length of 50 m.

What is the required heat duty (power requirement) for this section?

What is the maximum pipe wall temperature? (A thin-walled pipe is used.)

Comment on this result.

Plot the concentration of viable cells as a function of axial distance along the heater.

Immediately following this preheater is a section of well-insulated pipe having the same inside diameter as the preheater. What additional length of insulated pipe is required to reduce the viable cell population to the specified level?

502 Chapter 12 Convective Heat Transfer

**12.28** Using Eq. 12.8.16, find the distance downstream (measured in fiber diameters) at which the boundary layer thickness on a moving fiber just equals its radius.

**12.29** Using Eq. 12.8.19, present an analytical expression for $\alpha$ as a function of $vz/UR^2$ valid for $\alpha < 0.1$.

**12.30** Show that the Nusselt number, as defined in Eq. 12.8.24, is given by $Nu_D = 2/\beta$.

**12.31** Derive Eqs. 12.8.17 and 12.8.19.

**12.32** Derive Eq. 12.8.20.

**12.33** For the conditions given in Example 12.8.1, calculate and plot the fiber temperature as a function of position along the spinline.

**12.34** Argue that for the case Pr = 1, for convection from a fiber, $Nu_D$ (Pr = 1) = $2/\alpha$ where $\alpha$ is the function given in Fig. 12.8.2. With the result from Problem 12.29, give an analytical expression for $\overline{Nu}$ $(L)$ for values of $L$ such that $\alpha < 0.1$. Plot these results on Fig. 12.8.3.

**12.35** In Example 12.8.2 we evaluated the properties of the ambient air at 156°C, which is the average of the initial surface temperature of the fiber (290°C), and the ambient temperature. The fiber surface cools to 100°C in this example. Rework the problem, but evaluate the properties of air at the mean of the ambient temperature (22°C) and the mean surface temperature of (290 + 100)/2 = 195°C: that is, at 108.5°C.

**12.36** Liquid flows through a tube of circular cross section with a constant wall temperature. Plot the product $(w\Delta T)$ as a function of mass flowrate $w$ over the full range of laminar flowrates, for tube diameters of 0.1 and 1 cm. The fluid has all the properties of water, except that its viscosity is 50 times that of water. Inlet temperature is 10°C, and the tube wall is maintained uniformly at 20°C. For both tubes, $L/D = 25$.

**12.37** A correlation commonly recommended for predicting the heat transfer to a fluid in turbulent flow through a pipe is the Dittus–Boelter equation:

$$\overline{Nu}_D = 0.023\,Re_D^{0.8}\,Pr^n$$

$$n = 0.4 \text{ for heating the fluid}$$

$$n = 0.3 \text{ for cooling the fluid}$$

This correlation can be used down to Reynolds numbers as low as 2500, and for $L/D > 60$. How well does this equation fit the data of Problem 12.1?

**12.38** Give the derivation of Eq. 12.3.17. Assume that Pr > 1 so that $\delta/\delta_T > 1$.

**12.39 a.** Derive Eq. 12.2.28

**b.** Derive Eq. 12.3.22.

You will need the definition of the gamma function (see Appendix B, Section B.3.)

**12.40** Prepare a plot similar to Fig. 12.3.5, but for the case Pr = 1. What is the effect of the Prandtl number on the correction factor $\chi$?

**12.41** Find the relationship between $\gamma'$ of Eq. P12.19 and $\gamma$ of Eq. 12.3.12. Does the correction factor of Eq. 12.3.23 describe the theoretical solution presented in Fig. P12.20 for the case $\gamma' = 0.9$? Is this so over the whole range of the space variable $x^*$? Would you expect the type of model that leads to Eq. 12.3.23 to hold over the whole range of the space variable $x^*$? Explain your answer.

**12.42** When a liquid metal droplet is rapidly quenched and solidified, a very fine grain structure is frozen into the solid. This structure imparts favorable properties into the metal. One method of achieving rapid quenching is to form molten metal drops from the tip of a capillary and let the drops fall, under the influence of gravity, to impact on a very cold surface. Each drop cools as it falls, and we need a model for the temperature change of a drop as a function of the distance of fall. Develop such a model.

**12.43** Design a device for measuring velocities in gases, according to the following principles: A thin solid cylinder will be electrically heated internally to a uniform temperature. The cylinder will then be exposed to a cross-flow of the gas, and the heater will be turned off. A thermocouple within the cylinder will monitor the decay of temperature toward the free stream value. Your design should specify the following:

Cylinder dimensions and material of construction

Range of velocity for which your design is suitable

Time scale of the temperature decay

Using an appropriate analysis as support, present an equation from which you would convert some measure of the decay of temperature to the velocity of the gas.

**12.44** Suppose the device described in Problem 12.43 were available. Show how you could use it to measure the thermal conductivity of a gas.

**12.45** Dincer [*Int. J. Heat Mass Transfer*, **37**, 2781 (1994)] presents the following data for the heat transfer coefficient observed in the cooling, in air, of cucumbers. Compare the data to the correlation of Fig. 12.3.1.

| U(m/s) | h(W/m²·C) |
|--------|-----------|
| 1.0 | 18.2 |
| 1.25 | 19.9 |
| 1.5 | 21.3 |
| 1.75 | 23.1 |
| 2.0 | 26.6 |

The cucumbers have a length of 0.16 m and a diameter of 0.038 m. The physical properties are

$$\rho = 964 \text{ kg/m}^3 \qquad k = 0.62 \text{ W/m} \cdot {}^\circ\text{C}$$

$$\alpha_T = 1.5 \times 10^{-7} \text{ m}^2/\text{s}$$

**12.46** Return to Example 12.5.2. For the constraints specified in that example, including $L/D = 100$, find the required number of tubes $n_T$, and the tube diameter $D$, such that the pressure drop across the tube bank is 100 psi.

**12.47** For the conditions specified in Example 12.5.2 ($Q_{tot} = 0.002 \text{ m}^3/\text{s}$ and $\nu = 10^{-5} \text{ m}^2/\text{s}$ at 70°C) prepare a plot of $n_T L$ versus Pr, for $L/D$ in the range [50, 200], and for Pr in the range [10, 200].

**12.48** Rework Example 12.5.2, but make the following changes: the flowrate is increased to $Q_{tot} = 0.1 \text{ m}^3/\text{s}$, and the liquid kinematic viscosity is $\nu = 0.5 \times 10^{-6} \text{ m}^2/\text{s}$ at 70°C. Give the design parameters ($L$, $D$, and $n_T$) for $L/D = 100$.

**12.49** Make a "design chart" similar to Fig. 12.5.7 for Pr = 200.

**12.50** Rework Problem 12.13, but for the condition that the *surface* temperature reaches 35°C.

**12.51** Rework Problem 12.13, but for a disk that must be cooled to a mean temperature of 35°C in 100 s. The jet Reynolds number must be within the laminar regime. Suppose that the maximum possible Nusselt number that can be achieved in such a system is $Nu_D = 6000$. What would the quickest cooling time be?

**12.52** With reference to Section 12.9, show that the two Nusselt numbers for heat transfer in the limits of large and small values of $D/L$ may both be written in the form

$$\overline{Nu} = \frac{\overline{h}L}{k} = A\Pi^a \left(\frac{D}{L}\right)^b \qquad \text{(P12.52.1)}$$

where

$$\Pi \equiv \frac{\Delta P L^2}{\mu \alpha} \qquad \text{(P12.52.2)}$$

Give the coefficients $A$, $a$, and $b$ for the cases of large and small $D/L$.

On log–log coordinates, plot both Nusselt number expressions against $D/L$ in the range 0.5 $(D/L)_{opt}$ to $2(D/L)_{opt}$.

Examine the function

$$\overline{Nu} = \left[\frac{1}{(\overline{Nu})_{bl}^2} + \frac{1}{(\overline{Nu})_{fd}^2}\right]^{-1/2} \qquad \text{(P12.52.3)}$$

as an empirical model for Nu over the transition range from fully developed laminar flow to the boundary layer regime. We don't use the model for resistances in series:

$$\frac{1}{\overline{Nu}} = \frac{1}{(\overline{Nu})_{bl}} + \frac{1}{(\overline{Nu})_{fd}} \qquad \text{(P12.52.4)}$$

Why not?

**12.53** Repeat the derivation presented in Section 12.4.1, but for the case of a parallel plate heat exchanger with both planar surfaces maintained at temperature $T_R$. Let the separation between plates be $D$. Give the result analogous to Eq. 12.4.19. Write both Eq. 12.4.19 and your result for the parallel plate case in the general form

$$\frac{T_2 - T_R}{T_1 - T_R} = \exp\left(-\beta \, St \, \frac{L}{D_h}\right) \qquad \text{(P12.53.1)}$$

where the Stanton number is

$$St = \frac{h}{\rho C_p U} \qquad \text{(P12.53.2)}$$

and $D_h$ is the hydraulic or equivalent diameter, defined in general as

$$D_h \equiv \frac{4 \times \text{cross-sectional area}}{\text{wetted perimeter}} \qquad \text{(P12.53.3)}$$

What is the value of $\beta$ for each case?

**12.54** Repeat Problem 12.53, but for the case of an annular heat exchanger, with both the inner and outer tubes at the same temperature $T_R$. Let the radii of the inner and outer tubes be $R_1$ and $R_2$, and the heat transfer coefficients along the two surfaces be $h_1(z)$ and $h_2(z)$. Define clearly the average $h$ that appears in St in this case.

**12.55** Repeat Problem 12.53, but for the case that one of the plates is insulated.

**12.56** Air at 200°F is to be cooled to 70°F, at a mass flowrate of 70 lb/h. The heat exchanger is a copper pipe, 2-inch i.d., submerged in a stirred bath of ice and water. Find the required length of the pipe.

**12.57** Find $T_L$ for the system shown in Fig. P12.57.

**12.58** Cold water is heated in the system sketched in Fig. P12.58. We want to find the outlet temperature of the water. Give values for $T_{out}$, $Pr_{H_2O}$, $h_{inside}$, and $h_{outside}$. For airflow normal to the axis of a long cylinder, a correlation of experimental data leads to the expression

$$Nu_D = \frac{\overline{h}D}{k} = 0.3 + \frac{0.62\, Re_D^{1/2}\, Pr^{1/3}}{[1 + 0.4/(Pr^{2/3})]^{1/4}} \quad \textbf{(P12.58)}$$

$$\left[1 + \left(\frac{Re_D}{282,000}\right)^{5/8}\right]^{4/5}$$

which holds as long as $Re_D Pr > 0.2$. How does this correlation compare to Fig. 12.3.1?

**12.59** Water at 10°C enters a 2.54 cm i.d. pipe at a rate of 1.26 L/s. The pipe is 3.05 m long. If the wall temperature is held constant, at 99°C, and the water outlet temperature is 41°C, what is the Nusselt number for this flow?

What Nusselt number would you predict on the basis of the Colburn analogy?

**12.60** For laminar fully developed Newtonian flow in a circular pipe whose surface is at a constant temperature, the local Nusselt number can

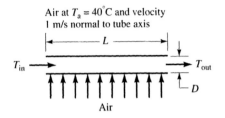

**Figure P12.58** System for heating water: mass flowrate, 0.1 kg/s; $T_{in}$, 20°C; $D$, 0.02 m; $L$, 1 m.

be found to be given by

$$Nu_D(z) = 1.076 \left(\frac{z}{D\, Re_D\, Pr}\right)^{-1/3} \quad \textbf{(P12.60)}$$

$$\text{for } \frac{z}{D\, Re_D\, Pr} \leq 0.01$$

Starting with this expression, find an expression for the average Nusselt number in a pipe of axial length $L$.

**12.61** Refer to the information presented in Problem 12.45. Estimate the time required to cool a single cucumber from 80°F to 40°F, using air at 35°F blowing across the cucumber at a speed of 2 m/s. The cucumber has a length of 0.16 m and a diameter of 0.038 m. How much faster will the cooling occur if the air speed is increased to 4 m/s? How much faster will the cooling occur if the air temperature is reduced to 33°F?

**12.62** You are the refreshment coordinator and backup drummer for the heavy metal group *Shriek and Mumble.* You have purchased liquid refreshments in a large pressurized vessel, and the liquid will be chilled and delivered into the hands of its consumers by passing it through a coil in an ice chest before it exits into a mug. Some thin-walled 3/8-inch i.d. copper tubing is available. What length of tubing is required to cool the liquid from 75°F to 40°F at a maximum flowrate of 2 gal/min?

**12.63** A circular disk of plastic, with a diameter of 0.2 m and a thickness of 0.02 m, is initially at a uniform temperature of 120°C. It is cooled from both sides by air, at 20°C, flowing parallel to the circular faces of the disk. The air speed is 0.2 m/s.

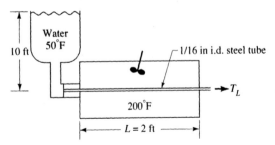

**Figure P12.57** A simple hot water heater.

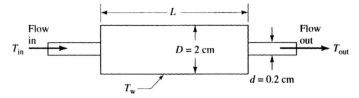

**Figure P12.64** Setup for experimental determination of $h$.

Find the time required to (nearly) cool the disk to the air temperature. For the plastic, use

$$\rho = 1200 \text{ kg/m}^3 \qquad k = 0.33 \text{ W/m} \cdot \text{K}$$

$$\alpha_T = 1.3 \times 10^{-7} \text{ m}^2/\text{s}$$

For the convective heat transfer coefficient, use Wedekind's correlation (Eq. P12.12). State clearly your definition of the final temperature of the disk ("near" thermal equilibirum with the air.)

**12.64** Design an experiment for the measurement of the heat transfer coefficient to a flow into and through a cylindrical tube. The geometry is as sketched in Fig. P12.64. The goal is to obtain an experimentally based correlation for the $j$-factor for heat transfer as a function of the Reynolds number for the flow. The desired Reynolds number range in the 2 cm diameter tube is 10 to 100.

Because the fluid enters the larger tube from a much smaller diameter tube, the flow will not likely be a simple Poiseuille flow over the length $L$. Hence our theory for the Nusselt number for laminar flow in a tube is not expected to be reliable. For this reason, an experimental program is necessary. However, we will use the simple laminar flow heat transfer model to estimate the behavior expected under any set of proposed design conditions. In other words, we use the available (though imperfect) model as an aid to design.

Make a choice for values of $T_w$ and $T_{in}$. What fluid(s) will you use? Give the required physical property values for the fluids you select. For one set of selected conditions, estimate and plot the outlet temperature as a function of the fluid mass flowrate. In addition, plot the estimated pressure drop (in units of psi) across the ends of the test section (across the length $L$), as a function of flowrate.

$L$ values of 10 and 40 cm are to be studied. You may need a separate set of design parameters for each case. Describe how you will impose a constant temperature $T_w$ on the surface of the tube, but not on the inlet and outlet tubes.

**12.65** Begin with Eq. 12.5.3. Introduce $\Theta$, $y^*$, and $x^*$ as in Eqs. 12.5.8 and 12.5.9. Show that the criterion for the approximation that leads to Eq. 12.5.10 is Pe $\gg 1$, where the Peclet number is defined as Pe $= UH/\alpha$.

**12.66** Water flows through a long tube of circular cross section at a mass flowrate of 0.01 kg/s. The wall flux is maintained uniformly along the axis at 2000 W/m². The water enters the tube at 20°C. The inside diameter is 6 cm.
**a.** Find the required length $L$ so that $T_{out} = 80°$C.
**b.** Find the tube wall temperature at the outlet $L$.

**12.67** Repeat Example 12.9.1, but for the case that $L = 0.1$ m and $D = 0.5$ mm.

**12.68** Repeat Example 12.9.1, but for the case that $L = 0.1$ m and $D = 0.5$ mm. In addition, assume that the cooling fan produces a pressure drop of 5000 Pa. Use Fig. 12.9.2 to find the value of the dimensionless heat transfer rate $Y$ under these conditions. By what factor is this below the maximum rate of heat transfer?

**12.69** Derive Eqs. 12.9.39 and 12.9.40.

**12.70** Repeat Example 12.9.2, but suppose that the heat transfer fluid is Freon 12 (a gas) instead of air. The properties of Freon 12 (at 300 K and 1 atm) may be taken as follows:

$$\rho = 4.9 \text{ kg/m}^3 \qquad \mu = 1.3 \times 10^{-5} \text{ Pa} \cdot \text{s}$$

$$k = 0.01 \text{ W/m} \cdot \text{K} \qquad \text{Pr} = 0.8$$

In particular, how does the optimum $D$ change, and by what factor is the allowable heat dissipation flux changed?

# Chapter 13

# Simple Heat Exchangers

**I**n this chapter we present and illustrate the principles of design of heat exchangers of very simple types, using the background developed in preceding chapters. Many commercial heat exchangers are far more complex than those illustrated here. Nevertheless, the ideas described will enable you to study these more complex systems when the need arises.

## 13.1  INTRODUCTION

Heat exchangers are used commonly throughout the chemical process industries as a means of heating and cooling process and product streams. Their analysis and design represent one of the most common practical applications of the fundamental principles we have studied so far. Most often, a heat exchanger is a system that provides a means for heat exchange between two fluids across a solid barrier that separates them. In such a system the two fluids are not in direct contact. *Direct contact* heat exchangers do exist: for example, a hot gas may be bubbled through a cold liquid for the purpose of raising the temperature of the liquid, or cooling the gas.

One of the simplest heat exchangers consists of a long circular pipe inside a second pipe. Heat is exchanged between the inner fluid and the outer fluid that flows through the annular region. Both fluids are usually flowing, and it is necessary to be able to estimate the heat transfer coefficients in both streams, using methods such as we have already discussed. We begin with an analysis of the so-called double-pipe heat exchanger.

## 13.2  DOUBLE-PIPE HEAT EXCHANGER ANALYSIS

Figure 13.2.1 shows a schematic of the double-pipe heat exchanger. The flows can be in the same direction along the pipe axes, or in opposite directions. The former case is called "cocurrent" flow, and the latter is "countercurrent" flow. The analyses of the two cases differ somewhat. We will analyze the cocurrent case, leaving the details of countercurrent heat exchange for a homework assignment.

We begin with a differential heat balance on the cold stream, which takes the form

$$dQ_h = (wC_p)_C \, dT_C = C_C \, dT_C \qquad (13.2.1)$$

where $C \equiv wC_p$

In this expression we define $dQ_h$ as the differential rate at which heat is transferred into the cold fluid, from the hot fluid, across a differential section of the exchanger. It

506

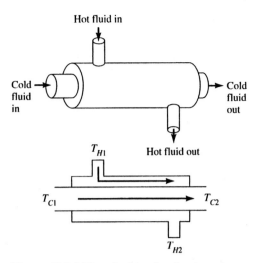

**Figure 13.2.1** The double-pipe heat exchanger with cocurrent flow.

is a matter of convenience to define the product term $C$ as above, since it appears throughout the following equations. We use $w$ for the *mass* flowrate, and so $C_p$ is the heat capacity per unit *mass*—not the *molar* heat capacity. If there is no heat loss from the exchanger, then a similar balance on the hot stream takes the form

$$dQ_h = -(wC_p)_H \, dT_H = -C_H \, dT_H \qquad (13.2.2)$$

Heat is transferred from the hot stream to the cold stream across a series of thermal resistances. There are convective resistances at the surfaces of the pipe that separates the two streams, and a conductive resistance associated with the pipe wall itself. These series resistances add (as described in an earlier chapter, Eq. 10.2.33) and we define an overall heat transfer coefficient $U$ such that

$$dQ_h = U \, dA(T_H - T_C) \qquad (13.2.3)$$

Here, $dA$ is the differential area that separates the two streams. The definition of $U$ includes the choice of whether we use the inner surface or the outer surface of the inside pipe in defining $dA$.

We now define a temperature difference between the hot and cold side as

$$d(T_H - T_C) = d(\Delta T) \qquad (13.2.4)$$

From Eqs. 13.2.1 and 13.2.2 we may write

$$dT_C = \frac{dQ_h}{C_C} \quad \text{and} \quad dT_H = -\frac{dQ_h}{C_H} \qquad (13.2.5)$$

When we subtract Eq. 13.2.1 from Eq. 13.2.2, we find

$$d(\Delta T) = dQ_h\left(-\frac{1}{C_H} - \frac{1}{C_C}\right) = -\frac{dQ_h}{C_H}\left(1 + \frac{C_H}{C_C}\right) \qquad (13.2.6)$$

The heat balances of Eq. 13.2.6 are differential, valid at any point along the axis of the exchanger. Now we write an *overall* heat balance, equating the heat lost by the hot fluid to that gained by the cold one, and find

$$Q_h = C_C(T_{C2} - T_{C1}) = C_H(T_{H1} - T_{H2}) \qquad (13.2.7)$$

This permits us to write Eq. 13.2.6 as

$$d(\Delta T) = \frac{dQ_h}{C_H} \frac{\Delta T_2 - \Delta T_1}{T_{H1} - T_{H2}} \quad \text{where} \quad \left( \begin{matrix} \Delta T_2 \equiv T_{H2} - T_{C2} \\ \Delta T_1 \equiv T_{H1} - T_{C1} \end{matrix} \right) \tag{13.2.8}$$

Note that $\Delta T_1$ and $\Delta T_2$ are the temperature differences between the two streams at the ends of the exchanger, while $\Delta T$ is the *local* temperature difference between the hot and cold streams. We may write Eq. 13.2.8 in the form

$$d(\Delta T) = \frac{dQ_h}{Q_h}(\Delta T_2 - \Delta T_1) = \frac{U \, dA \, \Delta T}{Q_h}(\Delta T_2 - \Delta T_1) \tag{13.2.9}$$

or

$$\frac{d\,\Delta T}{\Delta T} = \frac{U}{Q_h}(\Delta T_2 - \Delta T_1) \, dA \tag{13.2.10}$$

We may integrate this differential equation for $\Delta T$ if we assume $U$ is constant along the pipe axis, and find

$$\ln \frac{\Delta T_2}{\Delta T_1} = \frac{U}{Q_h}(\Delta T_2 - \Delta T_1)A \tag{13.2.11}$$

If we rearrange this equation, we may solve for the overall rate of heat transfer, with the result

$$Q_h = UA \frac{\Delta T_2 - \Delta T_1}{\ln(\Delta T_2/\Delta T_1)} = UA \, \Delta T_{\ln} \tag{13.2.12}$$

In Eq. 13.2.12 we have defined the so-called log–mean temperature difference as

$$\frac{\Delta T_2 - \Delta T_1}{\ln(\Delta T_2/\Delta T_1)} = \Delta T_{\ln} \tag{13.2.13}$$

This is the same definition we came upon in an earlier analysis of heat transfer in a single-tube heat exchanger (Eq. 12.4.24). For *countercurrent* flow we get the same final result (Eq. 13.2.12), though with some intermediate sign changes. (See Problem 13.1.)

We turn now to two example problems that give some insight into the differences between co- and countercurrent operation. We assume that the overall coefficient $U$ is known, and independent of the direction of the flow, for a given geometry and specified flowrates of the two streams.

### EXAMPLE 13.2.1  Maximum Cooling Capacity of an Exchanger of Fixed Area

Water at the rate of $w = 500$ lb$_m$/h at 100°F is available for use as a coolant in a double-pipe heat exchanger whose total surface area is 15 ft$^2$. The water is to be used to cool an oil ($C_p = 0.5$ Btu/lb $\cdot$ °F) from an initial temperature of 250°F. Because of other circumstances, an exit water temperature greater than 210°F cannot be allowed. The exit temperature of the oil must not be below 140°F. The overall heat transfer coefficient is 50 Btu/h $\cdot$ ft$^2$ $\cdot$ °F, and is assumed to be independent of flowrate of either stream in the range of interest.

Estimate the maximum flowrate of oil that may be cooled, assuming the flowrate of water is fixed at 500 lb$_m$/h.

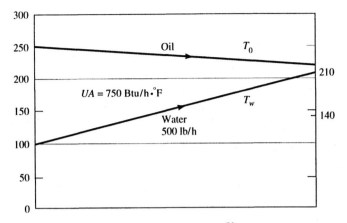

**Figure 13.2.2** Cocurrent temperature profiles.

*Solution*

1. First try *cocurrent* flow. As indicated in Fig. 13.2.2, the temperatures must satisfy

$$T_w \leq 210 \qquad T_0 \geq T_w$$
$$T_0 \geq 140$$

Hence $100 < T_w < 210$. The maximum flowrate of oil will be achieved if we allow $T_w$ to be at its maximum value of $T_w = 210°F$. This leaves $T_{oil}$ and $w_{oil}$ as unknowns, to be found from a heat balance:

$$Q_h = (wC_p)_{oil}(250 - T_{oil}) \qquad \text{(13.2.14)}$$

and a "performance" equation:

$$Q_h = UA\,\Delta T_{ln} = UA\frac{(250 - 100) - (T_0 - 210)}{\ln[150/(T_0 - 210)]} \qquad \text{(13.2.15)}$$

But for the water stream

$$Q_h = (wC_p)_w(210 - 100) = 500(1)110 = 55,000 \text{ Btu/h} \qquad \text{(13.2.16)}$$

so we find

$$\frac{360 - T_0}{\ln[150/(T_0 - 210)]} = \frac{55,000}{750} = 73.4 \qquad \text{(13.2.17)}$$

and it follows that

$$T_0 = 238.5°F \qquad \text{(13.2.18)}$$

Solving for the oil flow we find

$$w_{oil} = \frac{55,000}{0.5(11.5)} = 9560 \text{ lb/h} \qquad \text{(13.2.19)}$$

Cocurrent

The oil is cooled 11.5°F from its inlet value of 250°F.

2. Try countercurrent flow, referring to the temperature profiles in Fig. 13.2.3. We again let the water come to its maximum permissible temperature of 210°F. Then the oil

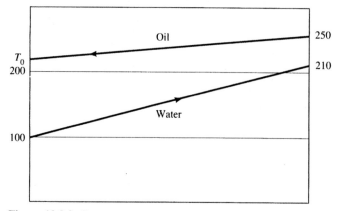

**Figure 13.2.3** Countercurrent temperature profiles.

outlet temperature is found from the heat balance. Equation 13.2.12 is valid regardless of the direction of the flows, so $\Delta T_{\ln}$ has the same value as before:

$$\Delta T_{\ln} = 73.4 = \frac{(T_0 - 100) - (250 - 210)}{\ln[(T_0 - 100)/40]} \tag{13.2.20}$$

Now, however, we find

$$T_0 = 221°F \tag{13.2.21}$$

and

$$w_{oil} = \frac{55,000}{0.5(29)} = 3800 \text{ lb/h} \tag{13.2.22}$$

Countercurrent

The oil is cooled 29°F.

Hence we find that the maximum oil flow is 9560 lb/h and it is under *cocurrent* conditions. There is a significant difference in the cooling capacity of co- and countercurrent exchangers, as this example illustrates. Note however that the degree of cooling differs in the two cases. We can cool more oil in the cocurrent configuration, but we cannot cool it as much as we do in the countercurrent configuration. For the problem specified here, it is not clear which configuration is favored, since we do not indicate the relative importance of oil flow versus degree of cooling. Example 13.2.2 gives a clearer look at the differences of the two configurations.

**EXAMPLE 13.2.2**    ***Find the Required Length of a Heat Exchanger with Specified Flows: Turbulent Flow in Both Streams***

The design constraints are given in the schematic of Fig. 13.2.4. We show this as a countercurrent configuration, but we will examine the cocurrent case as well. The benzene flow is specified as a mass flowrate (in pound *mass* units), and the water flow is given as a linear velocity. Heat transfer coefficients are not provided; we will have to calculate them based on discussions in Chapter 12.

The inside pipe is specified as "Schedule 40—1-1/4 inch steel." Pipe "schedules" are simply agreed-upon standards for pipe construction that specify the wall thickness of

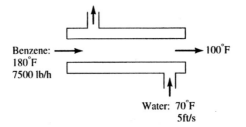

**Figure 13.2.4** Countercurrent flow.

the pipe. An engineering handbook gives us the information that for a Schedule 40—1-1/4 inch steel pipe the outside diameter is $D_o$ = 1.66 inches = 0.138 ft, the inside diameter is $D_i$ = 1.38 inches = 0.115 ft, and the cross-sectional area available to flow is $A_c$ = 0.0104 ft$^2$.

The outside pipe is stated to be "Schedule 40—2″ steel," which we find means that $D_i$ = 2.07 inches = 0.172 ft. We begin the analysis with the basic design equation

$$Q_h = UA \, \Delta T_{ln} \tag{13.2.23}$$

Note that the exchanger area $A$ is known only per unit length, since the pipe length is not specified. We must choose an *area basis* for the definition of $U$, since the exchanger surface is the wall of the inner pipe, whose inner and outer surfaces have different areas. We will base $U$ on the *outer* area (per unit of pipe length) $A_o/L = 2\pi r_o$ of the inner pipe. Hence we denote $U = U_o$ to remind us of this.

Our first major task is to estimate the overall heat transfer coefficient. We return to an earlier result (Eq. 10.2.33) for resistances in series, in a cylindrical system, rewritten here as

$$U_o = \frac{1}{r_o} \left[ \frac{1}{r_o h_o} + \frac{\ln r_o/r_i}{k} + \frac{1}{r_i h_i} \right]^{-1} \tag{13.2.24}$$

It is useful to write this in the form

$$U_o \frac{A_o}{L} = \left[ \frac{1}{h_o(A_o/L)} + \frac{\ln(r_o/r_i)}{2\pi k} + \frac{1}{h_i(A_i/L)} \right]^{-1} = (\Sigma R)^{-1} \tag{13.2.25}$$

We first calculate the inside coefficient (benzene), using the following data:

$T_b = 140°F$  $\qquad$  $k_b = 0.085 \text{ Btu/h} \cdot \text{ft} \cdot °F$

$\rho_b = 52.3 \text{ lb}_m/\text{ft}^3$  $\qquad$  $\mu_b = 0.39 \text{ cP} = \dfrac{0.39}{1000} \dfrac{1}{47.88} = 8.1 \times 10^{-6} \text{ lb}_f \cdot \text{s/ft}^2$

$\qquad\qquad\qquad\qquad$ $= 2.6 \times 10^{-4} \text{ lb}_m/\text{ft} \cdot \text{s}$

$C_{pb} = 0.45 \text{ Btu/lb}_m \cdot °F$  $\qquad$  $\text{Pr} = \dfrac{C_{pb}\mu}{k_b} = 5$

Note that we convert the viscosity to $\text{lb}_m/\text{ft} \cdot \text{s}$ units here. At a mass flowrate of $w$ = 7500 $\text{lb}_m/\text{h}$ = 2.08 $\text{lb}_m/\text{s}$, we find Re = $8.8 \times 10^4$.

For turbulent pipe flow we estimate the inside coefficient from the Dittus-Boelter equation (see Problem 12.37)

$$\text{Nu} = \frac{h_i D_i}{k_b} = 0.023 \, \text{Re}^{0.8} \, \text{Pr}^{0.3} = 337 \tag{13.2.26}$$

and find

$$h_i = 249 \text{ Btu/h} \cdot \text{ft}^2 \cdot {}^\circ\text{F} \tag{13.2.27}$$

The inside area per unit length is

$$\frac{A_i}{L} = \pi(0.115) = 0.361 \text{ ft}^2/\text{ft} \tag{13.2.28}$$

and so we find one of the terms in Eq. 13.2.25 above to be

$$\left(h_i \frac{A_i}{L}\right)^{-1} = [249\,(0.361)]^{-1} = 0.011 \text{ h} \cdot \text{ft} \cdot {}^\circ\text{F/Btu} \tag{13.2.29}$$

Now we turn to the water side (annular) flow. The geometrical parameters are given as

$$D_{oi} = 0.138 \text{ ft} \quad \text{and} \quad D_{io} = 0.172 \text{ ft}$$

for the outer diameter of the inside pipe and the inner diameter of the outside pipe, respectively. We find from experience that for turbulent flow through ducts of noncircular cross section we may use the available correlations for Nu as long as we replace the diameter in Nu and Re with an "equivalent" or "hydraulic" diameter. The "hydraulic diameter" is defined as four times the cross-sectional area divided by the "wetted perimeter." For the annular cross section this is

$$D_{eq} = 4\frac{\pi(D_{i,o}^2 - D_{o,i}^2)/4}{\pi(D_{i,o} + D_{o,i})} = D_{i,o} - D_{o,i} = 0.034 \text{ ft} \tag{13.2.30}$$

Next we need $T_{out}$ for the water side flow. From an overall energy balance we find

$$(wC_p \Delta T)_{benz} = (wC_p \Delta T)_{water} \tag{13.2.31}$$

The water flow, given as the linear velocity $V = 5$ ft/s, corresponds to a mass flowrate through the annulus of

$$w_{water} = 9300 \text{ lb}_m/\text{h} \tag{13.2.32}$$

(using $\rho_w = 62.4 \text{ lb}_m/\text{ft}^3$).

From Eq. 13.2.31 it follows that

$$7500\,(0.45)\,(100 - 180) = 9300\,(1)\,(70 - T_{out}) \tag{13.2.33}$$

or

$$T_{out} = 99{}^\circ\text{F} \tag{13.2.34}$$

We will evaluate the water properties at the average temperature of 84.7°F:

$$\mu = 0.8 \text{ cP} = 1.67 \times 10^{-5} \text{ lb}_f \cdot \text{s/ft}^2 \quad k = 0.34 \text{ Btu/h} \cdot \text{ft} \cdot {}^\circ\text{F}$$
$$\text{Pr} = 5.7$$

We need the Reynolds number on the annular side. We define and calculate the Reynolds number for flow through an annulus as[1]

$$\text{Re} \equiv \frac{\rho V D_{eq}}{\mu} = \frac{\dfrac{62.4 \text{ lb}_m/\text{ft}^3}{32.2 \text{ lb}_m \cdot \text{f/lb}_f \cdot \text{s}^2}\,(5 \text{ ft/s})\,0.034 \text{ ft}}{1.67 \times 10^{-5} \text{ lb}_f \cdot \text{s/ft}^2} = 2 \times 10^4 \tag{13.2.35}$$

---

[1] An alternative form of Re for flow through a *circular* pipe, in terms of the mass flowrate $w$, is Re $= 4w/\pi D\mu$, where $D$ is the pipe diameter. It does not follow that for flow through a pipe of annular cross section, Re $= 4w/\pi D_{eq}\mu$.

with $D_{eq}$ given by Eq. 13.2.30. We again calculate the heat transfer coefficient from the Dittus-Boelter equation, using $m = 0.4$ for the heated side:

$$\text{Nu} = 0.023 \text{ Re}^{0.8} \text{ Pr}^{0.4} = 127 \qquad \textbf{(13.2.36)}$$

The result is

$$h_o = 1270 \text{ Btu/h} \cdot \text{ft}^2 \cdot {}^\circ\text{F} \qquad \textbf{(13.2.37)}$$

Note $h_o \gg h_i$ for this system. We can now find the area (per unit pipe length) as

$$\frac{A_o}{L} = \pi(0.138) = 0.434 \text{ ft}^2/\text{ft} \qquad \textbf{(13.2.38)}$$

and

$$\left(h_o \frac{A_o}{L}\right)^{-1} = [1270\,(0.434)]^{-1} = 0.00181 \text{ h} \cdot \text{ft} \cdot {}^\circ\text{F/Btu} \qquad \textbf{(13.2.39)}$$

Finally we evaluate the thermal resistance of the pipe wall itself, although we expect it to be small relative to the convective resistances. From the geometry and the properties of steel we find

$$r_o - r_i = 0.012 \text{ ft} \qquad k_{\text{steel}} = 26 \text{ Btu/h} \cdot \text{ft} \cdot {}^\circ\text{F}$$

and

$$\frac{\ln r_o/r_i}{2\pi k} = 0.00116 \text{ h} \cdot \text{ft} \cdot {}^\circ\text{F/Btu} \qquad \textbf{(13.2.40)}$$

The total resistance/length is found as

$$\Sigma R = 0.00116 + 0.00181 + 0.011 = 0.014 \qquad \textbf{(13.2.41)}$$
$$\text{Wall} \qquad \text{Water} \qquad \text{Benzene}$$

The benzene side controls the heat transfer, but the other resistances are not negligible. Now we calculate the log–mean temperature difference and find

$$\Delta T_{\text{ln}} = \frac{(180 - 99) - (100 - 70)}{\ln (81/30)} = 51{}^\circ\text{F} \qquad \textbf{(13.2.42)}$$

The total heat transferred is found from a benzene side balance as

$$\begin{aligned} Q_h &= wC_p\,\Delta T \\ &= 7500\,(0.45)\,(180 - 100) = 2.7 \times 10^5 \text{ Btu/h} \end{aligned} \qquad \textbf{(13.2.43)}$$

The required length then follows from the basic design equation as

$$\frac{Q_h}{L} = \frac{UA}{L}\Delta T_{\text{ln}} = (\Sigma R)^{-1}\,\Delta T_{\text{ln}} = 3640 \text{ Btu/h} \cdot \text{ft} \qquad \textbf{(13.2.44)}$$

and so

$$L = \frac{Q_h}{3640} = \frac{2.7 \times 10^5}{3640} = 74 \text{ ft} \qquad \textbf{(13.2.45)}$$

What length do we get if the exchanger is run *cocurrent*? The only thing that changes is the log–mean temperature difference when we switch the direction of one of the streams. Figure 13.2.5 shows the temperature differences. We go from a log–mean $\Delta T$ of 51°F to a lower value of only 23.2°F. As a consequence, the required length of the

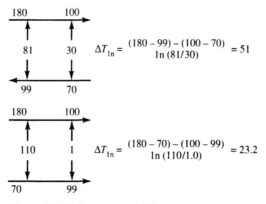

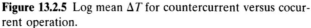

**Figure 13.2.5** Log mean $\Delta T$ for countercurrent versus cocurrent operation.

exchanger is greatly increased, from $L = 74$ ft to $L = 163$ ft. *The countercurrent exchanger requires less area for a fixed heat duty, at fixed flow rates.*

Note the need for precision in $\Delta T_{\text{out}}$ in the cocurrent case, since we have to calculate $100 - \Delta T_{\text{out}} = 100 - 99$. When $\Delta T_{\text{out}}$ is near 100, the log term is very sensitive to any error in $\Delta T_{\text{out}}$.

To summarize, we note that for a simple double-pipe heat exchanger the design equation is of the form

$$Q_h = UA\,\Delta T_{\text{ln}} \qquad (13.2.46)$$

We get the same *format* whether the system is co- or countercurrent, but $\Delta T_{\text{ln}}$ differs in each case, often appreciably.

In the example just worked, the length of the exchanger is unknown. Because both flows are turbulent, the heat transfer coefficients can be found independent of knowledge of the pipe length. If one of the flows is laminar, however, the heat transfer coefficient may be length dependent, and a trial-and-error procedure may be necessary. We illustrate this point in Example 13.2.3.

**EXAMPLE 13.2.3**  *Find the Required Length of a Heat Exchanger with Specified Flows: Laminar Flow on One Side*

It is necessary to heat an oil from 25°C to 30°C. The oil flowrate is specified as 20,000 gal/day. Hot water at 90°C is available, at a flowrate of 7000 gal/day. A decision has been made to build a double-pipe heat exchanger, using an inside pipe of inside diameter 6 inches and outside diameter 6.6 inches. The outside pipe will have an inside diameter of 8 inches. The oil flow is through the inside pipe.

The oil properties, available at 25°C, are

$$\rho = 53\ \text{lb}_m/\text{ft}^3 \qquad C_p = 0.46\ \text{Btu/lb}_m\cdot°\text{F} \qquad \mu = 150\ \text{cP} \qquad k = 0.11\ \text{Btu/h}\cdot\text{ft}\cdot°\text{F}$$

The pipe wall has a thermal conductivity of 25 Btu/h·ft·°F.

Let's begin by finding the Reynolds number for each stream. We will convert properties to SI units.

For the oil side, the volume flowrate is

$$Q = 2 \times 10^4 \text{ gal/day} \times \frac{1}{24 \times 3600 \text{ s/day}} \times 3.785 \times 10^{-3} \text{ m}^3/\text{gal} = 8.8 \times 10^{-4} \text{ m}^3/\text{s}$$

$$(13.2.47)$$

The oil density is

$$\rho_{\text{oil}} = 53 \text{ lb}_m/\text{ft}^3 \times 16.02 = 849 \text{ kg/m}^3 \qquad (13.2.48)$$

Hence the mass flowrate is

$$w = \rho_{\text{oil}} Q = 849 \times 8.8 \times 10^{-4} = 0.75 \text{ kg/s} \qquad (13.2.49)$$

The Reynolds number follows as

$$\text{Re}_{\text{oil}} = \frac{4w}{\pi D \mu} = \frac{4 \times 0.75}{\pi \times 6 \times 0.0254 \times 0.15} = 42 \qquad (13.2.50)$$

The flow is laminar on the oil side, and we can guess that the oil side will control the heat transfer.

For the water side, the volume flowrate is

$$Q = 7000 \text{ gal/day} \times \frac{1}{24 \times 3600 \text{ s/day}} \times 3.785 \times 10^{-3} \text{ m}^3/\text{gal} = 3 \times 10^{-4} \text{ m}^3/\text{s}$$

$$(13.2.51)$$

At 90°C the water density will be taken as 1000 kg/m³. (It is 4 or 5% below this value; since, however, we will be using heat transfer correlations that are accurate only to ±20%, great precision in calculating the physical properties that go into the Reynolds number is not necessary. For the same reason we do not bother to estimate a film temperature for the density calculation, given that density is a fairly weak function of temperature.) Hence the mass flowrate on the water side is

$$w = 0.3 \text{ kg/s} \qquad (13.2.52)$$

Because viscosity is a strong function of temperature, it is appropriate to calculate the viscosity at the film temperature. We need the outlet temperature of the water. This follows directly from an energy balance:

$$Q_h = (wC_p)_{\text{oil}}(30 - 25) = (wC_p)_{\text{water}}(90 - T_{w,\text{out}}) \qquad (13.2.53)$$

or

$$Q_h = (0.75 \times 0.46 \times 4190)(30 - 25) = (0.3 \times 1 \times 4190)(90 - T_{w,\text{out}}) = 7228 \text{ W}$$

$$(13.2.54)$$

Hence,

$$T_{w,\text{out}} = 84°C \qquad (13.2.55)$$

The film temperature for this system is just the arithmetic mean of the four terminal temperatures: $T_f = 57°C$. The viscosity of water at this temperature is $\mu = 0.48 \times 10^{-3}$ Pa·s. For the calculation of the Reynolds number in the annulus we need the equivalent diameter:

$$D_{\text{eq}} = D_{i,o} - D_{o,i} = (8 - 6.6)\, 0.0254 = 0.0356 \text{ m} \qquad (13.2.56)$$

or we can write the Reynolds number in terms of the mass flowrate as

$$\text{Re} = \frac{4w}{\pi(D_{i,o} + D_{o,i})\mu} = \frac{4 \times 0.3}{\pi(8 + 6.6)\,0.0254\,(0.48 \times 10^{-3})} = 2150 \qquad (13.2.57)$$

The flow is barely turbulent, and most correlations for turbulent heat transfer coefficients are obtained at Reynolds numbers higher than this. Still, we will use the Dittus–Boelter equation (Eq. 12.4.50):

$$\text{Nu} = 0.023\,\text{Re}^{0.8}\,\text{Pr}^{0.3} \qquad (13.2.58)$$

At the film temperature, the Prandtl number for water is $\text{Pr} = 3$, so we find

$$\text{Nu} = \frac{hD_{eq}}{k} = 0.023(2150)^{0.8}(3)^{0.3} = 15 \qquad (13.2.59)$$

and

$$h_{water} = \frac{15k}{D_{eq}} = \frac{15 \times 0.65 \text{ W/m}\cdot\text{K}}{0.0356 \text{ m}} = 274 \text{ W/m}^2\cdot\text{K} \qquad (13.2.60)$$

To find $h$ for the oil side, where the flow is laminar, we need an estimate for the length of the exchanger. How do we select a starting value? If we assume that $U \approx h_{water}$, we can estimate a length from

$$L = \frac{Q_h}{\pi D_{o,i} h_{water}\,\Delta T_{ln}} \qquad (13.2.61)$$

Now that we know $T_{w,out}$ we may find

$$\Delta T_{ln} = \frac{(84 - 25) - (90 - 30)}{\ln(59/60)} = 59.5°\text{C} \qquad (13.2.62)$$

This gives us an estimated length of

$$L = \frac{7228}{\pi(6.6 \times 0.0254)274 \times 59.5} = 0.84 \text{ m} \qquad (13.2.63)$$

We know this is an underestimate, because the oil side coefficient will be much lower than that of the water side, since the oil flow is laminar. We often find that oil side or laminar coefficients are an order of magnitude smaller than those on the water or turbulent side. Hence we expect $L$ to be as much as 10 times larger than 0.84 m. With these ideas in mind, we begin a trial-and-error calculation starting with $L = 10$ m.

We need a Prandtl number for the oil. We have the necessary data, and we find

$$\text{Pr} = \frac{\mu C_p}{k} = \frac{0.15(\text{Pa}\cdot\text{s})(0.46 \times 4190) \text{ W}\cdot\text{s/kg}\cdot\text{K}}{(0.11 \times 1.73) \text{ W/m}\cdot\text{K}} = 1520 \qquad (13.2.64)$$

Next we need to calculate

$$\frac{\text{Re Pr }D}{L} = \frac{41 \times 1520(6 \times 0.0254)}{10} = 950 \qquad (13.2.65)$$

Hence we use Eq. 12.5.49:

$$\overline{\text{Nu}}_{oil} = \frac{\overline{h}_{oil}D_{i,i}}{k} = 1.6\left(\frac{\text{Re Pr }D_{i,i}}{L}\right)^{1/3} = 16 \qquad (13.2.66)$$

and

$$\bar{h}_{oil} = \frac{16k}{D_{i,i}} = \frac{16(0.11 \times 1.73)}{6 \times 0.0254} = 20 \text{ W/m}^2 \cdot \text{K} \tag{13.2.67}$$

We may now calculate the overall heat transfer coefficient from

$$U_o = \left[ \frac{1}{h_{water}} + \frac{1}{\bar{h}_{oil}} \frac{D_{o,i}}{D_{i,i}} + \frac{D_{o,i}}{2k_{pipe}} \ln \left( \frac{D_{o,i}}{D_{i,i}} \right) \right]^{-1}$$

$$= \left[ \frac{1}{270} + \frac{1}{20} \frac{6.6}{6} + \frac{6.6 \times 0.0254}{2(25/1.73)} \ln \left( \frac{6.6}{6} \right) \right]^{-1} \tag{13.2.68}$$

$$= (0.0037 + 0.055 + 0.00055)^{-1} = 17 \text{ W/m}^2 \cdot \text{K}$$

Going back to Eq. 13.2.63 we obtain our new estimate of $L$:

$$L = \frac{7228}{\pi (6.6 \times 0.0254) 17 \times 59.6} = 13.5 \text{ m} \tag{13.2.69}$$

This is so close to the guessed value ($L = 10$ m) that another iteration would make only a small change in $L$, since, with the oil side in control, $U_o$ varies as $L^{-1/3}$ (according to Eq. 13.2.66).

Let's pause for a moment to look at the "structure" of the heat exchanger problems that we have been and will be examining. First, what are the "unknowns" (the design and operating parameters) in these problems? For a simple heat exchanger these are the *four* end temperatures, the *two* flowrates, the area of the exchanger, and the "heat duty" $Q_h$. Hence a simple exchanger is completely defined by these 8 variables. But these are constrained by only three equations: two heat balances of the form of Eq. 13.2.7, and the "performance" equation for the exchanger, of the form of Eq. 13.2.12.

It follows then that there are five "degrees of freedom" in these simple heat exchanger problems (8 variables minus 3 equations). A heat exchanger analysis is uniquely defined only if five design/operating parameters are specified. In the examples that follow, you will see that we must specify five such parameters to yield a unique design or performance.

## 13.3 OTHER HEAT EXCHANGER CONFIGURATIONS

Many heat exchangers are not simple double-pipe systems, but we retain the *format* of Eq. 13.2.46 with the introduction of an additional "correction" factor:

$$Q_h = UA \Delta T_{ln} F \tag{13.3.1}$$

The form of $F$ depends on the plumbing geometry of the exchanger. We can look at heat exchangers of several types, carry out the appropriate thermal balances, and then force the result for the heat duty $Q_h$ into the format of Eq. 13.3.1. We then, typically, present the function $F$ in a graphical format, as we shall see. By convention, $F$ is defined with $\Delta T_{ln}$ based on the equivalent *counter*-current flow.

**EXAMPLE 13.3.1**    *The 1 Shell Pass/2 Tube Pass Exchanger*

We now turn to a configuration in which a pipe "loops" through a larger container, or "shell," as shown in Fig. 13.3.1. We refer to this configuration as a "1 shell pass/2 tube pass" heat exchanger because the tube carries its stream along the length of the shell

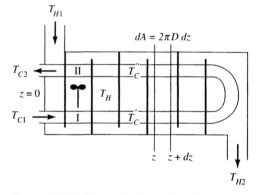

**Figure 13.3.1** The 1 shell pass/2 tube pass exchanger. The shell side is baffled in a way that ensures uniform shell side temperature in the direction normal to the $z$ axis in each baffled section.

of the exchanger twice, while the shell side fluid passes through once. The shell side is partitioned, or "baffled," in such a way that the shell side fluid is mixed by convection in the direction normal to the tubes, but not in the direction of the tube axes. Hence the shell side fluid temperature is a function of the $z$ axis shown in Fig. 13.3.1, but it is *assumed* that the shell side fluid has the same temperature surrounding both tubes I and II in any baffled compartment. If the baffles are close together in the $z$ direction, we may approximate the shell side fluid temperature by a *continuous* function of $z$. In reality, $T_H$ varies *discretely* from one baffled section to the next. This assumption of a continuous $T(z)$ permits us to derive differential equations for the axial temperature profiles of the hot and cold streams.

Our goal is to carry out a thermal analysis of this particular configuration and to derive the algebraic form of the factor $F$ defined in Eq. 13.3.1.

We begin by defining the ratio

$$R = \frac{(wC_p)_C}{(wC_p)_H} = \frac{C_C}{C_H} = \frac{T_{H1} - T_{H2}}{T_{C2} - T_{C1}} \tag{13.3.2a}$$

Equation 13.3.1 is written for the arbitrary choice that the cold fluid is on the tube side. In general, however, we define $R$ such that

$$R = \frac{(wC_p)_{tube}}{(wC_p)_{shell}} = \frac{C_{tube}}{C_{shell}} = \frac{\Delta T_{shell}}{\Delta T_{tube}} \tag{13.3.2b}$$

As noted before, we assume that the shell side fluid is well mixed in the direction normal to the pipe axes, so that at a specific position $z$ the shell side temperature $T_H$ is the same on the I loop and the II loop. *However, $T_H$ varies along the axis $z$.* A thermal balance over a differential length $dz$ on the I loop gives us

$$C_C(T'_{C_z} - T'_{C_{z+dz}}) = U\frac{dA}{2}(T'_C - T_H)_z \tag{13.3.3}$$

Similarly, a balance on the II loop over $dz$ yields

$$C_C(T''_{C_z} - T''_{C_{z+dz}}) = U\frac{dA}{2}(T_H - T''_C)_z \tag{13.3.4}$$

where prime and double prime denote the I and II loops, respectively. A balance on the shell (hot) side over the length $dz$ gives us

$$C_H(T_{H_z} - T_{H_{z+dz}}) = \frac{U\,dA(T_H - T'_C)_z}{2} + \frac{U\,dA(T_H - T''_C)_z}{2} \tag{13.3.5}$$

We divide both sides of Eqs. 13.3.3 and 13.3.4 by $dA$ and then take the limit as $dA \to 0$. The result is a pair of differential equations for $T_C$ in either loop:

$$\frac{C_C}{U}\frac{dT'_C}{dA} = \frac{T_H - T'_C}{2} \tag{13.3.6}$$

$$\frac{C_C}{U}\frac{dT''_C}{dA} = -\left(\frac{T_H - T''_C}{2}\right) \tag{13.3.7}$$

Similarly, Eq. 13.3.5 becomes

$$\frac{C_H}{U}\frac{dT_H}{dA} = \frac{T'_C - T_H}{2} + \frac{T''_C - T_H}{2} \tag{13.3.8}$$

With the following definition:

$$dn \equiv \frac{U\,dA}{C_C} \tag{13.3.9}$$

we may write these equations in the forms

$$\frac{dT'_C}{dn} = \frac{T_H - T'_C}{2} \tag{13.3.10}$$

$$\frac{dT''_C}{dn} = -\left(\frac{T_H - T''_C}{2}\right) \tag{13.3.11}$$

$$\frac{1}{R}\frac{dT_H}{dn} = \frac{T'_C - T_H}{2} + \frac{T''_C - T_H}{2} \tag{13.3.12}$$

Equations 13.3.10 to 13.3.12 are three differential equations in the three unknowns: $T'_C$, $T''_C$, and $T_H$. Our task now is to solve these equations.

We begin by making a thermal balance from any arbitrary value of $z$ to the end $z = L$:

$$C_H(T_H - T_{H2}) = C_C(T''_C - T'_C) \tag{13.3.13}$$

or

$$\frac{1}{R}(T_H - T_{H2}) = T''_C - T'_C \tag{13.3.14}$$

We can eliminate $T''_C$ first in terms of $T_H$ and $T'_C$:

$$T''_C = T'_C + \frac{1}{R}(T_H - T_{H2}) \tag{13.3.15}$$

From Eq. 13.3.12, and using Eqs. 13.3.10 and 13.3.11, we find

$$\frac{dT''_C}{dn} = \frac{dT'_C}{dn} + \frac{1}{R}\left(\frac{dT_H}{dn}\right) = -\frac{T_H}{2} + \frac{T'_C + (1/R)(T_H - T_{H2})}{2} \tag{13.3.16}$$

Also, we find

$$\frac{T_H - T_C'}{2} + \frac{1}{R}\frac{dT_H}{dn} = -\frac{T_H}{2} + \frac{T_C'}{2} + \frac{1}{2R}(T_H - T_{H2}) \qquad (13.3.17)$$

and

$$T_C' = T_H + \frac{1}{R}\frac{dT_H}{dn} - \frac{1}{2R}(T_H - T_{H2}) \qquad (13.3.18)$$

Putting this in Eq. 13.3.10 we obtain

$$\frac{dT_C'}{dn} = \frac{dT_H}{dn} + \frac{1}{R}\frac{d^2T_H}{dn^2} - \frac{1}{2R}\frac{dT_H}{dn} \qquad (13.3.19)$$

$$= \frac{T_H}{2} - \frac{1}{2}\left[T_H + \frac{1}{R}\frac{dT_H}{dn} - \frac{1}{2R}(T_H - T_{H2})\right]$$

We may finally write this as a differential equation with only a single dependent variable, $T_H$:

$$\frac{1}{R}\frac{d^2T_H}{dn^2} + \frac{dT_H}{dn} - \frac{1}{4R}(T_H - T_{H2}) = 0 \qquad (13.3.20)$$

Boundary conditions on $T_H$ are

$$T_H = T_{H1} \qquad \text{at} \quad n = 0 \qquad (13.3.21)$$

$$T_H = T_{H2} \qquad \text{at} \quad n = n_T \qquad (13.3.22)$$

where we have defined

$$n_T \equiv \frac{UA}{C_C} \qquad (13.3.23)$$

It is convenient to nondimensionalize Eq. 13.3.20, so we define

$$\theta = \frac{T_H - T_{H2}}{T_{H1} - T_{H2}} \qquad (13.3.24)$$

and find

$$\frac{d^2\theta}{dn^2} + R\frac{d\theta}{dn} - \frac{\theta}{4} = 0 \qquad (13.3.25)$$

with boundary conditions

$$\theta = 1 \qquad \text{at} \quad n = 0 \qquad (13.3.26)$$

$$\theta = 0 \qquad \text{at} \quad n = n_T \qquad (13.3.27)$$

The algebra from this point on is horrendous. We solve Eq. 13.3.25 for $\theta(T_H)$ first. Then we manipulate the other two differential equations (Eqs. 13.3.10 and 13.3.11, using the solution for $T_H$) so that we can find $T_C'$ and $T_C''$. We then write the overall heat balance and force it into the form of Eq. 13.3.1 so that we can find the form of the function $F$. The final result is expressible as

$$F = \frac{\dfrac{\sqrt{R^2 + 1}}{R - 1}\ln\left(\dfrac{1 - P}{1 - PR}\right)}{\ln\left(\dfrac{2/P - 1 - R + \sqrt{R^2 + 1}}{2/P - 1 - R - \sqrt{R^2 + 1}}\right)} \qquad (13.3.28)$$

where we define

$$P \equiv \frac{T_{C2} - T_{C1}}{T_{H1} - T_{C1}} \tag{13.3.29}$$

when the cold fluid is on the tube side.

In general, whether the cold fluid is on the shell side(s) or the tube side (t), $P$ is defined as follows:

$$P = \frac{T_{t,\text{out}} - T_{t,\text{in}}}{T_{s,\text{in}} - T_{t,\text{in}}} = \frac{\text{achieved heat transfer}}{\text{maximum possible heat transfer}} \tag{13.3.30}$$

The derivation of Eq. 13.3.30 is straightforward: it follows from the fact that $C_t$ $(T_{t,\text{out}} - T_{t,\text{in}})$ is the actual tube side heat exchanged. For the tube side, the maximum possible heat transfer would occur if $T_t$ could be changed from $T_{t,\text{in}}$ to $T_{s,\text{in}}$. Hence $P = C_t (T_{t,\text{out}} - T_{t,\text{in}})/C_t(T_{s,\text{in}} - T_{t,\text{in}})$ and Eq. 13.3.30 follows.

Note that in Eq. 13.3.28 the numerator of $F$ is indeterminate when $R$ is unity. It is useful to note, for computational purposes, that

$$\lim_{R \to 1} \frac{1}{R - 1} \ln \frac{1 - P}{1 - PR} = \frac{P}{1 - P} \tag{13.3.31}$$

Figure 13.3.2 shows the "correction factor" $F$ for this exchanger. Similar results can be derived, and are available, for other heat exchanger configurations. (See, e.g., Problem 13.5)

Examination of plots such as Fig. 13.3.2, for a variety of configurations, reveals a common feature: for a given value of $R$, there is a maximum value of $P$ possible. In view of the definition of $P$ (Eq. 13.3.30), we conclude that an initial choice of $R$ limits the performance of a particular exchanger. In particular, as we operate near the $P$ limit, the value of $F$ drops precipitously. Correspondingly, the exchanger area required to achieve the specified performance increases rapidly. Since real exchangers do not operate as ideally as the models that lead to these $F(R, P)$ graphs, we should avoid designing near the precipice—the region of high slope. We might, for example, find that nonidealities in mixing converted a marginal design to one that could not perform as demanded, at any exchanger length. A common recommendation is to avoid designs that lead to $F$ values less than about 0.8.

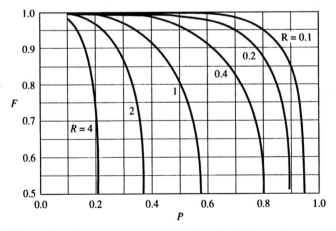

**Figure 13.3.2** Correction factor $F$ (Eq. 13.3.28) for heat exchanger with 1 shell pass and $2n$ tube passes.

## 13.4 EFFECTIVENESS CONCEPT FOR HEAT EXCHANGERS

If inlet or exit temperatures are to be evaluated in a heat exchanger analysis, trial and error is necessary because of the $(\ln \Delta T)$ term. An alternate approach to solving such problems is based on the concept of heat exchanger *effectiveness*.

We define the effectiveness $\varepsilon$ as

$$\varepsilon = \frac{\text{actual heat transfer}}{\text{maximum possible}} \tag{13.4.1}$$

The actual heat transfer is simply found as

$$\text{actual heat transfer} = (wC_p \Delta T)_h = (C \Delta T)_h = (wC_p \Delta T)_c = (C \Delta T)_c \tag{13.4.2}$$

The maximum possible $\Delta T$ is $(T_h - T_c)_{inlet}$. But, which fluid undergoes the maximum $\Delta T$? A little thought suggests that it is the one with the minimum $C$, since

$$C_{min} \Delta T_{max} = C_{max} \Delta T_{min} \tag{13.4.3}$$

Hence the maximum possible heat transfer is

$$Q_{h\,max} = C_{min}(T_h - T_c)_{inlet} \tag{13.4.4}$$

We have four possible cases for the double-pipe heat exchanger:

| **Parallel Flow** | **Counterflow** |
|---|---|
| Hot fluid = min | Hot fluid = min |
| Cold fluid = min | Cold fluid = min |

Equations 13.4.5 to 13.4.8 show this schematically. For parallel flow with the hot fluid as the minimum $C$ fluid, we write

$$\varepsilon_h = \frac{(wC_p)_h(T_{h1} - T_{h2})}{(wC_p)_h(T_{h1} - T_{c1})} = \frac{T_{h1} - T_{h2}}{T_{h1} - T_{c1}} \tag{13.4.5}$$

For the other cases:

$$\varepsilon_c = \frac{(wC_p)_c(T_{c2} - T_{c1})}{(wC_p)_c(T_{h1} - T_{c1})} = \frac{T_{c2} - T_{c1}}{T_{h1} - T_{c1}} \tag{13.4.6}$$

$$\varepsilon_h = \frac{(wC_p)_h(T_{h1} - T_{h2})}{(wC_p)_h(T_{h1} - T_{c2})} = \frac{T_{h1} - T_{h2}}{T_{h1} - T_{c2}} \tag{13.4.7}$$

$$\varepsilon_c = \frac{(wC_p)_c(T_{c1} - T_{c2})}{(wC_p)_c(T_{h1} - T_{c2})} = \frac{T_{c1} - T_{c2}}{T_{h1} - T_{c2}} \tag{13.4.8}$$

To obtain $\varepsilon$ we need the thermal energy balance, which at an intermediate stage gives, for the cocurrent case $\overset{\rightarrow}{\rightarrow}$ (go back to Eq. 13.2.11)

$$\ln \frac{T_{h2} - T_{c2}}{T_{h1} - T_{c1}} = -UA\left(\frac{1}{C_h} + \frac{1}{C_c}\right) = -\frac{UA}{C_c}\left(1 + \frac{C_c}{C_h}\right) \tag{13.4.9}$$

or

$$\frac{T_{h2} - T_{c2}}{T_{h1} - T_{c1}} = \exp\left[-\frac{UA}{C_c}\left(1 + \frac{C_c}{C_h}\right)\right] \tag{13.4.10}$$

We must now convert this temperature function to $\varepsilon$. Suppose the cold fluid is the minimum fluid. Then (Eq. 13.4.6)

$$\overset{\rightarrow}{\to} c \qquad \varepsilon = \frac{T_{c2} - T_{c1}}{T_{h1} - T_{c1}} \tag{13.4.11}$$

In any case, once $\varepsilon$ is known,

$$Q_h = \varepsilon Q_{max} = \varepsilon C_{min}(T_h - T_c)_{in} \tag{13.4.12}$$

From $Q_h = Q_c$ we find

$$T_{h2} = T_{h1} + \frac{C_c}{C_h}(T_{c1} - T_{c2}) \tag{13.4.13}$$

Then

$$T_{h2} - T_{c2} = T_{h1} + \frac{C_c}{C_h}(T_{c1} - T_{c2}) - T_{c2} \tag{13.4.14}$$

and so

$$\frac{T_{h2} - T_{c2}}{T_{h1} - T_{c1}} = \frac{(T_{h1} - T_{c1}) + (C_c/C_h)(T_{c1} - T_{c2}) + (T_{c1} - T_{c2})}{T_{h1} - T_{c1}} \tag{13.4.15}$$

$$= 1 + \frac{C_c}{C_h}(-\varepsilon) + (-\varepsilon) = 1 - \varepsilon\left(1 + \frac{C_c}{C_h}\right)$$

which, from before, (Eq. 13.4.10)

$$= \exp\left[-\frac{UA}{C_c}\left(1 + \frac{C_c}{C_h}\right)\right] \tag{13.4.16}$$

Therefore,

$$\varepsilon = \frac{1 - \exp\left[-\dfrac{UA}{C_c}\left(1 + \dfrac{C_c}{C_h}\right)\right]}{1 + \dfrac{C_c}{C_h}} \qquad \overset{\rightarrow}{\to} c \tag{13.4.17}$$

If we repeated for the case $\overset{\rightarrow}{\to} h$, we would find a similar result, and *both* could be written in a single form as

$$\overset{\rightarrow}{\to} \qquad \varepsilon = \frac{1 - \exp\left[-\dfrac{UA}{C_{min}}\left(1 + \dfrac{C_{min}}{C_{max}}\right)\right]}{1 + \dfrac{C_{min}}{C_{max}}} \tag{13.4.18}$$

For counterflow we find that both cases may be written in the single form

$$\overset{\leftarrow}{\to} \qquad \varepsilon = \frac{1 - \exp\left[-\dfrac{UA}{C_{min}}\left(1 - \dfrac{C_{min}}{C_{max}}\right)\right]}{1 - \dfrac{C_{min}}{C_{max}}\exp\left[-\dfrac{uA}{C_{min}}\left(1 - \dfrac{C_{min}}{C_{max}}\right)\right]} \tag{13.4.19}$$

We now define the dimensionless group (compare this to Eq. 13.3.23)

$$\frac{UA}{C_{min}} = NTU_{max} \qquad\qquad \text{(13.4.20)}$$

which is called "number of transfer units." For various heat exchanger configurations we can find graphs of $\varepsilon$ versus NTU, with $C_{min}/C_{max}$ as a parameter. Several such graphs are shown in Figs. 13.4.1 to 13.4.3. We turn now to an example of the utility of this

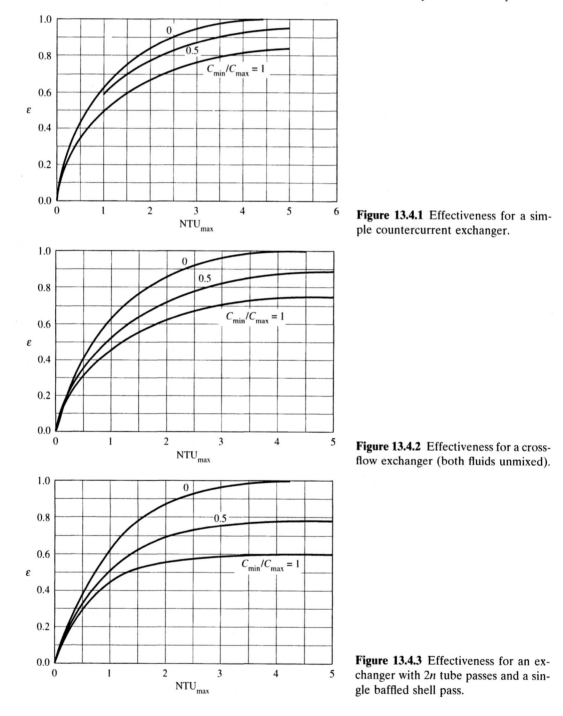

**Figure 13.4.1** Effectiveness for a simple countercurrent exchanger.

**Figure 13.4.2** Effectiveness for a crossflow exchanger (both fluids unmixed).

**Figure 13.4.3** Effectiveness for an exchanger with $2n$ tube passes and a single baffled shell pass.

effectiveness approach to heat exchanger problems in which one temperature and one flowrate are unknown.

EXAMPLE 13.4.1 *Performance of a "Cross-Flow" Heat Exchanger (One Temperature, One Flow Unknown)*

Consider a cross-flow heat exchanger with both fluids unmixed, as illustrated in Fig. 13.4.4. We wish to take 5000 ft$^3$/min of air at 1 atm from 60°F to 85°F (15.6°C to 29.4°C). We will use hot water at 180°F (82.2°C). For this exchanger we know that $UA = 4000$ Btu/h·°F = 2110 W/°C. Find the exit water temperature and the required water flowrate.

In this example, one temperature ($T_h$) and one flowrate ($w_{cold}$) are unknown. Trial-and-error manipulations are unavoidable, and we cannot use the $F(P,R)$ method because $R$ is unknown. Begin with an energy balance on the airstream. We find the air density from the ideal gas law:

$$\rho = \frac{P}{R_M T} = \frac{1.01 \times 10^5}{286(288.9)} = 1.22 \text{ kg/m}^3 \qquad \textbf{(13.4.21)}$$

Note that $\rho$ is mass (not molar) density. Hence we have used, in place of the gas constant $R_G$, the ratio of the gas constant (in SI units, 8314) to the molecular weight (29 kg/kg–mol):

$$R_M \equiv \frac{R_G}{M_{w,air}} = \frac{8314}{29} = 286 \qquad \textbf{(13.4.22)}$$

We convert the volumetric flowrate of the air to SI units and then find the mass flow as

$$w_c = 2.36 \text{ m}^3/\text{s} \times 1.22 \text{ kg/m}^3 = 2.9 \text{ kg/s} \qquad \textbf{(13.4.23)}$$

The heat balance on the cold (air) fluid is

$$Q_h = (wC_p \Delta T)_c = 2.9(1006)(29.4 - 15.6) = 40 \text{ kW} (= 1.4 \times 10^5 \text{ Btu/h})$$
$$= 4 \times 10^4 \text{ W} \qquad \textbf{(13.4.24)}$$

*We do not know which fluid is the minimum.* Assume air (for which we can calculate $C_c$) is the minimum fluid:

$$(wC_p)_c = C_c = 2.9 \, (1006) = 2900 \text{ W/°C} \qquad \textbf{(13.4.25)}$$

$$\text{NTU}_{max} = \frac{UA}{C_{min}} = \frac{2110}{2900} = 0.73 \qquad \textbf{(13.4.26)}$$

$$\varepsilon = \frac{\Delta T_{air}}{\Delta T_{max}} = \frac{29.4 - 15.6}{82.2 - 15.6} = 0.21 \qquad \textbf{(13.4.27)}$$

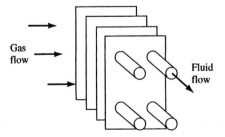

Gas
flow

Fluid
flow

**Figure 13.4.4** Cross-flow exchanger with both fluids unmixed.

Using the chart of $\varepsilon$ (versus NTU) for a cross-flow unmixed exchanger (Fig. 13.4.2), we are unable to find a point on any curves for $0 \leq C_{min}/C_{max} \leq 1$. *Hence the hot fluid must be minimum.* Thus we must do trial and error, assuming a value for $w$ for the water flow. We have the following relationships:

$$C_{max} = (wC_p)_c = 2900 \text{ W/°C} \tag{13.4.28}$$

$$\text{NTU} = \frac{UA}{C_{min}} = \frac{2110}{C_{min}} \tag{13.4.29}$$

$$\varepsilon = \frac{\Delta T_h}{\Delta T_{max}} = \frac{\Delta T_h}{82.2 - 15.6} = \frac{\Delta T_h}{66.6} \tag{13.4.30}$$

and, from a heat balance on the hot side,

$$\Delta T_h = \frac{4 \times 10^4 (\text{W})}{C_{min}(\text{W/°C})} = \frac{4 \times 10^4}{C_h} \tag{13.4.31}$$

We have *two unknowns:* $C_{min}$ (based on the water flowrate) and $\Delta T_h$ (which has the unknown exit water temperature). We will solve this set of equations by trial and error.

The procedure is as follows. Assume a value for $C_{min}$. Calculate NTU and $\Delta T_h$, based on the assumed value. Calculate $\varepsilon$ from Eq. 13.4.30, and check your result against the value found from Fig. 13.4.2, the graph for this exchanger. Continue this process until a converged value of $\varepsilon$ has been found. Table 13.4.1 shows a few iterations. Thus (after one final iteration) we estimate $C_{min} = (wC_p)_h = 645$ W/°C and so, with $C_p = 4180$,

$$\text{NTU} = \frac{UA}{C_{min}} = \frac{2110}{645} = 3.27 \qquad w_h = \frac{645}{4180} = 0.15 \text{ kg/s (1220 lb/h)} \tag{13.4.32}$$

$$T_{w,exit} = 82.2 - \frac{4 \times 10^4}{645} \qquad [\text{from } (wC_p \Delta T)_h = 4 \times 10^4] \tag{13.4.33}$$

$$= 20°C$$

In Example 13.4.2 we take a problem that involves unknown temperatures and compare the ease of solution by two methods: use of the $F(R, P)$ graphs and use of the effectiveness [NTU($\varepsilon$)] method.

---

**EXAMPLE 13.4.2** *Comparison of F(R,P) and NTU($\varepsilon$) (Both Flowrates Specified; Two Unknown Temperatures)*

Oil ($C_p = 0.5$ Btu/lb·°F) enters an exchanger at 200°F and 4000 lb/h. Cooling water at 50°F is available. An exchanger with one baffled shell pass and two tube passes is

**Table 13.4.1** Trial and Error Solution for $\varepsilon$

| $\dfrac{C_{min}}{C_{max}}$ | $C_{min}$ | NTU | $\Delta T_h$ | $\varepsilon$ From Fig. 13.4.2 | From Eq. 13.4.30: $\Delta T_h/\Delta T_{max}$ |
|---|---|---|---|---|---|
| 0.5 | 1450 | 1.45 | 28 | 0.65 | 0.42 |
| 0.25 | 725 | 2.9 | 56 | 0.89 | 0.84 |
| 0.22 | 640 | 3.3 | 63 | 0.93 | 0.95 |

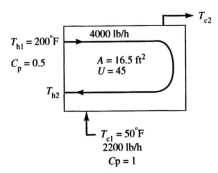

**Figure 13.4.5** Specifications for Example 13.4.2 (the units for $U$ are Btu/h · ft$^2$ · °F).

available, having an area of 16.5 ft$^2$. If the water flowrate is 2200 lb/h, find the exit temperatures. Take $U = 45$ Btu/h · ft$^2$ · °F. In this example the two flowrates are specified in Fig. 13.4.5, but we have two unknown temperatures.

Using the $F(R,P)$ method, trial and error is required. We begin with a guess for $T_{c2}$. With this initial guess:

1. Find $Q^*_{hc} = wC_p(T_{c2} - T_{c1})$
2. Find $T_{h2} = T_{h1} - (wC_p)_c/(wC_p)_h (T_{c2} - T_{c1})$
3. Calculate $R$ and $P$.
4. Find $F$ from Fig. 13.3.2.
5. Calculate $Q_h = UA \, \Delta T_{ln} F$.
6. Compare $Q_h$ to $Q^*_{hc}$.
7. Make a new guess on $T_{c2}$ until $Q_h = Q^*_{hc}$.

The result of carrying out these steps is:

$$T_{c2} = 87°F$$
$$T_{h2} = 160°F$$

A better way to do this type of design problem uses the (NTU, $\varepsilon$) method but does not require trial and error. We begin with the given data, from which we find

$$(wC_p)_{oil} = 4000 \, (0.5) = 2000 \qquad (wC_p)_{water} = 2200 \, (1) = 2200$$

We see that oil is the minimum fluid.

$$\text{NTU} = \frac{AU}{C_{min}} = \frac{16.5 \, (45)}{2000} = 0.37 \qquad \textbf{(13.4.34)}$$

$$\frac{C_{min}}{C_{max}} = 0.91 \qquad \textbf{(13.4.35)}$$

Then taking $\varepsilon$ from Fig. 13.4.3, we write:

$$\varepsilon = 0.25 = \frac{\Delta T_h}{\Delta T_{max}} = \frac{200 - T_{h2}}{200 - 50} \qquad \textbf{(13.4.36)}$$

$$T_{h2} = 200 - 0.25 \, (150) = 162.5°F \qquad \textbf{(13.4.37)}$$

$$T_{c2} = T_{c1} + \frac{C_{min}}{C_{max}} (T_{h1} - T_{h2}) = 50 + 0.91 \, (37.5) = 84.1°F \qquad \textbf{(13.4.38)}$$

The differences in values compared to the preceding method arise from inaccuracies in reading the graphs. Note that when both flowrates are specified, trial and error is not required with the (NTU, $\varepsilon$) method, as it would be with the $F(P,R)$ method.

---

**EXAMPLE 13.4.3**   *Performance of a Heat Exchanger*

From a performance test on a well-baffled, single-shell, two-tube-pass heat exchanger, the following data are available: oil ($C_p$ = 2100 J/kg·K) in turbulent flow inside the tubes entered at 340 K at the rate of 1.00 kg/s and left at 310 K; water flowing at an unknown rate on the shell side entered at 290 K and left at 300 K. A change in service conditions requires the cooling of a similar oil from an initial temperature of 370 K but at three-fourths of the oil flowrate used in the performance test. Estimate the outlet temperature of the oil for the same water rate and inlet conditions as before.

**Solution**

The test data may be used to determine the heat capacity rate of the water and the overall conductance (the $UA$ product) of the exchanger. The *heat capacity rate* (the value of $C$) of the water is, from a heat balance,

$$(wC_p)_c = C_c = C_h \frac{T_{h,in} - T_{h,out}}{T_{c,out} - T_{c,in}} = (1.00)(2100)\frac{340 - 310}{300 - 290} \qquad \textbf{(13.4.39)}$$
$$= 6300 \text{ W/K}$$

and the temperature ratio $P$ is

$$P = \frac{T_{t,out} - T_{t,in}}{T_{s,in} - T_{t,in}} = \frac{310 - 340}{290 - 340} = 0.6 \qquad \textbf{(13.4.40)}$$

The capacity ratio (not *rate*) is

$$R = \frac{300 - 290}{340 - 310} = 0.33 \qquad \textbf{(13.4.41)}$$

From Fig. 13.3.2, $F$ = 0.94 and

$$(F)(\Delta T_{ln}) = (0.94)\frac{(340 - 300) - (310 - 290)}{\ln[(340 - 300)/(310 - 290)]} = 27.1 \text{ K} \qquad \textbf{(13.4.42)}$$

From the design equation, the overall conductance $UA$ is

$$UA = \frac{Q_h}{F \Delta T_{ln}} = \frac{(1.00)(2100)(340 - 310)}{27.1} = 2525 \text{ W/K} \qquad \textbf{(13.4.43)}$$

Since the thermal resistance on the oil side is probably controlling, a decrease in velocity to 75% of the original value will increase the thermal resistance roughly by the velocity ratio raised to the 0.8 power. This can be verified by reference to the Dittus–Boelter equation (Eq. 12.4.50). Under the new conditions the conductance, the NTU, and the heat capacity rate ratio will therefore be approximately as follows:

$$UA \approx (2325)(0.75)^{0.8} = 1850 \text{ W/K} \qquad \textbf{(13.4.44)}$$

$$NTU = \frac{UA}{C_{oil}} = \frac{1850}{(0.75)(1.00)(2100)} = 1.17 \qquad \textbf{(13.4.45)}$$

$$\frac{C_{oil}}{C_{water}} = \frac{C_{min}}{C_{max}} = \frac{(0.75)(1.00)(2100)}{6300} = 0.25 \qquad \textbf{(13.4.46)}$$

From Fig. 13.4.3 the effectiveness is equal to 0.61. Hence from the definition of $\varepsilon$, the oil outlet temperature is

$$T_{\text{oil out}} = T_{\text{oil in}} - \varepsilon \, \Delta T_{\text{max}} = 370 - [0.61 \, (370 - 290)] = 321 \text{ K} \qquad \textbf{(13.4.47)}$$

### EXAMPLE 13.4.4  *Heat Exchanger with Condensation*

A heat exchanger (condenser) using saturated steam from the exhaust of a turbine at a pressure of 4.0 in.Hg abs. is to be used to heat 25,000 lb/h of seawater ($C_p = 0.95$ Btu/lb·°F) from 60°F to 110°F. The exchanger is to be sized for one shell pass and four tube passes with 60 parallel tube circuits of 0.995 inch-i.d. and 1.125-inch-o.d. brass tubing ($k = 60$ Btu/h·ft·°F). For the exchanger the average heat transfer coefficients on the steam and water sides are estimated to be 600 and 300 Btu/h·ft²·°F, respectively. Calculate the tube length required.

The exchanger, described as 1 shell pass/4 tube passes with 60 parallel tubes, is configured as in Fig. 13.4.6. Note the definition of the length of a tube, $L$.

**Solution**

At 4.0-in.Hg abs the temperature of condensing steam (we need thermodynamic data here, such as are found in Steam Tables) will be 125.4°F, so that the required effectiveness of the exchanger is

$$\varepsilon = \frac{T_{\text{c,out}} - T_{\text{c,in}}}{T_{\text{h,in}} - T_{\text{c,in}}} = \frac{110 - 60}{125.4 - 60} = 0.765 \qquad \textbf{(13.4.48)}$$

For a condenser, with the condensate on the hot side, we can show (see Problem 13.19) that $C_{\text{min}}/C_{\text{max}} = 0$ and from Figure 13.4.3, NTU = 1.4. The overall heat transfer coefficient per unit outside area of tube is (see Eq. 10.2.33)

$$U_{\text{o}} = \left[ \frac{1}{600} + \frac{1.125}{2 \times 12 \times 60} \ln\left(\frac{1.125}{0.995}\right) + \frac{1.125}{300 \times 0.995} \right]^{-1} = 180 \text{ Btu/h·ft}^2\text{·°F}$$

$$\textbf{(13.4.49)}$$

The total area $A_{\text{o}}$ is $60\pi D_{\text{o}} L$ and, since $U_{\text{o}} A_{\text{o}}/C_{\text{min}} = \text{NTU} = 1.4$, the length of the tube pass is

$$L = \frac{1.4 \times 25,000 \times 0.95 \times 12}{60 \times \pi \times 1.125 \times 180} = 10 \text{ ft} \qquad \textbf{(13.4.50)}$$

Each tube bank has four 10-foot tubes in series.

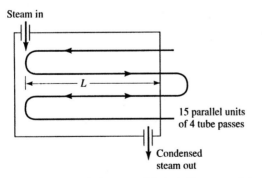

**Figure 13.4.6** Exchanger with 1 shell pass and 4 tube passes; 60 parallel tubes are arranged in 15 banks of 4 passes each.

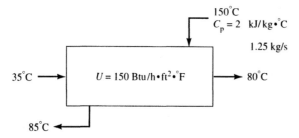

**Figure 13.4.7** A counterflow double-pipe heat exchanger.

### *EXAMPLE 13.4.5*   *Series/Parallel Exchangers*

A counterflow double-pipe heat exchanger (see Fig. 13.4.7) is used to heat 1.25 kg/s of water from 35°C to 80°C by cooling an oil ($C_p$ = 2 kJ/kg·°C) from 150°C to 85°C. The overall heat transfer coefficient is 150 Btu/h·ft²·°F. A similar arrangement is to be built at another plant location, but it is desired to compare the performance of the single counterflow heat exchanger with two smaller counterflow heat exchangers connected in series on the water side and in parallel on the oil side, as shown in Fig. 13.4.8. The oil flow is split equally between two exchangers, and it may be assumed that the overall heat transfer coefficient for the smaller exchangers is the same as for the large exchanger. If the smaller exchangers cost 20% more per unit surface area, which would be the more economical arrangement—the one large exchanger or the two equal-sized small exchangers?

### *Analysis of a Single Exchanger*

First we convert the heat transfer coefficient to SI units, using the factor 5.68 to multiply $h$ in Btu/h·ft²·°F to yield $h$ = 850 W/m²·°C. We write an overall heat balance to find the hot side flowrate. The heat duty is given by

$$
\begin{aligned}
Q_h &= (wC_p\,\Delta T)_c = (wC_p\,\Delta T)_h \\
&= (1.25)(4180)(80 - 35) = (wC_p)_h\,(150 - 85) \\
&= 2.35 \times 10^5 \text{ W}
\end{aligned} \tag{13.4.51}
$$

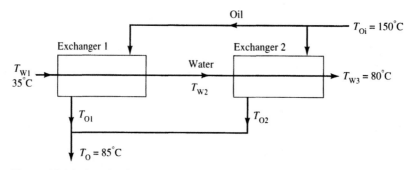

**Figure 13.4.8** A pair of exchangers connected in series on the cold side and in parallel on the hot side.

Then we find

$$(wC_p)_h = 3617 \text{ W/°C} \tag{13.4.52}$$

and

$$(wC_p)_c = 5222 \text{ W/°C} \tag{13.4.53}$$

Hence oil (the hot fluid) is the minimum fluid, and we need

$$\varepsilon_h = \frac{\Delta T_h}{150 - 35} = \frac{150 - 85}{150 - 35} = 0.57 \tag{13.4.54}$$

and

$$C_{min}/C_{max} = 0.7 \tag{13.4.55}$$

Using an $\varepsilon$(NTU) chart for a counterflow double-pipe exchanger, we find

$$\text{NTU} = 1.1 \tag{13.4.56}$$

Hence the area is

$$A = (\text{NTU})\frac{C_{min}}{U} = \frac{(1.1)(3620)}{850} = 4.7 \text{ m}^2 \text{ (50 ft}^2\text{)} \tag{13.4.57}$$

### Analysis of the Two Exchangers

We still have the same $C_{min}$ but it is split in two.
For each exchanger, $C_{min} = 1810$ W/°C. As before, $C_{max} = 5222$ W/°C. Now we find

$$C_{min}/C_{max} = 0.35 \tag{13.4.58}$$

NTU is the same for each exchanger of this configuration, but different from that for the single exchanger configuration. Hence, $\varepsilon_1 = \varepsilon_2$.

$$\varepsilon_1 = \frac{150 - T_{O1}}{150 - 35} = \varepsilon_2 = \frac{150 - T_{O2}}{150 - T_{W2}} \tag{13.4.59}$$

The constraint of an 85°C outlet is now applied to the mixture of *equal* flows from the two exchangers, with temperatures $T_{O1}$ and $T_{O2}$. Hence the mixed temperature is the mean, and $T_{O1}$ and $T_{O2}$ are connected by

$$\frac{T_{O1} + T_{O2}}{2} = 85 \tag{13.4.60}$$

An *energy balance* around exchanger 2 (see Fig. 13.4.8) gives

$$5230\,(80 - T_{W2}) = 1810\,(150 - T_{O2}) \tag{13.4.61}$$

We have three unknowns: $T_{O1}$, $T_{O2}$, $T_{W2}$, and they are found from these three last equations with the results:

$$T_{O1} = 77°C \qquad T_{O2} = 93°C \qquad T_{W2} = 60°C \tag{13.4.62}$$

Then we find the following:

$$\varepsilon_1 = \varepsilon_2 = 0.64 \tag{13.4.63}$$

$$\text{NTU} = 1.3 \text{ (from Fig. 13.4.1)} \tag{13.4.64}$$

$$A = \text{NTU}\frac{C_{min}}{U} = \frac{1.3(1810)}{850} = 2.8 \text{ m}^2 \tag{13.4.65}$$

for *each* exchanger. The total required area is 5.6 m², which is greater than that found for the single exchanger (4.7 m²). Hence the single exchanger is more economical.

### EXAMPLE 13.4.6    *Exchanger with a Bleed-Off Stream*

A counterflow double-pipe heat exchanger is used to heat 20,000 lb$_m$/h of water from 80°F to 150°F by cooling an oil ($C_p = 0.5$ Btu/lb$_m \cdot$°F) from 280°F to 200°F. It is desired to "bleed off" 5000 lb$_m$/h of water at 120°F, so the single exchanger is to be replaced by a two-exchanger arrangement that will permit this. The overall heat transfer coefficient is 80 Btu/h$\cdot$ft²$\cdot$°F for the single exchanger, and the same value may be taken for each of the smaller exchangers. The same total oil flow is used for the two-exchanger arrange-ment, except that the flow is split (but not equally) between the two exchangers. Deter-mine the areas of each of the smaller exchangers and the oil flowrates through each. Assume that the water flows in series through the two exchangers, with the "bleed-off" taking place between them. Assume the smaller exchangers have equal areas. Figure 13.4.9 gives the details.

Our set of working equations is the following:

$$Q_{h1} = 20,000 \,(1)\, 40 = 800,000 = w_1(0.5)(280 - T_1) \qquad \textbf{(13.4.66)}$$

$$Q_{h2} = 15,000 \,(1)\, 30 = 450,000 = w_2(0.5)(280 - T_2) \qquad \textbf{(13.4.67)}$$

$$w_1 + w_2 = 35,000 \qquad \textbf{(13.4.68)}$$

$$Q_{h1} = UAF\,\Delta T_{\ln 1} \qquad \textbf{(13.4.69)}$$

$$Q_{h2} = UAF\,\Delta T_{\ln 2} \qquad \textbf{(13.4.70)}$$

($F = 1$ if both exchangers are run as simple double-pipe heat exchangers.)

We have five equations with five unknowns ($w_1$, $w_2$, $T_1$, $T_2$, $A$). We note that $Q_{h1} \approx 2Q_{h2}$, and thus we expect to find $w_1 \approx 2w_2$. We will use this estimate to start a trial-and-error method:

Pick $w_1 = 23,000$ and $w_2 = 12,000$. (They must sum to 35,000. Otherwise, we have just made a guess of two values that are *about* a factor of 2 apart.) With this guess we find

$$\left. \begin{matrix} C_{\text{water 1}} = 20,000 & C_{\text{oil 1}} = 11,500 \\ C_{\text{water 2}} = 15,000 & C_{\text{oil 2}} = 6,000 \end{matrix} \right\} \text{min fluid} \qquad \textbf{(13.4.71)}$$

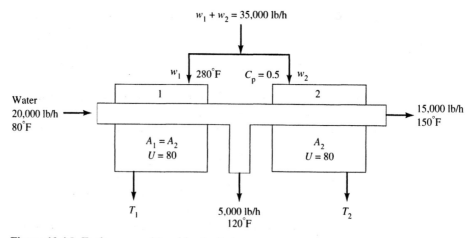

**Figure 13.4.9** Exchangers with a bleed-off stream.

and oil is the minimum fluid. From their definitions we have

$$\varepsilon_1 = \frac{280 - T_1}{280 - 80} \qquad (13.4.72)$$

$$\varepsilon_2 = \frac{280 - T_2}{280 - 120} \qquad (13.4.73)$$

From the definitions

$$\mathrm{NTU}_1 = \frac{UA}{C_{\min 1}} \qquad (13.4.74)$$

and

$$\mathrm{NTU}_2 = \frac{UA}{C_{\min 2}} \qquad (13.4.75)$$

we find

$$\frac{\mathrm{NTU}_1}{\mathrm{NTU}_2} = \frac{w_2}{w_1} = \frac{35{,}000}{w_1} - 1 \qquad (13.4.76)$$

The trial-and-error procedure now follows along these lines: having selected a guess for $w_1$, we use Eq. 13.4.66 to find $T_1$. From Eq. 13.4.68 we find $w_2$, and use Eq. 13.4.67 to get $T_2$. Now we can calculate $\varepsilon$ values from Eqs. 13.4.72 and 13.4.73. With these $\varepsilon$ values we use Fig. 13.4.1 to find the two NTUs. If we guessed the right value for $w_1$ in the first place, the calculated values of NTU will satisfy Eq. 13.4.76. If they don't, we make a new guess on $w_1$ and continue the process until we converge to a satisfactory value of $w_1$. Then the other unknowns follow from the equations above.

When this procedure is carried out, the final results are

$$w_1 = 26{,}700 \text{ lb/h} \qquad (13.4.77)$$
$$A_1 = A_2 = 67 \text{ ft}^2 \qquad (13.4.78)$$
$$T_1 = 220°F \qquad (13.4.79)$$
$$T_2 = 172°F \qquad (13.4.80)$$

## 13.5 OPTIMUM OUTLET TEMPERATURE

In many heat exchanger problems we deal not with the design of the exchanger, but rather with the specification of operating conditions. The exchanger may already be in place, and the task is to specify operating conditions (typically) for one stream. For example, we may have a required flowrate of one stream, and the temperature change of that stream may be specified by the details of the application. Since the heat duty ($Q_h$ of the specified stream) is fixed for one stream, a heat balance demands that the heat duty be the same for the second stream. However, this only constrains the *product* of flowrate and temperature change of the second stream. We raise the question now: On what basis might we make a rational choice of operating conditions of the second stream? Usually this choice is dictated by economics.

In this section we define and solve a simple optimization problem for a heat exchanger. For the sake of example we choose a simple countercurrent double-pipe heat exchanger, and we label the variables as in Fig. 13.5.1.

We assume that operating conditions on one stream are fixed and that the three unknowns are as follows: the flowrate and outlet temperature of the second stream,

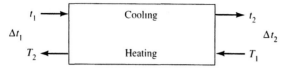

**Figure 13.5.1** A simple double-pipe heat exchanger.

and the area of the exchanger. These three unknowns (or "degrees of freedom") must satisfy two equations: a heat balance

$$Q_h = wc_p(t_2 - t_1) = WC_p(T_1 - T_2) \qquad (13.5.1)$$

and a design or performance equation

$$Q_h = UA \frac{(T_1 - t_2) - (T_2 - t_1)}{\ln\left[(T_1 - t_2)/(T_2 - t_1)\right]} \qquad (13.5.2)$$

(Note the change in notation from the examples of Section 13.4.)

In prior examples a third equation was supplied, in the form of a specification of one more operating condition—temperature, flowrate, or area. (We assume that $U$ is known in the range of acceptable operating conditions; otherwise, to keep the equations in balance, we add a relationship for $U$ as a function of flowrates of the two streams.) In this example the additional equation (constraint) enters in the form of a set of statements regarding the *economics* of the operation. We suppose that the "unknown" stream is the cooling water for this exchanger. One economic factor is the cost of this water, which includes the price paid to purchase the water, as well as the cost of any required pretreatment (e.g., to remove minerals that could deposit and foul the exchanger), and the "sewerage" costs of disposing of used water. We will assume that these costs are proportional to the amount of water used, and write a model for the annual cost of water (ACW) in the form

$$ACW = C_w w T_A \; (\$/\text{yr}) \qquad (13.5.3)$$

where

$$C_w = \text{cost of water/lb} \qquad (13.5.4)$$

and

$$T_A = \text{operating time/year} \qquad (13.5.5)$$

Note that $C_w$ is not a heat capacity, and $T_A$ is not a temperature!

The capital cost of a heat exchanger is taken to be in the form

$$ACC = C_A A \; (\$/\text{yr}) \qquad (13.5.6)$$

(We assume that this cost is linear in $A$ over a small range in $A$. A more realistic model would be a power function.) In this expression $C_A$ is the so-called *annualized* capital cost (per area). A capital cost is an expenditure to purchase an item. It is normally a "one-time" cost in the sense that when the item is received, we pay the bill. Hence it is measured in units of dollars. But we need a way to add capital costs to the (nearly) *continuous* costs of operating a process. These operating costs are measured and reported as a rate: $/yr.

It is common to *annualize* a capital cost. It is (roughly) the cost in dollars per year as if we were purchasing the capital item through continuous payments. (A crude example would be "buying" a house or a car through a loan agreement. We can then add the monthly payments to the monthly costs for power, maintenance, insurance, and replacement, and in that way state the annualized capital cost of the house or car in units of dollars per year.)

The additional costs of maintenance and depreciation of the equipment, including peripheral equipment such as pumps and valves and sensors and controllers) are usually taken as a fraction of annualized capital costs.

$$C_{MD} = f\, ACC \tag{13.5.7}$$

The total fixed costs (TFC) of the equipment (annualized) may be written, then, as

$$TFC \equiv (1+f)\, C_A \tag{13.5.8}$$

Note that this is written on a *per area* basis. The total annual cost model then takes the form

$$TAC = C_w w T_A + (1+f) C_A A \tag{13.5.9}$$

We now eliminate the flowrate $w$ and the area $A$ through the heat balance and design equations (Eqs. 13.5.1 and 13.5.2) with the result that our economic model takes the form

$$TAC = \frac{(C_w T_A) Q_h}{c_p (t_2 - t_1)} + \frac{TFC\, Q_h}{U \dfrac{(T_1 - t_2) - (T_2 - t_1)}{\ln\left[(T_1 - t_2)/(T_2 - t_1)\right]}} \tag{13.5.10}$$

Our goal is to minimize the cost of running this system. There will be a cooling water flowrate that leads to minimum cost. In the model, as we have written it, the minimum flowrate will be reflected in an optimum outlet water temperature. To find the optimum outlet water temperature, we set $\partial\,TAC/\partial t_2 = 0$. Note that the area does not appear in the algebra that follows. After some algebraic rearrangement, we have an expression for $t_2$ in terms of the parameters of the system:

$$0 = \frac{C_w T_A U}{c_p TFC} \frac{-1}{(t_2 - t_1)^2} + \frac{-1}{(T_1 - t_2) - \Delta t_1} \frac{\Delta t_1}{T_1 - t_2} + \ln\left(\frac{T_1 - t_2}{\Delta t_1}\right) \frac{-1}{(T_1 - t_2 - \Delta t_1)^2} \tag{13.5.11}$$

where we have defined

$$\Delta t_2 = T_1 - t_2 \tag{13.5.12}$$
$$\Delta t_1 = T_2 - t_1$$

We define the following dimensionless groups:

$$G_1 = \frac{C_w U T_A}{TFC\, c_p} \tag{13.5.13}$$

$$G_2 = \frac{\Delta t_2}{\Delta t_1} \tag{13.5.14}$$

$$G_3 = \frac{T_1 - T_2}{\Delta t_1} \tag{13.5.15}$$

Finally we may write Eq. 13.5.11 as

$$G_1 \left(\frac{G_2 - 1}{G_2 - 1 - G_3}\right)^2 = \ln G_2 - \left(1 - \frac{1}{G_2}\right) \tag{13.5.16}$$

Figure 13.5.2 shows $G_2$ versus $G_1$, with $G_3$ as a parameter. This graph may then be used to find the optimum operating conditions, as Example 13.5.1 shows.

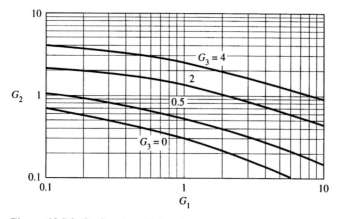

**Figure 13.5.2** Optimum outlet water temperature.

### EXAMPLE 13.5.1   *Calculation of the Optimum Outlet Water Temperature*

A viscous fluid is cooled from 175°F to 150°F with water available at 100°F. What is the optimum outlet water temperature? Using Fig. 13.5.1 as a guide, we find

$$175 - t_2 = \Delta t_2 \tag{13.5.17}$$

$$150 - 100 = \Delta t_1 = 50 \tag{13.5.18}$$

It will first be necessary to assume a value of $U$. Since the material is viscous, assume a low value for the overall coefficient. For the sake of example we take an arbitrary value of $U = 15$ Btu/ft$^2 \cdot$h$ \cdot$°F. To evaluate $G_1$ we need the following "data."

Assume we know that

$T_A$ = 8000 operating hours annually
$C_w$ is computed at $0.01/1000 gal = 0.01/8300 \$/lb

Take the annualized capital cost factor (Eq. 13.5.6) as

$$C_A = 0.92 \ \$/\text{ft}^2 \cdot \text{yr} \tag{13.5.19}$$

For additional annual charges, assume 20% repair and maintenance and 10% depreciation. The annualized fixed charge is then

$$\text{TFC} \equiv (1 + f)C_A = 1.3\,(0.92) = \$1.20/\text{ft}^2 \cdot \text{yr} \tag{13.5.20}$$

The specific heat of water is taken as 1.0 Btu/lb. We find

$$G_1 = \frac{15 \times 8000}{1.20 \times 1.0}\left(\frac{0.01}{8300}\right) = 0.12 \tag{13.5.21}$$

For the conditions specified, we find

$$G_3 = \frac{T_1 - T_2}{\Delta t_1} = \frac{175 - 150}{150 - 100} = 0.5 \tag{13.5.22}$$

From Fig. 13.5.2, we find

$$G_2 \equiv \frac{\Delta t_2}{\Delta t_1} = 1 \tag{13.5.23}$$

Then

$$\Delta t_2 \equiv T_1 - t_2 = 1 \times 50 = 50°F \qquad (13.5.24)$$

and the optimum outlet temperature is

$$t_2 = 175 - 50 = 125°F \qquad (13.5.25)$$

If we specify a particular flowrate for the viscous fluid, we can go on to find the required exchanger area $A$ for this optimum $t_2$, and the water flowrate as well.

## 13.6  ANALYSIS AND DESIGN OF A SYSTEM FOR HEATING A GELATIN SOLUTION

We end this chapter with a design problem that illustrates the integration of theoretical analysis with empirical correlations to produce design and operating recommendations for a specific thermal system. Figure 13.6.1 shows the system of interest. A batch of a viscous aqueous solution must be heated from some initial storage temperature $T_o$ to a final process temperature $T_f$. Hot water is available, at temperature $t_o$, and the hot stream is pumped through a heat exchanger to raise the temperature of the solution. The aqueous solution is heated in a batch mode, which means that the aqueous solution is circulated continuously through the heat exchanger. The heating stream makes a single pass through the exchanger and leaves the system at a temperature $t_f(t)$. Our notation reflects the expectation that as the process stream heats up, the outlet temperature of the heating fluid will change with time.

### 13.6.1  Transient Analysis of the Batch System

We can begin with a simple unsteady state thermal balance on the process batch.

$$MC_p \frac{dT}{dt} = WC_p(T_2 - T_1) \qquad (13.6.1)$$

Here, $M$ is the mass of solution in the batch, $W$ is the mass flowrate of the solution, and $C_p$ is the heat capacity (per unit mass) of the solution; $T_2$ is the temperature of the solution as it returns from the outlet of the heat exchanger, and $T_1$ is the temperature of the solution leaving the process vessel and entering the heat exchanger. We neglect all thermal losses in the process lines themselves. We will assume that the batch is very

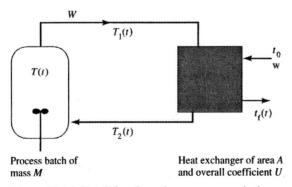

Figure 13.6.1 Batch heating of an aqueous solution.

well mixed, and therefore we will take

$$T(t) = T_1(t) \tag{13.6.2}$$

The heat exchanger equation will be written in the form

$$FUA \, \Delta T_{\text{ln}} = w c_{\text{p}}(t_{\text{o}} - t_{\text{f}}) \tag{13.6.3}$$

(Since we do not specify the type of exchanger here, $F$, as defined in Eq. 13.3.1, accounts for the particular choice that might be made.) Finally, with the assumption that heat exchange occurs between the two fluids only (i.e., there are no losses to the surrounding environment), we know that

$$WC_{\text{p}}(T_2 - T_1) = w c_{\text{p}}(t_{\text{o}} - t_{\text{f}}) \tag{13.6.4}$$

These equations are sufficient for us to solve for the temperature $T(t)$. We use the initial condition

$$T = T_{\text{o}} \qquad \text{at} \quad t = 0 \tag{13.6.5}$$

In the simplest case, the hot fluid outlet $t_{\text{f}}$ remains very near the inlet temperature $t_{\text{o}}$. This could be achieved, for example, by using condensing steam on the hot side, or by operating under conditions that $w c_{\text{p}} \gg WC_{\text{p}}$, so that a very small temperature drop occurs on the hot side.

Under this assumption, we replace Eqs. 13.6.3 and 13.6.4 by

$$WC_{\text{p}}(T_2 - T) = FUA \, \Delta T_{\text{ln}} = FUA \, \frac{(t_{\text{o}} - T) - (t_{\text{o}} - T_2)}{\ln\left[(t_{\text{o}} - T)/(t_{\text{o}} - T_2)\right]} \tag{13.6.6}$$

Note that in writing $\Delta T$ we use the equivalent countercurrent definition, consistent with our discussion of the factor $F$ in Section 13.3.

Equation 13.6.6 permits us to find $T_2$ in terms of $T$ as

$$T_2 = t_{\text{o}} - \frac{t_{\text{o}} - T}{K} \tag{13.6.7}$$

where

$$K \equiv \exp\left(\frac{FUA}{WC_{\text{p}}}\right) \tag{13.6.8}$$

We see from Eq. 13.6.7 that $K$ is a measure of the efficiency of the exchanger, since the approach of $T_2$ to $t_{\text{o}}$ gets small as $K$ increases. This is not surprising, since $FUA/WC_{\text{p}}$ is similar to NTU, defined earlier.

Upon elimination of $T_2$, we write the differential equation for $T$ in the form

$$MC_{\text{p}} \frac{dT}{dt} = \frac{(K - 1)WC_{\text{p}}}{K}(t_{\text{o}} - T) \tag{13.6.9}$$

The resulting solution for $T(t)$ may be found as

$$-\ln \Theta \equiv -\ln\left(\frac{t_{\text{o}} - T}{t_{\text{o}} - T_{\text{o}}}\right) = \frac{W(K - 1)}{MK}t \tag{13.6.10}$$

Consequently we expect a plot of $\ln \Theta$ versus time on arithmetic coordinates to yield a straight line, the slope of which will give a value for $K$, from which $FUA$ may then be obtained. The value of this model can be clarified in Example 13.6.1, where we examine data for an exchanger for which $F$ is unknown and $A$ is not well defined.

**EXAMPLE 13.6.1** *Performance of a Plate Heat Exchanger*

A plate heat exchanger is available for use in a pilot plant, but we know nothing about the internal area for transfer or the magnitude of the overall heat transfer coefficient or its dependence on flowrate. The exchanger is old, and the outside appearance suggests that it might be a mistake to try to determine the internal area by disassembling the unit. The seals are old, and the possibility of leaks in the reassembled system is a concern. It is decided to carry out a few basic experiments that will provide data on the exchanger's overall characteristics. In particular, we would like to determine its effective $UA$ product—actually, $FUA$—at various flowrates of each stream.

A schematic of a plate heat exchanger is shown in Fig. 13.6.2. The heat transfer surfaces are parallel plates, and plumbing and gasketing are designed to provide counter-current flow of the hot and cold streams. Usually the plates have an embossed chevron pattern, which promotes heat transfer.

An initially cold batch of water ($M = 189.3$ kg, $C_p = 4180$ J/kg) is circulated through the plate heat exchanger at a flowrate of 1 gal/min. Hot water is pumped rapidly (3.8 gal/min) through the hot side of the exchanger, and it is observed that the hot water outlet temperature is very close to the hot water inlet temperature, under these conditions. The temperature $T(t)$ of the cold water at the cold water outlet is measured over a short period of time, and the data are plotted in Fig. 13.6.3, in accordance with Eq. 13.6.10. A least-squares fit of the data yields a slope of $1.47 \times 10^{-4}$ s$^{-1}$, so

$$\frac{(K-1)W}{MK} = 1.47 \times 10^{-4}\text{s}^{-1} \tag{13.6.11}$$

Converting the flowrate of the cold water from 1 gal/min to $W = 0.063$ kg/s, we can solve this equation for $K$ and find

$$K = 1.79 \tag{13.6.12}$$

From the definition of $K$ we then find

$$FUA = WC_p(\ln K) = 0.063(4180)(0.58) = 153 \text{ J/s} \cdot \text{K} \tag{13.6.13}$$

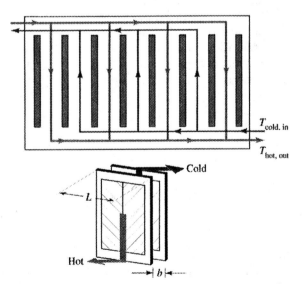

**Figure 13.6.2** Schematic of a plate heat exchanger (parallel counterflow).

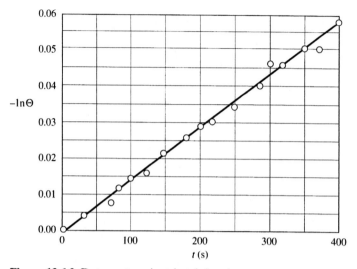

**Figure 13.6.3** Data on transient batch heating.

With this important characteristic of the exchanger established at this one pair of flowrates, and with additional such experiments at other flowrates, we could analyze a proposed batch heating process. What is most useful to us in the long term is the determination of the individual hot and cold side coefficients as functions of the flowrates of each stream. Hence we turn next to that task.

### 13.6.2 Determination of Individual Side Coefficients

For a plate heat exchanger, the relationship among the individual resistances is given by

$$\frac{1}{U} = \frac{1}{h_h} + \frac{1}{h_c} \tag{13.6.14}$$

where we denote the individual hot and cold side coefficients as $h_h$ and $h_c$. We neglect the usually small thermal resistance of the plates themselves, relative to the convective resistances. The information we need is the individual coefficients $h_h$ and $h_c$, as functions of flow conditions, for various fluids. From Eq. 13.6.14 we see that there are two relatively simple ways to gain this information. One is to operate one side (say, the hot side) at such a high flow rate that $h_h$ is very large. Then it follows that

$$\frac{1}{U} \approx \frac{1}{h_c} \tag{13.6.15}$$

Alternatively, if the flow conditions are identical on both the hot and cold sides, and if the plate surfaces are also identical on each side, then the heat transfer coefficients are identical on each side, and

$$\frac{1}{U} = \frac{1}{h_h} + \frac{1}{h_c} = \frac{2}{h_c} \tag{13.6.16}$$

Data have been obtained using the first of these methods (Eq. 13.6.15). The procedure described in Example 13.6.1 yielded values for $FU$ over a range of flow rates. (The area $A$ of the exchanger was known.) The hot side fluid was water, and the cold side fluid was an aqueous gelatin solution. Various concentrations of gelatin produced a set of solutions for which the viscosity varied over a threefold range. The data are shown in

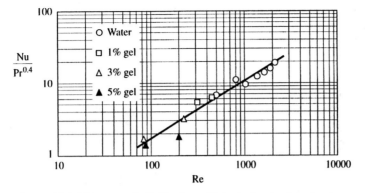

**Figure 13.6.4** Data on individual coefficients for a plate heat exchanger.

Fig. 13.6.4, nondimensionalized in the usual manner. The Reynolds number for the cold fluid is defined as (see Problem 13.33)

$$\text{Re} = \frac{4W}{(n_T - 1)\mu L}$$ (13.6.17)

where $L$ is the width of a plate in the direction *normal* to the flow, and $W$ is the total mass flowrate of the cold fluid. (For the hot fluid, $W$ is replaced by $w$.) The total number of plates is denoted $n_T$. The Nusselt number is defined as

$$\text{Nu} = \frac{2Fhb}{k}$$ (13.6.18)

where $b$ is the separation between plates ($2b$ is the hydraulic diameter, defined as 4 times the cross-sectional area, divided by the wetted perimeter). In general, the factor $F$ must be included to account for the specific flow configuration. In the case considered here, of parallel counterflow, it is a simple matter to determine the appropriate value of $F$. For the plate heat exchanger in parallel countercurrent flow, $F = 1$ (see Problem 13.30).

In designing a system for the heating of gelatin solutions, it is necessary to have data for the viscosity of the solution as a function of temperature and gelatin concentration. These data are shown in Fig. 13.6.5. We are now in position to illustrate a design procedure based on these ideas.

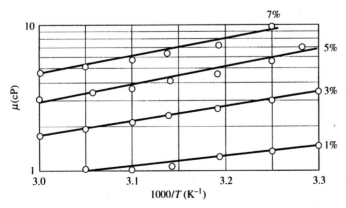

**Figure 13.6.5** Viscosity data for gelatin solutions as a function of temperature and weight percent gelatin in water.

**EXAMPLE 13.6.2** *Design of a Heating System for Gelatin*

We have to heat 500 L of a 7% gelatin solution from an initial temperature of 25°C to a final temperature of 40°C. Hot water is available as the heat transfer fluid, at a temperature of 70°C. A plate heat exchanger is available with an area of 5000 cm², with $n_T = 11$ plates, $L = 0.12$ m, and $b = 3$ mm. Specify flowrates of the hot and cold streams that will permit this batch to be taken through the required temperature increase in 15 min. The exchanger configuration is parallel counterflow, as illustrated in Fig. 13.6.2.
From Eq. 13.6.10 we find

$$-\ln \Theta \equiv -\ln\left(\frac{70-40}{70-25}\right) = 0.406 = \frac{W(K-1)}{(500\ \text{kg})K}(900\ \text{s}) \qquad (13.6.19)$$

or

$$\frac{W(K-1)}{K} = 0.225\ \text{kg/s} \qquad (13.6.20)$$

A second relationship between $W$ and $K$ follows from the definition of $K$ (Eq. 13.6.8):

$$K = \exp\frac{U(0.5\ \text{m}^2)}{(4180\ \text{J/kg})\ W} = \exp\left(\frac{1.2\times10^{-4}U}{W}\right) \qquad (13.6.21)$$

This is a relationship between $W$ and $K$ because $U$ depends on flowrate $W$, for a given exchanger and fluid. That dependence is given in Fig. 13.6.4, but in nondimensional terms. We will need estimates for physical properties of the gelatin solution. The average gelatin temperature is $(25 + 40)/2 = 33°C$. Assuming that the water side remains near 70°C, we will evaluate physical properties at the mean of 33 and 70°C, or 52°C = 325 K ($1000/T = 3.08$). From Fig. 13.6.5 we find the viscosity of a 7% solution as 5.4 cP = 0.0054 Pa·s. This is about 10 times greater than the viscosity of water at this temperature. Assuming $C_p$ and $k$ for gelatin are the same as for water (thermal properties should not be much different from water since this is a dilute solution—viscosity, however, is a very strong function of concentration), we will take the Prandtl number to be $\text{Pr}_{\text{gelatin}} = 10 \times \text{Pr}_{\text{water}} = 10 \times 3.3 = 33$. We may now write the following working equations, using consistent SI units:

$$\text{Re}_{\text{gel}} = \frac{4W}{(11-1)0.0054(0.12)} = 617W \qquad (13.6.22)$$

$$\text{Nu}_{\text{gel}} = \frac{2hb}{k} = \frac{2(0.003)h}{0.62} = 0.0097h \qquad (13.6.23)$$

We may now proceed to the solution of Eqs. 13.6.20 and 13.6.21 by a simple trial-and-error method. We pick a Reynolds number for the cold (gelatin) stream, and find the Nusselt number from Fig. 13.6.4. Using Eq. 13.6.23 we calculate $h_{\text{cold}}$. With the assumption that $h_{\text{cold}} \ll h_{\text{hot}}$, we replace $U$ in Eq. 13.6.21 with $h_{\text{cold}}$. With the selected Re, Eq. 13.6.22 gives us a value of $W$. From Eq. 13.6.20, $K$ follows. With $U$ and $W$, we use Eq. 13.6.21 to calculate a second value for $K$, from the chosen $W$ value. Trial and error continues until both equations yield the same $K$. A few such calculations produce the results presented in Table 13.6.1, with $h$ as well as $W$ in SI units.
Since Fig. 13.6.4 cannot be read with any great accuracy, the results of Table 13.6.1 are subject to some uncertainty. Nevertheless, we conclude that a gelatin flowrate in the neighborhood of 0.75 kg/s will yield the desired operation, on the assumption that

**Table 13.6.1** Results of Trial-and-Error Attempts to Find a Value for $K$

| Re | Nu/Pr$^{0.4}$ | Nu | $W$ (kg/s) | $h$ | $K$ Eq. 13.6.21 | $K$ Eq. 13.6.20 |
|---|---|---|---|---|---|---|
| 400 | 5.0 | 20 | 0.65 | 2090 | 1.47 | 1.53 |
| 500 | 5.8 | 23 | 0.81 | 2400 | 1.43 | 1.39 |
| 600 | 6.6 | 27 | 0.96 | 2760 | 1.41 | 1.3 |
| 1000 | 10 | 41 | 1.6 | 4200 | 1.37 | 1.16 |

the water side coefficient is large compared to the gelatin side coefficient. Hence we must examine the water side design.

For the water side we should operate under conditions that $h_{hot} = 10h_{cold} = 23,000$ J/s·m²·K. (We estimate $h_{cold}$ from Table 13.6.1.) For the hot water side we use Pr = 2.5 at 70°C, giving

$$\text{Nu}_{water} = \frac{2hb}{k} = \frac{2(23,000)(0.003)}{0.65} = 210 \tag{13.6.24}$$

and

$$\frac{\text{Nu}_{water}}{\text{Pr}^{0.4}} = \frac{210}{(2.5)^{0.4}} = 146 \tag{13.6.25}$$

From Fig. 13.6.4 (we must make a long extrapolation here) we find

$$\text{Re} = 23,000 = \frac{4w}{(11-1)(0.001)0.12} \tag{13.6.26}$$

and

$$w = 6.9 \text{ kg/s} \tag{13.6.27}$$

From Eq. 13.6.4 we can find that the water side cools only about 2°C. Hence the assumption that the hot side temperature remains nearly constant is validated for these conditions.

## SUMMARY

The design or predicted performance of a commercial heat exchanger begins with a "generic" design equation of the form

$$Q_h = FUA \Delta T_{ln}$$

This must be coupled with information from which the overall heat transfer coefficient $U$ may be estimated. The methods of Chapter 12 come into play here. The "correction factor" $F$ is specific to the design of the exchanger and depends on two important characteristics of the exchanger: the factors $R$ and $P$, defined in Eqs. 13.3.2b and 13.3.30, respectively. When sufficient information with respect to inlet and outlet temperatures is not available, trial-and-error methods are required to solve the design equations. Alternatively, a graphical method, based on the concept of exchanger *effectiveness,* can be used. This approach introduces the concept of the number of transfer units (NTU), and the method is referred to as the effectiveness–NTU method.

Heat exchanger design involves economic decisions at several levels. An example of economic optimization is presented in Section 13.5. We end the chapter with an example of a design problem for a plate heat exchanger, calling for the integration of much of what we have learned to this point.

## PROBLEMS

**13.1** Carry out the details of the analysis of the *countercurrent* double-pipe heat exchanger, parallel to the example of the cocurrent case given in the beginning of Section 13.2, and show that we still obtain Eq. 13.2.12. How are the differences $\Delta T_1$ and $\Delta T_2$ defined in this case? Compare to the definitions given in Eq. 13.2.8.

**13.2** Rework Example 13.2.1 and account for the effect of flowrate on the overall heat transfer coefficient. Drop the assumption that $U = 50$ Btu/h·ft²·°F, and instead calculate $U$ using the Dittus–Boelter equation (Eq. 13.2.26, with a 0.4 power on Pr if the stream is being heated instead of cooled). Take the pipe diameters to be 0.12 and 0.14 ft for the inner and outer diameters, respectively, of the inside pipe, and 0.17 ft as the inner diameter of the outside pipe. The inner pipe is made of copper. Compare the maximum flows attainable for cocurrent and countercurrent operation. Use physical property data given in Chapter 12, Problem 12.18.

**13.3** For the exchanger specified in Example 13.2.2, show that there is a minimum cooling flowrate below which it is not possible to achieve the required heat duty. Then prepare a plot of the required length of the exchanger as a function of the flowrate of the cooling stream, for flowrates in the range $2w_{min} < w < 20{,}000$ lb$_m$/h. What conclusion do you draw about the value of a strategy of enhancing exchanger efficiency by increasing the cooling flow? What is the maximum cooling flow worth using, and what is the minimum exchanger length achievable?

**13.4** For the exchanger specified in Example 13.2.2, find the required length if the inside pipe is changed so that the outside diameter is $D_o = 0.07$ ft, the inside diameter is $D_i = 0.06$ ft, with all other conditions remaining unchanged.

**13.5** Derive an analytical expression for $F(P,R)$ for the case of a tubular heat exchanger consisting of a set of parallel identical tubes contained within a shell, the shell side being *completely mixed* by some internal mixing device. Calculate $F(P)$ for $R = 2$ and check your values against Fig. P13.5.

**13.6** Rework Example 13.2.2, but assume that the shell fluid (the water side) is completely mixed by some mechanism. By what factor is the required area changed, with respect to the countercurrent case?

**13.7** Carry out an analysis of an exchanger with the shell side completely mixed (see Problem 13.5) and put the result in the form of a plot of efficiency ($\varepsilon$) versus NTU.

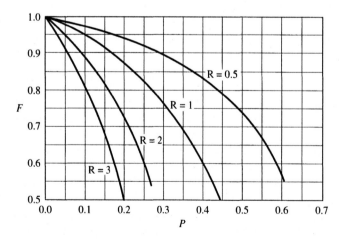

**Figure P13.5** $F(P, R)$ plot for an exchanger with a set of tubes passing through a well-mixed shell.

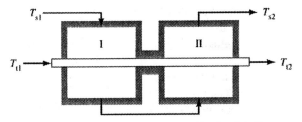

**Figure P13.9** Exchanger with two well-mixed shells in series.

**13.8** Continue Example 13.5.1, and at the optimum $t_2$, find the required exchanger area as a function of flowrate of the oil, for oil flows in the range $2000 < w < 8000$ lb/h. Take the oil viscosity to be 10 cP and the specific heat to be 0.5 Btu/lb·°F. Do not assume a value for $U$, but instead use an appropriate method of estimating heat transfer coefficients for the two streams. Neglect the conductive resistance of the pipe wall. Assume that the inner and outer pipe diameters are 0.5 and 0.75 inch, respectively. Plot the exchanger area at the optimum $t_2$, the water flowrate, and $U$, as functions of $w_{oil}$.

**13.9** A heat exchanger is designed as shown in Fig. P13.9. Derive an expression for $F(P,R)$. The two shells are well insulated and perfectly mixed, and have the same $UA$ product.

**13.10** Prepare an $F(P,R)$ plot for a heat exchanger designed according to the schematic of Fig. P13.10. Both sides of the heat exchange surface are perfectly mixed. State clearly your definition of $P$ and $R$ in terms of the temperatures in the figure.

**13.11** Prove the assertion in the footnote associated with Eq. 13.2.35.

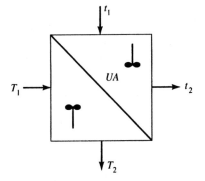

**Figure P13.10** Exchanger with two well-mixed shells in contact.

**13.12** Find the length $L$ required to heat a mineral oil from 20°C to 50°C in a circular tube heat exchanger of inside diameter $D_i = 1$ cm. The oil flowrate is 10 mL/s. The tube wall is maintained at $T = 100°C$.

For physical properties, use the following:

$$C_p = 2000 \text{ J/kg} \cdot \text{K} \qquad k = 0.147 \text{ W/m} \cdot \text{K}$$
$$\mu(\text{Fig. P12.18}) \qquad \rho = 830 \text{ kg/m}^3$$

State clearly how you find a heat transfer coefficient, and justify the method.

Give values for the following: $Re_D$, Pr, $h$(W/ m²·K), $L$(m).

**13.13** The oil described in Problem 13.12 is to be heated in a double-pipe heat exchanger. $D_i = 1$ cm as before. The outer pipe has $D_o = 2$ cm, and the heating fluid is water at a flowrate of 100 mL/s. (Use $C_p = 4200$ J/kg·K for water.) The inlet temperature of the water is 100°C.

What is the required exchanger length? Do both the cocurrent and countercurrent cases.

**13.14** The oil described in Problem 13.12 is to be heated in a double-pipe heat exchanger. The oil flowrate is increased (compared to Problem 13.12) to 200 mL/s, and $D_i = 1$ cm as before. The outer pipe has $D_o = 2$ cm, and the heating fluid is water at a flowrate of 200 mL/s. (Use $C_p = 4200$ J/kg·K for water.) The inlet temperature of the water is 100°C.

What is the required exchanger length? Do both the cocurrent and countercurrent cases.

**13.15** Rework Example 13.2.3, but do the cocurrent flow case.

**13.16** Go back to Example 13.2.2. The benzene stream is to be heated as specified. Instead of a double-pipe heat exchanger, use a shell-and-tube heat exchanger, with water on the shell side.

Assume that $U$ is given by $h_{benzene}$ in this case. The shell side is baffled.

Examine the following choices of tube side design: a single tube makes one pass (Fig. P13.16a), a single tube makes two passes (Fig. P13.16b), and a pair of tubes makes two passes (Fig. P13.16c). In all cases, use thin-walled copper tubing with a 1-inch inside diameter. In each case give the required area of tubing and the exchanger length $L$.

**13.17** A double-pipe heat exchanger is to receive 6.93 kg/s of 95% ethanol ($C_p$ = 3810 J/kg·K), to be cooled from 65.6°C to 39.4°C. Cooling water at 10°C is available at a flowrate of 6.3 kg/s ($C_p$ = 4187 J/kg·K). An overall heat transfer coefficient of $U$ = 568 W/m²·K is measured.

Find the outlet water temperature and the required exchanger area for cocurrent and countercurrent arrangements.

**13.18** A pair of double-pipe heat exchangers, of equal area, are arranged in series, as shown in Fig. P13.18. Water at 5000 lb/h flows through the inside and undergoes a temperature change from 40°C to 100°C.

Heat is supplied by an oil ($C_p$ = 0.5 Btu/lb·°F) available at 150°C. Oil goes through exchangers 1 and 2 at the rates of 3000 and 7000 lb/h, respectively. The hot oil streams exiting each exchanger are then mixed. What is the mixed temperature? Do both the cocurrent and countercurrent cases.

**13.19** For a heat exchanger in which the hot side uses a condensing vapor, the temperature of the hot side fluid remains approximately constant across the exchanger. Hence the temperature distributions would appear as shown in Fig. P13.19.

Argue that $C_{hot}$ = $(\dot{m}C_p)_{hot} \rightarrow \infty$ for this case.

**13.20** The condenser of a large power plant is a shell-and-tube exchanger with 30,000 tubes inside a single shell. *Each* tube executes *two* passes, as shown in Fig. P13.20.

| Tube Data | Steam Data |
|---|---|
| $D$ = 0.025 m | Condensation |
| Thin wall | temperature = 50°C |
| 30,000 tubes | Outside coefficient |
| $L$ = length/pass | $h_o$ = 11,000 W/m²·K |

The exchanger must exchange heat at a rate of $q = 2 \times 10^9$ W. This will be accomplished using 20°C water at 30,000 kg/s (1 kg/s per tube).

What is the temperature of the cooling water leaving the condenser?

What is the required tube length $L$ per pass?

**13.21** A shell-and-tube heat exchanger must be designed to heat 2.5 kg/s of water from 15°C to 85°C. Hot oil ($C_p$ = 2350 J/kg·K) at 160°C passes through the shell side, and the convective coefficient on the oil side is $h_o$ = 400 W/m²·K.

Ten tubes are used, of diameter $D$ = 25 mm, and each tube makes eight passes through the shell, as shown in Fig. P13.21.

**(a)** What flow rate will cool the oil down to 100°C?

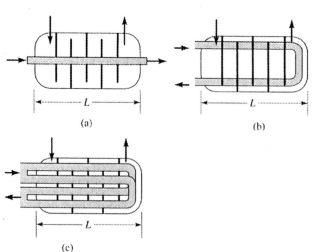

(a)

(b)

(c)

**Figure P13.16**

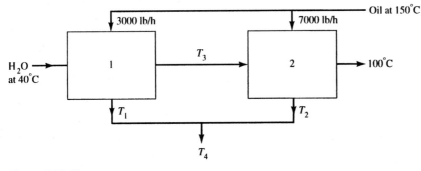

**Figure P13.18**

**(b)** What is the tube length under these conditions? (We define length in this case as the distance traveled by the water from entrance to exit, within a single tube).

**13.22** A simple heat exchanger is designed as shown in Fig. P13.22. The shell side is perfectly mixed. Find the unknown temperatures.

**13.23** Find the relationship between $NTU_{max}$, as defined in Eq. 13.4.20, and the Stanton number, defined earlier by Eq. 12.4.17.

**13.24** Oil flows through a long tube of circular cross section. The wall temperature is maintained uniformly at 40°C. The oil enters the tube at 20°C. The tube length is 10 cm, and the inside diameter is 0.2 cm. The Prandtl number of the oil (at 30°C) is 10. Find $T_{out}$ at oil flows such that Re = 20, 200, and 2000.

**13.25** A heat exchanger has two tube passes inside a baffled shell. The tube side fluid is an oil that enters at 200°F at a flowrate of 5000 lb/h ($C_p = 0.5$ Btu/lb·°F). The shell side fluid is water, which must be heated from 50°C to 80°F ($C_p = 1$ Btu/lb·°F). The $UA$ product for this exchanger is taken to be 488 Btu/h·°F.

What is the mass flowrate of the water? What is the oil exit temperature?

**13.26** Rework Example 13.4.2, but with the following changes. Suppose $UA$ were unknown, and instead we had

$T_{h1} = 200°F \qquad T_{h2} = 160°F \qquad T_{c1} = 50°F$
$w_h = 4000$ lb/h $\quad w_c = 2200$ lb/h
$C_{ph} = 0.5$ Btu/lb·°F $\quad C_{pc} = 1.0$ Btu/lb·°F

The task is to find values for $T_{c2}$, $UA$, and $Q_h$.

**13.27** Rework Example 13.4.2, but with the following changes:

$T_{h1} = 200°F \qquad T_{h2} = 160°F \qquad T_{c1} = 50°F$
$w_h = 4000$ lb/h $\qquad UA = 743$ Btu/h·°F
$C_{ph} = 0.5$ Btu/lb·°F $\quad C_{pc} = 1.0$ Btu/lb·°F

The task is to find values for $T_{c2}$, $w_c$, and $Q_h$.

**13.28** A countercurrent double pipe heat exchanger is available for the purpose of cooling an oil flow from 110 to 50°C. The oil flowrate is 2 kg/s ($C_p = 2.25$ kJ/kg·K). The coolant is water, which enters the exchanger at 20°C, at a flowrate of 1 kg/s. The $U$ value for this exchanger is known to be 400 W/m²·K.

What is the required area of the exchanger? What is the water exit temperature?

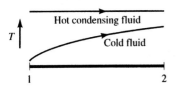

**Figure P13.19** Temperature profiles for a heat exchanger in which the hot side uses a condensing vapor.

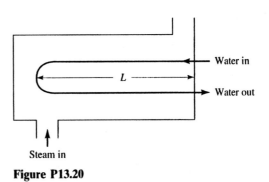

**Figure P13.20**

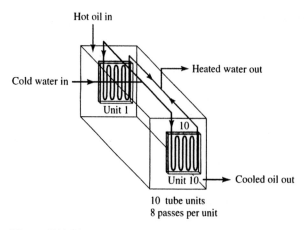

10 tube units
8 passes per unit

**Figure P13.21**

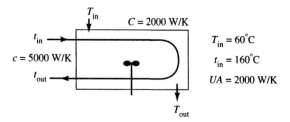

$C = 2000$ W/K

$T_{in} = 60°C$

$t_{in} = 160°C$

$UA = 2000$ W/K

**Figure P13.22**

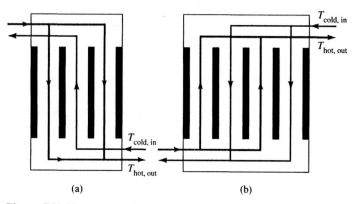

(a)

(b)

**Figure P13.32** Two parallel plate heat exchanger configurations.

**13.29** A mineral oil ($C_p = 2$ kJ/kg·K) flows through a circular tube of inside diameter $D_i = 1$ cm and is heated from 20°C to 50°C. The oil flowrate is 100 mL/s. The tube passes through a well-stirred vessel through which hot water is pumped at a flowrate of 100 g/s. The water inlet temperature is 100°C. What is the outlet temperature of the water?

**13.30** For the parallel counterflow plate heat exchanger, determine the factor $F$ used in Eq.

13.6.3 and subsequently in Eq. 13.6.18. Assume that each channel carries the same flowrate $Q_n$, and neglect the temperature dependence of the convective coefficients. Assume that the number of plates is large compared to unity.

**13.31** With reference to Example 13.6.2, suppose we double the gelatin flowrate from the value determined in the example. How long will it take to heat the gelatin through the specified range?

**13.32** Consider the two parallel plate heat exchanger configurations shown in Fig. P13.32. Analyze the performance of these two systems, and give an expression for the $F$ value for each exchanger. Assume that all flows are distributed equally among the individual channels.

**13.33** Show that Eq. 13.6.17 is consistent with the usual definition of the Reynolds number:

$$\text{Re} = \frac{\bar{u} D_{\text{h}} \rho}{\mu} \qquad \text{(P13.33.1)}$$

where $\bar{u}$ is the average velocity in an individual flow channel between two parallel plates, and $D_{\text{h}}$ is the hydraulic diameter, defined as

$$D_{\text{h}} \equiv \frac{4 \times \text{cross-sectional area for flow}}{\text{wetted perimeter}}$$

$$\text{(P13.33.2)}$$

# Chapter 14

# Natural Convection Heat Transfer

In many situations, heat is transferred from the surface of a body to a surrounding fluid by "natural convection," the motion set up in the surrounding fluid by the density variation between the fluid near the surface and the ambient fluid. When temperature differences between the surface and the ambient are large, these density variations can be large, and significant flows can be generated. Hence we have convective heat transfer, but from a flow field that is not imposed independently of the thermal process itself. In this chapter we will examine some fundamental features of natural convection (or, as it is often called, "free" convection) to see what important dimensionless groups control this process. Through a series of examples, we will learn how empirical correlations of convective heat transfer coefficients are used in the design and analysis of natural convection systems.

## 14.1  INTRODUCTION

In fluids, both gases and liquids, spatial temperature variations give rise to corresponding variations in the density of the fluid. For a single-component fluid, or for a mixture of uniform composition, the density may be written as a function of temperature in the form of a Taylor series about some reference temperature, such as the mean temperature $\overline{T}$:

$$\rho(T) = \rho = \rho(\overline{T}) + \left.\frac{\partial \rho}{\partial T}\right|_{\overline{T}} (T - \overline{T}) + \cdots \qquad (14.1.1)$$

If we define a thermodynamic property $\beta$ as

$$\beta = -\frac{1}{\rho}\left(\frac{\partial \rho}{\partial T}\right)_p \qquad (14.1.2)$$

and if we truncate the Taylor series after the linear term, we may write the density as

$$\rho(T) = \overline{\rho} - \overline{\rho}\beta(T - \overline{T}) \qquad (14.1.3)$$

The coefficient $\beta$ is an important physical property of the fluid, since its magnitude determines the degree to which a temperature variation in the fluid can establish a density variation.

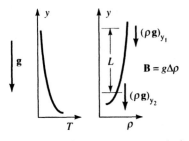

**Figure 14.1.1** Temperature variation in a fluid gives rise to a buoyancy force (per volume) **B**.

The role of a spatial density variation in a fluid can be appreciated by referring to Fig. 14.1.1. If there is a gravitational field acting on the fluid, then at any local region of the fluid there is a gravitational force (per unit volume) given by $\rho\mathbf{g}$, where $\rho$ is the local density. If there is a variation of temperature in the direction of gravity, then there is a corresponding variation in the body force. The difference in body force over some vertical distance in the fluid has the capacity to induce motion in that direction since, according to our principle of conservation of momentum, an unbalanced force can induce acceleration. Any induced motion in the fluid would be opposed by viscous friction. Thus a state of steady motion could ensue in which the body force is balanced by the viscous force, just as is the case when a solid body of density greater than that of the surrounding fluid falls through that fluid at a terminal (i.e., steady) velocity. We refer to the motion induced by temperature variations (really, by *density* variations) as "natural convection," or "free convection," in distinction to "forced convection."

Suppose we imagine a column of fluid of height $L$, with a cross sectional area $A$ normal to the $y$ axis. Then the body force acting across the ends of that column is $\mathbf{B}AL$. If the body force were unopposed by frictional forces, we would expect an increase in momentum in the fluid of the order of $\rho(\Delta u)^2 A$ (the momentum flux times the area across which the flux occurs). Hence we would expect that a buoyancy-induced velocity variation would be no greater than of the order of

$$u_b = O\left[\left(\frac{gL|\Delta\rho|}{\rho}\right)^{1/2}\right] \tag{14.1.4}$$

If we define a Reynolds number based on this buoyancy-induced velocity, and using $L$ as the appropriate length scale, we find a new dimensionless group appearing, of the form

$$(Re)_b = \frac{u_b L}{\nu} = \left(\frac{gL^3|\Delta\rho|}{\rho\nu^2}\right)^{1/2} = \left[\left(\frac{gL^3}{\nu^2}\right)\frac{|\Delta\rho|}{\rho}\right]^{1/2} \tag{14.1.5}$$

Normally we regard a Reynolds number as a ratio of inertial forces to viscous forces. In this case we may regard Eq. 14.1.5 as defining a ratio of buoyancy forces to viscous forces. Traditionally, this grouping is given the name "Grashof number":

$$Gr \equiv \left(\frac{gL^3}{\nu^2}\right)\frac{|\Delta\rho|}{\rho} \tag{14.1.6}$$

Hence we may think of the Grashof number as a Reynolds number for buoyancy-induced flow. Equation 14.1.3 may be written in the form

$$\frac{\Delta\rho}{\rho} = -\bar{\beta}\Delta T \tag{14.1.7}$$

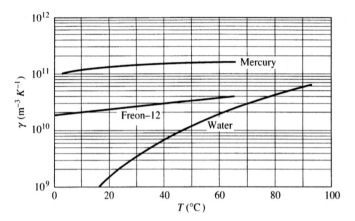

**Figure 14.1.2** Values of the parameter $\gamma$ as a function of temperature—liquids.

Thus we may write the Grashof number in the form

$$\mathrm{Gr} \equiv \left(\frac{g\beta}{\nu^2}\right) L^3 \Delta T \tag{14.1.8}$$

We note that this particular grouping of these terms separates the Grashof number into a product of a physical property group (*which is not dimensionless*) that we define as

$$\gamma \equiv \frac{g\beta}{\nu^2} \tag{14.1.9}$$

times a term that accounts for the temperature variation $\Delta T$ in the vertical direction, and a measure $L$ of the extent of the vertical distance over which that temperature difference is imposed. Figures 14.1.2 and 14.1.3 show some typical values of $\gamma$ as a function of temperature.

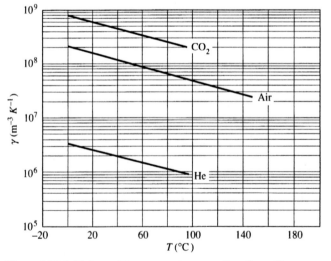

**Figure 14.1.3** Values of the parameter $\gamma$ as a function of temperature—gases (at p = 1 atm).

The analysis of natural convection is, generally, more complex than the fluid dynamic analyses we have met so far, because the flow field is intimately coupled with the temperature field. Hence, in general, we must solve both the momentum *and* energy equations to develop a model of natural convection. In the next section we look at some relatively simple natural convection problems.

## 14.2   BUOYANCY-INDUCED FLOW: NATURAL CONVECTION IN A CONFINED REGION

In this first example, the temperature and the flow fields are easily obtained by analytical methods because of the simplifying assumptions that we introduce.

**EXAMPLE 14.2.1**   *Analysis of Natural Convection between Infinite Parallel Plates*

Figure 14.2.1 shows the geometry and the boundary conditions for this model. Two very large plates are oriented in the vertical direction, parallel to each other with a separation distance given by $2b$. We begin with a *thermal energy balance:*

$$v_y \frac{\partial T}{\partial y} + v_z \frac{\partial T}{\partial z} = \alpha \left( \frac{\partial^2 T}{\partial y^2} + \frac{\partial^2 T}{\partial z^2} \right) \tag{14.2.1}$$

We have assumed no flow in the $x$ direction. The system has reached its steady state, and the viscosity and thermal diffusivity are taken as independent of temperature. Assume that the plates are long in the $z$ direction so that $T \neq T(z)$.

We write the velocity field in the form

$$\mathbf{v} = (v_z(y), 0) \tag{14.2.2}$$

Then Eq. 14.2.1 simplifies to

$$\frac{d^2 T}{dy^2} = 0 \tag{14.2.3}$$

with boundary conditions given by

$$\begin{aligned} T = T_1 \quad &\text{at} \quad y = b \\ T = T_2 \quad &\text{at} \quad y = -b \end{aligned} \tag{14.2.4}$$

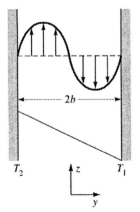

**Figure 14.2.1** Parallel plates at different temperatures induce flow.

The solution to this differential equation is easily found in the form

$$T = T_{\mathrm{m}} - \frac{1}{2} \Delta T \left( \frac{y}{b} \right) \tag{14.2.5}$$

where we have defined $\Delta T = T_2 - T_1$ and $T_{\mathrm{m}} = \frac{1}{2}(T_1 + T_2)$.

We now write the Navier–Stokes equation (the *momentum balance*) in the $z$ direction:

$$\mu \frac{d^2 v_z}{dy^2} = \frac{dp}{dz} + \rho g \tag{14.2.6}$$

In developing this model we will assume $\mu = $ constant but $\rho = \rho(T)$. Why is it reasonable to assume $\mu \neq \mu(T)$ but $\rho = \rho(T)$ when for most fluids the opposite is a better approximation? We need to look at the *essence* of this problem, which lies in the absence of any buoyancy-driven flow *unless* $\rho = \rho(T)$. Whether $\mu$ is constant is immaterial. When we write $\rho(T)$ in a Taylor series about some reference temperature $\overline{T}$, as in Eq. 14.1.1 earlier, Eq. 14.2.6 becomes

$$\mu \frac{d^2 v_z}{dy^2} = \frac{dp}{dz} + \bar{\rho} g - \bar{\rho} \bar{\beta} g (T - \overline{T}) \tag{14.2.7}$$

Buoyancy-driven flows are slow in some sense. We will *hope* (another word for "assume") that it is reasonable to neglect the effect of flow on $dp/dz$, so that the hydrostatic law holds:

$$\frac{dp}{dz} \approx -\bar{\rho} g \tag{14.2.8}$$

Then

$$\mu \frac{d^2 v_z}{dy^2} = -\bar{\rho} \bar{\beta} g (T - \overline{T}) \tag{14.2.9}$$

$$\underset{\text{Viscous}}{\underbrace{\phantom{xxxx}}} \quad \underset{\text{Buoyancy}}{\underbrace{\phantom{xxxx}}}$$
Viscous        Buoyancy
stresses      stresses

Since the two *dependent* variables, $v_z$ and $T$, appear in Eq. 14.2.9, we say that $v_z$ and $T$ are "coupled." However, in the simplified case we are examining, we have "uncoupled" the variables because we found $T(y)$ from an analysis that did not involve the flow field. Using the temperature distribution from above (Eq. 14.2.5) we now find

$$\mu \frac{d^2 v_z}{dy^2} = -\bar{\rho} \bar{\beta} g \left[ (T_{\mathrm{m}} - \overline{T}) - \frac{1}{2} \Delta T \left( \frac{y}{b} \right) \right] \tag{14.2.10}$$

Boundary conditions on $v_z$ are

$$v_z = 0 \quad \text{at} \quad y = \pm b \tag{14.2.11}$$

The solution to Eq. 14.2.10 is found by integration with respect to $y$, twice, and the result is

$$v_z = \frac{\bar{\rho} \bar{\beta} g \, b^2 \, \Delta T}{12 \, \mu} \left[ \left( \frac{y}{b} \right)^3 - A \left( \frac{y}{b} \right)^2 - \frac{y}{b} + A \right] \tag{14.2.12}$$

where we define

$$A = \frac{6(T_{\mathrm{m}} - \overline{T})}{\Delta T} \tag{14.2.13}$$

Note that we have not yet defined the choice of the reference temperature $\overline{T}$, although we earlier called it the mean temperature. These two temperatures are indeed the same, since the net flow must vanish. This implies that

$$\int_{-b}^{b} v_z dy = 0 \tag{14.2.14}$$

This yields $A = 0$ or $\overline{T} = T_m$, so we find

$$v_z = \frac{\overline{\rho}\,\overline{\beta}\,g\,b^2\,\Delta T}{12\,\mu}\left[\left(\frac{y}{b}\right)^3 - \frac{y}{b}\right] \tag{14.2.15}$$

This may be written in the format of a dimensionless velocity as

$$
\begin{aligned}
v_z \frac{b\overline{\rho}}{\mu} &= \frac{1}{12}\left[\frac{\overline{\rho}^2\,\overline{\beta}\,g\,b^3\,\Delta T}{\mu^2}\right]\left[\left(\frac{y}{b}\right)^3 - \frac{y}{b}\right]\\
&= \frac{1}{12}\left[\frac{\overline{\rho}\,g\,b^3\,\Delta\rho}{\mu^2}\right]\left[\left(\frac{y}{b}\right)^3 - \frac{y}{b}\right]\\
&= \frac{1}{12}\,\text{Gr}\left[\left(\frac{y}{b}\right)^3 - \frac{y}{b}\right]
\end{aligned}
\tag{14.2.16}
$$

where Gr is the Grashof number, but we see that in this case the length scale is the half-width $b$. There is no *vertical* length scale, since the plates are infinite in the vertical direction. Hence we use the only available length scale for the definition of Gr.

It is useful to calculate the magnitude of buoyancy-induced velocity, to obtain an idea of the degree to which a temperature variation can drive a flow. We will consider a temperature difference of 50°C, with hot and cold walls at 75 and 25°C, respectively, in air and in water, with a spacing between the plates given by $b = 1$ cm.

**EXAMPLE 14.2.2** *Buoyancy-Driven Convective Velocities in Water and Air*

From Eq. 14.2.16 we see that there is a distribution of velocities across the gap between the plates. We will calculate the *maximum* velocity that occurs in the region $[-b, b]$. All we need do is differentiate Eq. 14.2.16 with respect to $y$, and set the derivative to zero. This yields the position at which the maximum occurs. When that value of $y$ is substituted into Eq. 14.2.16, we find the maximum velocity. It is not difficult to show that the result may be written in the form

$$v_z^{\max} \frac{b\overline{\rho}}{\mu} = 0.032\,\text{Gr} \tag{14.2.17}$$

**For water** we find (Fig. 14.1.2)

$$\gamma = 1.2 \times 10^{10}\ \text{m}^{-3} \cdot \text{K}^{-1} \tag{14.2.18}$$

and (from Eq. 14.1.8)

$$\text{Gr} = \gamma b^3\,\Delta T = 6 \times 10^5 \tag{14.2.19}$$

We evaluate $\overline{\rho}$ and $\mu$ at $T = 50$°C, and we find

$$\nu = \frac{\mu}{\overline{\rho}} = 0.55 \times 10^{-6}\ \text{m}^2/\text{s} \tag{14.2.20}$$

Use of Eq. 14.2.17 yields the result

$$v_z^{\max} = 0.032 \, \frac{\mu}{b \, \bar{\rho}} \, \mathrm{Gr} = 0.032 \, \frac{0.55 \times 10^{-6}}{0.01} \, (6 \times 10^5) = 1.06 \text{ m/s} \qquad \textbf{(14.2.21)}$$

**For air** we find (Fig. 14.1.3)

$$\gamma = 10^8 \text{ m}^{-3} \text{ K}^{-1} \qquad \textbf{(14.2.18')}$$

and (from Eq. 14.1.8)

$$\mathrm{Gr} = \gamma b^3 \, \Delta T = 5000 \qquad \textbf{(14.2.19')}$$

We evaluate $\bar{\rho}$ and $\mu$ at $T = 50°\text{C}$, at which

$$\nu = 1.8 \times 10^{-5} \text{ m}^2/\text{s} \qquad \textbf{(14.2.20')}$$

From Eq. 14.2.17 we obtain

$$v_z^{\max} = 0.032 \, \frac{\mu}{b \, \bar{\rho}} \, \mathrm{Gr} = 0.032 \, \frac{1.8 \times 10^{-5}}{0.01} \, 5000 = 0.29 \text{ m/s} \qquad \textbf{(14.2.21')}$$

Even though water is by far the more viscous fluid, the buoyancy-induced flow is much greater in the water, because the key parameter is $\gamma$, which is much greater for water than for air.

We might call the flow field we have just examined a "confined" flow, and clearly the velocity depends strongly on the separation parameter $b$. In the next section we look at an unconfined flow driven by buoyancy. As we shall see, this flow has a boundary layer character.

## 14.3 BUOYANCY-INDUCED FLOW: NATURAL CONVECTION IN AN UNCONFINED REGION

In this section we examine a problem that does not permit us to solve for the flow and temperature fields by a very simple analytical method. Despite the complexity of the problem, however, it is possible to write an approximate mathematical model for this system, and obtain a solution to the model that is in very good agreement with experience.

Recall that a mathematical model is a quantitative statement of the relationships among the dependent and independent variables and the relevant parameters that describe a physical phenomenon. We sometimes use the term "model" to refer to the equations that *constrain* the relationships. Usually, these are the conservation equations of heat, mass, and momentum transfer. More commonly, we refer to a *solution* of the equations as the "model."

Here we introduce some modeling concepts through an example that illustrates the development of a model that may be obtained *without solving the equations* that constrain or describe the physics.

A flat plate, maintained at a constant temperature $T_0$, is held vertically in a large body of fluid maintained at a temperature $T_1$, far from the plate. In the neighborhood of the surface there is a buoyancy-driven flow of fluid that rises (assuming the plate is hotter than the surrounding fluid) roughly parallel to the plate surface. We seek a model for the heat loss from the plate, as a function of relevant parameters.

The plate is of finite height $H$, but it is infinite in width. Figure 14.3.1 serves to define a coordinate system for this plate, and the relevant variables. Because we assume that

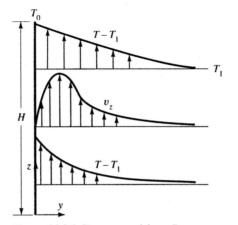

**Figure 14.3.1** Buoyancy-driven flow near an unconfined vertical plate.

the plate is very wide in the $x$ direction, we assume that there is no flow of heat, mass, or momentum in the $x$ direction.

The convective flow (driven by the buoyancy of the hot fluid near the hot surface) must satisfy the equations of fluid dynamics. These are the continuity equation (conservation of mass)

$$\frac{\partial v_y}{\partial y} + \frac{\partial v_z}{\partial z} = 0 \qquad (14.3.1)$$

and the Navier–Stokes equations (conservation of momentum for a Newtonian fluid). We assume that the major flow component is in the $z$ direction, and considering steady state behavior, we write

$$\rho\left(v_y \frac{\partial}{\partial y} + v_z \frac{\partial}{\partial z}\right) v_z = \mu\left(\frac{\partial^2}{\partial y^2} + \frac{\partial^2}{\partial z^2}\right) v_z - \frac{\partial p}{\partial z} - \rho g \qquad (14.3.2)$$

So far, we have written *two* constraining equations for the flow field. Note that we have *three* unknowns: $v_y$, $v_z$, and $p$, in only two equations. We need one more equation, or one less unknown.

We can eliminate $p$ by assuming that the buoyancy-driven flow field is so slow that the pressure gradient in the $z$ direction is given by the hydrostatic law:

$$\frac{\partial p}{\partial z} = -\rho g \qquad (14.3.3)$$

If we introduce Eq. 14.3.3 into Eq. 14.3.2, we see that the gravitational term disappears from the problem. While this seems like a simplification, in fact it renders the problem nonsensical. The only reason there is a buoyant flow field arising from the heated surface is that the density variations, due to a temperature-dependent fluid density, cause the hot fluid near the plate to rise. If we throw out the gravity term from the equations of motion, we have removed just the physical mechanism whose action we wish to describe. This form of intellectual carelessness, known as "throwing away the baby with the bathwater," is a common modeling error that arises when we allow the mathematics to obscure the physical picture.

The resolution of the problem is simple, and indeed we have already dealt with such equations in Section 14.1. As earlier, we replace Eq. 14.3.3 with

$$\frac{\partial p}{\partial z} = -\bar{\rho}g \tag{14.3.4}$$

where $\bar{\rho}$ is the *mean* density, evaluated at some arbitrary reference temperature $\bar{T}$.

Now we must introduce an equation of state that relates fluid density to temperature: $\rho = \rho(T)$. We assume that under conditions of slow flow, $\rho$ is independent of pressure. We write $\rho(T)$ as a Taylor series in $T$ as we did earlier (Eqs. 14.1.1 through 14.1.3), and we find Eq. 14.3.2 becomes

$$\rho\left(v_y\frac{\partial}{\partial y} + v_z\frac{\partial}{\partial z}\right)v_z = \mu\left(\frac{\partial^2}{\partial y^2} + \underline{\frac{\partial^2}{\partial z^2}}\right)v_z + \bar{\rho}\bar{\beta}g(T - \bar{T}) \tag{14.3.5}$$

We will choose the reference temperature to be $\bar{T} = T_1$ and drop the bars over $\rho$ and $\beta$ to simplify the notation.

We have two equations (Eqs. 14.3.1 and 14.3.5), and we still have three unknowns: $v_y$, $v_z$, and $T$. Now we must introduce an energy equation: the conduction/convection equation.

$$\rho\hat{C}_p\left(v_y\frac{\partial}{\partial y} + v_z\frac{\partial}{\partial z}\right)T = k\left(\frac{\partial^2}{\partial y^2} + \underline{\frac{\partial^2}{\partial z^2}}\right)T \tag{14.3.6}$$

We will evaluate $\rho\hat{C}_p$, and $k$ in Eq. 14.3.6 at the temperature $(T_1 + T_2)/2$.

At this stage we have a model defined by Eqs. 14.3.1, 14.3.5, and 14.3.6, a coupled set of nonlinear partial differential equations. These equations are too complex to solve analytically.

A common, and usually valid, approximation is to note that in this type of physical situation the variation of $v_z$ and $T$ is likely to be much stronger in the $y$ direction than in the $z$ direction. Hence we commonly drop the underlined terms in Eqs. 14.3.5 and 14.3.6. This still does not give us a set of equations solvable by analytical methods.

Let us avoid the solution for a moment by writing the appropriate boundary conditions. (In fact, we cannot really consider the solution process *until* we have written the boundary conditions.) We invoke the following physical statements:

| | | | |
|---|---|---|---|
| at $y = 0$ | $v_y = v_z = 0$ | $T = T_0$ | (14.3.7) |
| at $y = \infty$ | $v_z = 0$ | $T = T_1$ | (14.3.8) |
| at $z \to -\infty$ | $v_y = v_z = 0$ | $T = T_1$ | (14.3.9) |

This does not seem to help much, but we now have a complete formulation of the physical problem. Thus we have a mathematical model, but not a solution, that describes the process. If we cannot solve the equations, what else can we do? We could do dimensional analysis, which would give us some information about which dimensionless groups were important in this problem. In fact, however, we can go beyond dimensional analysis.

We do so by nondimensionalizing the equations by referring to reference values, the variables $v_y$, $v_z$, $T$, and $y$ and $z$. The trick is to make the right choices of reference values. We shall see shortly that there are some rational principles for doing so.

First, we choose reference values that simplify the numerical values of the dimensionless variables on the boundaries of the system. For example, a logical choice for nondimensionalizing $T$ is the following:

$$\text{define} \quad \Theta = \frac{T - T_1}{T_0 - T_1} \tag{14.3.10}$$

With this choice, $\Theta$ takes on the value $\Theta = 1$ on $y = 0$ and $\Theta = 0$ on $y = \infty$ and $z = -\infty$. Further, $\Theta$ is bounded between 0 and 1.

Usually we nondimensionalize the space variables $y$ and $z$ by means of characteristic length variables, if they exist. In this problem $H$ is a characteristic variable, or scale, in the $z$ direction. Hence it seems reasonable to

$$\text{define} \quad \zeta = \frac{z}{H} \tag{14.3.11}$$

There is no characteristic length scale in the $y$ direction. There are no characteristic velocity scales in this problem, either, so we do not know how to nondimensionalize $v_y$ and $v_z$ at this stage. For the moment,

$$\text{define} \quad \begin{cases} \phi_z = \dfrac{v_z}{V_z} & \text{(14.3.12)} \\[2mm] \phi_y = \dfrac{v_y}{V_y} & \text{(14.3.13)} \\[2mm] \eta = \dfrac{y}{Y} & \text{(14.3.14)} \end{cases}$$

where $V_y$, $V_z$, and $Y$ are as yet undefined. Equations 14.3.1, 14.3.5, and 14.3.6 are now rewritten as

$$\frac{HV_y}{V_z Y}\frac{\partial \phi_y}{\partial \eta} + \frac{\partial \phi_z}{\partial \zeta} = 0 \tag{14.3.15}$$

$$\frac{V_z}{\beta g \Delta T}\left(\frac{V_y}{Y}\frac{\partial}{\partial \eta} + \frac{V_z}{H}\frac{\partial}{\partial \zeta}\right)\phi_z = \frac{\mu V_z}{\rho \beta g \Delta T Y^2}\frac{\partial^2 \phi_z}{\partial \eta^2} + \Theta \tag{14.3.16}$$

(We have dropped the underlined term in Eq. 14.3.5 and defined $\Delta T = T_0 - T_1$.)

$$\left(\frac{V_y Y}{\alpha}\frac{\partial}{\partial \eta} + \frac{V_z Y^2}{H\alpha}\frac{\partial}{\partial \zeta}\right)\Theta = \frac{\partial^2 \Theta}{\partial \eta^2} \tag{14.3.17}$$

where the thermal diffusivity of the fluid is $\alpha = k/\rho C_p$.

We now observe that there are five dimensionless groups appearing in Eqs. 14.3.15, 14.3.16, and 14.3.17:

$$N_1 = \frac{V_y Y}{\alpha} \tag{14.3.18}$$

$$N_2 = \frac{V_z Y^2}{H\alpha} \tag{14.3.19}$$

$$N_3 = \frac{\mu V_z}{\rho \beta g \Delta T Y^2} \tag{14.3.20}$$

$$N_4 = \frac{V_y V_z}{\beta g \Delta T Y} = N_1 N_3 \frac{\alpha \rho}{\mu} = \frac{N_1 N_3}{\text{Pr}} \tag{14.3.21}$$

(where $\text{Pr} = \mu/\rho \alpha$ is the Prandtl number)

$$N_5 = \frac{V_z^2}{\beta g \Delta T H} = \frac{N_2 N_3}{\text{Pr}} \tag{14.3.22}$$

Thus we have *four independent* groups: the Prandtl number plus three other (strange) groups: $N_1$, $N_2$, and $N_3$.

But we still have not specified $V_y$, $V_z$, and $Y$. Is it possible that we can simplify the model by judicious choice of these three scaling factors? Since they are *arbitrary* scaling factors, in the sense that there are no natural parameters in the physics of the system with units of length (in the $y$ direction) or velocity, let us seek values for $V_y$, $V_z$, and $Y$ that eliminate $N_1$, $N_2$, and $N_3$ from the problem. We do this very simply, by making $N_1 = N_2 = N_3 = 1$. This then yields three equations for $V_y$, $V_z$, and $Y$.

Thus we set

$$\frac{V_y Y}{\alpha} = 1 \tag{14.3.23}$$

$$\frac{V_z Y^2}{H\alpha} = 1 \tag{14.3.24}$$

$$\frac{\mu V_z}{\rho g \beta \Delta T Y^2} = 1 \tag{14.3.25}$$

From Eq. 14.3.25 we find (defining $B = \rho g \beta \Delta T$)

$$\frac{Y^2}{V_z} = \frac{\mu}{\rho g \beta \Delta T} = \frac{\mu}{B} \tag{14.3.26}$$

From Eq. 14.3.24

$$Y^2 V_z = H\alpha \tag{14.3.27}$$

Equation 14.3.26 gives the result

$$Y^4 = \frac{H\alpha\mu}{B} \tag{14.3.28}$$

or

$$Y = \left(\frac{H\alpha\mu}{B}\right)^{1/4} \tag{14.3.29}$$

With Eq. 14.3.24 we find

$$V_z = \frac{H\alpha}{Y^2} = \frac{H\alpha}{(H\alpha\mu/B)^{1/2}} \tag{14.3.30}$$

or

$$V_z = \left(\frac{BH\alpha}{\mu}\right)^{1/2} \tag{14.3.31}$$

Finally, Eq. 14.3.24 yields

$$V_y = \frac{\alpha}{Y} = \left(\frac{B\alpha^3}{\mu H}\right)^{1/4} \tag{14.3.32}$$

Our model (i.e., the set of differential equations) now takes the form

$$\frac{\partial \phi_y}{\partial \eta} + \frac{\partial \phi_z}{\partial \zeta} = 0 \tag{14.3.33}$$

$$\frac{1}{Pr}\left(\phi_y \frac{\partial}{\partial \eta} + \phi_z \frac{\partial}{\partial \zeta}\right)\phi_z = \frac{\partial^2 \phi_z}{\partial \eta^2} + \Theta \tag{14.3.34}$$

$$\left(\phi_y \frac{\partial}{\partial \eta} + \phi_z \frac{\partial}{\partial \zeta}\right)\Theta = \frac{\partial^2 \Theta}{\partial \eta^2} \tag{14.3.35}$$

with boundary conditions

$$\text{at } \eta = 0 \qquad \phi_y = \phi_z = 0 \qquad \Theta = 1 \qquad \text{(14.3.36)}$$

$$\text{at } \eta = \infty \qquad \phi_y = \phi_z = 0 \qquad \Theta = 0 \qquad \text{(14.3.37)}$$

$$\text{at } \zeta = -\infty \qquad \phi_y = \phi_z = 0 \qquad \Theta = 0 \qquad \text{(14.3.38)}$$

What have we really achieved? The equations are still coupled and nonlinear, and no easier to solve just because they are nondimensional.

Let us examine the heat loss from the plate, a measure of the behavior of the system. We might have hoped for a model that would tell us how the heat loss would depend on $T_0 - T_1$, $H$, and the fluid properties ($\rho, \beta, \mu$, etc.).

The heat loss $q'$ (per unit width of plate) is given by

$$q' = \int_0^H -k \left.\frac{\partial T}{\partial y}\right|_{y=0} dz = -\frac{k\Delta T H}{Y} \int_0^1 \left.\frac{\partial \Theta}{\partial \eta}\right|_{\eta=0} d\zeta \qquad \text{(14.3.39)}$$

By inspection of Eqs. 14.3.33 to 14.3.35, and the boundary conditions, we can infer that

$$\Theta = \Theta(\eta, \zeta, \text{Pr}) \qquad \text{(14.3.40)}$$

Hence the integrand in Eq. 14.3.39, $\partial\Theta/\partial\eta$ at $\eta = 0$, depends only on $\zeta$ and Pr. Since the integration over $\zeta$ is a definite integral, the integral depends only on Pr. We will denote the value of the integral as $-C(\text{Pr})$. Then we find (using Eq. 14.3.29 to eliminate $Y$)

$$q' = Ck\Delta T(\text{Gr Pr})^{1/4} \qquad \text{(14.3.41)}$$

where

$$\text{Gr} = \frac{\rho^2 \beta g H^3 \Delta T}{\mu^2} \qquad \text{(14.3.42)}$$

is the Grashof number based on $H$.

We now have a model that predicts the following features for the heat loss from a plate:

$$q' \sim H^{3/4} \qquad \text{(14.3.43)}$$

$$q' \sim \Delta T^{5/4} \qquad \text{(14.3.44)}$$

Both predictions are borne out by experiments. The only thing the model fails to do is yield the dependence of $C$ on Prandtl number. But since Pr is a fluid property, a single experiment in the fluid of interest would serve to establish $C$.

We might guess that the dependence of $C$ on Pr is weak, since Pr appears in the problem as a factor multiplying the set of terms on the left-hand side of Eq. 14.3.34. If we trace back to the physics of this equation, we can note that these terms account for inertial forces in the momentum balance. Since free convection flows are in some sense slow flows, we would expect the inertial terms to be relatively unimportant. Thus we might infer that the solution of Eqs. 14.3.33 to 14.3.35 is a weak function of variations in those terms. This leads us to another inference: namely, that $C$ is a weak function of Pr. This dependence is confirmed by numerical solutions of the full equations, which give $C = 0.52$ for Pr $= 0.73$ (air) and $C = 0.61$ and $0.65$ at Pr $= 10$ and $100$, respectively.

Note how much information we obtained without solving any equations. Note, as well, how much more information we get from this procedure in comparison to what we obtain from dimensional analysis. This technique, in which we inspect the governing equations and reduce them to a nondimensional form containing the minimum number of dimensionless groups, is known as inspectional analysis.

Now that we have a "solution" for heat transfer in this system, we may define and calculate a heat transfer coefficient. We will define the convective coefficient $h$ as the coefficient in the expression

$$\frac{q'}{H} = h\,\Delta T \tag{14.3.45}$$

Since $q'$ is the rate of transfer *per unit width* in the $x$ direction, $q'/H$ is the convective flux from the plate into the fluid. From inspection of our model, we find that

$$\text{Nu} = \frac{hH}{k} = \frac{q'}{k\,\Delta T} = C(\text{Gr Pr})^{1/4} \tag{14.3.46}$$

The product of the Grashof and Prandtl numbers appears often in free convection models, and it is called the Rayleigh number (Ra):

$$\text{Gr Pr} = \frac{\rho^2 \beta\, g H^3\, C_p \Delta T}{\mu\, k} = \gamma\, \text{Pr}\, H^3\, \Delta T = \text{Ra} \tag{14.3.47}$$

In summary, while dimensional analysis could have told us that the Rayleigh number is an important group in free convection heat transfer, inspectional analysis tells us that the Nusselt number is expected to increase like the fourth root of the Rayleigh number—a much stronger result, and one confirmed by experience.

## 14.4 CORRELATIONS OF DATA FOR FREE CONVECTION HEAT TRANSFER COEFFICIENTS

Because there are relatively few simple analytical models of buoyancy-driven flows, we have very few analytical models from which heat transfer coefficients may be predicted and tested against data. Consequently, most design in the free convection regime is based on empirical correlations. What we do know, at this point of our discussion, is that the Grashof number is a key dimensionless group, and we should not be surprised to find correlations of Nusselt number as a function of the Grashof number. As in any convective heat transfer problem, the ratio of "diffusivities" of momentum and heat should play a role, and of course the Prandtl number is just that ratio.

We find that many heat transfer correlations are presented in the simple algebraic form

$$\text{Nu} = C(\text{Gr Pr})^n = C\,\text{Ra}^n \tag{14.4.1}$$

where $\text{Ra} = \text{Gr Pr}$ is the Rayleigh number defined in Eq. 14.3.47, and $C$ and $n$ are constants only in the sense that values are stated which depend on the range of Ra to which Eq. 14.4.1 is applied.

Physical properties are usually evaluated at the so-called film temperature, defined as

$$T_f = \frac{T_s + T_\infty}{2} \tag{14.4.2}$$

where $T_s$ and $T_\infty$ are the surface and ambient temperatures, respectively. We must always choose a characteristic length scale $L$ to use in defining Nu and Gr. Correlations based on experimental observations usually are, and always should be, presented with an accompanying restriction on the range of Grashof or Rayleigh number over which each is valid. Table 14.4.1 gives some examples.

Since Eq. 14.4.1 holds over a limited range of Ra, it may be better to avoid the restriction that this form of correlation implies and to seek correlations valid over a wider range of parameters. Churchill and Chu [*Int. J. Heat Mass Transfer*, **18**, 1049 (1975)] have tested and presented an equation that fits experimental data over a wide

**Table 14.4.1** Natural Convection Correlations of the Form Nu $= C$ Ra$^n$

| Shape | Ra | L | C | n |
|---|---|---|---|---|
| Vertical plane | $10^4$–$10^9$ | Vertical height of the surface | 0.58 | 0.25 |
| Vertical cylinder | $10^9$–$10^{13}$ | Axial length | 0.02 | 0.4 |
| Horizontal cylinders | $10^4$–$10^9$ | Diameter | 0.53 | 0.25 |
| Spheres and rectangular solids | | Use horizontal cylinder correlations, with $L = 0.5 D$ for a sphere $L_{rect} = (1/L_{horiz} + 1/L_{vert})^{-1}$ for rectangles | | |
| Horizontal surfaces | | | | |
|   Rectangular ($A \times B$) | | $L = 0.5 (A + B)$ | Different correlations | |
|   Disk | | $L = 0.9D$ | depending on | |
|   Others (not cylinders) | | $L =$ area/perimeter | whether transfer is from the upper or lower surface, and whether it is hot or cold | |
| Upper surface hot or lower surface cold | $10^4$–$10^7$ $10^7$–$10^{11}$ | | 0.54 0.15 | 0.25 0.33 |
| Lower surface hot or upper surface cold | $10^5$–$10^{11}$ | | 0.58 | 0.20 |

range of Rayleigh numbers, Prandtl numbers, and geometries. One must select the proper (i.e., the recommended) length scale $L$ that Ra and Nu are defined in terms of, for each geometry. Churchill and Chu's expression is

$$\text{Nu}_L^{1/2} = \text{Nu}_o^{1/2} + \left( \frac{\text{Ra}_L/300}{[1 + (0.5/\text{Pr})^{9/16}]^{16/9}} \right)^{1/6} \qquad (14.4.3)$$

Table 14.4.2 defines the recommended choices of $L$ for each geometry listed, along with the values of the coefficient Nu$_o$. Note that these choices of $L$ differ in some cases from those of Table 14.4.1. It is emphasized that the entries in Tables 14.4.1 and 14.4.2 are guidelines for quick estimates. It is far better to have experimental data available in the form of a dimensionless correlation. Several examples are presented in Figs. 14.4.1 to 14.4.3. Note the choice of length scale for Ra and Nu in each caption. We may now apply these correlations to some simple design problems.

**Table 14.4.2** Natural Convection Correlations of the Form of Eq. 14.4.3

| Geometry | L | Nu$_o$ |
|---|---|---|
| Vertical surface | L | 0.68 |
| Vertical cylinder | L | 0.68 |
| Horizontal cylinder | $\pi D$ | $0.36\pi$ |
| Sphere | $\pi D/2$ | $\pi$ |
| Inclined disk[a] | $9D/11$ | 0.56 |

[a] If $\theta$ is the angle made by the plane of the disk to the vertical, replace $g$ in Ra by $g \cos \theta$.

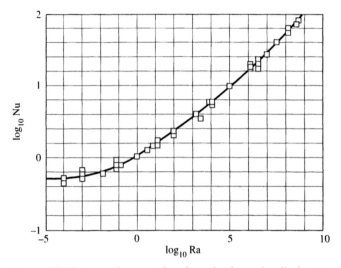

**Figure 14.4.1** Natural convection from horizontal cylinders to gases and liquids: use diameter $D$ for the length scale $L$. After McAdams, *Heat Transmission*, 3rd ed, McGraw-Hill, 1954.

---

**EXAMPLE 14.4.1**   *Heat Loss from a Horizontal Pipe*

An uninsulated pipe, 5 cm in outside diameter, runs horizontally across an open laboratory that is maintained at a temperature of 30°C. Air enters the pipe at a rate of 300 kg/h, at a temperature of 80°C, and the length of the pipe run is 20 m. The *downstream* pressure of the air *inside* the pipe is measured and found to be $1.0 \times 10^5$ Pa above atmospheric pressure. Find the temperature drop of the air flowing in the pipe due to natural convection losses to the room.

We will assume that the overall heat transfer coefficient is dominated by the *outside* convective resistance. Hence our primary focus is on external conditions. We will estimate the uninsulated external pipe wall temperature to be 70°C (just slightly cooler

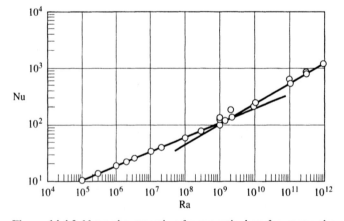

**Figure 14.4.2** Natural convection from vertical surfaces: use the vertical height of the surface for $L$ in Ra and Nu. After Eckert and Jackson, NACA RFM 50 D25, July 1950.

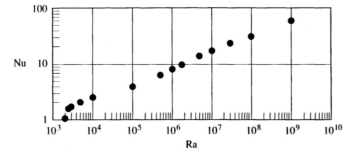

**Figure 14.4.3** Natural convection across a horizontal layer of water bounded by parallel rigid isothermal planes. The *lower* plane is hot. Use the vertical thickness of the layer for the length scale $L$ in Ra and Nu. After Hollands *et al.*, *Int. J. Heat Mass Transfer,* **18,** 879 (1975).

than the flowing stream within). At these assumed conditions we find (using the film temperature of 50°C)

$$\rho = 1 \text{ kg/m}^3 \qquad \mu = 2 \times 10^{-5} \text{ kg/m} \cdot \text{s} \qquad \text{Pr} = 0.7 \qquad k = 0.029 \text{ W/m} \cdot \text{K}$$
$$C_p = 1005 \text{ J/kg} \cdot \text{K}$$

Since air properties are not such a strong function of temperature, this set should yield a quick and reasonable approximation which can be improved by a simple trial-and-error procedure. From Fig. 14.1.3 we find $\gamma = 10^8 \text{ (m}^3 \cdot \text{K)}^{-1}$. We must now choose a correlation for Nu for a horizontal pipe. We will use Fig. 14.4.1. The length scale in Ra and Nu is the pipe diameter. Hence we find, for $\Delta T = 40$ K,

$$\text{Ra} = \text{Gr Pr} = \gamma L^3 \, \Delta T \, \text{Pr} = (10^8)(0.05^3)(40)(0.7) = 3.5 \times 10^5 \qquad \textbf{(14.4.4)}$$

From Fig. 14.4.1 the Nusselt number is approximately

$$\text{Nu} = 10 \qquad \textbf{(14.4.5)}$$

and we find the external heat transfer coefficient to be

$$h = \frac{k \, \text{Nu}}{D} = \frac{0.029 \times 10}{0.05} = 5.8 \text{ W/m}^2 \cdot \text{K} \qquad \textbf{(14.4.6)}$$

The heat balance now takes the form

$$Q_h = wC_p \, \Delta T_{\text{inside}} = hA \, \Delta T_{\text{outside}} \qquad \textbf{(14.4.7)}$$

or

$$\Delta T_{\text{inside}} = \frac{hA \, \Delta T_{\text{outside}}}{wC_p} \qquad \textbf{(14.4.8)}$$

The pipe area (the external surface area) is $A = \pi D L = 3.14 \text{ m}^2$, so we find

$$\Delta T_{\text{inside}} = \frac{5.8 \times 3.14 \times 40}{(300/3600) \, 1005} = 8.3 \text{ K} \qquad \textbf{(14.4.9)}$$

According to this result, the air exits at a temperature of about 72°C. Hence the use of 70°C for the pipe wall temperature seems reasonable, if we can confirm that the inside heat transfer coefficient is much greater than that outside. This is not difficult to do, and we leave it as an exercise (Problem 14.7).

---

**EXAMPLE 14.4.2**   *Heat Transfer across a Horizontal Bounded Air Film*

A horizontal air film separates two surfaces that are maintained at different temperatures. The lower surface is at 100°C, and the upper surface is at 50°C. We want to find the heat flux due to natural convection as a function of the film thickness. Assume that the air is at atmospheric pressure.

As a basis for comparison, let us first calculate the heat flux associated with *conduction* across a stagnant air film. From Fourier's law we simply have

$$q = \frac{k\,\Delta T}{L} = \frac{0.03 \times 50}{L} = \frac{1.5}{L} \tag{14.4.10}$$

in SI units. (We have evaluated the thermal conductivity of air at the mean air temperature, 75°C.) When we find a model for $q$ versus $L$ under natural convection, Eq. 14.4.10 will provide a basis for evaluating the degree to which natural convection enhances heat transfer across this "insulating" film.

We need a model for the convective coefficient in this system. We will use Fig. 14.4.3, noting that the original reference (Hollands et al.) indicates that the correlation presented works quite well for air as well. From Fig. 14.1.3 we find, at 75°C, $\gamma = 7 \times 10^7$ in SI units. The Rayleigh number is

$$\text{Ra} = \text{Gr Pr} = \gamma L^3\,\Delta T\,\text{Pr} = (7 \times 10^7)(L^3)(50)(0.7) = 2.5 \times 10^9\,L^3 \tag{14.4.11}$$

and, from Fig. 14.4.3, for $L$ values of 0.001, 0.005, 0.01, and 0.1 m, we find Nu values of 1, 1, 1.5, and 10, respectively. The flux is written as

$$q = h\,\Delta T = \text{Nu}\,\frac{k\,\Delta T}{L} \tag{14.4.12}$$

In this form, and noting Eq. 14.4.10, we see that the Nusselt number is the factor by which the flux is enhanced by natural convection. Hence we may present our results in the form of Fig. 14.4.4. We learn from this example that natural convection can severely degrade the insulating capability of a "stagnant" layer of low conductivity fluid, such as air, if $L$ is large.

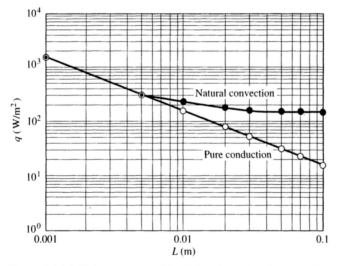

**Figure 14.4.4** Enhancement of heat flux by natural convection (Example 14.4.2).

**EXAMPLE 14.4.3**   *Temperature of a Surface*

A *net* radiant energy flux[1] of 800 W/m² is incident on a vertical black metal surface 3.5 m high and 2 m wide. The metal is insulated on the back side so that at equilibrium, the net incoming radiation is delivered by free convection to the surrounding air at 30°C. What average temperature will be attained by the plate?

Is 800 W/m² a large radiant energy flux? Solar radiation on the surface of the earth is approximately 1400 W/m², on a clear sunny summer day. Hence we can expect the surface to be similar to a black metal sheet in the sun. We assume uniform heat flux on the surface. Since we do not know the surface temperature, we must make an estimate. If the metal is not reflective, it could be very hot. Assume the surface temperature is 110°C. Evaluate air properties at a film temperature of (30 + 110)/2 = 70°C.

From Fig. 14.1.3 we estimate

$$\gamma = 7 \times 10^7 \text{ m}^{-3} \text{ K}^{-1} \qquad (14.4.13)$$

For $L = 3.5$ m,

$$\text{Gr}_L = \gamma \, \Delta T \, L^3 = (7 \times 10^7)(110 - 30)(3.5)^3 = 2.4 \times 10^{11} \qquad (14.4.14)$$

and

$$\text{Ra} = \text{Gr Pr} = 1.7 \times 10^{11} \qquad (14.4.15)$$

Using Fig. 14.4.2 we estimate

$$\text{Nu}_L = 600 \qquad (14.4.16)$$

$$h = \frac{k}{L} \text{Nu}_L = \frac{0.03}{3.5}(600) = 5.1 \text{ W/m}^2 \cdot {}^\circ\text{C} \qquad (14.4.17)$$

From $q = h \, \Delta T$ we find

$$\Delta T = \frac{q}{h} = \frac{800}{5.1} = 156°\text{C} = T_s - T_a = T_s - 30$$

$$T_s = 186°\text{C} \qquad (14.4.18)$$

This corresponds to a film temperature of

$$T_f = \frac{30 + 186}{2} = 108°\text{C} \qquad (14.4.19)$$

To get a more accurate result, we should "correct" $T_f$ and $\Delta T$ in Gr and do another iteration.

## 14.5   NATURAL CONVECTION IN AN ENCLOSED SPACE

In Example 14.2.1 we considered natural convection in a confined region between a pair of vertical parallel planes. If the region is also closed at the top and bottom, as suggested in Fig. 14.5.1, the hot fluid rising must turn around when it gets to the upper boundary, and flow down the cold surface. Hence the hot fluid transfers its heat to

---

[1] By "net" radiation we mean the difference between the absorbed radiation and the amount of energy reradiated away from the surface, by virtue of the temperature of the surface.

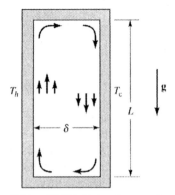

**Figure 14.5.1** Natural convection induces circulation in an enclosure.

the cold surface. Empirical relations exist for estimating the Nusselt number for such enclosures, and we review them briefly, and then look at an example.

We define the Grashof and Nusselt numbers for an enclosed fluid as follows:

$$\text{Gr}_\delta = \frac{g\,\beta(T_h - T_c)\,\delta^3}{\nu^2} \tag{14.5.1}$$

$$\text{Nu}_\delta = \frac{h\delta}{k} \tag{14.5.2}$$

Note that $\text{Gr}_\delta$ and $\text{Nu}_\delta$ are based on $\delta$ as the length scale, even though we normally use a vertical length as a characteristic scale for natural convection problems.

For liquids in an enclosure oriented as in Fig. 14.5.1 a recommended correlation for $\text{Nu}_\delta$ is

$$\text{Nu}_\delta = 0.42(\text{Gr}_\delta\,\text{Pr})^{0.25}\,\text{Pr}^{0.012}\left(\frac{L}{\delta}\right)^{-0.3} \tag{14.5.3}$$

in the parameter range

$$10^4 < \text{Ra} < 10^7$$
$$10 < \frac{L}{\delta} < 40 \tag{14.5.4}$$

For gases, a different correlation is usually used:

$$\text{Nu}_\delta = 0.2(\text{Gr}_\delta\,\text{Pr})^{0.25}\left(\frac{L}{\delta}\right)^{-0.11} \tag{14.5.5}$$

for

$$6 \times 10^3 < \text{Ra} < 2 \times 10^5$$
$$10 < \frac{L}{\delta} < 40 \tag{14.5.6}$$

Example 14.5.1 illustrates the use of such simple correlations.

**EXAMPLE 14.5.1**     *Natural Convection Heat Transfer across an Enclosed Air Space*

Air at atmospheric pressure is trapped between two sealed sheets of glass. Find the heat flux in the horizontal direction when the temperatures are (see Fig. 14.5.1) $T_h = 100°C$, $T_c = 40°C$, $L = 0.5$ m, and $\delta = 15$ mm. We will evaluate physical properties at the average "film" temperature, defined by

$$T_f = \frac{100 + 40}{2} = 70°C = 343 \text{ K} \tag{14.5.7}$$

We may use the ideal gas law for the air density at this temperature:

$$\rho = \frac{p}{RT} = \frac{10^5}{286(343)} = 1 \text{ kg/m}^3 \tag{14.5.8}$$

where, because we are calculating the *mass* density in SI units, we use $R = R_G/M_{w,air} = 8314 \text{ (J/kg–mol} \cdot \text{K)}/29\text{(kg/kg–mol)} = 286$.

For an ideal gas, $\beta$ is simply the inverse of the absolute temperature:

$$\beta = \frac{1}{T_f} = \frac{1}{343} = 2.9 \times 10^{-3} \text{ K}^{-1} \tag{14.5.9}$$

Other physical properties are found to be

$$\mu = 2 \times 10^{-5} \text{ kg/m} \cdot \text{s} \qquad k = 0.03 \text{ W/m} \cdot °C \tag{14.5.10}$$

$$\text{Pr} = 0.7 \tag{14.5.11}$$

We calculate the Grashof number and find

$$\text{Gr} = \frac{g \beta (T_h - T_c) \delta^3}{\nu^2} = \frac{9.8(2.9 \times 10^{-3})(100 - 40)(0.015)^3}{(2 \times 10^{-5}/1)^2} = 1.44 \times 10^4 \tag{14.5.12}$$

Hence the Rayleigh number is

$$\text{Ra} = \text{Gr Pr} = 10^4 \tag{14.5.13}$$

Since the enclosed fluid is a gas, we use Eq. 14.5.5 (noting that the restrictions of Eq. 14.5.6 hold):

$$\text{Nu}_\delta = \frac{h\delta}{k} = 0.2(\text{Gr}_\delta \text{Pr})^{0.25} \left(\frac{L}{\delta}\right)^{-0.11} = 1.35 \tag{14.5.14}$$

The heat flux is simply (cf. Eq. 14.4.12)

$$q = h \Delta T = \frac{k \, \text{Nu}_\delta \, \Delta T}{\delta} = 162 \text{ W/m}^2 \tag{14.5.15}$$

As in Example 14.4.2, the Nusselt number may be regarded as the factor by which natural convection enhances heat transfer, relative to conduction. Hence convection, in this example, increases the heat flux by 1.35 or 35%.

## 14.6   COMBINED NATURAL AND FORCED CONVECTION

We may encounter problems where both free and forced convection occur. The interaction of the two flows may be positive or negative in the sense suggested in Fig. 14.6.1. In interior flows, such as flow through a horizontal pipe, the interaction distorts the

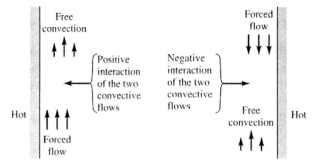

**Figure 14.6.1** The interaction of free and forced convection.

forced-flow profile significantly, and this effect may alter the total convection at the pipe wall. That is, the effects are not simply additive because of this "coupling" of the two flows.

For flow inside a horizontal pipe or tube, an important measure of the relative importance of forced and free convection is the ratio involving the Grashof and Reynolds numbers. Specifically, it is found that free convection often dominates when

$$\frac{Gr}{Re^2} \gg 1 \tag{14.6.1}$$

In general, we do not expect free convection to be very significant in a highly turbulent pipe flow, where forced convection is normally very strong. (See Problem 14.4 for a graphical method of evaluating the relative roles of free and forced convection.)

An empirical correlation for the Nusselt number in combined *laminar* pipe flow/free convection in *horizontal tubes* is of the form

$$Nu = 1.75 \left(\frac{\mu_b}{\mu_w}\right)^{0.14} [Gz + 0.012(Gz\ Gr^{1/3})^{4/3}]^{1/3} \tag{14.6.2}$$

The Graetz number is defined here as

$$Gz = Re\ Pr\ \frac{D}{L} \tag{14.6.3}$$

The term $\mu_b/\mu_w$ is the ratio of fluid viscosities at the bulk fluid temperature (this is simply the average fluid temperature) and the temperature at the inside surface of the tube. When raised to the 0.14 power, this ratio often yields a "correction" factor near unity, except where there is a very large temperature difference between the fluid and the wall. Many correlations of this form can be found in the literature, based on various sets of data. Example 14.6.1 illustrates the use of this correlation.

---

**EXAMPLE 14.6.1**  *Effect of Natural Convection on Laminar Flow Heat Transfer in a Horizontal Tube*

Air at atmospheric pressure and 25°C flows through a horizontal tube, 25 mm in diameter and 0.4 m long. The tube wall is maintained at 140°C. Plot the heat transfer coefficient as a function of velocity, for $\overline{U}$ in the range 1 to 100 cm/s.

To assess the need to account for natural convection, we must evaluate Re and Gr. Pick a value of $\overline{U}$ in the middle of the range of interest, say:

$$\overline{U} = 30\ cm/s \tag{14.6.4}$$

and evaluate physical properties at the film temperature:

$$T_f = \frac{140 + 25}{2} = 82.5°C = 355.5 \text{ K} \tag{14.6.5}$$

At this temperature

$$\rho_f = \frac{p}{RT} = \frac{10^5}{286(355.5)} = 1 \text{ kg/m}^3 \tag{14.6.6}$$

For gases, a good estimate of $\beta$ is obtained from the inverse of the absolute temperature (see Problem 14.3.):

$$\beta = \frac{1}{T_f} = 0.0028 \tag{14.6.7}$$

Other relevant properties are

$$\mu_f = 2.1 \times 10^{-5} \text{ kg/m} \cdot \text{s} \qquad \mu_w = 2.3 \times 10^{-5} \text{ kg/m} \cdot \text{s} \tag{14.6.8}$$

$$k_f = 0.03 \text{ W/m} \cdot °C \tag{14.6.9}$$

$$\text{Pr} = 0.7 \tag{14.6.10}$$

In Eq. 14.6.2, $\mu_b$ is evaluated at the bulk entrance temperature of 25°C:

$$\mu_b = 2 \times 10^{-5} \text{ kg/m} \cdot \text{s} \tag{14.6.11}$$

Then we find

$$\text{Re} = \frac{\rho \bar{U} D}{\mu} = \frac{(1)(0.3)(0.025)}{2.1 \times 10^{-5}} = 357 \tag{14.6.12}$$

$$\text{Gr} = \frac{\rho^2 g \beta (T_w - T_b) D^3}{\mu^2} = \frac{(1)^2(9.8)(0.0028)(140 - 25)(0.025)^3}{(2.1 \times 10^{-5})^2} = 1.1 \times 10^5 \tag{14.6.13}$$

$$\frac{\text{Gr}}{\text{Re}^2} = \frac{1.1 \times 10^5}{350^2} = O(1) \tag{14.6.14}$$

According to the rule of thumb stated earlier, free or natural convection will dominate only at velocities lower than this. Hence we might expect that a pure forced convection model would be appropriate at $\bar{U} = 30$ cms. But let's see what happens if we use Eq. 14.6.2 above. We need

$$\text{Gz} = \text{Re Pr} \frac{D}{L} = 350(0.7) \frac{0.025}{0.4} = 15 \tag{14.6.15}$$

so

$$\text{Nu} = 1.75 \left( \frac{2 \times 10^{-5}}{2.3 \times 10^{-5}} \right)^{0.14} [15 + 0.012[(15)(1.1 \times 10^5)^{1/3}]^{4/3}]^{1/3} = 7.8 \tag{14.6.16}$$

Then

$$h = \frac{k}{D} \text{Nu} = \frac{0.03(7.8)}{0.025} = 9.5 \text{ W/m}^2 \cdot °C \tag{14.6.17}$$

For laminar flow convection alone, using Fig. 12.5.6, we find

$$\text{Nu} = 4.3 \tag{14.6.18}$$

Hence the contribution of free convection is in fact comparable to that of laminar forced convection, at this flowrate. The determination of Nu at other velocities is left as a homework exercise. This example emphasizes the difficulty of using general rules regarding the relative importance of free and forced convection.

Turbulent flows can be altered by free convection only at very high Grashof numbers. An empirical expression recommended for use under such conditions is (see Fig. P14.4 in the Problems section for an indication of conditions under which this equation is appropriate.)

$$\text{Nu} = 4.69 \, \text{Re}^{0.27} \, \text{Pr}^{0.21} \, \text{Gr}^{0.07} \left(\frac{D}{L}\right)^{0.36} \tag{14.6.19}$$

For convection in *vertical* pipes and tubes the situation is more complex, since the effect of buoyancy is either to aid or oppose the flow, depending on the direction of the mean flow. Hence one finds empirical correlations for the two cases, and for each case a variety of correlations is available. We do not present any of them here, however.

## 14.7  AN INTEGRAL BOUNDARY LAYER ANALYSIS OF CONVECTION FROM A VERTICAL HEATED PLANE

We return to the problem described in Section 14.3 and illustrate the application of an integral boundary layer analysis. We imagine the existence of a boundary layer of thickness $\delta(z)$ across which the velocity and temperature fields change. For simplicity, we assume that the momentum and thermal boundary layers are identical. This is so only if the diffusivities for momentum and heat transfer are nearly the same (i.e., if the Prandtl number is near unity). This will be a reasonable approximation for gases. Figure 14.7.1 shows the control volume for the integral analysis.

The *net* rate of flow of $z$-directed momentum across horizontal planes at positions $z$ and $z + dz$ is (per unit width in the $x$ direction)

$$\frac{d}{dz}\left[\int_0^\delta \rho v_z^2 \, dy\right] \tag{14.7.1}$$

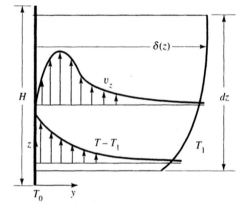

**Figure 14.7.1** Control volume for an integral boundary layer analysis.

The pressure difference that arises from buoyancy exerts a force on the control volume given by

$$\rho g \beta \left[ \int_0^\delta (T - T_1) dy \right] \tag{14.7.2}$$

A viscous shear force acts along the solid boundary of the control volume, and is given by

$$\mu \left[ \frac{\partial v_z}{\partial y} \right]_{y=0} \tag{14.7.3}$$

There is no comparable force along the boundary $\delta(z)$ because, by definition, this is the boundary at which the shear stress vanishes and the buoyancy-induced velocity vanishes. When we add these momentum and stress terms, we find

$$\frac{d}{dz} \left[ \int_0^\delta \rho v_z^2 \, dy \right] - \rho g \beta \left[ \int_0^\delta (T - T_1) dy \right] + \mu \left[ \frac{\partial v_z}{\partial y} \right]_{y=0} = 0 \tag{14.7.4}$$

This equation may also be obtained simply by averaging the $z$-directed momentum equation across the boundary layer. (See Problem 14.9.)

The net convective flow of thermal energy across the horizontal boundaries of the control volume is

$$\frac{d}{dz} \left[ \int_0^\delta \rho \, C_p \, v_z (T - T_1) dy \right] \tag{14.7.5}$$

and the contribution of conduction to the thermal balance is

$$k \left[ \frac{\partial T}{\partial y} \right]_{y=0} \tag{14.7.6}$$

so the energy balance takes the form

$$\frac{d}{dz} \left[ \int_0^\delta \rho \, C_p \, v_z (T - T_1) dy \right] + k \left[ \frac{\partial T}{\partial y} \right]_{y=0} = 0 \tag{14.7.7}$$

This single equation contains two unknowns: $v_z(z, y)$ and $T(z, y)$. The essence of the integral method is that we choose functional representations for these two dependent variables, as functions of the space variables $z$ and $y$, and then determine the coefficients of these representations that satisfy Eq. 14.7.7, if possible. The functional forms themselves must satisfy the boundary conditions on $v_z(z, y)$ and $T(z, y)$. One of the simplest choices we can make is a polynomial representation:

$$v_z = V(z) \frac{y}{\delta(z)} \left( 1 - \frac{y}{\delta(z)} \right)^2 \tag{14.7.8}$$

and

$$\frac{T - T_1}{T_0 - T_1} = \left( 1 - \frac{y}{\delta(z)} \right)^2 \tag{14.7.9}$$

These equations contain two unknown functions: $V(z)$ and $\delta(z)$. Equation 14.7.8 satisfies the condition that the induced velocity vanish at the solid surface, as well as at some (unknown) distance $y = \delta(z)$ from the surface. In addition, this choice of polynomial is "smooth," in the sense that the velocity gradient vanishes at $\delta(z)$. Similarly, Eq.

14.7.9 takes on the required values $T_o$ and $T_1$ at $y = 0$, and $y = \delta(z)$, respectively. The temperature gradient vanishes at $\delta(z)$, making this functional form smooth at that boundary.

When Eqs. 14.7.8 and 14.7.9 are substituted into Eqs. 14.7.4 and 14.7.7, a pair of ordinary differential equations for $V(z)$ and $\delta(z)$ is obtained:

$$\frac{d}{dz}\left(\frac{V^2 \delta}{105}\right) = \frac{g\beta (T_o - T_1) \delta}{3} - \frac{\nu V}{\delta} \tag{14.7.10}$$

and

$$\frac{d}{dz}\left(\frac{V \delta}{30}\right) = \frac{2\alpha}{\delta} \tag{14.7.11}$$

where

$$\alpha = \frac{k}{\rho C_p} \tag{14.7.12}$$

is the thermal diffusion coefficient.

Note what we have achieved at this point. Instead of working with (and attempting to solve) a pair of coupled, nonlinear partial differential equations for $v_z(z, y)$ and $T(z, y)$, we have a pair of coupled, nonlinear *ordinary* differential equations for $V$ and $\delta$. It is not difficult to confirm that the solutions are simple power functions in the forms

$$V = az^m \tag{14.7.13}$$

and

$$\delta = bz^n \tag{14.7.14}$$

When these expressions are substituted into Eqs. 14.7.10 and 14.7.11, we may equate the coefficients and the powers of $z$ and find the following solution functions:

$$V = 5.17 \left(\text{Pr} + \frac{20}{21}\right)^{-1/2} [g\beta (T_o - T_1)z]^{1/2} \tag{14.7.15}$$

and

$$\delta = 3.93 \,\text{Pr}^{-1/2} \left(\text{Pr} + \frac{20}{21}\right)^{1/4} \left[\frac{\nu^2 z}{g\beta (T_o - T_1)}\right]^{1/4} \tag{14.7.16}$$

While these solutions permit us to evaluate the detailed velocity and temperature profiles, our main interest is in the heat flux from the vertical surface. This is calculated from

$$q'(z) = -k \left(\frac{\partial T}{\partial y}\right)_{y=0} = \frac{2k \,\Delta T}{\delta(z)} \tag{14.7.17}$$

This local heat flux may be averaged over a vertical length $L$ with the result

$$\overline{q}' = \frac{1}{L} \int_0^L \frac{2k \,\Delta T}{bz^{1/4}} \, dz = \frac{8 \, k \,\Delta T \, L^{-1/4}}{3b} = \overline{h} \,\Delta T \tag{14.7.18}$$

where the coefficient $b$ may be obtained by comparing Eqs. 14.7.14 and 14.7.16. An average Nusselt number then follows as

$$\overline{\text{Nu}} = \frac{\overline{h}L}{k} = 0.68 \left(1 + \frac{20}{21 \,\text{Pr}}\right)^{-1/4} (\text{Pr Gr})^{1/4} \tag{14.7.19}$$

Upon inspection of Eq. 14.3.47, we see that the model presented there is quite close to the result of this integral analysis.

We note that the assumption that the velocity and thermal boundary layers are identical (hence there is only a single $\delta$ in this analysis) can be valid only if the Prandtl number is near unity. While this would appear to restrict the validity of this model to gases, it turns out that the coefficient of 0.68 is quite insensitive to Pr, and Eq. 14.7.19 gives a fairly good representation of data for both gases and liquids.

## 14.8   OPTIMUM SPACING FOR HEAT DISSIPATION

It is often necessary to rely on natural convection for the cooling of surfaces that are experiencing heating from internal sources. Applications range from motor housings to arrays of electronic circuit boards in computers and other devices. Some aspects of this problem can be understood by consideration of a simple model of natural convection in the region between a pair of parallel vertical smooth surfaces. Figure 14.8.1 shows the geometry and boundary conditions appropriate to the analysis that follows.

To find the average velocity induced by natural convection, we may write a simple momentum balance that equates the viscous losses due to laminar flow between parallel plates to the pressure gradient available to drive the flow. The fluid dynamic analysis, assuming constant properties and steady, fully developed flow with a mean velocity $U$, corresponds to a pressure gradient

$$\frac{dp}{dx} = -\frac{12\,\mu\,U}{b^2} - \rho g \tag{14.8.1}$$

If we treat the pressure gradient as in Section 14.3 (note the right-hand term of Eq. 14.3.5), we may write

$$\rho g + \frac{dp}{dx} = -\bar{\rho} g \beta\,(\overline{T}_f - T_o) \tag{14.8.2}$$

where we have introduced the mean fluid temperature $\overline{T}_f$ and the outside (ambient) temperature $T_o$. Hence we find an estimate of the average velocity based on this simple force balance as

$$U = \frac{\bar{\rho} g \beta\, b^2 (\overline{T}_f - T_o)}{12\,\mu} \tag{14.8.3}$$

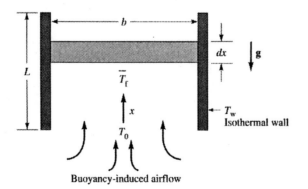

**Figure 14.8.1** Natural convection between isothermal parallel plates.

A heat balance over a differential section of "height" $dx$ takes the form

$$\rho b U C_p \, dT = 2h(T_w - T_f)dx \tag{14.8.4}$$

If we take $h$ to be constant along the length of the channel, we may integrate Eq. 14.8.4 directly to find the average fluid temperature along the channel:

$$\overline{T}_f = T_w - (T_w - T_o) \exp\left(\frac{-2hx}{\rho b U C_p}\right) \tag{14.8.5}$$

We will define a Nusselt number based on the temperature difference between the wall and the ambient fluid:

$$\mathrm{Nu}_o = \frac{h_o b}{k} = \frac{qb}{k(T_w - T_o)} \tag{14.8.6}$$

Note that this Nusselt number uses $b$ as the length scale. In this equation, $q$ is the heat flux from the isothermal surfaces into the fluid, averaged over the length $L$ in the vertical direction. This is obtained from an overall heat balance:

$$2q L = \overline{\rho} b U C_p [\overline{T}_f(L) - T_o] \tag{14.8.7}$$

We now use Eq. 14.8.5, at $x = L$, which leads to the following result:

$$2qL = \overline{\rho} b U C_p (\overline{T}_w - T_o)\left[1 - \exp\left(\frac{-2\,\mathrm{St}\,L}{b}\right)\right] \tag{14.8.8}$$

(Note the definition of the Stanton number in Chapter 12, Eq. 12.4.17.)

In applications of practical interest, it is often the case that the vertical length $L$ is sufficiently large compared to the spacing $b$ that $\mathrm{St}\,L/b \gg 1$. Then Eq. 14.8.8 simplifies to

$$2qL = \overline{\rho} b U C_p(\overline{T}_w - T_o) \tag{14.8.9}$$

and we find

$$\mathrm{Nu}_o = \frac{h_o b}{k} = \frac{\overline{\rho} b^2 U C_p}{2kL} \tag{14.8.10}$$

Consistent with the approximation for large $L$, the mean fluid temperature is approximately $T_w$ (note Eq. 14.8.5), and this permits us to write the mean velocity $U$ as

$$U = \frac{\overline{\rho} g \beta \, b^2(T_w - T_o)}{12\,\mu} \tag{14.8.11}$$

Finally, we find that Eq. 14.8.10 becomes, upon eliminating $U$,

$$\mathrm{Nu}_o = \frac{\overline{\rho}^2 g \beta \, b^4 C_p(T_w - T_o)}{24\,\mu k L} = \frac{\mathrm{Ra}'}{24} \tag{14.8.12}$$

where a modified Rayleigh number $\mathrm{Ra}'$ is defined as

$$\mathrm{Ra}' = \frac{\overline{\rho}^2 g \beta \, b^3 C_p(T_w - T_o)}{\mu k}\frac{b}{L} = \mathrm{Ra}_b \frac{b}{L} \tag{14.8.13}$$

This is a very different result from that obtained previously for an *isolated* vertical surface, Eq. 14.7.19, which we may write in the form

$$\mathrm{Nu}_\infty = 0.55(\mathrm{Ra}')^{1/4} \tag{14.8.14}$$

(In converting Eq. 14.7.19 we must note that the Nusselt number in that equation is based on $L$ as the length scale, as is the Grashof number. We also introduce the value

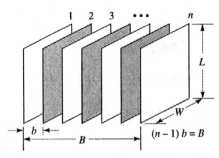

**Figure 14.8.2** A set of parallel equally spaced plates, each at $T_w$.

of Pr = 0.72 for air in the term following the coefficient 0.68, but leave the Prandtl number unspecified in the Pr Gr product, writing that product as the Rayleigh number.)

We expect that Eq. 14.8.12 provides a poor model for the Nusselt number as $b$ grows comparable to the length $L$ and that Eq. 14.8.14 becomes more appropriate. It is possible to combine Eqs. 14.8.12 and 14.8.14 into a format that fits data well over a wide range of values of $b$. Examination of experimental data suggests the following form:

$$ \text{Nu} = \left[ \frac{576}{(\text{Ra}')^2} + \frac{3.3}{\sqrt{\text{Ra}'}} \right]^{-1/2} \qquad \textbf{(14.8.15)} $$

We may regard Eq. 14.8.15 as an empirical relationship based on some simple theoretical models of natural convection. For large and small Ra' we recover Eqs. 14.8.12 and 14.8.14. Equation 14.8.15 will now serve as a basis for optimization with respect to the spacing between plates. Figure 14.8.2 will aid the discussion.

We assume that the plates are of negligible thickness and that each is somehow maintained at the same temperature $T_w$. The total width of the "stack" of plates is fixed at the value $B$. For any number of plates $n$, the interplate spacing is given by the expression

$$ b = \frac{B}{n-1} \qquad \textbf{(14.8.16)} $$

The goal of the analysis to follow is to find an optimum spacing $b^*$ that maximizes the total heat loss, within the constraint that $B$ is fixed. We assume that both surfaces of each plate behave according to Eq. 14.8.15. This will not be true for the outer surfaces of the two end plates, each of which will behave as an isolated plate. We assume, however, that $n$ is large enough that we may ignore this detail. Then the total heat loss is

$$ Q_{h,\text{total}} = 2nWLh\,\Delta T \qquad \textbf{(14.8.17)} $$

where $\Delta T = T_w - T_o$. We approximate $n$ as

$$ n = \frac{B}{b} \qquad \textbf{(14.8.18)} $$

and we write $h$ as

$$ h = \text{Nu}\frac{k}{b} \qquad \textbf{(14.8.19)} $$

Then Eq. 14.8.17 becomes

$$Q_{h.total} = 2 \frac{B}{b} WL \frac{k}{b} Nu \, \Delta T \tag{14.8.20}$$

It will be convenient to write the Rayleigh number in the form Ra'(b)

$$Ra' = \frac{\bar{\rho}^2 g\beta \, C_p(T_w - T_o)}{\mu kL} b^4 = \left(\frac{b}{M}\right)^4 \tag{14.8.21}$$

where $M$ is a combination of the parameters characteristic of this system, *other than b*. This permits us to write Eq. 14.8.20 in the form

$$F' \equiv \frac{M^2 \, Q_{h.total}}{2 \, WBLk \, \Delta T} = \left(\frac{M}{b}\right)^2 \left[\frac{576}{(b/M)^8} + \frac{3.3}{(b/M)^2}\right]^{-1/2} \tag{14.8.22}$$

To find the optimum spacing $b*$ that maximizes the heat loss, we differentiate Eq. 14.8.22 with respect to $b$, set the derivative to zero, and solve for $b = b*$. The algebra is simplified if we define

$$\phi = \left(\frac{b}{M}\right)^2 \tag{14.8.23}$$

and write Eq. 14.8.22 in the form

$$(F')^{-2} \equiv F = \frac{576}{\phi^2} + 3.3\phi \tag{14.8.24}$$

We now easily find

$$\frac{\partial F}{\partial \phi} = -\frac{1152}{\phi^3} + 3.3 = 0 \tag{14.8.25}$$

from which we can obtain the optimum we seek as

$$\phi* = 7.03 = \left(\frac{b*}{M}\right)^2 \tag{14.8.26}$$

or

$$Ra'* = 49.4 \tag{14.8.27}$$

With Eq. 14.8.21 we may find the optimum $b*$ for any given set of conditions. Note that the optimum $b$ depends on $T_w$.

### EXAMPLE 14.8.1    *Design and Performance of a Convective Heater*

A space heater is designed along the lines of the discussion of this section. It consists of a set of parallel heated plates of vertical height $L = 0.4$ m and width $W = 0.5$ m. A mechanism permits the plates to be maintained at a uniform temperature of $T_w = 80°C$. Optimize the design with respect to the total heat that can be transferred by natural convection to air at 20°C and 1 atm total pressure. The plates can be stacked in a region of width $B = 0.2$ m. Give the optimum plate spacing, and the expected rate of heat transfer to the room.

The optimum design is based on Eq. 14.8.27. We begin by finding values for physical properties of air at the mean temperature of 60°C. These are

$$\rho = 1.03 \text{ kg/m}^3 \qquad C_p = 1007 \text{ J/kg} \cdot \text{K} \qquad k = 0.028 \text{ W/m} \cdot \text{K}$$

$$\mu = 2 \times 10^{-5} \text{ Pa} \cdot \text{s} \qquad \beta = \frac{1}{T} = \frac{1}{323} \text{ K}^{-1}$$

From Eq. 14.8.21 we find (noting Eq. 14.8.27 for Ra' at the optimum)

$$\text{Ra}' = 8.68 \times 10^9 \, b^4 = 49.4 \qquad (14.8.28)$$

Hence the optimum spacing is

$$b = 0.0087 \text{ m} = 8.7 \text{ mm} \qquad (14.8.29)$$

From Eq. 14.8.22 we find

$$F' = \frac{M^2 Q_{h,\text{total}}}{2WBL \, k \, \Delta T} = \frac{1}{49.4}\left[\frac{576}{(49.4)^2} + \frac{3.3}{7.03}\right]^{-1/2} = 0.024 \qquad (14.8.30)$$

or

$$Q_{h,\text{total}} = \frac{2WBL \, k \, \Delta T}{M^2} F' = \frac{2(0.5)(0.2)(0.4)(0.028)(60)(0.024)}{(8.68 \times 10^9)^{-1/2}} = 302 \text{ W} \qquad (14.8.31)$$

## SUMMARY

Natural convection, sometimes called "free convection," arises from the density difference between the fluid in contact with a surface and the surrounding fluid. If the temperature difference between the surface and its surrounding fluid is great enough, the resulting buoyancy force can sustain a significant convective flow. Some very simple models of natural convection can be developed, and they show that the Grashof number (Eq. 14.1.6) is a key dimensionless parameter. In addition, simple models suggest that the Nusselt number for natural convection is expected to be related to the product of the Grashof number times the Prandtl number. This product is called the Rayleigh number (Eq. 14.3.47). Experimental data agree broadly with such simple models, and Nusselt number data are often correlated empirically by an expression of the form

$$\text{Nu} = C \, \text{Ra}^n$$

The coefficients $C$ and $n$ may depend on the Rayleigh number, and Tables 14.4.1 and 14.4.2 summarize many experimental observations. Over large ranges of the Rayleigh number, one finds graphical correlations of data, such as given in Figs. 14.4.1 to 14.4.3, to be useful for design purposes.

Often when there is *forced* convection, natural convection can contribute significantly to the overall heat transfer. Various suggestions are available for deciding whether one or the other mechanism dominates, and therefore what Nusselt number correlation to use, in specific situations. One suggestion is that natural convection dominates when $\text{Gr}/\text{Re}^2 \gg 1$. Alternatively, a "map" is shown in Fig. P14.4 which suggests that the choice of empirical correlation depends in a complicated way on the Reynolds number of the forced flow, and on the Rayleigh number and the geometry.

In Section 14.7 we demonstrate that the modeling tools we have acquired allow us to use an integral boundary layer analysis to derive a Nusselt number model that mimics experimental observations. In Section 14.8 the ideas introduced in this chapter permit us to carry out an optimal design of a cooling system dominated by natural convection.

## PROBLEMS

**14.1** Return to Chapter 10, Section 10.2.1, where we found that provision of a stagnant air film can significantly reduce the heat loss through a window pane. To what extent is the efficiency reduced by natural convection in the air space? Assume that the height of the window is 1 m.

**14.2** In addition to Eq. 14.6.2, several other recommended empirical correlations are available. For example, for laminar flow with a constant wall temperature, one finds

$$\text{Nu} = 1.75 \left(\frac{\mu_b}{\mu_w}\right)^{0.14}$$
$$\left[\text{Gz} + 12.6 \left(\frac{\text{Gr Pr } D}{L}\right)^{0.4}\right]^{1/3} \quad \textbf{(P14.2.1)}$$

and, for gases only,

$$\text{Nu} = 0.6(\text{Re Ra})^{0.2} \text{ for Pr} = 0.72 \quad \textbf{(P14.2.2)}$$

Use these two equations, and Eq. 14.6.2, and plot Nu versus Re for [1 < Re < 100] for air flowing through a horizontal pipe of diameter 3 cm and length 1 m. The air enters at 25°C, and the pipe wall is at 100°C. Repeat for a pipe of diameter 10 cm and length 10 m.

**14.3** Finish Example 14.6.1 by plotting the heat transfer coefficient as a function of velocity, for $\bar{U}$ in the range 1 to 100 cm/s. On the same plot show the result that is obtained by neglecting free convection.

**14.4** A rule of thumb is that free (natural) convection is negligible in pipe flow heat transfer as long as $\text{Gr}/\text{Re}^2 < 1$. Metais and Eckert (*J. Heat Transfer, Trans. ASME*, May 1964, p. 295) presented an empirical "roadmap" to be used in selecting a "route" to estimation of a Nusselt number when free and forced convection interact. Figure P14.4 is based on their recommendations. Is this consistent with the results of Problem 14.3?

**14.5** Using Fig. P14.4, would you anticipate that free convection would be significant in Example 14.6.1?

**14.6** Prove that $\beta$, defined in Eq. 14.1.2, is simply

$$\beta = \frac{1}{T} \quad \textbf{(P14.6)}$$

for an ideal gas, where $T$ is absolute temperature.

**14.7** Return to Example 14.4.1 and estimate the *internal* heat transfer coefficient; compare your estimate to that found external to the pipe. Is the neglect of the internal resistance reasonable under the conditions of the problem? Would this be true if the internal airflow were only 30 kg/h?

**14.8** Return to Example 14.5.1. What is the effect of reducing the pressure of the gas enclosed by the sheets of glass? Plot the heat flux as a function of absolute pressure in the range [0.1 < p < 1] atm.

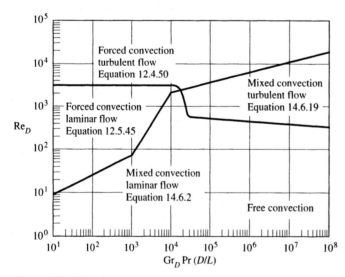

**Figure P14.4** Regimes of mixed free and forced convection, for flow through a *horizontal* pipe. Region boundaries are approximate.

**14.9** Show that Eq. 14.7.4 may be obtained by starting with the $z$-directed momentum equations in boundary layer form and then integrating each term with respect to $y$ over the region $[0, \delta]$. (This is tricky algebraically. At one point, to get the convective terms in the right form, you will need the continuity equation.)

**14.10** Use Eq. 14.7.17 to obtain the local Nusselt number along the vertical surface.

**14.11** A thin rectangular sheet is suspended vertically from a wire attached at the top of the sheet. The vertical length of the sheet is 20 cm, and the width is 10 cm. The sheet thickness is 100 $\mu$m. The density of the sheet is 6 g/cm³. The sheet is heated by radiation, and it has a uniform temperature of 150°C. The ambient medium is air at atmospheric pressure and 50°C. Is the shear stress of the thermally induced flow sufficient to lift the sheet?

**14.12** Circuit boards in electronic devices often depend on free convection cooling to maintain low enough temperatures to avoid thermal damage of the circuits. The geometry of these systems is often complex. Several experimental and theoretical studies relevant to this application have been carried out (see, e.g., Sefcik et al., *Int. J. Heat Mass Transfer*, **34**, 3037 (1991); Kato et al., *J. Chem. Eng. Japan*, **24**, 568 (1991); Bar-Cohen and Roshenow, *Trans. ASME, J. Heat Transfer*, **106**, 116 (1984)]. A geometrical model that has some of the important features is shown in Fig. P14.12.1. For the case of unobstructed flow ($G/W = 1$) there is an empirical correlation of the form

$$\overline{\mathrm{Nu}} = \frac{hW}{k} = \left[ \frac{144}{(\mathrm{Ra}\ W/H)^2} + \frac{2.8}{(\mathrm{Ra}\ W/H)^{1/2}} \right]^{-1/2}$$

**(P14.12)**

where the Rayleigh number is based on the width $W$, and not on the height $H$. Note that this correlation is for the case of asymmetric heating, where one surface is adiabatic and the other is isothermal.

Sefcik et al. solve the Navier–Stokes and convective energy equations numerically and find the average Nusselt number as a function of Grashof number, for varying combinations of geometrical parameters. One of their results is shown in Fig. P14.12.2. Compare the empirical correlation of Eq. P14.12 to this theoretical result. What is the Grashof number for the case $W = 2$ cm, $H = 10$ cm, $G/W = 0.33$, $T_w = 320$ K, and $T_{air} = 290$ K? (The Grashof number is based on the width $W$.)

**14.13** It is proposed to enhance the heat transfer by natural convection from a structure through the use of a set of vertical parallel fins. Use the analysis of Section 14.8 to provide an estimate of the optimum spacing. The ambient temperature is 320 K, and the fins are expected to be at a uniform temperature of 380 K. The ambient fluid is air at atmospheric pressure. The fin height is $L = 5$ cm. What spacing do you suggest?

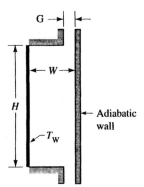

**Figure P14.12.1** Two-dimensional heated enclosure.

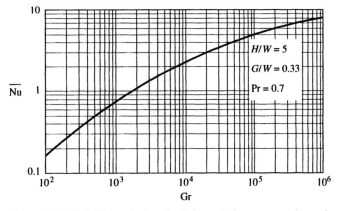

**Figure P14.12.2** Numerical calculation of the average Nusselt number along the isothermal surface for the geometry of Fig. P14.12.1.

**14.14** Equation P14.12 is quite different from Eq. 14.8.15, particularly with regard to the coefficient of the first term in brackets. Explain this difference.

**14.15** Rework Problem 14.13. Take into account the possibility that the fins are not 100% efficient, and so are not at 380 K uniformly along the fin length. Assume that the fins are of thickness 1 mm and are made of copper.

**14.16** A stack of silicon wafers, at 300°C, is placed in a still-air environment at 25°C, as shown in Fig. P14.16. Wafers are 6 inches in

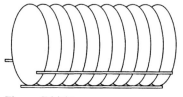

**Figure P14.16**

diameter, 1 mm thick, spaced 1 cm apart. Estimate the time required to bring the wafers to room temperature.

**14.17** A silicon wafer, 6 inches in diameter and 1 mm thick, is initially at 300°C. The wafer is sitting in a vertical orientation in air at 25°C.

**a.** If the air were blowing rapidly past the wafer, normal to the face of the wafer, how fast would the wafer cool to the air temperature?

**b.** In fact, the air is not being blown. The wafer is observed to cool to room temperature in 10 min. Estimate the heat transfer coefficient that corresponds to this rate of cooling.

**14.18** Data on natural convection heat transfer from a vertical flat plate in air were presented by Saunders [*Proc. R. Soc.*, **A157**, 278 (1936)]. The data were obtained using plates with heights in the range $0.33 < L < 23$ cm, and over a range of air pressures from 0.04 to 68 atm. A selection of these data is replotted in Fig. P14.18. Test the ability of Eq. 14.3.46 to fit these data.

**14.19** Data presented (see Fig. P14.19) by Hanzawa et al. [*J. Chem. Eng. Japan*, **23**, 378 (1990)] represent the natural convection heat transfer from the lower disk-shaped heated surface ($T = T_H$) of a cylindrical can of height $H$ and radius $r_w$. The upper disk-shaped surface was insulated, and the curved sidewall of the can was maintained at a constant temperature $T_w$. Test the ability of Eq. 14.4.1 to fit these data. Give a value for the coefficient $n$ of that equation.

**14.20** Shiina et al. [*Int. J. Heat Mass Transfer*, **37**, 1605 (1994)] studied natural convection within a hemispherical enclosure heated on the lower horizontal circular face and cooled on the hemispherical surface. They measured the heat flux across the hemispherical surface and calculated the average Nusselt number on that surface. Some of their data obtained with a variety of liquids are shown in Fig. P14.20. Test the ability of Eq. 14.4.1 to fit these data. Give a value for the coefficient $n$ of that equation.

**14.21** A pan of water 8 cm deep is placed on a stove-top burner. The burner element is thermostatically controlled and maintains the bottom of the pan at 100°C. Assuming that the top sur-

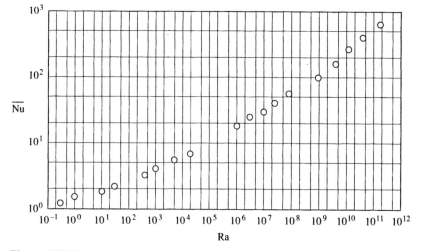

**Figure P14.18** Data on natural convection from a vertical flat plate.

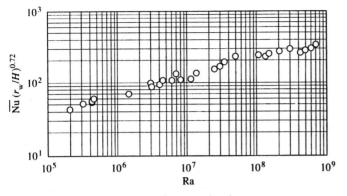

**Figure P14.19** Data on natural convection in a can.

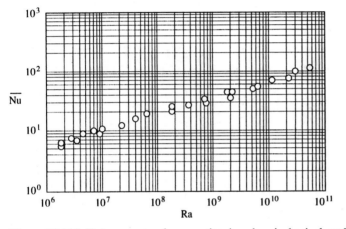

**Figure P14.20** Data on natural convection in a hemispherical enclosure.

face of the water is initially at 20°C, what is the initial rate of heat transfer from the burner to the water? The pan is circular, with a 15 cm diameter.

**14.22** Following the lines of Example 14.8.1, plot the optimum spacing and the expected heat transfer at each optimum as a function of $T_w$ in the range [40, 95°C].

**14.23** Following the lines of Example 14.8.1, plot the expected heat transfer as a function of $T_w$ in the range [40, 95°C], assuming a constant (hence, nonoptimum) plate spacing of 8.7 mm.

**14.24** Data are available from the manufacturer for the properties of a high temperature heat transfer fluid, Dow Syltherm 800. From the following information, plot $\beta$ versus $T$ (see Eq. 14.1.2) and $\gamma$ versus $T$ (see Eq. 14.1.9) over the range $0 < T < 300$°C.

$$\rho(kg/m^3) = 954 - 0.919T + (4.25 \times 10^{-4} T^2)$$
$$- (1.67 \times 10^{-6} T^3) \quad \text{with } T \text{ in } °C$$

$$\ln \mu(Pa \cdot s) = -14.03 + \frac{5324}{T} - \frac{1.178 \times 10^6}{T^2}$$

$$+ \frac{1.253 \times 10^8}{T^3} \quad \text{with } T \text{ in } K$$

**14.25** Confirm that Eqs. 14.7.13 and 14.7.14 satisfy Eqs. 14.7.10 and 14.7.11, and in doing so derive Eqs. 14.7.15 and 14.7.16.

# Chapter 15

# Heat Transfer by Radiation

In some respects the physics of radiation is simpler than that of convective heat transfer. Details of the flow field at the radiating surface are irrelevant, and we need only concern ourselves with the surface itself, as well as any other surfaces that are radiating to the surface of interest. Certain physical, and especially geometrical, aspects of radiation are complex, however, and as a consequence we have to choose between an exact approach that is quite detailed and a grossly oversimplified approximate approach. It is the latter—simplified—approach that we deal with in this chapter. In many cases of technological importance this will provide a model of the heat transfer that is accurate enough to be an aid to analysis and design.

## 15.1 INTRODUCTION

Radiation is the energy emitted by any material object as a consequence of its temperature. The transmission of this form of energy does not require an intervening *material* medium. Radiation can be transmitted through a vacuum. Whether radiation is transmitted through a *material* body—solid, liquid, or gas—depends on the nature of the body and the wavelength or frequency of the radiation.

Radiation that impinges on a material object can be *absorbed, reflected,* and/or *transmitted.* Figure 15.1.1 shows this situation schematically. We denote the fraction of the incident flux that is reflected, absorbed, and transmitted by the terms $\rho$, $a$, and $\tau$, respectively. Although $\rho$, $a$, and $\tau$ depend on material properties, they may vary with the wavelength or frequency of the incident radiation. We can at least say that the fractions sum to unity:

$$\rho + a + \tau = 1 \tag{15.1.1}$$

We begin our discussion of radiation by treating hypothetical idealized situations, and then we apply these ideas—with modifications—to real problems.

When a hot object *emits* radiation, measured by an energy flux $q^e$, some of that flux may be *incident* on another surface. We denote the incident flux by $q^i$. For *opaque* solids, by definition, the incident flux is not transmitted through the body. Hence

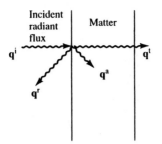

**Figure 15.1.1** Fate of an incident radiant flux impinging on matter.

$\tau = 0$. From Eq. 15.1.1 we see that the incident flux may be absorbed and reflected. We define the *absorptivity a* by

$$a = \frac{q^a}{q^i} \tag{15.1.2}$$

Absorptivity may be frequency (wavelength) dependent, so we can define

$$a_\nu = \frac{q^a_\nu}{q^i_\nu} \tag{15.1.3}$$

as the ratio of energy absorbed at a specific frequency, $\nu$, to the incident energy of that same frequency.

We define three idealized "bodies" with respect to their absorption of radiation. A **black body** absorbs *all* incident radiation of any frequency. Hence

$$a_\nu \equiv 1 \qquad \text{for all } \nu \tag{15.1.4}$$
**Black body**

A **gray body** does not absorb all incident radiation. It reflects some fraction $\rho = 1 - a$ of the incident radiation. However, its absorptivity is independent of frequency. Hence

$$a_\nu \neq f(\nu) \tag{15.1.5}$$
$$a_\nu = a < 1$$
**Gray body**

A **real body** is anything other than gray or black.

$$a_\nu = a_\nu(\nu) < 1 \tag{15.1.6}$$
**Real body**

With these introductory definitions in hand, we turn immediately to models of radiation exchange.

## 15.2 EMISSION AND EXCHANGE OF RADIATION

A black body emits radiation according to the so-called Planck distribution law:

$$q^e_{b\lambda} = \frac{2\pi c^2 h}{\lambda^5} \left[ \exp\left(\frac{ch}{\lambda \kappa T}\right) \right]^{-1} \tag{15.2.1}$$

We call $q^e_{b\lambda}$ the "emissive power" of the radiation. It is actually a power *flux*—it is a *rate* of energy (hence power) transmitted by radiation, per unit of radiating area. Note that $q^e_{b\lambda}$ depends on the wavelength of the radiation, $\lambda$. (Wavelength and frequency are related by the velocity of the wave, which in this case is the speed of light, denoted $c$.) In Eq. 15.2.1 we find the speed of light $c$, Planck's constant $h$, and the Boltzmann constant $\kappa$. In SI units these constants have the values

$$c = 3 \times 10^8 \text{ m/s} \qquad h = 6.63 \times 10^{-34} \text{ J} \cdot \text{s} \qquad \kappa = 1.38 \times 10^{-23} \text{ J/K}$$

If we integrate $q^e_{b\lambda}$ over all wavelengths, we find

$$q^e_b = \int_0^\infty q^e_{b\lambda} \, d\lambda = \sigma T^4 \tag{15.2.2}$$

where $\sigma$, the Stefan–Boltzmann constant, has the value $\sigma = 5.67 \times 10^{-8} \text{ W/m}^2 \cdot \text{K}^4$.

Real bodies emit less radiation than the idealized black body, and we define the *emissivity* of a real body as

$$e = \frac{q^e}{q^e_b} \qquad \text{and} \qquad e_\nu = \frac{q^e_\nu}{q^e_{b\nu}} \tag{15.2.3}$$

Typical values for $e$ are as follows:

| | |
|---|---|
| Polished metals | 0.03–0.08 |
| Nonmetals | 0.65–0.95 |

Kirchhoff's law states that for black bodies at thermal equilibirum, $e = a$. We use this statement as an assumption if no specific information is available on the relationship of $e$ to $a$ under *nonequilibrium* conditions. Since most heat transfer problems of interest to us do take place under nonequilibrium conditions, this approximation is frequently used.

It follows from the foregoing definitions that a real body emits radiation according to

$$q^e = e\sigma T^4 \tag{15.2.4}$$

The first thing we note is the very strong dependence of the magnitude of the flux of radiation power on absolute temperature. Thus, in practice, we can expect to find that radiation is largely a high temperature phenomenon.

### EXAMPLE 15.2.1   *Heat Losses from a Plate*

A large square metal plate, one meter in length and width, forms a section of the floor of a room. The room air is stationary (i.e., there is no forced convection) and is at 300 K (27°C). Plot the heat flux from the plate due to free convection, and radiation, as a function of plate temperature in the range [350 K $< T <$ 700 K]. Assume that the plate does not receive radiation from any source in its surroundings. Take the emissivity to be $e = 0.3$.

Begin with the lowest plate temperature: $T = 350$ K. For the loss by free convection we will use Eq. 14.1.1 and Table 14.4.1. We evaluate physical properties at the mean "film" temperature, which is the mean between the surface and the surroundings. For air at 325 K (52°C) we find (Fig. 14.1.3) $\gamma = 0.9 \times 10^8$. We calculate the Grashof number using $L = 1$ m and $\Delta T = 50$ K.

$$\text{Gr} = \gamma L^3 \Delta T = 0.9 \times 10^8 (1)^3 (50) = 4.5 \times 10^9 \tag{15.2.5}$$

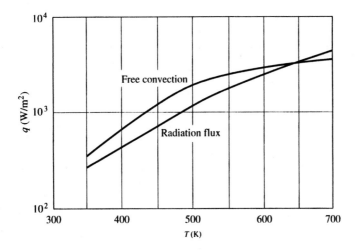

**Figure 15.2.1** Fluxes due to radiation and free convection (Example 15.2.1).

With a Prandtl number for air of Pr = 0.7, we find

$$Ra = 3.2 \times 10^9 \qquad (15.2.6)$$

From Table 14.4.1 we find $C = 0.15$ and $n = 0.33$. Hence

$$Nu = C\,Ra^{0.33} = 0.15\,(3.2 \times 10^9)^{0.33} = 219 \qquad (15.2.7)$$

The heat transfer coefficient is

$$h = \frac{k}{L}Nu = \frac{0.028}{1}219 = 6.1\ \text{W/m}^2\cdot\text{K} \qquad (15.2.8)$$

This yields a heat flux of

$$q_{\text{conv}} = h\,\Delta T = 6.1 \times 50 = 305\ \text{W/m}^2 \qquad (15.2.9)$$

The radiation flux follows from Eq. 15.2.4, and is given by

$$q_{\text{rad}} = e\sigma T^4 = 0.3(5.67 \times 10^{-8})(350)^4 = 255\ \text{W/m}^2 \qquad (15.2.10)$$

We see that at this low temperature, the radiation heat flux is smaller than the flux due to free convection. This situation reverses itself as plate temperature increases, since the emitted radiative flux increases with the *fourth* power of temperature of the plate, and the free convection flux increases roughly linearly with the temperature difference between the plate and the surrounding air. Figure 15.2.1 shows the results of subsequent calculations.

### 15.2.1   Radiation Exchange Between Black Bodies

Now we want to consider *two* bodies that are *exchanging* radiation. Only a fraction of the radiation emitted by one body is intercepted by the other, and vice versa. We should begin, then, with a discussion of the *geometry* of radiation.

Figure 15.2.2 shows the coordinate system in which we analyze the geometry of radiation. We consider a "beam" of radiation emitted from some point on a "black" planar surface. This point defines the origin of a spherical coordinate system. At some

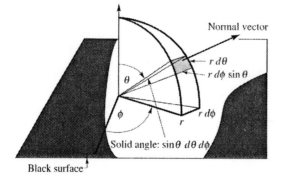

**Figure 15.2.2** The geometry of radiation from a surface.

distance $r$ from the surface of origin (the black surface), the radiation vector (the "normal" vector of Fig. 15.2.2) passes through a differential area of magnitude

$$dA = r^2 \sin \theta \, d\theta \, d\phi \tag{15.2.11}$$

By analogy to planar geometry, we may define a *solid* angle $d\Omega$ by

$$dA = r^2 d\Omega \tag{15.2.12}$$

Then it follows that

$$d\Omega = \sin \theta \, d\theta \, d\phi \tag{15.2.13}$$

For a hemisphere, by this definition of $\Omega$, we see that $A = r^2\Omega$. But the area of a hemisphere is $A = 2\pi r^2$, so we see that $\Omega = 2\pi$ for a hemisphere, and $4\pi$ for a sphere. Thus we find that a hemispherical surface corresponds to a *solid* angle of $2\pi$. (In the planar analog, a semicircle corresponds to an angle $\pi$.)

The *intensity* of radiation $I_b$ that is emitted from a planar surface is *defined* so that the flux crossing a hemisphere surrounding the surface is given by

$$q_b^e = \int_0^{2\pi} \int_0^{\pi/2} I_b \, d\Omega \cos \theta = \int_0^{2\pi} \int_0^{\pi/2} I_b \sin \theta \cos \theta \, d\theta \, d\phi \tag{15.2.14}$$

$$= I_b \int_0^{2\pi} \int_0^{\pi/2} \sin \theta \cos \theta \, d\theta \, d\phi$$

We have used the fact that for a black body the radiation intensity is independent of direction. Such emissions are called *diffuse* radiation. Hence the flux is related to the intensity by

$$q_b^e = \pi I_b = \sigma T^4 \tag{15.2.15}$$

Now we return to the issue of *exchange* of radiation between two bodies, as indicated in Fig. 15.2.3. The *intensity* of emitted radiation from 1 is $I_{b1}$ (energy per area of emitting surface *per unit solid angle*).

The emitted radiant *flux* is (see Eq. 15.2.15)

$$q_{b1} = \pi I_{b1} \tag{15.2.16}$$

The *rate* at which energy is *received* by the differential area $dA_2$ of body 2, denoted by $dQ_{12}$, is

$$dQ_{12} = (I_{b1})(\cos \theta_1 \, dA_1) \, d\Omega \tag{15.2.17}$$

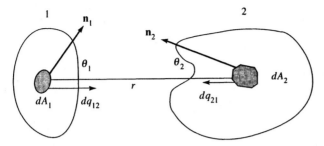

**Figure 15.2.3** Radiant fluxes with directions $n_1$ and $n_2$ are emitted from two bodies.

The receiving area $dA_2$ is related to $d\Omega$ by

$$dA_2 = r^2 d\Omega \qquad (15.2.18)$$

But all of the area $dA_2$ is not "seen" (i.e., it is not normal to the flux $dq_{12}$) because of the angle $\theta_2$. Hence

$$\text{the "seen area"} = \cos\theta_2 dA_2 \qquad (15.2.19)$$

and so

$$dQ_{12} = I_{b1}\cos\theta_1 \, dA_1 \left(\cos\theta_2 \frac{dA_2}{r^2}\right) \qquad (15.2.20)$$

$$= q_{b1} \, dA_1 \left\{\frac{\cos\theta_1 \cos\theta_2 \, dA_2}{\pi r^2}\right\}$$

Also, body 1 receives radiation from body 2 according to a similar expression:

$$dQ_{21} = q_{b2} dA_2 \left\{\frac{\cos\theta_2 \cos\theta_1 \, dA_1}{\pi r^2}\right\} \qquad (15.2.21)$$

The *net* radiation is simply the difference

$$dQ_{net} = (q_{b1} - q_{b2}) \frac{\cos\theta_1 \cos\theta_2 \, dA_1 \, dA_2}{\pi r^2} \qquad (15.2.22)$$

This gives the net exchange between two surfaces of arbitrary orientation on each body. The *total* net exchange between the bodies is given by an integration over the total surface of each body:

$$Q_{net} = \int_{A_2} \frac{dQ_{12}}{dA_2} dA_2 - \int_{A_1} \frac{dQ_{21}}{dA_1} dA_1 \qquad (15.2.23)$$

($A_1$, $A_2$ are total surface areas of the bodies.)

$$Q_{net} = (q_{b1} - q_{b2}) A_1 \left[\frac{1}{A_1}\int\int \frac{\cos\theta_1 \cos\theta_2 \, dA_1 \, dA_2}{\pi r^2}\right] \qquad (15.2.24)$$

Note that we have multiplied and divided both sides by one of the areas, $A_1$. Thus we may write the *rate* of radiation exchange in terms of one of the areas:

$$Q_{net} = (q_{b1} - q_{b2}) A_1 F_{12} \qquad (15.2.25)$$

Equation 15.2.25 defines the "view factor," $F_{12}$. It is the fraction of the area of body 1 seen by body 2, or, equivalently, the fraction of the radiation leaving 1 that is intercepted by 2. Thus a black body (1) with a convex surface, completely surrounded by a second black body (2), has a view factor $F_{12}$ of unity, since 1 is completely surrounded by 2, and therefore all the radiation leaving 1 reaches 2. However, the view factor $F_{21}$ is not unity, since some of the radiation leaving 2 is intercepted by another region of itself, if 2 completely surrounds 1. The restriction that body 1 must be convex implies that 1 does not intercept any of its own radiation. A simple example of these ideas can be worked out to give analytical expressions for the view factors without any complex geometrical calculations.

**EXAMPLE 15.2.2**　**View Factors for Black Body Radiation between a Hemisphere and a Plane**

Figure 15.2.4 shows the geometry. Surface 1 is the *interior* of the hemisphere of radius $R$, and 2 is the planar disk of radius $R$. There are four view factors:

$F_{11}$ is the fraction of radiation leaving 1 that is intercepted by itself.
$F_{12}$ is the fraction of radiation leaving 1 that is intercepted by 2.
$F_{21}$ is the fraction of radiation leaving 2 that is intercepted by 1.
$F_{22}$ is the fraction of radiation leaving 2 that is intercepted by itself.

Since *all* the radiation leaving 2 is intercepted by 1, $F_{21}$ is unity. Since *none* of the radiation leaving 2 is seen by 2, $F_{22}$ is zero. All the radiation leaving 1 is intercepted either by itself or by 2. Hence

$$F_{11} + F_{12} = 1 \qquad (15.2.26)$$

The net exchange of black body radiation is given by Eq. 15.2.24. Had we multiplied and divided the right-hand side by $A_2$, instead of $A_1$, we would have defined the bracketed term as $F_{21}$. For two surfaces that see only each other, then, Eq. 15.2.25 may be written as

$$Q_{net} = (q_{b1} - q_{b2})A_1F_{12} = (q_{b1} - q_{b2})A_2F_{21} \qquad (15.2.27)$$

It follows then that if two black surfaces see only each other,

$$A_1F_{12} = A_2F_{21} \qquad (15.2.28)$$

For the specific geometry at hand, we have

$$\frac{A_1}{A_2} = \frac{2\pi R^2}{\pi R^2} = 2 \qquad (15.2.29)$$

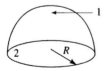

**Figure 15.2.4** Radiation between a hemisphere 1 and the plane 2.

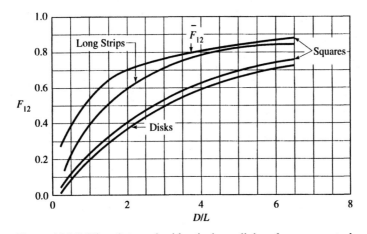

**Figure 15.2.5** View factors for identical parallel surfaces separated by a distance $L$, where $D$ is the diameter or side length. The "strip" is of infinite length and of width $D$. $\overline{F}$ is called the *black body interchange factor* (see text).

Hence

$$F_{12} = \tfrac{1}{2} F_{21} = \tfrac{1}{2} \qquad (15.2.30)$$

and, from Eq. 15.2.26,

$$F_{11} = \tfrac{1}{2} \qquad (15.2.31)$$

View factors have been calculated for a variety of geometries and are usually presented in graphical form. Some examples are given in Figs. 15.2.5 through 15.2.7. In addition, some other important cases, involving one or more surfaces within an enclosure, have been worked out, and the results are presented without derivation. Note that 1 refers to the inner surface of the outer cylinder in Fig. 15.2.6. $F_{12}$ is not unity because the

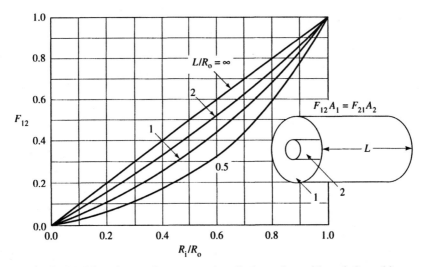

**Figure 15.2.6** View factors for concentric cylinders of equal length $L$, and inner and outer radii $R_i$ and $R_o$.

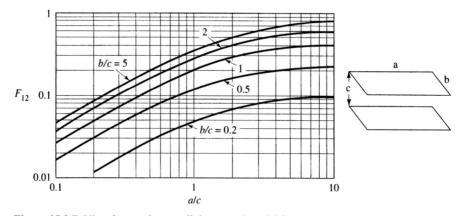

**Figure 15.2.7** View factors for parallel rectangles of sides $a \times b$ separated by a distance $c$.

outer cylinder has a view that is partially of a portion of itself and partially of the inner cylinder. $F_{21}$ is not unity because some of the radiation that leaves the inner cylinder "escapes" through the ends normal to the cylinder axes. For large values of $L/R_o$, $F_{21}$ would approach unity.

## 15.2.2   Radiation Exchange Between Black Bodies in an Adiabatic Enclosure

We are often interested in problems that involve radiation between a pair of surfaces that are themselves connected by other surfaces. One of the simplest examples is that of a pair of parallel walls of an enclosure, exchanging radiation, and connected by *adiabatic* walls. An adiabatic wall is defined as one for which there is no net heat flux—the wall is in thermal equilibrium with its surroundings. Figure 15.2.8 shows the geometry. If all the surfaces radiate as black bodies, the heat exchange between any one surface and *all* the others is given by the sum

$$Q_i = \sigma A_i \sum_{j=1}^{n} F_{ij}(T_i^4 - T_j^4) \tag{15.2.32}$$

This is the net heat loss (or gain) from surface $i$. The summation is over $j$ for all surfaces, so $j$ takes on the values 1 through 6 for the example shown. The term for $j = i$ will vanish, however: since all the surfaces in this example are planar, no surface exchanges radiation with itself.

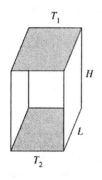

**Figure 15.2.8** Radiation between two walls of an enclosure.

We suppose, now, that we want to calculate the net heat flow between surfaces 1 and 2, and that surfaces 3 to 6 (the sidewalls) are adiabatic. Then we may write a set of heat equations of the form of Eq. 15.2.32, but we set $Q = 0$ for the sidewalls. Thus we write

$$Q_1 = \sigma A_1 [F_{12}(T_1^4 - T_2^4) + F_{13}(T_1^4 - T_3^4) + F_{14}(T_1^4 - T_4^4) + F_{15}(T_1^4 - T_5^4) + F_{16}(T_1^4 - T_6^4)]$$

$$(15.2.33)$$

or, factoring the $T_1$ terms, and noting that $\Sigma F_{1j} = 1$,

$$Q_1 = \sigma A_1 \left[ T_1^4 - \sum_{j=1}^{6} F_{1j} T_j^4 \right] \qquad (15.2.34)$$

Similarly we find

$$Q_2 = \sigma A_2 \left[ T_2^4 - \sum_{j=1}^{6} F_{2j} T_j^4 \right] \qquad (15.2.35)$$

For the adiabatic walls Eq. 15.2.32 becomes, in turn,

$$Q_3 = 0 = \sigma A_3 \left[ T_3^4 - \sum_{j=1}^{6} F_{3j} T_j^4 \right] \qquad (15.2.36)$$

$$Q_4 = 0 = \sigma A_4 \left[ T_4^4 - \sum_{j=1}^{6} F_{4j} T_j^4 \right] \qquad (15.2.37)$$

$$Q_5 = 0 = \sigma A_5 \left[ T_5^4 - \sum_{j=1}^{6} F_{5j} T_j^4 \right] \qquad (15.2.38)$$

$$Q_6 = 0 = \sigma A_6 \left[ T_6^4 - \sum_{j=1}^{6} F_{6j} T_j^4 \right] \qquad (15.2.39)$$

We have six equations and as many unknowns: $Q_1$ and $Q_2$, and the four sidewall temperatures. We regard $T_1$ and $T_2$ as given temperatures, not as unknowns. Hence we may eliminate the four unknown temperatures and find $Q_1$ and $Q_2$ in terms of $T_1$ and $T_2$. The result is complex, algebraically, and involves the individual view factors $F_{ij}$. (See Problem 15.21.) We may write the final expression for the net heat transfer in the form

$$Q_1 = \sigma A_1 [\bar{F}_{12}(T_1^4 - T_2^4)] \qquad (15.2.40)$$

where $\bar{F}_{12}$ is called the *black body interchange factor* for the enclosure. Since $\bar{F}_{12}$ depends on the individual view factors, which are given in Fig. 15.2.7 for the parallel surfaces and are available (but not shown here) for pairs of planar walls at right angles to each other, we may find $\bar{F}_{12}$ for any rectangular enclosure. The result for the factor $\bar{F}_{12}$ between a pair of square parallel planes is shown on Fig. 15.2.5, as a function of the distance between the planes.

**EXAMPLE 15.2.3**  *Design of a Solar Heater (Black Bodies: No Convection)*

Solar energy has the capacity to raise the temperature of a surface to a level significantly above the ambient temperature. On a clear summer day on the earth's *surface*, the solar radiation flux is approximately 1140 W/m² = 360 Btu/h·ft². Let's examine some features of solar radiation that are relevant to the design of a solar heating system. First, we can find the temperature a planar surface might reach if it were subject to solar radiation incident upon, and normal to, one side. We assume that the surface behaves as a black body and that it does not lose heat to its surroundings by convection or conduction. (This will give us an upper limit on the surface temperature.) A simple heat balance equates the incoming (solar) radiation to the radiative emission from both sides of the hot surface. The sky is taken as a black body at absolute zero temperature (0 K). Then the surface temperature follows as a solution to

$$q_{\text{solar}} - 2\sigma T_s^4 = 0 \tag{15.2.41}$$

With $\sigma$ in SI units we find

$$T_s = \left(\frac{1.14 \times 10^3}{2 \times 5.67 \times 10^{-8}}\right)^{1/4} = 316 \text{ K} = 43°\text{C} \tag{15.2.42}$$

Now suppose this surface is the upper surface of an enclosure, and the lower surface is receiving radiation from the upper surface while losing heat from its other side only by convection to a heat exchange fluid. We assume that there is *no convection within the enclosure*, however. We assume that the lower surface loses heat by convection to some fluid at a temperature of 300 K, and such that the heat transfer coefficient is 0.003 W/cm$^2$·K. The geometry is such that the enclosure is very narrow in the vertical direction, in comparison to the dimensions of the surfaces, and (note Fig. 15.2.5) we may take the interchange factor $\bar{F}_{12}$ to be nearly unity.

Now a heat balance takes the form of a pair of equations:

*For the upper surface*

$$q_{\text{solar}} - \sigma(2T_1^4 - T_2^4) = 0 \tag{15.2.43}$$

*For the lower surface*

$$\sigma(T_1^4 - T_2^4) - h(T_2 - T_f) = 0 \tag{15.2.44}$$

The first equation is a balance on the upper surface, which *gains* heat on one side by radiation from the sun and on the other side by radiation from the lower surface at $T_2$, and *loses* heat by radiation to the sky (from its upper surface) and also to the colder lower surface. Hence the factor of 2 appears in the $T_1$ term, since there are radiative losses from both sides of that surface. The second equation is a heat balance on the lower surface, again simply equating the heat input (radiation from the upper surface, at $T_1$) to the loss (radiation *to* the upper surface and convection to the fluid at temperature $T_f$). We take $T_f$ as given in this example. This pair of nonlinear algebraic equations may be easily solved by trial and error to yield the solutions

$$T_1 = T_s = 347 \text{ K} = 44.4°\text{C} \tag{15.2.45a}$$
$$T_2 = 309 \text{ K} = 36°\text{C} \tag{15.2.45b}$$

Note that, as expected, the upper surface is slightly warmer (cf. Eq. 15.2.42) because it absorbs some thermal radiation from the (colder) lower surface. We shall see that in a real solar heating system it is necessary to account for convective losses from the surface exposed to the atmosphere, and these losses significantly reduce the efficiency of the system.

### 15.2.3   Radiation Exchange Between Gray Bodies

The assumption that a body radiates and absorbs as a black body is not always in keeping with the physics of real surfaces, so we turn now to an analysis of radiation between *gray* bodies. Gray bodies have emissivities and absorptivities that are not equal to unity, but they are assumed to be equal to each other—that is, $a = e$—and these coefficients are assumed to be independent of wavelength.

---

**EXAMPLE 15.2.4**   *Analysis of the Net Radiation Flux between Parallel Gray Surfaces*

We consider the case of infinite parallel *gray* surfaces at temperatures $T_1$ and $T_2$, as in Fig. 15.2.9, setting $e_1 = a_1$ and $e_2 = a_2$ as is usually assumed. Surface 1 emits energy by radiation. The radiant *flux* is

$$q_1 = e_1 \sigma T_1^4 \tag{15.2.46}$$

A fraction of that flux is reflected back to surface 1. We may write this reflected radiation flux, assuming no transmission of radiation, so that $a + e = 1$, in the form

$$e_1(1 - e_2)\sigma T_1^4 \tag{15.2.47}$$

but a fraction of that flux is reflected away from surface 1:

$$e_1(1 - e_2)(1 - e_1)\sigma T_1^4 \tag{15.2.48}$$

If we look at the *net* rate of energy *loss* from surface 1 by primary emission and multiple reflection/absorption, we may write the net loss (considering only radiation associated with surface 1) as

$$\sigma T_1^4[e_1 - e_1(1 - e_2) + e_1(1 - e_1)(1 - e_2) - e_1(1 - e_1)(1 - e_2)^2 + \cdots] \tag{15.2.49}$$

We can factor this to the form

$$\sigma T_1^4[e_1 e_2][1 + (1 - e_1)(1 - e_2) + (1 - e_1)^2(1 - e_2)^2 + \cdots] \tag{15.2.50}$$

In the same way, the loss from surface 2 that arises from its primary radiation at temperature $T_2$ and subsequent reflection and absorption of that radiation is

$$\sigma T_2^4[e_1 e_2][1 + (1 - e_2)(1 - e_1) + (1 - e_2)^2(1 - e_1)^2 + \cdots] \tag{15.2.51}$$

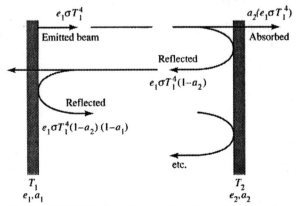

**Figure 15.2.9** Radiation from surface 1 is reflected and reradiated from surface 2.

Thus the net flux of energy at surface 1 is the difference between the sums in Eqs. 15.2.50 and 15.2.51:

$$q_{12} = \sigma(T_1^4 - T_2^4)e_1e_2 \sum_{p=0}^{\infty} [(1 - e_1)(1 - e_2)]^p \qquad (15.2.52)$$

This can be simplified if we note that

$$\sum x^p = \frac{1}{1 - x} \qquad \text{if } x < 1 \qquad (15.2.53)$$

This lets us write Eq. 15.2.52 as

$$q_{12} = \frac{\sigma(T_1^4 - T_2^4)e_1e_2}{1 - (1 - e_1)(1 - e_2)} = \frac{\sigma(T_1^4 - T_2^4)}{\dfrac{1 - (1 - e_1 - e_2 + e_1e_2)}{e_1e_2}} \qquad (15.2.54)$$

or

$$q_{12} = \frac{\sigma(T_1^4 - T_2^4)}{\dfrac{1}{e_1} + \dfrac{1}{e_2} - 1} \qquad (15.2.55)$$

For a black body, $e_1 = e_2 = 1$, and $q_{12} = \sigma(T_1^4 - T_2^4)$, as expected.

This example is very simple because the surfaces are infinite and parallel, and so the view factors are unity, since any point on one surface sees all points on the other. For other shapes we can find similar expressions, but we must account for the view factors.

In Fig. 15.2.10 we see a surface at temperature $T_1$ completely surrounded by an isothermal enclosure at temperature $T_2$. The net radiation exchange between the two surfaces is given by Eq. 15.2.56, where $F_{12}$ is the view factor appropriate to the specific geometry. For example, for concentric pipes, Fig. 15.2.6 provides the view factor. For concentric spheres, $F_{12} = 1$ and $A_1/A_2 = (r_1/r_2)^2$.

$$q_{12} = \frac{\sigma(T_1^4 - T_2^4)}{\dfrac{1}{F_{12}} + \left(\dfrac{1}{e_1} - 1\right) + \dfrac{A_1}{A_2}\left(\dfrac{1}{e_2} - 1\right)} \qquad (15.2.56)$$

In Fig. 15.2.11 we see a pair of *gray* surfaces exchanging radiation with each other. However, the pair is completely surrounded by an *adiabatic* surface, defined earlier. Although this surface emits radiation according to its temperature $T_{ad}$, that temperature

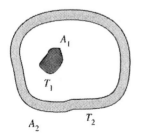

**Figure 15.2.10** Surface in an enclosure.

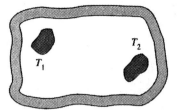

**Figure 15.2.11** Two surfaces exchange heat with each other, surrounded by an adiabatic surface.

drops out of the analysis because it can be related to the temperatures $T_1$ and $T_2$. The result of the analysis for the heat flux from body 1 to body 2 is found to be

$$q_{12} = \frac{\sigma(T_1^4 - T_2^4)}{\dfrac{A_1 + A_2 - 2A_1F_{12}}{A_2 - A_1F_{12}^2} + \left(\dfrac{1}{e_1} - 1\right) + \dfrac{A_1}{A_2}\left(\dfrac{1}{e_2} - 1\right)} \qquad (15.2.57)$$

To use this expression, it is necessary to know the view factor $F_{12}$ that appears in the denominator.

## EXAMPLE 15.2.5   *Radiation Effect on Temperature Measurement*

A temperature-measuring device is placed in a hot gas flowing through a pipe, as shown in Fig. 15.2.12. The device is simply a thermocouple embedded in a metallic cylinder. The cylinder is made of a high conductivity metal, to ensure that the measured temperature is essentially the surface temperature of the device. The pipe walls are within a furnace, and the whole system is a heater for the gas. The pipe wall temperature is measured and is found to be 800°F. The thermocouple measures a temperature of 300°F. To what extent does radiation affect the reading? In other words, how close is the actual gas temperature to the thermocouple reading?

This is simply a problem in heat balancing: the heat lost by the thermocouple by convection must balance the net heat exchange by radiation between the thermocouple and the pipe wall. For a gray body, this heat balance takes the form

$$e\sigma(T_t^4 - T_w^4) = h(T_g - T_t) \qquad (15.2.58)$$

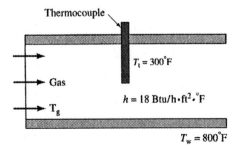

**Figure 15.2.12** Temperature measurement in a flowing gas.

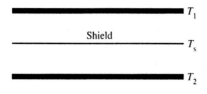

**Figure 15.2.13** A radiation shield.

Note that we use a view factor of unity here. For a small cylinder (the thermocouple) in a long pipe, we may assume that the thermocouple "sees" only the pipe walls. By definition, then, its view factor is unity.

In the British units selected here we use[1] $\sigma = 0.1712 \times 10^{-8}$ Btu/h·ft$^2$·R$^4$. We choose an emissivity value of $e = 0.98$. Then Eq. 15.2.58 becomes

$$0.98(0.1712 \times 10^{-8})((759)^4 - (1259)^4) = 18(T_g - 759) \qquad (15.2.59)$$

and we solve directly for the gas temperature and find

$$T_g = 556 \text{ R} = 97°\text{F} \qquad (15.2.60)$$

The thermocouple (which reads 300°F) is in error to a very large extent because it is absorbing so much radiant energy emitted from the very hot wall of the pipe. This is an unacceptable error.

Suppose we select a thermocouple with a *highly reflective surface*, enabling the reduction of the absorptivity (and emissivity) to a very small value. Then we can reduce the error due to radiation from the hot pipe. For example, if $e = 0.03$, we find

$$0.03(0.1712 \times 10^{-8})(T_t^4 - (1259)^4) = 18 (556 - T_t) \qquad (15.2.61)$$

Note that now we know (or assume) the gas temperature (556 R = 97°F), and we are solving for the temperature reading. We find, in this case,

$$T_t = 104°\text{F} \qquad (15.2.62)$$

The error (7 F°) has been reduced to an acceptable level for many applications.

**EXAMPLE 15.2.6** *Effectiveness of a Radiation Shield*

From Example 15.2.5, we see that one means of controlling the radiation heat transfer between two bodies is through the choice of surface material. A second obvious method to reduce radiation transfer to or from a surface is to introduce an intervening surface that blocks the passage of radiation. For example, suppose we wish to reduce the temperature of a surface that is being heated by radiation, as in Fig. 15.2.13. We attempt to do this by placing a reflective surface between the transmitting and receiving surfaces. Unless the radiation "shield" is *completely* reflective, however, the shield will heat up

---

[1] Absolute temperature in British units is degrees Rankine (R), where R = °F + 459. Absolute zero temperature is −459°F.

and radiate to the protected surface. The issue is with respect to the degree to which such a shield is effective. We can develop a simple model of this system for the special case of surfaces that are large parallel planes.

Between the receiving surface (1) and the shield, the net radiative exchange is given by Eq. 15.2.55:

$$q_{1s} = \frac{\sigma(T_1^4 - T_s^4)}{\dfrac{1}{e_1} + \dfrac{1}{e_s} - 1} \qquad (15.2.63)$$

The same expression holds for the net exchange between the transmitting surface (2) and the shield:

$$q_{s2} = \frac{\sigma(T_s^4 - T_2^4)}{\dfrac{1}{e_s} + \dfrac{1}{e_2} - 1} \qquad (15.2.64)$$

From a simple heat balance, it follows that

$$q_{s2} = q_{1s} \equiv q_{12} \qquad (15.2.65)$$

If we eliminate the temperature $T_s$ between these two equations, we find

$$q_{12} = \frac{\sigma(T_1^4 - T_2^4)}{\left(\dfrac{1}{e_1} + \dfrac{1}{e_s} - 1\right) + \left(\dfrac{1}{e_2} + \dfrac{1}{e_s} - 1\right)} \qquad (15.2.66)$$

The efficiency of the shield may be characterized by the ratio of the fluxes with and without the shield. Specifically, we define an "effectiveness factor," $\eta$, as that ratio, with the result

$$\eta = 1 - \frac{\dfrac{1}{e_1} + \dfrac{1}{e_2} - 1}{\left(\dfrac{1}{e_1} + \dfrac{1}{e_s} - 1\right) + \left(\dfrac{1}{e_2} + \dfrac{1}{e_s} - 1\right)} \qquad (15.2.67)$$

### 15.2.4 Radiation Flux Plots

When we are doing quick estimates of radiative heat transfer, using the simple models presented in the preceding examples, there is a lot of repetitive calculation of the term $\sigma T^4$. Figures 15.2.14 and 15.2.15 will aid in quick estimates of the radiation flux emitted by a black body. They are based on

$$\text{flux} = \sigma T^4 \qquad (15.2.68)$$

with

$$\sigma = 5.67 \times 10^{-8}\,\text{W/m}^2 \cdot \text{K}^4 \qquad \text{or} \qquad \sigma = 0.1712 \times 10^{-8}\,\text{Btu/h} \cdot \text{ft}^2 \cdot \text{R}^4 \qquad (15.2.69)$$

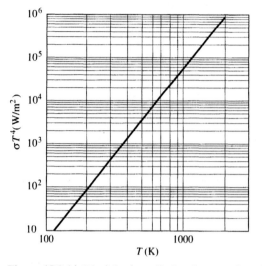

**Figure 15.2.14** Black body radiation flux as a function of Kelvin temperature.

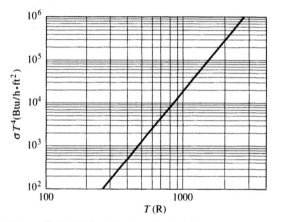

**Figure 15.2.15** Black body radiation flux as a function of Rankine temperature.

## 15.3  RADIATION AND CONVECTION

In most practical engineering settings, radiation heat transfer occurs in the presence of additional heat transfer mechanisms such as conduction and convection. We have already seen a simple example of this situation—Example 15.2.5. No new ideas are introduced in considering such problems, so we simply offer a few additional examples here.

**EXAMPLE 15.3.1**    *Design of a Solar Heater (Accounting for Convection)*

We return to the solar heater example (Example 15.2.3), but account for the role of convective heat losses. Consider a system designed as shown in Fig. 15.3.1. Tubes carrying a heat exchange fluid are enclosed in a conductive slab, which is exposed to

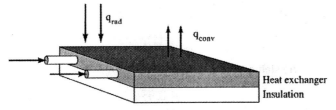

**Figure 15.3.1** Elements of a solar heater.

solar radiation. The slab is raised to an elevated temperature by the radiation flux, and heat conduction through the slab to the tube surfaces is very efficient. Heat is then removed from the slab by convection to the fluid. Since the goal is to maximize the heat transfer to the tubes, we design the system so that the slab temperature is raised well above that of the ambient air. As a consequence, it is quite likely that free convection will reduce the maximum attainable temperature of the slab, thereby reducing the efficiency of this system as a heat exchanger. We want to explore two questions here:

1. To what extent is free convection effective in reducing the slab temperature? (Do we really need to be concerned about this factor?)
2. How do we reduce convective losses without also reducing the input by radiation?

Let's suppose, as in Example 15.2.3, that the solar radiation flux is 1140 W/m². In the absence of convection and if the lower surface is insulated, we can find[1] that the surface could be raised to 103°C. Now we will account for free convection from a hot horizontal surface. We take the ambient air temperature to be 20°C. The slab is in the form of a rectangular panel of dimensions 1 m × 2 m. We need a model for the convective loss.

We will use the model of Eq. 14.4.1 and the coefficients of Table 14.4.1:

$$Nu = C \, Ra^n \qquad (15.3.1)$$

The length scale $L$ in the Nusselt and Rayleigh numbers is (according to Table 14.4.1)

$$L = 0.5(A + B) = 1.5 \text{ m} \qquad (15.3.2)$$

To estimate the Rayleigh number, we use a film temperature for physical properties. With a guess that the slab surface is at 80°C, and with an ambient air temperature of 20°C, the film temperature (the mean of these two temperatures) is 50°C. At this temperature (from Fig. 14.1.3) $\gamma = 9 \times 10^7$ m⁻³K⁻¹. The Prandtl number of air is 0.7. This permits us to estimate the Rayleigh number as (see Eq. 14.3.47)

$$Ra = \gamma Pr L^3 \, \Delta T = (9 \times 10^7)(0.7)(1.5)^3(80 - 20) = 1.3 \times 10^{10} \qquad (15.3.3)$$

Then, from Table 14.4.1 we use $C = 0.15$ and $n = 0.33$ in Eq. 15.3.1. This gives us

$$Nu = 324 \qquad (15.3.4)$$

and (with the air thermal conductivity of $k = 0.03$ W/m·K)

$$h = \frac{k \, Nu}{L} = \frac{0.03 \times 324}{1.5} = 6.5 \text{ W/m}^2 \cdot °C \qquad (15.3.5)$$

To estimate the effect of natural convection, we will ignore the heat transfer to the exchanger fluid and first calculate the reduction in slab temperature due to the convective

---

[1] We solve Eq. 15.2.41, but without the factor of 2, since loss is from one side only.

loss. Thus our heat balance is

$$q_{solar} - h_{conv}(T_s - T_{air}) - \sigma T_s^4 = 0 \qquad \textbf{(15.3.6)}$$

(Note that we treat the slab as a black body.) Solving for $T_s$ we find

$$T_s = 345 \text{ K} = 72°C \qquad \textbf{(15.3.7)}$$

The value is 30 C° below that obtained when free convection is neglected. This solution for $T_s$ is based on the use of a film temperature of 50°C (for calculation of physical properties), which is so close to the average of this estimate of $T_s$ and the ambient temperature that it is not necessary to do an iterative correction. The point of the example is established: free or natural convection can significantly reduce the effectiveness of solar heating. Hence we turn to the next issue: How do we reduce free convection?

One of the simplest techniques for dealing with this problem is to provide an enclosure above the exchanger surface, as suggested in Fig. 15.3.2. A transparent glass plate provides an air space that may act as insulation with respect to convective losses from the slab to the ambient air. Glass will reduce the incoming radiation flux to some degree, since the glass will have a finite reflectance and absorbance. By the same token, the glass will reduce the radiation loss from the slab back to the ambient (as in a greenhouse) so there will be some compensation of these effects. Instead of carrying out a detailed analysis, for which we do not have enough information at this stage, and since our goal is to examine some broad features of this system, we will assume that the glass is perfectly transparent to radiation (in both directions) and focus instead on the degree to which we can reduce the free convective loss from the slab.

For the example, we will choose to place the glass plate in a way that creates a 1 cm air gap. Our first task is to estimate the convective loss from the slab with this air enclosure above it. Some elements of the problem at hand have already been considered in Example 14.4.2, where we estimated the convective heat flux across a horizontal confined air gap. However, we no longer know the temperatures of the upper and lower surfaces—they are part of the thermal problem that we must solve. Again we ignore heat transfer to the heat exchange fluid, to be able to focus on the effect of the glass enclosure on the temperature of the slab.

We assume that the system has come to a steady state. We must write two heat balances. For the slab we write the heat balance as

$$q_{solar} - h_{int}(T_s - T_{glass}) - \sigma T_s^4 = 0 \qquad \textbf{(15.3.8)}$$

where $h_{int}$ is the convective coefficient within the enclosed air layer.

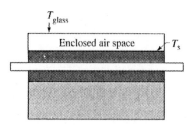

**Figure 15.3.2** An enclosure to reduce free convection.

To calculate $h_{int}$, we need an estimate for the Nusselt number for the trapped air layer, but this requires an estimate of the temperatures that appear in Eq. 15.3.8. We will assume the following values in estimating the Rayleigh number:

$$T_s = 90°C \quad \text{and} \quad T_{glass} = 30°C \tag{15.3.9}$$

To calculate thermal properties of the air layer, we need a "film" temperature (the average between the solid surface and the bulk fluid above it). For a "film" temperature, we estimate that the air layer temperature is 60°C, and thus the "film" temperature is 75°C. From Fig. 14.1.3 we find $\gamma = 6.4 \times 10^7$ m$^{-3}$K$^{-1}$. The Rayleigh number is (using our estimates just above)

$$\text{Ra} = \gamma \text{Pr} L^3 \, \Delta T = (6.4 \times 10^7)(0.7)(0.01)^3 (90 - 30) = 2700 \tag{15.3.10}$$

(Note that since this is the air layer, the length scale is $L = 1$ cm $= 0.01$ m.) According to Fig. 14.4.3, the free convection is suppressed almost completely. We find Nu $= 1.5$ and so

$$h_{int} = \frac{k}{L} \text{Nu} = \frac{0.03}{0.01} \times 1.5 = 4.5 \text{ W/m}^2 \cdot \text{K} \tag{15.3.11}$$

Now we turn to the *exterior* convection. The heat balance around the glass plate (which is assumed to be transparent to radiation) is simply

$$h_{int}(T_s - T_{air}) - h_{ext}(T_g - T_a) = 0 \tag{15.3.12}$$

where $T_a$ is the ambient temperature. With the assumed glass temperature given in Eq. 15.3.9, we find an external "film" temperature of 25°C, at which $\gamma = 1.4 \times 10^8$ m$^{-3}$ K$^{-1}$. The Rayleigh number is

$$\text{Ra} = \gamma \text{Pr} L^3 \, \Delta T = (1.4 \times 10^8)(0.7)(1.5)^3 (30 - 20) = 3.3 \times 10^9 \tag{15.3.13}$$

(Again, note which $L$ is used here.) Then, from Table 14.4.1 we use $C = 0.15$ and $n = 0.33$ in Eq. 15.3.1. This gives us

$$\text{Nu} = 208 \tag{15.3.14}$$

and (with the air thermal conductivity of $k = 0.03$ W/m $\cdot$ K)

$$h_{ext} = \frac{k \, \text{Nu}}{L} = \frac{0.03 \times 208}{1.5} = 4.2 \text{ W/m}^2 \cdot \text{K} \tag{15.3.15}$$

The two unknown temperatures must be found from simultaneous solution of Eqs. 15.3.8 and 15.3.12:

$$1140 - 4.5(T_s - T_g) - 5.67 \times 10^{-8} T_s^4 = 0 \tag{15.3.16}$$

and

$$4.5(T_s - T_g) - 4.2(T_g - 293) = 0 \tag{15.3.17}$$

We may solve Eqs. 15.3.16 and 15.3.17 by trial and error, and find

$$T_s = 360 \text{ K} = 87°C \quad \text{and} \quad T_g = 318 \text{ K} = 45°C \tag{15.3.18}$$

We have significantly reduced the free convection heat loss, and the slab temperature is much closer to the value it would have in the absence of any convective losses. We note that this estimate of the two temperatures differs somewhat from the values assumed (Eq. 15.3.9) for evaluation of physical properties. A second round of calculations, using these results for parameter estimation, will improve the estimate. Since our point is to illustrate the extent of improvement possible through the use of an enclosure, we do not carry the calculation further.

In treating this example we have assumed that all the incoming radiation is rejected either by reradiation or by natural convection. The absorbing plate is insulated below, and so it is passive in the sense that it transfers no heat to a heat exchanger. We have also assumed that the absorbing plate acts as a black body. If such a system were to be used as a solar heater, its efficiency would be very poor. We can see this immediately with a crude calculation along the following lines.

If we are to transfer heat to a heat exchanger, using the absorber plate as the hot surface, it is essential that the hot surface be *significantly* hotter than the fluid to which heat is transferred. We certainly want a temperature difference of some tens of degrees kelvin. Let's assume that the transfer fluid is at 300 K, and the absorber is at 330 K. For a black body radiating at this temperature, the heat loss by reradiation will be

$$q_{loss} = \sigma T_s^4 = 5.67 \times 10^{-8}(330)^4 = 672 \text{ W/m}^2 \tag{15.3.19}$$

This is a substantial portion of the expected solar radiation available. The value used in this example (1140 W/m$^2$) is really an upper limit under ideal conditions. Even so, the efficiency of the system would be no better than

$$\eta_{eff} = 100 \frac{1140 - 672}{1140} = 41\% \tag{15.3.20}$$

Under more realistic conditions that account for convective losses and use a more realistic estimate of the solar radiation flux (a value of 750 W/m$^2$ is often used for a seasonal average in regions where solar heating is worth considering), the efficiency would be very small. For example, at this lower solar flux, and assuming an absorber temperature of only 320 K, the efficiency falls to 21%.

This thermal problem is transformed, at this point, to a challenging *materials* problem. It is possible to create surfaces for the absorber that have an absorptivity near that of a black body, but with an emissivity that is much reduced compared to unity. In other words, these surfaces are not gray bodies. It is also possible to find coatings for the upper glass plate that transmit solar radiation but have a finite reflectivity for the thermal radiation from the absorber. Transmission and reflectance can be functions of the wavelength of the radiation, and we can take advantage of this property. Solar radiation lies largely in the visible portion of the spectrum, while the hot surface of the absorber plate (which is, after all, not glowing) radiates largely in the infrared. We will take these points into account in Example 15.3.2.

**EXAMPLE 15.3.2** *Design of a Solar Heater for Domestic Hot Water*

We want to supply hot water at a temperature of 140°F (60°C or 333 K) using a system designed along the lines of Figs. 15.3.1 and 15.3.2. We want to be able to meet a water demand of 0.5 gal/min. The feed water enters the heat exchanger at a temperature of 50°F (10°C or 283 K). Carry out a rough design of such a system.

The first task is to find the thermal demand for this system. The water demand corresponds to a flow of 3.2 × 10$^{-5}$ m$^3$/s, and we must raise the temperature of this water by 50 kelvin degrees. Hence the heat load is

$$Q_h = \rho Q C_p \Delta T = 1000(3.2 \times 10^{-5})4180 \times 50 = 6690 \text{ W} \tag{15.3.21}$$

With an input from the sun of the order of 1000 W/m², which is reduced by various convective and reradiative losses to perhaps a third of this value, we would require something like 20 m² of surface area to meet this demand. This means that we would need 10 panels in parallel.

Let's examine the performance of a single panel whose characteristics are sketched in Fig. 15.3.3. The panel is identical to that considered in Example 15.3.1, except that a set of parallel copper tubes is integral with the absorbing plate. The tubes have an inside diameter of 1.5 cm, and there are 40 such tubes in parallel across the 1 m width of the panel. In our analysis, we will assume a temperature for the absorber plate, make a heat balance on the plate, and determine how much heat can be removed from the plate by transfer to the heat exchanger. By specifying the desired outlet temperature of the water, we can then find the water flowrate that this system will support. With this result as a "base case," we can estimate the performance of the system, in the form of a curve of outlet water temperature as a function of water flowrate.

We make the following assumptions about the characteristics of the system:

The average solar flux is 750 W/m².
The glass cover plate is transparent to the solar radiation.
The absorbing plate has an absorbance for solar radiation of 0.9.
The absorbing plate is at a uniform temperature of $T_s = 350$ K.
At this temperature, the absorbing plate has an emissivity of 0.3.
The heat loss by free convection from the absorbing surface corresponds to an overall
   heat transfer coefficient $U$ of 2 W/m²·K (see Problem 15.9).
The ambient air is at $T_a = 293$ K.

The heat balance on the absorbing plate takes the form

$$a_{solar}q_{solar}A = e\sigma T_s^4 A + U(T_s - T_a)A + Q_{h,exch} \qquad (15.3.22)$$

From this we find that the available useful heat delivery is

$$0.9 \times 750 \times 2 - 0.3(5.67 \times 10^{-8})(350^4)2 - 2(350 - 293)2$$
$$= Q_{h,exch} = 1350 - 511 - 228 = 611 \text{ W} \quad (15.3.23)$$

Using a heat capacity of $C_p = 4180$ W·s/kg·K we may write the heat delivery in terms of the mass flowrate of the water as

$$Q_{h,exch} = wC_p \Delta T \qquad (15.3.24)$$

If we require an increase in the water temperature of 50 kelvin degrees, we find

$$w = \frac{Q_{h,exch}}{C_p \Delta T} = \frac{611}{4180 \times 50} = 2.9 \times 10^{-3} \text{ kg/s} \qquad (15.3.25)$$

The specified demand of 0.5 gal/min corresponds to a mass flowrate of $3.2 \times 10^{-2}$ kg/s. This is 10 times the flow we can deliver from a single panel, with the specified temperature increase and with the assumed absorber plate temperature. Hence to meet the demands for hot water, we would need 10 such panels in parallel.

On the basis of these rough calculations, it appears that we could design a solar heater to meet this demand. The important constraints on the design are the total flowrate and the outlet water temperature, which both must satisfy the demand stated earlier. We do not require that the absorber plate rise to a specified temperature.

Suppose the system consists of 10 panels, designed as shown in Fig. 15.3.3 and as stated in the text. Each panel then carries $3.2 \times 10^{-3}$ kg/s of water. Suppose that each panel has 40 tubes, for a flowrate in each tube of $w' = 8 \times 10^{-5}$ kg/s. We may write

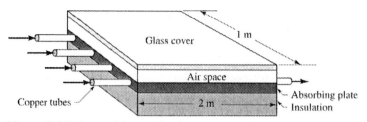

**Figure 15.3.3** A panel for a solar heater.

the Reynolds number, evaluating physical properties at the mean of the inlet and outlet temperatures, as follows:

$$\text{Re} = \frac{4w'}{\pi D \mu} = \frac{4 \times (8 \times 10^{-5})}{\pi \times 0.015 \times (0.72 \times 10^{-3})} = 9.4 \tag{15.3.26}$$

Thus we see that we have a laminar flow heat exchanger. The Nusselt number for this system is obtained from Fig. 12.5.5. More generally, we may find the relationship of temperature rise, flowrate, and tube wall temperature from Eq. 12.4.19 in the form

$$\frac{T_{\text{out}} - T_s}{T_{\text{in}} - T_s} = \text{esp}(-4 \text{ ST } L/D) \tag{15.3.27}$$

where St = Nu/Re Pr, as usual. In this example, we find Nu = 3.66.

The design problem is seen to be overspecified at this point, because we have fixed $T_{\text{in}} = 10°C$, $T_{\text{out}} = 60°C$, and $w' = 8 \times 10^{-5}$ kg/s. This would yield a specific solution for the panel temperature $T_s$, without consideration of Eq. 15.3.22. Hence we would no longer satisfy the heat balance that matches $Q_{\text{h,exch}}$ to the heat demand per panel. We don't really care about the temperature $T_s$ as long as we can satisfy the overall heat demand. Hence we relax the specification on $T_s$. We relax the specification on $w'$ as well (equivalent to not specifying the number of tubes per panel, or the number of panels) and we solve for the two unknowns, $w'$ (per tube) and $T_s$ from Eqs. 15.3.27 and 15.3.22 (with 15.3.24):

$$\alpha_{\text{solar}} q_{\text{solar}} A = \varepsilon \sigma T_s^4 A + U(T_s - T_a)A + n_t w' C_p (T_{\text{out}} - T_{\text{in}}) \tag{15.3.28}$$

or

$$0.9 \times 750 \times 2 - 0.3(5.67 \times 10^{-8})(T_s^4)2 - 2(T_s - 293)2 = n_t w' 4180(50) \tag{15.3.29}$$

Equation 15.3.27 becomes (see Prob. 15.10)

$$\frac{333 - T_s}{283 - T_s} = \exp\left(-\frac{0.004}{w'}\right) \tag{15.3.30}$$

(We used Pr = 4 at an assumed $T_f = 45°C$. We will find the results insensitive to this choice.)

We require that the total flow per panel (assuming we use $N_p$ panels) be

$$N_p n_t w' = 3.2 \times 10^{-2} \text{ kg/s} \tag{15.3.31}$$

With an assumed value for $N_p$, this set of three equations is sufficient to yield values for $w'$ (per tube), $n_t$, and $T_s$. We then must check to confirm that the required number of tubes per panel is consistent with good engineering design, and the available space. We must also be sure that the absorber surface temperature $T_s$ is not too high, from a safety or materials point of view.

If we select $N_p = 10$ we find that the absorber plate temperature falls to 333K = 60°C, and we require $n_t = 35$ tubes per panel to meet the thermal demand ($\Delta T = 50°C$ and $T_{out} = 60°C$ for 0.5 gal/min). Alternatively, if we maintain $n_t = 40$ tubes per panel, 9 panels will suffice.

Other aspects of this type of design may be explored through this model. Problems 15.10 and 15.11 provide an opportunity to do this.

## 15.4 RADIATION AND CONDUCTION

Sometimes a body is heated by radiation to its surface, followed by conduction into the interior of the body. An example of the simplest such problem is illustrated in Fig. 15.4.1, which shows a large planar slab of thickness $B$, insulated on one surface and receiving radiation on the other parallel surface. Assuming that the radiation flux is uniform across the exposed surface and that the body is opaque to radiation, the transient heat conduction equation and boundary conditions take the forms

$$\frac{\partial T}{\partial t} = \alpha \frac{\partial^2 T}{\partial x^2} \tag{15.4.1}$$

with the initial condition

$$T = T_o \quad \text{at} \quad t = 0 \tag{15.4.2}$$

and boundary conditions

$$\frac{\partial T}{\partial x} = 0 \quad \text{at} \quad x = 0 \tag{15.4.3}$$

$$-k\frac{\partial T}{\partial x} = \sigma \bar{F}_{12}(T^4 - T_e^4) \quad \text{at} \quad x = B \tag{15.4.4}$$

Here $\bar{F}_{12}$ is the radiation interchange factor between the slab and its surroundings. The radiating (emitting) surface is taken to be at the steady temperature $T_e$. Black body radiation is assumed. We begin the analysis by nondimensionalizing this boundary value problem. Define a dimensionless temperature as

$$\theta = \frac{T - T_e}{T_o - T_e} \tag{15.4.5}$$

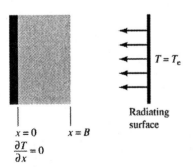

$x = 0$     $x = B$

$\frac{\partial T}{\partial x} = 0$

**Figure 15.4.1** A planar slab of thickness $B$, insulated at $x = 0$, receives radiation on the surface $x = B$.

and introduce the Fourier number $X_{\mathrm{Fo}} = \alpha t/B^2$ as the time variable and $\xi = x/B$ as the space variable. Then the boundary value problem transforms to

$$\frac{\partial \theta}{\partial X_{\mathrm{Fo}}} = \frac{\partial^2 \theta}{\partial \xi^2} \tag{15.4.6}$$

$$\theta = 1 \qquad \text{at} \quad X_{\mathrm{Fo}} = 0 \tag{15.4.7}$$

$$\frac{\partial \theta}{\partial \xi} = 0 \qquad \text{at} \quad \xi = 0 \tag{15.4.8}$$

and the boundary condition on the surface exposed to radiation is

$$\frac{\partial \theta}{\partial \xi} = \left( \frac{\sigma B \overline{F}_{12} T_{\mathrm{e}}^3}{k} \right) \frac{1 - [1 - \theta(1 - T_{\mathrm{o}}/T_{\mathrm{e}})]^4}{1 - T_{\mathrm{o}}/T_{\mathrm{e}}} \qquad \text{at} \quad \xi = 1 \tag{15.4.9}$$

We conclude that the solution depends on the values of a dimensionless temperature ratio $T_{\mathrm{o}}/T_{\mathrm{e}}$ and a new dimensionless group

$$\Sigma = \frac{\sigma B \overline{F}_{12} T_{\mathrm{e}}^3}{k} \tag{15.4.10}$$

Because of the nonlinear terms in $\theta$ that appear in the radiation boundary condition (Eq. 15.4.9), it is necessary to solve this boundary value problem numerically. In Fig. 15.4.2 we show solutions for the surface temperature $\theta_1$ as a function of time, with $\Sigma$ as a parameter.

An analytical solution to this problem is possible if we assume that the heat conduction is very efficient, in some sense, yielding a nearly flat temperature profile across the slab. This would correspond to small values of the parameter $\Sigma$. In that case a heat balance would take the form

$$\frac{d}{dt}(\rho C_{\mathrm{p}} TAB) = A \overline{F}_{12} \sigma (T_{\mathrm{e}}^4 - T^4) \tag{15.4.11}$$

where $A$ is the area normal to the $x$ axis, assumed to be large compared to $B^2$. (Keep in mind that absolute temperatures are used in radiative models.) It is not difficult to show that this equation may be written in the form

$$-\frac{d\theta}{d\Sigma X_{\mathrm{Fo}}} = \frac{1 - [1 - \theta(1 - T_{\mathrm{o}}/T_{\mathrm{e}})]^4}{1 - T_{\mathrm{o}}/T_{\mathrm{e}}} \tag{15.4.12}$$

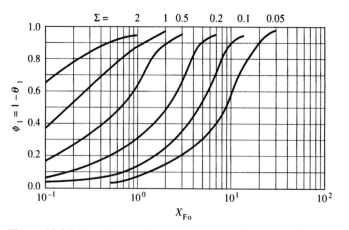

**Figure 15.4.2** Transient surface temperature due to a radiant flux: $T_{\mathrm{o}}/T_{\mathrm{e}} \ll 1$. (Zerkle and Sunderland, *J. Heat Transfer, Trans. ASME*, February 1965, p. 117).

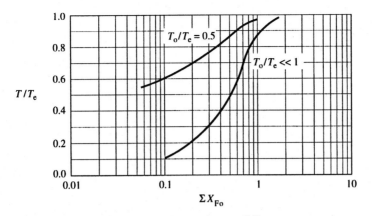

**Figure 15.4.3** Transient temperature for small $\Sigma$.

Equation 15.4.12 can be integrated analytically. For example, for the case $T_o/T_e \ll$ 1, the solution may be written in the form

$$-\frac{1}{4}\ln\frac{1-\phi}{1+\phi} + \frac{1}{2}\tan^{-1}\phi = \Sigma X_{Fo}$$  (15.4.13)

where

$$\phi = 1 - \theta = \frac{T - T_o}{T_e - T_o}$$  (15.4.14)

Figure 15.4.3 shows this solution, in terms of $T/T_e$, as well as a solution for the case $T_o/T_e = 0.5$. Note that for the case of $T_o/T_e \ll 1$, $\phi$ is just $T/T_e$.

**EXAMPLE 15.4.1**   *Radiative Heat Rise in a Solid Panel*

A large square metal panel, of thickness $B = 1$ cm and side length $L = 1$ m, is at a uniform temperature of 27°C. It is suddenly exposed to radiation from an environment at an equivalent black body temperature of 600 K. If there are no convective losses from the panel, how long will the panel take to achieve equilibrium with this radiative source? The panel radiates as a black body. Assume a thermal conductivity of the panel of 100 Btu/h·ft·°F, a thermal diffusivity of 3 ft²/h, and a radiation interchange factor of $\bar{F}_{12} = 0.5$.

We begin by calculating a value of the parameter $\Sigma$. To do so, we require a value in SI units for the thermal conductivity $k$:

$$k = 100 \, \text{Btu/h} \cdot \text{ft} \cdot °\text{F} = 173 \, \text{W/m} \cdot \text{K}$$  (15.4.15)

It follows that

$$\Sigma = \frac{\sigma B \bar{F}_{12} T_e^3}{k} = \frac{5.67 \times 10^{-8}(0.01)(0.5)(600)^3}{173} = 3.5 \times 10^{-4}$$  (15.4.16)

For this case, $T_o/T_e = 0.5$. From Fig. 15.4.3 (valid for this small $\Sigma$ case), equilibrum occurs at (approximately) $\Sigma X_{Fo} = 1.5$, or

$$X_{Fo} = \frac{\alpha t}{B^2} = \frac{1.5}{3.5 \times 10^{-4}} = 4240$$  (15.4.17)

With

$$\alpha = 3 \text{ ft}^2/h = 7.7 \times 10^{-5} \text{ m}^2/\text{s} \qquad (15.4.18)$$

we find the time to reach equilibrium with the radiation source to be

$$t = \frac{4240(0.01)^2}{7.7 \times 10^{-5}} = 5480 \text{ s} = 1.52 \text{ h} \qquad (15.4.19)$$

## SUMMARY

To develop models of radiative heat transfer, we must understand the factors that control radiant emission of energy from a body, as well as the absorption of radiation by that, or another, body. Emission is a temperature-dependent phenomenon and is most simply described by Eq. 15.2.4: $q^e = e\sigma T^4$. The emissivity $e$ of the surface is a material property. The black body corresponds to the case $e = 1$. Absorption of radiation is much more complex, since the absorption depends on the relative orientations (as described by view factors) of the emitting and absorbing surfaces. View factors have been calculated for a variety of pairs of surfaces, and examples of view factors are presented in Section 15.2. With these basic concepts, a number of simple expressions may be derived which serve to provide models for the radiation interchange between surfaces of simple shape.

In many cases of engineering interest, radiation and convection interact to determine the thermal equilibrium of a system, such as a solar heater. While real engineering systems are quite complex, it is often possible to simplify the geometrical and physical features to the point of using the methods of this and earlier chapters to derive models than can serve as useful starting points for quick and approximate design calculations.

We end the chapter with a simple example of combining our knowledge of transient conduction with radiation to produce a model of the radiative heating of a solid surface.

## PROBLEMS

**15.1** Repeat Example 15.2.1, but for a surface with an emissivity of 0.8. At what temperature are the radiant and convective fluxes equal?

**15.2** A 75 W incandescent lightbulb is approximated as a sphere of diameter 55 mm, with a gray surface that has an emissivity of 0.75. If the ambient air temperature is 20°C, what is the surface temperature of the bulb? Assume that the bulb actually radiates power at the rated value of 75 W.

**15.3** The walls of a furnace are maintained at a temperature of 1500 K, and the surfaces may be assumed to behave as black bodies. A square copper plate of thickness 1 mm and side 0.5 m is at $T_o = 300$ K when it is placed in the furnace. The plate has an emissivity of 0.5. What is the steady state temperature of the plate, and how long will it take to achieve its steady temperature?

**15.4** Liquid oxygen is stored in a spherical container that is surrounded by a second concentric sphere. The air between the two spheres is evacuated to reduce the convective heat transfer to the oxygen container. The sphere diameters are 0.5 and 1 m, respectively, and all surfaces have an emissivity of 0.03. Assume that the inner sphere surface is at the temperature of boiling oxygen (90 K), and the outer surface is at 270 K. Because of the heat transfer into the oxygen, boiling occurs continuously. To prevent pressure buildup, a vent permits the oxygen vapor to escape to the atmosphere. If the latent heat of vaporization of oxygen is $2 \times 10^5$ J/kg, what is the rate of loss of mass of oxygen from the container?

**15.5** To gain an understanding of the importance of controlling the surface properties of cryogenic fluid containers, return to Problem 15.4 and plot the rate of loss of oxygen as a function of emissivity in the range $0.02 < e < 1$.

**15.6** A cryogenic liquid flows through a thin-walled pipe of diameter 0.1 m. The liquid is at 70 K at the pipe entrance. The pipe crosses a large open room whose walls may be regarded as black surfaces at a temperature of 300 K, the same as the ambient air temperature. The outer surface of the pipe has an emissivity of 0.6. Estimate the net heat transfer to the pipe from free convection, and radiation, per unit length of pipe. Which mode of heat transfer is the more significant under the stated conditions?

**15.7** A heat exchanger for a space vehicle is designed along the lines suggested in Fig. P15.7. Find the temperature of the heat exchange fluid as a function of the separation between the surfaces 1 and 2. Take the surrounding space as a black body at 0 K. Disks 1 and 2 have diameters of 1 m.

**15.8** A silicon wafer with a diameter of 150 mm and a thickness of 1 mm is initially at a uniform temperature of 800 K. It is suddenly transferred to an evacuated chamber that completely surrounds the wafer. The chamber surface has an emissivity $e = 0.9$ and is maintained at $T = 300$ K. How long does it take for the wafer to cool to $T = 350$ K? Use the following physical properties of silicon:

$$e = 0.6 \qquad k = 0.15 \text{ W/cm} \cdot \text{K}$$
$$C_p = 0.7 \text{ J/g} \cdot \text{K} \qquad \rho = 2.3 \text{ g/cm}^3$$

**15.9** In Example 15.3.1 we found internal and external convective heat transfer coefficients of 4.5 and 4.2 W/m$^2 \cdot$K, respectively. In Example 15.3.2 we used an "overall" heat transfer coefficient of 2 W/m$^2 \cdot$K. What is the relationship of this number to the internal and external coefficients?

**15.10** Repeat Example 15.3.2 with these constraints: $n_t = 40$ tubes per panel, $T_{in} = 10°C$, and a total flow of 0.5 gal/min. Find $T_s$ and $T_{out}$ for $N_p = 6$ panels. What is Re for the tube flow? What is Nu?

**15.11** (a) Repeat Problem 15.10, but for $1 < N_p < 20$, and then present a plot of $T_s$ and $T_{out}$ as functions of $N_p$. (b) Do the same, but for $n_t = 10$ tubes per panel.

**15.12** Consider a single solar panel designed and operating according to Example 15.3.2. Assume that the panel has 20 tubes of diameter 1.5 cm within the absorbing plate. Prepare a model of the performance of this system. In particular, plot the outlet water temperature as a function of the water flowrate.

**15.13** Go back to Example 14.4.3 and include the radiation loss from the surface. Use $e = 0.7$. Let the incident radiation flux be 800 W/m$^2$.

**15.14** As an alternative to Eq. 15.4.5, define a dimensionless temperature as $\Phi = T/T_e$. Give the equations that correspond to Eqs. 15.4.6 through 15.4.9 in this case. Which nondimensionalization is preferable? Why? Do the same for Eq. 15.4.12 (i.e., write it in terms of $\Phi$). What is the initial condition?

**15.15** Using the nondimensionalization described in Problem 15.14, give the solution corresponding to Eq. 15.4.13, for the case of arbitrary $T_o/T_e$. Put a line on Fig. 15.4.3 for the case $T_o/T_e = 0.25$ and 0.75.

**15.16** For the system described and modeled in Section 15.4, find the time (as the Fourier number) at which the slab reaches steady state, as a function of the parameter $\Sigma$. State clearly your definition of steady state. Is there any limitation on the values of $\Sigma$ for which your result is valid?

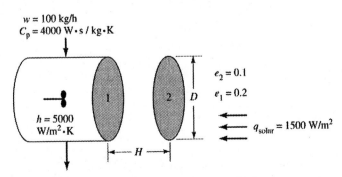

$w = 100$ kg/h
$C_p = 4000$ W $\cdot$ s / kg $\cdot$ K

$e_2 = 0.1$
$e_1 = 0.2$

$h = 5000$ W/m$^2 \cdot$K

$q_{solar} = 1500$ W/m$^2$

**Figure P15.7** Heat exchanger for a space vehicle.

**15.17** Modify the analysis in Section 15.4 to account for convective loss from the exposed surface of the slab, $x = B$. What form does the boundary condition at the exposed surface (Eq. 15.4.4) take in this case? Give an analytical expression for the equilibrium (steady state) temperature of the slab, under these conditions.

Assume that $\Sigma$ is small enough to permit the neglect of internal temperature gradients, and present the differential equation that must be solved to yield the thermal history subsequent to the onset of radiation to the panel. Take $T = T_a$ as the initial condition, and give the solution for the thermal history in an appropriate nondimensional format.

**15.18** For the system described in Section 15.4, show that if the slab is initially at temperature $T_o$, and cools by radiation to a medium at zero absolute temperature, the cooling rate is given by

$$\frac{T}{T_o} = (1 + 3\Sigma_o X_{Fo})^{-1/3} \quad \textbf{(P15.18.1)}$$

where

$$\Sigma_o = \frac{\sigma B \overline{F}_{12} T_o^3}{k} \quad \textbf{(P15.18.2)}$$

**15.19** Rework Problem 15.17, but let the initial temperature be $T_o$, which differs from $T_a$. Give the solution for the thermal history in an appropriate nondimensional format.

**15.20** In examining the approach to Problem 11.18, we note that radiation was not accounted for. Is this a reasonable assumption? What additional dimensionless group should have been considered? What criterion, in terms of this group, determines the role of radiation?

**15.21** Write an explicit expression for $\overline{F}_{12}$ that appears in Eq. 15.2.40, in terms of $F_{12}$. Give the numerical value of $\overline{F}_{12}$ for two cases:
**a.** All six surfaces are squares.
**b.** Surfaces 1 and 2 are squares of side length $L$, and the four sides are rectangles of sides $L$ by $H$.

**15.22** Find the radiation interchange factor $\overline{F}_{12}$ for an enclosure whose surfaces are those of a right circular cylinder, in terms of $F_{12}$. Give the numerical value of $\overline{F}_{12}$ for two cases:
**a.** $H = D$
**b.** $H = 10D$
surfaces 1 and 2 are the two circular ends of the cylinder.

**15.23** Find the view factor for the two surfaces shown in Fig. P15.23, using only the summation

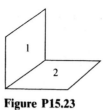

**Figure P15.23**

relationship for an enclosure: $\Sigma_j F_{ij} = 1$, and the reciprocity relationship $A_1 F_{12} = A_2 F_{21}$. Surfaces 1 and 2 are squares at right angles to each other.

**15.24** The fireball of a hydrogen bomb radiates as a black body at a temperature $T = 7200$ K. Take the diameter of the fireball as 1 mile. Assume that air is transparent to this radiation. Calculate the radiant energy flux incident on an area of a vertical wall of a house 25 miles from the center of a blast occurring at an altitude of 1 mile. Use an emissivity of $e = 0.7$ for the wall.

**15.25** Liquid oxygen, which boils at $-297°F$, is to be stored in a spherical container of 1 ft diameter. The system is insulated by an evacuated space between the inner sphere and a surrounding 1.5 ft i.d. concentric sphere. Both spheres are made of polished aluminum ($e = 0.03$), and the temperature of the outer sphere is maintained at 30°F. At what rate is heat transferred to the oxygen?

**15.26** A study of turbulent heat transfer in a duct of square cross section has been completed, and some concern has been expressed about possible temperature-measuring errors due to radiation effects. See Fig. P15.26. Estimate the percentage error in absolute temperature under the following conditions:

The temperature measuring device is a thermocouple within a 3 mm diameter polished aluminum rod that is oriented normal to the gas flow. Assume good thermal contact between the thermocouple junction and the rod.

The duct walls are at a uniform temperature of 1000 K. The temperature reading is 900 K. The square duct has sides of 1 meter in length, and the Reynolds number (based on the side of

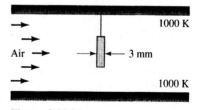

**Figure P15.26**

the duct) is $10^7$. The gas is air at 1 atm pressure.

**15.27** Copper disks, 1 mm thick and 10 cm in diameter, are stacked with an air gap of 1 cm between each one. The plane of the disk surfaces is vertical. The disks are initially at 500°C and are suddenly exposed to an ambient air environment at 25°C. Write an ordinary differential equation, and boundary conditions, that (if you could solve it) would permit you to find the time required to cool the center disk of a 101-disk stack from 500°C to 50°C.

At the initial instant of exposure of the disks to the ambient, what are the relative contributions of convection and radiation to the heat loss?

**15.28** A flat plate solar collector operates as shown in Fig. P15.28.

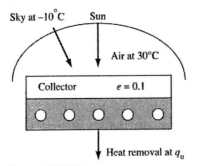

Sky at $-10°$C    Sun

Air at 30°C

Collector     $e = 0.1$

Heat removal at $q_u$

**Figure P15.28**

The collector absorbs radiation from the sky, and from the sun. It loses heat by radiant emission, and by free convection. The following facts are available:

| | |
|---|---|
| Sky | Behaves like a blackbody at $-10°$C. (This is called the "effective sky temperature.") The sky radiation is in the same spectral region as the collector surface emission. Hence use $a_{sky} \approx e = 0.1$, where $a_{sky}$ means the *collector* absorptivity for sky radiation. |
| Sun | The insolation is 750 W/m². The solar absorptivity is $a_{solar} = 0.95$. |
| Collector | Horizontal planar surface. At an air temperature of 30°C the collector surface is observed to be at 120°C when heat is removed at a rate $q_u$. On a calm day only free convection occurs—no forced convection. |

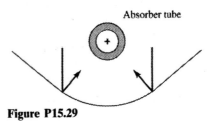

Absorber tube

**Figure P15.29**

Under these conditions, what is the efficiency of the collector, defined as the ratio

$$\varepsilon = \frac{\text{useful heat removal rate } q_u}{\text{insolation}}$$

**15.29** A solar water heater is designed as in Fig. P15.29. Water flows through a copper tube of diameter $D = 60$ mm at a flowrate of $w = 0.01$ kg/s. The water enters the tube at $T_i = 20°$C, and we want to heat the water to $T_o = 80°$C. The tube is aligned with the focal point of a parabolic reflector. As a consequence, solar radiation is reflected and concentrated on the tube. From independent measurements, we know that the concentrated heat flux *absorbed* by the tube is $q_s = 2000$ W/m². (This is a value averaged

**b.** What is the surface temperature of the *tube* at the outlet? Ignore radiation losses for this calculation.

**15.30** The temperature of air flowing inside a tube is being measured with a thermocouple in the system sketched in Fig. P15.30. The tube walls are at a uniform temperature of 440°F. The

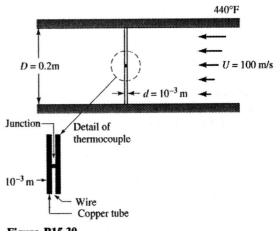

440°F

$D = 0.2$m

$U = 100$ m/s

$d = 10^{-3}$ m

Junction

Detail of thermocouple

$10^{-3}$ m

Wire
Copper tube

**Figure P15.30**

thermocouple indicates that the gas temperature is 940°F. The thermocouple reading is in error because the thermocouple actually radiates some energy to the cold tube walls. Take the emissivity of the thin solid copper tube that contains the thermocouple wires and junction to be $e = 0.7$. Define the temperature difference between the gas and the apparent gas temperature, $\Delta T = T_g - 940$, as the "thermocouple error." Estimate $\Delta T$.

**15.31** Prove that in the situation represented in Fig. P15.31, a black radiation shield reduces the heat transfer from surface 1 to surface 2 by a factor of 2, if the two surfaces behave as black bodies themselves. All surfaces are parallel planes that are large in comparison to the separation between them.

**Figure P15.31**

**15.32** Estimate the temperature at a point 10 cm back from the front edge of the roof of an automobile traveling at 55 mph in air at 30°C. The solar radiation flux at the time is 800 W/m². The roof may be regarded as flat, 1 m wide and 1.5 m long. Take the absorptivity for solar radiation as $a = 0.8$, and the emissivity (at the surface temperature) as $e = 0.4$. What is the rooftop temperature if the automobile is not moving?

**15.33** A heat exchanger is designed as shown in Fig. P15.33. Find the fluid outlet temperature.

**15.34** A black horizontal disk, with a diameter of 1 m, in contact with the ground, is being heated by solar radiation ($q_{solar} = 1000$ W/m²). Heat is lost to the surrounding air at 25°C by natural convection. There is no forced convection. The surface opposite the sun is insulated from the ground. Find the surface temperature.

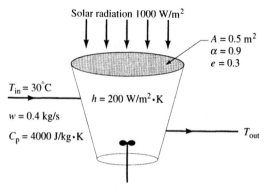

**Figure P15.33**

**15.35** Give the derivations of Eqs. 15.2.66 and 15.2.67.

**15.36** The enclosure in Fig. P15.36 has a triangular cross section and is very wide in the direction normal to the page. The lengths are related by $L_1 = 0.1 L_2 = 0.1 L_3$. Solar radiation heats surface

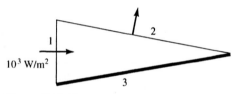

**Figure P15.36**

1, and the net flow of heat across that surface, into the enclosure, corresponds to a flux of 1000 W/m². Surface 3 is insulated, and no heat transfer occurs across it. Surface 2 is a black surface with respect to radiation. The enclosure can be considered to be completely surrounded by a "black" sky at 10 K.

**a.** Find the temperature $T_2$.

**b.** Give values for the view factors $F_{13}, F_{23}, F_{21}$, and $F_{12}$.

**c.** There is insufficient information given to permit us to find $T_1$. Derive an algebraic relationship for $T_3$ as a function of $T_1$ and $T_2$.

# Chapter 16

# Simultaneous Heat and Mass Transfer

**H**eat and mass transfer often occur simultaneously under conditions of engineering importance. Such problems can be quite difficult to analyze because of the interaction of the two phenomena. In this chapter we introduce a few examples of this class of problem and illustrate some simple models that mimic the observed physics.

## 16.1 INTRODUCTION

In a number of technologically significant problems, heat and mass transfer occur simultaneously, with one dependent on the other. The simplest example is provided by a liquid that evaporates from the surface of a droplet into a surrounding vapor. Evaporation requires the transfer of energy to supply the latent heat of vaporization of the liquid. The rate of evaporation could be limited by the rate of supply of energy to the surface of the liquid. The evaporation rate is also determined by the rate of mass transfer from the surface into the ambient medium. This would depend in part on the partial pressure of the evaporating species at the liquid/gas interface. But vaporization can be accompanied by a significant cooling effect, and vapor pressure is a strong function of temperature. As a result, the rate of vaporization is controlled by the "coupling" between the heat and mass transfer phenomena occurring at the liquid/vapor interface. Such problems can be very complex, and we will examine some simple cases to gain an understanding of the nature of simultaneous heat and mass transfer.

## 16.2 EVAPORATIVE COOLING OF A LIQUID DROPLET

With reference to Fig. 16.2.1 we consider the evaporation of a pure liquid droplet surrounded by a gas. The rate of evaporation corresponds to a molar flux that we denote $N_A$, and *for low evaporation rates* we may write $N_A$ as

$$N_A = k_y(y_{A,R} - y_{A\infty}) \tag{16.2.1}$$

Note our use of a mass transfer coefficient $k_y$ based on the mole fraction "driving force" in the gas phase (denoted $y_A$), in writing Eq. 16.2.1.

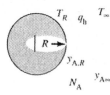

**Figure 16.2.1** Evaporation of a droplet.

The evaporative flux corresponds to (and requires the supply of) heat transferred at a rate that matches the heat of vaporization of the liquid:

$$q_h = -N_A \Delta \tilde{H}_{v,A} \tag{16.2.2}$$

(We specify a molar heat of vaporization here, since we are using a molar evaporative flux.) Note the sign: the thermal flux is opposite in direction to the molar flux.

The heat flux is expressed in terms of a heat transfer coefficient in the form

$$q_h = h(T_R - T_\infty) \tag{16.2.3}$$

We make the assumption that the droplet is small enough to permit internal heat conduction to produce a uniform temperature $T_R$ throughout the droplet. Upon combining these three equations we find

$$\frac{y_{A,R} - y_{A\infty}}{T_\infty - T_R} = \frac{h}{k_y \Delta \tilde{H}_{v,A}} \tag{16.2.4}$$

This equation expresses the necessary coupling between the composition and temperature at the droplet surface.

The nature of the convective heat and mass transfer processes is such that in most cases of interest to us the Chilton–Colburn analogy between the so-called $j$-factors holds (see Eqs. 6.2.16 and 12.1.19):

$$j_H = j_D \tag{16.2.5}$$

or

$$\frac{h}{k_y} = \tilde{C}_{pf} \left( \frac{Sc}{Pr} \right)_f^{2/3} \tag{16.2.6}$$

Quite a bit of manipulation separates Eqs. 16.2.5 and 16.2.6. For example, we must note our earlier definition of the Sherwood number for mass transfer as

$$Sh = \frac{kL}{D_{AB}} \tag{16.2.7}$$

where $L$ is some appropriate length scale, $D_{AB}$ is the diffusion coefficient in the surrounding fluid, and $k$ is the mass transfer coefficient based on *molar concentration* as the driving force:

$$k = \frac{N_A}{\Delta c} = \frac{N_A}{C \Delta y} = \frac{k_y}{C} \tag{16.2.8}$$

where $C$ is the molar density of the surrounding fluid. (For an ideal gas, $C = P/R_G T$.) Note that the *molar* heat capacity of the fluid appears in Eq. 16.2.6. With these substitutions, we may write Eq. 16.2.4 in the form

$$\frac{y_{A,R} - y_{A\infty}}{T_\infty - T_R} = \frac{\tilde{C}_{pf}}{\Delta \tilde{H}_{v,A}} \left( \frac{Sc}{Pr} \right)_f^{2/3} \tag{16.2.9}$$

The right-hand side of Eq. 16.2.9 involves only physical properties of the system. Hence, for various liquids and surrounding gases, there will be a unique relationship between the surface temperature $T_R$ and the surface composition $y_{A,R}$. A second relationship between these two parameters exists, based on vapor/liquid equilibrium data. This is most commonly expressed in the form

$$y_{A,R} = \frac{p_{vap,A}}{P} = f(T_R) \tag{16.2.10}$$

where we note that the vapor pressure of the liquid is a function of the liquid temperature.

This model permits us to estimate the temperature of an evaporating droplet, as Example 16.2.1 illustrates.

**EXAMPLE 16.2.1** *Temperature of an Evaporating Droplet of Water*

A small droplet of water falls slowly through ambient air that is at one atmosphere pressure and 300 K. The relative humidity of the ambient air is 20%. We wish to find the surface temperature of the droplet. The first thing we need is a curve for vapor pressure versus temperature. This is provided in Fig. 16.2.2.

We need, as well, the following physical properties:

$$\hat{C}_p = 1 \text{ kJ/kg} \cdot \text{K (for air)} \qquad \Delta \hat{H}_v = 2260 \text{ kJ/kg (for water)}$$
$$\text{Sc} = 0.6 \qquad \text{Pr} = 0.7$$

Note that while Eq. 16.2.9 has the heat capacity and the heat of vaporization *per mole,* we have found these properties *per mass.* On a kilogram-mole basis, we may convert these values, using 29 for the molecular weight of air, and 18 for water. Equation 16.2.9 becomes

$$\frac{y_{A,R} - y_{A\infty}}{T_\infty - T_R} = \frac{1 \times 29}{2260 \times 18}\left(\frac{0.6}{0.7}\right)^{2/3} = 6.4 \times 10^{-4} \tag{16.2.11}$$

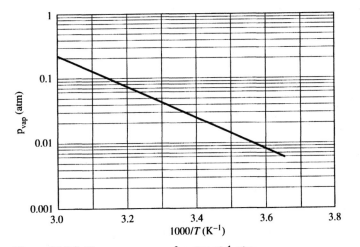

**Figure 16.2.2** Vapor pressure of water at 1 atm.

Since the relative humidity of the surrounding air is specified as 20%, we find, from Fig. 16.2.2, at $T_\infty = 300$ K,

$$y_{A\infty} = 0.2 \frac{p_{vap,A}}{P} = 0.2 \frac{0.035 \text{ atm}}{1 \text{ atm}} = 0.007 \qquad (16.2.12)$$

We now find $T_R$ by trial and error, using Eq. 16.2.11 and Fig. 16.2.2. The result is

$$T_R = 287 \text{ K } (14°C) \qquad (16.2.13)$$

We find that the drop is cooled, relative to the surrounding air, by 13°C. This significant degree of cooling can be used in the design of a cooling system for water. Homework Problem 16.2 addresses this possibility.

## 16.3 EVAPORATIVE BLOWING EFFECT ON HEAT TRANSFER

In the preceding section we examined a simplified model of the effect of evaporative cooling. The heat transferred to support evaporation must lead to a reduced temperature, the consequence of which is to reduce the vapor pressure, hence the driving force for mass transfer. In this section we examine a very simple model of the interaction of heat and mass transfer, but in this case we focus on another property of evaporation: namely, that it produces an outward directed flow of vapor (what we call "blowing") that retards the flux of heat to the surface.[1] To simplify the algebra and to avoid obscuring the physics, we will look at transfer to and from a *planar* surface through a convective resistance equivalent to a uniform film of thickness $\delta_T$ for heat transfer and $\delta_{AB}$ for mass transfer. Figure 16.3.1 shows the geometry. We use $y$ to denote the position coordinate normal to the surface, and now $x_A$ is the mole fraction of species A in the gas phase.

We will consider a binary gas system in which species A is transported by diffusion and convection only in the direction $y$ normal to the planar interface between liquid A and the binary vapor AB. Species B is assumed to be insoluble in liquid A. We will first neglect the effect of evaporative cooling on the dynamics of this system, in the sense that we will take the temperature of the liquid surface to be already known and given as $T_0$. There is a temperature profile across the thermal film thickness $\delta_T$, but we will ignore any effect of temperature variation on physical properties such as diffusion coefficients and molar density. With these assumptions we may write the steady state species balance as

$$\frac{dN_{A,y}}{dy} = 0 \qquad (16.3.1)$$

where $N_{A,y}$ is the molar flux of species A in the direction normal to the surface $y = 0$, relative to fixed coordinates. We invoke Fick's law, noting the assumption that species B is insoluble in the liquid and so cannot cross the surface, and write (see Eq. 2.1.18)

$$N_{A,y} = -cD_{AB} \frac{dx_A}{dy} + x_A N_{A,y} \qquad (16.3.2)$$

---

[1] We treated a problem with a similar character in Chapter 5 (Section 5.5), where the "blowing" was the consequence of chemical reactions that produced more moles of product gas than incoming reactant species.

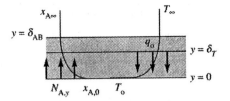

**Figure 16.3.1** Definition sketch for consideration of the effect of "blowing."

or

$$N_{A,y_o} = \frac{-cD_{AB}}{1 - x_A}\frac{dx_A}{dy} \tag{16.3.3}$$

We have added a subscript to the flux $N_{A,y}$ to reflect what Eq. 16.3.1 tells us about the flux: that it is a constant, independent of $y$. The solution for the mole fraction profile is easily found in the form

$$\frac{1 - x_A}{1 - x_{A,0}} = \exp\left(\frac{N_{A,y_o}}{cD_{AB}}y\right) \tag{16.3.4}$$

For a boundary condition, we take the mole fraction of species A to be known and have the value $x_{A\infty}$ for $y = \delta_{AB}$ and beyond. Then it follows that

$$\frac{1 - x_{A\infty}}{1 - x_{A,0}} = \exp\left(\frac{N_{A,y_o}}{cD_{AB}}\delta_{AB}\right) \tag{16.3.5}$$

This equation provides a means of connecting the film thickness $\delta_{AB}$, which is not really measurable, to the flux $N_{A,y_o}$, which we can measure.

Let's turn now to the energy equation for this system. The key issue is how to write the energy flux to account for the flow normal to the surface $y = 0$, which evaporation produces. This evaporative mass flux is the equivalent of a convective velocity against which heat must be conducted. Hence evaporation, when we account for it as we do here, turns this problem into a convection/conduction problem. The starting point is back in Chapter 11, in Eq. 11.2.4, where we wrote the energy flux across a boundary as[2] (but in different notation)

$$e_y = -u_y\rho\hat{C}_pT + k_T\frac{dT}{dy} \tag{16.3.6}$$

The *mass* average velocity $u_y$ can be related to the *molar* flux $N_{A,y}$ through

$$u_y\rho = N_{A,y_o}M_{w,A} \tag{16.3.7}$$

Using the fact that

$$u_y\rho\hat{C}_{pA} = N_{A,y_o}M_{w,A}\hat{C}_{pA} = N_{A,y_o}\tilde{C}_{pA} \tag{16.3.8}$$

---

[2] Note in particular that we use $k_T$ for the thermal conductivity here, to avoid confusion later with a mass transfer coefficient $k_x$.

(where $\tilde{C}_{pA}$ is the *molar* heat capacity of the *vapor*), we may write the energy equation for this system as

$$\frac{d}{dy}\left[N_{A.y_o}\tilde{C}_{pA}(T - T_o) - k_T\frac{dT}{dy}\right] = 0 \tag{16.3.9}$$

or simply

$$N_{A.y_o}\tilde{C}_{pA}(T - T_o) - k_T\frac{dT}{dy} = q_o \tag{16.3.10}$$

where $q_o$ is the constant heat flux to the surface. This equation is easily solved to yield

$$1 - \frac{N_{A.y_o}\tilde{C}_{pA}(T - T_o)}{q_o} = \exp\left(\frac{N_{A.y_o}\tilde{C}_{pA}}{k_T}y\right) \tag{16.3.11}$$

With the boundary condition that

$$T = T_\infty \quad \text{at} \quad y = \delta_T \tag{16.3.12}$$

we may find

$$T_\infty - T_o = \frac{q_o}{N_{A.y_o}\tilde{C}_{pA}}\left[1 - \exp\left(\frac{N_{A.y_o}\tilde{C}_{pA}}{k_T}\delta_T\right)\right] \tag{16.3.13}$$

Now we need to eliminate the thermal film thickness $\delta_T$ from this analysis, and likewise we must eliminate $\delta_{AB}$ from Eq. 16.3.5. To begin, we define the heat transfer coefficient for this system, in the absence of "blowing," as

$$h^o \equiv \frac{k_T}{\delta_T} \tag{16.3.14}$$

Of course this is simply the result we would expect for pure conduction across the film $\delta_T$ if we were unaware that blowing was affecting the heat transfer. The actual heat transfer coefficient would be defined as

$$h^\oplus \equiv \frac{q_o}{T_o - T_\infty} \tag{16.3.15}$$

This lets us write Eq. 16.3.13 as

$$1 + \frac{(T_\infty - T_o)N_{A.y_o}\tilde{C}_{pA}}{q_o} = \exp\left(\frac{N_{A.y_o}\tilde{C}_{pA}}{h^o}\right) \tag{16.3.16}$$

In the same manner, we can define a mass transfer coefficient for this system, in the absence of "blowing," as

$$k_x^o \equiv \frac{cD_{AB}}{\delta_{AB}} \tag{16.3.17}$$

The actual mass transfer coefficient would be defined as

$$k_x^\oplus \equiv \frac{N_{A.y_o}(1 - x_{A,o})}{x_{A,o} - x_{A\infty}} \tag{16.3.18}$$

This lets us write Eq. 16.3.5 without the film thickness $\delta_{AB}$, and we write it in the form

$$1 + \frac{x_{A,o} - x_{A\infty}}{1 - x_{A,o}} = \exp\left(\frac{N_{A.y_o}}{k_x^o}\right) \tag{16.3.19}$$

Between Eqs. 16.3.18 and 16.3.19, we may eliminate the flux $N_{A,y_0}$ and find the ratio of the mass transfer coefficient with blowing to that in the absence of blowing, with the result

$$\frac{k_x^{\oplus}}{k_x^o} = \frac{\ln(1 + R_A)}{R_A} \qquad (16.3.20)$$

where $R_A$ is defined as

$$R_A \equiv \frac{x_{A,o} - x_{A\infty}}{1 - x_{A,o}} \qquad (16.3.21)$$

In the same manner we may introduce Eq. 16.3.15 into Eq. 16.3.16 and find the ratio of the heat transfer coefficient with blowing to that in the absence of blowing, with the result

$$\frac{h^{\oplus}}{h^o} = \frac{\ln(1 + R_T)}{R_T} \qquad (16.3.22)$$

where $R_T$ is defined as

$$R_T \equiv \frac{N_{A,y_0} \check{C}_{pA}(T_\infty - T_o)}{q_o} \qquad (16.3.23)$$

Figure 16.3.2 plots the function $[\ln(1 + R)]/R$ versus $R]$. From Eq. 16.3.21 we see that if we have an estimate of the surface temperature $T_o$, and vapor pressure data in the form $p_{vap}(T_o)$, we can estimate a value for $R_A$. To evaluate the correction factor to the *heat* flux (the left-hand side of Eq. 16.3.22), we must have an estimate of $R_T$. Hence we must be able to estimate heat and mass transfer fluxes in the absence of blowing: $N_{A,y_0}$ and $q_o$, or equivalently, the transfer coefficients $k_x^o$ and $h^o$. It is these latter coefficients that are usually available to us through empirical correlations. The procedure for evaluating the importance of this blowing phenomenon on simultaneous heat and mass transfer is best understood by following through the details of an example problem.

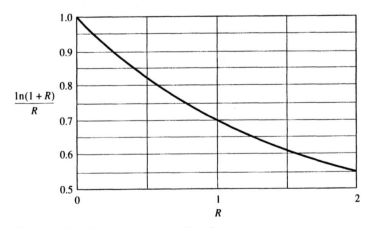

**Figure 16.3.2** The blowing correction factor.

**EXAMPLE 16.3.1**   *Evaporation of a Droplet Falling through Air*

We begin with the estimation of the heat and mass transfer coefficients in the absence of blowing. Then we will calculate the surface temperature of the drop, much as we did in the beginning of this chapter. With this information, and vapor pressure data, we can estimate the factor $R_A$. From this we can evaluate the degree to which blowing alters the rate of evaporation of the droplet. If there is a significant effect on the rate of evaporation, we will have to correct the heat transfer rate used to estimate the surface temperature to obtain an improved estimate of the evaporation rate.

In the case of interest we consider a small droplet of butane ($C_4H_{10}$) of initial diameter $D_{drop} = 100 \ \mu$m, falling through still air at 325 K. The initial drop temperature is 250 K. Butane is a gas at the air temperature, at atmospheric pressure. It boils at 272.7 K. We will find that the drop temperature remains below the boiling point because of the evaporative cooling effect.

We first calculate the terminal velocity $U$ of the drop in air so that we can estimate the heat and mass transfer coefficients for the drop. For the terminal velocity we will assume, because the drop is so small, that Stokes' law holds:

$$U = \frac{D_{drop}^2 \Delta \rho g}{18 \ \mu_{air}} \tag{16.3.24}$$

The density of liquid butane is $\rho = 580$ kg/m³. At 325 K the viscosity of air is $2 \times 10^{-5}$ Pa·s, and the density is 1.1 kg/m³. We estimate the terminal velocity to be

$$U = \frac{(100 \times 10^{-6})^2(580 - 1.1)9.8}{18(2 \times 10^{-5})} = 0.158 \text{ m/s} \tag{16.3.25}$$

This corresponds to a Reynolds number

$$\text{Re} = \frac{U D_{drop} \rho_{air}}{\mu_{air}} = \frac{0.158(100 \times 10^{-6})1.1}{2 \times 10^{-5}} = 0.87 \tag{16.3.26}$$

This is a little large for the use of Stokes' law, but we will use it anyway. We will see that the calculations we are making are tolerant to errors of this order.

To estimate the heat and mass transfer coefficients, we need models for the Nusselt and Sherwood numbers for a falling drop. At low Reynolds numbers, a recommended expression is

$$\text{Nu} = \frac{h^\circ D_{drop}}{k_{T,air}} = [4 + 1.21(\text{Re Pr})^{2/3}]^{1/2} \tag{16.3.27}$$

with an identical expression for the Sherwood number (see Eq. 6.6.6), but with Pr replaced by Sc. We should calculate a Prandtl number for air, but we have seen in earlier chapters that Pr = 0.7 is a good approximation at the air temperature of interest and that Pr is a weak function of temperature. Again, we will see that great accuracy is not required here. The Nusselt number is calculated to be

$$\text{Nu} = [4 + 1.21(0.87 \times 0.7)^{2/3}]^{1/2} = 2.2 \tag{16.3.28}$$

This is very close to the expected lower limit of Nu = 2 for a stationary drop. We can see that small errors in Re and Pr would not change this number very much. The heat transfer coefficient then follows from

$$h^\circ = \frac{k_{T,air}}{D_{drop}} \text{Nu} = \frac{6.7 \times 10^{-3}}{100 \times 10^{-6}}(2.2) = 147 \text{ g·cal/s·m}^2\text{·K} \tag{16.3.29}$$

Note that we use a value for the thermal conductivity of air ($k_{T,air}$) in units of *gram-calories/per second-meter-kelvin*.

In the same way we find a value for the mass transfer coefficient, beginning with

$$\text{Sh} = \frac{k_c^o D_{drop}}{D_{AB,butane/air}} = [4 + 1.21(\text{Re Sc})^{2/3}]^{1/2} \tag{16.3.30}$$

where $k_c$ is the mass transfer coefficient based on concentration as the driving force. The relationship between $k_c$ and $k_x$ is simply

$$k_x = k_c \frac{P}{R_G T} \tag{16.3.31}$$

where $P$ is the total pressure on the system (one atmosphere). For the diffusion coefficient of butane in air at 325 K we use the Chapman–Enskog theory (Chapter 2) with the following parameters:

$$\sigma_{butane} = 4.69 \text{ Å} \qquad (\varepsilon/k_B)_{butane} = 531 \text{ K} \qquad M_{w,butane} = 58$$
$$\sigma_{air} = 3.71 \text{ Å} \qquad (\varepsilon/k_B)_{air} = 78.6 \text{ K} \qquad M_{w,air} = 29$$

For the air/butane mixture we find

$$\sigma_{AB} = \frac{\sigma_A + \sigma_B}{2} = \frac{4.69 + 3.71}{2} = 4.2 \text{ Å} \tag{16.3.32}$$

and

$$\frac{\varepsilon_{AB}}{k_B} = \frac{1}{k_B}(\varepsilon_A \times \varepsilon_B)^{1/2} = (531 \times 78.6)^{1/2} = 204 \text{ K} \tag{16.3.33}$$

At $T = 325$ K we find

$$\frac{k_B T}{\varepsilon_{AB}} = \frac{325}{204} = 1.6 \tag{16.3.34}$$

From Fig. 2.2.1 we find $\Omega_D = 1.2$. With these parameters, the Chapman–Enskog theory yields the diffusion coefficient as

$$D_{AB} = 1.86 \times 10^{-3} T^{3/2} \frac{[(M_{w,A} + M_{w,B})/M_{w,A}M_{w,B}]^{1/2}}{P\sigma_{AB}^2 \Omega_D} \tag{16.3.35}$$

$$= 1.86 \times 10^{-3}(325)^{3/2} \frac{[(29 + 58)/29 \times 58]^{1/2}}{(1)(4.2)^2(1.2)} = 0.12 \text{ cm}^2/\text{s}$$

The Schmidt number then follows as

$$\text{Sc} = \frac{\mu_{air}}{\rho_{air} D_{AB}} = \frac{2 \times 10^{-5} \text{ Pa} \cdot \text{s}}{1.1(\text{kg/m}^3)(0.12 \times 10^{-4})(\text{m}^2/\text{s})} = 1.5 \tag{16.3.36}$$

Finally we can return to the calculation of the mass transfer coefficient. From Eq. 16.3.30 we find

$$\text{Sh} = \frac{k_c^o D_{drop}}{D_{AB,butane/air}} = [4 + 1.21(0.87 \times 1.5)^{2/3}]^{1/2} = 2.3 \tag{16.3.37}$$

and so

$$k_x^o = \frac{P}{R_G T} k_c^o = \frac{P}{R_G T} \text{Sh} \frac{D_{AB}}{D_{drop}} = \frac{1}{82.06 \times 325} 2.3 \frac{0.12 \,(\text{cm}^2/\text{s})}{100 \times 10^{-4} \text{ cm}} = 10^{-3} \text{ g–mol/cm}^2 \cdot \text{s}$$

$$= 10 \text{ g–mol/m}^2 \cdot \text{s} \tag{16.3.38}$$

(Since our gas constant is 82.06, we use cgs units in this calculation.)

Our next step is the estimation of the temperature of the droplet, accounting for the evaporative cooling effect. Making an energy balance, equating the evaporative heat loss to the heat gain $q_o$, we find

$$k_x^o(x_{A,o} - x_{A\infty})\pi D_{\text{drop}}^2 \Delta \tilde{H}_{\text{vap}} = h^o(T_\infty - T_{\text{drop}})\pi D_{\text{drop}}^2 \quad (16.3.39)$$

or

$$T_\infty - T_{\text{drop}} = \frac{\text{Sh}}{\text{Nu}} \frac{cD_{AB} \Delta \tilde{H}_{\text{vap}}}{k_{T,\text{air}}}(x_{A,o} - x_{A\infty}) \quad (16.3.40)$$

For butane we use a heat of vaporization of 5352 g·cal/g–mol. As before, we calculate the molar density of air at 325 K from $P/R_G T_\infty$. Then Eq. 16.3.40 becomes

$$T_\infty - T_{\text{drop}} = \frac{2.3}{2.2} \frac{1}{82.06 T_\infty} \frac{0.12 \times 5352}{6.7 \times 10^{-5}}(x_{A,o} - x_{A\infty}) = 1.2 \times 10^5 \frac{x_{A,o} - x_{A\infty}}{T_\infty}$$
$$(16.3.41)$$

(Again, since our gas constant is 82.06, we use cgs units in this calculation.)

Equation 16.3.41 provides one relationship between the drop temperature and the mole fraction of butane in the gas phase at the surface of the drop. The second relationship is the vapor pressure as a function of temperature. We use the Antoine equation with the coefficients given in Reid, Prausnitz, and Poling for $n$-butane:

$$\ln p_{\text{vap}} = 15.68 - \frac{2155}{T - 34.4} \quad (16.3.42)$$

with $T$ in kelvins and $p_{\text{vap}}$ in millimeters of mercury. The vapor pressure–temperature curve is plotted in Fig. 16.3.3. Instead of the vapor pressure, we plot the mole fraction, using

$$x_{A,o} = \frac{p_{\text{vap}}}{P} \quad (16.3.43)$$

If we plot Eq. 16.3.41 on Fig. 16.3.3, for the case $T_\infty = 325$ K, and $x_{A\infty} = 0$, the intersection gives the drop temperature and the mole fraction of butane in the vapor phase at the drop surface. In this case we find

$$x_{A,o} = 0.23 \quad \text{and} \quad T_{\text{drop}} = 240 \text{ K} \quad (16.3.44)$$

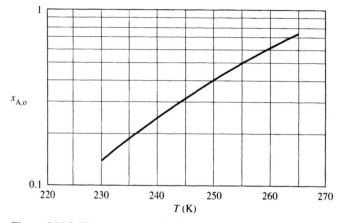

**Figure 16.3.3** Vapor pressure (mole fraction) versus temperature for $n$-butane.

The drop is well below its boiling temperature, even though the ambient temperature is about 50°C above the boiling temperature.

Now we are finally in a position to assess the effect of blowing. (See Problem 16.12.) If we find that blowing has significantly reduced the rates of heat and mass transfer, we may have to recalculate the drop temperature. By an iterative process, we could then ultimately arrive at an accurate estimate of the rate of evaporation of the drop. In this example we find

$$R_A = \frac{x_{A,o}}{1 - x_{A,o}} = \frac{0.23}{0.77} = 0.3 \tag{16.3.45}$$

Hence

$$\frac{k_x^\oplus}{k_x^o} = \frac{\ln(1.3)}{0.3} = 0.88 \tag{16.3.46}$$

Thus the effect of the evaporative flux (this is what "blowing" is) is to reduce the rate of mass transfer to some extent. However, this calculation is based on the solution for the drop surface temperature, in the absence of blowing. We must now find the correction factor to the heat transfer coefficient. This requires calculation of $R_T$ (see Eq. 16.3.23), which begins with a calculation of the molar flux $N_{A,y_o}$. From Eq. 16.3.19,

$$N_{A,y_o} = k_x^o \ln(1 + R_A) = 10 \ln(1.3) = 2.6 \text{ g–mol/m}^2 \cdot \text{s} \tag{16.3.47}$$

It is not difficult to show that this is an extremely large molar flux, relative to the number of moles in a 100 $\mu$m drop (see Problem 16.16).

$R_T$ now follows from Eq. 16.3.23 as[3]

$$R_T = \frac{N_{A,y_o}\tilde{C}_{pA}(T_o - T_\infty)}{q_o} = \frac{k_x^o\tilde{C}_{pA}\ln(1 + R_A)}{h^o} \tag{16.3.48}$$
$$= \frac{10 \times 19 \times \ln(1.3)}{147} = 0.34$$

and we find

$$\frac{h^\oplus}{h^o} = \frac{\ln(1 + R_T)}{R_T} = \frac{\ln(1.34)}{0.34} = 0.86 \tag{16.3.49}$$

The supply of energy to the droplet is reduced by 14%, relative to simple convection in the absence of blowing.

We should now return to Eq. 16.3.39 and replace the heat and mass transfer coefficients by their corrected values. However, since the *ratio* of these two coefficients appears, and the correction factors are nearly identical for each, there will be no change in Eq. 16.3.41, and our estimates of the evaporative flux and the droplet temperature are accurate.

**EXAMPLE 16.3.2**   *Evaporation of Alkane Fuel Droplets*

Combustion systems that utilize liquid fuels usually involve a spray chamber that disperses the fuel into small droplets. There is an important design issue with respect to

---

[3] The molar heat capacity of butane vapor is estimated to be 19 g · cal/g-mol · K, using methods described in Reid, Prausnitz, and Poling.

the lifetime of a fuel droplet—the time required for the droplet to completely evaporate. We can analyze droplet evaporation now, taking into account the effect of evaporative cooling. We simplify the analysis by ignoring combustion, dealing only with the evaporation process. We also ignore the contribution of radiation to the heat transfer process.

We imagine a single spherical droplet, initially with a radius $R_o$, suddenly injected into a hot gas maintained at a temperature $T_\infty$. A simple mass balance on the liquid drop takes the form

$$-\frac{dm_L}{dt} = \frac{d}{dt}\left(\frac{4\pi R^3 \rho_L}{3}\right) \tag{16.3.50}$$

If we assume that the drop is very quickly brought to its vaporization temperature, then a heat balance takes the form

$$h(4\pi R^2)(T_\infty - T_R) = -4\pi R^2 C_L \Delta\tilde{H}_v \frac{dR}{dt} \tag{16.3.51}$$

where $C_L$ is the molar density of the liquid.

Fuel droplets are often so small that the rate of heat transfer to the surrounding gas is limited by conduction into the gas. In other words, convection is largely suppressed. For a single spherical drop losing heat by conduction into a surrounding infinite fluid, we know that the limiting value of the Nusselt number, corresponding to heat conduction, is $\mathrm{Nu}_D = 2$, or

$$h = \frac{k_{T,f}}{R} \tag{16.3.52}$$

where $k_{T,f}$ is the fluid thermal conductivity. Hence Eq. 16.3.51 becomes

$$-R\frac{dR}{dt} = \frac{k_{T,f}(T_\infty - T_R)}{C_L \Delta\tilde{H}_v} \tag{16.3.53}$$

Next we must account for the mass transfer process at the drop surface. If we use a mass transfer coefficient based on the mole fraction "driving force" in the gas phase, as we did in writing Eq. 16.2.1, we find

$$k_x(4\pi R^2)(x_{A,R} - x_\infty) = -4\pi R^2 C_L \frac{dR}{dt} \tag{16.3.54}$$

where $C_L$ is the molar density of the liquid.

Consistent with the assumption that $\mathrm{Nu}_D = 2$, we assume also that $\mathrm{Sh}_D = 2kR/D_{AB} = 2$ so that

$$k = \frac{k_x}{C} = \frac{D_{AB}}{R} \tag{16.3.55}$$

Then Eq. 16.3.54 becomes

$$-R\frac{dR}{dt} = \frac{CD_{AB}(x_{A,R} - x_\infty)}{C_L} \tag{16.3.56}$$

Either Eq. 16.3.53 or 16.3.56 may be used to find $R(t)$. If we assume that the surrounding vapor is free of fuel ($x_\infty = 0$), we obtain a simple differential equation for $R(t)$:

$$-R\frac{dR}{dt} = \frac{CD_{AB}x_{A,R}}{C_L} = -\frac{k_{T,f}(T_R - T_\infty)}{C_L \Delta\tilde{H}_v} = \frac{K}{2} \tag{16.3.57}$$

where $K$ is a constant. To find $R(t)$, we must couple this equation with a vapor pressure–temperature curve

$$x_{A,R} = \frac{p_{vap,A}}{P} = f(T_R) \tag{16.3.58}$$

First we integrate Eq. 16.3.57 to find

$$R^2 = R_0^2 - Kt \tag{16.3.59}$$

This is an explicit solution for $R(t)$, but the coefficient $K$ (which is called the *evaporation constant*) is unknown, since the temperature $T_R$ and the mole fraction $x_{A,R}$ are not known at this point.

We wish to predict the rate of evaporation of small droplets of pure alkane fuels suspended in an airstream at 1000 K. Since some data are available for $n$-heptane and hexadecane, we will examine these two fluids as examples. We need the following physical property information. Unless otherwise noted, we use Reid, Prausnitz, and Poling (*The Properties of Gases and Liquids*) for physical property data, or for estimation methods.

To calculate the evaporation constant $K$ we must know the drop temperature $T_R$. We expect that the drop will be at a temperature below the normal boiling point ($T_{bp} = 372$ K). We will *assume* $T_R = 370$ K, and we could confirm that the value for $K$ that follows does not change much if $T_R$ is really 10 or 15°C lower than this.

**Heptane**                     **Hexadecane**

$$\Delta \tilde{H}_v = 3.2 \times 10^7 \text{ J/kg} \cdot \text{mol} \quad \Delta \tilde{H}_v = 5.1 \times 10^7 \text{ J/kg} \cdot \text{mol}$$

For air at 370 K, and $P = 1$ atm, $k_{T,f} = k_{T,air} = 0.031$ W/m·K (see Fig. C10.2)
We will calculate $K$ from

$$K = \frac{2k_{T,f}(T_\infty - T_R)}{C_L \Delta \tilde{H}_v} \tag{16.3.60}$$

For heptane ($C_7H_{16}$) we get (using the Le Bas method) $V_b = 7 \times 14.8 + 16 \times 3.7$ = 162.8 cm³/g-mol or

$$C_L = \frac{1000}{V_b} = 6.1 \frac{\text{kg-mol}}{\text{m}^3}$$

From Eq. 16.3.60 the evaporation constant is found to be

$$K = \frac{2 \times 0.031(1000 - 370)}{6.1(3.2 \times 10^7)} = 2 \times 10^{-7} \text{ m}^2/\text{s} \tag{16.3.61}$$

Thus we predict that the droplet radius (in units of meters) changes according to

$$R^2 = R_0^2 - 2 \times 10^{-7} t \tag{16.3.62}$$

An important characteristic of this system is the time required for the droplet to completely evaporate. This follows from Eq. 16.3.62 as

$$t_\infty = \frac{R_0^2}{2 \times 10^{-7}} \tag{16.3.63}$$

Equation 16.3.62 predicts that the drop radius will decrease quadratically with time. Figure 16.3.4 shows data for two alkane fuels. For the case of heptane, the line corresponding to Eq. 16.3.62 is labeled "simple model," and we see that it fits the data quite well. For hexadecane, the data for radius versus time show a slow initial period, followed by a linear decrease of $R^2$, as predicted. Departures from our simple model, not shown

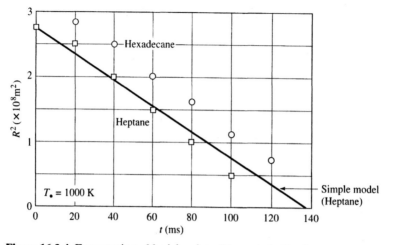

**Figure 16.3.4** Evaporation of fuel droplets. Wang et al., *Comb. Flame*, **56**, 175 (1984).

on Fig. 16.3.4 for the case of hexadecane, are due in part to our neglect of any "heat-up" time for the droplet, which is longer for hexadecane because it has a much higher boiling temperature than heptane. We have assumed that the droplet reaches its equilibrium temperature instantaneously. We have also neglected any contribution to heat transfer due to radiation. While these shortcuts might seem suspect in view of the high temperatures in a combustion system, the alkane fuels have a relatively low absorptivity for radiation, hence are relatively unaffected by radiation.

## 16.4 ANALYSIS OF A LIQUID SOURCE DELIVERY SYSTEM

Chemical vapor deposition is a common means of growing solid films of some element or compound onto another solid surface. This technology is of special importance in semiconductor fabrication and in the production of high performance composite materials. High purity source materials are essential to the success of such processes. One means of promoting the required level of purity is to use a *liquid* source as the feed material. If the liquid itself is very pure chemically, and if the container and "plumbing" are also free of mobile particles and volatile contaminants, then the vapor evolved from such a liquid source should satisfy the requisite constraints on purity.

Figure 16.4.1 illustrates some of the physical events occurring at and near the liquid/gas interface of an evaporator. If the pressure in the headspace above the liquid is comparable to the vapor pressure of the liquid, substantial evaporation will occur at the gas/liquid interface. The heat required to support the evaporation is withdrawn from the liquid, and evaporative cooling occurs in a thin layer just below the interface. The resultant reduction in temperature reduces the vapor pressure and, therefore, suppresses the rate of vaporization. Unless heat is supplied from the ambient medium, the liquid temperature will fall continuously until (at least in principle) the freezing point is reached. Of course, the design of any commercial evaporator includes an active heating unit to sustain evaporation and produce a steady rate of mass flow.

Heat transfer from the ambient medium is enhanced within the liquid by natural convection. In addition, vapor bubbles may evolve from nucleation sites (microscopic cracks and roughness in the container surface). Liquid flow is then induced by these

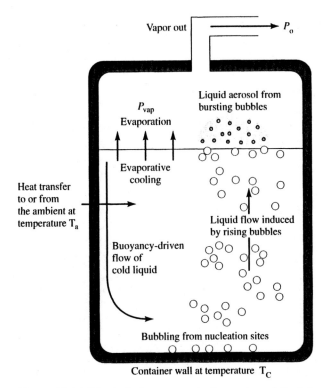

**Figure 16.4.1** Physical phenomena in the region of the interface of an evaporator.

rising bubbles, which further enhances heat transfer within the liquid. In this section, we consider the case of an evaporator operating under steady thermal conditions and develop a model with which the vapor delivery rate may be estimated as a function of the design and operating characteristics of the system, and as a function of the thermodynamic properties of the liquid.

Steady state evaporation requires that the heat input to the evaporator just balance the heat losses due to the generation of vapor and any heat losses to the surroundings. The mass carryover rate may be written in the form

$$\dot{m} = \rho q \qquad (16.4.1)$$

where $q$ is the *volumetric* (not molar) flowrate and $\rho$ is the vapor mass density. In a steady state system, $\dot{m}$ is constant.

An energy balance may be written as

$$\frac{d}{dt}(M\hat{C}_{p,L}T) = -\rho q \Delta\hat{H}_v + UA(T_C - T) \qquad (16.4.2)$$

We let $M$ represent the mass of liquid in the evaporator at any time. For true steady state operation, $M$ is constant, and liquid must be fed into the evaporator at the same rate at which it evaporates. If this feed stream is at a temperature $T_f$ different from the steady operating temperature $T$, an additional heat exchange must take place to maintain steady thermal conditions. Normally we would add fresh liquid at a temperature very near to the steady operating temperature $T$. Hence this sensible heat term would be small in comparison to the heat loss due to vaporization. We will not consider it here.

The heat capacity and heat of vaporization of the liquid (*per mass*) are $\hat{C}_{p,L}$ and $\Delta\hat{H}_v$, respectively. Heat transfer from an external heat source is determined by the overall heat transfer coefficient $U$ and the area $A$ across which this heat is transferred. Heat losses due to poor insulation of the vessel are neglected in this simple model. At steady state, then, since $T$ is constant, Eq. 16.4.2 simplifies to

$$\rho q \Delta\hat{H}_v + UA(T_C - T) = 0 \tag{16.4.3}$$

In the simple model presented here, we assume that the evaporator is immersed in a well-stirred, constant temperature bath. The temperature of the container wall, $T_C$, is assumed to be constant and is specified. Equations 16.4.1 and 16.4.3 can be combined to yield a single relationship between the mass carryover rate $\rho q$, and the liquid temperature $T$. One more relationship between $\rho q$ and $T$ is required. It is provided by the relationship of flowrate to pressures in the system. While normally the derivation would entail the introduction of a knowledge of the friction factors for flow through the straight tubes and energy loss factors $e_v$ for the bends, expansions, and contractions characteristic of the plumbing, this information enters our model through the definition of something called the *conductance*, $F$, which is also characteristic of the series of flow elements (tubes, orifices, valves, etc.) that connect the evaporator to the downstream pump.

For a flow $q_v$ driven by a pressure difference $P_{vap} - P_o$, $F$ is defined as

$$F = \frac{P_{vap}q_v}{P_{vap} - P_o} \tag{16.4.4}$$

The vapor pressure of the liquid is $P_{vap}$, which is a strong function of liquid temperature $T$. The downstream pressure is $P_o$. The volumetric flowrate $q_v$ is subscripted to indicate that it is evaluated at the pressure $P_{vap}$. (Keep in mind that the vapor is a compressible fluid.) With the assumption of ideal gas behavior, we may write

$$\rho q = \frac{M_{w,v}}{R_G T} P_{vap}q_v \tag{16.4.5}$$

where $M_{w,v}$ is the vapor molecular weight. For a given plumbing configuration, we may write the conductance as a function of the pressures $P_{vap}$ and $P_o$. For the sake of example, we will take the conductance to be limited by *viscous* flow of a compressible gas through a small capillary of radius $R$ and length $L$, for which the conductance is given by

$$F = \frac{\pi R^4}{8\mu L}\overline{P} = \frac{\pi R^4}{16\mu L}(P_{vap} + P_o) \tag{16.4.6}$$

We note that the conductance itself depends on the operating temperature we are trying to calculate, since the vapor pressure $P_{vap}$ is strongly temperature dependent. (The viscosity is a weak function of temperature, for a vapor.)

For the required vapor pressure–temperature relationship, we will use the Antoine equation in the form

$$\ln P_{vap} = A - \frac{B}{T + C} \tag{16.4.7}$$

where $A$, $B$, and $C$ are thermodynamic coefficients for the liquid.

Given sufficient information about the system, we may use the equations presented above to calculate the temperature of the liquid and the flowrate $q$. If we choose to report $q$ in units of *standard* conditions (such as standard cubic centimeters per time) then

$$q_s = \frac{P_{vap} q_v}{760} \frac{273}{T} \qquad (16.4.8)$$

when $P_{vap}$ is in units of millimeters of mercury, or torr, and $T$ is in kelvins.

In a typical engineering design problem, the process pressure $P_o$ and the desired delivery rate $\dot{m} = \rho q$ will be specified. We would like to determine the operating temperature that will yield the required flowrate. This turns out to be a problem in which the evaporative heat transfer and the fluid dynamics are strongly coupled.

If we combine Eqs. 16.4.4 to 16.4.6, we may write a *design equation* in the form

$$P_{vap}^2 = P_o^2 + \frac{16 \dot{m} R_G T \mu L}{M_{w,v} \pi R^4} \qquad (16.4.9)$$

When this is combined with the *thermodynamic equation* (the Antoine equation), we have a pair of equations in the two unknowns, $P_{vap}$ and $T$. The solution yields the required design temperature $T$. Such a solution is easily found by graphical or numerical procedures. We illustrate the graphical procedure here.

**EXAMPLE 16.4.1**  *Performance of a Vapor Delivery System*

Silicon dioxide films are grown at low temperature and pressure from any of several organometallic compounds, of which tetraethoxysilane (TEOS) is a common example. Our goal is to present a performance curve that indicates the rate at which we can supply TEOS as a function of the bubbler temperature. The following conditions and data are given to us:

$$P_o = 10^{-3} \text{ torr} \quad M_{w,v} = 208 \text{ g/mol}$$
$$R = 0.1 \text{ cm} \quad \mu(T) \text{ (see Fig. 16.4.2)}$$
$$L = 5 \text{ cm} \quad P_{vap}(T) \text{ (see Fig. 16.4.2)}$$

Figure 16.4.2 shows the graphical solution to this problem for several values of $\dot{m}$. From Eq. 16.4.7 (and the coefficients $A$, $B$, and $C$ and the critical temperature for TEOS),

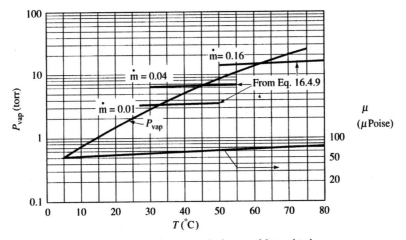

**Figure 16.4.2** Graphical solution to a design problem. ($\dot{m}$ in units of g/s.)

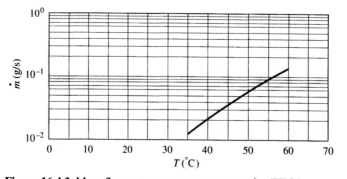

**Figure 16.4.3** Mass flowrate versus temperature for TEOS.

we get a curve for vapor pressure versus $T$. From Eq. 16.4.9, for a given value of $\dot{m}$ (mass flow), we get another curve of $P_{vap}$ versus $T$. For each mass flow, the intersection of the $\dot{m}$ curve with the $P_{vap}/T$ curve yields the temperature required to achieve a specific mass flowrate for a *fixed* flow-restricting element. Figure 16.4.3 shows such a curve, with the downstream pressure $P_o$ at a value of $10^{-3}$ torr. Different solutions would be obtained for other liquids, and for conductance functions other than the simple one adopted here.

In fact, the conductance function in this example is treated as if the primary flow controlling element (the capillary) is fixed, hence *passive*. In such a case the only means to adjust the mass flowrate is to change the liquid temperature or the process pressure. Normally, commercial systems utilize an *active* flow resistance, a so-called mass flow controller (MFC). In an MFC, upstream variations that give rise to a change in delivery rate ($\dot{m}$) are sensed, and the conductance is altered to bring $\dot{m}$ back to its design value. An MFC is a *variable conductance* device.

The model illustrated here seems, then, to be artificial or unrealistic in the sense that one does not normally control temperature as a means of providing mass flow control. What the model *does* provide is a method of predicting the *maximum possible* mass flow under the condition of the maximum conductance of the system (i.e., a wide-open valve in the MFC) subject to the constraint of the upper limit on temperature that the MFC can withstand, given materials or electronics limitations.

We can see this point more clearly if we write the conductance as

$$F = \frac{K}{\mu}\overline{P} \qquad \text{for } viscous \text{ flow} \tag{16.4.10}$$

where $K$ is a characteristic geometrical parameter for the MFC. Then Eq. 16.4.9 takes the form

$$\dot{m} = \frac{M_{w,v}}{R_G T}\frac{K}{2\mu}(P_{vap}^2 - P_o^2) \qquad \text{for } viscous \text{ flow} \tag{16.4.11}$$

If a wide-open valve has a parameter value $K_{max}$, Eq. 16.4.11 gives the maximum mass flow as a function of temperature for a specific vapor, and for a specified downstream pressure $P_o$. If certain considerations provide an upper limit on temperature, $T_{max}$, this effectively yields the maximum $\dot{m}$ as

$$\dot{m}_{max} = \frac{M_{w,v}}{R_G T_{max}}\frac{K_{max}}{2\mu_{max}}(P_{vap,max}^2 - P_o^2) \tag{16.4.12}$$

where $\mu$ and $P_{vap}$ are evaluated at $T = T_{max}$.

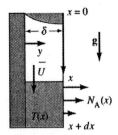

**Figure 16.5.1** Heat transfer and evaporation from a falling liquid film.

## 16.5   EVAPORATIVE COOLING OF A FALLING LIQUID FILM

In our analysis of the falling film evaporator (Section 5.1 in Part I) we assumed that the flow was isothermal and that all physical properties were independent of temperature. In fact, there can be a significant evaporative *cooling* effect. The film will cool, with a corresponding (significant) reduction in vapor pressure at the interface. The analysis of such a system is quite complicated, but it is possible to develop a simple model that reflects the essential physics. We begin with a heat balance on a section of the liquid, as depicted in Fig. 16.5.1.

We assume that for a thin liquid film there will be very little resistance to heat conduction across the film. Hence we assume that the liquid film temperature is uniform in the $y$ direction, varying only in the direction of flow. We assume, also, that the convection of heat in the $x$ direction is much greater than conduction in the same direction. Then a simple heat balance leads us to

$$-\frac{dT}{dx} = \frac{M_w \Delta \hat{H}_v}{\rho \hat{C}_p \overline{U} \delta} N_A = \frac{M_w \Delta \hat{H}_v}{\rho \hat{C}_p \overline{U} \delta} \left(\frac{p_{vap}}{R_G T}\right) \left(\frac{3 D_{AB} \overline{U}}{2\pi x}\right)^{1/2} \tag{16.5.1}$$

where $M_w$ is the molecular weight (kg/kg–mole) of the evaporative species and $\hat{H}_v$ and $\hat{C}_p$ are on a per unit *mass* basis. (Since the *ratio* appears here, we could use molar values as well.) This equation follows from Eq. 5.1.25, but we use the mean velocity $\overline{U}$ instead of the surface velocity $v_m$ (noting $\overline{U} = 2v_m/3$), and we write the concentration difference in terms of the vapor pressure at the interface. In the simplest possible model, we can assume that the concentration ($p_{vap}/R_G T$, in kg–mol/m³) is constant along the flow direction, thereby ignoring the effect of cooling on this term. While this is not a good assumption, it does permit an immediate approximate solution, since we can integrate Eq. 16.5.1 to yield the solution

$$T(x) = T_{x=0} - 1.382 \frac{M_w \hat{H}_v D_{AB}^{1/2}}{\rho \hat{C}_p \overline{U}^{1/2} \delta} \left(\frac{p_{vap}}{R_G T}\right) x^{1/2} \tag{16.5.2}$$

It remains for us to evaluate the extent to which this simplified model provides any reasonable approximation of reality. We do this in Example 16.5.1.

**EXAMPLE 16.5.1**   *Prediction of Experimental Cooling Data*

Tsay et al. [*Int. J. Multiphase Flow*, **16**, 853 (1990)] present the experimental data shown in Fig. 16.5.2 on the temperature at the *solid*/liquid interface for a falling liquid film.

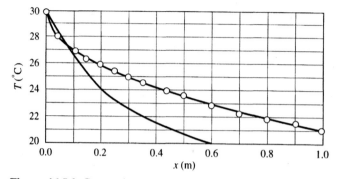

**Figure 16.5.2** Comparison of our simple model (the lower curve: Eq. 16.5.4) to data for cooling of a liquid film of ethanol.

The liquid used was ethanol. The model we developed above assumes no temperature gradients in the $y$ direction, and no heat transfer across the solid/liquid boundary. In the experiments performed, the solid boundary was a very poor conductor of heat. We will first examine the ability of Eq. 16.5.2 to describe these data.

The following information can be obtained from the reference cited.

$$\dot{m} = 0.01 \text{ kg/m} \cdot \text{s} \qquad \delta = 1.67 \times 10^{-4} \text{ m} \qquad T_{in} = 303 \text{ K}$$

where $\dot{m}$ is the mass flowrate of the liquid, per unit of width, $\delta$ is the inlet liquid film thickness, and $T_{in}$ is the inlet liquid temperature. The ambient air is also at 303 K, and it is free of ethanol vapor. Physical properties are as follows:

$$\rho = 788 \text{ kg/m}^3 \text{ (liquid)} \quad \hat{C}_p = 2512 \text{ J/kg} \cdot \text{K} \quad \hat{H}_v = 9.42 \times 10^5 \text{ J/kg} \quad M_w = 46 \text{ kg–mol/kg}$$

$$D_{ethanol/air} = 1.85 \times 10^{-5} \text{ m}^2/\text{s}$$

With these numbers it is not difficult to find (note Eq. 5.1.2) that the mean velocity of the film is $\bar{U} = 0.076$ m/s.

The vapor pressure of ethanol is shown in Fig. 16.5.3; it is obviously a strong function of temperature in the range of temperatures observed. We have assumed a constant vapor pressure in our model, and this is likely to be the major source of error.

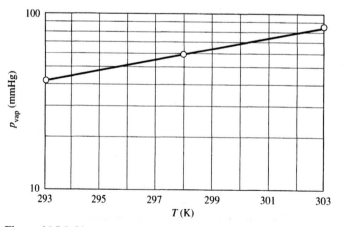

**Figure 16.5.3** Vapor pressure data for ethanol.

We first use the model given by Eq. 16.5.2 with the vapor pressure calculated at the initial (inlet) temperature. From the vapor pressure at 303 K (87 mmHg), we find

$$\frac{p_{vap}}{R_G T} = \frac{87/760}{82.06\,(303)} \times 10^6 = 4.6\ \text{g–mol/m}^3 = 0.0046\ \text{kg–mol/m}^3 \qquad (16.5.3)$$

Note carefully the manipulation of the units here, and especially that we are using SI units with *kilogram*–moles. Equation 16.5.2 takes the form

$$T = 30 - 13\sqrt{x} \qquad (16.5.4)$$

This is the lower curve on Fig. 16.5.2. As expected, this simple model is not very accurate. It overpredicts the temperature drop because it uses a vapor pressure that is too high. As the film cools, the vapor pressure will fall and the rate of evaporation will decrease. These effects, in turn, decrease the rate of evaporative cooling. With this prediction as a starting point, however, we can choose a vapor pressure at the predicted average temperature over the 1-meter length of the film. From Eq. 16.5.4 we have predicted $T = 17°C$ at $x = 1$ m. Thus the average temperature is $(30 + 17)/2 = 23.5°C = 297$ K. For the next approximation, we will use a vapor pressure of 60 mmHg. This yields the curve (shown in Fig. 16.5.2) passing through the experimental data. To some extent the apparent success of this simplified model is due to our use of data for which the temperature change is relatively small. Hence the vapor pressure data can be *linearized* over a small range of temperature, and the use of an average temperature yields a reasonable approximation. A more accurate, but still simple and analytical, model is possible if an exponential curve is fit to the vapor pressure–temperature data. (See Problem 16.11.)

## SUMMARY

In this final chapter we examine some situations in which heat and mass transfer are strongly coupled. Through several examples we illustrate the approach to developing models for the effect of evaporative mass transfer on the thermal history of an interface, and the effect of evaporative cooling on the evaporation rate itself. Thermodynamic information is essential to the development of these models. In particular, we must have data on the temperature dependence of vapor pressure, as well as heat of vaporization data. These ideas are illustrated first through some simple drop cooling and drop evaporation examples. In Section 16.3 we combine what we know about fluid dynamics, heat, and mass transfer to produce a model of the effect of evaporative mass transfer ("blowing") on heat and mass transfer across the boundary of an interface with its surroundings. In Section 16.4 we apply these ideas to the design of a system for delivering vapor to a reactor. We see that our knowledge of fluid dynamics is also brought to bear here.

## PROBLEMS

**16.1** Suppose, instead of adopting the Chilton–Colburn analogy (Eq. 16.2.5), we assumed that Nu = Sh. How would Eq. 16.2.9 change, and would that be a significant change?

**16.2** Water droplets are sprayed into the top of a tall tower. The drops fall through air that has a temperature of 280 K and a relative humidity of 40%. The mean drop diameter is 2 mm. Find the temperature of the water at the bottom of the tower for these conditions. For arbitrary air conditions, prepare a graph of water temperature as a function of relative humidity and air temperature.

**16.3 a.** Find the surface temperature $T_R$ of a small ethanol drop ($D_{drop} = 100 \ \mu$m) suspended in air at 600 K.

**b.** Find the evaporation constant (Eq. 16.3.57) for ethanol in air at 600 K.

We have the following physical property information for ethanol:

$$\rho_L = 760 \ \text{kg/m}^3 \quad C_L = (760/46) \ \text{kg--mol/m}^3$$
$$\Delta \hat{H}_v = 8.46 \times 10^5 \ \text{J/kg}$$

For air at 600 K, and $P = 1$ atm

$$k_{T,f} = k_{T,air} = 0.047 \ \text{W/m} \cdot \text{K}$$

For ethanol diffusing in air, at 600 K, $D_{AB} = 3.8 \times 10^{-5} \ \text{m}^2/\text{s}$.

Vapor pressure data may be written in the form of the Antoine equation:

$$\ln p_{vap}(\text{mmHg}) = 18.9 - \frac{3804}{T_R - 41.7} \quad \textbf{(P16.3.1)}$$

where $T_R$ is in kelvins.

**c.** Is the "blowing" effect (Section 16.3) significant under these conditions?

**16.4** Return to Example 16.3.2 and plot drop radius (squared) versus time for hexadecane.

**16.5** Experimental studies of evaporative losses from an outdoor swimming pool have been carried out, and the data [Smith et al., Solar Energy, **53**, 3 (1994)] have been correlated in the following form:

$$\frac{\dot{m}}{A} = \frac{(30.6 + 32.1 \ U_w)(P_w - P_a)}{\Delta H_v} \quad \textbf{(P16.5)}$$

where

$\dot{m}/A$ = evaporative flux (kg/m$^2 \cdot$h)
$U_w$ = wind speed measured 0.3 m above the water surface (m/s)
$P_w$ = vapor pressure of water at the pool temperature (mmHg)
$P_a$ = vapor pressure of water at the air dewpoint temperature (mmHg)
$\Delta H_v$ = latent heat of vaporization of water at the pool temperature (kJ/kg)

Using models developed in earlier chapters of this book, predict the evaporative loss as a function of wind speed, and compare your prediction to this correlation.

The shape of the pool was not specified, but its surface area is 383 m$^2$. Wind speeds were measured in the range of 0.1 to 3 m/s.

**16.6** A drop of water, of initial diameter $D = 2$ mm, falls through still air. The air is at temperature $T = 90°$F and has a relative humidity of 60%. Find the equilibrium surface temperature of the drop and the initial percentage rate of change in drop diameter.

**16.7** Give the derivation of the following model for calculation of the relative humidity (RH) from measurements of wet bulb ($T_{wb}$) and dry bulb ($T_{db}$) temperatures:

$$\text{RH} = \frac{p_{wb}}{p_{db}} - \frac{(T_{db} - T_{wb})\tilde{C}_p}{\Delta \tilde{H}_{vap}} \left( \frac{Sc}{Pr} \right)^{2/3} \frac{P}{p_{db}} \quad \textbf{(P16.7)}$$

where $P$ is atmospheric pressure, $p_{wb}$ and $p_{db}$ are equilibrium vapor pressures of water at $T_{wb}$ and $T_{db}$. Does the validity of this model require the assumption that the airflow is *turbulent* around the thermometer bulb? Explain your answer.

**16.8** Repeat the analysis of Example 16.4.1, but for a vapor with the following properties:

$M_w = 250$ g/g-mol     $P_{vap} = 10$ torr at 35°C
$P_{vap} = 1$ torr at 0°C     $\mu = 70 \ \mu P$ at 80°C
$\mu = 50 \ \mu P$ at 0°C     $P_{vap} = 100$ torr at 80°C

**16.9** Repeat the analysis of Example 16.4.1, but for water vapor, over the temperature range [5, 25°C]. At what rate must heat be added to the system to maintain constant temperature, over this range of temperatures? Assume that the liquid reservoir and all tubing are well insulated.

**16.10** Extend the analysis of Section 16.5 to predict the *total* change in temperature as a function of mass flowrate (per unit width) for an ethanol liquid film over a 1-meter vertical length. The inlet liquid temperature is 30°C, which is the same as the ambient air temperature. Tsay et al. [*Int. J. Multiphase Flow*, **16**, 853 (1990)] present the experimental data shown below. Compare your predictions to these data.

| $\dot{m}$ (kg/m·s) | $\delta (\times 10^{-4}$ m) | $\Delta T$ (°C) |
|---|---|---|
| 0.01 | 1.67 | −9 |
| 0.02 | 2.11 | −6 |
| 0.04 | 2.66 | −3 |

**16.11** The vapor pressure data shown in Fig. 16.5.3 suggest that a reasonable analytical expression for the $p_{vap}(T)$ relationship is

$$p_{vap} = A \exp(bT) \quad \textbf{(P16.11)}$$

Introduce this relationship into Eq. 16.5.1 and solve for $T(x)$. Use the resulting model to predict the data of Fig. 16.5.2.

**16.12** In Section 16.3 we carried out an analysis of the effect of blowing on heat and mass transfer to a planar surface. Show that the results obtained, especially Eqs. 16.3.20 and 16.3.22, hold for the spherically symmetric case of evaporation from a drop.

**16.13** Plot Eq. 16.3.41 on Fig. 16.3.3, for the case $T_\infty = 325$ K, and $x_{A\infty} = 0$, and confirm the results given in Eq. 16.3.44.

**16.14** Repeat Example 16.3.1, but for an ambient temperature of 350 K.

**16.15** How sensitive is the result for $T_o$ found in Example 16.3.1 to errors or uncertainties in physical properties?

**a.** Suppose, for example, that the diffusion coefficient were really $D_{AB} = 0.24$ cm$^2$/s.

**b.** What if the heat of vaporization were really half the value used in the example?

**c.** What if the heat of vaporization were really twice the value used in the example?

**16.16** If the butane droplet evaporated at the initial rate given by the flux in Eq. 16.3.47, how long would be required for complete evaporation?

**16.17** Review the analysis in Section 16.3 that deals with the effect of "blowing" on heat and mass transfer at a gas/liquid interface. Argue the following points:

**a.** Equation 16.3.23 may be written as

$$R_T = \frac{\tilde{C}_{pA}(T_\infty - T_o)}{\Delta \tilde{H}_v} \quad \text{(P16.17.1)}$$

**b.** If the Prandtl and Schmidt numbers are equal,

$$R_T = R_A \quad \text{(P16.17.2)}$$

**c.** Under conditions corresponding to Pr = Sc, the drop temperature during evaporation may be found as the intersection of the curves

$$R_A(T) = \frac{x_{A,o}(T) - x_{A\infty}}{1 - x_{A,o}(T)} \quad \text{(P16.17.3)}$$

and $R_T(T)$, where $x_{A,o}(T)$ is found from Eqs. 16.3.42 and 16.3.43. Rework Example 16.3.1 using this procedure.

# Appendix A

# Solutions to the Diffusion Equation

In many cases of one dimensional transient heat conduction and diffusion, the equation

$$\frac{\partial Y}{\partial X} = \frac{1}{s^n} \frac{\partial}{\partial s} \left( s^n \frac{\partial Y}{\partial s} \right)$$

arises upon nondimensionalization. The cases $n = 0$, 1, and 2 correspond to Cartesian, cylindrical, and spherical coordinates, respectively. For simple initial and boundary conditions, analytical solutions can be found. In this appendix we present a selection of examples of such solutions. We use the word "diffusion" in the general sense to refer to mass diffusion or heat diffusion (conduction). The word "concentration" can be replaced by "temperature," as well.

## A1  PLANAR DIFFUSION IN ONE DIMENSION

1. Diffusion within a slab of thickness $2b$, initially at a uniform concentration $C_o$ throughout (see Fig. A1.1.) For time $t > 0$ both surfaces are held at a uniform concentration $C_s$.

$$Y = \frac{4}{\pi} \sum_{n=0}^{\infty} (-1)^n \frac{\cos[(2n+1)/2]\pi\eta}{2n+1} \exp\left[ -\frac{(2n+1)^2}{4} \pi^2 X_D \right] \qquad \textbf{(A1.1)}$$

where

$$Y = \frac{C - C_s}{C_o - C_s} \qquad X_D = \frac{D_{AB}t}{b^2} \qquad \eta = \frac{y}{b} \qquad \textbf{(A1.2)}$$

2. Diffusion within a slab of thickness $2b$, initially at a uniform concentration $C_o$ throughout (see Fig. A1.2). For time $t > 0$, both surfaces are exposed to an ambient

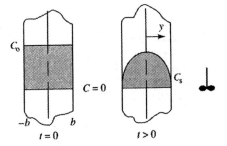

**Figure A1.1** Slab with symmetric transfer to a well-stirred external medium.

medium of uniform concentration $C^{f\infty}$. Transfer across the boundaries $y = \pm b$ occurs by convection with a finite Biot number.

$$Y = \sum_{n=1}^{\infty} \frac{4 \sin \lambda_n}{2\lambda_n + \sin 2\lambda_n} \cos (\lambda_n \eta) \exp(-\lambda_n^2 X_D) \qquad \text{(A1.3)}$$

where

$$Y = \frac{C - C^{f\infty}/\alpha}{C_0 - C^{f\infty}/\alpha} \qquad X_D = \frac{D_{AB}t}{b^2} \qquad \eta = \frac{y}{b} \qquad \text{(A1.4)}$$

The eigenvalues $\lambda_n$ are the roots to (see Fig. A1.3)

$$\lambda_n \tan \lambda_n = \text{Bi} \qquad \text{(A1.5)}$$

with the Biot number Bi defined as

$$\text{Bi} = \frac{\alpha k_c b}{D_{AB}} \qquad \text{(A1.6a)}$$

for mass transfer and

$$\text{Bi} = \frac{hb}{k_T} \qquad \text{(A1.6b)}$$

for heat transfer.[1]

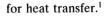

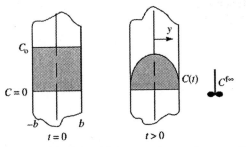

**Figure A1.2** Slab with symmetric transfer to a stirred external medium with a finite Biot number.

---

[1] In addition, for heat transfer, replace all $C$ by T, and set $\alpha = 1$ in the definition of $Y$. Replace $D_{AB}$ by the thermal diffusivity $\alpha_T$ in $X_D$. ($X_D$ is then the Fourier number $X_{Fo}$.)

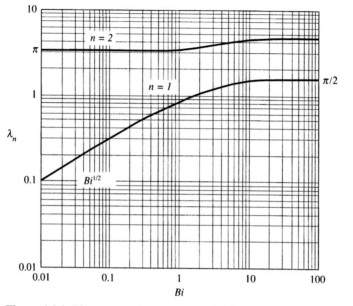

**Figure A1.3** Eigenvalues (roots) of Eq. A1.5.

For mass transfer, the equilibrium solubility coefficient $\alpha$ is defined as

$$\alpha = \frac{C^{f\infty}}{C_{eq}^s} \qquad \text{(A1.7)}$$

where $C_{eq}^s$ is the concentration that would be observed within the slab if it reached equilibirum with the ambient medium.

3. Diffusion within a semi-infinite region $y > 0$ bounded by the infinite plane $y = 0$ (see Fig. A1.4). The region is initially at the uniform concentration $C_o$ throughout $y > 0$. For time $t > 0$, the surface at $y = 0$ is held at a uniform concentration $C_s$.

$$Y = \text{erf}\left(\frac{y}{2\sqrt{D_{AB}t}}\right) \qquad \text{(A1.8)}$$

where "erf" is the error function and $Y$ is defined as in Eq. A1.2. For heat transfer, replace $D_{AB}$ by the thermal diffusivity $\alpha_T$.

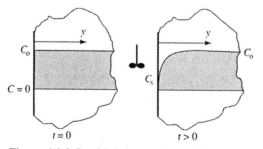

**Figure A1.4** Semi-infinite medium with uniform initial concentration. Transfer is at the planar face $y = 0$ to a well-stirred ambient medium.

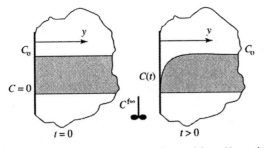

**Figure A1.5** Semi-infinite medium with uniform initial con-centration. Transfer is at the planar face $y = 0$ to a stirred external medium with a finite Biot number.

4. Diffusion within a semi-infinite region $y > 0$ bounded by the plane $y = 0$ (see Fig. A1.5). The region is initially at the uniform concentration $C_o$ throughout $y > 0$. For time $t > 0$, the surface at $y = 0$ is exposed to an ambient medium of uniform concentration $C^{f\infty}$. Transfer across the boundary $y = 0$ occurs by convection with a finite Biot number.

$$Y = \mathrm{erf}\left(\frac{y}{2\sqrt{D_{AB}t}}\right) + \exp\left[2\tau^{1/2}\left(\frac{y}{2\sqrt{D_{AB}t}} + 2\tau^{1/2}\right)\right]\mathrm{erfc}\left(\frac{y}{2\sqrt{D_{AB}t}} + 2\tau^{1/2}\right)$$

$$\text{(A1.9)}$$

where erf is the error function introduced earlier and plotted in Fig. B2.1, and erfc( ) = 1 − erf( ). $Y$ is defined as in Eq. A1.4. The dimensionless time $\tau$ is defined for mass transfer as

$$\tau = \frac{(\alpha k_c)^2 t}{D_{AB}} \qquad \text{(A1.10a)}$$

with $\alpha$ defined as in Eq. A1.7.
  For heat transfer

$$\tau = \frac{h^2 t}{\rho \hat{C}_p k_T} \qquad \text{(A1.10b)}$$

5. Diffusion within a semi-infinite region $y > 0$ bounded by the plane $y = 0$. The region is initially at the uniform concentration $C_o$ within a finite thickness $0 < y < a$, and otherwise at the concentration $C_o = 0$ for $y > a$ (see Fig. A1.6). For

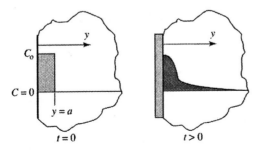

**Figure A1.6** Semi-infinite medium with uniform initial con-centration over a finite region $[0, a]$, and zero concentration elsewhere. At time $t > 0$ the planar face $y = 0$ is impermeable to mass or heat transfer.

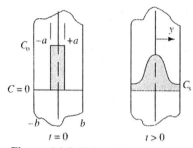

**Figure A1.7** Slab with symmetric transfer to a well-stirred external medium at $C_s$. Initially the concentration is uniformly distributed over the finite region $[-a, +a]$.

time $t > 0$, the surface at $y = 0$ is impermeable. The solution is

$$Y = \frac{C(y,t)}{C_o} = \frac{1}{2}\left(\text{erf}\frac{a - y}{2\sqrt{D_{AB}t}} + \text{erf}\frac{a + y}{2\sqrt{D_{AB}t}}\right) \qquad \textbf{(A1.11)}$$

It is helpful to know that $\text{erf}(-x) = -\text{erf}(x)$.

6. Diffusion from a slab of thickness $2b$, initially at a uniform concentration $C_o$ in the region $(-a < y < a)$ (see Fig. A1.7). For time $t > 0$, the surfaces at $y = \pm b$ are held at a uniform concentration $C_s$.

$$Y = \frac{4}{\pi}\sum_{n=0}^{\infty}\sin\left(\frac{2n + 1}{2}\right)\pi\frac{a}{b}\frac{\cos[(2n + 1)/2]\pi\eta}{2n + 1}\exp\left(-\frac{(2n + 1)^2}{4}\pi^2 X_D\right)$$

$$\textbf{(A1.12)}$$

where

$$Y = \frac{C - C_s}{C_o - C_s} \qquad X_D = \frac{D_{AB}t}{b^2} \qquad \eta = \frac{y}{b} \qquad \textbf{(A1.13)}$$

## A2  RADIAL DIFFUSION IN A LONG CYLINDER

7. Diffusion within a cylinder of radius $R$, initially at a uniform concentration $C_o$ throughout. For time $t > 0$, the surface at $r = R$ is exposed to an ambient medium of uniform concentration $C^{f\infty}$. Transfer across the boundary $r = R$ occurs by convection with a finite Biot number.

$$Y = 2\,\text{Bi}\sum_{n=1}^{\infty}\frac{J_0(\lambda_n s)}{(\lambda_n^2 + \text{Bi}^2)J_0(\lambda_n)}\exp(-\lambda_n^2 X_D) \qquad \textbf{(A2.1)}$$

where

$$X_D = \frac{D_{AB}t}{R^2} \qquad s = \frac{r}{R} \qquad Y = \frac{C - C^{f\infty}/\alpha}{C_o - C^{f\infty}/\alpha} \qquad \textbf{(A2.2)}$$

and the eigenvalues $\lambda_n$ are the roots to

$$\lambda_n J_1(\lambda_n) - \text{Bi}\,J_0(\lambda_n) = 0 \qquad \textbf{(A2.3)}$$

with Bi defined as

$$\text{Bi} = \frac{\alpha k_c R}{D_{AB}} \qquad \textbf{(A2.4a)}$$

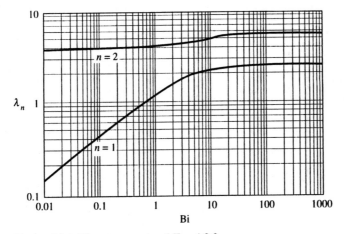

**Figure A2.1** First two roots of Eq. A2.3.

for mass transfer and

$$Bi = \frac{hR}{k_T} \qquad \text{(A2.4b)}$$

for heat transfer.

The first two roots $\lambda_1$ and $\lambda_2$ are plotted as functions of Bi in Fig. A2.1. In the limit of Bi $\to \infty$, Eq. A2.1 reduces to

$$Y = 2 \sum_{n=1}^{\infty} \frac{J_0(\lambda_n s)}{\lambda_n J_1(\lambda_n)} \exp(-\lambda_n^2 X_D) \qquad \text{(A2.5)}$$

The $\lambda_n$ for Bi $\to \infty$ are obtained from Fig. A2.1. Their values are $\lambda_1 = 2.405$ and $\lambda_2 = 5.52$.

8. Diffusion within a hollow cylinder of inner radius $\kappa R$, and outer radius $R$, initially at a uniform concentration $C_o$ throughout. For time $t > 0$, both surfaces are held at a uniform concentration $C_s$.

$$Y = \pi \sum_{n=1}^{\infty} \frac{J_0(\kappa\lambda_n)U_0(\lambda_n s)}{J_0(\kappa\lambda_n) + J_0(\lambda_n)} \exp(-\lambda_n^2 X_D) \qquad \text{(A2.6)}$$

where

$$Y = \frac{C - C_s}{C_o - C_s} \qquad X_D = \frac{D_{AB}t}{R^2} \qquad s = \frac{r}{R} \qquad \text{(A2.7)}$$

and the function $U_0$ is defined as

$$U_0(\lambda_n s) = J_0(\lambda_n s)Y_0(\lambda_n) - J_0(\lambda_n)Y_0(\lambda_n s) \qquad \text{(A2.8)}$$

The $\lambda_n$ are the roots to

$$J_0(\kappa\lambda_n)Y_0(\lambda_n) - J_0(\lambda_n)Y_0(\kappa\lambda_n) = 0 \qquad \text{(A2.9)}$$

The first two $\lambda_n$ are obtained from Fig. A2.2 as a function of $\kappa$.

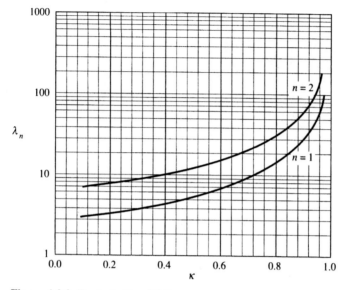

**Figure A2.2** Roots to Eq. A2.9.

## A3 RADIAL DIFFUSION IN A SPHERE

**9.** Diffusion within a sphere of radius $R$, initially at a uniform concentration $C_0$ throughout. For time $t > 0$, the surface at $r = R$ is exposed to an ambient medium of uniform concentration $C^{f\infty}$. Transfer across the boundary $r = R$ occurs by convection with a finite Biot number.

$$Y = 2\,\text{Bi}\,\frac{1}{s}\sum_{n=1}^{\infty} \frac{[\lambda_n^2 + (\text{Bi}-1)^2]\sin\lambda_n}{\lambda_n^2[\lambda_n^2 + \text{Bi}(\text{Bi}-1)]}\sin(\lambda_n s)\exp(-\lambda_n^2 X_D) \qquad \textbf{(A3.1)}$$

where

$$X_D = \frac{D_{AB}t}{R^2} \qquad s = \frac{r}{R} \qquad Y = \frac{C - C^{f\infty}/\alpha}{C_0 - C^{f\infty}/\alpha} \qquad \textbf{(A3.2)}$$

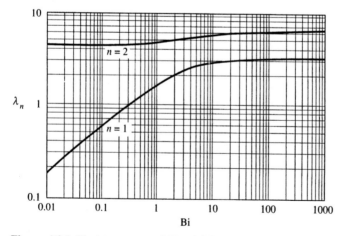

**Figure A3.1** First two roots of Eq. A3.3.

and the eigenvalues $\lambda_n$ are the roots to

$$\lambda_n \cot \lambda_n + \mathrm{Bi} - 1 = 0 \qquad \textbf{(A3.3)}$$

with Bi defined as

$$\mathrm{Bi} = \frac{\alpha k_c R}{D_{AB}} \qquad \textbf{(A3.4a)}$$

for mass transfer and

$$\mathrm{Bi} = \frac{hR}{k_T} \qquad \textbf{(A3.4b)}$$

for heat transfer.

The first two roots $\lambda_1$ and $\lambda_2$ are plotted as functions of Bi in Fig. A3.1. In the limit of $\mathrm{Bi} \to \infty$, Eq. A3.1 reduces to

$$Y = \frac{2}{s} \sum_{n=1}^{\infty} \frac{(-1)^{n+1}}{n\pi} \sin(n\pi s) \exp(-n^2\pi^2 X_D) \qquad \textbf{(A3.5)}$$

# Appendix B

# Some Useful Functions

**M**athematical models of relatively simple problems in fluid dynamics, or heat and mass transfer, often lead to differential equations whose solutions are not simple algebraic functions. When these equations occur frequently, it is useful to have the solutions available in either tabulated or graphical format. In this section we present a number of such functions.

## B1 BESSEL FUNCTIONS

The ordinary differential equation

$$\frac{d^2y}{dx^2} + \frac{1}{x}\frac{dy}{dx} + \left(\lambda^2 - \frac{n^2}{x^2}\right)y = 0 \tag{B1.1}$$

is called Bessel's equation of order $n$, and it has the general solution

$$y = aJ_n(\lambda x) + bY_n(\lambda x) \tag{B1.2}$$

when $n$ is an integer and $\lambda$ is some constant. We call the functions $J_n$ and $Y_n$ the Bessel functions of the first and second kind, of order $n$.

We often need to write derivatives or integrals of $J_n$. The following relations are useful to us:

$$\frac{dJ_n(\lambda x)}{dx} = \frac{n}{x}J_n(\lambda x) - \lambda J_{n+1}(\lambda x) \tag{B1.3}$$

for integer values of $n$, and especially

$$\frac{dJ_0(\lambda x)}{dx} = -\lambda J_1(\lambda x) \tag{B1.4}$$

The most useful integral relationship is

$$\int x^{n+1}J_n(\lambda x)\,dx = \frac{1}{\lambda}x^{n+1}J_{n+1}(\lambda x) \tag{B1.5}$$

and especially

$$\int xJ_0(\lambda x)\,dx = \frac{x}{\lambda}J_1(\lambda x) \tag{B1.6}$$

646

It is helpful to know that for small arguments the Bessel functions $J_0$ and $J_1$ can be approximated by

$$J_0(x) = 1 - \left(\frac{x}{2}\right)^2 + \frac{1}{4}\left(\frac{x}{2}\right)^4 - \cdots \tag{B1.7}$$

and

$$J_1(x) = \left(\frac{x}{2}\right) - \frac{1}{2}\left(\frac{x}{2}\right)^3 + \cdots \tag{B1.8}$$

while for the function $Y_0(x)$ a good approximation for small arguments is

$$Y_0(x) = \frac{2}{\pi}\ln x \tag{B1.9}$$

A plot of the functions $J_0$, $J_1$, and $Y_0$ is given in Fig. B1.1. The functions $J_0$ and $J_1$ oscillate about $y = 0$ with a decreasing amplitude, and eventually they both become periodic with a period equal to $\pi$. We often need the first few roots of $J_0$ and $J_1$, that is, the solutions to $J_0 = 0$ and $J_1 = 0$. These are:

$J_0 = 0: x = 2.405, 5.520, 8.654, 11.792$ (after this, add $\pi$ for successive roots)

$J_1 = 0: x = 3.832, 7.016, 10.173, 13.324$ (after this, add $\pi$ for successive roots)

The ordinary differential equation

$$\frac{d^2y}{dx^2} + \frac{1}{x}\frac{dy}{dx} - \left(\lambda^2 + \frac{n^2}{x^2}\right)y = 0 \tag{B1.10}$$

is called the modified Bessel's equation of order $n$, and it has the general solution

$$y = aI_n(\lambda x) + bK_n(\lambda x) \tag{B1.11}$$

when $n$ is an integer and $\lambda$ is some constant. We call the functions $I_n$ and $K_n$ the modified Bessel functions of the first and second kind, of order $n$.

Useful relationships include the following:

$$\frac{dI_n(\lambda x)}{dx} = \frac{n}{x}I_n(\lambda x) + \lambda I_{n+1}(\lambda x) \quad \text{and} \quad \frac{dK_n(\lambda x)}{dx} = \frac{n}{x}K_n(\lambda x) - \lambda K_{n+1}(\lambda x) \tag{B1.12}$$

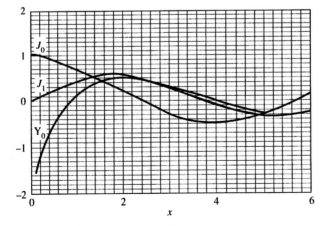

**Figure B1.1** Bessel functions $J_0$, $J_1$, and $Y_0$.

and especially

$$\frac{dI_0(\lambda x)}{dx} = \lambda I_1(\lambda x) \qquad \text{and} \qquad \frac{dK_0(\lambda x)}{dx} = -\lambda K_1(\lambda x) \qquad \textbf{(B1.13)}$$

It is helpful to know that for small arguments the modified Bessel functions can be represented by

$$I_0(x) = 1 + \left(\frac{x}{2}\right)^2 + \frac{1}{4}\left(\frac{x}{2}\right)^4 - \cdots \qquad \text{and} \qquad K_0(x) \approx -\ln x \qquad \textbf{(B1.14)}$$

and

$$I_1(x) = \left(\frac{x}{2}\right) + \frac{1}{2}\left(\frac{x}{2}\right)^3 + \cdots \qquad \text{and} \qquad K_1(x) \approx \frac{1}{2x} \qquad \textbf{(B1.15)}$$

A plot of these functions is given in Fig. B1.2. A double logarithmic scale is used for plotting, since the values grow very large or very small as the arguments increase.

For large arguments, good approximations are

$$I_n(x) = \frac{e^x}{\sqrt{2\pi x}}\left(1 - \frac{4n^2 - 1}{8x} + \cdots\right)$$

and

$$\textbf{(B1.16)}$$

$$K_n(x) = \frac{e^{-x}}{\sqrt{2x/\pi}}\left(1 + \frac{4n^2 - 1}{8x} + \cdots\right) \qquad \text{for} \quad x > 10$$

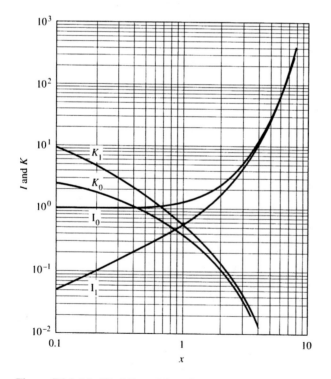

**Figure B1.2** Modified Bessel functions.

Sometimes differential equations arise that are not as simple as Eqs. B1.1 or B1.10 but have solutions that are Bessel functions. A generic Bessel equation is the following:

$$x^2 \frac{d^2y}{dx^2} + x(a + 2bx^r)\frac{dy}{dx} + [c + dx^{2s} - b(1 - a - r)x^r + b^2x^{2r}]y = 0 \quad \textbf{(B1.17)}$$

The solution of this equation is

$$y = x^{(1-a)/2} \exp\left(-\frac{bx^r}{r}\right)\left\{A_1 Z_p\left[\frac{\sqrt{|d|}}{s}x^s\right] + A_2 Z_{-p}\left[\frac{\sqrt{|d|}}{s}x^s\right]\right\} \quad \textbf{(B1.18)}$$

where the order of the Bessel function is

$$p = \frac{1}{s}\left[\left(\frac{1-a}{2}\right)^2 - c\right]^{1/2} \quad \textbf{(B1.19)}$$

Depending on the values of the coefficients $a, b, c, r$, and $s$ of the differential equation, $Z_p$ stands for one of the Bessel functions:

If $\sqrt{d}/s$ is real and $p$ is zero or an integer, $Z_p = J_p$ and $Z_{-p} = Y_p$.
If $\sqrt{d}/s$ is imaginary and $p$ is zero or an integer, $Z_p = I_p$ and $Z_{-p} = K_p$.

For example, although it is not obvious on inspection, the equation

$$\frac{d}{dx}\left(x\frac{dy}{dx}\right) - \beta y = 0 \quad \textbf{(B1.20)}$$

has the solution

$$y = A_1 I_0(2\sqrt{\beta}x) + A_2 K_0(2\sqrt{\beta}x) \quad \textbf{(B1.21)}$$

## B2   THE ERROR FUNCTION

A partial differential equation of the form

$$\frac{\partial y}{\partial t} = \alpha \frac{\partial^2 y}{\partial x^2} \quad \textbf{(B2.1)}$$

arises commonly in our study of transport phenomena. We really can't talk about a solution to this equation without first specifying the initial and boundary conditions. Of particular interest is the set of conditions

$$y = 1 \quad \text{at} \quad t = 0 \quad \textbf{(B2.2a)}$$
$$y = 0 \quad \text{at} \quad x = 0 \quad \textbf{(B2.2b)}$$
$$y = 1 \quad \text{at} \quad x \to \infty \quad \textbf{(B2.2c)}$$

Equation B2.1 arises, for example, in the case of heat or mass transfer in a semi-infinite medium, when the boundary at $x = 0$ is held at a constant temperature or concentration different from the uniform initial condition.

It is not very difficult to show that the solution to Eq. B2.1 with these initial and boundary conditions is expressible as an integral in the form

$$y = \frac{2}{\sqrt{\pi}}\int_0^{x'/\sqrt{4\alpha t}} e^{-\eta^2}\, d\eta \quad \textbf{(B2.3)}$$

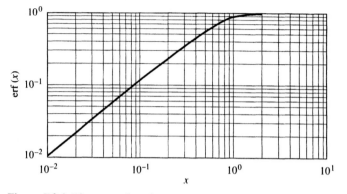

**Figure B2.1** The error function.

Because this integral occurs so often in mathematical models, its value is tabulated and plotted as a function of the upper limit of the integral. It is called the error function and denoted *erf*:

$$y = \text{erf}\left(\frac{x}{\sqrt{4\alpha t}}\right) = \frac{2}{\sqrt{\pi}} \int_0^{x'/\sqrt{4\alpha t}} e^{-\eta^2} d\eta \qquad \text{(B2.4)}$$

The error function is plotted in Fig. B2.1. Among the useful properties of the error function are the following:

$$\frac{d}{dx}\text{erf}(x) = \frac{2}{\sqrt{\pi}} e^{-x^2} \qquad \text{(B2.5)}$$

and

$$\text{erf}(-x) = -\text{erf}(x) \qquad \text{(B2.6)}$$

For small arguments, the exponential in the argument of Eq. B2.4 can be expanded in a series, using

$$e^x = 1 + x + \frac{x^2}{2} + \frac{x^3}{6} + \cdots \qquad \text{(B2.7)}$$

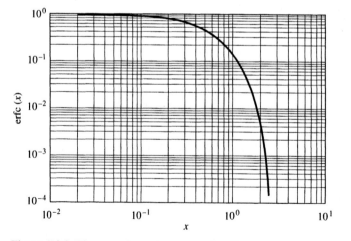

**Figure B2.2** The complementary error function.

After integrating term by term, one finds

$$\operatorname{erf} x = \frac{2}{\sqrt{\pi}} x \left( 1 - \frac{x^2}{3} + \cdots \right) \qquad \text{for} \quad x \ll 1 \tag{B2.8}$$

Sometimes it is useful to have what is called the complementary error function, which is simply

$$\operatorname{erfc}(x) = 1 - \operatorname{erf}(x) \tag{B2.9}$$

and is plotted in Fig. B2.2.

## B3   THE GAMMA FUNCTION

The integral

$$\Gamma(x) = \int_0^\infty s^{x-1} e^{-s} \, ds \qquad x > 0 \tag{B3.1}$$

arises often in the analysis of problems of interest to us. It is called the gamma function. One of its important properties is that

$$\Gamma(1 + x) = x \Gamma(x) \tag{B3.2}$$

The gamma function is plotted in Fig. B3.1 on the range $[1, 2]$. We simply use Eq. B3.2 for values of $x$ outside that range.

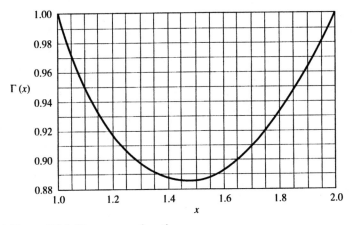

**Figure B3.1** The gamma function.

## REFERENCE

For an extensive discussion of and compilation of these and many other functions, a primary reference is A. Abramowitz, and I. A. Stegun, *Handbook of Mathematical Functions*, Dover, New York, 1965.

# Appendix C

# Physical Property Values

To implement mathematical models as part of the analysis and design activity, it is necessary to have access to the values of the physical properties that appear in the models. In this appendix we review some of the methods available for the calculation of the physical properties that appear in models of the types developed in the text.

We can categorize the physical properties of most interest to us into two broad groups: transport properties and thermodynamic properties. The transport properties include viscosity, the diffusion coefficient, and the thermal conductivity. Among the thermodynamic properties, we are interested in the vapor pressure of pure liquids (and sometimes solids), the partial pressure of solutes above their (usually) aqueous solutions, the solubility of a species between two phases, the heat of vaporization and the heat of fusion of liquids, the specific heat (heat capacity) of fluids and solids, and the interfacial tension between two immiscible fluid phases. If these properties have been measured, we obviously need to know where to look for their values. If we cannot find data, we must be able to estimate their values from fundamental physical parameters such as molecular weight and chemical structure, the critical constants, and the normal boiling point (for liquids). Fortunately a great deal of effort has gone into the development of extensive data-bases and a variety of methods for property estimation.

## C1  VAPOR PRESSURE OF PURE COMPOUNDS

The most common correlation for pure-component vapor pressure as a function of temperature is the Antoine equation

$$\ln p_{\text{vap}} = A - \frac{B}{T + C} \tag{C1.1}$$

where $A$, $B$, and $C$ are empirical constants determined by fitting data. $T$ is *absolute temperature* $(K)$ and $p_{\text{vap}}$ will be in units of millimeters of mercury with the Antoine constants listed in Table C1.1. An extensive table of these constants is given by Reid et al.

**Table C1.1** Antoine Constants for Calculation of Vapor Pressure of Liquids

| Compound | Formula | Constants | | |
|---|---|---|---|---|
| | | A | B | C |
| Silicon tetrachloride | $SiCl_4$ | 15.8 | 2634 | −43.2 |
| Water | $H_2O$ | 18.3 | 3816 | −46.1 |
| Carbon tetrachloride | $CCl_4$ | 15.9 | 2808 | −46.0 |
| Chloroform | $CHCl_3$ | 16.0 | 2697 | −46.2 |
| Methanol | $CH_4O$ | 18.6 | 3626 | −34.2 |
| Ethylene | $C_2H_4$ | 15.5 | 1347 | −18.2 |
| Ethanol | $C_2H_6O$ | 18.9 | 3804 | −41.7 |
| Ethylene glycol | $C_2H_6O_2$ | 20.3 | 6022 | −28.3 |
| Acetone | $C_3H_6O$ | 16.7 | 2940 | −35.9 |
| Isopropyl alcohol | $C_3H_8O$ | 18.7 | 3640 | −53.5 |
| Methyl ethyl ketone | $C_4H_8O$ | 16.6 | 3150 | −36.7 |
| n-Butane | $C_4H_{10}$ | 15.7 | 2155 | −34.4 |
| n-Pentane | $C_5H_{12}$ | 15.8 | 2477 | −39.9 |
| Benzene | $C_6H_6$ | 15.9 | 2789 | −52.4 |
| Phenol | $C_6H_6O$ | 16.4 | 3491 | −98.6 |
| Cyclohexane | $C_6H_{12}$ | 15.7 | 2766 | −50.5 |
| Methyl isobutyl ketone | $C_6H_{12}O$ | 15.7 | 2894 | −70.8 |
| n-Hexane | $C_6H_{14}$ | 15.8 | 2698 | −48.8 |
| Trimethyl gallium | $Ga(CH_3)_3$ | 8.1 | 1703 | 0 |
| TEOS | $Si(OC_2H_5)_4$ | 8.3 | 2381 | 0 |

## C2  VAPOR/LIQUID EQUILIBRIUM AND SOLUBILITY (HENRY'S LAW)

We often need to know the compositions in equilibrium when some species partitions itself between air and water. When the species is relatively dilute in the aqueous phase, this information is often presented in the form of Henry's law:

$$p_A = H_{A,x} x_A \qquad \text{(C2.1)}$$

The composition in the gas phase is given in units of partial pressure in this specific form of Henry's law, with the liquid phase composition in units of mole fraction. Hence $H_{A,x}$ has (typically) units of atmospheres per mole fraction. The Henry's law constant is a function of temperature. Figure C2.1 shows $H_{A,x}(T)$ behavior for some common *gases*, ranging from highly soluble ($NH_3$) to sparingly soluble (CO).

Most *organic liquids* that are immiscible in water actually have a limited solubility, and so there is an equilibrium established between the liquid and vapor phases. Because of the low solubility, Henry's law is usually observed to be a good approximation. Figure C2.2 shows some data for the $H_{A,x}(T)$ behavior for some common *organic liquids*, in water.

An extensive tabulation of *solubility* data for 400 organic compounds in water at atmospheric pressure and at a single temperature (largely, at 25°C) is presented by Yaws et al. Such information is also obtainable directly from Henry's law. When a pure gas is in equilibrium with water, the vapor pressure of the dissolved gas is one atmosphere, so

$$x_A = \frac{p_a}{H_{A,x}} = \frac{1 \text{ atm}}{H_{A,x}} \equiv S_A \qquad \text{(C2.2)}$$

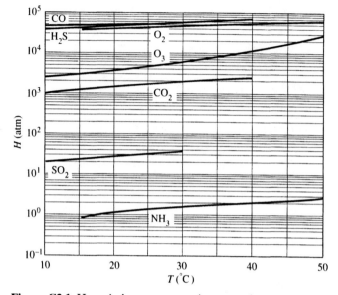

**Figure C2.1** Henry's law constants, in atmospheres per mole fraction, for several common gases with respect to their aqueous solutions. Small $H$ corresponds to high solubility.

and $S_A$ is the equilibrium solubility (as a mole fraction) in water. Hence we can use Henry's law constants as a function of temperature to calculate the solubility as a function of temperature.

The solubility of water in organics may also be of interest. A table of such values is given for about 20 compounds in Chapter 21 of Lyman et al.

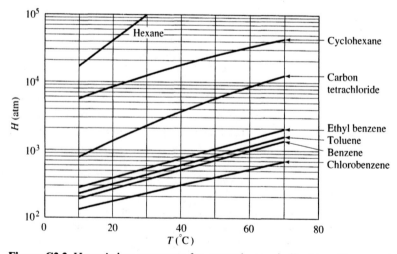

**Figure C2.2** Henry's law constants for several organic liquids with respect to their aqueous solutions. Small $H$ corresponds to high solubility.

## C3   MASS DENSITY

### C3.1   Gases

The density of *gases* is most easily estimated by using the ideal gas law:

$$\rho_{A,gas} = \frac{PM_{w,A}}{R_G T} \tag{C3.1}$$

where $P$ is the total pressure, $M_{w,A}$ is the molecular weight, $R_G$ is the gas constant, and $T$ is absolute temperature. Of course the ideal gas law will not be a good approximation of the behavior near the critical conditions, so one must check this point in using Eq. C3.1.

### C3.2   Liquids

To estimate the density of *liquids,* the method of Grain is recommended by Lyman et al. It takes the form

$$\rho_{A,liq} = \frac{M_{w,A}}{v_{bp\,A,liq}} \left[ 3 - 2\frac{T}{T_{bp\,A}} \right]^n \tag{C3.2}$$

where $v_{bp\,A,liq}$ is the molar volume of the liquid at its boiling point, and $T_{bp\,A}$ is the normal boiling point. The exponent $n$ depends on the chemical class of the species. Recommended values are as follows:

$n = 0.25$ for alcohols
$n = 0.29$ for hydrocarbons
$n = 0.31$ for other organics

The molar volume of the liquid at its normal boiling point can be estimated using the Le Bas additive volume method described in Chapter 2, using Table 2.2.3. The additional information required is the temperature at the normal boiling point. Reid et al. give this information for several hundred compounds, as do Yaws et al.

### C3.3   Solids

For calculating the *mass* densities of *solids* the method of Immirzi and Perini is suggested and discussed by Lyman et al. It is based on the equation

$$\rho_{A,s} = \frac{1.66\,M_{w,a}}{V_{A,cry}} \tag{C3.3}$$

The term $V_{A,cry}$ is the so-called *crystal volume* of the solid, in units of cubic angstroms. It is estimated by a volume increment method in which values are assigned to the unit volumes of the atomic elements in the compound. The crystal volume is then calculated from

$$V_{A,cry} = \sum_i n_i v_i \tag{C3.4}$$

where $n_i$ is the number of times each atomic element appears in the molecule. We need a table of unit volumes $v_i$ for each atomic element, and we need to know the molecular structure of the compound, since the atomic volumes depend on the bonding. For the latter, the *Merck Index* is a good source. Some selected values of $v_i$ are given in Table C3.1. A complete table is presented by Immirzi and Parini.

**Table C3.1** Unit Volumes of Atomic Elements

| Element | $v_i(\text{Å}^3)$ |
|---|---|
| —H | 6.9 |
| =C= | 15.3 |
| =C< | 13.7 |
| —C— | 11.0 |
| =O | 14.0 |
| —O— | 9.2 |
| Six carbons in the benzene structure | 75.2 |
| Ten carbons in the naphthalene structure | 123.7 |

**EXAMPLE** *Calculate the Mass Density of Solid Benzoic Acid, $C_7H_6O_2$*

First we need the structure of benzoic acid, which is

The molecular weight is 122. From Table C3.1 we find the following units volumes:

75.2 for the benzene structure itself
13.7 for the double-bonded carbon in the acid (COOH) group
14 for the double-bonded oxygen
9.2 for the single-bonded oxygen
6.9 for each of the six hydrogens: five on the benzene ring plus the one in the OH group

From Eq. C3.4 we find $V_{A,cry} = 153.5$ Å$^3$, and from Eq. C3.3 it follows that $\rho_{A,s} = 1.32$ g/cm$^3$. This is exactly the value stated in the *Merck Index*.

## C4 MOLAR HEAT OF VAPORIZATION OF LIQUIDS

We often need the molar heat of vaporization of a pure liquid at its normal boiling point, $\Delta \tilde{H}_v$. Among the variety of methods available, the method of Chen is simple to implement and is accurate. It takes the form

$$\Delta \tilde{H}_v = R_G T_c T_{br} \frac{3.98 T_{br} - 3.94 + 1.56 \ln P_c}{1.07 - T_{br}} \qquad \text{(C4.1)}$$

where $T_c$ and $P_c$ are the critical temperature and pressure, respectively, and $T_{br}$ is the reduced normal boiling point temperature, $T_{br} = T_b/T_c$. All temperatures are in kelvins. We use $R_G = 1.987$ cal/g–mol·K and find $\Delta \tilde{H}_v$ in units of cal/g–mol. Obviously we need critical constants to implement this method. Reid et al. provide critical constant data for several hundred compounds.

## C5   MOLAR HEAT CAPACITIES

### C5.1   Gases

For a monatomic ideal gas, the molar heat capacity is simply

$$\tilde{C}_p^o = \frac{5}{2} R_G = \frac{5 \times 1.987}{2} = 4.97 \text{ g} \cdot \text{cal/g-mol} \cdot \text{K} \tag{C5.1}$$

No such simple relationship exists for polyatomic gases. While a variety of methods exist for estimating the heat capacity of gases, the simplest is based on a polynomial fit of data for the ideal gas heat capacity $\tilde{C}_p^o(T)$ in the form

$$\tilde{C}_p^o = A + BT + CT^2 + DT^3 \tag{C5.2}$$

Value of the constants $A$, $B$, $C$, and $D$ are given by Reid et al. for several hundred compounds. For gases for which these constants are not listed, group contribution methods exist based on the molecular structure of the gas. These methods, which are reviewed in Reid et al., are not described here. If the heat capacity is required under conditions for which the ideal gas assumption is poor, correction methods exist, also reviewed in Reid et al.

### C5.2   Liquids

For liquids, the method of Lyman and Danner is recommended by Reid et al. This method of estimating heat capacities is quite complicated in the sense that one must have data for the critical constants of the molecule, as well as its radius of gyration. The details are described in Chapter 5 of Reid et al., including tables of the required molecular parameters. A simple rule of thumb is that near the normal boiling point, most *organic* liquids have heat capacities *per unit mass* between 0.4 and 0.5 cal/g·K. Data are reported in many of the handbooks listed in the references to this appendix.

### C5.3   Solids

As is usually the case, it is very difficult to estimate the heat capacity of solids with any accuracy. Usually, one must seek data in the various handbooks listed.

## C6   VISCOSITY OF GASES AND LIQUIDS

### C6.1   Gases

For gases, we usually use the Chapman–Enskog theory, in the form

$$\mu = 2.67 \times 10^{-6} \frac{\sqrt{M_w T}}{\sigma^2 \Omega_v} \text{ Pa} \cdot \text{s} \tag{C6.1}$$

where $M_w$ and $T$ are molecular weight and absolute temperature (in K). The parameters $\sigma$ and $\Omega_v$ have molecular interpretations, but they are actually determined by fitting Eq. C6.1 to viscosity data. The "collision diameter" $\sigma$ is sometimes referred to as the molecular diameter. We use angstrom units for $\sigma$ in Eq. C6.1. (1 Å = $10^{-10}$ m). To estimate the "viscosity collision integral," $\Omega_v$, it is necessary to have one more empirical parameter, the "characteristic energy" $\varepsilon$. Tables of $\sigma$ and $\varepsilon$ are available. (See, e.g.,

**Table C6.1** Parameters for use in Eq. C6.1

| Compound | Formula | $\sigma$ (Å) | $\varepsilon/k_B$(K) | $M_w$ |
|---|---|---|---|---|
| Air | | 3.71 | 78.6 | 29 |
| Carbon tetrachloride | $CCl_4$ | 5.95 | 323 | 154 |
| Chloroform | $CHCl_3$ | 5.39 | 340 | 119 |
| Methanol | $CH_3OH$ | 3.63 | 482 | 32 |
| Methane | $CH_4$ | 3.76 | 149 | 16 |
| Carbon monoxide | CO | 3.69 | 91.7 | 28 |
| Carbon dioxide | $CO_2$ | 3.94 | 195 | 44 |
| Ethane | $C_2H_6$ | 4.44 | 216 | 30 |
| Propane | $C_3H_8$ | 5.12 | 237 | 44 |
| $n$-Butane | $C_4H_{10}$ | 4.69 | 531 | 58 |
| Acetone | $CH_3COCH_3$ | 4.6 | 560 | 58 |
| Benzene | $C_6H_6$ | 5.35 | 412 | 78 |
| Hydrogen | $H_2$ | 2.83 | 59.7 | 2 |
| Water | $H_2O$ | 2.64 | 809 | 18 |
| Ammonia | $NH_3$ | 2.9 | 558 | 17 |
| Nitrogen | $N_2$ | 3.8 | 71.4 | 28 |
| Oxygen | $O_2$ | 3.47 | 107 | 32 |
| Silane | $SiH_4$ | 4.08 | 208 | 18 |
| Helium | He | 2.58 | 10.2 | 4 |
| Argon | Ar | 3.54 | 93.3 | 40 |
| Neon | Ne | 2.82 | 32.8 | 20 |

*Source:* Reid et al., 1987.

Reid et al. An abbreviated table is given here as Table C6.1.) With a value for $\varepsilon$ one calculates a dimensionless temperature $T^*$ as

$$T^* = \frac{k_B T}{\varepsilon} \qquad \textbf{(C6.2)}$$

where $k_B$ is Boltzmann's constant ($= 1.38 \times 10^{-23}$ N·m/K). Figure C6.1 plots $\Omega_v$ versus $T^*$.

## C6.2  Liquids

It is difficult to predict the viscosity of pure liquids from a small number of well-defined parameters such as the molecular weight of the liquid. There are empirical correlations

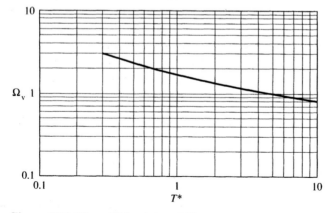

**Figure C6.1** The collision integral $\Omega_v$.

**Table C6.2** Pseudocritical Viscosity $\mu^+$ for Several
Organic Liquids

| Liquid | $\mu^+$ (Pa·s) |
|---|---|
| Hydrocarbons | $8.75 \times 10^{-5}$ |
| Halogenated hydrocarbons | $14.8 \times 10^{-5}$ |
| Benzene derivatives | $8.95 \times 10^{-5}$ |
| Halogenated benzene derivatives | $12.3 \times 10^{-5}$ |
| Alcohols | $8.19 \times 10^{-5}$ |
| Organic acids | $11.7 \times 10^{-5}$ |
| Ethers, ketones, aldehydes, and acetates | $9.6 \times 10^{-5}$ |
| Phenols | $1.26 \times 10^{-5}$ |

*Source:* Reid et al., 1987, Table 9.9.

available, many of them based on parameters that characterize the molecular structure. A detailed discussion of various methods of predicting viscosities is found in Reid, Prausnitz, and Poling. One method given in that reference, which offers a compromise between accuracy and simplicity, is that of Morris, who suggested

$$\log \frac{\mu}{\mu^+} = J\left(\frac{1}{T_r} - 1\right) \tag{C6.3}$$

This model involves two parameters, a so-called pseudocritical viscosity $\mu^+$ and a structure function $J$. The "reduced temperature," $T_r$ is the ratio $T/T_c$ of the temperature to the critical temperature of the liquid (both are absolute temperatures, in K). Thus it is necessary to have data for $T_c$ for various liquids. Some selected values for $\mu^+$ are given in Table C6.2.

The structure function $J$ is calculated from

$$J = \left(0.0577 + \sum_i b_i n_i\right)^{1/2} \tag{C6.4}$$

where the $b_i$ are group contributions given in Table C6.3, taken from Reid *et al.*, and $n_i$ is the number of times that group appears in the molecule.

**Table C6.3** Group Contributions to $J$

| Group | $b_i$ | Group | $b_i$ |
|---|---|---|---|
| $CH_3$, $CH_2$, CH | 0.0825 | $CH_3$, $CH_2$, CH adjoining ring | 0.0520 |
| Halogen-substituted $CH_3$ | 0 | $NH_2$ adjoining ring | 0.7645 |
| Halogen-substituted $CH_2$ | 0.0893 | F, Cl adjoining ring | 0 |
| Halogen-substituted CH | 0.0667 | OH for alcohols | 2.0446 |
| Halogen-substituted C | 0 | COOH for acids | 0.8896 |
| Br | 0.2058 | C=O for ketones | 0.3217 |
| Cl | 0.147 | O=C—O for acetates | 0.4369 |
| F | 0.1344 | OH for phenols | 3.4420 |
| I | 0.1908 | —O— for ethers | 0.1090 |
| Double bond | −0.0742 | | |
| $C_6H_4$ benzene ring | 0.3558 | | |
| Additional H in ring | 0.1446 | | |

*Source:* Reid et al., 1987.

With an accompanying table of critical temperature data (see Appendix A of Reid et al.), one may estimate liquid viscosities by this method. Reid et al. show that the average error in using this method for about 40 common organic liquids is about 15%, but errors as large as 50–70% are observed for some liquids.

## C7 THERMAL CONDUCTIVITIES OF GASES AND LIQUIDS

### C7.1 Monatomic Gases

For monatomic gases the thermal conductivity $k_T$ may be calculated from

$$k_T = \frac{1.99 \times 10^{-4}(T/M_w)^{1/2}}{\sigma^2 \Omega_v} \text{ cal/cm} \cdot \text{s} \cdot \text{K} \tag{C7.1}$$

As usual, $T$ is absolute temperature (K) and $M_w$ is the gram-molecular weight. Values of $\sigma$ and $\varepsilon/k_B$ (needed to find $\Omega_v$) may be found for monatomic gases in Table C6.1. Figure C6.1 plots $\Omega_v$ versus $T^*$.

### C7.2 Polyatomic Gases

For polyatomic gases, the thermal conductivity $k_T$ may be calculated from a variety of semiempirical models discussed by Reid et al. However, these methods require input parameters that themselves are often no more accessible than the conductivity data. Hence one should examine the experimental literature for polyatomic gases, such as the compilations of data by Touloukian and by Vines and Bennett in the list of references. Figure C7.1 gives data for a few selected polyatomic gases at atmospheric pressure.

### C7.3 Liquids

For organic liquids one of the simplest methods to implement is that of Sato and Reidel, given in Reid et al. in the form

$$k_T = \frac{2.64 \times 10^{-3}}{M_w^{1/2}} \frac{3 + 20(1 - T_r)^{2/3}}{3 + 20(1 - T_{rb})^{2/3}} \tag{C7.2}$$

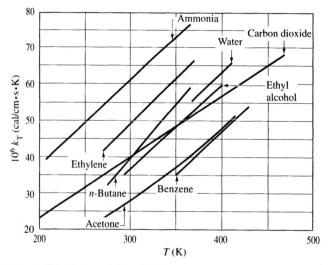

**Figure C7.1** Selected data for $k_T$ versus $T$ for polyatomic gases at atmospheric pressure. Note that all values of $k_T$ were multiplied by $10^6$ before being plotted.

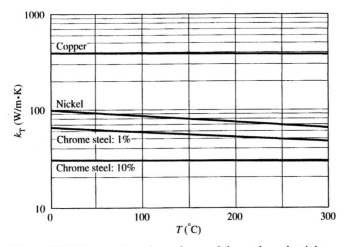

**Figure C7.2** Temperature dependence of thermal conductivity
for metals.

with $k_T$ in units of calories per centimeter-second-kelvin. The reduced normal boiling
point temperature $T_{rb} = T_b/T_c$, and $T_r = T/T_c$, where as usual $T_c$ is the critical temperature. The method has been tested for organic liquids, and errors as large as 20% are
observed. Nevertheless, this is a useful suggestion when a quick estimate will serve
the purpose.

## C7.4   Solids

The thermal conductivities of solids vary widely, and there is no simple method for
prediction that can be applied to metals, ceramics, plastics, and so on. Figure C7.2 shows
examples of data that can be found in various handbooks. Table C7.1 gives a selection

**Table C7.1** Thermal Properties of a Few Solids[a]

| Solid | $k_T$ (W/m · K) | $\hat{C}_p$ (J/kg · K) | $\rho$ (kg/m³) |
|---|---|---|---|
| Aluminum | 150 | 900 | 2700 |
| Asphalt | 0.7 | 920 | 2100 |
| Brick | 0.45 | 840 | 1800 |
| Chrome steel | | | |
| 1% Cr | 61 | 460 | 7870 |
| 10% Cr | 31 | 460 | 7800 |
| Copper | 390 | 380 | 8950 |
| Diamond | 1350 | 510 | 3250 |
| Fat | 0.17 | 1900 | 910 |
| Nickel | 90 | 446 | 8900 |
| Ice (0°C) | 2.2 | 2000 | 915 |
| Paper | 0.12 | 1200 | 700 |
| Polyethylene | 0.35 | 2300 | 920 |
| Polyvinylchloride (PVC) | 0.15 | 960 | 1400 |
| Quartz glass | 1.4 | 730 | 2200 |
| Silicon | 153 | 700 | 2300 |
| Window glass | 0.8 | 800 | 2800 |

[a] Values are nominal, at "room" temperature. Substances belonging to individual categories such as "brick," "paper," or "silicon" may differ in composition.

of thermal conductivity values at 20°C. Also listed are the heat capacities (*per mass*) and the mass densities.

## C8  DIFFUSION COEFFICIENTS IN GASES AND LIQUIDS

For gases and liquids we illustrated the Chapman–Enskog theory and the Wilke–Chang correlation, respectively, as discussed in Chapter 2. More extensive databases for the required physical/molecular parameters are given in Reid et al. For the case of diffusion coefficients in *water*, Reid et al. recommend an alternative to the Wilke–Chang equation—the method of Hayduk and Laudie, which is based on the equation

$$D_{A,w}(cm^2/s) = \frac{13.26 \times 10^{-5}}{\mu_w^{1.14} v_A'^{0.589}} \tag{C8.1}$$

where $\mu_w$ is the viscosity of water, at the temperature of interest, in units of centipoise, and $v_A'$ is the Le Bas molar volume (cm³/mol). A table of values for the calculation of $v_A'$ is given in Chapter 2 (Table 2.2.3).

## C9  INTERFACIAL TENSION

The interfacial tension of a pure liquid in equilibirum with its own vapor is called the "surface tension." Methods of estimating surface tension from molecular structure are described in detail by Reid et al. An extensive compilation of data is given by Jasper. Some selected data are shown in Fig. C9.1. Most organic liquids have surface tensions in the range of 20 to 40 dyn/cm.

Data for the interfacial tensions of various organic compounds with water are available, for example, in Donahue and Bartell, and in Good. As in the case of pure liquids, the values lie in a narrow range—about 15 to 40 dyn/cm. Interfacial tensions of some typical organic compounds with water are given in Table C9.1.

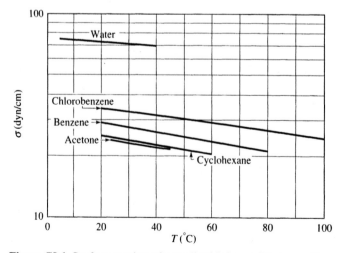

**Figure C9.1** Surface tension of pure liquids in equilibrium with their vapor.

**Table C9.1** Interfacial Tensions
of a Few Organics with
Water (20°C)

| Compound | $\sigma$ (dyn/cm) |
|---|---|
| Benzene | 29 |
| n-Butanol | 25 |
| Cyclohexanol | 34 |
| Hexane | 19 |
| n-Octane | 22 |
| Toluene | 28 |

## C10   PROPERTIES OF WATER AND AIR

Since water and air are the two most common fluids whose properties we need, we
provide Figs. C10.1 and C10.2 for quick reference.

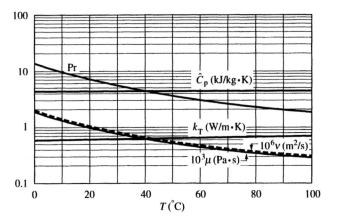

**Figure C10.1** Properties of water.

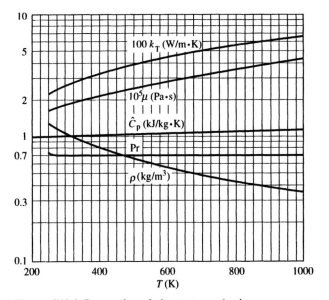

**Figure C10.2** Properties of air at atmospheric pressure.

## C11 REFERENCES

Many "handbooks" are available which contain large collections of physical property data for selected materials. For common materials, these references often offer the quickest starting point. A few such handbooks are listed below. If data are not readily available, the best resource for methods of estimation is Reid, Prausnitz, and Poling. In addition Lyman, Reehl, and Rosenblatt provide a summary of much of what is available in Reid et al., along with additional recommendations, with applications specifically to organic compounds of environmental interest. Only a brief summary of methods and data is given in this appendix. For more detail and extensive referencing, see the books by Reid et al. and Lyman et al.

## REFERENCES

Dean, J. A., Ed., *Lange's Handbook of Chemistry*, 13th ed., McGraw-Hill, New York, 1985.

Donahue, J. D., and F. E. Bartell, "The Boundary Tension at Water–Organic Liquid Interfaces," *J. Phys. Chem.*, **56,** 480 (1952).

Good, R. J., "Generalization of Theory for Estimation of Interfacial Energies," *Ind. Eng. Chem.*, **62,** 55 (1970).

Immirzi, A., and B. Perini, "Prediction of Density in Organic Crystals," *Acta Crystallogr., Sect. A*, **33,** 216 (1977).

Jasper, J. J., *J. Phys. Chem. Ref. Data*, **1,** 841 (1972).

Lyman, W. J., W. F. Reehl, and D. H. Rosenblatt, *Handbook of Chemical Property Estimation Methods*, McGraw-Hill, New York, 1982.

National Research Council, *International Critical Tables*, E. W. Washburn, Ed., McGraw-Hill, New York, 1926.

Perry, R. H., and C. H. Chilton, *Chemical Engineers' Handbook*, 5th ed., McGraw-Hill, New York, 1973.

Perry, R. H., and D. Green, *Chemical Engineers' Handbook*, 6th ed., McGraw-Hill, New York, 1984.

Reid, R. C., J. M. Prausnitz, and B. E. Poling, *The Properties of Gases and Liquid,* 4th ed., McGraw-Hill, New York, 1987.

Rossini, F. D., D. D. Wagman, W. H. Evans, S. Levine, and I. Jaffe, *Selected Values of Chemical Thermodynamic Properties*, Parts I and II, NBS Circular 500, U.S. National Bureau of Standards, Washington, DC 1952.

Touloukian, Y. S. (Series Ed.), *Thermophysical Properties of Matter*, see especially Vol. 3, *Thermal Conductivity of Nonmetallic Liquids and Gases*, IFI/Plenum Data Corp., New York, 1970.

Vargaftik, N. B., *Tables on the Thermophysical Properties of Liquids and Gases*, 2nd ed., Hemisphere, Washington DC, 1975.

Vines, R. G., and L. A. Bennett, *J. Chem. Phys.*, **22,** 360 (1954).

Weast, R. C., Astle, M. C., and W. H. Beyer, Eds., *CRC Handbook of Chemistry and Physics*, 68th ed., CRC Press, Boca Raton, FL, 1987.

Windholz, M., Ed., *The Merck Index*, 10th ed., Merck & Co., Rahway, NJ, 1983.

Yaws, C. L., H-C. Yang, J. R. Hopper, and K. C. Hansen, "Hydrocarbons: Water Solubility Data," *Chem. Eng.*, April 1990, p. 177; "Organic Chemicals: Water Solubility Data," *Chem. Eng.*, July 1990, p. 115.

# Appendix D

# Unit Conversions

**W**hile the International System (SI) of units is being adopted by most journals and texts, much of the existing and important literature uses other units. It is necessary to be able to convert from one system to another. This is especially the case with physical property data, which may be available in several units depending on the sources used.

---

The fundamental mechanical units of the SI system are the meter (m) for length, the kilogram (kg) for mass, and the second (s) for time. From these basic "measures" we can derive units for commonly measured mechanical and thermochemical quantities. These units, in turn, are often given special names in honor of pioneers in the fields of physics, mechanics, and thermodynamics. For example, the following units appear commonly throughout this textbook:

| | | |
|---|---|---|
| Force | $kg \cdot m/s^2$ | newton (N) |
| Pressure | $kg/m \cdot s^2 = N/m^2$ | pascal (Pa) |
| Energy, work | $kg \cdot m^2/s^2 = N \cdot m$ | joule (J) |
| Power | $kg \cdot m^2/s^3 = J/s$ | watt (W) |

The dominant system of units of the second half of the twentieth century has been the metric, or cgs, system. In that system we commonly find

| | | |
|---|---|---|
| Force | $g \cdot cm/s^2$ | dyne |
| Pressure | $g/cm \cdot s^2$ | $dyn/cm^2$ |
| Energy, work | $g \cdot cm^2/s^2$ | erg |

The British system of units (ft, lb, s) is largely being replaced in modern technical literature, but there are a number of commonplace uses of the British system (e.g., gallons and cubic feet for volume measure; pounds force for weight), and one must be able to convert them to SI units.

Unit Conversion Table

| To convert from | to | multiply by |
|---|---|---|
| **Area** | | |
| in.$^2$ | m$^2$ | $6.45 \times 10^{-4}$ |
| ft$^2$ | m$^2$ | 0.0929 |
| cm$^2$ | m$^2$ | $10^{-4}$ |
| **Density** | | |
| g/cm$^3$ | kg/m$^3$ | 1000 |
| lb/ft$^3$ | kg/m$^3$ | 16.02 |
| **Diffusion Coefficient, Species or Thermal** | | |
| cm$^2$/s | m$^2$/s | $10^{-4}$ |
| ft$^2$/s | m$^2$/s | 0.0929 |
| ft$^2$/h | m$^2$/s | $2.58 \times 10^{-5}$ |
| **Energy** | | |
| Btu (thermochemical) | $kg \cdot m^2/s^2 = N \cdot m = joule\ (J)$ | 1054 |
| calorie (thermochemical) | joule (J) | 4.184 |
| kilocalorie (thermochemical) | joule (J) | 4184 |
| erg | joule (J) | $10^{-7}$ |
| ft $\cdot$ lb$_f$ | joule (J) | 1.356 |
| kW $\cdot$ h | joule (J) | $3.6 \times 10^6$ |
| **Force** | | |
| dyne | $kg \cdot m/s^2 = N$ | $10^{-5}$ |
| lb$_f$ | N | 4.448 |
| poundal | N | 0.138 |
| **Heat Capacity (per mass)** | | |
| Btu $\cdot$ lb$_m$ $\cdot$ °F | $J/kg \cdot K = W \cdot s/kg \cdot K$ | 4190 |
| cal/g $\cdot$ K | $W \cdot s/kg \cdot K$ | 4190 |
| **Heat Transfer Coefficient** | | |
| g/s$^3$ $\cdot$ K | $kg/s^3 \cdot K = W/m^2 \cdot K$ | $10^{-3}$ |
| cal/cm$^2$ $\cdot$ s $\cdot$ K | $W/m^2 \cdot K$ | $4.184 \times 10^4$ |
| lb/s$^3$ $\cdot$ °F | $W/m^2 \cdot K$ | 0.816 |
| Btu/ft$^2$ $\cdot$ h $\cdot$ °F | $W/m^2 \cdot K$ | 5.68 |
| **Length** | | |
| angstrom (Å) | m | $10^{-10}$ |
| ft | m | 0.3048 |
| inch | m | 0.0254 |
| micrometer ($\mu$m) | m | $10^{-6}$ |
| **Mass** | | |
| pound (lb) | kg | 0.4536 |
| ton (2000 lb) | kg | 907 |

Unit Conversion Table

| To convert from | to | multiply by |
|---|---|---|
| **Power** | | |
| Btu/s | $kg \cdot m^2/s^3 = J/s = watt\ (W)$ | 1054 |
| cal/s | W | 4.184 |
| ft $\cdot$ lb$_f$/s | W | 1.356 |
| horsepower | W | 745.7 |
| kcal/s | W | 4184 |
| **Pressure** | | |
| atmosphere (atm) | $kg/m \cdot s^2 = N/m^2 = pascal\ (Pa)$ | $1.013 \times 10^5$ |
| bar | Pa | $10^5$ |
| cmHg | Pa | 1333 |
| cmH$_2$O | Pa | 98.06 |
| dyn/cm$^2$ | Pa | 0.1 |
| inch H$_2$O | Pa | 249 |
| lb$_f$/ft$^2$ | Pa | 47.88 |
| lb$_f$/in.$^2$ (psi) | Pa | 6895 |
| mmHg | Pa | 133.3 |
| torr | Pa | 133.3 |
| **Thermal Conductivity** | | |
| ergs/s $\cdot$ cm $\cdot$ K | $kg \cdot m/s^3 \cdot K = W/m \cdot K$ | $10^{-5}$ |
| lb$_f$/s $\cdot$ K | W/m $\cdot$ K | 8.007 |
| cal/s $\cdot$ cm $\cdot$ K | W/m $\cdot$ K | 418.4 |
| Btu/h $\cdot$ ft $\cdot$ °F | W/m $\cdot$ K | 1.731 |
| **Viscosity** | | |
| centipoise (cP) | $kg/m \cdot s = Pa \cdot s$ | 0.001 |
| poise | Pa $\cdot$ s | 0.1 |
| lb$_f$ $\cdot$ s/ft$^2$ | Pa $\cdot$ s | 47.88 |
| lb$_m$/ft $\cdot$ s | Pa $\cdot$ s | 1.49 |
| **Volume** | | |
| ft$^3$ | m$^3$ | 0.0283 |
| gallon (U.S.) | m$^3$ | $3.785 \times 10^{-3}$ |
| in.$^3$ | m$^3$ | $1.639 \times 10^{-5}$ |
| liter (L) | m$^3$ | 0.001 |

# Index

Printed in the United States
54670LVS00001B/77-78